Computational Fluid Dynamics and Heat Transfer

Series in Computational and Physical Processes in Mechanics and Thermal Sciences

Series Editors

W. J. Minkowycz
Mechanical and Industrial Engineering
University of Illinois at Chicago
Chicago, Illinois

E. M. Sparrow
Mechanical Engineering
University of Minnesota, Twin Cities
Minneapolis, Minnesota

The Finite Element Method
Basic Concepts and Applications, Second Edition
Darrell W. Pepper and Juan C. Heinrich

Advances in Numerical Heat Transfer, Volume 3
W. J. Minkowycz

Heat Conduction Using Green's Functions
Kevin D. Cole, James Beck, A. Haji-Sheikh, and Bahman Litkouhi

Computer Methods for Engineering with MATLAB® Applications, Second Edition
Yogesh Jaluria

Computational Fluid Mechanics and Heat Transfer, Third Edition
Richard Pletcher, John C. Tannehill, and Dale Anderson

Nanoparticle Heat Transfer and Fluid Flow
W. J. Minkowycz, E.M. Sparrow, and J.P. Abraham

The Finite Element Method
Basic Concepts and Applications with MATLAB®, MAPLE®, and COMSOL®, Third Edition
Darrell W. Pepper and Juan C. Heinrich

Numerical Simulation of Heat Exchangers: Advances in Numerical Heat Transfer Volume V
W. J. Minkowycz, E.M. Sparrow, J.P. Abraham, and J.M. Gorman

Computational Fluid Mechanics and Heat Transfer, Third Edition
Dale Anderson, John C. Tannehill, Richard Pletcher, and Ramakanth Munipalli

Computational Fluid Dynamics and Heat Transfer, Second Edition
Pradip Majumdar

For more information about this series, please visit: https://www.routledge.com/Computational-and-Physical-Processes-in-Mechanics-and-Thermal-Sciences/book-series/TFSE00230

Computational Fluid Dynamics and Heat Transfer

Second Edition

Pradip Majumdar

CRC Press
Taylor & Francis Group
Boca Raton London New York

CRC Press is an imprint of the
Taylor & Francis Group, an **informa** business

Second edition published 2022
by CRC Press
6000 Broken Sound Parkway NW, Suite 300, Boca Raton, FL 33487-2742

and by CRC Press
4 Park Square, Milton Park, Abingdon, Oxon, OX14 4RN

First edition published by CRC Press 2005

CRC Press is an imprint of Taylor & Francis Group, LLC

ISBN: 978-1-498-70374-1 (hbk)
ISBN: 978-1-032-04094-3 (pbk)
ISBN: 978-0-429-18300-3 (ebk)

Typeset in Times
by codeMantra

Access the Support Materials: https://www.routledge.com/9781498703741.

To my late Parents

Snehalata and Rati Ranjan,

my Wife

Srabani,

and my Children

Diya and Ishan

Contents

PART III Finite Element Method

Preface

With the advances and availability of current and emerging high-end computing hardware such as the GPUs and cloud computing infrastructures, computational fluid dynamics (CFD) and heat transfer methods are now reaching new heights in its increased use for the analysis and design of processes and products for a variety complex engineering problems. CFD and heat transfer codes are extensively used as a simulation tool for the virtual or digital prototype development of products and devices involving complex transport and multiphysic phenomena. It is an essential element of the agile product development environment in all sectors of manufacturing for developing products in a very accelerated time cycle.

There is a continuous demand from the industry for graduating engineers with the basic knowledge and skills of computational methods for fluid dynamics and heat transfer. This book introduces fundamentals of two important computational techniques for solving fluid flow and heat transfer: *Finite Difference/control volume method* and *Finite Element Method.* The theory and procedures are presented in a simple and straight forward manner with detail derivations and illustrations through numerous classical example problems from fluid flows, heat transfer, and mass transfer. Students and industrial developers will be able to develop their basic skills and build on their abilities to use commercial Computational Fluid Dynamics (CFD) and Finite Element Method (FEM) codes in a more efficient manner. The primary objective is to help the students understand the concepts and procedures thoroughly and be able to implement the computational methodology into a customized computer code for solving more complex problems on their own.

Book chapters are grouped into three parts. Part I presents reviews of basic equations of heat transfer, mass transfer and fluid dynamics in Chapter 1; basic concepts of numerical approximations and errors in Chapter 2; numerical solution techniques for systems of linear algebraic equations in Chapter 3 and numerical integrations and quadrature formulas in Chapter 4. Part II presents finite difference/control volume method in Chapters 5–10. Two different approaches for discretization procedures, one based on finite difference formulas derived using truncated Taylor series expansions and the other based on taking integration over a control volume, are presented. Part III presents finite element method in Chapters 11–16. Both Galerkin and variational methods of finite element formulation are presented. Basic concepts of the calculus of variations and variational method of approximations are also presented for clear understanding of the finite element method. The topics on finite difference/control volume and finite element method are presented in several chapters with the first one devoted to basic theory and examples dealing with one-dimensional steady-state diffusion problems, followed by multidimensional problems, transient or unsteady state problems, and convection problems. Solution procedures are demonstrated through several worked-out example problems using hand calculations and through implementation of the solution algorithms into a computer codes. Since it is expected that students may use different kind of computer programming languages, no computer programs are included. Rather, the algorithms are presented in the form of pseudo codes for students to implement them into computer codes using their choice of programming languages. At the end of chapters problems are included to help students get further understanding of the techniques by working out the problems in hand calculations and by developing computer codes, implementing the solution algorithms.

This book is primarily written for an introductory textbook for a senior undergraduate and first-year graduate-level course. The book is also comprehensive enough to be considered as a reference book for independent study by students and practicing engineers and scientist who are involved in developing their own codes as well as those who use commercial codes for the analysis and design of the processes and products involving fluid dynamics, heat, and mass transfer.

The book is intended as a textbook on CFD and heat transfer for an undergraduate course and/or for a first-year graduate course. It is not intended that all 16 chapters be covered in the first course at senior undergraduate or first-year graduate level. Subject materials covered in the book could be used in a second advanced level course on CFD. A review of basic governing equations and boundary conditions is given in Chapter 1. This chapter also includes presentations of all fluid flow and transport equation including a basic set of two-equation turbulent flow model in a generalized form. This chapter could be assigned to the students as a self-study reading assignment to review basic concepts, laws and governing equations, and boundary conditions. Chapters 2 and 4 are included primarily for students who have had no prior course on numerical analysis. The materials presented in Chapter 3 are quite comprehensive on solvers for system of linear equations that arise from CFD discretization methods and may be reviewed before proceeding with Finite Difference/Control volume and Finite Element CFD methods.

The title of the book has been changed to *Computational Fluid Dynamics and Heat Transfer* to better reflect the content of the book. While all chapters are reviewed for better clarity in concepts and worked-out examples, additional topics on mesh generation and turbulent fluid flows are included based on practical applications. A list of major revisions and new additions is described below:

Chapter 1 includes additional sections: Section 1.2.5 includes a brief description on turbulent flow and modeling and Section 1.7 includes a general form of the governing equations that will be used to introduce the computational discretization schemes in latter chapters.

Chapter 3 includes an additional section on Generalized Minimal Residual Method, which is one of the most popular iterative solvers used in most commercial CFD codes. The iterative refinement scheme is also revised to show its applications in Chapter 9 for deriving correction equations for velocity and pressure.

Chapter 5 includes a new section on higher order approximation schemes for convective terms such as Second-Order Upwind Scheme, Third-Order Quick Scheme, and MUSCL Scheme – Monatomic Upstream – Centered Scheme for Conservation Law.

Chapter 8 includes an additional very commonly used *pressure-correction algorithm* known as the *Pressure Implicit with Splitting of Operators* algorithm.

Two new chapters: Chapter 9 "Additional Features in Computational Model and Mesh Generations" and Chapter 10 "Computational Model for Turbulence Flow" are added to expand description of CFD and heat transfer for practical applications. In Chapter 9, some of the essential features and guidelines for developing CFD and heat transfer codes implementing discretization procedure for control volume/finite difference method and finite element are discussed. These important features include guidelines for mesh generation and quality; adapting meshing; multigrid methods; and initial, inlet, and boundary conditions. A few case study example problems are also included to demonstrate important aspects of CFD and heat transfer.

Chapter 10 starts with a brief review of the physical description and some major characteristics of turbulence. Further, computational modeling of turbulence flow is discussed with a primary focus on the Reynolds Averages Navier Stokes turbulence model. Descriptions of various turbulence closure models and some guidelines for choosing a turbulence model are also given. Furthermore, some important aspects related to the turbulent fluid flow computations such as wall function treatments; selection of $y+$ values; and wall and boundary conditions for turbulence scalar variables are outlined.

I would like to express my sincere gratitude and thanks to Professor E. M. Sparrow and Professor W. J. Minkowycz for their extremely valuable comments, advice, and encouragements. I thank all reviewers for their constructive comments. I would like to express my deep appreciation to all editors, managers, designers, and staff members at Taylor & Francis for their efforts, supports, understanding, and patience during the production of this book. A special thanks to my children Diya and Ishan for helping me select the cover page of the book.

The book evolved from several years of teaching a course on computational heat transfer and fluid dynamics. I believe that the content of the book with additional subject matters will help students build a stronger grip about computational simulation and modeling. I thank many students and readers of the book in the industry from different parts of the world for theirs comments and suggestions for the enhancement of the book. I welcome suggestions from interested readers of the book.

I appreciate immensely the unlimited support of my wife, Srabani, and children, Diya and Ishan, during the long hours that spent in completing the book. They were the continuous source of my motivation to continue and complete the book.

Pradip Majumdar

Access the Support Materials: https://www.routledge.com/9781498703741.

Author

Pradip Majumdar earned an MS and a PhD in Mechanical Engineering at Illinois Institute of Technology. He was a professor and the chair in the Department of Mechanical Engineering at Northern Illinois University. He is recipient of the 2008 Faculty of the Year Award for Excellence in Undergraduate Education. Dr. Majumdar has been the lead investigator for numerous federal and industrial projects. Dr. Majumdar authored numerous papers on fluid dynamics, heat and mass transfer, energy systems, fuel cell, Li-ion battery storage, electronics cooling and electrical devices, engine combustion, nano-structured materials, advanced manufacturing, and transport phenomena in biological systems. Dr. Majumdar is the author of three books, including *Computational Methods for Heat and Mass Transfer*; *Fuel Cells – Principles, Design and Analysis*; and *Design of Thermal Energy Systems*. Dr. Majumdar is an editor of the *International Communications in Heat and Mass Transfer*. He has previously served as the associate editor of *ASME Journal of Thermal Science and Engineering*. Dr. Majumdar has been making keynote and plenary presentations on Li-ion battery storage, fuel cell, electronics cooling, nanostructure materials at national/international conferences and workshops. Dr. Majumdar has participated as an international expert in GIAN lecture series on fuel cell and Li-ion battery storage. Dr. Majumdar is a fellow of the American Society of Mechanical Engineers (ASME).

Part I

Basic Equations and Numerical Analysis

1 Review of Basic Laws and Equations

The computational analysis of fluid flow and heat and mass transfer processes starts from the basic rate and conservation equations for these processes. In this chapter, we will give a brief review of the fundamental concepts, principles and basic equations in fluid dynamics, heat and mass transfer. Knowledge and understanding of the basic concepts and principles are essential to analyze these processes using computational techniques.

1.1 BASIC EQUATIONS

Analysis of fluid flow and heat transfer starts with the statement of the basic conservation laws. This includes (a) conservation of mass, (b) Newton's second law of motion or conservation of momentum, and (c) first law of thermodynamics. Also, we need additional relations such as the equation of state or constitutive relations. These basic laws are the same as those in mechanics and thermodynamics. However, these laws are reformulated in a manner suitable for fluid flows and heat and mass transfer.

We can formulate the basic laws in terms of infinitesimal or finite systems and control volumes. Formulation based on infinitesimal system and control volumes leads to **differential formulation**. Solution of the differential formulation gives detailed knowledge or point-to-point behavior of fluid flows and heat transfer, whereas the **integral formulation** that results from the finite system and control volume gives gross or average behavior. In many fluid flow and heat transfer problems, we may additionally apply the integral formulation to check the accuracy of the computational solution of the differential equations by observing the overall balance of mass, momentum, and energy.

1.2 FLUID FLOW

1.2.1 Fluid Properties

Fluid properties are classified into four basic types: (a) *Thermodynamic properties* such as temperature, pressure, density, enthalpy, and entropy; (b) *Transport properties* such as viscosity, thermal conductivity, and mass diffusivity; (c) *Kinematics properties* such as velocity, acceleration, angular velocity, vorticity, and deformation or strain rate; and (d) other miscellaneous properties such as surface tension. Among these, *kinematics properties* are truly the properties of the fluid motion itself rather than that of the fluid.

1.2.1.1 Kinematics of Fluid and Kinematic Properties

Velocity Field The velocity at a point in a flow field is defined as the velocity of a fluid particle that passes through that point. The velocity of a fluid particle is defined in the same way as in dynamics, i.e., as the rate of change of the particle's position vector with respect to time. So, velocity at any point of a flow field is written as

$$\vec{V} = \frac{d\tilde{x}}{dt} \tag{1.1}$$

In Cartesian co-ordinate system, where we define the position vector as $\tilde{x} = \hat{i}x + \hat{j}y + \hat{k}z$, the velocity vector can be written as

$$\vec{V}(x,y,z,t) = \hat{i}\frac{dx}{dt} + \hat{j}\frac{dy}{dt} + \hat{k}\frac{dz}{dt} \tag{1.2a}$$

or

$$\vec{V}(x,y,z,t) = \hat{i}u + \hat{j}v + \hat{k}w \tag{1.2b}$$

where $u = dx/dt$, $v = dy/dt$, and $w = dz/dt$ are the x, y, and z component velocities, respectively.

Similarly, the velocity vectors in cylindrical and spherical co-ordinate systems are defined in Figure 1.1.

Cylindrical

$$\vec{V}(r,\theta,z,t) = \hat{i}u_r + \hat{j}u_\theta + \hat{k}u_z$$

Spherical

$$\vec{V}(r,\theta,\phi,t) = \hat{i}u_r + \hat{j}u_\theta + \hat{k}u_\phi$$

Volume and Mass Flow Rates Once the velocity field is known, we can calculate the volume flow rate, mass flow rate, and average velocity for flow through any fixed or arbitrary control surface as shown in Figure 1.2.

Volume Flow Rate Volume flow rate through the differential area dA is

$$d\dot{\forall} = (\vec{V}\cdot\hat{n})dA = \vec{V}\cdot d\vec{A} \tag{1.3}$$

Integrating over the whole area, we get the total flow rate

$$\dot{\forall} = \int_{CS}(\vec{V}\cdot\hat{n})dA = \int\vec{V}\cdot d\vec{A} \tag{1.4}$$

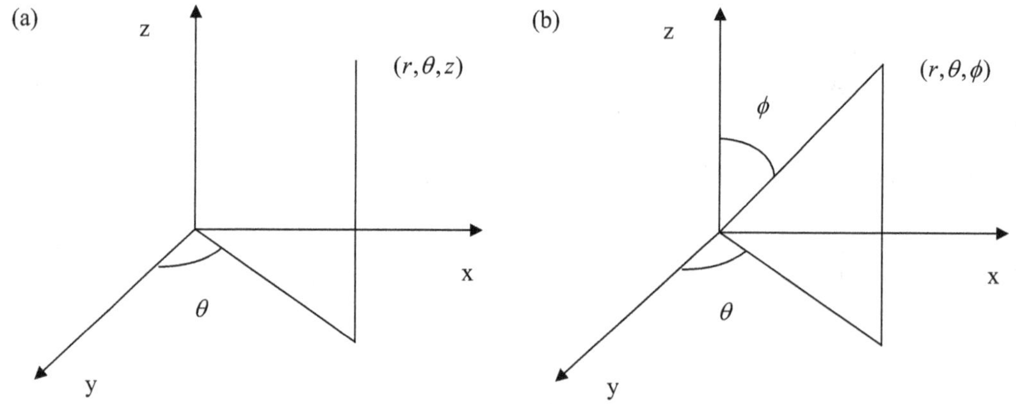

FIGURE 1.1 Co-ordinate systems: (a) cylindrical and (b) spherical.

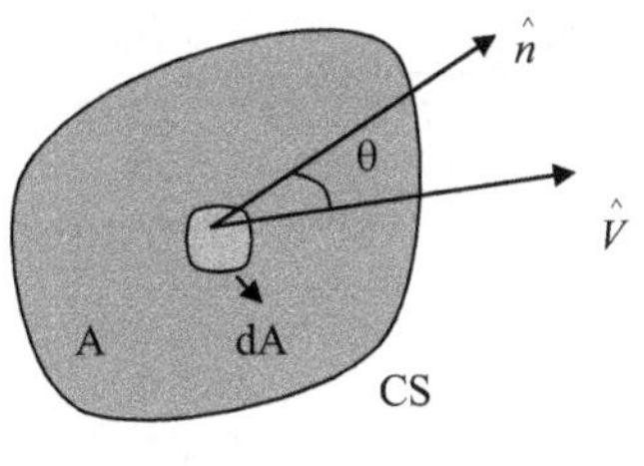

FIGURE 1.2 Arbitrary control volume surface.

Mass Flow Rate

$$\dot{m} = \int_{CS} \rho\left(\vec{V}\cdot\hat{n}\right)\mathrm{d}A \tag{1.5}$$

Average Velocity

$$\bar{V} = \frac{\dot{\forall}}{A} = \frac{\int_{CS} \vec{V}\cdot\vec{\mathrm{d}A}}{A} \tag{1.6}$$

Acceleration The acceleration of a fluid particle at any point on the flow field is defined as the rate of change of velocity of the particle with time. For a velocity field, the ***acceleration*** can be derived as

$$\vec{a} = \frac{D\vec{V}}{Dt} = \frac{\partial \vec{V}}{\partial t} + u\frac{\partial \vec{V}}{\partial x} + v\frac{\partial \vec{V}}{\partial y} + w\frac{\partial \vec{V}}{\partial z} \tag{1.7}$$

The term $D\vec{V}/Dt$ on the left-hand side is called total derivative or material derivative. It represents the total acceleration of the fluid particles. It is evident from the right-hand side of the expression that the total acceleration has two major components: the first term $\partial\vec{V}/\partial t$ is the local acceleration, and the remainder is convective acceleration given by

$$u\frac{\partial \vec{V}}{\partial x} + v\frac{\partial \vec{V}}{\partial y} + w\frac{\partial \vec{V}}{\partial z}$$

It can be mentioned here that these nonlinear convective terms pose considerable mathematical and numerical difficulties. The expression for acceleration can be written in vector notation as

$$\vec{a} = \frac{D\vec{V}}{Dt} = \frac{\partial \vec{V}}{\partial t} + \left(\vec{V}\cdot\nabla\right)\vec{V} \tag{1.8}$$

The scalar component equations are written as

Cartesian Coordinate (x, y, z)

$$a_x = \frac{Du}{Dt} = \frac{\partial u}{\partial t} + u\frac{\partial u}{\partial x} + v\frac{\partial u}{\partial y} + w\frac{\partial u}{\partial z} \tag{1.9a}$$

$$a_y = \frac{Dv}{Dt} = \frac{\partial v}{\partial t} + u\frac{\partial v}{\partial x} + v\frac{\partial v}{\partial y} + w\frac{\partial v}{\partial z} \tag{1.9b}$$

$$a_z = \frac{Dw}{Dt} = \frac{\partial w}{\partial t} + u\frac{\partial w}{\partial x} + v\frac{\partial w}{\partial y} + w\frac{\partial w}{\partial z} \tag{1.9c}$$

Cylindrical Coordinate (r, θ, z)

$$a_r = \frac{Du_r}{Dt} = \frac{\partial u_r}{\partial t} + u_r\frac{\partial u_r}{\partial r} + \frac{u_\theta}{r}\frac{\partial u_r}{\partial \theta} - \frac{u_\theta^2}{2} + u_z\frac{\partial u_r}{\partial z} \tag{1.10a}$$

$$a_\vartheta = \frac{Du_\theta}{Dt} = \frac{\partial u_\theta}{\partial t} + u_r\frac{\partial u_\theta}{\partial r} + \frac{u_\theta}{r}\frac{\partial u_\theta}{\partial \theta} + \frac{u_r u_\theta}{r} + u_z\frac{\partial u_\theta}{\partial z} \tag{1.10b}$$

$$a_z = \frac{Du_z}{Dt} = \frac{\partial u_z}{\partial t} + u_r\frac{\partial u_z}{\partial r} + \frac{u_\theta}{r}\frac{\partial u_z}{\partial \theta} + u_z\frac{\partial u_z}{\partial z} \tag{1.10c}$$

Spherical Coordinates (r, θ, φ)

$$a_r = \frac{Du_r}{Dt} = \frac{\partial u_r}{\partial t} + u_r\frac{\partial u_r}{\partial r} + \frac{u_\phi}{r}\frac{\partial u_r}{\partial \phi} + \frac{u_\theta}{r\sin\phi}\frac{\partial u_r}{\partial \theta} - \frac{u_\theta^2 + u_\phi^2}{r} \tag{1.11a}$$

$$a_\theta = \frac{Du_\theta}{Dt} = \frac{\partial u_\theta}{\partial t} + u_r\frac{\partial u_\theta}{\partial r} + \frac{u_\phi}{r}\frac{\partial u_\theta}{\partial \phi} + \frac{u_\theta}{r\sin\phi}\frac{\partial u_\theta}{\partial \theta} + \frac{u_r u_\theta}{r} - \frac{u_\theta u_\phi \cot\phi}{r} \tag{1.11b}$$

$$a_\phi = \frac{Du_\phi}{Dt} = \frac{\partial u_\phi}{\partial t} + u_r\frac{\partial u_\phi}{\partial r} + \frac{u_\phi}{r}\frac{\partial u_\phi}{\partial \phi} + \frac{u_\theta}{r\sin\phi}\frac{\partial u_\phi}{\partial \phi} + \frac{u_r u_\phi}{r} - \frac{u_\theta^2 \cot\phi}{r} \tag{1.11c}$$

Motion or Kinematics of Fluid Element In order to describe the motion or kinematics of fluid, let us consider a differential control volume of a fluid element of volume dv in Cartesian co-ordinate as shown in Figure 1.3.

Let us assume that a fluid element initially at a point **P** at time t then moves to a new position P' at time $t+dt$. As the fluid element moves in the flow field, it may experience several things. These include (a) translation, (b) rotation, (c) linear deformation or extensional strain, and (d) angular deformation or shear strain. These are discussed briefly by considering a motion on an x–y plane. Similar motions can also be illustrated for y–z and z–x planes.

Translation The **translation** is defined by the displacements u dt and v dt of the point P to P' as shown in Figure 1.4.

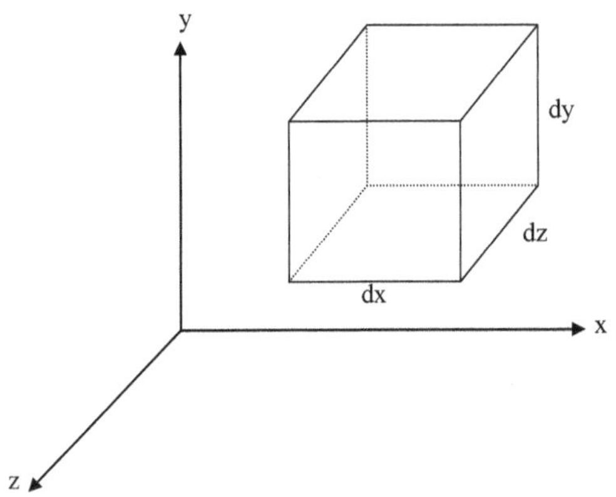

FIGURE 1.3 Differential control volume in Cartesian co-ordinates.

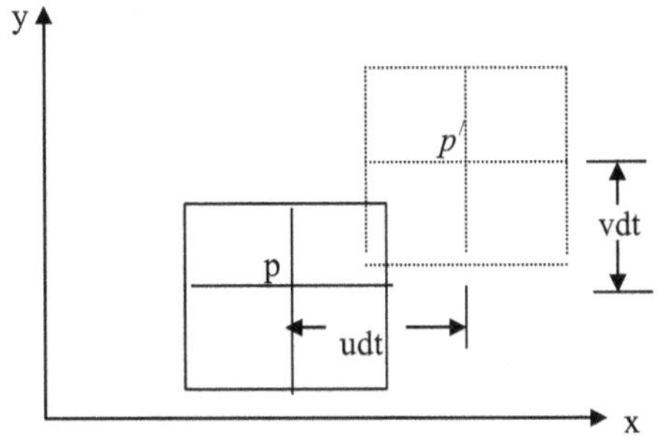

FIGURE 1.4 Translation in the x–y plane.

The rates of translations are u and v. For three-dimensional fluid flows, the translations are defined by velocity components u, v, and w.

Rotation The fluid **rotation**, $\vec{\omega}$, is a vector quantity with rotation about all three coordinate axes, and it is written as

$$\vec{\omega} = \hat{i}\omega_x + \hat{j}\omega_y + \hat{k}\omega_z \tag{1.12}$$

where ω_x, ω_y, and ω_z are rotations about x, y, and z axes, respectively.

Each of these rotational components is defined as the average angular velocity of any two mutually perpendicular lines in a differential element. Figure 1.5 shows the rotation of such an element in the x–y plane.

The rotation ω_z of this element is defined as the average angular velocity of two mutually perpendicular lines PA and PB.

The angular velocity of line PA is given as

$$\omega_{\text{PA}} = \lim_{\Delta t \to 0} \frac{\Delta \theta}{\Delta t} = \lim_{\Delta t \to 0} \frac{\Delta y'/\Delta x}{\Delta t} = \frac{\partial v}{\partial x} \tag{1.13}$$

and the angular velocity of line PB is given as

$$\omega_{\text{PB}} = \lim_{\Delta t \to 0} \frac{\Delta \phi}{\Delta t} = \lim_{\Delta t \to 0} \frac{\Delta x'/\Delta y}{\Delta t} = -\frac{\partial u}{\partial y} \tag{1.14}$$

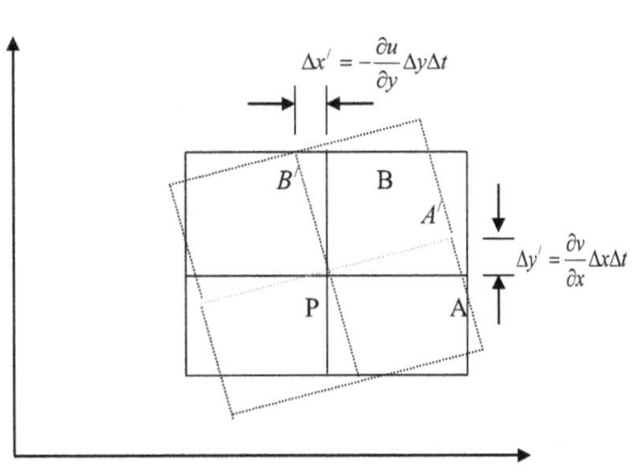

FIGURE 1.5 Rotation of a fluid element in the x–y plane.

The rotation of the fluid element about the z-axis is expressed as

$$\omega_z = \frac{1}{2}\left(\frac{\partial v}{\partial x} - \frac{\partial u}{\partial y}\right) \tag{1.15a}$$

Similarly, the rotation of the differential fluid element about the x-axis and the y-axis is given as

$$\omega_x = \frac{1}{2}\left(\frac{\partial w}{\partial y} - \frac{\partial v}{\partial z}\right) \tag{1.15b}$$

and

$$\omega_y = \frac{1}{2}\left(\frac{\partial u}{\partial z} - \frac{\partial w}{\partial x}\right) \tag{1.15c}$$

Substituting, the complete expression of rotation is written as

$$\vec{\omega} = \frac{1}{2}\left[\hat{i}\left(\frac{\partial w}{\partial y} - \frac{\partial v}{\partial z}\right) + \hat{j}\left(\frac{\partial u}{\partial z} - \frac{\partial w}{\partial x}\right) + \hat{k}\left(\frac{\partial v}{\partial x} - \frac{\partial u}{\partial y}\right)\right] \tag{1.15d}$$

We can express this in a compact form using vector notation as

$$\vec{\omega} = \frac{1}{2}\left(\nabla \times \vec{V}\right) \tag{1.15e}$$

Another important quantity of great interest in fluid mechanics is the ***vorticity***, which is defined as

$$\vec{\zeta} = 2\vec{\omega} = \nabla \times \vec{V} \tag{1.16}$$

It can be mentioned here that a flow is irrotational if vorticity $\vec{\zeta} = 0$ or angular velocity $\vec{\omega} = 0$. Also, an initial irrotational flow field can only become rotational with a simultaneous angular deformation caused by viscous or shear stresses.

Linear Deformation or Extensional Strain The **linear deformation** or **extensional strain** is defined as the fractional increase in length of one side of the element (see Figure 1.6).

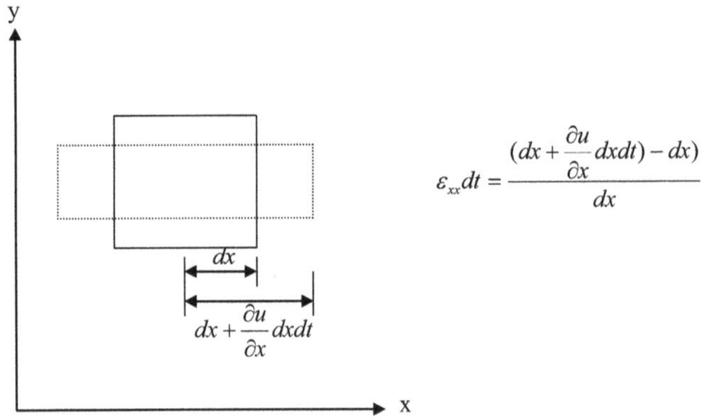
y
x
dx
$dx + \frac{\partial u}{\partial x} dxdt$
$\varepsilon_{xx} dt = \frac{(dx + \frac{\partial u}{\partial x} dxdt) - dx)}{dx}$

FIGURE 1.6 Linear deformation of the fluid element in the x–y plane.

We can express the extensional strain or linear deformation of side dx as

$$\varepsilon_{xx}\,\mathrm{d}t = \frac{\left[dx + \left(\partial u/\partial x\right)\mathrm{d}x\mathrm{d}t\right] - \mathrm{d}x}{\mathrm{d}x}$$

Thus, the extensional strain rate in the x-direction is

$$\varepsilon_{xx} = \frac{\partial u}{\partial x} \tag{1.17a}$$

Similarly, the y and z components of extensional strain rates are

$$\varepsilon_{yy} = \frac{\partial v}{\partial y} \tag{1.17b}$$

and

$$\varepsilon_{zz} = \frac{\partial w}{\partial z} \tag{1.17c}$$

Angular Deformation or Shear Strain The two-dimensional **angular deformation** or **shear strain** is defined as the average decrease of the angle between two lines PA and PB as shown in Figure 1.7.

Thus the shear strain rate for the x–y plane is

$$\varepsilon_{xy} = \frac{1}{2}\left(\frac{\mathrm{d}\theta}{\mathrm{d}t} + \frac{\mathrm{d}\phi}{\mathrm{d}t}\right) \tag{1.18}$$

Now, noting that $\mathrm{d}\theta/\mathrm{d}t = \partial v/\partial x$ and $\mathrm{d}\phi/\mathrm{d}t = \partial u/\partial y$, we get

$$\varepsilon_{xy} = \frac{1}{2}\left(\frac{\partial v}{\partial x} + \frac{\partial u}{\partial y}\right) \tag{1.19a}$$

Similarly, the shear strain rates for the y–z and z–x planes are

$$\varepsilon_{yz} = \frac{1}{2}\left(\frac{\partial w}{\partial y} + \frac{\partial v}{\partial z}\right) \tag{1.19b}$$

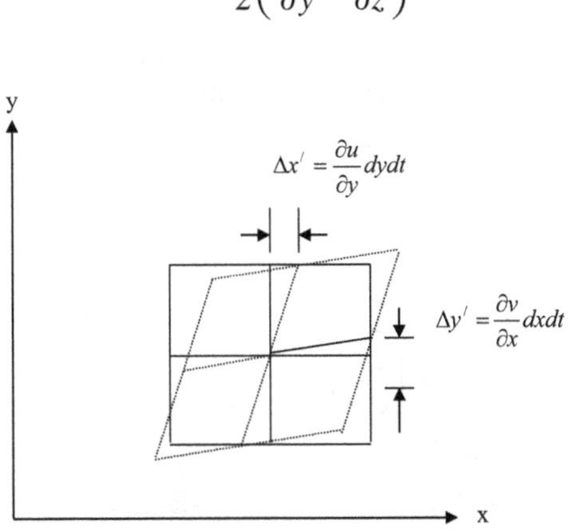

FIGURE 1.7 Angular deformation of fluid element in the x–y plane.

and

$$\varepsilon_{zx} = \frac{1}{2}\left(\frac{\partial u}{\partial z} + \frac{\partial w}{\partial x}\right) \tag{1.19c}$$

Also, the shear strain rates are symmetric; that is, $\varepsilon_{xy} = \varepsilon_{yx}, \varepsilon_{yz} = \varepsilon_{zy}$ and $\varepsilon_{zx} = \varepsilon_{xz}$.

We can write all the components of extensional and shear strains as a symmetric second-order strain tensor

$$\varepsilon_{ij} = \begin{pmatrix} \varepsilon_{xx} & \varepsilon_{xy} & \varepsilon_{xz} \\ \varepsilon_{yx} & \varepsilon_{yy} & \varepsilon_{yz} \\ \varepsilon_{zx} & \varepsilon_{zy} & \varepsilon_{zz} \end{pmatrix} \tag{1.20}$$

Type of Forces in a Fluid Forces in a fluid can be categorized into **surface forces** and **body forces**. The **surface forces** are exerted by direct contact at normal as well as tangential to the surfaces. These forces include pressure force and viscous force. **Body forces** on the other hand are external forces developed without any body contact and are distributed over the entire volume of the fluid. These forces include gravitational forces, magnetic forces, etc. These forces are a direct function of mass. For example, the gravitational force that acts on a fluid element of volume $d\forall$ is

$$d\vec{F}_B = \rho \vec{g} d\forall \tag{1.21}$$

where

ρ = density of the fluid
g = acceleration due to gravity

Stress Field Stresses in a medium result from surface forces acting on some portion of the medium. The stress is defined as a force per unit area. Both force and area are vector quantities, and stress is a tensor quantity of second order containing nine components. Let us consider an area dA through a point A as shown in Figure 1.8.

The unit vector, $\hat{n}$, gives the orientation of the area. The force acting on the area can be resolved into two components: a normal component, δF_n, and a tangential component, δF_t. Normal and shear stresses now can be defined as

$$\text{Shear stress} = \lim_{\delta A_n \to 0} \frac{\delta F_n}{\delta A_n} \tag{1.22}$$

$$\text{Shear stress} = \lim_{\delta A_n \to 0} \frac{\delta F_t}{\delta A_n} \tag{1.23}$$

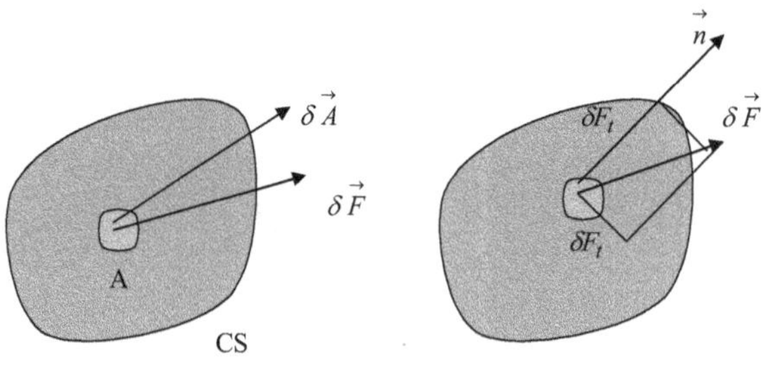

FIGURE 1.8 Forces on an arbitrary area.

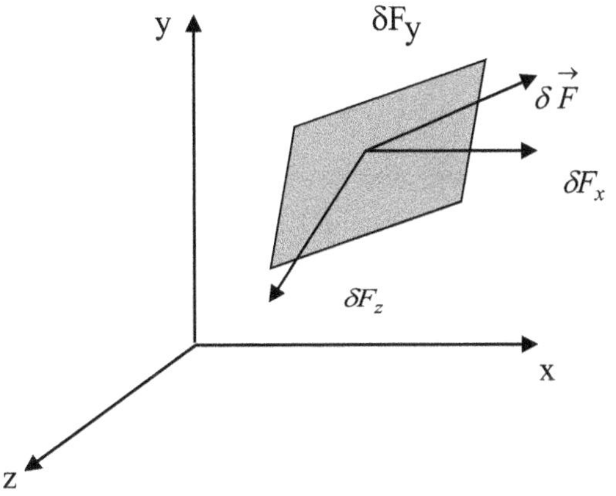

FIGURE 1.9 Forces on a positive x-plane.

Stress on a Plane Let us consider a differential force $\delta \vec{F}$ that acts on a positive x-plane of area δA_x as shown in Figure 1.9.

If we resolve the force δF into three components δF_x, δF_y, and δF_z in x-, y-, and z-directions, respectively, then the three stress components on the x-plane can be defined as

$$\sigma_{xx} = \lim_{\delta A_x \to 0} \frac{\delta F_x}{\delta A_x} = \text{normal stress on the } x \text{ plane in the } x \text{ direction} \tag{1.24a}$$

$$\tau_{xy} = \lim_{\delta A_x \to 0} \frac{\delta F_y}{\delta A_x} = \text{tangential or shear stress on the } x \text{ plane in the } y \text{ direction} \tag{1.24b}$$

$$\tau_{xz} = \lim_{\delta A_x \to 0} \frac{\delta F_z}{\delta A_x} = \text{tangential or shear stress on the } x \text{ plane in the } z \text{ direction} \tag{1.24c}$$

Similarly, we may have three component stresses σ_{yy}, τ_{yx}, and τ_{yz} on the y-plane of area δA_y, and σ_{zz}, τ_{zx}, and τ_{zy} on the z-plane of area δA_z.

Stress at a Point Stress at a point is given by nine stress components on three mutually perpendicular planes and can be expressed as a nine-component stress tensor

$$T_{ij} = \lim_{\delta A_i \to 0} \frac{\delta F_j}{\delta A_i}, \quad \text{where, } i = x, y \text{ and } z;\ j = x, y \text{ and } z \tag{1.25a}$$

or

$$T_{ij} = \begin{bmatrix} \sigma_{xx} & \tau_{xy} & \tau_{xz} \\ \tau_{yx} & \sigma_{yy} & \sigma_{yz} \\ \tau_{zx} & \tau_{zy} & \sigma_{zz} \end{bmatrix} \tag{1.25b}$$

Stresses on a Differential Element For a differential element, there are six surfaces (two x-planes, two y-planes, and two z-planes) on which stresses act, as shown in Figure 1.10.

All stresses are shown as positive. A stress component is positive when the direction of the stress component and the plane on which it acts are either both positive or both negative. For example, τ_{yx} represents shear stress on a positive y-plane acting in a positive x-direction or shear stress on a negative y-plane acting in a negative x-direction.

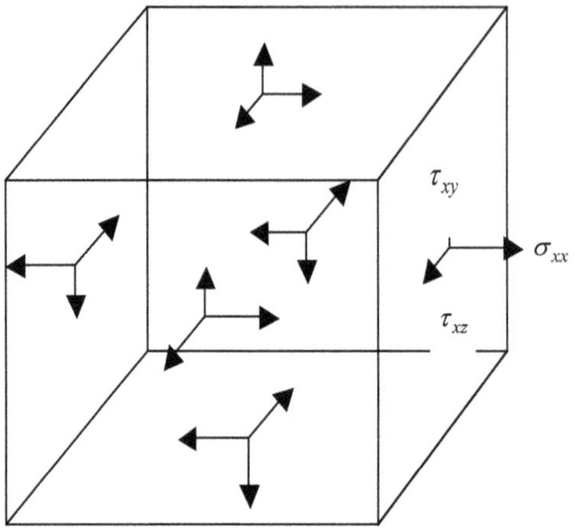

FIGURE 1.10 Stresses on a differential element.

1.2.2 Basic Equations in Integral Form

There are two basic approaches for analyzing fluid flow and heat transfer problems: **system** and **control volume**. In the following sections we will present basic equations for a system and for a control volume.

1.2.2.1 Basic Equations for a System

Conservation of Mass The conservation of mass for a system states that mass M of the system is constant, so we have

$$\left.\frac{\mathrm{d}M}{\mathrm{d}t}\right|_{\text{System}} = 0 \tag{1.26a}$$

where

$$M_{\text{System}} = \int_m \mathrm{d}m = \int_{\forall} \rho \mathrm{d}\forall \tag{1.26b}$$

Conservation of Momentum Newton's second law states that the time rate of change of momentum of a system is equal to the sum of all externally applied forces and is expressed as

$$\left.\frac{\mathrm{d}\vec{L}}{\mathrm{d}t}\right|_{\text{System}} = \vec{F} \tag{1.27a}$$

where the linear momentum $\left.\vec{L}\right|_{\text{System}}$ is given by

$$\left.\vec{L}\right|_{\text{System}} = \int_m \vec{V} \mathrm{d}m = \int_{\forall} \vec{V} \rho \mathrm{d}\forall \tag{1.27b}$$

and $\vec{F}$ includes all surface and body forces that act on the system and expressed as

$$\vec{F} = \vec{F}_S + \vec{F}_B \tag{1.27c}$$

The First Law of Thermodynamics – Conservation of Energy The first law of thermodynamics for a system undergoing a process is stated as

$$\left.\frac{dE}{dt}\right|_{System} = \pm\ \dot{Q} \pm\ \dot{W} + \dot{Q}_{gen} \tag{1.28a}$$

where $\dot{Q}$ is the heat transfer rate across the system boundary, $\dot{W}$ is the work done by the system on the surrounding or vice-versa, E is the total energy content of the system, and $\dot{Q}_{gen}$ is the energy generation in the system. The total energy is expressed as

$$E_{System} = \int_m e\,dm = \int_{\forall} e\rho\,d\forall \tag{1.28b}$$

and the specific energy e is a sum of specific internal energy (i), kinetic energy, and potential energy, and written as

$$e = i + \frac{V^2}{2} + gZ \tag{1.28c}$$

It can be noted that all conservation statements, given by Equations 1.26–1.28, involve a time derivative of an extensive property of the system (such as mass M, linear momentum $\vec{L}$ and energy E). These extensive properties can be represented by a single symbol, X, which can be expressed in a general form

$$X = \int_m x\,dm = \int_{\forall} x\rho\,d\forall \tag{1.29}$$

where x is the extensive property per unit mass, and equals 1 for the mass balance Equation 1.26, equals $\vec{V}$ for the momentum balance Equation 1.27, and equals e for the energy Equation 1.18.

The basic equation for the system analysis is converted into a control volume analysis by simply using the **Reynolds' transport equation** (Bennett and Myers, 1982), which is stated as

$$\left.\frac{dX}{dt}\right|_{System} = \left.\frac{\partial X}{\partial t}\right|_{CV} + \int_{CS} x\rho\vec{V}\cdot d\vec{A} \tag{1.30}$$

where the first term on the right-hand side of the equation represents the rate of change of the extensive property X within the system. The second term represents the net rate of flow of the extensive property through the control volume surface. Using **Reynold's transport equation**, we can get all the basic conservation equations for a control volume as summarized below.

1.2.2.2 Basic Equations for a Control Volume

Conservation of Mass Using $X = M$ and $x = 1$, we get the conservation of mass equation for a control volume as

$$\frac{\partial}{\partial t}\left(\int_{CV} \rho\,d\forall\right) + \int_{CS} \rho\vec{V}\cdot d\vec{A} = 0 \tag{1.31a}$$

For steady flow

$$\int_{CS} \rho\vec{V}\cdot d\vec{A} = 0 \tag{1.31b}$$

For steady flows, the net mass flow rates into a control volume must be equal to the net mass flow out of the control volume.

For incompressible flow in a nondeformable control volume

$$\int_{CS} \vec{V}\cdot d\vec{A} = 0 \tag{1.31c}$$

In incompressible flows, the net volume flow rate into a control volume must be equal to the net volume flow rate out of the control volume.

Conservation of Momentum Substituting $X = \vec{L}$ and $x = \vec{V}$ into Equation 1.29, we obtain the conservation of momentum for a control volume from Equations 1.27 and 1.30 as

$$\frac{\partial}{\partial t}\left(\int_{\forall} \vec{V}\rho d\forall\right) + \int_{CS} \vec{V}\rho\vec{V}\cdot d\vec{A} = \sum \vec{F}_S + \sum \vec{F}_B \tag{1.32}$$

where

$$\sum \vec{F}_S = \text{sum of all surface forces}$$

$$\sum \vec{F}_B = \text{sum of all body forces}$$

The momentum equation can be written in three component scalar equations in different coordinate systems. For example, in Cartesian coordinate system, we can write

$$x\text{-component}: \quad \frac{\partial}{\partial t}\left(\int_{\forall} u\rho d\forall\right) + \int_{CS} u\rho\vec{V}\cdot d\vec{A} = \sum \vec{F}_{Sx} + \sum \vec{F}_{Bx} \tag{1.33a}$$

$$y\text{-component}: \quad \frac{\partial}{\partial t}\left(\int_{\forall} v\rho d\forall\right) + \int_{CS} v\rho\vec{V}\cdot d\vec{A} = \sum \vec{F}_{Sy} + \sum \vec{F}_{By} \tag{1.33b}$$

$$z\text{-component}: \quad \frac{\partial}{\partial t}\left(\int_{\forall} w\rho d\forall\right) + \int_{CS} w\rho\vec{V}\cdot d\vec{A} = \sum \vec{F}_{Sz} + \sum \vec{F}_{Bz} \tag{1.33c}$$

The first term on the left-hand side represents the rate of change of momentum inside the control volume. The second term is the net rate of momentum flux through the control volume. The sum of these two terms is equal to the sum of all surface and body forces acting on the control volume.

The First Law of Thermodynamics Substituting $X = E$ and $x = e$, we obtain the conservation of energy for a control volume from Equations 1.28 1.30 as

$$\frac{\partial}{\partial t}\left(\int_{\forall} e\rho d\forall\right) + \int_{CS} e\rho\vec{V}\cdot d\vec{A} = \pm\, \dot{Q} \pm \dot{W} + \dot{Q}_{gen} \tag{1.34}$$

1.2.3 Differential Analysis of Fluid Motion

To obtain detailed point-to-point knowledge, we need to have the basic equations in differential form. These basic differential equations can be obtained by considering infinitesimal systems and control volume, and applying conservation principles. A good understanding of these equations is important because they give many insights or explain physics clearly. Detailed discussions on differential analysis of fluid motions are given in the fluid mechanics books by Fox and McDonald (1998), White (1988), and Schlichting (2000). A brief summary of these equations is given in the following section.

1.2.3.1 Conservation of Mass

Cartesian (*x*, *y*, *z*) Coordinate System The differential equation for conservation of mass, or the continuity equation, is derived by considering a cubic infinitesimal control volume ΔV surrounding the point p, as shown in Figure 1.11. At point p, the density is assumed to be ρ and the velocity is given by $\vec{V} = \hat{i}u + \hat{j}v + \hat{k}w$.

Considering the mass balance over a control volume of infinitesimal size, the conservation of mass or continuity equation is given as

$$\frac{\partial \rho}{\partial t} + \frac{\partial(\rho u)}{\partial x} + \frac{\partial(\rho v)}{\partial y} + \frac{\partial(\rho w)}{\partial z} = 0 \tag{1.35a}$$

or

$$\frac{\partial \rho}{\partial t} + \nabla.\rho\vec{V} = 0 \tag{1.35b}$$

Equation 1.35 is simplified for special cases as follows.

Incompressible Flow For incompressible flow, density is constant, i.e., ρ = constant, and Equation 1.35 reduces to

$$\frac{\partial u}{\partial x} + \frac{\partial v}{\partial y} + \frac{\partial w}{\partial z} = 0 \tag{1.36a}$$

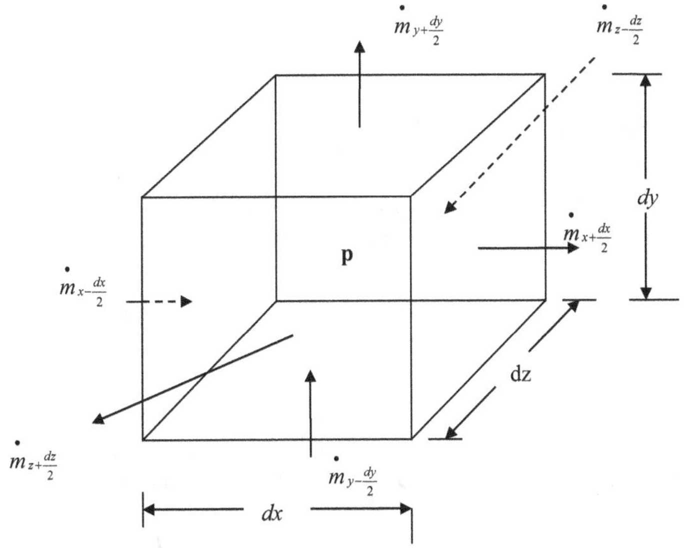

FIGURE 1.11 Differential control volume in Cartesian co-ordinate.

or

$$\nabla \cdot \vec{V} = 0 \tag{1.36b}$$

We have indicated before that the three derivative terms of Equation 1.36a represent the extensional strain in the x-, y-, and z-directions. So, the conservation of mass equation simply states that the net dilation of the fluid element as zero.

Steady Flow For steady flow the density is independent of time, and so conservation of mass equation transforms into

$$\frac{\partial(\rho u)}{\partial x} + \frac{\partial(\rho v)}{\partial y} + \frac{\partial(\rho w)}{\partial z} = 0 \tag{1.37a}$$

or

$$\nabla \cdot \rho \vec{V} = 0 \tag{1.37b}$$

Cylindrical Coordinate System In a cylindrical coordinate system, we consider a cylindrical infinitesimal control volume where the mass conservation equation is given as

$$\frac{\partial \rho}{\partial t} + \frac{\rho u_r}{r} + \frac{\partial(\rho u_r)}{\partial r} + \frac{1}{r}\frac{\partial(\rho u_\theta)}{\partial \theta} + \frac{\partial(\rho u_z)}{\partial z} = 0 \tag{1.38a}$$

or

$$\frac{\partial \rho}{\partial t} + \frac{1}{r}\frac{\partial(\rho r u_r)}{\partial r} + \frac{1}{r}\frac{\partial(\rho u_\theta)}{\partial \theta} + \frac{\partial(\rho u_z)}{\partial z} = 0 \tag{1.38b}$$

Incompressible Flow

$$\frac{1}{r}\frac{\partial(r u_r)}{\partial r} + \frac{1}{r}\frac{\partial(u_\theta)}{\partial \theta} + \frac{\partial(u_z)}{\partial z} = 0 \tag{1.39}$$

Steady Flow

$$\frac{1}{r}\frac{\partial(r\rho u_r)}{\partial r} + \frac{1}{r}\frac{\partial(\rho u_\theta)}{\partial \theta} + \frac{\partial(\rho u_z)}{\partial z} = 0 \tag{1.40}$$

Spherical Coordinate System In a spherical coordinate system, we consider an infinitesimal control volume where the mass conservation equation is given as

$$\frac{\partial \rho}{\partial t} + \frac{1}{r^2}\frac{\partial(\rho r^2 u_r)}{\partial r} + \frac{1}{r\sin\phi}\frac{\partial(\rho u_\phi \sin\phi)}{\partial \phi} + \frac{1}{r\sin\phi}\frac{\partial(\rho u_\theta)}{\partial \theta} = 0 \tag{1.41}$$

Conservation of Momentum The conservation of momentum equation is derived from Newton's second law of motion, which expresses proportionality between applied force and resulting acceleration of a particle. So, the conservation of momentum for an infinitesimal fluid mass is

$$\Delta\vec{F} = \Delta m\vec{a} = \Delta m\frac{D\vec{V}}{Dt} \tag{1.42}$$

or

$$\Delta\vec{F}_B + \Delta\vec{F}_S = \rho\frac{D\vec{V}}{Dt}\Delta\forall \tag{1.43}$$

where $\Delta\vec{F}$ is the net applied force on the fluid element composed of two types of force, net body force $\Delta\vec{F}_B$ and net surface force $\Delta\vec{F}_S$. $\vec{a}$ is the acceleration of the fluid element and is derived in the Eulerian system as

$$\vec{a} = \frac{D\vec{v}}{Dt} = \frac{\partial\vec{V}}{\partial t} + u\frac{\partial\vec{V}}{\partial x} + v\frac{\partial\vec{V}}{\partial y} + w\frac{\partial\vec{V}}{\partial z} \tag{1.44}$$

The body force has three components written as

$$\Delta\vec{F}_B = \text{bodyforce} = \hat{i}\Delta F_{Bx} + \hat{j}\Delta F_{By} + \hat{j}\Delta F_{Bz} \tag{1.45}$$

Assuming that the body force is due to gravity only, we can express as

$$\Delta\vec{F}_B = \left(\hat{i}\rho g_x + \hat{j}\rho g_y + \hat{k}\rho g_z\right)\Delta\forall = \rho\vec{g}\Delta\forall \tag{1.46}$$

The surface force $\Delta\vec{F}_s$ is computed from the symmetric stress tensor given by Equation 1.25. For example, the estimation of net surface force in the x-direction is obtained by considering all x-component stresses on six surfaces of the fluid element as shown in Figure 1.12. If the x-component stresses at the center of the differential element are taken to be σ_{xx}, τ_{yx}, and τ_{zx}, then stresses acting in the x-direction on each face of the element are derived from a truncated Taylor series expansion as demonstrated in Figure 1.12.

Adding all forces in the x-direction, we get the net x-component force

$$\mathrm{d}F_{sx} = \left(\frac{\partial\sigma_{xx}}{\partial x} + \frac{\partial\tau_{yx}}{\partial y} + \frac{\partial\tau_{zx}}{\partial z}\right)\mathrm{d}x\mathrm{d}y\mathrm{d}z \tag{1.47a}$$

Similarly, the net y-component and z-component surface forces are given as

$$\mathrm{d}F_{sy} = \left(\frac{\partial\tau_{yx}}{\partial x} + \frac{\partial\sigma_{yy}}{\partial y} + \frac{\partial\tau_{yz}}{\partial z}\right)\mathrm{d}x\mathrm{d}y\mathrm{d}z \tag{1.47b}$$

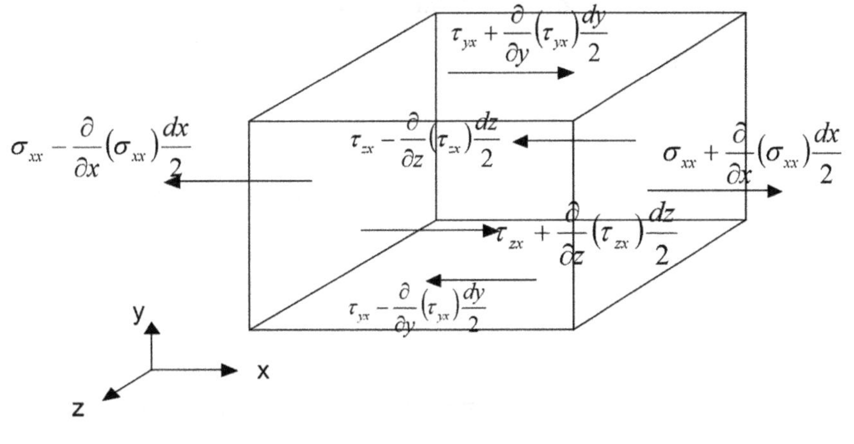

FIGURE 1.12 x-component stresses on a fluid element.

$$\mathrm{d}F_{sz} = \left(\frac{\partial \tau_{zx}}{\partial x} + \frac{\partial \tau_{zy}}{\partial y} + \frac{\partial \sigma_{zz}}{\partial z} \right) \mathrm{d}x\mathrm{d}y\mathrm{d}z \tag{1.47c}$$

Substituting Equations 1.45–1.47 into Equation 1.44 and dividing through by the volume $\mathrm{d}\forall = \mathrm{d}x\mathrm{d}y\mathrm{d}z$, we obtain the equation of motion in component form as

x-momentum

$$\rho \left(\frac{\partial u}{\partial t} + u\frac{\partial u}{\partial x} + v\frac{\partial u}{\partial y} + w\frac{\partial u}{\partial z} \right) = \rho g_x + \left(\frac{\partial \sigma_{xx}}{\partial x} + \frac{\partial \tau_{yx}}{\partial y} + \frac{\partial \tau_{zx}}{\partial z} \right) \tag{1.48}$$

y-momentum

$$\rho \left(\frac{\partial v}{\partial t} + u\frac{\partial v}{\partial x} + v\frac{\partial v}{\partial y} + w\frac{\partial v}{\partial z} \right) = \rho g_y + \left(\frac{\partial \tau_{yx}}{\partial x} + \frac{\partial \sigma_{yy}}{\partial y} + \frac{\partial \tau_{yz}}{\partial z} \right) \tag{1.48b}$$

z-momentum

$$\rho \left(\frac{\partial w}{\partial t} + u\frac{\partial w}{\partial x} + v\frac{\partial w}{\partial y} + w\frac{\partial w}{\partial z} \right) = \rho g_z + \left(\frac{\partial \tau_{zx}}{\partial x} + \frac{\partial \tau_{zy}}{\partial y} + \frac{\partial \sigma_{zz}}{\partial z} \right) \tag{1.48c}$$

Equation 1.48 is applicable to the motion of any fluid that satisfies the continuum assumption. In order to solve this equation for the flow field or flow properties, it is necessary to express the stresses in terms of measurable quantities such as velocity vector $\vec{V}$ and its derivatives. This is done through an established relationship between stress (T_{ij}) and deformation rate or strain rate (ε_{ij}), which is known as **Stokes viscous deformation law** (Schlichting and Gersten, 2000; White, 1988).

Stokes Viscous Deformation Law The viscous deformation law for Newtonian fluids was first postulated by Stokes in 1845. Basic assumptions made in deriving this relationship are: (a) fluid is isotropic and continuous, (b) the shear stress–strain rate relation is **linear**, and (c) the stress equals the hydrostatic pressure *P* for cases with zero strain rate. Stokes postulates lead to the following deformation law.

Normal Stresses

$$\sigma_{xx} = -p + 2\mu\varepsilon_{xx} - \frac{2}{3}\mu\nabla\cdot\vec{V} \tag{1.49a}$$

$$\sigma_{yy} = -p + 2\mu\varepsilon_{yy} - \frac{2}{3}\mu\nabla\cdot\vec{V} \tag{1.49b}$$

$$\sigma_{zz} = -p + 2\mu\varepsilon_{zz} - \frac{2}{3}\mu\nabla\cdot\vec{V} \tag{1.49c}$$

Shear Stresses

$$\tau_{yx} = \tau_{xy} = 2\mu\varepsilon_{xy} \tag{1.50a}$$

$$\tau_{zx} = \tau_{xz} = 2\mu\varepsilon_{xz} \tag{1.50b}$$

$$\tau_{yz} = \tau_{zy} = 2\mu\varepsilon_{yz} \tag{1.50c}$$

Using Equations 1.18 and 1.19 for strain rates, we can write the deformation law as

$$\tau_{xy} = \tau_{yx} = \mu\left(\frac{\partial v}{\partial x} + \frac{\partial u}{\partial y}\right) \tag{1.51a}$$

$$\tau_{yz} = \tau_{zy} = \mu\left(\frac{\partial w}{\partial y} + \frac{\partial v}{\partial z}\right) \tag{1.51b}$$

$$\tau_{zx} = \tau_{xz} = \mu\left(\frac{\partial u}{\partial z} + \frac{\partial w}{\partial x}\right) \tag{1.51c}$$

and

$$\sigma_{xx} = -p - \frac{2}{3}\mu\nabla\cdot\vec{V} + 2\mu\frac{\partial u}{\partial x} \tag{1.51d}$$

$$\sigma_{yy} = -p - \frac{2}{3}\mu\nabla\cdot\vec{V} + 2\mu\frac{\partial v}{\partial y} \tag{1.51e}$$

$$\sigma_{zz} = -p - \frac{2}{3}\mu\nabla\cdot\vec{V} + 2\mu\frac{\partial w}{\partial z} \tag{1.51f}$$

The Navier–Stokes Equations Substituting deformation law (1.51) into equation of motion (1.48), we get the **Navier–Stokes equations**, which represent the momentum or equation of motion for Newtonian viscous fluid

$$\rho\frac{Du}{Dt} = \rho g_x - \frac{\partial p}{\partial x} + \frac{\partial}{\partial x}\left[\mu\left(2\frac{\partial u}{\partial x} - \frac{2}{3}\nabla\cdot\vec{V}\right)\right] + \frac{\partial}{\partial y}\left[\mu\left(\frac{\partial u}{\partial y} + \frac{\partial v}{\partial x}\right)\right] + \frac{\partial}{\partial z}\left[\mu\left(\frac{\partial w}{\partial x} + \frac{\partial u}{\partial z}\right)\right] \tag{1.52a}$$

$$\rho\frac{Dv}{Dt} = \rho g_y - \frac{\partial p}{\partial y} + \frac{\partial}{\partial x}\left[\mu\left(\frac{\partial u}{\partial y} + \frac{\partial v}{\partial x}\right)\right] + \frac{\partial}{\partial y}\left[\mu\left(2\frac{\partial v}{\partial y} - \frac{2}{3}\nabla\cdot\vec{V}\right)\right] + \frac{\partial}{\partial z}\left[\mu\left(\frac{\partial v}{\partial z} + \frac{\partial w}{\partial y}\right)\right] \tag{1.52b}$$

$$\rho\frac{Dw}{Dt} = \rho g_z - \frac{\partial p}{\partial z} + \frac{\partial}{\partial x}\left[\mu\left(\frac{\partial w}{\partial x} + \frac{\partial u}{\partial z}\right)\right] + \frac{\partial}{\partial y}\left[\mu\left(\frac{\partial v}{\partial z} + \frac{\partial w}{\partial y}\right)\right] + \frac{\partial}{\partial z}\left[\mu\left(2\frac{\partial w}{\partial z} - \frac{2}{3}\nabla\cdot\vec{V}\right)\right] \tag{1.52c}$$

Incompressible Fluids For incompressible fluids, density is constant, and conservation of mass equation simplifies to

$$\frac{\partial u}{\partial x} + \frac{\partial v}{\partial y} + \frac{\partial w}{\partial z} = 0 \tag{1.53a}$$

or

$$\nabla\cdot\vec{V} = 0 \tag{1.53b}$$

and substituting $\nabla\cdot\vec{V} = 0$ into Equation 1.52, momentum equations reduces to

$$\rho\left(\frac{\partial u}{\partial t} + u\frac{\partial u}{\partial x} + v\frac{\partial u}{\partial y} + w\frac{\partial u}{\partial z}\right) = \rho g_x - \frac{\partial p}{\partial x} + \frac{\partial}{\partial x}\left(\mu\frac{\partial u}{\partial x}\right) + \frac{\partial}{\partial y}\left(\mu\frac{\partial u}{\partial y}\right) + \frac{\partial}{\partial z}\left(\mu\frac{\partial u}{\partial z}\right) \tag{1.54a}$$

$$\rho\left(\frac{\partial v}{\partial t}+u\frac{\partial v}{\partial x}+v\frac{\partial v}{\partial y}+w\frac{\partial v}{\partial z}\right)=\rho g_y-\frac{\partial p}{\partial y}+\frac{\partial}{\partial x}\left(\mu\frac{\partial v}{\partial x}\right)+\frac{\partial}{\partial y}\left(\mu\frac{\partial v}{\partial y}\right)+\frac{\partial}{\partial z}\left(\mu\frac{\partial v}{\partial z}\right) \quad (1.54b)$$

$$\rho\left(\frac{\partial w}{\partial t}+u\frac{\partial w}{\partial x}+v\frac{\partial w}{\partial y}+w\frac{\partial w}{\partial z}\right)=\rho g_z-\frac{\partial p}{\partial y}+\frac{\partial}{\partial x}\left(\mu\frac{\partial w}{\partial x}\right)+\frac{\partial}{\partial y}\left(\mu\frac{\partial w}{\partial y}\right)+\frac{\partial}{\partial z}\left(\mu\frac{\partial w}{\partial z}\right) \quad (1.54c)$$

Further assuming a fluid with constant viscosity (μ), Equation 1.52 reduces to

$$\rho\left(\frac{\partial u}{\partial t}+u\frac{\partial u}{\partial x}+v\frac{\partial u}{\partial y}+w\frac{\partial u}{\partial z}\right)=\rho g_x-\frac{\partial p}{\partial x}+\mu\left(\frac{\partial^2 u}{\partial x^2}+\frac{\partial^2 u}{\partial y^2}+\frac{\partial^2 u}{\partial z^2}\right) \quad (1.55a)$$

$$\rho\left(\frac{\partial v}{\partial t}+u\frac{\partial v}{\partial x}+v\frac{\partial v}{\partial y}+w\frac{\partial v}{\partial z}\right)=\rho g_y-\frac{\partial p}{\partial y}+\mu\left(\frac{\partial^2 v}{\partial x^2}+\frac{\partial^2 v}{\partial y^2}+\frac{\partial^2 v}{\partial z^2}\right) \quad (1.55b)$$

$$\rho\left(\frac{\partial w}{\partial t}+u\frac{\partial w}{\partial x}+v\frac{\partial w}{\partial y}+w\frac{\partial w}{\partial z}\right)=\rho g_z-\frac{\partial p}{\partial z}+\mu\left(\frac{\partial^2 w}{\partial x^2}+\frac{\partial^2 w}{\partial y^2}+\frac{\partial^2 w}{\partial z^2}\right) \quad (1.55c)$$

and in vector notation

$$\rho\frac{D\vec{V}}{dt}=\rho\vec{g}-\nabla p+\mu\nabla^2\vec{V} \quad (1.55d)$$

It can be mentioned here that Equation 1.54 is not suitable for viscous-incompressible nonisothermal flow, particularly for liquids for which viscosity is a strong function temperature, and hence a function of space co-ordinates.

Frictionless Flow or Inviscid Flow For frictionless flow, viscosity $\mu = 0$ and flow is defined as inviscid flow. Substituting $\mu = 0$, Equation 1.54 gives the momentum equation for inviscid flow as

$$\rho\frac{D\vec{V}}{Dt}=\rho\vec{g}-\nabla p \quad (1.56)$$

Equation 1.55 is also known as the **Euler equation.**

General Form Incompressible Momentum Equations Let us also present here a general form of the incompressible momentum equation to facilitate our presentation of the finite difference/control volume method and finite element method in later chapters.

x-momentum

$$\rho\left(\frac{\partial u}{\partial t}+u\frac{\partial u}{\partial x}+v\frac{\partial u}{\partial y}+w\frac{\partial u}{\partial z}\right)=\frac{\partial}{\partial x}\left(\mu\frac{\partial u}{\partial x}\right)+\frac{\partial}{\partial y}\left(\mu\frac{\partial u}{\partial y}\right)+\frac{\partial}{\partial x}\left(\mu\frac{\partial u}{\partial z}\right)+S_u \quad (1.57a)$$

y-momentum

$$\rho\left(\frac{\partial v}{\partial t}+u\frac{\partial v}{\partial x}+v\frac{\partial v}{\partial y}+w\frac{\partial v}{\partial z}\right)=\frac{\partial}{\partial x}\left(\mu\frac{\partial v}{\partial x}\right)+\frac{\partial}{\partial x}\left(\mu\frac{\partial v}{\partial y}\right)+\frac{\partial}{\partial z}\left(\mu\frac{\partial v}{\partial z}\right)+S_v \quad (1.57b)$$

z-momentum

$$\rho\left(\frac{\partial w}{\partial t}+u\frac{\partial w}{\partial x}+v\frac{\partial w}{\partial y}+w\frac{\partial w}{\partial z}\right)=\frac{\partial}{\partial x}\left(\mu\frac{\partial w}{\partial x}\right)+\frac{\partial}{\partial x}\left(\mu\frac{\partial w}{\partial y}\right)+\frac{\partial}{\partial z}\left(\mu\frac{\partial w}{\partial z}\right)+S_w \quad (1.57c)$$

where $S_u = \frac{\partial P}{\partial x} + \rho g_x + S_{u0}$, $S_v = \frac{\partial P}{\partial y} + \rho g_y + S_{v0}$ and $S_w = \frac{\partial P}{\partial z} + \rho g_z + S_{w0}$ represents the source terms in the momentum equations taking in account of the pressure force term, gravitational force term, and any additional force terms.

Turbulent Fluid Flow Equations

While a comprehensive discussion of turbulent fluid flow modeling and computations is given in Chapter 10, we present here transport equations for one of the widely used turbulence flow model known as the two-equation $\kappa - \varepsilon$ model. This two-equation turbulence model includes **two transport equations**: a transport equation for the **turbulent kinetic energy**, κ, and the transport equation for the **turbulent dissipation**, ε. The transport equations for the $\kappa - \varepsilon$ turbulence model are given as follows:

Turbulence Kinetic Energy Transport

$$\rho\left(\frac{\partial \kappa}{\partial t} + u\frac{\partial \kappa}{\partial x} + v\frac{\partial \kappa}{\partial y} + w\frac{\partial \kappa}{\partial z}\right) = \frac{\partial}{\partial x}\left(\mu + \frac{\mu_t}{\sigma_\kappa}\frac{\partial \kappa}{\partial x}\right) + \frac{\partial}{\partial y}\left(\mu + \frac{\mu_t}{\sigma_\kappa}\frac{\partial \kappa}{\partial y}\right) + \frac{\partial}{\partial z}\left(\mu + \frac{\mu_t}{\sigma_\kappa}\frac{\partial \kappa}{\partial z}\right) + S_\kappa \quad (1.58a)$$

where S_κ is the turbulence kinetic energy source given as

$$S_\kappa = P - \varepsilon \quad (1.58b)$$

$$\mathrm{P} = 2\mu_t\left[\left(\frac{\partial u}{\partial x}\right)^2 + \left(\frac{\partial v}{\partial y}\right)^2 + \left(\frac{\partial w}{\partial z}\right)^2\right] + \mu_t\left[\left(\frac{\partial u}{\partial y} + \frac{\partial v}{\partial x}\right)^2 + \left(\frac{\partial v}{\partial z} + \frac{\partial w}{\partial y}\right)^2 + \left(\frac{\partial w}{\partial x} + \frac{\partial u}{\partial z}\right)^2\right] \quad (1.58c)$$

Energy Dissipation Rate Transport

$$\rho\left(\frac{\partial \varepsilon}{\partial t} + u\frac{\partial \varepsilon}{\partial x} + v\frac{\partial \varepsilon}{\partial y} + w\frac{\partial \varepsilon}{\partial z}\right) = \frac{\partial}{\partial x}\left(\mu + \frac{\mu_t}{\sigma_\varepsilon}\frac{\partial \varepsilon}{\partial x}\right) + \frac{\partial}{\partial y}\left(\mu + \frac{\mu_t}{\sigma_\varepsilon}\frac{\partial \varepsilon}{\partial y}\right) + \frac{\partial}{\partial z}\left(\mu + \frac{\mu_t}{\sigma_\varepsilon}\frac{\partial \kappa}{\partial z}\right) + S_\varepsilon \quad (1.59a)$$

where S_ε is the source term for the turbulence energy dissipation rate and expressed as

$$s_\varepsilon = \frac{\varepsilon}{\kappa}(C_{\varepsilon 1}P - C_{\varepsilon 2}\varepsilon) \quad (1.59b)$$

The turbulent viscosity, μ_t, is computed by combining k and ε as

$$\mu_t = \rho C_\mu \frac{k^2}{\varepsilon} \quad (1.60)$$

1.2.4 Boundary Conditions for Flow Field

Solution of conservation of mass and momentum equations require mathematically tenable and physically realistic boundary conditions, which represent the physical conditions at extreme values of the independent variables. A detailed description of hydrodynamic boundary conditions is given in Goldstein (1965), White (1988), Panton (1984), and Schlichting and Gersten (2000). The following are the most commonly used boundary conditions.

1.2.4.1 Solid–Fluid Interface

Tangential Component Velocity We normally assume that at the solid–fluid interface there is no relative speed difference or finite discontinuities of velocities between the solid surface and the fluid in contact. So, the fluid velocity equals the velocity of the solid surface with which it moves. Such a condition is known as **no-slip condition**. This is expressed mathematically as

$$u = V_{wx} \quad \text{and} \quad w = V_{wz} \tag{1.61a}$$

where V_{wx} and V_{wz} are the x- and z-components velocity of the solid wall.

For more discussions on the validity of no-slip conditions and use of slip conditions, refer to White (1988).

For situations where fluid is moving adjacent to a stationary solid surface, all tangential velocity components are assumed as zero and written as

$$u = 0 \text{ and } w = 0 \tag{1.61b}$$

This boundary condition with assigned known velocity component is mathematically referred to as the **Dirichlet condition.**

Normal Component Velocity The velocity component v normal to the wall becomes zero for an impermeable wall. However, it becomes nonzero for a permeable or porous wall through which fluid is sucked in or blown out. For example, for a wall with flow suction, this condition is stated mathematically as

$$v = v_0 \tag{1.62}$$

where v_0 is the normal velocity at the surface.

1.2.4.2 Fluid–Fluid Interface

Liquid–Liquid Interface Consider two liquids in contact as shown in Figure 1.13.

The velocity as well the shear stress are continuous across the interface and these are stated as follows.

No-slip condition for axial component:

$$u_1 = u_2 \tag{1.63a}$$

Continuity in normal component velocity:

$$v_1 = v_2 \tag{1.63b}$$

Continuity in shear stress:

$$\tau_1 = \tau_2 \tag{1.63c}$$

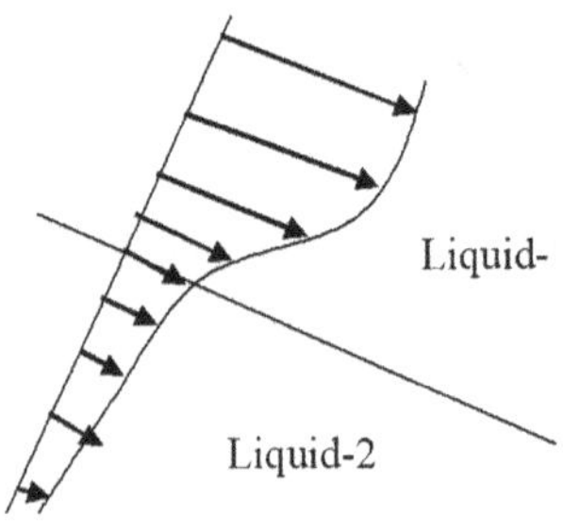

FIGURE 1.13 Liquid–liquid interface.

or

$$\mu_1 \frac{\partial u_1}{\partial n} = \mu_2 \frac{\partial u_2}{\partial n} \tag{1.63d}$$

It can be noted that, for dissimilar viscosity, there exists a discontinuity in the velocity gradients.

Liquid–Gas Interface At liquid–gas interfaces the momentum flux and hence the velocity gradient in the liquid are negligibly small compared to that in the gas phase since the viscosity of the liquid is much greater than that of the gas (Figure 1.14).

So, the interface boundary condition for the liquid phase is stated as

$$\frac{\partial u_l}{\partial n} = 0 \tag{1.64}$$

This zero-velocity gradient condition is mathematically referred to as the **Newmann boundary condition**, which prescribes the gradient normal of a variable at the boundary as constant.

Deformable Solid–Fluid Interface. There are many problems in which the solid is coupled to the fluid through deformation and dynamic interaction. In such cases the equation of motion for solving the fluid and that of the solid is solved simultaneously while satisfying continuity of velocity and stress at the interface, which is defined as

$$\tau_f = \tau_s \tag{1.65}$$

where τ_f is the fluid stress given, for example, by Newtonian fluid behavior and τ_s is the solid stress given, for example, by Hook's law.

Inlet Boundary Conditions

Inlet boundaries are specified at some parts of the boundary where the fluid enters the solution domain. At the inlet boundaries, known velocity and scalar quantities such as temperature, mass species concentration, and turbulence quantities can be specified.

Outlet Boundary Conditions

Outlet boundaries are specified at some parts of the boundary where the fluid leaves the solution domain.

Free-Stream Boundary Conditions

Free-stream boundary represents the conditions at far-field regions.

Pressure Boundary Conditions

Pressure boundary conditions can be specified at boundaries where the pressure values or pressure distributions are known.

For more discussions on boundary and inlet conditions, details are given in Chapter 9.

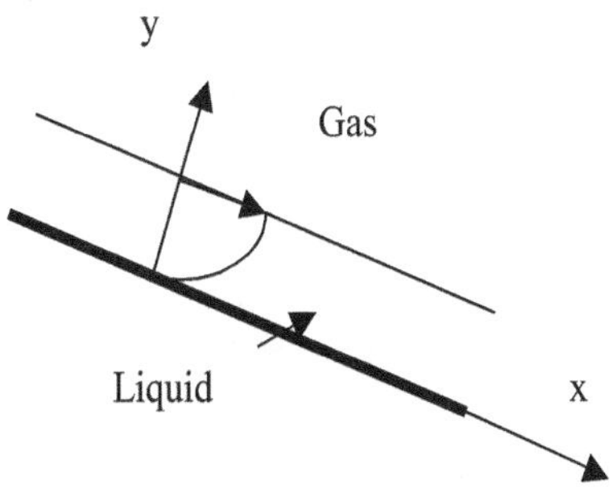

FIGURE 1.14 Liquid–gas interfaces.

1.3 HEAT TRANSFER

Heat transfer is a science that determines how and at what rate heat energy is transferred as a result of a temperature gradient or difference. There are three basic modes of heat transfer: conduction, convection, and radiation.

1.3.1 Basic Modes and Transport Rate Equation

Conduction Heat Transfer This mode is important primarily for heat transfer in solid and stationary fluid.

Conduction Rate Equation The conduction rate equation is governed by Fourier's law, which states that, in a homogeneous substance, the local heat flux is proportional to the local temperature gradient. For one-dimensional heat flow in x-direction, such as in a plane slab (Figure 1.15), this law is written as

$$\frac{q}{A} = -\frac{\mathrm{d}T}{\mathrm{d}x} \tag{1.66a}$$

Introducing the constant of proportionality

$$q'' = -k\frac{\mathrm{d}T}{\mathrm{d}x} \tag{1.66b}$$

where $q'' = q/A$ = heat flow per unit area per unit time or heat flux, and k is the thermal conductivity of the material. The negative sign is included to ensure that heat flows in the direction of decrease in temperature.

Similarly, we can write the heat conduction rate equation in y-and z-directions. In general, the heat flux is a vector quantity and is expressed as

$$\vec{q} = -k\nabla T \tag{1.67a}$$

where Equation 1.65 can be expressed in different coordinate systems as follows.

Cartesian Coordinate

$$\vec{q} = -k\nabla T = \hat{i}q''_x + \hat{j}q''_y + \hat{k}q''_z \tag{1.67b}$$

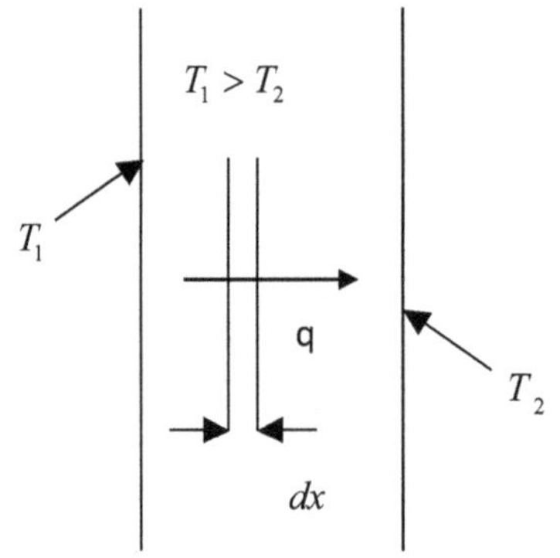

FIGURE 1.15 Heat conduction through a plane wall.

where

$$q_x'' = -k\frac{\partial T}{\partial x},\quad q_y'' = -k\frac{\partial T}{\partial y}\quad \text{and}\quad q_z'' = -k\frac{\partial T}{\partial z}$$

Cylindrical Coordinate

$$\vec{q} = -k\nabla T = \hat{i}q_r'' + \hat{j}q_\theta'' + \hat{k}q_z'' \tag{1.67c}$$

where

$$q_r'' = -k\frac{\partial T}{\partial r},\quad q_\theta'' = -k\frac{1}{r}\frac{\partial T}{\partial \theta}\quad \text{and}\quad q_z'' = -k\frac{\partial T}{\partial z}$$

Spherical Coordinate

$$\vec{q} = -k\nabla T = \hat{i}_r q_r'' + \hat{i}_\phi q_\phi'' + \hat{i}_\theta q_\theta'' \tag{1.67d}$$

where

$$q_r'' = -k\frac{\partial T}{\partial r},\quad q_\phi'' = -k\frac{1}{r}\frac{\partial T}{\partial \phi}\quad \text{and}\quad q_\theta'' = -k\frac{1}{r\sin\phi}\frac{\partial T}{\partial \theta}$$

We can also express the heat flux normal to any arbitrary surface (Figure 1.16) as

$$q_n'' = \vec{q}''\cdot\hat{n} = -k\frac{\partial T}{\partial n} \tag{1.68}$$

Also, the total heat flow through a surface area A_n is estimated by integrating the local heat transfer rate over the entire area as

$$q_n = \int_{A_n} q_n'' \mathrm{d}A_n = -\int_{A_n} k\frac{\partial T}{\partial n}\mathrm{d}A_n \tag{1.69}$$

For uniform heat flux over the area, Equation 1.70 reduces to

$$q_n = -kA_n\frac{\partial T}{\partial n} \tag{1.70}$$

The **Fourier** law given by Equation 1.70 is valid for an **isotropic** material for which the thermal conductivity is directionally independent. In an **anisotropic** medium, thermal conductivity depends on coordinate directions. One special category of this is the orthotropic medium such as in laminated composites, and the heat flux vector given by Equation 1.71 for Cartesian co-ordinate system is written as

$$\vec{q} = -\left(\hat{i}k_x\frac{\partial T}{\partial x} + \hat{j}k_y\frac{\partial T}{\partial y} + \hat{k}k_z\frac{\partial T}{\partial z}\right) \tag{1.71}$$

where the heat flux components are

$$q_x' = -k_x\frac{\partial T}{\partial x},\quad q_y' = -k_y\frac{\partial T}{\partial y}\quad \text{and}\quad q_z' = -k_z\frac{\partial T}{\partial z} \tag{1.72}$$

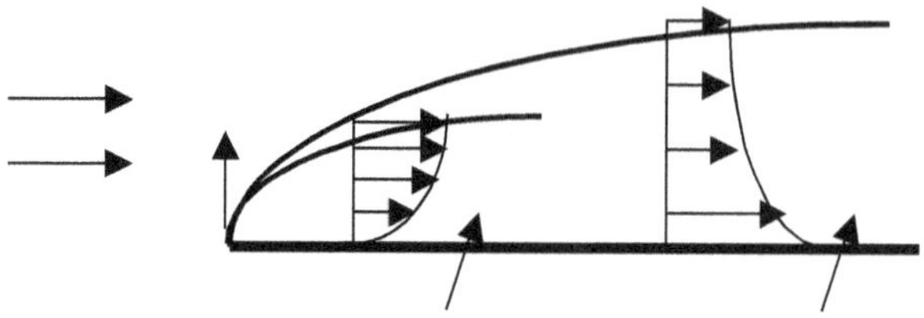

FIGURE 1.16 Hydrodynamic and thermal boundary layers for flow over a solid surface.

Convection Heat Transfer The convection heat transfer occurs between a moving fluid and an exposed solid surface. Let us consider the fluid flow over a solid surface at a temperature T_S as shown in Figure 1.16. The fluid upstream temperature and velocity are T_∞ and u_∞, respectively.

Due to the effect of the viscosity or no-slip condition, a thin fluid region develops, known as the hydrodynamic boundary layer, inside which velocity varies from the solid surface velocity to the outer stream velocity, u_∞. Similarly, there is a thermal boundary layer inside which fluid temperature changes from surface temperature T_S to outer fluid temperature T_∞. Since the fluid is stationary at the solid surface, the heat is transferred by conduction normal to the surface and is expressed by the conduction rate equation 1.64 as

$$q_S'' = -k_f \left.\frac{\partial T}{\partial y}\right|_{y=0} \tag{1.73}$$

where k_f is the thermal conductivity of the fluid.

In order to estimate the heat flux by convection, the temperature gradient or the temperature distribution in the thermal boundary layer needs to be known. This temperature distribution depends on the surface geometry, transport and thermo-physical properties, and the nature of the fluid motion or the velocity field, and is determined by solving the energy equation along with the mass and momentum equations. Based on the nature of the flow field, the convection heat transfer is classified as **forced convection, free or natural convection,** or **phase change heat transfer**, such as in condensation and boiling. In **forced convection** some external forces generated by pumps, fans, or winds induce the flow field. On the other hand, for **free or natural convection**, the flow is induced by natural forces such as buoyancy or **Marangoni** forces. The buoyancy force arises from density variations, which is caused by temperature variations. In both forced and free convections, energy being transferred is in the form of sensible heat of the fluid. On the other hand, in **phase change heat transfer**, the energy transfer is in the form of latent heat of the fluid, and the flow field is created due to the formation of vapor bubbles, as in boiling heat transfer, or due to the condensation of vapor on a solid surface, as in condensation heat transfer.

Irrespective of this classification of convection heat transfer, the overall effect is given by a **convection rate equation**, governed by **Newton's law of cooling**, expressed as

$$q_c'' \propto (T_S - T_\infty) \tag{1.74}$$

We introduce a proportionality constant

$$q_c'' = h_c(T_S - T_\infty) \tag{1.75}$$

where h_c, the constant of proportionality, is called the **convection heat transfer coefficient** or **film coefficient**. The defining equation for the convection heat transfer coefficients is obtained by recasting Equation 1.75 as

$$h_c = \frac{q''_c}{(T_S - T_\infty)} \tag{1.76}$$

Combining Equations 1.73 and 1.75 we obtain the alternate defining equation for the convection heat transfer coefficient

$$h_c = \frac{-k_f(\partial T/\partial y)\big|_{y=0}}{(T_S - T_\infty)} \tag{1.77}$$

It is important to note that the temperature gradient inside the thermal boundary layer varies along the surface, and hence the convection heat transfer coefficient varies locally from point to point. The local convection coefficient is defined based on local heat transfer rate, as

$$\mathrm{d}q = h_c \mathrm{d}A(T_S - T_\infty) \tag{1.78}$$

The total heat transfer rate can either be calculated by integrating the local heat transfer rate over the area or using Equation 1.75 with an average heat transfer coefficient $\bar{h}_c$, which is defined as

$$\bar{h}_c = \frac{1}{A}\iint_A h_c \mathrm{d}A \tag{1.79}$$

Convection heat transfer coefficients are derived for many flow conditions in the form of a correlation. For forced convections the correlations are of the form

$$\mathrm{Nu} = f(\mathrm{Re}, \mathrm{Pr})$$

where

$$\mathrm{Nu} = \text{Nusselt number} = \frac{h_c L_c}{k}$$

$$\mathrm{Re} = \text{Reynolds number} = \frac{\rho U_c L_c}{\mu}$$

$$\mathrm{Pr} = \text{Prandtl number} = \frac{\mu c_P}{k} = \frac{\nu}{\alpha}$$

L_c and U_c represent the characteristics length and velocity in the problem. A list of many correlations for different flow geometries and conditions can be found in Incropera and DeWitt (2002), Holman (1997), and Mills (1999).

Thermal Radiation Thermal radiation is the intermediate portion (0.1–100 μm) of the electromagnetic radiation emitted by a substance as a result of its temperature. Thermal radiation heat transfer involves transmission and exchange of electromagnetic waves or photon particles as a result of temperature difference.

The thermal radiation emitted by a substance encompasses a range of wavelength (λ), and it is referred to as spectral distribution. The total black-body emissive power is obtained by integrating the spectral emissive power over the entire range of wavelengths and is derived as

$$E_b = \sigma T^4 \tag{1.80}$$

where σ is the Stefan–Boltzmann constant $\left(5.6697 \times 10^{-8}\ \mathrm{W/m^2{\cdot}K^4}\right)$ and T is the temperature in absolute scale.

Equation 1.80 is referred to as the **Stefan–Boltzmann law**, which gives total emissive power of the **black** or **ideal body** at a given temperature. It is called black-body radiation because a body which behaves according to this law appears black, since there is no reflection of any incident radiation from it. For a nonblack body, the emitted power is less than the black body emissive power, and it is given as

$$E = \varepsilon E_b \tag{1.81a}$$

or

$$E_b = \varepsilon \sigma T^4 \tag{1.81b}$$

where ε is the emissivity factor of a **nonblack surface** and is defined as the ratio of its emissive power to that of a black body, i.e., $\varepsilon = E/E_b$. In addition to the emissivity factor, other material optical properties such as absorptivity, transmissivity, and reflectivity are also essential for heat transfer analysis involving radiation.

Radiation Heat Exchange Based on the Stefan–Boltzmann law, **the radiation heat transfer between two black surfaces** through a nonparticipating medium (see Figure 1.17) is expressed as

$$q_R = \sigma A_1 F_{ij}\left(T_i^4 - T_j^4\right) \tag{1.82a}$$

Using a reciprocity relation $A_i F_{ij} = A_j F_{ji}$, Equation 1.82 can also be written as

$$q_R = \sigma A_j F_{ji}\left(T_i^4 - T_j^4\right) \tag{1.82b}$$

where A_i, A_j are areas of two surfaces, T_i, T_j are temperatures of two surfaces, and F_{ij}, F_{ji} are the shape factors or view factors.

The shape factor F_{ij} is defined as the fraction of energy leaving surface A_i that is intercepted by surface A_j. The shape factor depends on the size, shape, and orientations of the two surfaces i and j.

Shape factors are computed for many common geometries and configurations and presented in the form of formulas and graphs (Sparrow and Cess, 1978; Siegel and Howell, 1992).

Equation 1.82 can also be written in an electrical network analogy as

$$q = \frac{E_{bi} - E_{bj}}{1/A_i F_{ij}} = \frac{E_{bi} - E_{bj}}{R_{ij}} \tag{1.83}$$

where $R_{ij} = 1/A_i F_{ij}$ is termed as the **radiation space resistance** between the two surfaces i and j.

The radiation heat exchange between two diffuse gray surfaces that form an enclosure is derived as

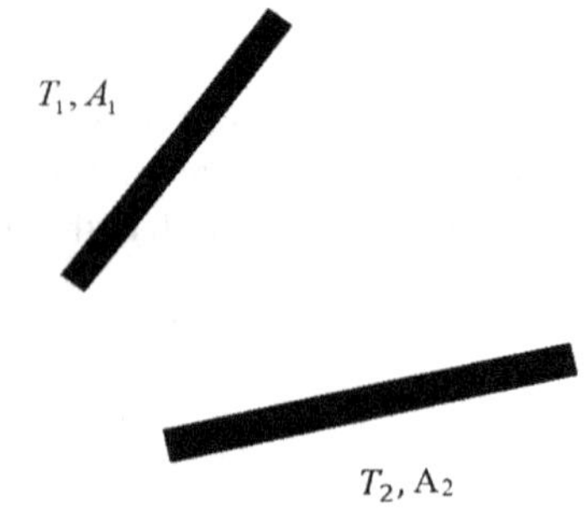

FIGURE 1.17 Radiation exchange between two black bodies.

$$q = \frac{E_{bi} - E_{bj}}{\frac{1-\varepsilon_i}{\varepsilon_i A_i} + \frac{1}{A_i F_{ij}} + \frac{1-\varepsilon_j}{\varepsilon_j A_j}}$$

$$= \frac{\sigma T_i^4 - T_j^4}{\frac{1-\varepsilon_i}{\varepsilon_i A_i} + \frac{1}{A_i F_{ij}} + \frac{1-\varepsilon_j}{\varepsilon_j A_j}} \tag{1.84}$$

Equation 1.83 can also be written in an electrical network analogy as

$$q = \frac{E_{bi} - E_{bj}}{R_{ij}} \tag{1.85}$$

where

$$R_{ij} = \frac{1-\varepsilon_i}{\varepsilon_i A_i} + \frac{1}{A_i F_{ij}} + \frac{1-\varepsilon_j}{\varepsilon_j A_j}$$

is the total radiation resistance between two gray surfaces i and j. The term $(1-\varepsilon)/\varepsilon A$ is defined as the **radiation surface resistance**.

For many applications, the expression Equation 1.82 for radiation heat transfer is given in a linearized form as

$$q = h_r A\left(T_i - T_j\right) \tag{1.86}$$

where

$$h_r = \sigma\left(T_i + T_j\right)\left(T_i^2 + T_j^2\right)/R_{ij} \tag{1.87}$$

1.3.2 The First Law of Thermodynamics and Heat Equation

In order to estimate the heat transfer rates in a medium, we need to determine the temperature distribution or temperature field resulting from the physical conditions or thermal boundary conditions that exist at the boundary. The temperature field is determined by solving the heat equation, which is a statement of conservation of energy or the first law of thermodynamics. Let us consider a differential control volume in a Cartesian coordinate, and various forms of energy transfer across the control volume surfaces as shown in Figure 1.18.

The **first law of thermodynamics** for a differential control volume can be derived from Equation 1.34 stated for a control volume as

$$\pm \dot{Q} \pm \dot{W} + \dot{Q}_{\text{gen}} = \frac{\partial}{\partial t}\int_{\forall} e\rho d\forall + \int_{\text{CS}} e\rho \vec{V}\cdot d\vec{A} \tag{1.88}$$

The **first term** $\dot{Q}$ represents heat transfer across the control volume surfaces by conduction (Q_C) and by radiation heat transfer (Q_R). The net conduction heat transfer per unit area per unit time across the differential control volume surfaces is derived as

$$\dot{Q}_C = -\left(\frac{\partial q_x}{\partial x} + \frac{\partial q_y}{\partial y} + \frac{\partial q_z}{\partial z}\right)dxdydz = -\left(\frac{\partial q_x}{\partial x} + \frac{\partial q_y}{\partial y} + \frac{\partial q_z}{\partial z}\right)d\forall \tag{1.89}$$

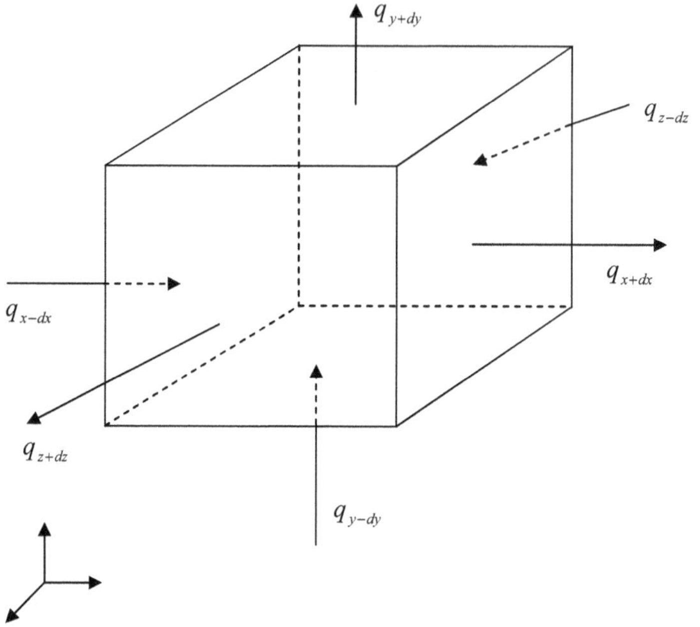

FIGURE 1.18 Differential control volume for energy balance.

or

$$\dot{Q}_C = -(\nabla \cdot \vec{q})\mathrm{d}\forall \tag{1.90}$$

The net radiation heat transfer across the control volume surface is given as

$$\dot{Q}_R = (\nabla \cdot \vec{q}_R)\mathrm{d}\forall \tag{1.91}$$

Combining Equations 1.90 and 1.91, we get total heat transfer rate through the control volume surface as

$$\dot{Q} = -(\nabla \cdot \vec{q})\mathrm{d}\forall + (\nabla \cdot \vec{q}_R)\mathrm{d}\forall \tag{1.92}$$

The **second term** $\dot{W}$ in Equation 1.88 represents the energy transfer across the control volume surfaces due to the net work done by the surface forces on the differential elements. The surface forces include the pressure force and the viscous normal and shear forces. The three components of this work are given as

$$\dot{W}_x = \left[\frac{\partial}{\partial x}\{u(\sigma_{xx} - p)\} + \frac{\partial}{\partial y}(u\tau_{yx}) + \frac{\partial}{\partial z}(u\tau_{zx})\right]\mathrm{d}\forall \tag{1.93a}$$

$$\dot{W}_y = \left[\frac{\partial}{\partial x}(v\tau_{xy}) + \frac{\partial}{\partial y}\{v(\sigma_{yy} - p)\} + \frac{\partial}{\partial z}(v\tau_{zy})\right]\mathrm{d}\forall \tag{1.93b}$$

$$\dot{W}_z = \left[\frac{\partial}{\partial x}(w\tau_{xz}) + \frac{\partial}{\partial y}(w\tau_{yz}) + \frac{\partial}{\partial x}\{w(\sigma_{zz} - p)\}\right]\mathrm{d}\forall \tag{1.93c}$$

The total work done by all surface stresses is given as

$$\dot{W} = \left[\frac{\partial}{\partial x}\left\{ u\sigma_{xx} + v\tau_{xy} + w\tau_{xz} \right\} + \frac{\partial}{\partial y}\left\{ u\tau_{yx} + v\sigma_{yy} + w\tau_{yz} \right\} + \frac{\partial}{\partial z}\left\{ u\tau_{zx} + v\tau_{zy} + w\sigma_{zz} \right\} \right] \mathrm{d}\forall - \left[\frac{\partial}{\partial x}(up) + \frac{\partial}{\partial y}(vp) + \frac{\partial}{\partial z}(wp) \right] \mathrm{d}\forall \tag{1.94}$$

The **third term** $\dot{Q}_{\text{gen}}$ in Equation 1.88 represents the total heat generation within the differential element and is expressed as

$$\dot{Q}_{\text{gen}} = \dot{Q}'''\mathrm{d}\forall \tag{1.95}$$

where $\dot{Q}'''$ is the volumetric heat generation rate. Energy generation may occur as a result of chemical reaction, absorption of nuclear radiation, electrical heat generation, or absorption of solar or laser energy in a semitransparent medium.

The **fourth term** in Equation 1.88 represents the rate of change of energy within the volume, which for the case of negligible potential energy is the rate of change of internal energy and kinetic energy given as

$$\frac{\partial}{\partial t}\int_{\forall} e\rho \mathrm{d}\forall = \frac{\partial}{\partial t}\left(\rho i + \frac{1}{2}\rho V^2 \right)\mathrm{d}\forall \tag{1.96}$$

where i is the internal energy per unit mass of the fluid in the differential element and V is the magnitude of the local velocity defined at the center of the element.

The **fifth term** in Equation 1.88 represents the net energy flux across the control volume surfaces due to fluid motion, and expressed as

$$\int_{\text{CS}} e\rho\vec{V}\cdot\mathrm{d}\vec{A} = \left[\frac{\partial}{\partial x}\left(\left(i + \frac{V^2}{2} \right)\rho u \right) + \frac{\partial}{\partial y}\left(\left(i + \frac{V^2}{2} \right)\rho v \right) + \frac{\partial}{\partial z}\left(\left(i + \frac{V^2}{2} \right)\rho w \right) \right] \mathrm{d}x\mathrm{d}y\mathrm{d}z$$

or

$$\int_{\text{CS}} e\rho\vec{V}\cdot\mathrm{d}\vec{A} = \left[\nabla\left(i + \frac{V^2}{2} \right)\rho\vec{V} \right]\mathrm{d}\forall \tag{1.97}$$

Substituting Equations 1.92 and 1.94–1.97 into Equation 1.88 we have

$$\frac{\partial}{\partial t}\left(\rho\left(i + \frac{V^2}{2} \right) \right) = -\nabla\cdot\vec{q} + \left(\nabla\cdot\vec{q}_R\right) - \nabla\left[\rho\left(i + \frac{V^2}{2} \right)\vec{V} \right] - \nabla\cdot\left(p\vec{V}\right) + \dot{Q}''' + \nabla\cdot\left(\tau\vec{V}\right) \tag{1.98}$$

Equation 1.98 is further simplified and regrouped as

$$\rho\frac{D}{Dt}\left(i + \frac{V^2}{2} \right) = -\nabla\cdot\vec{q} + \nabla\cdot\vec{q}_R - \nabla\left(p\vec{V}\right) + \dot{Q}''' + \nabla\cdot\left(\tau\vec{V}\right) \tag{1.99}$$

Equation 1.99 is referred to as the total energy balance equation in a strictly conservative form. A more popular form in heat transfer analysis, known as the thermal or internal energy balance, is obtained by subtracting the mechanical energy equation

$$\rho \frac{D}{Dt}\left(\frac{V^2}{2}\right) = -\vec{V}\nabla p + \vec{V}\nabla \cdot \tau \tag{1.100}$$

from Equation 1.99, and written as

$$\rho \frac{Di}{Dt} = -\nabla \cdot \vec{q} + \nabla \cdot \vec{q}_R - p\nabla \cdot \vec{V} + \dot{Q}''' + \tau\nabla \cdot \vec{V} \tag{1.101}$$

The thermal energy equation can also be written in terms of enthalpy by substituting $h = i + (p/\rho)$ into Equation 1.101 as

$$\rho \frac{Dh}{Dt} = -\nabla \cdot \vec{q} + \nabla \cdot \vec{q}_R + \frac{dp}{dt} + \dot{Q}''' + \tau\nabla \cdot \vec{V} \tag{1.102}$$

With the substitution of Equation 1.51a–f for normal stresses and shear stresses, and Equation 1.65 for conduction transport rate equation in Equations 1.101 and 1.102, we get

$$\rho \frac{Di}{Dt} = \nabla \cdot (k\nabla T) + \nabla \cdot \vec{q}_R - p\nabla \cdot \vec{V} + \dot{Q}''' + \Phi$$
$$\rho \frac{Di}{Dt} = \nabla \cdot (k\nabla T) + \nabla \cdot \overrightarrow{q_R} - p\nabla \cdot \vec{V} + \dot{Q}''' + \Phi \tag{1.103}$$

and

$$\rho \frac{Dh}{Dt} = \nabla \cdot (k\nabla T) + \nabla \cdot \overrightarrow{q_R} + \frac{Dp}{Dt} + \dot{Q}''' + \Phi \tag{1.104}$$

where Φ is called the **viscous heat dissipation** term given as

$$\Phi = \mu\left\{\left[\left(\frac{\partial u}{\partial y} + \frac{\partial v}{\partial x}\right)^2 + \left(\frac{\partial v}{\partial z} + \frac{\partial w}{\partial y}\right)^2 + \left(\frac{\partial w}{\partial x} + \frac{\partial u}{\partial z}\right)^2\right] + 2\left[\left(\frac{\partial u}{\partial x}\right)^2 + \left(\frac{\partial v}{\partial y}\right)^2 + \left(\frac{\partial w}{\partial z}\right)^2\right]\right\}$$
$$-\mu\left[\frac{2}{3}\left(\frac{\partial u}{\partial x} + \frac{\partial v}{\partial y} + \frac{\partial w}{\partial z}\right)^2\right] \tag{1.105}$$

The **viscous heat dissipation term** represents the rate at which mechanical work is irreversibly converted to thermal heat energy due to the viscous effects of the fluid. The **first term** of the viscous heat dissipation given by Equation 1.105 is contributed by the viscous stresses. The **second** and the **third terms** are the contribution from the viscous normal stresses. In most heat transfer problems, thermal energy equations 1.103 and 1.104 are written solely in terms of primary variables such as temperature by expressing internal energy and enthalpy in terms of primary variables using thermodynamic relations (Sonntag et al., 1998).

The temperature-based energy equation is given in two different forms

$$\rho c_v \frac{DT}{Dt} = \nabla \cdot (k\nabla T) + \nabla \cdot \vec{q}_R - T\left(\frac{\partial p}{\partial T}\right)_v (\nabla \cdot \vec{V}) + \dot{Q}''' + \Phi \tag{1.106}$$

and

$$\rho c_p \frac{DT}{Dt} = \nabla \cdot (k\nabla T) + \nabla \cdot \vec{q}_R + \beta T \frac{DP}{Dt} + \dot{Q}''' + \Phi \tag{1.107}$$

The energy equation is simplified for some commonly used cases as follows.

1. **Incompressible Liquid** For incompressible fluid $\nabla \cdot \vec{V} = 0$, $\beta = 0$, and Equations 1.106 and 1.107 reduce to

$$\rho c \frac{DT}{Dt} = \nabla \cdot (k \nabla T) + \nabla \cdot \vec{q}_R + \dot{Q}''' + \Phi \tag{1.108}$$

and the viscous heat dissipation is given as

$$\Phi = \mu \left\{ \left[\left(\frac{\partial u}{\partial y} + \frac{\partial v}{\partial x} \right)^2 + \left(\frac{\partial v}{\partial z} + \frac{\partial w}{\partial y} \right)^2 + \left(\frac{\partial w}{\partial x} + \frac{\partial u}{\partial z} \right)^2 \right] + 2 \left[\left(\frac{\partial u}{\partial x} \right)^2 + \left(\frac{\partial v}{\partial y} \right)^2 + \left(\frac{\partial w}{\partial z} \right)^2 \right] \right\}$$

2. **Ideal Gas** For an ideal gas $P\nu = RT$ it gives $(\partial p/\partial T)_v = R/v$, and this reduces the internal energy-based energy Equation 1.103 to

$$\rho c_v \frac{DT}{Dt} = \nabla \cdot (k \nabla T) + \nabla \cdot \vec{q}_R - p(\nabla \cdot \vec{V}) + \dot{Q}''' + \Phi \tag{1.109}$$

Also, for the ideal gas it can be shown that $\beta = 1/T$ and the enthalpy-based energy equation 1.107 reduces to

$$\rho c_p \frac{DT}{Dt} = \nabla \cdot (k \nabla T) + \nabla \cdot \vec{q}_R + \frac{DP}{Dt} + \dot{Q}''' + \Phi \tag{1.110}$$

3. **Simplified Heat Equation** A common form of energy or heat equation is very often used in many applications for (a) negligible viscous dissipation Φ; (b) negligible compressibility effect; and (c) negligible radiation heat transfer rate, i.e., $\vec{q}_R = 0$. The heat equation is given as

$$\rho c_p \left(\frac{\partial T}{\partial t} + u \frac{\partial T}{\partial x} + \upsilon \frac{\partial T}{\partial y} + w \frac{\partial T}{\partial z} \right) = \frac{\partial}{\partial x}\left(k \frac{\partial T}{\partial x} \right) + \frac{\partial}{\partial y}\left(k \frac{\partial T}{\partial y} \right) + \frac{\partial}{\partial z}\left(k \frac{\partial T}{\partial z} \right) + Q''' \tag{1.111a}$$

or

$$\rho c_p \frac{DT}{Dt} = \nabla \cdot (k \nabla T) + Q''' \tag{1.111b}$$

For problems with constant thermal conductivity, the equation is reduced to the following form:

$$\rho c_p \frac{DT}{Dt} = k \nabla^2 T + \dot{Q}''' \tag{1.112a}$$

and with no volume heat generation as

$$\rho c_p \frac{DT}{Dt} = k \nabla^2 T \tag{1.112b}$$

In terms of specific coordinate systems, we can write

Cartesian (x, y, z)

$$\rho c_{\mathrm{p}}\left(\frac{\partial T}{\partial t}+u\frac{\partial T}{\partial x}+\upsilon\frac{\partial T}{\partial y}+w\frac{\partial T}{\partial z}\right)=k\left(\frac{\partial^2 T}{\partial x^2}+\frac{\partial^2 T}{\partial y^2}+\frac{\partial^2 T}{\partial z^2}\right) \tag{1.112a}$$

Cylindrical (r, θ, z)

$$\rho c_{\mathrm{p}}\left(\frac{\partial T}{\partial t}+u_r\frac{\partial T}{\partial r}+\frac{u_\theta}{r}\frac{\partial T}{\partial \theta}+u_z\frac{\partial T}{\partial z}\right)=k\left[\frac{1}{r}\frac{\partial}{\partial r}\left(r\frac{\partial T}{\partial r}\right)+\frac{1}{r^2}\frac{\partial^2 T}{\partial \theta^2}+\frac{\partial^2 T}{\partial z^2}\right] \tag{1.112b}$$

Spherical (r, θ, ϕ)

$$\rho c_{\mathrm{p}}\left(\frac{\partial T}{\partial t}+u_r\frac{\partial T}{\partial r}+\frac{u_\phi}{r}\frac{\partial T}{\partial \phi}+\frac{u_\theta}{r\sin\phi}\frac{\partial T}{\partial \theta}\right)$$
$$=k\left[\frac{1}{r^2}\frac{\partial}{\partial r}\left(r^2\frac{\partial T}{\partial r}\right)+\frac{1}{r^2\sin\phi}\frac{\partial}{\partial \phi}\left(\sin\phi\frac{\partial T}{\partial \phi}\right)+\frac{1}{r^2\sin^2\phi}\frac{\partial^2 T}{\partial \theta^2}\right] \tag{1.112c}$$

While dealing with high viscous flows with large velocity gradients, the model above is modified by adding the viscous dissipation term Φ, which considers the internal heating due to viscous dissipation

$$\rho C_p\left(\frac{\partial T}{\partial t}+u\frac{\partial T}{\partial x}+v\frac{\partial T}{\partial y}+w\frac{\partial T}{\partial z}\right)=\frac{\partial}{\partial x}\left(k\frac{\partial T}{\partial x}\right)+\frac{\partial}{\partial y}\left(k\frac{\partial T}{\partial y}\right)+\frac{\partial}{\partial z}\left(k\frac{\partial T}{\partial z}\right)+\Phi \tag{1.113a}$$

Such problems are encountered in lubrication, thermo-mechanical analysis of metal forming processes, and convective heat transfer involving high viscous fluids and/or high velocities.

A generalized form of the heat equation that will be considered in later chapters for developing discretized equations based on finite difference/control volume and finite element method is given as

$$\rho C_p\left(\frac{\partial T}{\partial t}+u\frac{\partial T}{\partial x}+v\frac{\partial T}{\partial y}+w\frac{\partial T}{\partial z}\right)=\frac{\partial}{\partial x}\left(k\frac{\partial T}{\partial x}\right)+\frac{\partial}{\partial y}\left(k\frac{\partial T}{\partial y}\right)+\frac{\partial}{\partial z}\left(k\frac{\partial T}{\partial z}\right)+S_T \tag{1.113b}$$

where $S_T=\Phi+\dot{Q}'''$.

4. **Heat Conduction in Solids – Heat Diffusion Equation** For solids, $\vec{V}=0$, $\Phi=0$, and the energy equation 1.113, along with the inclusion of the volumetric heat generation term, transform into a form known as the **heat diffusion equation** for solids

$$\rho c\frac{\partial T}{\partial t}=\nabla\cdot\left(k\nabla T\right)+Q''' \tag{1.114}$$

and for constant thermal conductivity it reduces to

$$\frac{1}{\alpha}\frac{\partial T}{\partial t}=\nabla^2 T+Q''' \tag{1.115}$$

where $\alpha=k/\rho c$ is called the thermal diffusivity of the material.

Equations 1.114 and 1.115 are written in a specific co-ordinate system as given in Table 1.1.

TABLE 1.1
Heat Diffusion Equation in Specific Coordinate Systems

Cartesian Coordinate

$$\rho c\frac{\partial T}{\partial t}=\frac{\partial}{\partial x}\left(k\frac{\partial T}{\partial x}\right)+\frac{\partial}{\partial y}\left(k\frac{\partial T}{\partial y}\right)+\frac{\partial T}{\partial z}\left(k\frac{\partial T}{\partial z}\right)+\dot{Q}''' \tag{1.116a}$$

$$\frac{1}{\alpha}\frac{\partial T}{\partial t}=\frac{\partial^2 T}{\partial x^2}+\frac{\partial y^2 T}{\partial y}+\frac{\partial^2 T}{\partial y^2}+\frac{\dot{Q}'''}{k} \tag{1.116b}$$

Cylindrical Coordinate

$$\rho c\frac{\partial T}{\partial t}=\frac{1}{r}\frac{\partial}{\partial r}\left(kr\frac{\partial T}{\partial r}\right)+\frac{1}{r^2}\frac{\partial}{\partial \theta}\left(k\frac{\partial T}{\partial \theta}\right)+\frac{\partial}{\partial z}\left(k\frac{\partial T}{\partial z}\right)+\dot{Q}''' \tag{1.116c}$$

$$\frac{1}{\alpha}\frac{\partial T}{\partial t}=\frac{1}{r}\frac{\partial}{\partial r}\left(r\frac{\partial T}{\partial r}\right)+\frac{1}{r^2}\frac{\partial^2 T}{\partial \theta^2}+\frac{\partial^2 T}{\partial z^2}+\frac{\dot{Q}'''}{k} \tag{1.116d}$$

Spherical Coordinate

$$\rho c\frac{\partial T}{\partial t}=\frac{1}{r^2}\frac{\partial}{\partial r}\left(kr^2\frac{\partial T}{\partial r}\right)+\frac{1}{r^2\sin\phi}\frac{\partial}{\partial \phi}\left(k\sin\phi\frac{\partial T}{\partial \phi}\right)+\frac{1}{r^2\sin^2\phi}\frac{\partial}{\partial \theta}\left(k\frac{\partial T}{\partial \theta}\right)+\dot{Q}''' \tag{1.116e}$$

$$\frac{1}{\alpha}\frac{\partial T}{\partial t}=\frac{1}{r^2}\frac{\partial}{\partial r}\left(r^2\frac{\partial T}{\partial r}\right)+\frac{1}{r^2\sin\phi}\frac{\partial}{\partial \phi}\left(\sin\phi\frac{\partial T}{\partial \phi}\right)+\frac{1}{r^2\sin^2\phi}\frac{\partial^2 T}{\partial \theta^2}+\frac{\dot{Q}'''}{k} \tag{1.116f}$$

Equations 1.114 and 1.115 can further be simplified in some special cases as follows.

1. *Unsteady heat conduction without heat source*

$$\rho c\frac{\partial T}{\partial t}=\nabla\cdot\left(k\nabla T\right) \tag{1.117a}$$

$$\frac{1}{\alpha}\frac{\partial T}{\partial t}=\nabla^2 T \tag{1.117b}$$

2. *Steady-state heat conduction with heat source*

$$\nabla\cdot\left(k\nabla T\right)+\dot{Q}'''=0 \tag{1.118a}$$

$$\nabla^2 T+\frac{\dot{Q}'''}{k}=0 \tag{1.118b}$$

3. *Steady-state heat conduction without heat source*

$$\nabla\cdot\left(k\nabla T\right)=0 \tag{1.119a}$$

$$\nabla^2 T=0 \tag{1.119b}$$

1.3.3 Initial and Boundary Conditions for Heat Transfer

In order to determine the temperature distribution and heat transfer rate in the medium, we need to solve the appropriate form of the differential heat diffusion equation using the physical condition in time, known as the **initial condition**, and physical conditions in space, known as the **boundary conditions**.

Initial Condition The initial condition is needed for problems that are transient or time-dependent. Since the heat diffusion equation is first-order in time, only one initial condition is needed. The initial condition specifies the known temperature distribution in the medium at an instant of time, usually at the origin of time, and it is stated as

$$\text{At time } t = 0, \quad T(\vec{x},0) = f(\vec{x}) \quad \text{or} \quad T_1 \tag{1.120}$$

Boundary Conditions Temperature boundary conditions are derived from the physical condition that exists at the boundary. Some of the most common boundary conditions are described as follows:

1. Boundary condition of the first kind or *Dirichlet boundary condition* (Figure 1.19). The surface temperature is specified as a constant temperature or a function of space and time, i.e.,

$$T = T_S(\vec{x},t) \quad \text{on the surface } S \tag{1.121}$$

   where $T_S(\vec{x},t)$ is a function of time and position or it could be a constant surface temperature, T_S. For the special case of $T_S = 0$, the boundary condition is referred to as the *homogeneous boundary condition of the first kind.*
2. The boundary condition of the second kind or *Newmann condition* (Figure 1.20). The heat flux at the surface is specified to be constant or a function of space and time, i.e.,

$$q_n'' = \pm q_S''(\vec{r},t) \quad \text{or} \quad \pm q_S'' \tag{1.122a}$$

   or

$$-k\frac{\partial T}{\partial n} = \pm q_S''(\vec{r},t) \quad \text{or} \quad \pm q_S'' \tag{1.122b}$$

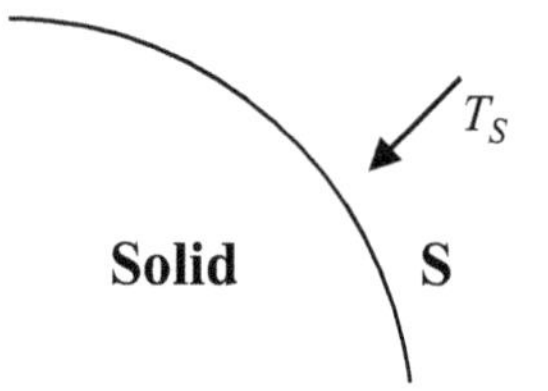

FIGURE 1.19 Boundary condition of the first kind.

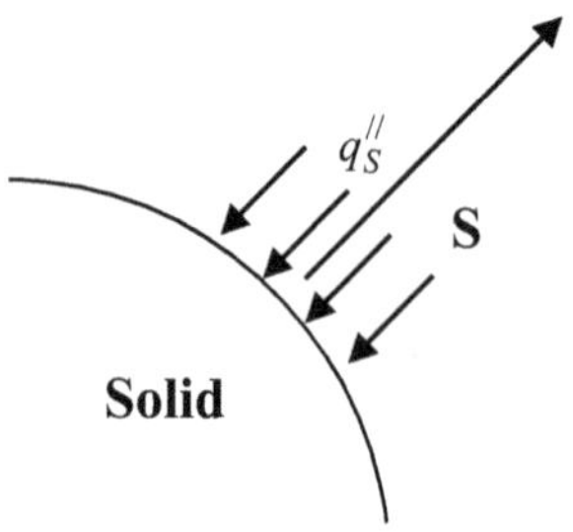

FIGURE 1.20 Boundary condition of the second kind.

where $q_S(\vec{x},t)$ is a function of time and position or it could be a constant surface heat flux q_S'', and q_n'' is the conduction heat transfer rate per unit area as given by the **Fourier's** law of heat conduction equation 1.68, and shown as positive in the outward normal direction to the surface. For a heat transfer rate in an inward normal direction to the surface, the heat transfer rate q_n'' will be considered as negative, and the boundary condition will be stated as

$$-q_n'' = \pm q_S''(\vec{r},t) \quad \text{or} \quad \pm q_S'' \tag{1.123a}$$

or

$$k\frac{\partial T}{\partial n} = \pm q_S''(\vec{r},t) \quad \text{or} \quad \pm q_S'' \tag{1.123b}$$

For the special case of

$$q_S'' = 0 \quad \text{or} \quad \left.\frac{\partial T}{\partial n}\right|_S = 0 \tag{1.124}$$

the boundary condition is referred to as the homogeneous boundary condition of the second kind (Figure 1.21). This is also referred as adiabatic condition, or perfectly insulated condition or symmetric condition.

3. The boundary condition of the third kind or *convective surface condition* (Figure 1.22)

$$\pm q_n'' = h_c(T_s - T_\infty) \tag{1.125a}$$

or

$$\pm k\left.\frac{\partial T}{\partial x}\right|_s = h_c(T_s - T_\infty) \tag{1.125b}$$

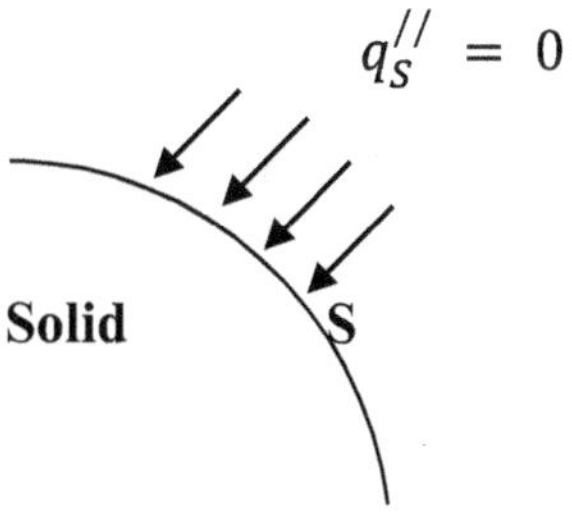

FIGURE 1.21 Homogeneous boundary condition of the second kind.

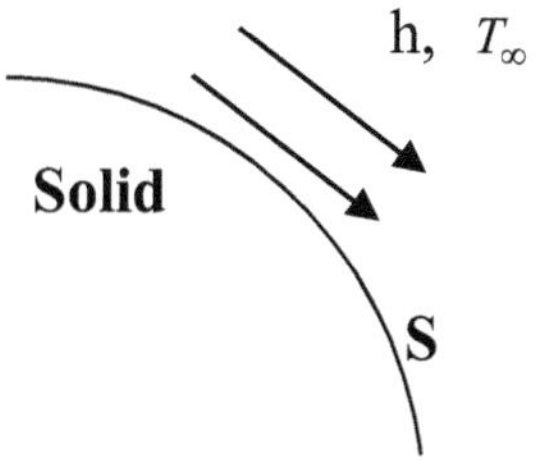

FIGURE 1.22 Boundary condition of the third kind.

For combined heat losses by convection and radiation from the surface, an energy balance leads a boundary condition as

$$\pm k \left.\frac{\partial T}{\partial x}\right|_{\mathrm{s}} = h_{\mathrm{c}}(T_{\mathrm{s}} - T_{\infty}) + h_{\mathrm{r}}(T_{\mathrm{s}} - T_{\mathrm{sur}}) \tag{1.126}$$

where h_r is the radiation film coefficient given by Equation 1.87 as

$$h_r = \sigma\left(T_s + T_{\mathrm{sur}}\right)\left(T_s^2 + T_{\mathrm{sur}}^2\right)/R_{s,\mathrm{sur}} \tag{1.127}$$

1.4 MASS TRANSFER

Mass transfer is a science that describes the relative motion of species in a mixture of chemical species as a result of mass concentration gradient or difference. There are two basic modes of mass transfer: **mass transfer by diffusion** and **mass transfer by convection.** There is a close similarity between heat transfer and mass transfer in terms of the transport rate equation and the transport conservation equation. Let us briefly define some of the basic mixture properties before describing mass transport rate and conservation equations in the following section. For more comprehensive discussion, refer to Sherwood et al. (1975), Bird et al. (2002), Incropera and DeWitt (2002), Kays and Crawford (1993), Mills (1999), Mills (2001), and Treybal (1980).

The amount of any species i in a mixture of chemical species is given in terms of mass density or concentration $\rho_i\left(\mathrm{kg/m^3}\right)$ on mass basis. This is also defined on a molar basis as molar density or concentration $C_i\left(\mathrm{kmol/m^3}\right)$. The mass density and concentration are related through molecular weight M_i as

$$C_i = \frac{\rho_i}{M_i} \tag{1.128}$$

The total mass density $\rho\left(\mathrm{kg/m^3}\right)$ and total molar density $C\left(\mathrm{mole/m^3}\right)$ are defined as

$$\rho = \sum_i \rho_i \tag{1.129}$$

and

$$C = \sum_i C_i \tag{1.130}$$

The quantity of a species can also be expressed in terms of mass fraction m_i or mole fraction x_i, which are defined as

$$m_i = \frac{\rho_i}{\rho} \tag{1.131}$$

and

$$x_i = \frac{C_i}{C} \tag{1.132}$$

1.4.1 BASIC MODES AND TRANSPORT RATE EQUATION

Diffusion Mass Transfer   This mode is primarily important for mass transfer in a stationary medium such as in a solid and stationary fluid. The **diffusion rate equation** is given by **Fick's law**, which expresses the transfer of a species 1 in a binary mixture of 1 and 2 as

$$\vec{j}_1 = -\rho D_{12} \nabla m_1 \tag{1.133a}$$

or

$$\vec{\bar{j}}_1 = -c D_{12} \nabla x_1 \tag{1.133b}$$

where
$\vec{j}_1$ = diffusive mass flux of species 1 $\left(kg/m^2 \cdot s \right)$
$\vec{\bar{j}}_1$ = diffusive molar mass flux of species 1 $\left(kmol/m^2 \cdot s \right)$
m_1 = mass fraction of species 1
x_1 = mole fraction of species 1
ρ = local mixture density $\left(kg/m^3 \right)$
D_{12} = binary diffusion coefficient or mass diffusivity $\left(m^2/s \right)$

In scalar form, the equation can be written in three components and in a Cartesian coordinate system as

$$j_{x_1} = -\rho D_{12} \frac{\partial m_1}{\partial x} \tag{1.134a}$$

$$j_{x_1} = -\rho D_{12} \frac{\partial m_1}{\partial y} \tag{1.134b}$$

$$j_{x_1} = -\rho D_{12} \frac{\partial m_1}{\partial z} \tag{1.134c}$$

Mass Diffusion Coefficient   As we can see **Fick's law** is analogous to **Fourier's law** for heat conduction and introduces in an analogous way the **mass density** or **binary diffusion coefficient** as a material transport property. Empirical correlations for a binary diffusion coefficient are derived based on kinetic theory and using experimental data. A widely used such correlation is given as

$$D_{12} = \frac{0.00100 T^{1.75} \left(\dfrac{1}{M_1} + \dfrac{1}{M_2} \right)^{1/2}}{P \left[\left(\sum v \right)_1^{1/3} + \left(\sum v \right)_2^{1/3} \right]^2} \tag{1.135}$$

where M_1 and M_2 are the molecular weights of species 1 and 2, respectively, and the quantity $\left(\sum v \right)$ represents the summation of atomic-diffusion volumes for each species of the binary. A list is given in Sherwood et al. (1975).

A commonly used theoretical equation for the binary diffusion coefficient at low pressure is given as

$$D_{12} = \frac{0.001858T^{1.5}\left(\left(1/M_1\right)+\left(1/M_2\right)\right)^{1/2}}{P\sigma_{12}\Omega_D} \tag{1.136}$$

where
T = temperature in K
P = pressure in atm
$\Omega_D = f\left(kT/\varepsilon_{12}\right)$
σ_{12}, ε_{12} = Lennard–Jones force constants
k = Boltzmann constant

The binary values of σ_{12} and ε_{12} are computed from the values of component species based on the following rules

$$\sigma_{12} = \frac{1}{2}(\sigma_1 + \sigma_2) \tag{1.137a}$$

and

$$\frac{\varepsilon_{12}}{k} = \left(\frac{\varepsilon_1}{k}\frac{\varepsilon_2}{k}\right)^{1/2} \tag{1.137b}$$

Values of $\Omega_D = f\left(kT/\varepsilon_{12}\right)$, ε/k, and σ for many common pure substances are given in Sherwood et al. (1975).

In a porous media the diffusion mechanism can be of three different types: ordinary diffusion, Knudsen diffusion, and surface diffusion. If the pores are much larger than the mean free path length, then the molecules collide with each other more frequently than with the pore walls, and ordinary diffusion is assumed to be the dominant diffusion mechanism. Knudsen diffusion is encountered in smaller pores and/or at lower pressure or density. In this case, molecules collide more frequently with the walls than with each other. The Knudsen diffusion coefficient is given, based on kinetic theory, as

$$D_{1k} = 2/3a\bar{v}_1 \tag{1.138}$$

where a is the effective pore radius in meters and $\bar{v}_1$ is the average molecular speed of species, and given as

$$\bar{v}_1 = \left(\frac{8RT}{\pi M_1}\right) \tag{1.139}$$

Substituting the value for the gas constant and combining Equations 1.138 and 1.139, the expression for Knudsen diffusion co-efficient is given as

$$D_{1k} = 97a\left(\frac{T}{M_1}\right)^{1/2}, \quad m^2/s \tag{1.140}$$

where T is temperature in Kelvin and M_1 is the molecular weight.

In an intermediate range of pore sizes, both ordinary and Knudsen diffusions contribute to the transfer of the species in the media. In this range, the combined ordinary and Knudsen diffusion can be represented by assuming parallel resistances expressed as

$$D_1 = \left(\frac{1}{D_{12}} + \frac{1}{D_{1k}} \right)^{-1} \tag{1.141}$$

For even smaller pores and for species adsorbed on pore surfaces, the dominant mechanism is surface diffusion in which molecules diffuse along the surface in the direction of decrease in concentration.

Other important factors that have to be considered for porous media are the presence of the tortuous path and changes and reduction in the cross-sectional area of pore channels. The effective diffusivity is expressed as

$$D_{1,e} = \frac{\varepsilon}{\tau} D_1 \tag{1.142}$$

where ε is the **porosity** of the media that accounts for the reduction of the free area for diffusion due to the presence of the solid phase and τ is the **tortuosity factor** that accounts for the increase in the diffusional path due to the tortuous path of the pores.

Convection Mass Transfer The convection mass transfer is analogous to convection heat transfer and occurs between a moving mixture fluid species and an exposed solid surface. Like hydrodynamic and thermal boundary layers, a concentration boundary layer forms over the surface if the free stream concentration of a species 1, $C_{1\infty}$, differs from species concentration at the surface, C_{1S}.

Since the fluid is stationary at the solid surface, the mass of species 1 is transferred by diffusion normal to the surface, and is expressed by the diffusion rate equation (1.33) as

$$m''_{1S} = -D_{12} \left. \frac{\partial C_1}{\partial y} \right|_{y=0} \tag{1.143}$$

Like convection heat transfer, the mass flux of the species 1 is given by a **convection rate equation** as

$$m''_{1S} = h_m \left(C_{S1} - C_{\infty 1} \right) \tag{1.144}$$

where h_m is called the **convection mass transfer coefficient** or **mass film coefficient**. The defining equation for the convection mass transfer coefficients is obtained as

$$h_m = \frac{m''_{1S}}{\left(C_{1S} - C_{1\infty} \right)} \tag{1.145}$$

and

$$h_m = \frac{-D_{12} \left(\partial C_1 / \partial y \right) \Big|_{y=0}}{\left(C_{IS} - C_{1\infty} \right)} \tag{1.146}$$

The correlations for convective mass transfer coefficient are given in a similar manner as in convections heat transfer, but in terms of mass transfer parameters as

$$\text{Sh} = f(\text{Re}, \text{Sc}) \tag{1.147}$$

where

$$\text{Sh} = \text{Sherwood number} = \frac{h_m L_c}{D_{12}} \tag{1.148}$$

$$\text{Sc} = \text{Schmidt number} = \frac{\nu}{D_{12}} \tag{1.149}$$

Relation between Heat and Mass Transfer Examination of the governing equations and boundary conditions shows that mass transfer phenomena are closely related to heat transfer. So, the correlations for heat and mass transfer can be used interchangeably. The relation between heat and mass transfer convection coefficients is given by

$$\frac{h}{h_m} = \rho c_P \text{Le}^{1/2} \tag{1.150}$$

where $\text{Le} = \alpha/D_{12}$ is defined as the Lewis number.

1.4.2 Conservation of Mass Species and Mass Concentration Equation

Like heat transfer rates, the species mass flux can be determined from the mass concentration field by solving the species mass concentration equation, which is a statement of conservation of mass species. The species mass concentration equation is derived following the procedure outlined for conservation of energy as

$$\frac{DC_1}{Dt} = \nabla\cdot\left(D_{12}\nabla C_1\right) + \dot{n}_1 \quad \text{in molar form} \tag{1.151a}$$

and

$$\frac{D\rho_1}{Dt} = \nabla\cdot\left(D_{12}\nabla \rho_1\right) + \dot{m}_1 \quad \text{in mass basis} \tag{1.151b}$$

where $\dot{n}_1$ and $\dot{m}_1$ are the volumetric generation or consumption of the species 1 per unit volume in molar basis and mass basis, respectively.

In Cartesian coordinate systems, we can write the equation as

Cartesian (x, y, z)

$$\left(\frac{\partial C_1}{\partial t} + u\frac{\partial C_1}{\partial x} + \upsilon\frac{\partial C_1}{\partial y} + w\frac{\partial C_1}{\partial z}\right) = \frac{\partial}{\partial x}\left(D_{12}\frac{\partial C_1}{\partial x}\right) + \frac{\partial}{\partial y}\left(D_{12}\frac{\partial C_1}{\partial y}\right) + \frac{\partial}{\partial z}\left(D_{12}\frac{\partial C_1}{\partial z}\right) + \dot{n}_1 \tag{1.152a}$$

or

$$\left(\frac{\partial \rho_1}{\partial t} + u\frac{\partial \rho_1}{\partial x} + \upsilon\frac{\partial \rho_1}{\partial y} + w\frac{\partial \rho_1}{\partial z}\right) = \frac{\partial}{\partial x}\left(D_{12}\frac{\partial \rho_1}{\partial x}\right) + \frac{\partial}{\partial y}\left(D_{12}\frac{\partial \rho_1}{\partial y}\right) + \frac{\partial}{\partial z}\left(D_{12}\frac{\partial \rho_1}{\partial z}\right) + \dot{m}_1 \tag{1.152b}$$

The second, third and fourth terms on the left-hand side of Equation 1.152 represent the net transport of the species 1 by convective motion in x-, y-, and z-directions, respectively. The first three terms on the right-hand side of the equation represent the net transport by diffusion. The last term on the right-hand side represents the generation or consumption of the species 1 by processes such as absorption, adsorption/desorption, and other chemical or electro-chemical reactions.

Mass Diffusion Equation in Solids For solids, $\vec{V} = 0$ and the mass equation 1.151a transforms into a form known as the **mass diffusion equation** for solids

$$\frac{\partial C_1}{\partial t} = \nabla\cdot\left(D_{12}\nabla C_1\right) + \dot{n}_1 \tag{1.153}$$

and for constant mass diffusivity, it reduces to

$$\frac{1}{D_{12}}\frac{\partial C_1}{\partial t} = \nabla^2 C_1 + \frac{\dot{n}_1}{D_{12}} \tag{1.154}$$

In Cartesian coordinates, this takes the form

$$\frac{1}{D_{12}}\frac{\partial C_1}{\partial t} = \frac{\partial^2 C_1}{\partial x^2} + \frac{\partial^2 C_1}{\partial y^2} + \frac{\partial^2 C_1}{\partial z^2} + \frac{\dot{n}_1}{D_{12}} \tag{1.155}$$

1.4.3 Initial and Boundary Conditions for Mass Transfer

Initial Condition The initial condition is specified in terms of known concentration distribution in the medium at an instant of time. For example, it can be stated as

$$\text{at time } t = 0, \quad C(\vec{x},0) = f(\vec{x}) \quad \text{or} \quad C_t \tag{1.156}$$

Boundary Conditions Concentration boundary conditions are specified in the same manner as in temperature based on the physical condition that exists at the boundary. Some of the most common boundary conditions are described as follows:

1. **Specified Surface Concentration** The surface concentration is specified as a constant concentration or a function of space and time, i.e.,

$$C = C_S(\vec{x},t) \quad \text{on the surface } S \tag{1.157}$$

   where $C_S(\vec{x},t)$ is a function of time and position or it could be a constant surface concentration, C_S.
2. **Surface Mass Generation and Consumption** The mass flux at the surface is specified to be constant or a function of space and time, i.e.,

$$-D_{12}\left.\frac{\partial C}{\partial x}\right|_S = \pm m_S''(\vec{x},t) \tag{1.158}$$

   where $m_S(\vec{x},t)$ is a function of time and position or it could be a constant surface heat flux, m_S''.

   For an **impermeable surface** $m_S'' = 0$ and the boundary condition reduces to

$$\left.\frac{\partial C}{\partial x}\right|_S = 0 \tag{1.159}$$

3. The convective surface condition

$$\pm D_{12}\left.\frac{\partial C}{\partial x}\right|_s = h_m(C_s - C_\infty) \tag{1.160}$$

1.5 GENERALIZED FORM OF TRANSPORT EQUATION

The governing equations for momentum, temperature, and mass species transport can be represented in a general transport equation given as

$$C\left(\frac{\partial \phi}{\partial t}+u\frac{\partial \phi}{\partial x}+\upsilon\frac{\partial \phi}{\partial y}+w\frac{\partial \phi}{\partial z}\right)=\frac{\partial}{\partial x}\left(\Gamma\frac{\partial \phi}{\partial x}\right)+\frac{\partial}{\partial y}\left(\Gamma\frac{\partial \phi}{\partial y}\right)+\frac{\partial}{\partial z}\left(\Gamma\frac{\partial \phi}{\partial z}\right)+S_{\phi} \tag{1.161}$$

where the symbol ϕ represents the velocity components u, v, and w for flow field given by Equation 1.57; temperature T given by the heat Equation 1.113, and mass species concentration C given by Equation 1.152. Further it can be shown in later chapter on turbulence flow modeling that Equation (.) also represents two equations for turbulence kinetic energy, κ, and turbulence dissipation rete, ε. The term $S_{\varnothing}$ represents source term in the respective equations. Table 1.2 below gives a summary of the dependent variables and the source terms in all the transport equations:

Mass Continuity

$$\frac{\partial u}{\partial x}+\frac{\partial v}{\partial y}+\frac{\partial w}{\partial z}=0 \tag{1.162a}$$

x-momentum

$$\rho\left(\frac{\partial u}{\partial t}+u\frac{\partial u}{\partial x}+v\frac{\partial u}{\partial y}+w\frac{\partial u}{\partial z}\right)=-\frac{\partial p}{\partial x}+\frac{\partial}{\partial x}\left(\mu\frac{\partial u}{\partial x}\right)+\frac{\partial}{\partial y}\left(\mu\frac{\partial u}{\partial y}\right)+\frac{\partial}{\partial x}\left(\mu\frac{\partial u}{\partial z}\right)+S_u \tag{1.162b}$$

y-momentum

$$\rho\left(\frac{\partial v}{\partial t}+u\frac{\partial v}{\partial x}+v\frac{\partial v}{\partial y}+w\frac{\partial v}{\partial z}\right)=-\frac{\partial p}{\partial y}+\frac{\partial}{\partial x}\left(\mu\frac{\partial v}{\partial x}\right)+\frac{\partial}{\partial x}\left(\mu\frac{\partial v}{\partial y}\right)+\frac{\partial}{\partial z}\left(\mu\frac{\partial v}{\partial z}\right)+S_v \tag{1.162c}$$

z-momentum

$$\rho\left(\frac{\partial w}{\partial t}+u\frac{\partial w}{\partial x}+v\frac{\partial w}{\partial y}+w\frac{\partial w}{\partial z}\right)=\frac{\partial}{\partial x}\left(\mu\frac{\partial w}{\partial x}\right)+\frac{\partial}{\partial x}\left(\mu\frac{\partial w}{\partial y}\right)+\frac{\partial}{\partial z}\left(\mu\frac{\partial w}{\partial z}\right)+S_w \tag{1.162d}$$

TABLE 1.2
Description of Dependent Variable and Source Terms for the General Transport Equation

Equation	C	ϕ	Γ	Source Term, S_ϕ
x-momentum	ρ	u	μ	$-\frac{\partial P}{\partial x}+\rho g_x+S_{u0}$
y-momentum	ρ	v	μ	$-\frac{\partial P}{\partial y}+\rho g_y+S_{v0}$
z-momentum	ρ	w	μ	$-\frac{\partial P}{\partial z}+\rho g_z+S_{w0}$
Energy	ρC_p	T	K	$S_T=\Phi+\dot{Q}'''$
Mass concentration	1	C	D	S_C
Turbulence kinetic energy	ρ	K	$\mu+\mu_t$	$S_\kappa=P-\varepsilon$
Turbulence energy dissipation rate	ρ	ε	$\mu+\mu_t$	$S_\varepsilon=\frac{\varepsilon}{\kappa}(C_{\varepsilon1}P-C_{\varepsilon2}\varepsilon)$

Heat Transport Equation

$$\rho c_p\left(\frac{\partial T}{\partial t}+u\frac{\partial T}{\partial x}+\upsilon\frac{\partial T}{\partial y}+w\frac{\partial T}{\partial z}\right)=\frac{\partial}{\partial x}\left(k\frac{\partial T}{\partial x}\right)+\frac{\partial}{\partial y}\left(k\frac{\partial T}{\partial y}\right)+\frac{\partial}{\partial z}\left(k\frac{\partial T}{\partial z}\right)+S_T \quad (1.162e)$$

Mass Species Transport

$$\rho\left(\frac{\partial C_i}{\partial t}+u\frac{\partial C}{\partial x}+\upsilon\frac{\partial C}{\partial y}+w\frac{\partial C}{\partial z}\right)=\frac{\partial}{\partial x}\left(D\frac{\partial C}{\partial x}\right)+\frac{\partial}{\partial y}\left(D\frac{\partial C}{\partial y}\right)+\frac{\partial}{\partial z}\left(D\frac{\partial c}{\partial z}\right)+S_C \quad (1.162f)$$

Turbulence Kinetic Energy Transport

$$\rho\left(\frac{\partial\kappa}{\partial t}+u\frac{\partial\kappa}{\partial x}+v\frac{\partial\kappa}{\partial y}+w\frac{\partial\kappa}{\partial z}\right)=\frac{\partial}{\partial x}\left(\mu+\frac{\mu_t}{\sigma_\kappa}\frac{\partial\kappa}{\partial x}\right)+\frac{\partial}{\partial y}\left(\mu+\frac{\mu_t}{\sigma_\kappa}\frac{\partial\kappa}{\partial y}\right)+\frac{\partial}{\partial z}\left(\mu+\frac{\mu_t}{\sigma_\kappa}\frac{\partial\kappa}{\partial z}\right)+S_\kappa \quad (1.162g)$$

where S_κ is the turbulence kinetic energy source given as

$$S_\kappa = P - D$$

$$P=2\mu_t\left[\left(\frac{\partial u}{\partial x}\right)^2+\left(\frac{\partial v}{\partial y}\right)^2+\left(\frac{\partial w}{\partial z}\right)^2\right]+\mu_t\left[\left(\frac{\partial u}{\partial y}+\frac{\partial v}{\partial x}\right)^2+\left(\frac{\partial v}{\partial z}+\frac{\partial w}{\partial y}\right)^2+\left(\frac{\partial w}{\partial x}+\frac{\partial u}{\partial z}\right)^2\right]$$

Energy Dissipation Rate Transport

$$\rho\left(\frac{\partial\varepsilon}{\partial t}+u\frac{\partial\varepsilon}{\partial x}+v\frac{\partial\varepsilon}{\partial y}+w\frac{\partial\varepsilon}{\partial z}\right)$$

$$=\frac{\partial}{\partial x}\left(\mu+\frac{\mu_t}{\sigma_\varepsilon}\frac{\partial\varepsilon}{\partial x}\right)+\frac{\partial}{\partial y}\left(\mu+\frac{\mu_t}{\sigma_\varepsilon}\frac{\partial\varepsilon}{\partial y}\right)+\frac{\partial}{\partial z}\left(\mu+\frac{\mu_t}{\sigma_\varepsilon}\frac{\partial\kappa}{\partial z}\right)+S_\varepsilon\frac{\partial}{\partial t}(\rho\varepsilon)+\frac{\partial}{\partial x_j}\left[\rho u_j\varepsilon\right]$$

$$=\frac{\partial}{\partial x_j}\left[\left(\mu+\frac{\mu_t}{\sigma_\varepsilon}\right)\frac{\partial\varepsilon}{\partial x_j}\right]+S_\varepsilon \quad (1.162h)$$

where S_ε is the source term for the turbulence energy dissipation rate and expressed as

$$s_\varepsilon=\frac{\varepsilon}{\kappa}\left(C_{\varepsilon 1}P-C_{\varepsilon 2}\varepsilon\right)$$

Table 1.2 gives summary of the different dependent variables and source terms for the general transport Equation 1.161.

1.6 MATHEMATICAL CLASSIFICATION OF GOVERNING EQUATIONS

The governing differential equations for fluid flow and heat transfer can be transformed into a standard form depending on the physical nature under consideration. We will give a brief discussion on the mathematical classification of the different forms of these governing differential equations. This will help in selecting appropriate discretization and solution procedures. Let us consider a general two-dimensional second-order partial differential equation of the form

$$a(x,y)\frac{\partial^2\phi}{\partial x^2}+b(x,y)\frac{\partial\phi}{\partial x\partial y}+c(x,y)\frac{\partial^2\phi}{\partial y^2}+F\left(x,y,\phi,\frac{\partial\phi}{\partial x},\frac{\partial\phi}{\partial y}\right)=0 \quad (1.163)$$

The equation is linear for

$$F\left(x, y, \phi, \frac{\partial \phi}{\partial x}, \frac{\partial \phi}{\partial y}\right) = d\frac{\partial \phi}{\partial x} + e\frac{\partial \phi}{\partial y} + f\phi + g \tag{1.164}$$

where coefficients a, b, c, d, e, and f are arbitrary functions of independent variables x and y. The dependent variable φ and its normal derivatives are known on a smooth curve Γ in the x–y plane. A combination of these coefficients leads to different standard types of partial differential equations such as **elliptic**, **parabolic**, and **hyperbolic**. Results arise from the existence of a unique solution for all second derivatives at a point P on the curve. The necessary and sufficient conditions for such a solution state that if $a(\mathrm{P}) \neq 0$ the curve will be characteristic at the point P if and only if the slope of the curve at the point satisfies

$$\frac{\mathrm{d}y}{\mathrm{d}x} = \frac{b \pm \sqrt{b^2 - 4ac}}{a} \tag{1.165}$$

Three special cases can be highlighted.

1. If $b^2 - 4ac = 0$ at point P, there is only one characteristic direction at P and Equation 1.161 is parabolic at P. For example, the diffusion equation $\partial\phi/\partial t = \alpha\,\partial^2\phi/\partial x^2$ is parabolic in the x–t plane.
2. If $b^2 - 4ac < 0$, no real curve can satisfy Equation 1.163 and is not characteristic at point P, and Equation 1.161 is said to be **elliptic** at P. For the two-dimensional Laplace equation, $\partial^2\phi/\partial x^2 + \partial^2\phi/\partial x^2 = 0$, we get $b^2 - 4ac = -1$, so this equation is elliptic in the x–y plane.
3. If $b^2 - 4ac > 0$, there are two characteristic directions at P, and Equation 1.161 is said to be **hyperbolic**. For the wave equation of the form $\partial^2\phi/\partial t^2 = \beta\,\partial^2\phi/\partial x^2$, we get $b^2 - 4ac = \beta$, i.e., $b^2 - 4ac > 0$ for a positive value of β. So, the wave equation is hyperbolic in the x–t plane. More discussions about this classification are given by Anderson et al. (1984), Courant and Hilbert (1962), and Stakgold (1979).

This mathematical classification is often tied with the concepts of one-way and two-way coordinates (Patankar, 1980). A one-way coordinate is defined as the one in which the properties at a given point are influenced by changes in properties on only one side of that point, i.e., either upstream or downstream side of that point. In two-way coordinates, the properties at a point are influenced by changes in properties on both sides, i.e., both upstream and downstream of that point. While the term parabolic represents a one-way coordinate, the elliptic represents a two-way concept. Time is always one-way, based on the fact that the flow or temperature variable at a given time is only influenced by changes at a previous time. Space coordinates, on the other hand, are normally two-way co-ordinates. However, in some fluid flow and convection heat transfer problems, it is assumed to be one-directional or can be treated as parabolic. This situation arrives when there is a strong unidirectional flow in the coordinate direction so that the significant influences travel only from upstream points to downstream points. It should be remembered that diffusion processes have two-way characteristics. However, for strong unidirectional flow, convection overpowers diffusion and makes the space coordinate one-way.

PROBLEMS

1.1 Start from the Navier–Stokes equations and derive the governing equations in Cartesian coordinates for low Reynolds number flow, referred to as **creep flow**.

1.2 Make appropriate assumptions and drive the two-dimensional boundary-layer equations for flow over a flat plate in Cartesian coordinates.

1.3 Consider the flow of liquid film in steady and laminar motion, falling down an inclined surface with slope θ from the vertical direction of the gravity. The thickness, h, of the film is constant along the length. Assume the flow as fully developed and at zero-pressure gradient, i.e., $dP/dx = 0$.

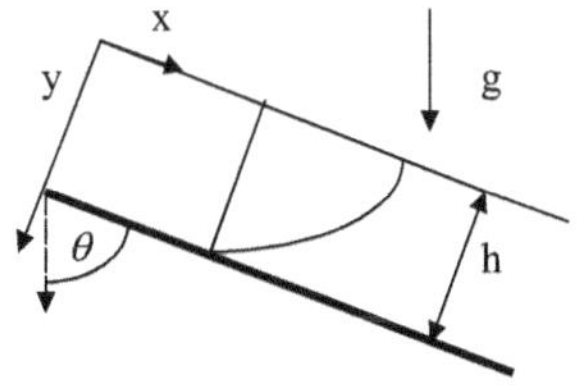

Derive the mathematical statement of the problem from mass and momentum equations of fluid flow for determining velocity distribution and mass flow rate of the liquid film.

1.4 Consider the flow of a viscous incompressible fluid in a square cavity shown in the figure below. The flow is induced by the motion of the top wall at a constant velocity U_w.

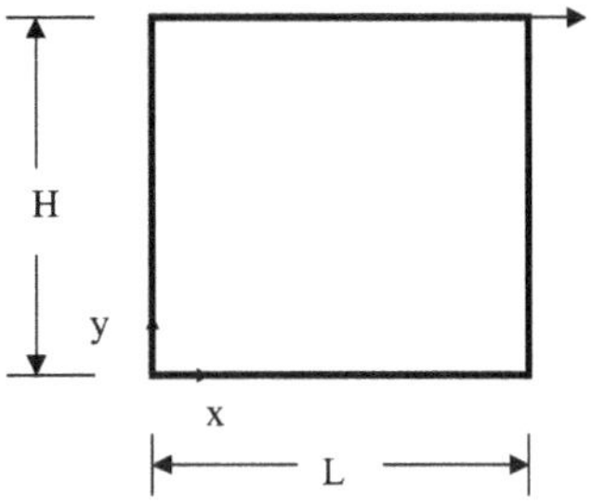

Derive the governing equation and boundary conditions.

1.5 Consider the fully developed laminar steady flow of Newtonian viscous fluid through a two-dimensional channel formed by two infinite parallel plates as shown below. The lower plate is stationary and the upper plate is moving in the x-direction with a constant speed, U_w, in the positive x-direction.

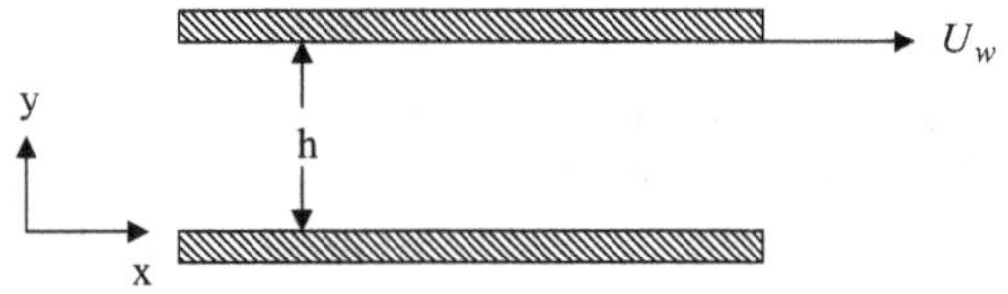

Derive the mathematical statement of the problem for determining from mass and momentum equations of fluid flow, referred to as Couette flow.

1.6 Consider the problem of developing laminar steady flow of Newtonian viscous fluid flow in a two-dimensional channel formed by two infinite parallel stationary plates as shown below. The upper plate is maintained at a constant temperature T_w and the lower plate is adiabatic.

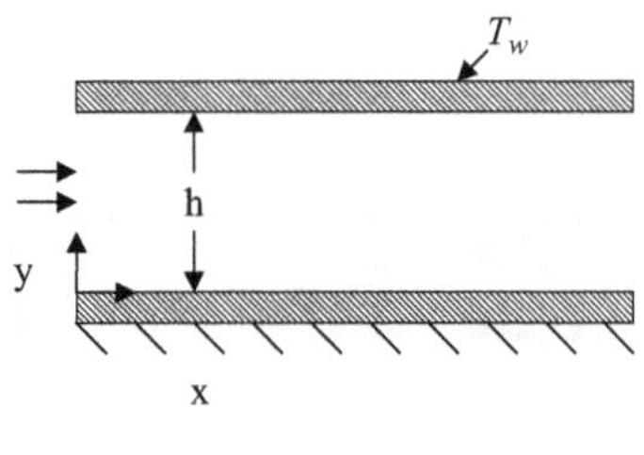

What are the continuity and x and y momentum equations for this pressure-driven $\left(dp/dx < 0 \right)$ channel flow? What is the appropriate energy equation with significant viscous heat dissipation?

1.7 Consider one-dimensional unsteady-state heat conduction in a plane slab of thickness L. Initially the slab is at a uniform temperature of T_I. At time $t > 0$, the left side of the slab is subjected to uniform constant surface heat flux q_s'' and the right surface is maintained at a constant temperature, T_R.

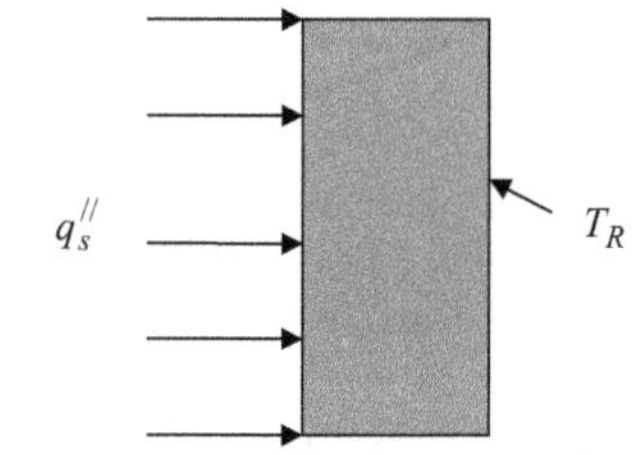

State the mathematical statement (governing equation, initial conditions, and boundary conditions) for determining the transient temperature distribution in the slab.

1.8 Consider two-dimensional steady-state conduction in a rectangular slab with the top surface maintained at a high temperature. All other surfaces are subjected to convection condition with h_c and T_∞.

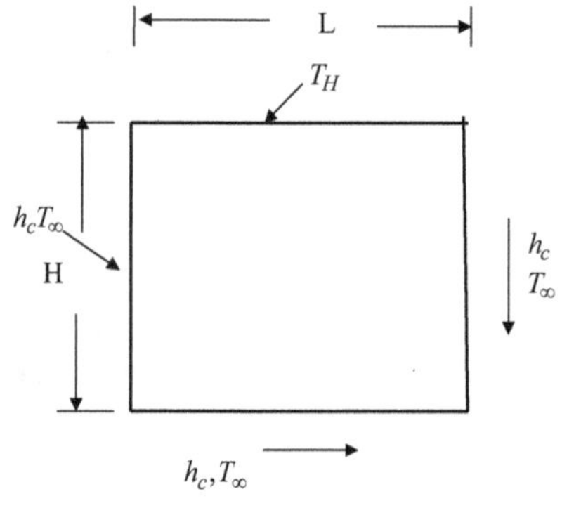

State the mathematical statement (governing equation and boundary conditions) for determining the steady-state temperature distribution in the slab.

1.9 Consider the diffusion of moisture from an air stream in a two-dimensional channel to the porous adsorbing material felt lined on the bottom of the air flow channel as shown. The felt is supported at the bottom by an aluminum plate, thus making it impermeable to mass diffusion. Air enters the channel at uniform velocity u_{in} and at uniform concentration C_{in}. Initial concentration distribution in the felt is uniform at C_0. Assume a constant rate of moisture adsorption m_{ad}''' and state the governing equations and boundary conditions for determining the transient moisture concentration distribution in the channel and in the solid felt.

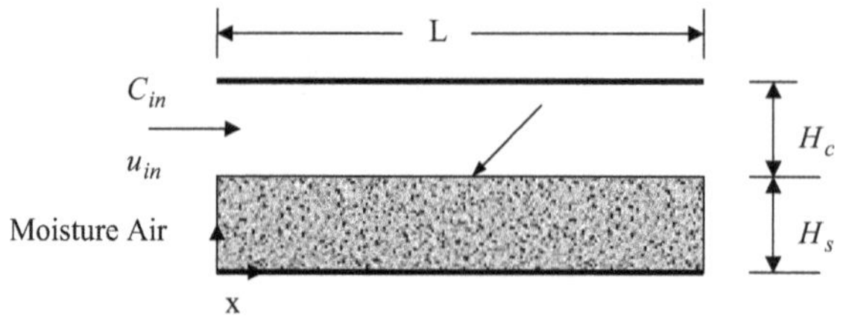

1.10 The figure below shows the schematic diagram of laser heating of a material. The incident laser beam is absorbed by the material and rapidly heats up the thin layer of the material surface and forms a thin layer of molten pool.

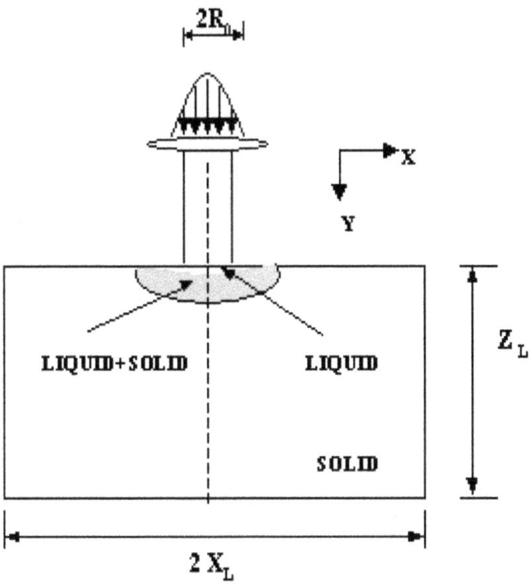

What are the governing equations in terms of continuity and x- and z-component momentum equations and boundary condition for the flow field in the liquid pool as developed by the surface tension force? What are the heat equations in the solid and liquid region and the associated boundary conditions?

2 Approximations and Errors

Understanding the concept of error is important, not only for the effective use of numerical methods but also to get an insight into improving the mathematical model and considering other variables and parameters. Numerical solutions always deviate from exact solutions, giving rise to errors since inaccuracies are introduced while approximating mathematical formulas and representing a quantity or a data by a computer numbering system. Inaccuracies are also introduced during arithmetic operations using data. Exact solutions are also subject to errors as computations are performed retaining only a finite number of digits. One of the major challenges of numerical methods is to retain the error within certain acceptable limits. There are two basic categories of error, **round off error** and **truncation error**, which are related directly to numerical methods and termed as computational errors. In addition, there are other sources of error that are not related to the computational methods themselves such as **gross blunders**, **formulation or modeling errors**, and **data uncertainty**. In this chapter we will discuss primarily the computational errors.

2.1 TRUNCATION ERROR

Truncation errors are those that result from using an approximation in place of exact mathematical operations and quantities. The mathematical statement of fluid flow and heat transfer problems usually involves derivatives or integrals of different orders. Approximations to such quantities are obtained by retaining a certain number of terms in a power series instead of using an infinite number of terms to obtain exact results. Let us consider the problem in which a quantity is approximated by a truncated power series. If a function $\varphi(x)$ and its $n+1$ derivatives are continuous on an interval $[x_{i+1}, x_i]$, then the value of the function at x_{i+1} is given in terms of the function value and its derivatives at another close point x_i by the Taylor series

$$\phi(x_{i+1}) = \phi(x_i) + \phi'(x_i)h + \frac{\phi''(x_i)}{2!}h^2 + \frac{\phi'''(x_i)}{3!}h^3 + \cdots + \frac{\phi^{(n)}(x_i)}{n!}h^n + R_n \tag{2.1}$$

where h is the step size defined by $h = \Delta x = x_{i+1} - x_i$. R_n is called the **reminder term** that represents the error in approximating the function by the sum of the first $n+1$ terms.

Accuracy of the approximation increases with increase in number of terms retained in the truncated series. The rest of the terms are grouped as a reminder term, which basically represents the **truncation error**, and it is expressed in integral form as

$$\mathrm{R_n} = \int_{x_i}^{x_{i+1}} \frac{(x_{i+1} - x_i)^{n+1}}{(n+1)!} \phi^{(n+1)}(x')\mathrm{d}x' \tag{2.2}$$

An alternative form of the reminder term is obtained by using the second mean value theorem for the integral (Equation 2.2) as

$$R_n = \frac{\phi^{n+1}(\xi)}{(n+1)!} h^{(n+1)} \tag{2.3}$$

where ξ is a value of x that lies between x_{i+1} and x_i.

The major difficulty in evaluating the truncation error exactly by using Equations 2.2 or 2.3 is that the function $\varphi(x)$ and its $(n + 1)$ derivatives are not known for problems for which a numerical solution is sought. Additionally, the value of ξ is not known exactly. However, these expressions for truncation error give a relative estimate of the truncation error. In order to express the relative error, the reminder term given by Equations 2.2 or 2.3 is also expressed as $R_n = O\left(h^{n+1}\right)$, and this means that the truncation error is of h^{n+1} order.

It can be seen that the **truncation error** is directly proportional to the step size, h, and so gives an idea about comparative errors of computational methods based on the Taylor series. For example, if the error is $O(h)$, reducing the step size by half will reduce the error by half. If the error is of $O(h^2)$, then reducing the step size by half will reduce the error by quarter. The important conclusion that we can make here is that retaining additional terms in the truncated Taylor series decreases the truncation error. In many cases, if h is sufficiently small, the first- and second-order terms will account for a disproportionately high percentage of the error. In such cases only a few terms are retained to retain an adequate level of approximation, and the desired accuracy of the solution is obtained by reducing the step size progressively. However, in many fluid flow and heat transfer problems we may reach a limit to decrease the step size any further. With decrease in step size, number calculation steps increase and, hence, it will increase the round-off error and will require increased computer memory. In such cases we will have to use higher-order approximation. In general, higher-order approximations converge faster, but require additional computations.

The physical interpretation of the order of such approximations can further be demonstrated in Figure 2.1 and considering the following cases.

In the case of a **zero-order approximation**, only the first term is retained in the estimation of the function $\phi(x_{i+1})$ as

$$\phi(x_{i+1}) \approx \phi(x_i) \tag{2.4a}$$

with error of the order $O(h)$. Such an approximation is good when the unknown function being approximated has a small variation or is close to constant over the interval x_{i+1} and x_i, as shown in Figure 2.1.

In the case of a **first-order approximation**, the first two terms are retained in the estimation of the function $\phi(x_{i+1})$ as

$$\phi(x_{i+1}) \approx \phi(x_i) + \phi'(x_i)h \tag{2.4b}$$

with error of the order $O(h^2)$. In such an approximation, the unknown function is assumed to vary in a linear straight-line manner over the interval x_{i+1} and x_i as shown in the figure.

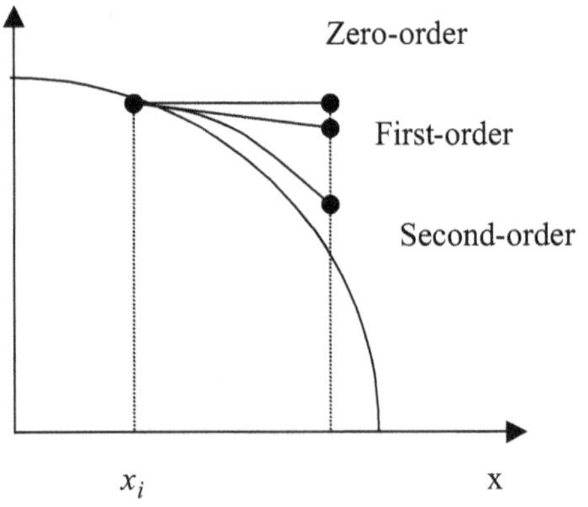

FIGURE 2.1 Order of approximation.

In the case of a **second-order approximation**, the first three terms are retained in the estimation of the function $\phi(x_{i+1})$ as

$$\phi(x_{i+1}) \approx \phi(x_i) + \phi'(x_i)h + \frac{\phi''(x_i)}{2!}h^2 \tag{2.4c}$$

with error of the order $O(h^3)$. In such an approximation, the unknown function is assumed to vary in a quadratic manner over the interval x_{i+1} and x_i as shown in the figure.

It can be noted that, with increased order of approximations, the function is more closely approximated.

Example 2.1: Truncation Error

Estimate the value of the function $\phi(x) = \sin x$ at $x = \pi/6$, based on the values of the function $\phi(x)$ and its derivatives at $x = \pi/5$ by considering one to six terms in the Taylor series.

Solution

In this problem, the step size is

$$h = \frac{\pi}{6} - \frac{\pi}{5} = -\frac{\pi}{30}$$

The zero-order approximation is computed as

$$\phi\left(\frac{\pi}{6}\right) \approx \phi\left(\frac{\pi}{5}\right)$$

or

$$\sin\left(\frac{\pi}{6}\right) \approx \sin\left(\frac{\pi}{5}\right) = 0.58778525229247$$

The first-order approximation is computed as

$$\phi\left(\frac{\pi}{6}\right) \approx \phi\left(\frac{\pi}{5}\right) + \phi'\left(\frac{\pi}{5}\right)h$$

or

$$\sin\left(\frac{\pi}{6}\right) \approx \sin\left(\frac{\pi}{5}\right) + \cos\left(\frac{\pi}{5}\right) \times \left(-\frac{\pi}{30}\right) = 0.50306519075389$$

The second-order approximation is computed as

$$\phi\left(\frac{\pi}{6}\right) \approx \phi\left(\frac{\pi}{5}\right) + \phi'\left(\frac{\pi}{5}\right)h + \frac{\phi''(\pi/5)}{2!}h^2$$

or

$$\sin\left(\frac{\pi}{6}\right) \approx \sin\left(\frac{\pi}{5}\right) + \cos\left(\frac{\pi}{5}\right) \times \left(-\frac{\pi}{30}\right) + \frac{-\sin(\pi/5)}{2!}\left(-\frac{\pi}{30}\right)^2 = 0.49984229746893$$

The procedure is repeated for the third- to fifth-order approximations and the results are summarized in the table below.

Order (n)	$\phi(x) = \sin(\pi/6)$
0	0.587785252
1	0.503065190
2	0.499842297
3	0.499997140
4	0.500000086
5	0.500000011
∞	0.5

We note that, with the addition of a greater number of terms in the Taylor series, the truncation error decreases and the approximate value approaches the exact solution.

2.2 ROUND-OFF ERROR

Round-off errors originate in all numerical calculations because actual numbers are approximated while storing and performing arithmetic operations on a computer. This is because that the computers can only represent quantities with a finite number of digits, i.e., retain only a fixed number of significant figures during calculations. Numbers such as π, e, $\sqrt{15}$, etc., cannot be expressed by a finite number of significant figures, and hence cannot be represented exactly. This discrepancy introduced by the omission of **significant figures** is called the **round-off error**, and it depends primarily on the **computer number system**. To obtain a better understanding of the round-off error, a clear understanding of the concept of significant figures and computer number systems is essential.

2.2.1 Significant Figures or Digits

The **significant digits** or **figures** of a number are those that can be used with confidence. They correspond to some certain digits plus one estimated digit. We must also remember that zeros are not always used as significant figures. For example, heat flux data 26 400 W/m^2 may have **three**, **four**, or **five** significant digits depending on whether the zeros are known with confidence. This uncertainty can be resolved by using scientific notation such as 2.64×10^4, 2.640×10^4, and 2.6400×10^4. Significant figures are important to particularly express an *approximate computational solution* with confidence. We should decide that our approximate solution is correct to some significant digits.

Example 2.2: Significant Figure

Consider a temperature reading from a temperature gauge or a thermometer with a resolution of 1°C. A temperature reading of 62.25°C has three certain digits 62.2 and one estimated digit 0.05, which is one half of the smallest scale division. So, the significant figure for temperature is 62.25°C.

2.2.2 Computer Number System

The information that is stored in a computer may be numerical data such as numbers or nonnumeric data that consist of alphabets, digits, and symbols. Digital computers store all information in binary form defined by different fundamental units such as **bit**, **byte**, and **word**. A bit is a binary digit, i.e., a zero (0) or one (1). A **byte** is a larger unit, which consists of 8 bits, and a *word* is an even bigger unit, which consists of several bytes or a string of bits. The word lengths could be 16 bits, 32 bits or 64 bits, depending on the type of computer. For example, a 32-bit word consists of 4 bytes or 32 bits. Numbers are stored in one or more words.

A number system is a way to represent a number. Some of the number systems that are commonly being used are the decimal or the 10-base system, octal or 8-base system, and the binary or 2-base system. Each computer uses a specific number system to express numbers. For example, integer numbers are whole numbers with no fractional part, and they are represented by a fixed-point numbering system such as

$$(I_n)_b = \pm(d_{n-1}d_{n-2}\ldots d_2d_1d_0)_b \tag{2.5}$$

where b is the base of the number system, n is the number of digits, and the digits d_j represent any integer values from $j = 0, 1, 2, \ldots (b-1)$.

For example, in a decimal system the number is evaluated as

$$\begin{aligned}(I_n)_{10} = (45\,682)_{10} &= 10^4(4)+10^3(5)+10^2(6)+10^1(8)+10^0(2)\\ &= 40\,000+5000+600+80+2\\ &= 45\,682\end{aligned}$$

and the decimal value of a binary number is evaluated as

$$\begin{aligned}(I_n)_2 = (11\,010\,1)_2 &= 2^5(1)+2^4(1)+2^3(0)+2^2(1)+2^1(0)+2^0(1)\\ &= 32+16+0+4+0+1=(53)_{10}\\ &= 32+16+0+4+0+1=(53)_{10}\\ &= (53)_{10}\end{aligned}$$

A real number is expressed using a **floating-point system** of the form

$$(R_n)_b = \pm d_1d_2d_3\ldots d_n \times b^e \tag{2.6}$$

where b is the base of the computer number system, digits $d_1d_2d_3\ldots d_n$ are defined as the **mantissa** (m) of the number, e is the exponent, and n is the finite length of the mantissa. Each d is an integer between 0 and $b-1$, and $d_1 \neq 0$. The mantissa is usually normalized in order to make sure that $d_1 \neq 0$ and this gives $1/b \leq m \leq 1$.

For example, a number such as 0.0004657891 is stored as 0.0004657×10^0 in a floating-point number system that allows only seven significant digits to be stored and without any normalization. However, with the use of normalization, the floating-point number system will store the number as 0.4657891×10^{-3}.

Another important item to be remembered is that the size of a **word** in terms of number of binary digits or bits limits the number of significant digits that can be retained in storing and representing a number. For example, in a typical 32-bit computer with 1 bit for the sign, 1 bit for the sign of the exponent, 7 bits for exponent (e), and 23 bits for the mantissa (n), the computer will represent a number with about seven decimal places or seven significant decimal digits because the least significant bit in the mantissa represents 2^{-23} or approximately 10^{-7}. The accuracy can be improved by using an additional **word** in representing the floating-point number. In a **double-precision** or **extended-precision** computation, floating point numbers are represented by two **words** with approximately twice as many bits in the mantissa, and this leads to double the number of significant digits or significant decimal places. However, double-precision computations require additional computer storage and are slower in computation than that in a single-precision computation.

Round-off error is introduced since a computer approximates a number consisting of more digits than the number of significant digits. The actual approximation is achieved in one of two ways: **chopping** or **rounding**. **Chopping** means that any quantity falling within an interval of length Δx will be retained as a quantity with the lower bound of the interval, and this leads to an upper bound of the error as Δx. On the other hand, in **rounding** any quantity falling within an interval Δx will be retained as the nearest allowable number, i.e., a number given either by the lower bound or the upper bound of the interval, whichever is the nearest. This leads to an upper bound of the error as $\Delta x/2$.

2.2.3 Machine Epsilon

Often, error bounds are expressed in normalized form to consider the fact that the interval Δx increases as the magnitude of the number itself increases. For **chopping**, the error bound is expressed as

$$\frac{|\Delta x|}{|x|} \le \varepsilon \tag{2.7}$$

and for rounding the error it is expressed as

$$\frac{|\Delta x|}{|x|} \le \frac{\varepsilon}{2} \tag{2.8}$$

where ε is defined as the **machine epsilon** that is computed for a specific computer as

$$\varepsilon = b^{1-n} \tag{2.9}$$

and where b is the number base and n represent the number of significant digits in the mantissa.

The value of the machine epsilon (ε) varies from computer to computer depending on the *word* length, number of words used (i.e., one or two), *base* (b), and type of number-approximation (*rounding* or *chopping*) used.

Example 2.3: Machine Epsilon

In a computer the value of $\pi = 3.141592654$ is approximated as 3.14159 by chopping. What is the number of significant digits retained in the computer?

Solution

The error bound is expressed as

$$\left|\frac{3.141592654 - 3.14159}{3.141\,592654}\right| \le b^{1-n}$$

or

$$0.845 \times 10^{-6} < 10^{1-7}$$

Hence, the number of significance digits for the approximation of π is 7.

Aside from the **computer number system in representing and storing a number**, there could be **round-off** error due to mathematical operations such as *Addition*, *Subtraction*, *Multiplication*, *Division*, etc. The loss of significant numbers during *subtraction* is the greatest source of **round-off**

error. Quite often we can use double precision to reduce the effect of round-off errors. However, extra price is paid in terms of more memory and execution time.

The inequality in Equations 2.7 and 2.8 specifies the worst possible cases. In the following section we will see the significance of the absolute magnitude of the **round-off** error as well as the normalized form of the error, and the *machine epsilon* in the convergence and stopping criteria in an iterative numerical scheme.

2.3 ERROR DEFINITIONS

As mentioned before, numerical errors arise from the use of an approximation to represent exact mathematical operations and quantities. The relationship between true or exact value and the approximate value is

$$\text{true value} = \text{approximate value} + \text{error} \tag{2.10}$$

Note that in the above equation the error term represents the true error, and this is expressed as

$$\text{true error}\ (E_t) = \text{true value}, \phi_{\text{tv}} - \text{approximate value}, \tag{2.11}$$

Normalizing the error with respect to the true value, we express the **true relative error** as

$$\varepsilon_t = \frac{E_t}{\phi_{\text{tv}}} \tag{2.12}$$

and **true percent relative error** as

$$\varepsilon_t = \frac{E_t}{\phi_{\text{tv}}} \times 100\% \tag{2.13}$$

Example 2.4: Error Estimation

Let us consider the true values of temperature and heat transfer rate in a heat conduction problem as T = 100°C and q = 10 000 W. Estimate the error and percent relative error if the approximate computational solutions are T = 99°C and q = 9999 W.

Solution

Error for temperature is

$$E_t = 100 - 99 = 1°\text{C}$$

$$\varepsilon_t = \frac{1}{100} \times 100\% = 1\%$$

Error for heat transfer rate is

$$E_t = 10\,000 - 9999 = 1\,\text{W}$$

$$\varepsilon_t = \frac{1}{10\,000} \times 100\% = 0.01\%$$

It can be concluded that the computational solution is adequate for predicting the heat transfer rate. However, the computational solution for temperature could be improved.

2.4 APPROXIMATE ERROR

Computational errors cannot be estimated exactly for problems in which there are no analytical solutions. In such cases we must rely on approximate errors. In cases where true values are not known, we estimate the error using the best available estimate of the true value, i.e., the approximate value itself, and express the *approximate percent relative error* as

$$\varepsilon_a = \frac{E_a}{\phi_{\text{av}}} \times 100\% \tag{2.14}$$

where

E_a = approximate error
ϕ_{av} = approximate value

In these cases, the major challenge is to determine the error in the absence of the true value. In most computational methods for fluid flows and heat transfer we employ an iterative scheme to compute answers. In this approach the present approximate value $\left(\phi^{k+1}\right)$ is estimated based on the previous approximate value $\left(\phi^{k}\right)$. This process is repeated or iterated to compute better and better approximation. In such cases we estimate **approximate error** and **approximate percent relative error** as

$$E_a = \phi^{k+1} - \phi^{k} \tag{2.15}$$

and

$$\varepsilon_a = \frac{\phi^{k+1} - \phi^{k}}{\phi^{k}} \tag{2.16}$$

2.5 CONVERGENCE CRITERIA

Signs of the approximate error can be positive or negative. However, we are interested in the absolute value of the error, i.e., $|\varepsilon_a|$. In an iterative scheme, convergence is assumed to have been reached when the absolute value of the error is less than a pre-specified tolerance limit, ε_s. In such a case computation is repeated or iterated until

$$|\varepsilon_a| \le \varepsilon_s \tag{2.17a}$$

or

$$\left|\frac{\phi^{k+1} - \phi^{k}}{\phi^{k}}\right| \le \varepsilon_s \tag{2.17b}$$

The prespecified tolerance limit is selected based on the number of significant digits desired in the numerical solution. It should be noted that the smallest numerical error that can be achieved is limited by the fact that the pre-assigned tolerance limit has to be greater than the error bound given by Equations 2.7 or 2.8, which relates the error to the number of significant figures or digits in the approximation. If this tolerance limit is met, then we can be assured that the computational result is correct to at least n significant figures or digits.

In the case of a vector array, $\{\phi\} = \{\phi_1, \phi_2 \ldots \phi_n\}$ or ϕ_i, Equation 2.17 is applied successively to all elements of the vector as

$$|\varepsilon_{ia}| \le \varepsilon_s \tag{2.18a}$$

or

$$\left|\frac{\phi_i^{k+1} - \phi_i^k}{\phi_i^k}\right| \le \varepsilon_s, \quad i = 1,2,3\ldots n \tag{2.18b}$$

In problems where the number of unknowns, n, or vector size is large, an alternate convergence criterion based on **Euclidean norm** is used as

$$\left\|\varepsilon_{ia}^{k+1}\right\|_e \le \varepsilon_s \tag{2.19}$$

where the **Euclidean norm** is defined as

$$\left\|\varepsilon_{ia}^{k+1}\right\|_e = \sqrt{\sum_{i=1}^{n}\left(\varepsilon_{ia}^{k+1}\right)^2} = \sqrt{\left(\left(\varepsilon_{1a}^{k+1}\right)^2 + \left(\varepsilon_{2a}^{k+1}\right)^2 \cdots + \left(\varepsilon_{na}^{k+1}\right)\right)^2} \tag{2.20}$$

In problems where the unknowns are arranged in a two-dimensional vector array or matrix array ϕ_{ij}, the convergence criterion can be assigned in the form of a **Frobenius norm** as

$$\left\|\varepsilon_{ij}^{k+1}\right\|_e \le \varepsilon_s \tag{2.21}$$

where the **Frobenius norm** is defined as

$$\left\|\varepsilon_{ij}^{k+1}\right\|_e = \sqrt{\sum_{i=1}^{n}\sum_{j=1}^{n}\left(\varepsilon_{ij}^{k+1}\right)^2} \tag{2.22}$$

Another alternative convergence criterion is sometimes given in terms of uniform vector norm as follows. For a vector array

$$\left\|\varepsilon_{i}^{k+1}\right\|_\infty \le \varepsilon_s \tag{2.23a}$$

where

$$\left\|\varepsilon_{i}^{k+1}\right\|_\infty = \max_{1\le i\le n}\left|\varepsilon_i\right| \tag{2.23b}$$

and for a matrix array

$$\left\|\varepsilon_{ij}^{k+1}\right\|_\infty \le \varepsilon_s \tag{2.23c}$$

where

$$\left\|\varepsilon_{ij}^{k+1}\right\|_\infty = \max_{1\le i\le n}\sum_{j=1}^{n}\left|\varepsilon_{ij}\right| \tag{2.24}$$

Example 2.5: Total Error

Compute the numerical estimate of the derivative and the associated total numerical error for the function $y = x^4$ using an approximation scheme for the first derivative and using progressively smaller step size. Estimate the true percent relative error and approximate percent relative error. Identify the optimum step size and error.

Solution

Let us estimate the derivative at a point $x_i = 10$ and $x_{i+1} = x_i + h$, and with progressively decreasing the step size by a factor of 10.

The exact derivative of the function is given as

$$\text{Der}E = y'(x_i) = 4x_i^3 = 4000 \tag{E.2.5.1}$$

We can approximate the first derivative of the function using the first-order approximation given by Equation 2.4b

$$\text{Der}A = y'(x_i) \cong \frac{y(x_{i+1}) - y(x_i)}{h} \tag{E.2.5.2}$$

Using the first estimate of the derivative with a step size $h = 0.1$, we get

$$\text{Der}A = \frac{(10+0.1)^4 - 10^4}{0.1} = 4060.401000000$$

We repeat this procedure with decreasing step sizes. Results for the estimate of the derivative and associated true and approximate errors are given in the table below and shown as a plot in Figure 2.2.

Step Size	Estimated Derivative	True Percent Relative Error	Approximate Percent Relative Error
1.0E−01	4060.401000000	1.510025000	-
1.0E−02	4006.004001000	0.150100025	1.357886787
1.0E−03	4000.600039999	0.015001000	0.135078762
1.0E−04	4000.060000391	0.001500010	0.013500788
1.0E−05	4000.005999853	0.000149996	0.001350011
1.0E−06	4000.000597006	0.000014925	0.000135071
1.0E−07	4000.000035687	0.000000892	0.000014033
1.0E−08	4000.000337001	0.000008425	0.000007533
1.0E−09	4000.000330961	0.000008274	0.000000151
1.0E−10	4000.000330961	0.000008274	0.000000000
1.0E−11	3999.645059594	0.008873510	0.008882572
1.0E−12	4000.355602329	0.008890058	0.017761989
1.0E−13	3979.039320257	0.524016994	0.535714286
1.0E−14	4263.256414561	6.581410364	6.666666667
1.0E−15	7105.427357601	77.635683940	40.000000000

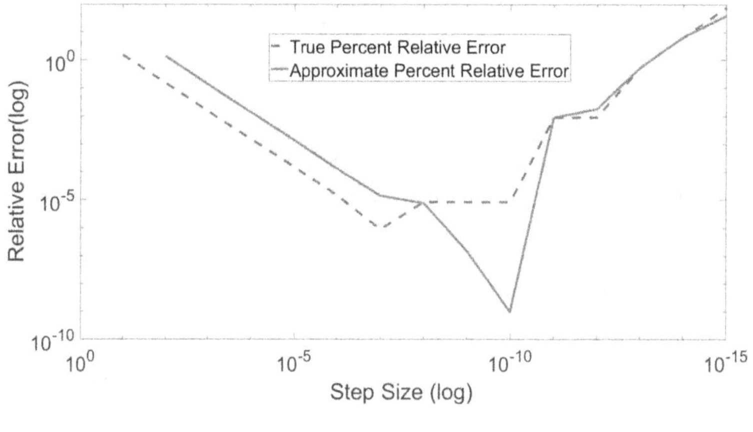

FIGURE 2.2 Plots for true and approximate relative error with decrease in step size.

We can see that the optimum size for the minimum total error is around h = 1.0E−09 for approximate percent relative error. With further decrease in step size, the total error starts increasing. The approximate percent relative error follows a similar pattern.

PROBLEMS

2.1 What would be the assigned tolerance limit, (ε_s), if convergence with four significant digits is required in an iterative scheme.

2.2 How many significant digits are there in the number 0.6843×10^{-3}?

2.3 Round the number 0.78654×10^3 to four significant digits.

2.4 Convert the binary number 11011011 to an equivalent decimal number in an 8-bit computer.

2.5 Represent the following binary numbers in decimal form: (a) 1011101 and (b) 110101.

2.6 Round the following numbers correctly to six significant digits: (a) 35.67457, (b) 0.00023145, and (c) 50 143.567.

2.7 In a computer the value of $\pi = 3.141592654$ is approximated as 3.14159 by chopping. What is the number of significant digits retained in the computer?

2.8 Write a computer program to estimate the value of the function $\phi(x) = \cos x$ at $x = \pi/3$ based on the values of the function $\phi(x)$ and its derivatives at $x = \pi/4$ by considering one to six terms in the Taylor series. Present your value of the function, true percent relative error, and approximate percent relative error at each step in a table.

2.9 The series expansion for sin x is given as

$$\sin x = x - \frac{x^3}{3!} + \frac{x^5}{5!} - \frac{x^7}{7!} + \cdots$$

Starting with the simplest version $\sin x \cong x$, add terms one at a time to estimate $\sin(\pi/3)$. Add terms until the absolute value of the approximate percent relative error falls below the assigned tolerance value of ε_s=0.5%.

2.10 Write a computer program to compute the numerical estimate of the derivative and total numerical error of the function $y = x^6$ using an approximation scheme for the first derivative and progressively smaller step size. Present your results for estimated derivate, true percent relative error and approximate percent relative error in a table. Decrease step sizes by a factor of 10, i.e., $\Delta x = 1/10^i$ and identify the optimum step size and error.

3 Numerical Solutions of Systems of Equations

In the subsequent chapters, we shall solve the system of simultaneous equations that are obtained by approximating the governing differential equations of heat and mass transfer problems, using discretization methods such as finite difference–control volume and finite element methods. The mathematical analysis of a linear physical system often results in a model consisting of simultaneous linear algebraic equations. The solution methods of nonlinear systems or differential equations use the technique of linearization of the models, thus requiring the repetitive solution of simultaneous linear algebraic equations.

These problems may range in complexity from a set of a few algebraic equations to a set of 1000 or more. For small- and medium-size problems, the time required for the solution of such a system of equations does not exceed that required for the formation of the system of equations using discretization methods. However, for large problems, the solution time for such systems dominates the total computing time. Also, the overall effectiveness of a discretization method depends to a large extent on the numerical procedure used for the solution of system of equations. As will be noted in subsequent chapters, the accuracy of the discretization method, in general, can be improved if a finer discretization mesh distribution is used. This means that the cost and effectiveness of a discretization method depend to a considerable extent on the solution algorithm for solving the resulting system of equations. Therefore, the use of inappropriate techniques for the solution of the system equations may lead to excessive computational time and even a numerically unstable solution algorithm.

The major objective of this chapter is to present various numerical methods for solving linear systems of simultaneous equations. Discretization of differential equations gives rise to various types of linear systems. Based on the physical problems under consideration, these systems pose special properties and structure, such as symmetric, nonsymmetric, banded and sparse, diagonally dominant, tridiagonal, block banded, block tridiagonal, and positive definite. The solution methods for solving these systems are described briefly in the following sections. For a detailed and further study of this topic, readers are suggested to refer to Ortega and Poole (1981), Datta (1995), Chapra and Canale (2002), Golub and Van Loan (1989), James et al. (1985), Hildebrand (1974), Forsythe et al. (1977), Saad (1996), and Rao (2002).

3.1 MATHEMATICAL BACKGROUND

3.1.1 Representation of the System of Equations

Consider the system of simultaneous linear algebraic equations represented as

$$\begin{aligned}
&a_{11}x_1 + a_{12}x_2 + \cdots + a_{1n}x_n = c_1 \quad &\text{(a)}\\
&a_{21}x_1 + a_{22}x_2 + \cdots + a_{2n}x_n = c_2 \quad &\text{(b)}\\
&\vdots\\
&a_{n1}x_1 + a_{n2}x_2 + \cdots + a_{nn}x_n = c_n \quad &\text{(n)}
\end{aligned} \tag{3.1}$$

where a_{ij} terms are the known coefficients of the equation, the x_j terms are the unknown variables, and the c_i terms are the known constants. Equation 3.1 represents a system of linear equations with n unknowns.

In developing a systematic approach for the solution of linear algebraic equations, we will make use of matrix-vector notations. For this reason, and for the benefit of the reader, a brief review of selected matrix and vector operations is given in Appendix A. In general, these systems of equations are often expressed in matrix notion as

$$[A]\{x\}=\{c\} \tag{3.2a}$$

or simply as

$$Ax = c \tag{3.2b}$$

where $[A]$ is the coefficient matrix written as

$$[A]=\begin{bmatrix} a_{11} & a_{12} & \dots & \dots & \dots & a_{1n} \\ a_{21} & a_{22} & \dots & \dots & \dots & a_{2n} \\ \vdots & & & & & \\ \vdots & & & & & \\ a_{n1} & a_{n2} & \dots & \dots & \dots & a_{nn} \end{bmatrix} \tag{3.3}$$

The elements of the coefficient matrix are the known coefficients of the system of equations and are designated by a_{ij}. The first subscript i designates the row number in which the element lies. The second subscript j designates the corresponding column number. $\{c\}$ is a $(n \times 1)$ column vector of constants or known values

$$\{c\}^{\mathrm{T}} = \begin{bmatrix} c_1 & c_2 \dots c_n \end{bmatrix} \tag{3.4}$$

and $\{x\}$ is the $(n \times 1)$ column vector of unknowns:

$$\{x\}^{\mathrm{T}} = \begin{bmatrix} x_1 & x_2 \dots x_n \end{bmatrix} \tag{3.5}$$

We will consider cases when the matrix A is square; that is, the number of unknowns equals the number of equations. Also, the nonhomogeneous set of linear equations (3.1) is assumed to be nonsingular; that is, it has linearly independent rows (and columns) or the determinant of the coefficient matrix is nonzero, for which a unique nontrivial solution exists.

3.1.2 Cramer's Rule and the Elimination of Unknowns

The solution of a set of a few linear algebraic equations can be obtained by **Cramer's rule** or by the **method of elimination of unknowns**. Cramer's rule is a solution technique that is used to calculate the solution of the nonhomogeneous linear algebraic equation (3.1) by expressing each unknown as the ratio of determinants of the coefficient matrix A and substituted matrix A_j as

$$x_j = \frac{|A_j|}{|A|}, \quad j = 1,2,\dots,n \tag{3.6}$$

The substituted matrix A_j is obtained by replacing jth columns of the matrix A with the known vector c as

$$[A_j] = \begin{bmatrix} a_{11} & a_{12} & \dots & c_1 & \dots & a_{1n} \\ a_{21} & a_{22} & \dots & c_2 & \dots & a_{2n} \\ \vdots & & & & & \\ \vdots & & & & & \\ a_{n1} & a_{n2} & \dots & c_n & \dots & a_{nn} \end{bmatrix} \tag{3.7}$$

The solution of a set of a few linear algebraic equations can be obtained easily by the application of Cramer's rule. However, for systems involving five or more equations, Cramer's rule requires a rapidly escalating number of arithmetic operations to evaluate the determinants, too large even for today's high-speed computers, and becomes impractical. The **method of elimination of unknowns** is an algebraic approach that involves multiplying the equations by constants in order that one of the unknowns will be eliminated when two equations are combined. This will result in a single equation that can be solved for the remaining unknown. The value of this unknown can then be substituted into one of the original equations to compute the other variables.

A solution procedure based on the elimination of unknowns appears to be simple to evaluate for systems with three or four equations. However, for larger systems, the elimination of unknown methods becomes too complex and extremely tedious to implement by hand calculations.

In the subsequent sections, we will discuss more efficient alternative solution methods. There are two general types of numerical methods available for solving a nonhomogeneous set of linear algebraic equations. Methods of the first type are called **direct methods** and the methods of the second type are called **iterative methods**.

3.2 DIRECT METHODS

The direct methods perform a finite number of steps to obtain the solution. Among these we will consider Gaussian elimination, the Gauss–Jordan elimination or factorization, LU decomposition, Cholesky decomposition, and the diagonally dominant system.

3.2.1 Gaussian Elimination

The Gaussian elimination method uses systematically the procedure utilized in the method of elimination of unknowns. This method employs the scheme or algorithm in two steps, which are referred to as the **elimination** and **back substitutions**. In the **elimination step**, a set of n equations in n unknowns are reduced to an equivalent upper triangular set, which is then easily solved in the **back-substitution step**. The procedure will be illustrated by solving the set of simultaneous equations given by Equation 3.1.

Elimination Step

1. This step is designed to reduce the set of equations to an upper triangular set. The initial step will be to eliminate the first unknown, x, from the second through the nth equation. To do this, divide Equation 3.1a by the coefficient of x_1, i.e., a_{11}, referred to as the **pivot element**, to get

$$x_1 + \frac{a_{12}}{a_{11}} x_2 + \frac{a_{13}}{a_{11}} x_3 + \dots + \frac{a_{1n}}{a_{11}} x_n = \frac{c_1}{a_{11}} \tag{3.8}$$

Next, multiply Equation 3.8 by the coefficient of x in Equation 3.1b, i.e., a_{21}, to get

$$a_{21}x_1 + \frac{a_{21}}{a_{11}}a_{12}x_2 + \frac{a_{21}}{a_{11}}a_{13}x_3 + \cdots + \frac{a_{21}}{a_{11}}a_{1n}x_n = \frac{a_{21}}{a_{11}}c_1 \tag{3.9}$$

Now this equation can be subtracted from Equation 3.1b to get

$$\left(a_{22} - \frac{a_{21}}{a_{11}}a_{12}\right)x_2 + \left(a_{23} - \frac{a_{21}}{a_{11}}a_{13}\right)x_3 + \cdots + \left(a_{2n} - \frac{a_{21}}{a_{11}}a_{1n}\right)x_n = \left(c_2 - \frac{a_{21}}{a_{11}}c_1\right) \tag{3.10}$$

or

$$a'_{22}x_2 + a'_{23}x_3 + \cdots + a'_{2n}x_n = c'_2 \tag{3.11}$$

In a similar manner, x_1 is eliminated from all Equations 3.1c–n of the set except Equation 3.1a, so that the set assumes the modified form

$$\begin{aligned}
a_{11}x_1 + a_{12}x_2 + a_{13}x_3 + \cdots + a_{1n}x_n &= c_n && \text{(a)}\\
a'_{22}x_2 + a'_{23}x_3 + \cdots + a'_{2n}x_n &= c'_2 && \text{(b)}\\
a'_{32}x_2 + a'_{33}x_3 + \cdots + a'_{3n}x_n &= c'_3 && \text{(c)}\\
&\vdots\\
&\vdots\\
a'_{n2}x_2 + a'_{n3}x_3 + \cdots + a'_{nn}x_n &= c'_n && \text{(n)}
\end{aligned} \tag{3.12}$$

In the foregoing steps, Equation 3.1a is used to eliminate the unknown in the subsequent equations and is called the **pivot equation**. In the pivot equation, a_{11}, the coefficient of the unknown, x_1, which is to be eliminated from the subsequent equations, is called the **pivot element**. The prime superscripts indicate that the coefficients are modified in the elimination process.

2. Repeat the above step in order to eliminate the second unknown from all equations following the pivot equation, that is, Equation 3.12b and using a'_{22} as the pivot element. This elimination results in

$$\begin{aligned}
a_{11}x_1 + a_{12}x_2 + a_{13}x_3 + \cdots + a_{1n}x_n &= c_n && \text{(a)}\\
a'_{22}x_2 + a'_{23}x_3 + \cdots + a'_{2n}x_n &= c'_2 && \text{(b)}\\
a''_{33}x_3 + \cdots + a''_{3n}x_n &= c''_3 && \text{(c)}\\
&\vdots\\
&\vdots\\
a''_{n3}x_3 + \cdots + a''_{nn}x_n &= c''_n && \text{(n)}
\end{aligned} \tag{3.13}$$

3. This procedure is continued until the original set of equations has been transformed to an upper triangular system in the $(n-1)$th elimination step

$$
\begin{aligned}
a_{11}x_1 + a_{12}x_2 + a_{13}x_3 + \cdots + a_{1n}x_n &= c_n && \text{(a)} \\
a'_{22}x_2 + a'_{23}x_3 + \cdots + a'_{2n}x_n &= c'_2 && \text{(b)} \\
a''_{33}x_3 + \cdots + a''_{3n}x_n &= c''_3 && \text{(c)} \\
&\vdots \\
a^{n-2}_{n-1,n-1}x_{n-1} + a^{n-2}_{n-1,n}x_n &= c^{n-2}_{n-1} && \text{(n}-1\text{)} \\
a^{n-1}_{nn}x_n &= c^{n-1}_n && \text{(n)}
\end{aligned}
\qquad (3.14)
$$

Back Substitution

Once the upper triangular system of equations is obtained, the unknown can be solved one at a time through back substitution. Equation 3.14n is solved directly for x_n as

$$x_n = \frac{c^{n-1}_n}{a^{n-1}_{nn}} \qquad (n)$$

This value is then substituted into Equation 3.14$n-1$ to solve for x_{n-1}

$$x_{n-1} = \frac{c^{n-2}_{n-1} - a^{n-2}_{n-1,n}x_n}{a^{n-2}_{n-1,n-1}} \qquad (n-1).$$

This back-substitution procedure is continued to solve the remaining unknown as

$$
\begin{aligned}
x_{n-2} &= \frac{c^{n-3}_{n-2} - a^{n-3}_{n-2,n-1}x_{n-1} - a^{n-3}_{n-2,n}x_n}{a^{n-3}_{n-2,n-2}} && (n-2) \\
&\vdots \\
x_2 &= \frac{c'_2 - a'_{23}x_3 \ldots a'_{2,n-1}x_{n-1} - a'_{2,n}x_n}{a'_{22}} && \text{(b)} \\
x_1 &= \frac{c_1 - a_{12}x_2 - a_{13}x_3 \ldots a_{1,n-1}x_{n-1} - a_{1n}x_n}{a_{11}} && \text{(a)}
\end{aligned}
\qquad (3.15)
$$

Generalized Formula The procedure outlined in the elimination and back substitution steps can be generalized to express the modified elements and the unknown as

Elimination

$$a^k_{ij} = a^{k-1}_{ij} - \frac{a^{k-1}_{kj}}{a^{k-1}_{kk}} a^{k-1}_{ik} \quad \text{for} \quad k+1 \le j \le m, \quad k+1 \le i \le n \quad \text{and} \quad k = 1,2\ldots n-1, \qquad (3.16)$$

where

i = index identifying the row number of the matrix

j = index identifying the column number of the augmented matrix

k = index identifying pivot row or the elimination step

n = number of rows in the matrix

m = number of columns in the augmented matrix = $n + 1$

It can be noted that the lower limit for j is $k+1$ instead of k. There is no need to calculate the anticipated initial zero values that appear during the elimination steps.

Back Substitution

$$x_n = \frac{a_{nm}}{a_{nn}} \tag{3.17a}$$

and

$$x_i = \frac{a_{im}^{i} - \sum_{j=i+1}^{n} a_{ij}^{i-1} x_j}{a_{ii}^{i-1}}, \quad \text{for} \quad i = n-1, n-2, \ldots, 1 \tag{3.17b}$$

Some problems arise in the Gauss elimination procedure when a pivot element is zero or close to zero. This leads to division by zero or to large round-off errors. To avoid these problems, a procedure known as **pivoting** is used in the elimination step. Each elimination step should be made by choosing a row with the largest pivot element as the pivot row. This is achieved by searching for the row with the largest element and switching the rows with the largest element. This procedure is known as **partial pivoting**. If columns as well as rows are searched and switched for the largest element, then the procedure is called **complete pivoting**. For solving a large system of equations, partial pivoting is always incorporated in the computer program. Complete pivoting is also desirable for some problems. However, it adds more complexity to the computer program.

As it has been mentioned before, there may be a considerable loss of accuracy in the results owing to the round-off error that accumulates in many arithmetic operations. This may be quite significant when solving a large system of equations. So, it is important to check the accuracy by substituting the answers back into the original equations and compute the residuals as

$$\begin{aligned} r_1 &= a_{11}x_1 + a_{12}x_2 \cdots a_{1n}x_n - c_1 && \text{(a)} \\ r_2 &= a_{21}x_1 + a_{22}x_2 + \cdots a_{2n}x_n - c_2 && \text{(b)} \\ &\vdots \\ r_n &= a_{n1}x_1 + a_{n2}x_2 + \cdots + a_{nn}x_n - c_n && \text{(c)} \end{aligned} \tag{3.18}$$

Example 3.1: Gauss Elimination Method

Use the Gauss elimination method to solve the following system of equations:

$$\begin{bmatrix} 4.855 & -4 & 1 & 0 \\ -4 & 5.855 & -4 & 1 \\ 1 & -4 & 5.855 & -4 \\ 0 & 1 & -4 & 4.855 \end{bmatrix} \begin{pmatrix} x_1 \\ x_2 \\ x_3 \\ x_4 \end{pmatrix} = \begin{pmatrix} -1.59 \\ 1 \\ 1 \\ -1.64 \end{pmatrix} \tag{E.3.1.1}$$

Solution

In this problem, we use the basic Gauss elimination algorithm given by Equation 3.16 for elimination steps and Equation 3.17 for the back-substitution step. Following this algorithm, we get

Elimination step: $k = 1$, pivot row = 1, pivot element: $a_{11} = 4.855$

$$\begin{bmatrix} 4.85500 & -4 & 1 & 0 \\ 0 & 2.55944 & -3.17610 & 1 \\ 0 & -3.17610 & 5.64902 & -4 \\ 0 & 1 & -4 & 4.85500 \end{bmatrix} \begin{pmatrix} x_1 \\ x_2 \\ x_3 \\ x_4 \end{pmatrix} = \begin{bmatrix} -1.59000 \\ -0.309980 \\ 1.32719 \\ -1.64000 \end{bmatrix} \quad \text{(E.3.1.2)}$$

Elimination step: $k = 2$, pivot row = 2, pivot element: $a'_{22} = 2.55944$

$$\begin{bmatrix} 4.85500 & -4 & 1 & 0 \\ 0 & 2.55944 & -3.17610 & 1 \\ 0 & 0 & 1.70771 & -2.75907 \\ 0 & 0 & -2.75907 & 4.46429 \end{bmatrix} \begin{pmatrix} x_1 \\ x_2 \\ x_3 \\ x_4 \end{pmatrix} = \begin{pmatrix} -1.59000 \\ -0.309980 \\ 0.942827 \\ -1.51888 \end{pmatrix} \quad \text{(E.3.1.3)}$$

Elimination step: $k = 3$, pivot row = 3, pivot element: $a''_{33} = 1.70771$

$$\begin{bmatrix} 4.85500 & -4 & 1 & 0 \\ 0 & 2.55944 & -3.17610 & 1 \\ 0 & 0 & 1.70771 & -2.75907 \\ 0 & 0 & & 0.006600 \end{bmatrix} \begin{pmatrix} x_1 \\ x_2 \\ x_3 \\ x_4 \end{pmatrix} = \begin{pmatrix} -1.59000 \\ -0.309980 \\ 0.942827 \\ 0.004390 \end{pmatrix} \quad \text{(E.3.1.4)}$$

Back Substitution Step

$$x = \begin{pmatrix} 0.686706 \\ 1.63768 \\ 1.62674 \\ 0.665151 \end{pmatrix} \quad \text{(E.3.1.5)}$$

To check the accuracy of the answer, we now evaluate the residuals by substituting the answer back into the original system of equation

$$r = \begin{pmatrix} 0.00002237 \\ 0.00001660 \\ 0.00005530 \\ -0.000028105 \end{pmatrix} \quad \text{(E.3.1.6)}$$

3.2.2 Gauss–Jordan Elimination Method

This method is a variation of the Gaussian elimination method. The major difference is that when an unknown is eliminated, it is eliminated from all the other equations preceding the pivot equation as well as those following it. Thus, the elimination steps reduce the coefficient matrix to an identity matrix rather than a triangular matrix. The solution is then obtained directly without requiring a back-substitution step. The solution vector becomes exactly equal to the modified right-hand side vector. The following steps illustrate the procedure for solving the system given by Equation 3.1.

Original System

$$\begin{aligned} a_{11}x_1 + a_{12}x_2 \cdots a_{1n}x_n &= c_1 \\ a_{21}x_1 + a_{22}x_2 \cdots a_{2n}x_n &= c_2 \\ &\vdots \\ a_{n1}x_1 + a_{n2}x_2 \cdots a_{nn}x_n &= c_n \end{aligned} \tag{3.19}$$

Elimination step = 1, pivot row = 1, pivot element = a_{11}

$$\begin{aligned} x_1 + a'_{12}x_2 + a'_{13}x_3 + \cdots + a'_{1n} &= c'_1 \\ a'_{22}x_2 + a'_{23}x_3 + \cdots + a'_{2n} &= c'_2 \\ &\vdots \\ a'_{n2}x_2 + a'_{n3}x_3 + \cdots + a'_{nn} &= c'_n \end{aligned} \tag{3.20}$$

Elimination step = 2, pivot row = 2, pivot element = a'_{22}

$$\begin{aligned} x_1 + 0x_2 + a''_{13}x_3 + \cdots + a''_{1n} &= c''_1 \\ 0x_2 + a''_{23}x_3 + \cdots + a''_{2n} &= c''_2 \\ &\vdots \\ 0x_2 + a''_{n3}x_3 + \cdots + a''_{nn} &= c''_n \end{aligned} \tag{3.21}$$

Elimination step = n, pivot row = n, pivot element = a_{nn}^{n-1}

$$\begin{aligned} x_1 \qquad\qquad &= c_1^n \\ x_2 \qquad &= c_2^n \\ &\vdots \\ x_n &= c_n^n \end{aligned} \tag{3.22}$$

The solution vector $\{x\}$ is given directly by the modified $\{c\}$ vector without requiring any back-substitution step.

Generalized Formula The procedure outlined in the Gauss–Jordan method can be generalized to express modified elements for all elimination steps as

$$a_{kj}^{k} = \frac{a_{kj}^{k-1}}{a_{kk}} \tag{3.23a}$$

$$a_{ij}^{k} = a_{ij}^{k-1} - a_{ik}^{k-1} a_{kj}^{k} \tag{3.23b}$$

for $k = 1, 2, \ldots, n$, $1 \le i \le n$ except $i = k$, and $k + 1 \le j \le n + 1$.

Example 3.2: Gauss–Jordan Method:

Use the Gauss–Jordan method to solve the following system of equations:

$$\begin{bmatrix} 3 & -6 & 7 \\ 9 & 0 & -5 \\ 5 & -8 & 6 \end{bmatrix} \begin{Bmatrix} x_1 \\ x_2 \\ x_3 \end{Bmatrix} = \begin{Bmatrix} 3 \\ 3 \\ -4 \end{Bmatrix} \qquad \text{(E.3.2.1)}$$

Solution

We combine the coefficient matrix and the right-hand side vector to write the augmented matrix, A, as

$$\begin{bmatrix} 3 & -6 & 7 & 3 \\ 9 & 0 & -5 & 3 \\ 5 & -8 & 6 & -4 \end{bmatrix} \qquad \text{(E.3.2.2)}$$

We then use the Gauss–Jordan algorithm given by Equation 3.23 that reduces the original coefficient matrix to an identity matrix. Following this algorithm, we have

Elimination step, $k = 1$, pivot row = 1, pivot element $a_{11} = 3$

Elements of row 1 are calculated using Equation 3.23a, which is $a'_{ij} = a_{ij}/a_{11}$ for this elimination step. This gives the modified matrix as

$$\begin{bmatrix} 1 & a'_{12} & a'_{13} & a'_{14} \\ a_{21} & a_{22} & a_{23} & a_{24} \\ a_{31} & a_{32} & a_{33} & a_{34} \end{bmatrix} = \begin{bmatrix} 1 & -2 & 2.33 & 1 \\ 9 & 0 & -5 & 3 \\ 5 & -8 & 6 & -4 \end{bmatrix} \qquad \text{(E.3.2.3)}$$

Equation 3.23b is then used to calculate elements of rows 2 and 3 as follows:

For $i = 2, j = 2, 3, 4$

$$\begin{bmatrix} 1 & a'_{12} & a'_{13} & a'_{14} \\ 0 & a'_{22} & a'_{23} & a'_{24} \\ a_{31} & a_{32} & a_{33} & a_{34} \end{bmatrix} = \begin{bmatrix} 1 & -2 & 2.33 & 1 \\ 0 & 18 & -26 & -6 \\ 5 & -8 & 6 & -4 \end{bmatrix} \qquad \text{(E.3.2.4)}$$

For $i = 3, j = 2, 3, 4$

$$\begin{bmatrix} 1 & a'_{12} & a'_{13} & a'_{14} \\ 0 & a'_{22} & a'_{23} & a'_{24} \\ 0 & a'_{32} & a'_{33} & a'_{34} \end{bmatrix} = \begin{bmatrix} 1 & -2 & 2.33 & 1 \\ 0 & 18 & -26 & -6 \\ 0 & 2 & -5.65 & -9 \end{bmatrix} \qquad \text{(E.3.2.5)}$$

Elimination step, $k = 2$, pivot row = 2, pivot element $a'_{22} = 18$

Elements of row 2 are calculated using Equation 3.23a, which reduces to $a''_{2j} = a'_{2j}/a'_{22}$. This gives the modified matrix as

$$\begin{bmatrix} 1 & a'_{12} & a'_{13} & a'_{14} \\ 0 & 1 & a''_{23} & a''_{24} \\ 0 & a'_{32} & a'_{33} & a'_{34} \end{bmatrix} = \begin{bmatrix} 1 & -2 & 2.33 & 1 \\ 0 & 1 & 1.44 & -0.33 \\ 0 & 2 & -5.65 & -9 \end{bmatrix} \qquad \text{(E.3.2.6)}$$

Equation 3.23b is then used to calculate elements of rows 3 and 1 as

For $i = 3, j = 3, 4$

$$\begin{bmatrix} 1 & 0 & a'_{13} & a'_{14} \\ 0 & 1 & a''_{23} & a''_{24} \\ 0 & 0 & a''_{33} & a''_{34} \end{bmatrix} = \begin{bmatrix} 1 & -2 & 2.33 & 1 \\ 0 & 1 & -1.44 & -0.33 \\ 0 & 0 & -2.77 & -8.33 \end{bmatrix} \quad \text{(E.3.2.7)}$$

For $i = 1, j = 3, 4$

$$\begin{bmatrix} 1 & 0 & a''_{13} & a''_{14} \\ 0 & 1 & a''_{23} & a''_{24} \\ 0 & 0 & a''_{33} & a''_{34} \end{bmatrix} = \begin{bmatrix} 1 & 0 & -0.55 & 0.33 \\ 0 & 1 & -1.44 & -0.33 \\ 0 & 0 & -2.77 & -8.33 \end{bmatrix} \quad \text{(E.3.2.8)}$$

Elimination step, $k = 3$, pivot row = 3, pivot element $a''_{33} = -2.77$
Elements of row 3 are calculated using $a'''_{3j} = a''_{3j}/a''_{33}$. This modifies the augmented matrix to

$$\begin{bmatrix} 1 & 0 & a''_{13} & a''_{14} \\ 0 & 1 & a''_{23} & a''_{24} \\ 0 & 0 & 1 & a'''_{34} \end{bmatrix} = \begin{bmatrix} 1 & 0 & -0.55 & 0.33 \\ 0 & 1 & -1.44 & -0.33 \\ 0 & 0 & 1 & 3.01 \end{bmatrix} \quad \text{(E.3.2.9)}$$

Equation 3.23b is then used to calculate elements of row 1 and 2 as
For $i = 1, j = 4$

$$\begin{bmatrix} 1 & 0 & 0 & a'''_{14} \\ 0 & 1 & a''_{23} & a''_{24} \\ 0 & 0 & 1 & a'''_{34} \end{bmatrix} = \begin{bmatrix} 1 & 0 & 0 & 1.99 \\ 0 & 1 & -1.44 & -0.33 \\ 0 & 0 & 1 & 3.01 \end{bmatrix} \quad \text{(E.3.2.10)}$$

For $i = 2, j = 4$

$$\begin{bmatrix} 1 & 0 & 0 & a'''_{14} \\ 0 & 1 & 0 & a'''_{24} \\ 0 & 0 & 1 & a'''_{34} \end{bmatrix} = \begin{bmatrix} 1 & 0 & 0 & 1.99 \\ 0 & 1 & 0 & 4.00 \\ 0 & 0 & 1 & 3.01 \end{bmatrix} \quad \text{(E.3.2.11)}$$

The solution vector is then given by the modified elements of the right-side vector as

$$\begin{Bmatrix} x_1 \\ x_2 \\ x_3 \end{Bmatrix} = \begin{Bmatrix} a'''_{14} \\ a'''_{24} \\ a'''_{34} \end{Bmatrix} = \begin{Bmatrix} 1.99 \\ 4.00 \\ 3.01 \end{Bmatrix} \quad \text{(E.3.2.12)}$$

All problems and improvements discussed for the Gauss elimination method are applicable to the Gauss–Jordan method. So, the method of pivoting is applicable to the Gauss–Jordan method to avoid zero division and to reduce round-off error. It can be mentioned here that the Gauss–Jordan method usually takes more operations than Gauss-elimination. Therefore, the Gauss-elimination method is usually the method of choice for obtaining the exact solutions of a system of linear equations. The Gauss–Jordan method is preferred for obtaining a matrix inverse (A^{-1}) that provides a convenient way of evaluating a set of simultaneous equations with multiple right-hand side force vectors, $\{c\}$.

The solution of the set of simultaneous equations given in Equation 2.1 can be obtained by using the matrix inverse A^{-1} as

$$\{x\} = [A]^{-1}\{c\} \tag{3.24}$$

Applications of such procedures occur when it is necessary to solve the same system of equations as in Equation 2.1 many times with different right-hand side vector $\{c\}$. The step for obtaining the matrix inverse is performed only once. This step involves augmenting the coefficient matrix $[A]$ with an identify matrix, $[I]$. Then the Gauss–Jordan algorithm is employed to reduce the coefficient matrix to an identify matrix. This procedure will provide the inverse matrix as the modified augmented unit matrix.

For example, let us append a 3 × 3 matrix

$$A = \begin{bmatrix} a_{11} & a_{12} & a_{13} \\ a_{21} & a_{22} & a_{23} \\ a_{31} & a_{32} & a_{33} \end{bmatrix}$$

with a 3 × 3 identity matrix,

$$I = \begin{bmatrix} 1 & 0 & 0 \\ 0 & 1 & 0 \\ 0 & 0 & 1 \end{bmatrix}$$

in the following manner:

$$\left[\begin{array}{ccc|ccc} & A & & & I & \\ a_{11} & a_{12} & a_{13} & 1 & 0 & 0 \\ a_{21} & a_{22} & a_{23} & 0 & 1 & 0 \\ a_{31} & a_{32} & a_{33} & 0 & 0 & 1 \end{array}\right]$$

The Gauss-Jordon algorithm is then employed. This procedure will transform the co-efficient matric into an identity matrix. During this process, the elements of the identity matrix will be modified and the **modified matrix** represents **inverse matrix** as

$$A^{-1} = B = \begin{bmatrix} b_{11} & b_{12} & b_{13} \\ b_{21} & b_{22} & b_{23} \\ b_{31} & b_{32} & b_{33} \end{bmatrix}$$

3.2.3 Decomposition or Factorization Methods

Another class of elimination methods that is sometimes more efficient than Gauss elimination or the Gauss–Jordan method is based on decomposition or factorization of the coefficient matrix. The main advantage of these methods is that a smaller number of arithmetic operations are needed, and this results in a smaller round-off error and less computing time.

Basic LU Decomposition

In the basic LU decomposition method, the system $Ax = c$ is solved in three steps. In the **first step**, the coefficient matrix, A, is decomposed into an upper triangular matrix, U, and a lower triangular matrix, L, such that

$$A = LU \tag{3.25}$$

where the upper triangular matrix U is given as

$$U = \begin{bmatrix} 1 & u_{12} & u_{13} & \cdots & \cdots & u_{1n} \\ 0 & 1 & u_{23} & \cdots & \cdots & u_{2n} \\ \vdots & & & & & \\ \vdots & & & & & \\ \vdots & & & & & \\ 0 & 0 & 0 & \cdots & \cdots & 1 \end{bmatrix} \tag{3.26}$$

and the lower triangular matrix is expressed as

$$L = \begin{bmatrix} l_{11} & 0 & 0 & \cdots & \cdots & 0 \\ l_{21} & l_{22} & 0 & \cdots & \cdots & 0 \\ \vdots & & & & & \\ \vdots & & & & & \\ \vdots & & & & & \\ l_{n1} & l_{n2} & & & & l_{n2} \end{bmatrix} \tag{3.27}$$

Once the decomposition or factorization of A is obtained, the solution of the system becomes equivalent to solving two systems $Ly = c$ and $Ux = y$.

In the **second step**, the lower triangular matrix, L, and known right-hand side vector, c, are used to solve

$$Ly = c \tag{3.28}$$

using back substitution given by the generalized formula

$$y_1 = \frac{c_1}{l_{11}} \tag{3.29a}$$

and

$$y_i = \frac{c_i - \sum_{j=1}^{i-1} l_{ij} y_j}{l_{ii}}, \quad \text{for } i = 2,3,\ldots,n \tag{3.29b}$$

In the **third step**, the upper triangular matrix U is used along with vector y as the known right-hand side vector to solve the system

$$Ux = y \tag{3.30}$$

for the unknown vector x using the generalized formula for back substitution as

$$x_n = y_n \tag{3.31a}$$

$$x_i = y_i - \sum_{j=i+1}^{n} u_{ij} x_j, \quad i = n-1, n-2, \ldots, 1 \tag{3.31b}$$

Decomposition Procedure There are two different LU decomposition methods: **Dolittle's** method and **Crout's** method. In **Dolittle's** method, the matrix L has all the diagonal elements equal to 1, and the decomposed matrices are obtained using the Gauss-elimination method. The upper triangular matrix U is obtained in the forward elimination step as

$$U = \begin{bmatrix} a_{11} & a_{12} & a_{13} & \cdots & \cdots & a_{1n} \\ 0 & a'_{22} & a'_{23} & \cdots & \cdots & a'_{2n} \\ \vdots & & & & & \\ \vdots & & & & & \\ \vdots & & & & & \\ 0 & 0 & 0 & \cdots & \cdots & a_{nn}^{n-1} \end{bmatrix} \tag{3.32}$$

The lower triangular matrix, L, is formed by retaining the multipliers, $m_{kj} = a_{kj}^{k-1} / a_{kk}^{k-1}$, for $j = k+1 \cdots n$, which are used during the elimination step and are used as elements l_{ij}

$$L = \begin{bmatrix} 1 & 0 & 0 & \cdots & \cdots & 0 \\ l_{21} = m_{21} & 1 & 0 & \cdots & \cdots & 0 \\ l_{31} = m_{31} & l_{32} = m_{32} & 1 & & & \\ \vdots & & & & & \\ \vdots & & & & & \\ l_{n1} = m_{n1} & l_{n2} = m_{n2} & l_{n3} = m_{n3} & \cdots & \cdots & 1 \end{bmatrix} \tag{3.33}$$

In **Crout's method**, the U matrix has 1s in the diagonal elements. This is one of the most efficient decomposition procedures without requiring the time-consuming elimination steps. This involves a direct matrix multiplication of $[L]$ $[U]$ and equating the resulting terms with the corresponding elements of the coefficient matrix A. This procedure results in the following generalized formula for the elements in the L and U matrices.

First column of L

$$l_{i1} = a_{i1}, \quad \text{for } i = 1,2,3,\ldots,n \tag{3.34a}$$

First row of U

$$u_{1j} = \frac{a_{1j}}{l_{11}}, \quad \text{for } j = 2,3,4,\ldots,n \tag{3.34b}$$

Rest of the columns in L

$$l_{ij} = a_{ij} - \sum_{k=1}^{j-1} l_{ik} u_{kj}, \quad \text{for } j = 2,3,n-1 \text{ and } i = j, j+1,\ldots,n \tag{3.34c}$$

Rest of the rows in U

$$u_{ij}=\frac{a_{ij}-\sum_{k=1}^{i-1}l_{ik}u_{kj}}{l_{ii}},\quad \text{for } i=2,3,4,\ldots,n-1 \text{ and } j=i+1,i+2,\ldots,n \tag{3.34d}$$

Last diagonal element in U

$$l_{nn}=a_{nn}-\sum_{k=1}^{n-1}l_{nk}u_{kn} \tag{3.34e}$$

Since there is no need to store the 0s and 1s of the L and U matrices, considerable reduction in computer storage is obtained by storing both U and L matrices in the same storage location for the original coefficient matrix A. The original a_{ij} elements of the coefficient matrix are stored first. Then, calling both the U_{ij} and L_{ij} elements as a_{ij}, the old a_{ij} element values are replaced with u_{ij} and l_{ij} element values as the new a_{ij} element values.

Example 3.3: *LU* Decomposition

Use Crout's *LU* decomposition to solve

$$\begin{bmatrix} -4 & 1 & 1 & 0 \\ 1 & -4 & 0 & 1 \\ 1 & 0 & -4 & 1 \\ 0 & 1 & 1 & -4 \end{bmatrix}\begin{Bmatrix} x_1 \\ x_2 \\ x_3 \\ x_4 \end{Bmatrix}=\begin{Bmatrix} 0 \\ -90 \\ 0 \\ -90 \end{Bmatrix} \tag{E.3.3.1}$$

Solution

In the first step of the solution, we use the generalized formula given by Equation 3.34 to determine elements of matrices L and U.

Equation 3.34a gives elements of the first column of L

$$\begin{aligned} &\text{For } j=1\,(\text{first column}) \text{ and} \\ &i=1,\quad l_{11}=a_{11}=-4 \\ &i=2,\quad l_{21}=a_{21}=1 \\ &i=3,\quad l_{31}=a_{31}=1 \end{aligned}$$

Equation 3.34b gives elements of the first row of U

$$\begin{aligned} &\text{For } i=1\,(\text{first row}) \text{ and} \\ &j=2,\quad u_{12}=\frac{a_{12}}{l_{11}}=\frac{1}{-4}=-0.25 \\ &j=3,\quad u_{13}=\frac{a_{13}}{l_{11}}=\frac{1}{-4}=-0.25 \\ &j=4,\quad u_{14}=\frac{a_{14}}{l_{11}}=\frac{0}{-4}=0.0 \end{aligned}$$

Equation 3.34c gives elements of the second column of L

For $j = 2(\text{column}-2)$ and

$$\begin{aligned} i = 2, \quad & l_{22} = a_{22} - l_{21}u_{12} = -4 - (1)(-0.25) = -3.75 \\ i = 3, \quad & l_{32} = a_{32} - l_{31}u_{12} = 0 - (1)(-0.25) = 0.25 \\ i = 4, \quad & l_{42} = a_{42} - l_{41}u_{14} = 1 - (0)(0) = 1 \end{aligned}$$

Equation 3.34d gives elements of the second row of U

For $i = 2(\text{row}-2)$ and

$$j = 3, \quad u_{23} = \frac{a_{23} - l_{21}u_{13}}{l_{22}} = \frac{0 - (1)(-0.25)}{-3.75} = -0.06667$$

$$j = 4, \quad u_{24} = \frac{a_{24} - l_{21}u_{14}}{l_{22}} = \frac{1 - (1)(0)}{-3.75} = -0.26667$$

Equation 3.34c gives elements of the third column of L

For $j = 3$ and

$$\begin{aligned} i = 3, \quad l_{33} &= a_{33} - (l_{31}u_{13} + l_{32}u_{23}) \\ &= -4 - [(1)(-0.25) + (0.25)(-0.06667)] = -3.73333 \\ i = 4, \quad l_{43} &= a_{43} - (l_{41}u_{13} + l_{42}u_{23}) \\ &= 1 - [(0)(-0.25) + (1)(-0.06667)] = 1.06667 \end{aligned}$$

Equation 3.34d gives elements of the third row of U

For $i = 3(\text{row}-3)$ and

$$\begin{aligned} j = 4, \quad u_{34} &= \frac{a_{34} - [l_{31}u_{14} + l_{32}u_{24}]}{l_{33}} \\ &= \frac{1 - [(1)(0) + (0.25)(-0.26667)]}{-3.73333} = -0.28571 \end{aligned}$$

Equation 3.34e gives the last diagonal element of L

For $j = 4$ and $i = 4$

$$\begin{aligned} l_{44} &= a_{44} - (l_{41}u_{14} + l_{42}u_{24} + l_{43}u_{34}) \\ &= -4 - [(0)(0) + (1)(-0.26667) + (1.06667)(-0.28571)] = -3.42857 \end{aligned}$$

Assembling the elements, we have the decomposition of A into L and U as

$$L = \begin{bmatrix} -4 & 0 & 0 & 0 \\ 1 & -3.75 & 0 & 0 \\ 1 & 0.25 & -3.73333 & 0 \\ 0 & 1 & 1.0666 & -3.42857 \end{bmatrix} \quad \text{(E.3.3.2)}$$

and

$$U = \begin{bmatrix} 1 & -0.25 & -0.25 & 0 \\ 0 & 1 & -0.06667 & -0.026667 \\ 0 & 0 & 1 & -0.28571 \\ 0 & 0 & 0 & 1 \end{bmatrix} \quad \text{(E.3.3.3)}$$

In the **second step**, the system of equation $Ly = c$ is solved for the vector y using back substitution (Equation 3.29) as

$$y_1 = \frac{c_1}{l_{11}} = \frac{0}{-4} = 0$$

$$y_2 = \frac{c_2 - l_{21}y_1}{l_{22}} = \frac{-90-(1)(0)}{-3.75} = 24$$

$$y_3 = \frac{c_3 - l_{31}y_1 - l_{32}y_2}{l_{33}} = \frac{0-[(1)(0)-(0.25)(24)]}{-3.73333} = 1.60714$$

$$y_4 = \frac{c_4 - [l_{41}y_1 + l_{42}y_2 + l_{43}y_3]}{l_{44}}$$

$$= -90 - \frac{[(0)(0)+(1)(24)+(1.0666)(1.60714)]}{-3.42857} = 33.75001$$

Thus, the vector y is

$$y = \begin{Bmatrix} 0 \\ 24 \\ 1.60714 \\ 33.75001 \end{Bmatrix} \quad \text{(E.3.3.4)}$$

In the **third step**, the system $Ux = y$ is solved for the solution vector x using back substitution (Equation 3.31) as

$$x_4 = y_4 = 33.75001$$

$$x_3 = y_3 - u_{34}x_4 = 1.60714 - (-0.28571)(33.75001) = 11.24986$$

$$x_2 = y_2 - [u_{23}x_3 + u_{24}x_4]$$

$$= 24 - [(-0.06667)(11.24986) + (-0.26667)(33.75001)] = 33.75014$$

$$x_1 = y_1 - [u_{12}x_2 + u_{13}x_3 + u_{14}x_4]$$

$$= 0 - [(-0.25)(33.5014) + (-0.25)(11.24986) + (0)(33.75001)]$$

$$= 11.25$$

The final solution vector x is

$$x = \begin{Bmatrix} 11.25 \\ 33.75014 \\ 11.24986 \\ 33.75001 \end{Bmatrix} \quad \text{(E.3.3.5)}$$

Cholesky Factorization Method

A symmetric square matrix often arises in many engineering problems. For such a matrix, the decomposition can be more efficiently performed using a recursive procedure known as the

Cholesky factorization method. The method is based on factorizing the symmetric matrix A into a lower triangular matrix L and its transpose matrix L^T such that

$$A = LL^T \tag{3.35}$$

where the lower triangular matrix L and its transpose matrix L^T are expressed as

$$L = \begin{bmatrix} l_{11} & 0 & 0 & \cdots & 0 \\ l_{21} & l_{22} & 0 & \cdots & 0 \\ \vdots & & & & \\ \vdots & & & & \\ l_{n1} & l_{n2} & 0 & \cdots & l_{nn} \end{bmatrix} \quad \text{and} \quad L^T = \begin{bmatrix} l_{11} & l_{21} & \cdots & \cdots & l_{n1} \\ & l_{22} & \cdots & \cdots & l_{n2} \\ & & & & \vdots \\ & & & & \vdots \\ & & & & l_{nn} \end{bmatrix}$$

The elements of the Cholesky factorization are obtained directly from Equation 3.35 and elements of factorized matrices are given by the generalized formulas

$$l_{11} = \sqrt{a_{11}} \tag{3.36a}$$

$$l_{i1} = \frac{a_{i1}}{l_{11}}, \quad \text{for } i = 2,3,\ldots,n \tag{3.36b}$$

$$l_{ii} = \sqrt{a_{ii} - \sum_{k=1}^{i-1} l_{ik}^2} \tag{3.36c}$$

$$l_{ij} = \frac{1}{l_{jj}}\left(a_{ij} - \sum_{k=1}^{j-1} l_{ik} l_{jk}\right), \quad \text{for } j = 2,3,\ldots,i-1, \text{ and } j < i \tag{3.36d}$$

Once the factorization is obtained, the solution to the linear system $Ax = c$ is obtained by first solving the lower triangular system $Ly = c$ and followed by solving the upper transpose system $L^T x = y$.

The Cholesky decomposition method offers computational advantages because only half the storage is needed and in most cases requires half the computational time as compared to the *LU* decomposition method.

Example 3.4: Cholesky Factorization Method

Use the Cholesky factorization method to decompose the symmetric matrix

$$A = \begin{bmatrix} 1 & -4 & 1 & 0 \\ -4 & 60 & -4 & 0 \\ 1 & -4 & 6 & -4 \\ 0 & 0 & -4 & 25 \end{bmatrix}$$

Solution

$$l_{11} = \sqrt{a_{11}} = \sqrt{1} = 1$$

The second row ($i = 2$)

$$j=1,\quad l_{21}=\frac{a_{21}}{l_{11}}=\frac{-4}{1}=-4$$

$$j=2,\quad l_{22}=\sqrt{a_{22}-\sum_{k=1}^{2-1} l_{2k}^2}=\sqrt{60-(-4)^2}=\sqrt{44}$$

Third row ($i = 3$)

$$j=1,\quad l_{31}=\frac{a_{31}}{l_{11}}=\frac{1}{1}=1$$

$$j=2,\quad l_{32}=\frac{\left(a_{32}-\sum_{k=1}^{2-1} l_{3k}l_{2k}\right)}{l_{22}}=\frac{\left[a_{32}-(l_{31}l_{21})\right]}{l_{22}}=\frac{\left[-4-(1)(-4)\right]}{\sqrt{44}}=0$$

$$j=3,\quad l_{33}=\sqrt{a_{33}-\sum_{k=1}^{3-1} l_{3k}^2}=\sqrt{6-\left(l_{31}^2+l_{32}^2\right)}=\sqrt{6-\left(1^2+0^2\right)}=\sqrt{5}$$

Fourth row ($i = 4$)

$$j=1,\quad l_{41}=\frac{a_{41}}{l_{11}}=\frac{0}{1}=0$$

$$j=2,\quad l_{42}=\frac{\left(a_{42}-\sum_{k=1}^{2-1} l_{4k}l_{2k}\right)}{l_{22}}=\frac{0-\left[(0)(-4)\right]}{\sqrt{44}}=0$$

$$j=3,\quad l_{43}=\frac{\left(a_{43}-\sum_{k=1}^{3-1} l_{4k}l_{3k}\right)}{l_{33}}$$

$$=\frac{\left[a_{43}-(l_{41}l_{31}+l_{42}l_{32})\right]}{l_{33}}=\frac{\left[-4-((0)(1)+(0)(0))\right]}{\sqrt{5}}=-\frac{4}{\sqrt{5}}$$

$$j=4,\quad l_{44}=\sqrt{a_{44}^2-\sum_{k=1}^{4-1} l_{4k}^2}=\sqrt{a_{44}-\left[l_{41}^2+l_{42}^2+l_{43}^2\right]}$$

$$=\sqrt{25-\left[(0)^2+(0)^2+\left(-\frac{4}{\sqrt{5}}\right)^2\right]}=\sqrt{\frac{109}{5}}$$

The matrix L and its transpose matrix L^{T} can now be written as

$$L=\begin{bmatrix} 1 & 0 & 0 & 0 \\ -4 & \sqrt{44} & 0 & 0 \\ 1 & 0 & \sqrt{5} & 0 \\ 0 & 0 & -\frac{4}{\sqrt{5}} & \sqrt{\frac{109}{5}} \end{bmatrix} \quad \text{and} \quad L^{\mathrm{T}}=\begin{bmatrix} 1 & -4 & 1 & 0 \\ 0 & \sqrt{44} & 0 & 0 \\ 0 & 0 & \sqrt{5} & -\frac{4}{\sqrt{5}} \\ 0 & 0 & 0 & \sqrt{\frac{109}{5}} \end{bmatrix}$$

Factorization by Householder Method

In the QR factorization method, the matrix A is factorized into the product of an orthogonal matrix Q and an upper triangular matrix R such that

$$A = QR \tag{3.37}$$

There are different choices for QR factorization techniques such as the one based on the **Householder method** or based on the **Given's method**. We will discuss here the QR factorization based on the Householder method in which the matrix A is transformed into a form for which the decomposition can be more rapidly obtained by the Householder method. One such desired form is the **Hessenberg matrix**, which has the form

$$A_{\mathrm{H}} = \begin{bmatrix} \overset{*}{a}_{11} & \overset{*}{a}_{12} & \overset{*}{a}_{13} & \cdots & \cdots & \overset{*}{a}_{1n} \\ \overset{*}{a}_{21} & \overset{*}{a}_{22} & \overset{*}{a}_{23} & \cdots & \cdots & \overset{*}{a}_{2n} \\ 0 & \overset{*}{a}_{32} & \overset{*}{a}_{33} & & & \\ 0 & 0 & \overset{*}{a}_{43} & \ddots & & \\ \vdots & \vdots & \vdots & \ddots & \ddots & \\ 0 & 0 & 0 & \cdots & \overset{*}{a}_{n-1,1} & \overset{*}{a}_{nn} \end{bmatrix} \tag{3.38}$$

Note that it has one nonzero diagonal below the main diagonal and the elements above the main diagonal are nonzero, in general. The reduction of the original matrix A to the Hessenberg form can be done by Householder transformations in $n-1$ steps. The procedure is outlined as follows:

Step 1 If we select a nonzero vector z_1 to be such that the Householder matrix can be defined as $H_1 = I - \left(2z_1z_1^{\mathrm{T}}/z_1^{\mathrm{T}}z_1\right)$, it can be used to create 0s in a vector as

$$H_1 \begin{Bmatrix} a_{11} \\ a_{21} \\ a_{31} \\ \vdots \\ \vdots \\ a_{n1} \end{Bmatrix} = \begin{Bmatrix} a_{11}^1 \\ 0 \\ 0 \\ \vdots \\ \vdots \\ 0 \end{Bmatrix} \tag{3.39}$$

We transform the matrix A to the form

$$A_{\mathrm{H}}^1 = H_1A = \begin{bmatrix} a_{11}^1 & a_{12}^1 & a_{13}^1 & \cdots & a_{1n}^1 \\ 0 & a_{22}^1 & a_{23}^1 & \cdots & a_{2n}^1 \\ 0 & a_{32}^1 & a_{33}^1 & \cdots & a_{3n}^1 \\ \vdots & \vdots & & & \\ \vdots & \vdots & & & \\ 0 & a_{n2}^1 & a_{n3}^1 & \cdots & a_{nn}^1 \end{bmatrix} \tag{3.40}$$

Note that the matrix $A_H^1 = H_1A$ has 0s below the diagonal element in the first column.

Step 2 We select a nonzero vector z_2 to be such that the Householder matrix can be defined as $H_2 = I - \left(2z_2z_2^{\mathrm{T}}/z_2^{\mathrm{T}}z_2\right)$, and it can be used to transform the matrix to the form

$$A_H^2 = H_2 A_H^1 = \begin{bmatrix} a_{11}^1 & a_{12}^1 & a_{13}^1 & \cdots & a_{1n}^1 \\ 0 & a_{22}^2 & a_{23}^2 & \cdots & a_{2n}^2 \\ 0 & 0 & a_{33}^2 & \cdots & a_{3n}^2 \\ \vdots & \vdots & & & \\ \vdots & \vdots & & & \\ 0 & 0 & a_{n3}^2 & \cdots & a_{nn}^2 \end{bmatrix} \tag{3.41}$$

Note that the matrix $A_H^2 = H_2 A_H^1 = H_2 H_1 A$ has 0s in the first two columns below the main diagonal.

This procedure is repeated until in the $(n-1)$th step the Householder transformation leads to

$$A_H^{n-1} = H_{n-1} A_H^{n-2} = \begin{bmatrix} a_{11}^1 & a_{12}^1 & a_{13}^1 & \cdots & & a_{1n}^1 \\ 0 & a_{22}^2 & a_{23}^2 & \cdots & & a_{2n}^2 \\ 0 & 0 & a_{33}^3 & \cdots & & a_{3n}^3 \\ \vdots & \vdots & & \ddots & & \\ \vdots & \vdots & & & a_{n-1,n-1}^{n-1} & a_{n-1,n}^{n-1} \\ 0 & 0 & 0 & \cdots & 0 & a_{nn}^{n-1} \end{bmatrix} \tag{3.42}$$

which represents an upper triangular matrix R as

$$R = H_{n-1} \ldots H_2 H_1 A \tag{3.43}$$

Now, if we define

$$Q^{\mathrm{T}} = H_{n-1} \ldots H_2 H_1 \tag{3.44}$$

then Equation 3.43 can be written as $Q^{\mathrm{T}} A = R$. Since each Householder matrix H is orthogonal, the product Q^{T} is also orthogonal. So, $Q^{-1} = Q^{\mathrm{T}}$, and Equation 3.43 is written as

$$A = QR \tag{3.45}$$

Once the QR factorization is obtained, the system of linear equations $Ax = c$ is solved in the following two steps.

In the **first step**, the vector y is obtained from

$$y = Q^{\mathrm{T}} c \tag{3.46a}$$

In the **second step**, the upper triangular matrix R is used along with y as the right-hand side vector to solve for unknown vector x as

$$Rx = y \tag{3.46b}$$

3.2.4 Banded Systems

In many heat transfer and fluid flow problems, the discretization equations form a system in which not all the unknowns are present in each equation and hence consist of many zero elements in the coefficient matrix. A banded matrix is one such matrix that has all elements equal to zero except for a band centered around the main diagonal as shown below.

$$\begin{bmatrix} a_{11} & a_{12} & a_{13} & 0 & 0 & 0 & \cdots & 0 \\ a_{21} & a_{22} & a_{23} & a_{24} & 0 & 0 & \cdots & 0 \\ a_{31} & a_{32} & a_{33} & a_{34} & a_{35} & 0 & \cdots & 0 \\ 0 & a_{42} & a_{43} & a_{44} & a_{45} & a_{46} & \cdots & 0 \\ 0 & 0 & a_{53} & a_{54} & & \leftarrow \text{hbw} & \rightarrow & 0 \\ \vdots & & & \leftarrow & \leftarrow & \text{bw} & \rightarrow & \rightarrow \\ \vdots & & & & & & & \\ 0 & 0 & 0 & 0 & 0 & a_{n6} & \cdots & a_{nn} \end{bmatrix} \tag{3.47}$$

The dimensions of such a banded system are quantified by two parameters: the bandwidth **bw** and the half bandwidth **hbw**, as shown. These parameters are related to each other by **bw** = 2 **hbw** + 1. In general, a system $Ax = c$ is defined as banded when elements of the coefficient matrix $a_{ij} = 0$ for $|i-j| >$ **hbw**.

Although all direct methods, such as the elimination and the factorization methods, can be used to solve such banded systems, they become quite inefficient for cases where pivoting is not required. A **diagonally dominant matrix** is one such case where no partial pivoting is required during triangularization procedures. For such matrices, the absolute value of the diagonal element in each of the equations must be larger than the sum of the absolute values of other elements in the equation. That is, $| a_{jj} | > \sum |a_{ij}|$, where the summation is taken from $i = 1$ ton, excluding the element for $i = j$. Considerable economics are achieved while solving such systems using alternate, more efficient algorithms that do not involve zero elements and require less expenditure of space and time in storing and manipulating zero elements.

Tridiagonal Systems and Tridiagonal-Matrix Algorithm (TDMA) One of the most encountered banded systems is the tridiagonal system, which has a bandwidth of **bw** = 3. Such a matrix has the form

$$A = \begin{bmatrix} a_1 & b_1 & 0 & 0 & 0 & \cdots & 0 \\ c_2 & a_2 & b_2 & 0 & 0 & \cdots & 0 \\ 0 & c_3 & a_3 & b_3 & 0 & \cdots & 0 \\ 0 & 0 & c_4 & a_4 & b_4 & \cdots & 0 \\ \vdots & & & & & & \\ & & & & c_{n-1} & a_{n-1} & b_{n-1} \\ & & & & & c_n & a_n \end{bmatrix} \tag{3.48}$$

Note that the notation for matrix elements is changed from a single two-dimensional matrix array, a_{ij}, to three one-dimensional vectors a_i, b_i, and c_i, that store only the elements in the three diagonals. Because of this simple form of the system, the elimination or factorization process leads into a convenient algorithm known as **tridiagonal matrix algorithm (TDMA)** or **Thomas algorithm**.

Thomas Algorithm – A Gauss Elimination Method for Tridiagonal System As with the conventional Gauss elimination method, the scheme utilized in this algorithm involves two steps: **elimination** and **back substitution**. The procedure will be illustrated by considering the set of equations written as

$$\begin{bmatrix} a_1 & b_1 & 0 & 0 & 0 & \cdots & 0 \\ c_2 & a_2 & b_2 & 0 & 0 & \cdots & 0 \\ 0 & c_3 & a_3 & b_3 & 0 & \cdots & 0 \\ 0 & 0 & c_4 & a_4 & b_4 & \cdots & 0 \\ \vdots & & & & & & \\ 0 & 0 & 0 & \cdots & c_{n-1} & a_{n-1} & b_{n-1} \\ 0 & 0 & 0 & \cdots & 0 & c_n & a_n \end{bmatrix} \begin{Bmatrix} x_1 \\ x_2 \\ x_3 \\ x_4 \\ \vdots \\ x_{n-1} \\ x_n \end{Bmatrix} = \begin{Bmatrix} d_1 \\ d_2 \\ d_3 \\ d_4 \\ \vdots \\ d_{n-1} \\ d_n \end{Bmatrix} \tag{3.49}$$

In the **elimination step**, as Gaussian elimination is applied to this system; only one of the c's is eliminated from the column containing the pivot element in each step, since the remaining elements below the diagonal element are zero. The original zero elements remain unchanged and the matrix reduces to the form

$$\begin{bmatrix} a_1 & b_1 & 0 & 0 & 0 & \cdots & 0 \\ 0 & a_2' & b_2' & 0 & 0 & \cdots & 0 \\ 0 & & a_3' & b_3' & 0 & \cdots & 0 \\ 0 & 0 & & a_4 & b_4 & \cdots & 0 \\ \vdots & & & & & & \\ 0 & 0 & 0 & \cdots & 0 & a_{n-1}' & b_{n-1}' \\ 0 & 0 & 0 & \cdots & 0 & 0 & a_n' \end{bmatrix} \begin{Bmatrix} x_1 \\ x_2 \\ x_3 \\ x_4 \\ \vdots \\ x_{n-1} \\ x_n \end{Bmatrix} = \begin{Bmatrix} d_1 \\ d_2' \\ d_3' \\ d_4' \\ \vdots \\ d_{n-1}' \\ d_n' \end{Bmatrix} \tag{3.50}$$

From this reduced system of equations, the unknowns x_i may easily be obtained by back substitution.

The **generalized formulas** that express the modified elements in the system during each elimination step and the solution during the substitution steps are given by

Elimination

$$c_i' = 0, \quad a_i' = a_i - b_{i-1}\frac{c_i}{a_{i-1}'} \tag{3.51a}$$

$$b_i' = b_i, \quad d_i' = d_i - d_{i-1}'\frac{c_i}{a_{i-1}'} \quad \text{for } i = 2,3,\ldots,n \tag{3.51b}$$

Back Substitution

$$x_n = \frac{d_n'}{a_n'} \tag{3.52a}$$

$$x_i = \frac{d_i' - b_{i+1}'}{a_i'} \quad \text{for } i = n-1, n-2, \ldots, 3, 2, 1 \tag{3.52b}$$

Thomas Algorithm – An *LU* Decomposition Method for Tridiagonal System As with the conventional *LU* decomposition method, this algorithm consists of three steps: **decomposition, forward**, and **backward substitution**. In the first step, the *LU* decomposition of the tridiagonal matrix *A* gives upper bidiagonal and lower bidiagonal matrices such that

$$A = LU \tag{3.53a}$$

or

$$\begin{bmatrix} a_1 & b_1 & 0 & 0 & 0 & \cdots & 0 \\ c_2 & a_2 & b_2 & 0 & 0 & \cdots & 0 \\ 0 & c_3 & a_3 & b_3 & 0 & \cdots & 0 \\ 0 & 0 & c_4 & a_4 & b_4 & \cdots & 0 \\ \vdots & & & & & & \\ & & & & c_{n-1} & a_{n-1} & b_{n-1} \\ & & & & & c_n & a_n \end{bmatrix}$$

$$= \begin{bmatrix} 1 & & & & & & \\ l_2 & 1 & & & & & \\ 0 & l_3 & 1 & & & & \\ 0 & 0 & l_4 & 1 & & & \\ \vdots & & & & & & \\ & & & & l_{n-1} & 1 & \\ & & & & & l_n & 1 \end{bmatrix} \begin{bmatrix} u_1 & b_1 & & & & & \\ & u_2 & b_2 & & & & \\ & & u_3 & b_3 & & & \\ & & & u_4 & b_4 & & \\ & & & & & & \\ & & & & & u_{n-1} & b_{n-1} \\ & & & & & & u_n \end{bmatrix} \tag{3.53b}$$

The elements of decomposed matrices are obtained directly by equating the corresponding elements of matrices on both sides of Equation 3.48 and expressed by the formulas

$$u_1 = a_1 \tag{3.54a}$$

and

$$l_i = \frac{c_i}{u_{i-1}} \tag{3.54b}$$

$$u_i = a_i - l_i b_{i-1} \quad \text{for } i = 2,3,\ldots,n \tag{3.54c}$$

Once the decomposition is obtained, the solution to the tridiagonal linear system $Ax = c$ is obtained by solving the lower bidiagonal system $Ly = c$ and then by solving the upper bidiagonal system $ux = y$.

TDMA Algorithm – A General Elimination Method for Tridiagonal System This is one of the most popular algorithms (Patankar, 1980) for solving a tridiagonal system. This algorithm employs a scheme in two steps: **forward substitution** and **backward substitution**. In the forward substitution step x_i is expressed in terms of x_{i+1} until x_n is obtained as a numerical value. In the subsequent back substitution step $x_{n-1}, x_{n-2}, \ldots, x_3, x_2, x_1$ are obtained. The procedure will be illustrated by solving the set of equations (3.49) written as

$$a_i x_i = -b_i x_{i+1} - c_i x_{i-1} + d_i \tag{3.55}$$

where $c_1 = 0$, $b_n = 0$, and $i = 1, 2, \ldots, n$.

The **forward substitution** step is designed to reduce the set of equations to the following form:

$$x_i = P_i x_{i+1} + Q_i \tag{3.56}$$

where P_i and Q_i are obtained by substituting $x_{i-1} = P_{i-1} x_i + Q_{i-1}$ into Equation 3.55 as

$$a_i x_i = -b_i x_{i+1} - c_i \left(P_{i-1} x_i + Q_{i-1} \right) + d_i \tag{3.57}$$

Rearranging Equation 3.57, we get

$$x_i = -\frac{b_i}{a_i + c_i P_{i-1}} x_{i+1} + \frac{d_i - c_i Q_{i-1}}{a_i + c_i P_{i-1}} \tag{3.58}$$

Comparison of Equations 3.56 and 3.58 leads to the recurrence relations

$$P_i = -\frac{b_i}{a_i + c_i P_{i-1}} \tag{3.59a}$$

$$Q_i = \frac{d_i - c_i Q_{i-1}}{a_i + c_i P_{i-1}} \tag{3.59b}$$

To start the recurrence process, P_1 and Q_1 are expressed by substituting $c_1 = 0$ into Equation 3.59, as

$$P_1 = -\frac{b_1}{a_1} \tag{3.60a}$$

$$Q_1 = \frac{d_1}{a_1} \tag{3.60b}$$

The recurrence process leads to $P_n = 0$ since $b_n = 0$, and hence from Equation 3.56

$$x_n = Q_n \tag{3.61}$$

In the subsequent back substitution step x_{n-1} is obtained from Equation 3.56 as

$$x_{n-1} = P_{n-1} x_n + Q_{n-1} \tag{3.62a}$$

This back-substitution procedure is continued to solve the remaining unknowns in the following manner:

```
        P1 = -b1 / a1
        Q1 = d1  / a1

for i = 2, 3 to n

        Pi = -bi / (ai + ci Pi-1)
        Qi = (di - ci Qi-1) / (ai + ci Pi-1)

enddo
        xn = Qn
        do for i = n-1, n-2 - - - 3, 2, 1
           xi = Pi xi+1 + Qi
        enddo
```

FIGURE 3.1 Pseudo code for TDMA.

$$
\begin{aligned}
x_{n-2} &= P_{n-2}x_{n-1} + Q_{n-2} && (b)\\
&\vdots\\
x_2 &= P_2x_3 + Q_2 && (n-1)\\
x_1 &= P_1x_2 + Q_1 && (n)
\end{aligned}
\qquad (3.62b)
$$

A pseudo code for the TDMA algorithm is presented in Figure 3.1.

Unlike general direct methods as applied to a dense coefficient matrix, the number of operations needed for solving a tridiagonal system using TDMA or the Thomas method is of $0(n)$ rather than $0(n^2)$ or $0(n^3)$. Therefore, much smaller times and round-off errors arise in the solutions of such systems while using these alternate algorithms.

Unfortunately, the Thomas algorithm based on the factorization procedure breaks down if any u_i is zero. In fact, the stability of the process in general cannot be guaranteed even if all u_i are nonzero. However, in many heat transfer and fluid flow problems, the discretization procedure leads to a system that is tridiagonal as well as symmetric positive definite. In such cases, the Thomas algorithm based on LU factorization is quite stable. In the general case, the TDMA algorithm based on the elimination method is preferred to ensure stability.

Example 3.5: Tridiagonal Matrix Algorithm

Use the TDMA to solve the following system of equations:

$$
\begin{bmatrix}
-2 & 1 & 0 & 0 & 0 & 0 & 0\\
1 & -2 & 1 & 0 & 0 & 0 & 0\\
0 & 1 & -2 & 1 & 0 & 0 & 0\\
0 & 0 & 1 & -2 & 1 & 0 & 0\\
0 & 0 & 0 & 1 & -2 & 1 & 0\\
0 & 0 & 0 & 0 & 1 & -2 & 1\\
0 & 0 & 0 & 0 & 0 & 1 & -2
\end{bmatrix}
\begin{Bmatrix} x_1\\ x_2\\ x_3\\ x_4\\ x_5\\ x_6\\ x_7 \end{Bmatrix}
=
\begin{Bmatrix} -240\\ -40\\ -40\\ -40\\ -40\\ -40\\ -60 \end{Bmatrix}
\qquad (E.3.5.1)
$$

Solution

We use Equation 3.60 for $i = 1$

$$P_1 = -\frac{b_1}{a_1} = -\frac{1}{-2} = 0.5, \quad Q_1 = \frac{d_1}{a_1} = \frac{-240}{-2} = 120$$

Use Equation 3.59 for $i > 1$

$$i = 2, \quad P_2 = \frac{-b_2}{a_2 + c_2 P_1} = \frac{-1}{-2 + 1 \times 0.5} = 0.66667,$$

$$Q_2 = \frac{d_2 - c_2 Q_1}{a_2 + c_2 P_1} = \frac{-40 - 1 \times 120}{-2 + 1 \times 0.5} = 106.66667$$

$$i = 3, \quad P_3 = \frac{-b_3}{a_3 + c_3 P_2} = \frac{-1}{-2 + 1 \times 0.66667} = 0.75,$$

$$Q_3 = \frac{d_3 - c_3 Q_2}{a_3 + c_3 P_2} = \frac{-40 - 1 \times 106.66667}{-2 + 1 \times 0.66667} = 110$$

$$i = 4, \quad P_4 = \frac{-b_4}{a_4 + c_4 P_3} = \frac{-1}{-2 + 1 \times 0.75} = 0.80,$$

$$Q_4 = \frac{d_4 - c_4 Q_3}{a_4 + c_4 P_3} = \frac{-40 - 1 \times 110}{-2 + 1 \times 0.75} = 120$$

$$i = 5, \quad P_5 = \frac{-b_5}{a_5 + c_5 P_4} = \frac{-1}{-2 + 1 \times 0.80} = 0.83333,$$

$$Q_5 = \frac{d_5 - c_5 Q_4}{a_5 + c_5 P_4} = \frac{-40 - 1 \times 120}{-2 + 1 \times 0.80} = 133.33333$$

$$i = 6, \quad P_6 = \frac{-b_6}{a_6 + c_6 P_5} = \frac{-1}{-2 + 1 \times 0.83333} = 0.85714,$$

$$Q_6 = \frac{d_6 - c_6 Q_5}{a_6 + c_6 P_5} = \frac{-40 - 1 \times 133.3333}{-2 + 1 \times 0.83333} = 148.57100$$

$$i = 7, \quad P_7 = 0,$$

$$Q_7 = \frac{d_7 - c_7 Q_6}{a_7 + c_7 P_6} = \frac{-40 - 1 \times 148.57100}{-2 + 1 \times 0.85714} = 148.57100$$

Use Equations 3.61 and 3.62 in the back-substitution step for the solution vector x as follows. For

$$i = 7, \quad x_7 = Q_7 = 182.5$$

$$i = 6, \quad x_6 = P_6 x_7 + Q_6 = 0.85714(182.5) + 148.571 = 305$$

$$i = 5, \quad x_5 = P_5 x_6 + Q_5 = 0.83333(305) + 133.33333 = 387.5$$

$$i = 4, \quad x_4 = P_4 x_5 + Q_4 = 0.8(387.5) + 120 = 430$$

$$i = 3, \quad x_3 = P_3 x_4 + Q_3 = 0.75(430) + 110 = 432.5$$

$$i = 2, \quad x_2 = P_2 x_3 + Q_2 = 0.66667(432.5) + 106.66667 = 395$$

$$i = 1, \quad x_1 = P_1 x_2 + Q_1 = 0.5(395) + 120 = 317.5$$

The solution vector is

$$\{x\} = \begin{Bmatrix} 317.5 \\ 395 \\ 432.5 \\ 430 \\ 387.5 \\ 305 \\ 182.5 \end{Bmatrix} \quad \text{(E.3.5.2)}$$

Block Tridiagonal System Many physical problems in fluid flow and heat transfer give rise to a special system such as the block tridiagonal system when numerical discretization schemes are applied to the governing equations. For example, application of a finite difference–control volume discretization scheme to a two-dimensional elliptic field equation in the form a *Poisson's* equation with *Dirichlet* boundary conditions leads to a system of linear equations of form

$$\begin{bmatrix} 4 & -1 & 0 & 0 & \cdots & -1 & 0 & 0 & 0 & 0 \\ -1 & 4 & -1 & 0 & \cdots & 0 & -1 & 0 & 0 & 0 \\ 0 & -1 & 4 & -1 & \cdots & & 0 & -1 & 0 & 0 \\ 0 & 0 & -1 & 4 & \cdots & 0 & & 0 & -1 & 0 \\ & & & & \vdots & & & & & \\ 0 & 0 & 0 & 0 & \cdots & 4 & -1 & 0 & 0 & 0 \\ -1 & 0 & & & \cdots & -1 & 4 & -1 & 0 & 0 \\ 0 & -1 & 0 & & \cdots & 0 & -1 & 4 & -1 & 0 \\ 0 & 0 & -1 & 0 & \cdots & & 0 & -1 & 4 & -1 \\ 0 & 0 & 0 & -1 & \cdots & & & 0 & -1 & 4 \end{bmatrix} \begin{Bmatrix} x_1 \\ x_2 \\ x_3 \\ x_4 \\ \vdots \\ x_n \end{Bmatrix} = \begin{Bmatrix} d_1 \\ d_2 \\ d_3 \\ d_4 \\ \vdots \\ d_n \end{Bmatrix} \quad (3.63)$$

Equation 3.63 can be written in a block matrix form as

$$\begin{bmatrix} A_N & -I_N & & & \\ -I_N & A_N & & & \\ & -I_N & & & \\ & & & & -I_N \\ & & & -I_N & A_N \end{bmatrix} \begin{Bmatrix} x_{1N} \\ x_{2N} \\ \vdots \\ x_{NN} \end{Bmatrix} = \begin{Bmatrix} d_{1N} \\ d_{2N} \\ \\ x_{NN} \end{Bmatrix} \quad (3.64)$$

where A_N is a $N \times N$ matrix given as

$$A_N = \begin{bmatrix} 4 & -1 & & & \\ -1 & 4 & & & \\ & & & & \\ & & & 4 & -1 \\ & & & -1 & 4 \end{bmatrix} \quad (3.65)$$

and I_N is the $N \times N$ identity matrix.

Equation 3.64 can also be written in a block tridiagonal matrix form like Equation 3.49 as

$$\begin{bmatrix} A_1 & B_1 & & & \\ C_2 & A_2 & B_2 & & \\ & C_3 & & & \\ & & & & B_{N-1} \\ & & & C_N & A_N \end{bmatrix} \begin{Bmatrix} x_{1N} \\ x_{2N} \\ \vdots \\ x_{N-1,N} \\ x_{NN} \end{Bmatrix} = \begin{Bmatrix} d_{1N} \\ d_{2N} \\ \\ x_{N-1,N} \\ x_{NN} \end{Bmatrix} \tag{3.66}$$

So, Equation 3.66 represents a block tridiagonal system, which is an $n \times n$ square matrix, and in which each element is a $N \times N$ matrix, each solution vector element is a vector, and each constant vector element is a vector. The algorithm for the solution of a block tridiagonal system follows the same procedure as in the case of scalar tridiagonal algorithm: factorization and substitution. The basic technique based on *LU* factorization discussed in Section 3.2.3 can be used for solving a block tridiagonal system. This is called **block *LU* factorization.**

Block *LU* Factorization In the **first step**, the block tridiagonal matrix A_N is factorized into upper and lower triangular matrices such that

$$A_N = LU \tag{3.67a}$$

or

$$\begin{bmatrix} A_1 & B_1 & & & \\ C_2 & A_2 & B_2 & & \\ & C_3 & & & \\ & & & & B_{N-1} \\ & & & C_N & A_N \end{bmatrix} = \begin{bmatrix} I & & & & \\ L_2 & I & & & \\ & L_3 & & & \\ & & & & \\ & & & L_N & I \end{bmatrix} \begin{bmatrix} U_1 & B_1 & & & \\ & U_2 & B_2 & & \\ & & & & \\ & & & & B_{N-1} \\ & & & & U_N \end{bmatrix} \tag{3.67b}$$

The elements of the block *L* and *U* matrices are obtained directly by equating the corresponding elements on both sides of Equation 3.67b, and expressed by the formulas

$$U_1 = A_1 \tag{3.68a}$$

and

$$L_i = \frac{C_i}{U_{i-1}} \tag{3.68b}$$

$$U_i = A_i - L_i B_{i-1} \quad \text{for } i = 2 \text{ to } n \tag{3.68c}$$

In the **second step**, the lower triangular matrix L and right-hand side vector d are used to solve the system $Ly = d$ by the block forward elimination algorithm as

$$y_1 = d_1 \tag{3.69a}$$

and

$$y_i = d_i - L_{i-1}y_{i-1}, \quad \text{for } i = 2 \text{ to } n \tag{3.69b}$$

In the **third step**, the upper triangular matrix U and vector y are used to solve the system $Ux = y$ using the block back substitution algorithm

$$U_n x_n = y_n \tag{3.70a}$$

and

$$U_i x_i = y_i - B_i x_{i+1}, \quad \text{for } i = n-1,\ldots,1 \tag{3.70b}$$

3.2.5 Error Equation and Iterative Refinement

Direct methods may lead to considerable error in the result due to round-off error. This is particularly significant in a large system of equations. In many cases, using an iterative refinement procedure that involves use of error equations can reduce this error. The procedure is outlined as follows.

Consider a system of the form

$$Ax = c \tag{3.2b}$$

If x^* is the computed solution vector of the system, then substitution of this result into Equation 3.2b gives

$$Ax^* = c^* \tag{3.71}$$

Now, suppose that $\hat{x}$ is the correction vector to the approximate solution vector x^*, then the exact solution vector x is expressed as

$$x = x^* + \hat{x} \tag{3.72}$$

Substituting Equation 3.72 into Equation 3.2b the following system results in

$$A\left(x^* + \hat{x}\right) = c \tag{3.73}$$

Now subtracting Equation 3.71 from Equation 3.73 results in

$$A\hat{x} = c - c^* \tag{3.74a}$$

or

$$A\hat{x} = e \tag{3.74b}$$

where e is the *residual error vector* and can be rewritten as

$$e = c - Ax^* \tag{3.75}$$

Equation 3.74 is a system of *error equations*, which can also be solved using the direct solver, and the correction vector, $\hat{x}$, can be applied to refine the solution as specified by Equation 3.72.

This refinement procedure is continued until a desired accuracy is achieved. The desired accuracy can be set by a criterion such as

$$\|\hat{x}\|_e \le \varepsilon_s \tag{3.76a}$$

or

$$\frac{\|x^*\|_e}{\|x\|_e} \le \varepsilon_s \tag{3.76b}$$

where ε_s is the assigned tolerance limit. It is relatively simple to implement this iterative refinement algorithm into computer programs for any direct methods. However, it is especially effective for those direct methods for which triangularization of the matrix exist, such as in *LU* decomposition or matrix inversion by the Gauss–Jordan elimination method.

Figure 3.2 shows a pseudo code to implement this iterative refinement procedure.

The error refinement algorithm discussed will be extensive used in the solution of fluid flow and heat transfer equations to be discussed in Chapter 8. Let us now consider here the use of this error refinement strategy while solving the generalized form of transport equation. As discussed in Chapter 1, the governing equations for momentum, temperature, and mass species transport can be represented in a general transport equation given as

$$C\left(\frac{\partial\phi}{\partial t} + u\frac{\partial\phi}{\partial x} + \upsilon\frac{\partial\phi}{\partial y} + w\frac{\partial\phi}{\partial z}\right) = \frac{\partial}{\partial x}\left(\Gamma\frac{\partial\phi}{\partial x}\right) + \frac{\partial}{\partial y}\left(\Gamma\frac{\partial\phi}{\partial y}\right) + \frac{\partial}{\partial z}\left(\Gamma\frac{\partial\phi}{\partial z}\right) + S_\phi \tag{3.77}$$

```
initialize xi = x*
do for
    compute residual vector e from e = c - A x*
    Solve the correction vector x̂ from A x̂ = e
    Refine solution vector x from x_{i+1} = x_i + x̂
    If ( ||x̂||_e < ε_s ) then
        Solution converged
        Stop
    Else
End do
```

FIGURE 3.2 Pseudo code for iterative refinement.

where the symbol ϕ represents the velocity components u, v, and w for flow field; temperature T; and mass species concentration C; and two-equations for turbulence kinetic energy, κ and turbulence dissipation rete, ε.

As we will demonstrate, the discretization procedure generally leads to the formation of the linear set of algebraic equations

$$a_i\phi_i^{m+1} = a_{i+1}\phi_{i+1}^{m+1} + a_{i-1}\phi_{i-1}^{m+1} + a_{j+1}\phi_{j-1}^{m+1} + a_{j-1}\phi_{j-1}^{m+1} + a_{k+1}\phi_{k+1}^{m+1} + a_{k-1}\phi_{k-1}^{m+1} + c \tag{3.78a}$$

Or, in the general matrix form as

$$A\,\{\phi\} = \{c\} \tag{3.78b}$$

Following the iterative procedure outlined, we can designate the solution of Equation 3.78 as the estimated values, ϕ^*, given as

$$A\,\{\phi^*\} = \{c^*\} \tag{3.79}$$

and the updated values as

$$\phi = \phi^* + \phi' \tag{3.80}$$

where the corrected values, ϕ', are obtained from the correction equations given as

$$A\,\{\phi'\} = \{c - c^*\} \tag{3.81a}$$

or

$$A\,\{\phi'\} = \{e\} \tag{3.81b}$$

where e is the *residual error vector* and can be rewritten

$$\{e\} = \{c\} - \{A\phi^*\} \tag{3.82}$$

Notice that error equation 3.81b is like the original linear system of equation except for the right-hand side residual vector.

3.3 ITERATIVE METHODS

The direct methods discussed in the preceding sections are usually effective for a small number of equations, typically of the order of several hundreds. This number can often be expanded if the system is well conditioned, the matrix is sparse, and iterative refinement is employed. However, because of the round-off error, computer time, and storage, direct methods become inadequate for larger systems. The discretization of governing equations in many fluid flows and heat transfer problems results in a system as large as several thousand or more. For these kinds of problems, **iterative methods** can be used more effectively.

The solution of a large matrix system requires a large memory as well as a large portion of solution time along with the assembly of the global matrix. Direct methods have been successfully applied to the solution of such systems of equations, running on high-speed computers including super computers. However, memory limitations on super computers, as well as high cost/performance ratios, severely limit the routine application of direct solvers for large-scale three-dimensional flow problems. On the other hand, because of their lower storage requirements and operation costs, iterative methods are adapted to engineering workstations, which are characterized by large memory, fast-scale performance, and low-cost/performance ratio.

Iterative methods are approximate methods, which start with an initial guess solution and iterate to converged solutions with some prespecified tolerance limits. There are many iterative methods that can be used to solve large systems. These include (a) Gauss–Seidel method, (b) Jacobi method, (c) successive over-relaxation (SOR) method, (d) conjugate gradient (CG) method, and (e) generalized minimal residual (GMRES) method.

The **Gauss–Seidel** method is the most used iterative method. It is a simple modification of the **Jacobi** method and is a special case of the SOR method. The Gauss–Seidel method converges for diagonally dominant matrices and for symmetric positive definite matrices. The CG method is highly efficient for symmetric problems. The choice of iterative methods for the system is, however, complicated by the nonsymmetric and nonpositive definitiveness of the matrix. Classical iterative methods such as Gauss–Seidel, SOR, and CG methods become inapplicable. Under such cases, several other methods look more attractive. These include (a) GMRES, (b) CG squared, (c) bi-conjugate gradient stabilized, and (d) transpose free quasi-minimal residual.

3.3.1 Jacobi Method

Let us consider the system of equations given by Equation 3.1, i.e.,

$$Ax = c \tag{3.2b}$$

This system of equations can be rewritten explicitly for the unknown x_i as

$$\begin{aligned} x_i &= \frac{c_1 - a_{12}x_2 - a_{13}x_3 \ldots a_{1n}x_n}{a_{11}} \\ x_2 &= \frac{c_2 - a_{21}x_1 - a_{23}x_3 \ldots a_{2n}x_n}{a_{22}} \\ &\vdots \\ &\vdots \\ x_n &= \frac{c_n - a_{n1}x_1 - a_{n2}x_2 \ldots a_{n,n-1}x_{n-1}}{a_{nn}} \end{aligned} \tag{3.83a}$$

Equation 3.77a can be written in a compact form as

$$x_i = \frac{c_i - \sum\limits_{j=1, j\neq i}^{n} a_{ij}x_j}{a_{ii}}, \quad \text{for} \quad i = 1,2,\ldots,n \tag{3.83b}$$

The sequence of steps constituting the Jacobi method is as given below.

In the **first step**, assign an initial guess value for each unknown appearing in the system. If no reasonable guess of these values can be made, then any arbitrarily chosen values such as zeros can be made, i.e., $x_1^0 = 0, x_2^0 = 0,\ldots, x_n^0 = 0$.

In the **second step**, these initial guess values are substituted into Equation 3.83 to compute a new value of x_i^1 as

$$x_1^1 = \frac{c_1 - a_{12}x_2^0 - a_{13}x_3^0 \ldots a_{1n}x_n^0}{a_{11}}$$
$$x_2^1 = \frac{c_2 - a_{21}x_1^0 - a_{23}x_3^0 \ldots a_{2n}x_n^0}{a_{22}}$$
$$\vdots \tag{3.84a}$$
$$x_n^1 = \frac{c_n - a_{n1}x_1^0 - a_{n2}x_2^0 \ldots a_{n,n-1}x_{n-1}^0}{a_{nn}}$$

and in a compact form as

$$x_i^1 = \frac{c_i - \sum_{j=1, j\neq i}^{n} a_{ij}x_j^0}{a_{ii}}, \quad \text{for} \quad i = 1,2,\ldots,n \tag{3.84b}$$

When all values are estimated by the above equations, the first iteration (k =1) is said to have been completed.

In **step 3**, the iteration procedure outlined in step two is continued by substituting the values of each unknown determined in an iteration, k, into the right-hand side of Equation 3.83 as

$$x_i^{k+1} = \frac{c_i - \sum_{j=1, j\neq i}^{n} a_{ij}x_j^k}{a_{ii}}, \quad \text{for} \quad i = 1,2,\ldots,n \tag{3.85}$$

The iteration process is continued until the unknown values obtained in the present iteration $k + 1$ differs from its respective values obtained in the preceding iteration k by a specified tolerance limit, ε_s. This convergence criterion is expressed as

$$\left| \frac{x_i^{k+1} - x_i^k}{x_i^{k+1}} \right| \leq \varepsilon_s \tag{3.86}$$

The convergence can also be met using an alternative criterion based on the Euclidean norm such as

$$\frac{\left\| x_i^{k+1} - x_i^k \right\|_e}{\left\| x_i^{k+1} \right\|_e} \leq \varepsilon_s \tag{3.87}$$

This alternate convergence criterion is particularly suitable for large systems to save computational time.

3.3.2 Gauss–Seidel Method

It can be observed in the **Jacobi method** that all unknown values are computed using values estimated in the previous iteration step, rather than using most recent values of the other unknowns. Considerable improvement in the storage requirements and the rate of convergence are achieved in many problems by employing an alternative iterative scheme known as the **Gauss–Seidel method**, in which as unknowns are computed in an iteration step; they are subsequently employed in the computation of the rest of the unknown in the same iteration step. This iterative scheme can be expressed as

$$x_i^{k+1} = \frac{c_i - \sum_{j=1}^{i-1} a_{ij} x_j^{k+1} - \sum_{j=i+1}^{n} a_{ij} x_j^k}{a_{ii}}, \quad \text{for} \quad i = 1, 2, \ldots, n \tag{3.88}$$

There are many problems where the Jacobi method converges faster. However, in most cases, the Gauss–Seidel method is the method of choice.

3.3.3 Convergence Criterion for Iterative Methods

One major disadvantage of the iteration method is that it does not always converge. There are problems in which the solution may oscillate and eventually diverge. A **sufficient** condition for the convergence of the Gauss–Seidel and Jacobi methods is that the solution is guaranteed to converge if the coefficient matrix A of the system of Equation 1.1 is diagonally dominant, that is,

$$|a_{ij}| > \sum_{j=1, j \neq i}^{n} |a_{ij}| \tag{3.89}$$

It is important to note that such a condition is a sufficient condition and not a **necessary** condition. That is, at times, the system may violate the criterion and still attain converged solutions. It is quite desirable to have a discretization scheme that leads to a diagonally dominant system.

3.3.4 The SOR Method

There are many problems in which the Gauss–Seidel method converges at a slow rate, especially when the system is large. However, the rate of convergence can be enhanced by using a parameter, Λ, and using a solution scheme given by

$$x_i^{k+1} = \Lambda x_i^{k+1} + (1 - \Lambda) x_i^k \tag{3.90}$$

where Λ is a constant and termed as the **relaxation parameter**, and x_i^{k+1} on the right-hand side of the equation is the value of x_i obtained in the $(k+1)$th iterative step by using the Gauss–Seidel iteration scheme given by Equation 3.88. Substituting Equation 3.88 into Equation 3.90, we obtain the relaxation scheme as

$$x_i^{k+1} = \Lambda \left[\frac{c_i - \sum_{j=1}^{i-1} a_{ij} x_j^{k+1} - \sum_{j=i+1}^{n} a_{ij} x_j^k}{a_{ii}} \right] + (1 - \Lambda) x_i^k \tag{3.91}$$

If $1 < \Lambda < 2$, then this modified Gauss–Seidel iterative scheme is known as the **SOR method**. It can be noted from Equation 3.91 that the SOR iteration method reduces to Gauss–Seidel method when

$\Lambda = 1$. If $0<\Lambda<1$, the iteration scheme is known as **successive under-relaxation (SUR) method**. A SUR method is generally employed to make a nonconvergent Gauss–Seidel iteration method to converge or to facilitate convergence by dampening out any small-scale oscillation. An optimum value of the relaxation parameter Λ is established through numerical experimentation.

Example 3.6: Gauss–Seidel Method

Use the Gauss–Seidel algorithm to solve the following system of equations:

$$\begin{aligned} 3x_1 + x_2 - 2x_3 &= 9 \\ -x_1 + 4x_2 - 3x_3 &= -8 \\ x_1 - x_2 + 4x_3 &= 1 \end{aligned} \tag{E.3.6.1}$$

Continue iteration until all unknowns are accurate to three decimal places, i.e., converged to specified tolerance of $\varepsilon_s = 0.0009$.

Solution

In the **first step**, we use Equation 3.77a to express each equation for its unknown on the diagonal in an explicit form as

$$x_1^{k+1} = \frac{9 - x_2^k + 2x_3^k}{3} \tag{E.3.6.2a}$$

$$x_2^{k+1} = \frac{-8 + x_1^{k+1} + 3x_3^k}{4} \tag{E.3.6.2b}$$

$$x_3^{k+1} = \frac{1 - x_1^{k+1} + x_2^{k+1}}{4} \tag{E.3.6.2c}$$

In the **second step**, we assume the initial guess values as

$$x_1^0 = 0 \tag{E.3.6.3a}$$

$$x_2^0 = 0 \tag{E.3.6.3b}$$

$$x_3^0 = 0 \tag{E.3.6.3c}$$

In the **third step**, we start the iteration process. The first iteration is completed by substituting the initial guess values into Equation E.3.6.2 as

Iteration, $k = 1$

$$x_1^1 = \frac{9 - x_2^0 + 2x_3^0}{3} = \frac{9-(0)+2(0)}{3} = 3.00000000 \tag{E.3.6.4a}$$

$$x_2^1 = \frac{-8 + x_1^1 + 3x_3^0}{4} = \frac{-8+(3.00000000)+3(0)}{4} = -1.25000000 \tag{E.3.6.4b}$$

$$x_3^1 = \frac{1 - x_1^1 + x_2^1}{4} = \frac{1-(3.00000000)+(-1.25000000)}{4} = -0.8125 \tag{E.3.6.4c}$$

Percent Relative Error

$$|\varepsilon_{a1}| = \left|\frac{x_1^1 - x_1^0}{x_1^1}\right| = 1.00000000 \tag{E.3.6.5a}$$

$$|\varepsilon_{a2}| = \left|\frac{x_2^1 - x_2^0}{x_2^1}\right| = 1.00000000 \tag{E.3.6.5b}$$

$$|\varepsilon_{a3}| = \left|\frac{x_3^1 - x_3^0}{x_3^1}\right| = 1.00000000 \tag{E.3.6.5c}$$

Iteration, $k = 2$

$$x_1^2 = \frac{9 - x_2^1 + 2x_3^1}{3} = \frac{9-(-1.25000000)+2(-0.8125)}{3} = 2.875 \tag{E.3.6.6a}$$

$$x_2^2 = \frac{-8 + x_1^2 + 3x_3^1}{4} = \frac{-8+(2.875)+3(-0.8125)}{4} = -1.890625 \tag{E.3.6.6b}$$

$$x_3^2 = \frac{1 - x_1^2 + x_2^2}{4} = \frac{1-(2.875)+(-1.890625)}{4} = -0.94140625 \tag{E.3.6.6c}$$

Percent Relative Error

$$|\varepsilon_{a1}| = \left|\frac{x_1^2 - x_1^1}{x_1^2}\right| = 0.043478261 \tag{E.3.6.7a}$$

$$|\varepsilon_{a2}| = \left|\frac{x_2^2 - x_2^1}{x_2^2}\right| = 0.338\,842975 \tag{E.3.6.7b}$$

$$|\varepsilon_{a3}| = \left|\frac{x_3^2 - x_3^1}{x_3^2}\right| = 0.136929461 \tag{E.3.6.7c}$$

For the subsequent iterations, a similar procedure is repeated until a solution vector converges to the specified tolerance of $\varepsilon_s = 0.00009$. The results are summarized in the table below.

Iteration	x_1	x_2	x_3	ε_{a1}	ε_{a2}	ε_{a3}
1	3.0000000	−1.2500000	−0.8125	1.0000000	1.0000000	1.0000000
2	2.875	−1.890625	−0.94140625	0.04347821	0.3388429	0.1369294
3	3.0026041	−1.95540364	−0.98950195	0.04249783	0.0331280	0.0486059
4	2.99213324	−1.99409315	−0.9965566	0.34648510	0.0194020	0.0070790
5	3.00032665	−1.99733578	−0.99941561	0.00273083	0.0032426	0.0028606
6	2.99950152	−1.99968632	−0.99979619	0.00027508	0.0011754	0.00038066
7	3.00003131	−1.99983932	−0.99996765	0.00017659	0.0000765	0.00017146
8	2.99996800	−1.99998374	−0.99998793	0.00002110	0.0000722	0.00002027

3.3.5 CG Method

One difficulty associated with the SOR method is that it is not always guaranteed to converge and often the convergence depends upon the relaxation parameter Λ that is quite difficult to choose without numerical experimentation. The **CG** method, on the other hand, is guaranteed to converge, particularly for a large and sparse symmetric positive definite system of equations. In this section, we present the basic CG method of Hestenes and Stiefel (1952) that has its origin in optimization theory. The CG method can be considered as an acceleration technique that does not require estimation of acceleration or relaxation parameter as in the case of other iteration techniques such as the SOR method.

As stated, the CG method can be viewed as an optimization problem. If A is a symmetric and positive definite, then it can be shown that the problem of solving a linear system $Ax = c$ is equivalent to minimizing a quadratic function

$$\Pi(x) = \frac{1}{2}x^{\mathrm{T}}Ax - x^{\mathrm{T}}c \tag{3.92}$$

The quadratic function, $\Pi(x)$, attains its minimum value of $-\frac{1}{2}c^{\mathrm{T}}\, A^{-1}c$ at $x = A^{-1}C$. Our objective is not only to decrease the function Π^{k+1} efficiently, but also to estimate x^{k+1} efficiently and attain a rapid convergence process for converging the quadratic function to a minimum.

In the basic CG method, the minimization problem is solved iteratively by the recursive scheme

$$x^{k+1} = x^k + \alpha_p^k p^k \tag{3.93}$$

where the scalar quantity α_p^k is chosen such as to minimize the function $\Pi(x)$, and it is expressed as

$$\alpha_p^k = \frac{r^{(k)^{\mathrm{T}}}\, r^k}{p^{(k)^{\mathrm{T}}}\, Ap^{(k)}} \tag{3.94}$$

The residual r^k is estimated as

$$r^{k+1} = r^k - \alpha_p^k Ap^k \tag{3.95}$$

where the vector p^k is called the direction vector and is linearly independent. The quadratic function is minimized in the space spanned by these vectors, which are generated in the preceding iterative step as

$$p^{k+1} = r^{k+1} + \beta_p^k p^k \tag{3.96}$$

where

$$\beta^k = \frac{r^{(k+1)^{\mathrm{T}}}\, r^k}{r^{(k)^{\mathrm{T}}}\, r^k} \tag{3.97}$$

The iterative process starts with an initial guess value of x^1 and by setting $p^1 = r^1 = c - Ax^1$ the iterative process is continued until convergence.

The basic CG algorithm can be summarized as follows:

Choose the initial guess x^1 (often a null vector)

Choose convergence tolerance ε_s

Calculate the residual $r^1 = c - Ax^1$

Set $p^1 = r^1$

for $k = 1, 2, \ldots,$calculate

$$\alpha_p^k = \frac{r^{(k)^{\mathrm{T}}} r^k}{p^{(k)^{\mathrm{T}}} A p^k}$$

$$x^{k+1} = x^k = \alpha_p^k p^k$$

$$r^{k+1} = r^k - \alpha_p^k A p^k$$

$$\beta_p^k = \frac{r^{(k+1)^{\mathrm{T}}} r^{k+1}}{r^{(k)^{\mathrm{T}}} r^k}$$

$$p^{(k+1)} = r^{k+1} + \beta_p^k p^k$$

Continue iteration until $\left\| r^k \right\|_e \le \varepsilon_s$

A convergence criterion on $\left| \dfrac{x^{k+1} - x^k}{x^{k+1}} \right|$ could also be used. A pseudo code for basic CG algorithm is illustrated in Figure 3.3.

The rate of convergence of the CG method depends on the condition number of the matrix A. The larger the condition number, the slower the convergence, and in cases where the matrix A is ill-conditioned, the convergence can be extremely slow. The convergence speed of the CG method can be increased using **preconditioning**. In fact, the CG method is rarely used without any preconditioning.

Example 3.7: CG Method

Use the CG method to solve the following system of equations. Continue iteration until converged to a tolerance limit of $\varepsilon_s = 0.0009$ using the Euclidean norm

$$\begin{bmatrix} 12 & 1 & 7 \\ 1 & 12 & 3 \\ 7 & 3 & 12 \end{bmatrix} \begin{Bmatrix} x_1 \\ x_2 \\ x_{31} \end{Bmatrix} = \begin{Bmatrix} 27 \\ -24 \\ 3 \end{Bmatrix}$$

Solution

Step 1

Choose initial guess as

Step 1

Choose x^1, ε_s and itmax

Choose k = 0 and err = 0

Calculate the residual $r^1 = c - Ax^1$

Set $p^1 = r^1$

Step 2

do while ((k < itmax) and err = 0)

err = 1

dofor k = 1, 2 . . .

$$\alpha_p^k = \frac{\mathrm{r}^{(k)^T}\mathrm{r}^k}{p^{(k)^T}\mathrm{A}\,p^k}$$

$$\mathrm{x}^{k+1} = \mathrm{x}^k + \alpha_p^k\, p^k$$

$$\mathrm{r}^{k+1} = \mathrm{r}^k - \alpha_p^k\,\mathrm{A}\,p^k$$

$$\beta_p^k = \frac{\mathrm{r}^{(k+1)^T}\mathrm{r}^{k+1}}{\mathrm{r}^{(k)^T}\mathrm{r}^k}$$

$$p^{(k+1)} = \mathrm{r}^{k+1} + \beta_p^k\, p^k$$

If ((err = 1) and $\mathrm{r}^{k+1} \neq 0$)

$\varepsilon_a = \left\|\mathrm{r}^k\right\|$

If ($\varepsilon_a > \varepsilon_s$) then err = 0

end if

enddo

FIGURE 3.3 A pseudo code for CG algorithm.

$$x^1 = \begin{Bmatrix} 0 \\ 0 \\ 0 \end{Bmatrix}$$

$$r^1 = C - Ax^1 = \begin{Bmatrix} 27 \\ -24 \\ 3 \end{Bmatrix} - \begin{bmatrix} 12 & 1 & 7 \\ 1 & 12 & 3 \\ 7 & 3 & 12 \end{bmatrix} \begin{Bmatrix} 0 \\ 0 \\ 0 \end{Bmatrix} = \begin{Bmatrix} 27 \\ -24 \\ 3 \end{Bmatrix}$$

$$p^1 = \begin{Bmatrix} 27 \\ -24 \\ 3 \end{Bmatrix},\quad p^{(1)^T} = \begin{Bmatrix} 27 & -24 & 3 \end{Bmatrix}$$

Step 2

$$\alpha_p^1 = \frac{r^{(k)^{\mathrm{T}}} r^k}{p^{(k)^{\mathrm{T}}} A p^k} = \frac{\left(\|r^1\|\right)^2}{p^{(1)^{\mathrm{T}}} A p^1} = 0.087$$

$$x^2 = x^1 + \alpha_p^1 p^1 = \begin{Bmatrix} 0 \\ 0 \\ 0 \end{Bmatrix} + 0.087 \begin{Bmatrix} 27 \\ -24 \\ 3 \end{Bmatrix}, \quad x^2 = \begin{pmatrix} 2.349 \\ -2.088 \\ 0.261 \end{pmatrix}$$

$$r^2 = r^1 - \alpha_p^1 A p^1 = \begin{Bmatrix} 27 \\ -24 \\ 3 \end{Bmatrix} - 0.087 \begin{bmatrix} 12 & 1 & 7 \\ 1 & 12 & 3 \\ 7 & 3 & 12 \end{bmatrix} \begin{Bmatrix} 27 \\ -24 \\ 3 \end{Bmatrix}, \quad r^2 = \begin{pmatrix} -0.927 \\ -2.076 \\ -10.311 \end{pmatrix}$$

$$\beta_p^1 = \frac{r^{(2)^{\mathrm{T}}} r^2}{r^{(1)^{\mathrm{T}}} r^1} = \frac{(-0.927)^2 + (-2.076)^2 + (-10.311)^2}{(27)^2 + (-24)^2 + (3)^2}, \quad \beta_p^1 = 0.084\,845$$

$$p^2 = r^2 + \beta_p^1 p^1 = \begin{pmatrix} -0.927 \\ -2.076 \\ -10.311 \end{pmatrix} + 0.084845 \begin{Bmatrix} 27 \\ -24 \\ 3 \end{Bmatrix}, \quad p^2 = \begin{pmatrix} 1.368 \\ -4.116 \\ -10.056 \end{pmatrix}$$

Check for convergence using the **Euclidean norm**, i.e., $\left\|r_i^{k+1}\right\|_e \le \varepsilon_s$:

$$\left(\|r^2\|\right)^2 = \left[(1.368)^2 + (-4.116)^2 + (-10.056)^2\right] = 111.485826$$

or

$$\|r^2\| = 10.558$$

Since $\left(\|r^2\|\right) > \varepsilon_s$, we continue the iteration steps and the results are summarized in the table below.

Iteration No.	r_1^k	r_2^k	r_3^k	x_1^k	x_2^k	x_3^k	$\left(\|r^2\|\right)$
1	1.368	−4.116	−10.056	2.349	−2.088	0.261	10.558
2	3.4299	3.7884	−1.0527	2.4516	−2.3967	0.4932	5.21769
3	0.036341	−0.07639	0.01714	2.997335	−1.99300	−1.00162	0.08631
4	0.0069231	−0.00326	−0.00322	3.000661	−1.99973	−1.00018	0.00830
5	−0.001065	0.000290	0.002092	3.000076	−2.00009	−1.00045	0.00223

Solution converged in five iteration steps.

3.3.6 Preconditioned CG Method

The basic CG method works poorly for an ill-conditioned system. Convergence speed of the CG method can be increased using preconditioning. Preconditioning means that we change variables to obtain a new system $\tilde{A}x = c$ whose coefficient matrix $\tilde{A}$ has a much-improved eigenvalue distribution or condition number, condition $(\tilde{A})$, compared to that of the original system $Ax = c$. The new system is obtained by substituting $x = L^{\mathrm{T}}\tilde{x}$ into the original system as

$$AL^{T}\tilde{x} = c \tag{3.98}$$

Premultiplying Equation 3.92 by L yields the equivalent system

$$LAL^{T}\tilde{x} = Lc$$

or

$$\tilde{A}\tilde{x} = \tilde{c} \tag{3.99}$$

where

$$\tilde{A} = LAL^{T} \tag{3.100}$$

and

$$\tilde{c} = Lc \tag{3.101}$$

The CG method is now applied to the new system and the original variable x is then recovered from the variable $\tilde{x}$ using $x = L^{T}\tilde{x}$.

The nonsingular matrix $A_p = (LL^{T})^{-1}$ is called the **preconditioning matrix or preconditioner**. The main objective with this transformation is to obtain a matrix $\tilde{A}$ that has a much-improved condition number. The preconditioned CG algorithm is expressed as follows.

Choose the initial guess x^1

Choose convergence tolerance ε_s

Calculate the residual $r^1 = c - Ax^1$

Calculate $z^1 = A_p^{-1} r^1$

Set $p^1 = z^1$

Calculate for $k = 2,3, \ldots$

$$\alpha_p^k = \frac{z^{(k)^{T}} r^k}{p^{(k)^{T}} A p^k}$$

$$x^{k+1} = x^k + \alpha_p^k p^k$$

$$r^{k+1} = r^k - \alpha_p^k A p^k$$

$$z^{k+1} = A_p^{-1} r^{k+1}$$

$$\beta_p^k = \frac{z^{(k+1)^{T}} r^{k+1}}{z^{(k)^{T}} r^k}$$

$$p^{(k+1)} = z^{k+1} + \beta_p^k p^k$$

Continue iteration until $\left\| r^k \right\|_e \leq \varepsilon_s$

It can be noted that in this algorithm an intermediate vector z^k has been introduced. If no preconditioning is used, i.e., if $A_p = I$, then this intermediate vector z^k is equal to r^k and the precondition CG algorithm reduces to the basic CG algorithm. Significant savings in computer storage and time are gained by using the precondition CG method.

One of the important steps that must be considered before using the precondition CG method is to determine a preconditioner, A_p. Many different preconditioners have been proposed and used in solving fluid flow and heat transfer problems. Among them, the most popular ones are the incomplete Cholesky factorization, the polynomial preconditioner, the diagonal scaling preconditioner, and the block factorization preconditioner. The choice of a good preconditioner can have a dramatic effect upon the rate of convergence. For a detailed description of preconditioners, see Golub and Van Loan (1989), Ortega and Poole (1981), and Saad (1996).

3.3.7 Generalized Minimal Residuals Method

The CG and preconditioned conjugate methods discussed in previous section are an efficient solver for large sparse symmetric and positive definite system of linear equations. However, in many fluid dynamics and heat transfer problem, the discretization process leads to the formation of nonsymmetric and indefinite matrix form. For such problems, GMRES iterative algorithm is found to be very efficient for large scale Computational Fluid Dynamics (CFD) problems. GMRES as an iterative algorithm based on the approach of minimizing the norm of the residual error vector over a Krylov space as demonstrated by Saad (Saad, 1996; Saad and Schultz, 1986).

The GMRES algorithm is designed as a generalization of the MINRES (Paige and Saunders, 1975) algorithm and based on the methodologies presented by Arnoldi (Arnoldi, 1951; Saad, 1980) for any nonsingular arbitrary sparse matrices. The Arnoldi's method derives an upper Hessenberg matrix using an orthonormal basis to minimize the residual norm over the Krylov subspace. The GMRES iterative algorithm proposed by Saad et al. (1986) implemented Arnoldi's method to generate an upper Hessenberg orthonormal basis to minimize the residual error norm.

The iterative GMRES method relies on the main concept that a smaller set consecutive iterative values exhibit linearity pattern and so the solution can be obtained solved as a least-square problem based on the property of minimizing the norm of the residual vector over a **Krylov subspace**, which is defined for an associated dimension r for an $n \times n$ matrix, A, as

$$k^r \left(A,x^0\right) = \text{span}\left[x^{(0)}, A\,x^0, A^2x^{(0)} \ldots k^{(r-1)}x^{(0)}\right] \tag{3.102}$$

Here x^0 is the initial guess value in each iteration steps of the GMRES algorithm for solving the least-square problem based on the property of minimizing the norm of residual vector. The matrix A is the coefficient matrix of the linear system of equation given as

$$Ax = c \quad \text{or} \quad a_{ij}x_j = c_i \tag{3.103}$$

One important point that can be noted here that to proceed with the solution of the least square problem, we need to determine the orthonormal basis of the Krylov subspace to construct a Hessenberg matrix H^{r+1} and followed by its reduction to an upper Hessenberg matrix H^r. There are a number of different approaches used in the implementation of GMRES algorithm. Then the most important and popular ones are the Gram-Schmidt method used by Arnoldi (1986); the Householder transformation (1964); and the Givens matrix rotation and transformation for reduction to an upper Hessenberg matrix. Let us demonstrate here the GMRES method and algorithm based on **Given rotation process** to derive the solution during iteration process.

Before proceeding with description of the GRMES method, let us briefly review the Given Matrix and rotation for Hessenberg matrix reduction.

A **Given's rotation matrix** is defined in the following form:

$$R_{ij}(c,s) = \begin{bmatrix} 1 & & & & & & & & & \\ & 1 & & & & & & & & \\ & & 1 & & & & & & & \\ & & & 1 & & & & & & \\ & & & & c_r & s_r & & & & \\ & & & & -s_r & c_r & & & & \\ & & & & & & 1 & & & \\ & & & & & & & 1 & & \\ & & & & & & & & 1 & \\ & & & & & & & & & 1 \end{bmatrix} \tag{3.104a}$$

where

$$c^2 + r^2 = 1 \tag{3.104b}$$

which satisfies $c = \cos\theta$ and $s = \sin\theta$ and this enables the Given matrix $R_{ij}(c,s)$ to rotate orthogonally a pair of unit vectors i and j through a given angle θ in the (x,y) or (i,j) plane. Also, the Givens matrix $R_{ij}(c,s)$ rotation is orthogonal as $R_{ij} \cdot R_{ij}^T = I$.

One of the important uses of the **Givens matrix** is to create zeros in a specific position of a vector or in specific position in column vector of a matrix. Using such an approach, **Givens rotations** can be applied successively over the column vector one at a time and transform a $n \times n$ matrix A into an upper triangular matrix like **Hessenberg** matrix form, H.

Before demonstrating this procedure to reduce a matric into Hessenberg form, let us show the use of the Givens matrix $\begin{bmatrix} c_r & s_r \\ -s_r & c_r \end{bmatrix}$ to create zero in a specific position row # 2 ($i = 2$) of the column vector ($j = 1$), $\begin{Bmatrix} x_1 \\ x_2 \end{Bmatrix}$ using Givens rotation in the following manner:

$$\begin{bmatrix} c_r & s_r \\ -s_r & c_r \end{bmatrix} \begin{Bmatrix} x_1 \\ x_2 \end{Bmatrix} = \begin{Bmatrix} x^* \\ 0 \end{Bmatrix} \tag{3.105}$$

Considering the values of the vector as $\begin{Bmatrix} x_1 \\ x_2 \end{Bmatrix} = \begin{Bmatrix} 1 \\ \frac{1}{3} \end{Bmatrix}$, we can estimate

$$c_r = \frac{x_1}{\sqrt{x_1^2 + x_2^2}} = \frac{1}{\sqrt{1 + \frac{1}{9}}} = \frac{3}{\sqrt{10}}$$

$$s_r = \frac{x_2}{\sqrt{x_1^2 + x_2^2}} = \frac{\frac{1}{3}}{\sqrt{1 + \frac{1}{9}}} = \frac{1}{\sqrt{10}}$$

Substituting these values into the left-hand side of Equation 3.105, we get the reduction of the vector as

$$\begin{bmatrix} \frac{3}{\sqrt{10}} & \frac{1}{\sqrt{10}} \\ -\frac{1}{\sqrt{10}} & \frac{3}{\sqrt{10}} \end{bmatrix} \begin{Bmatrix} x_1 \\ x_2 \end{Bmatrix} = \begin{Bmatrix} \frac{6}{\sqrt{10}} \\ 0 \end{Bmatrix}$$

The procedure can be repeated for any other positions within a matrix. Givens rotation method can also be applied for the QR factorization of a matrix A.

3.3.8 GMRES Method

The GMRES method falls under a special class of iterative methods that is based on the optimization of the residuals. For a linear system of equations $Ax = c$ $\left(\text{or } a_{ij}x_j = c_i\right)$, the GMRES scheme starts with an iterative scheme given as

$$x_j^{(0)} = x_j^{(0)} + \bar{x}_j \tag{3.106}$$

where $x_j^{(0)}$ is the guess value and $\bar{x}_j$ is an element of the Krylov space K of dimension r used in order to minimize the **residual error-norm**. The residual error with this initial guess is then

$$E = c - Ax^{(0)}. \tag{3.107}$$

The objective is to minimize the norm of the error vector E_i as

$$\text{Minimize}\, E_i = c_i - a_{ij}\left(x_j^{(0)} + \bar{x}_j\right) \tag{3.108}$$

The process starts with the selection of a smaller value of r for the Krylov space and iterate until convergence. To proceed with the minimization process in an orthonormal basis, another important step is performed to compute a new vector $\boldsymbol{y}$ which minimizes the residual norm. We can write the minimization statement given by Equation 3.108 in terms of this new vector as

$$\text{Min}\left\| c_i - a_{ij}\left(x_j^{(0)} + \bar{x}_j\right)\right\| = \text{Min}\left\| E_i^0 - a_{ij}\, H_{\text{jm}}^{(r,r)}\, y_m\right\| \tag{3.109}$$

where $H_{\text{jm}}^{(r,r)}$ is an Hessenberg matrix. Considering $\{h_1,\ h_2,\ \ldots\ h_r\}$ one such orthonormal basis vector, we can construct an orthogonal matrix H^r by placing each basis vector as column elements of the matrix.

In order to simplify the solution for the $\boldsymbol{y}$ vector, the Hessenberg matrix H_{jm}^{r+1} can be transformed into an **upper Hessenberg matrix,** $\bar{H}_{\text{jm}}^r$ using orthonormal rotation scheme such as Given rotation matrix, R so that $\boldsymbol{y}$ vector can be solve by simply using back substitution steps. The minimized error statement given by Equation 3.109 now can be written as

$$\text{Min}\left\| E_i^0 - a_{ij}H_{\text{jm}}^r y_m\right\| = \left\| \bar{e}_i - H_{\text{im}}^r y_m\ \right\| \approx 0 \tag{3.110}$$

Once $\boldsymbol{y}$ vector is solved from this equation directly, we can then start a new iteration step beginning with computing the new residual vector

$$E_i^r = \bar{E}_{ij} y_j \tag{3.111}$$

where $\bar{E}_{ij}$ matrix is formed by replacing each column of the matrix with the residual vectors $\bar{E}_i^r$ computed during Gram-Schmidt orthogonalization steps.

In the next step, the guess value is updated using residual error value as

$$x_j^{(1)} = x_j^{(0)} + E_j^r \tag{3.112}$$

The **GMRES algorithm** is summarized below:

Step 1 Choose $x_j^{(0)}$ as the initial guess

1. Evaluate residual error

$$E_i^{(0)} = c_i - a_{ij}\ x_j^{(0)}$$

Estimate the residual vector norm

$$E_i^{(0)} = \sqrt{\sum \left(E_i^{(0)}\right)^2}$$

and normalized error vector

$$\bar{E}_i^{(1)} = \frac{E_i^{(0)}}{E_i^{(0)}}$$

Step 2 Begin iterations to determine the residuals
Do for $k = 1, 2, \ldots r$

$$\tilde{E}_i^{(k)} = a_{ij}\ \bar{E}_j^{(k)}$$

Do for $l = 1, \ldots, k$

$$h^{(k,l)} = \tilde{E}_i^{(k)}\ \bar{E}_i^{(l)}$$

$$\tilde{E}_i^{(k)} = \tilde{E}_i^{(k)} - h^{(k,l)}\ \bar{E}_i^{(l)}$$

$$\tilde{E}_i^{(k)} = \sqrt{\sum \left(\tilde{E}_i^{(k)}\right)^2}$$

$$\bar{E}_i^{(k+1)} = \frac{E_i^{(k)}}{E_i^{(k)}} = \begin{Bmatrix} \bar{E}_1^{(k+1)} \\ \bar{E}_2^{(k+1)} \\ \vdots \\ \bar{E}_n^{(k+1)} \end{Bmatrix}$$

Continue
Continue

Step 3 Construct **Hessenberg Matrix**

$$H_{om}^{(r+1,\,r)} = \begin{bmatrix} h^{(1,1)} & h^{(2,1)} & h^{(3,1)} \\ \tilde{E}_i^{(1)} & h^{(2,2)} & h^{(3,2)} \\ 0 & \tilde{E}_i^{(2)} & h^{(3,3)} \\ 0 & 0 & \tilde{E}_i^{(3)} \end{bmatrix}$$

The system equation based on this Hessenberg matrix form is

$$\begin{bmatrix} h^{(1,1)} & h^{(2,1)} & h^{(3,1)} \\ \tilde{E}_i^{(1)} & h^{(2,2)} & h^{(3,2)} \\ 0 & \tilde{E}_i^{(2)} & h^{(3,3)} \\ 0 & 0 & \tilde{E}_i^{(3)} \end{bmatrix} \begin{Bmatrix} y1 \\ y2 \\ y3 \end{Bmatrix} = \begin{Bmatrix} E_i^{(0)} \\ 0 \\ 0 \end{Bmatrix}$$

Step 4 Construct reduced upper Hessenberg matrix (using **Givens rotation matrix** R_{io}^{r}). For example, in the first rotation, the Given matric elements can be computed as follows:

$$c_l = \frac{H_{ll}}{\sqrt{\left(H_{l,l}\right)^2 + \left(H_{l+1,l}\right)^2}} = \frac{h^{(1,1)}}{\sqrt{\left(h^{(1,1)}\right)^2 + \left(\tilde{E}_i^{(1)}\right)^2}}$$

$$s_l = \frac{H_{l+1,l}}{\sqrt{\left(H_{l,l}\right)^2 + \left(H_{l+1,l}\right)^2}} = \frac{\tilde{E}_i^{(1)}}{\sqrt{\left(H_{l,l}\right)^2 + \left(\tilde{E}_i^{(1)}\right)^2}}$$

The rotation matrix can be constructed with its elements given by the column of element of Hessenberg matrix obtained in Step 4 for orthonormalization and satisfying

$$\overline{H}_{\text{im}}^{(r,r)} = R_{\text{io}}^{(r,r+1)}\, H_{\text{om}}^{(r+1,\,r)}$$

where $H_{\text{om}}^{(r+1,\,r)}$ is an upper **Hessenberg matrix (obtained in Step 3)**

During the rotation process, the right-side vector will also be transformed to a new form.

$$\begin{bmatrix} \overset{=}{h}^{(1,1)} & \overset{=}{h}^{(2,1)} & \overset{=}{h}^{(3,1)} \\ 0 & \overset{=}{h}^{(2,2)} & \overset{=}{h}^{(3,2)} \\ & & \overset{=}{h}^{(3,3)} \end{bmatrix} \begin{Bmatrix} y_1 \\ y_2 \\ y_3 \end{Bmatrix} = \begin{Bmatrix} E_1^{(0)} \\ E_2^{(0)} \\ E_3^{(0)} \end{Bmatrix}$$

where

$$\bar{\bar{H}}_{\text{im}}^{(r,r)} = \begin{bmatrix} \bar{\bar{h}}^{(1,1)} & \bar{\bar{h}}^{(2,1)} & \bar{\bar{h}}^{(3,1)} \\ 0 & \bar{\bar{h}}^{(2,2)} & \bar{\bar{h}}^{(3,2)} \\ & & \bar{\bar{h}}^{(3,3)} \end{bmatrix}$$

and the right-hand side known vector $\left\{ \begin{matrix} [\![E_i^{(0)}]\!] \\ 0 \\ 0 \end{matrix} \right\}$ changes to

$$\left\{ \begin{matrix} [\![E_1^{(0)}]\!] \\ [\![E_2^{(0)}]\!] \\ [\![E_3^{(0)}]\!] \end{matrix} \right\}$$

Step 5 Solve Equation 3.110 by back substitution for the solution of the vector $\boldsymbol{y}$.

$$\left\{ \begin{matrix} y_1 \\ y_2 \\ y_3 \end{matrix} \right\} =$$

Step 6 Compute the residuals

$$\left\{ \begin{matrix} E_1^{(r)} \\ E_2^{(r)} \\ \vdots \\ E_n^{(r)} \end{matrix} \right\} = \begin{bmatrix} \bar{E}_1^{(1)} & \bar{E}_1^{(2)} & \cdots & \bar{E}_1^{(n)} \\ \bar{E}_2^{(1)} & \bar{E}_1^{(2)} & \cdots & \bar{E}_2^{(n)} \\ \vdots & \vdots & \cdots & \vdots \\ \bar{E}_n^{(1)} & \bar{E}_n^{(2)} & \cdots & \bar{E}_n^{(n)} \end{bmatrix} \left\{ \begin{matrix} y_1 \\ y_2 \\ \vdots \\ y_n \end{matrix} \right\}$$

Step 7 Update new values of x_i

$$\begin{Bmatrix} x_1 \\ x_2 \\ \vdots \\ x_n \end{Bmatrix} = \begin{Bmatrix} x_1^{(0)} \\ x_2^{(0)} \\ \vdots \\ x_n^{(0)} \end{Bmatrix} + \begin{Bmatrix} E_1^{(r)} \\ E_2^{(r)} \\ \vdots \\ E_n^{(r)} \end{Bmatrix}$$

PROBLEMS

3.1 Consider the following matrix:

$$[A] = \begin{bmatrix} 4 & -1 & 2 \\ 4 & -8 & 2 \\ -2 & 1 & 6 \end{bmatrix}$$

Determine (a) the determinant of the matrix $|A|$, (b) the inverse of the matrix A^{-1}, and (c) the condition number **Cond (*A*)** using the row-sum norm.

3.2 Develop a code **Gauss** to implement the Gauss elimination algorithm with partial pivoting to solve the following system of equations.

$$\begin{bmatrix} 4 & -1 & 0 & 0 & 0 & 0 \\ 0 & 4 & -1 & 0 & 0 & 0 \\ 0 & -1 & 4 & -1 & 0 & 0 \\ 0 & 0 & -1 & 4 & -1 & 0 \\ 0 & 0 & 0 & -1 & 4 & 0 \\ 0 & 0 & 0 & 0 & -1 & -4 \end{bmatrix} \begin{Bmatrix} x_1 \\ x_2 \\ x_3 \\ x_4 \\ x_5 \\ x_6 \end{Bmatrix} = \begin{Bmatrix} -90 \\ 90 \\ 0 \\ 0 \\ 90 \\ 90 \end{Bmatrix}$$

Print the augmented matrix and the pivot row and pivot element in every elimination step. Use the results to estimate the residual terms.

3.3 Develop a code **Gauss–Jordan** implementing the Gauss–Jordan algorithm to determine the inverse of the coefficient matrix and solve the following system of equations.

$$\begin{bmatrix} 2 & -1 & 0 & 0 & 0 & 0 \\ 0 & 2 & -1 & 0 & 0 & 0 \\ 0 & -1 & 2 & -1 & 0 & 0 \\ 0 & 0 & -1 & 2 & -1 & 0 \\ 0 & 0 & 0 & -1 & 2 & 0 \\ 0 & 0 & 0 & 0 & -1 & -2 \end{bmatrix} \begin{Bmatrix} x_1 \\ x_2 \\ x_3 \\ x_4 \\ x_5 \\ x_6 \end{Bmatrix} = \begin{Bmatrix} -30 \\ 30 \\ 0 \\ 0 \\ 30 \\ 30 \end{Bmatrix}$$

3.4 Determine the Euclidean norm $\|A\|_e$ and the row-sum norm $\|A\|_\infty$ for the coefficient matrix A of the system given in Problem 3.2. Determine the condition number Cond (A) for the system in Problem 3.2 using row-sum norm.

3.5 Develop a computer code ludecom based on *LU* decomposition algorithms to solve the following system of equations and present results in each elimination step.

$$\begin{bmatrix} -4 & 1 & 0 & 4 & 0 & 0 & 0 & 0 & 0 \\ 1 & -4 & 1 & 0 & 1 & 0 & 0 & 0 & 0 \\ 0 & 1 & -4 & 0 & 0 & 1 & 0 & 0 & 0 \\ 1 & 0 & 0 & -4 & 1 & 0 & 1 & 0 & 0 \\ 0 & 1 & 0 & 1 & -4 & 1 & 0 & 1 & 0 \\ 0 & 0 & 1 & 0 & 1 & -4 & 0 & 0 & 1 \\ 0 & 0 & 0 & 1 & 0 & 0 & -4 & 0 & 0 \\ 0 & 0 & 0 & 0 & 1 & 0 & 1 & -4 & 1 \\ 0 & 0 & 0 & 0 & 0 & 1 & 0 & 1 & -4 \end{bmatrix} \begin{Bmatrix} T_1 \\ T_2 \\ T_3 \\ T_4 \\ T_5 \\ T_6 \\ T_7 \\ T_8 \\ T_9 \end{Bmatrix} = \begin{Bmatrix} -420 \\ -400 \\ -420 \\ -20 \\ 0 \\ -20 \\ -40 \\ -20 \\ -40 \end{Bmatrix}$$

Use partial pivoting if necessary.

3.6 Solve the following system of equations using the TDMA algorithm.

$$\begin{bmatrix} 1.0835 & -0.04175 & 0 & 0 \\ -0.04175 & 1.0835 & -0.04175 & 0 \\ 0 & -0.04175 & 1.0835 & -0.04175 \\ 0 & 0 & -0.04175 & 1.0835 \end{bmatrix} \begin{Bmatrix} x_1 \\ x_2 \\ x_3 \\ x_4 \end{Bmatrix} = \begin{Bmatrix} 4.175 \\ 0 \\ 0 \\ 2.0875 \end{Bmatrix}$$

3.7 Develop a computer code tdma implementing the TDMA algorithm and solve the system of equations in Problem 3.1.

3.8 Use the Cholesky factorization method to solve the system of equations given in Problem 3.2.

3.9 Use the Gauss–Seidel method to solve the following system of equations.

$$\begin{aligned} 12x_1 + x_2 + 7x_3 &= 27 \\ x_1 + 12x_2 + 3x_3 &= -24 \\ 7x_1 + 3x_2 + 12x_3 &= 3 \end{aligned}$$

Show three iterations in hand calculations.

3.10 Consider the following system of equations:

$$\begin{aligned} 3x_1 - 0.1x_2 - 0.2x_3 &= 7.85 \\ 0.1x_1 + 7x_2 - 0.3x_3 &= -19.3 \\ 0.3x_1 - 0.2x_2 + 10x_3 &= 71.4 \end{aligned}$$

a. Determine whether the coefficient matrix is diagonally dominant.
b. Use the Gauss–Seidel method to obtain the solution of the system. Show three iterations and percent relative error in each step of hand calculations.
c. Use the SOR method with a relaxation factor $\lambda = 1.1$ to obtain the solution of the system. Show three iterations and percent relative error in each step of hand calculations.

3.11 Develop a computer code **sor** by using the SOR iterative algorithm and solve the system of equations given in Problem 3.3. Continue iteration until convergence is reached for all variables with a tolerance limit of $\varepsilon_s = 0.01\%$. Present your results in a summary table for unknown values and associated approximate percent relative error in each iteration.

3.12 Use the CG algorithm to solve the following system of equations.

$$\begin{bmatrix} 3 & -3 & 0 & 0 & 0 & 0 \\ -3 & 10 & -1 & -4 & -2 & 0 \\ 0 & -1 & 44 & -2 & 19 & 0 \\ 0 & -4 & -2 & 10 & -4 & 0 \\ 0 & -2 & 19 & -4 & 90 & -17 \\ 0 & 0 & 0 & 0 & -17 & 43 \end{bmatrix} \begin{Bmatrix} x_1 \\ x_2 \\ x_3 \\ x_4 \\ x_5 \\ x_6 \end{Bmatrix} = \begin{Bmatrix} 2 \\ 77 \\ 75 \\ 79 \\ 77 \\ 2 \end{Bmatrix}$$

a. Show hand calculations in two iterations.

b. Develop a computer code **cg** *implementing the CG algorithm and solve the system of equations.* Continue iterations until all unknowns are accurate to three decimal places, i.e., converged to a specified tolerance of $\varepsilon_s = 0.0009$. Present the results in a summary table showing values and percent relative error in each iteration step.

3.13 Consider a system of equations with a nonsymmetric coefficient given as

$$\begin{bmatrix} 5 & 3 & -2 \\ -4 & -1 & 1 \\ 3 & -2 & -1 \end{bmatrix} \begin{Bmatrix} \\ \\ \\ \end{Bmatrix} = \begin{Bmatrix} 1 \\ -4 \\ -3 \end{Bmatrix}$$

Solve the system of equations using GMRES method.

4 Numerical Integration

Quite often we will be confronted with problems of integrating functions for which no convenient integrals are available or which cannot be integrated in terms of elementary function. Some examples of such integrals that we will see in fluid flow and heat transfer problems are as follows.

For flow in any arbitrary flow section

Average velocity

$$\bar{V} = \frac{1}{A}\int_A \vec{V}\mathrm{d}\vec{A}$$

Mass flow rate

$$m = \int_A \rho\vec{V}\cdot\mathrm{d}\vec{A}$$

Average fluid temperature at a section of fluid flow in a circular pipe

$$\bar{T} = \frac{\int_A \rho c_p u T \mathrm{d}A}{\int_A \rho u \mathrm{d}A} = \frac{\int_0^R \rho c_p u T 2\pi r \mathrm{d}r}{\int_0^R \rho u 2\pi r \mathrm{d}r}$$

Average heat transfer coefficient for convective heat transfer from a surface

$$\bar{h} = \frac{1}{A}\int_A h \mathrm{d}A$$

Numerical integrations are also encountered in the evaluation element characteristic matrices and source vectors in finite element formulations. Some examples of such integrals are

a. $I = \int_{x_i}^{x_j} \Gamma(x)\frac{\mathrm{d}[N]^{\mathrm{T}}}{\mathrm{d}x}\frac{\mathrm{d}\phi}{\mathrm{d}x}\mathrm{d}x, \quad I = \int_{x_i}^{x_j} S(x)[N]^{\mathrm{T}}\mathrm{d}x, \quad I = \int_{x_i}^{x_j} q(x)[N]^{\mathrm{T}}[N]\mathrm{d}x$

where

$$[N]^{\mathrm{T}} = \begin{Bmatrix} N_i(x) \\ N_j(x) \end{Bmatrix} = \begin{Bmatrix} \dfrac{x_j - x}{x_j - x_i} \\ \dfrac{x - x_i}{x_j - x_i} \end{Bmatrix}, \quad \phi^* = N_i\phi_i + N_j\phi_j$$

b. $I = \int_{-1}^{1} q(x)[N]^{\mathrm{T}}[N]\mathrm{d}x$

where

$$[N]^{\mathrm{T}} = \begin{Bmatrix} N_i \\ N_j \\ N_k \end{Bmatrix} = \begin{Bmatrix} \frac{x}{2}(x-1) \\ -(x^2-1) \\ \frac{x}{2}(x+1) \end{Bmatrix}$$

Note that the following integral represents a matrix whose each element involves an integral given as

$$I = \int_{-1}^{1} q(x) \begin{bmatrix} N_i^2 & N_iN_j & N_iN_k \\ N_jN_i & N_j^2 & N_jN_k \\ N_kN_i & N_kN_j & N_k^2 \end{bmatrix} \mathrm{d}x$$

c. $I = \int_A \{D\}^{\mathrm{T}}[\Gamma]\{D\}dA$

where

$$[\Gamma] = \text{transport property matrix} = \begin{bmatrix} \Gamma_x & 0 \\ 0 & \Gamma_y \end{bmatrix}$$

$$\{D\} = \text{column vector with derivatives} = \begin{Bmatrix} \dfrac{\partial[N]}{\partial x} \\ \dfrac{\partial[N]}{\partial y} \end{Bmatrix}$$

Numerical integration, also referred to as **quadrature formulas**, involves approximation of the integrand by a function that can be evaluated exactly. Numerical integration essentially finds a close approximation to the area under a curve represented by the function $f(x)$

$$I = \int_a^b f(x)\mathrm{d}x \tag{4.1}$$

The function $f(x)$ could be a continuous function, as in Figure 4.1a, or a distribution of discrete values such as velocity and temperature at discrete points, as shown in Figure 4.1b. The integration methods can be classified into two basic groups, the Newton–Cotes formulas and the Gauss quadrature formula.

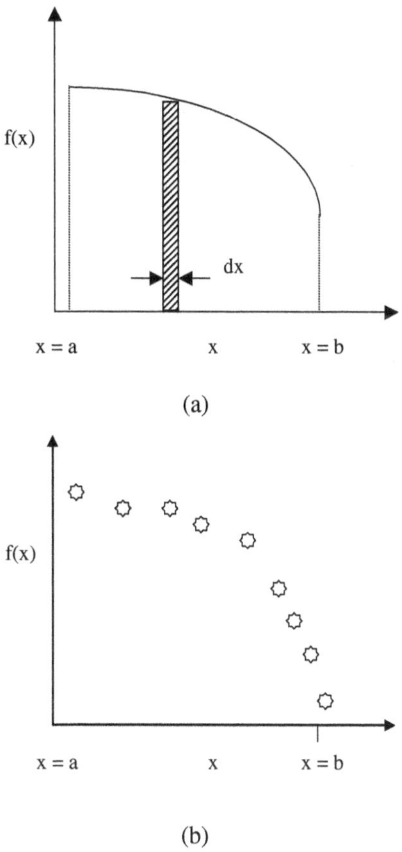

FIGURE 4.1 Function variation.

4.1 NEWTON–COTES INTEGRATION FORMULAS

This is the most common numerical integration scheme. The basic strategy is to replace a complicated function or distribution of data with some approximating function such as

$$I = \int_a^b f(x)\mathrm{d}x = \int_a^b f_n(x)\mathrm{d}x \tag{4.2}$$

where $f_n(x)$ is a polynomial of the form

$$f_n(x) = a_0 + a_1 x + a_2 x^2 + \cdots + a_n x^n \tag{4.3}$$

Here n represents the order of the polynomial or order of the approximation. For example:

$n = 1$ represents first-order approximation or a straight-line approximation given as $f_n(x) = a_0 + a_1 x$ and as depicted in Figure 4.2.

$n = 2$ represents second-order approximation that uses a single parabola requiring three consecutive points, as given by the function $f_n(x) = a_0 + a_1 x + a_2 x^2$ and as depicted in Figure 4.3.

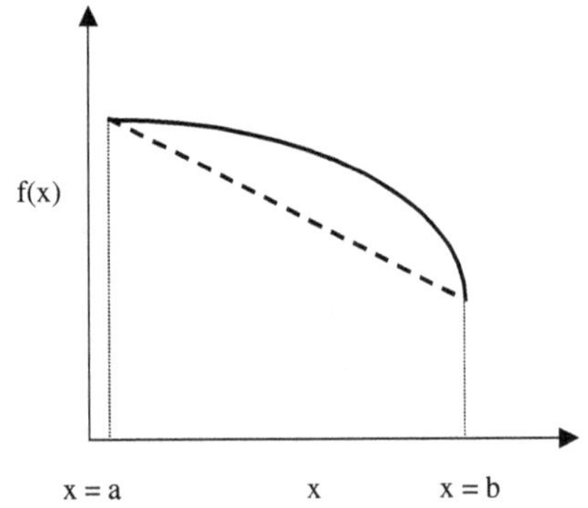

FIGURE 4.2 First-order linear approximation.

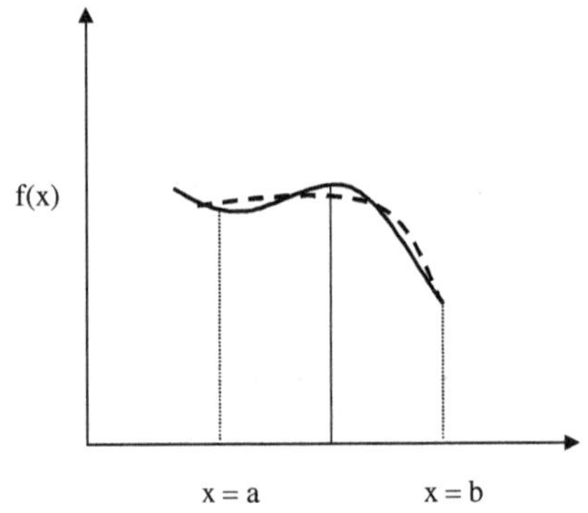

FIGURE 4.3 Second-order quadratic approximation.

4.1.1 The Trapezoidal Rule

This is the first of the **Newton–Cotes** integration formulas and is the first-order or straight-line approximation shown in Figure 4.2. The integration is approximated by

$$I = \int_a^b f(x)\mathrm{d}x \cong \int_a^b f_1(x)\mathrm{d}x \tag{4.4}$$

where the function $f_1(x) = a_0 + a_1 x$ is expressed in terms of the function values at the two limits as

$$f_1(x) = f(x_i) + \frac{f(x_{i+1}) - f(x_i)}{h}(x - x_i) \tag{4.5}$$

Substituting Equation 4.5 into Equation 4.4, we obtain the first-order estimate of the integral as

$$I \cong \frac{h}{2}\left[f(x_i) + f(x_{i+1})\right] \tag{4.6}$$

Equation 4.6 also represents the trapezoidal area under the straight line, which approximates the area under the curve. So, Equation 4.6 is referred to as the trapezoidal rule.

In order to obtain greater accuracy, the area under the curve is divided into multiple areas, as shown in Figure 4.4, and the trapezoidal rule is applied many times to estimate the total area under the curve.

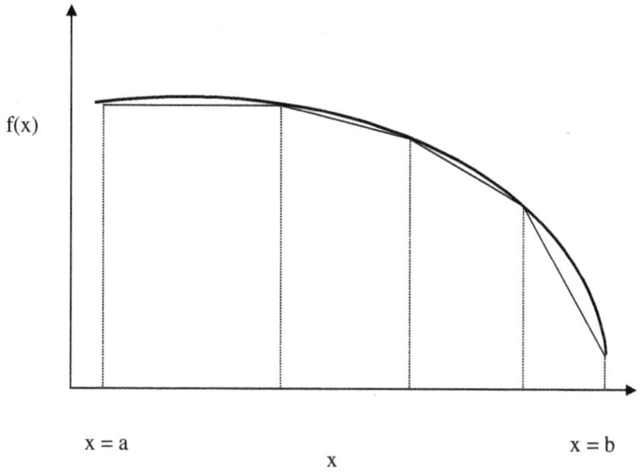

FIGURE 4.4 Multiple uses of the trapezoidal rule with multiple strips.

Multiple application of the trapezoidal rule leads to a general **trapezoidal quadrature formula**

$$I \cong \frac{h}{2}\left[f(x_0)+2\sum_{i=2}^{n-1} f(x_i)+f(x_n)\right] \tag{4.7}$$

where

n = number of strips

$h = \frac{b-a}{n}$ = step size

It can be shown by using the Taylor series expansion that the truncation error associated with the single strip trapezoidal rule is given as

$$E_{t1} = -\frac{h^3}{12} f''(\xi_1) \tag{4.8}$$

where ξ_1 is a point in the interval between the two limiting points a and b.

Evaluating the truncation errors for all the strips in a similar manner, the total truncation error for the general trapezoidal formula can be written as

$$E_t = -\frac{h^3}{12}\left[f''(\xi_1)+f''(\xi_2)+\cdots+f''(\xi_n)\right] \tag{4.9}$$

Equation 4.9 is further simplified by introducing an average value for the second derivative of the function as

$$E_t = -\frac{1}{12}\left(nf''(\xi)_{\text{av}}\, h^3\right) \tag{4.10}$$

Substituting $h = (b - a)/n$, Equation 4.10 reduces to

$$E_t = -\frac{1}{12}(b-a) f''(\xi)_{\text{av}}\, h^2 \tag{4.11}$$

Since the point ξ is difficult to estimate and $f''(\xi)_{\text{av}}$ may not be known, Equation 4.11 is written in an alternate form as

$$E_t \cong ch^2 \tag{4.12a}$$

or

$$E_t \cong O\left(h^2\right) \tag{4.12b}$$

Equation 4.12 is particularly useful for comparative evaluation of integral estimate using different step sizes, h, or number of strips, n.

Example 4.1: Trapezoidal Rule

Numerical results for the axial velocity distribution in a circular pipe are given at discrete points at a cross-section in the table below. The pipe is 1 ft in diameter. Compute the volume flow rate and the average velocity in the pipe using the trapezoidal rule.

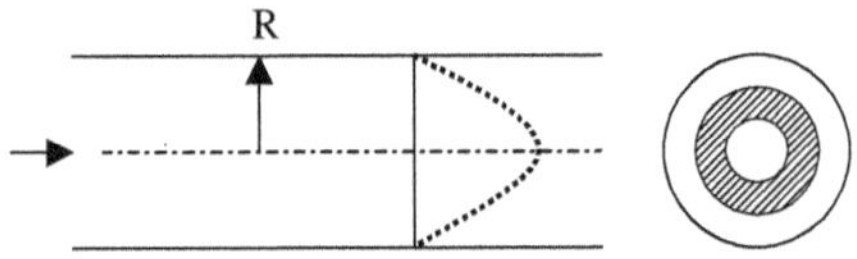

I	*r* (ft)	u_z (fps)
1	0	10.000
2	1/12	9.722
3	1/6	8.889
4	1/4	7.500
5	1/3	5.556
6	5/12	3.056
7	1/2	0

Solution

The volume flow rate of an incompressible fluid is given by the integral

$$\forall = \int_A \vec{V}\cdot d\vec{A} \tag{E.4.1.1}$$

If the axial velocity distribution at any cross-section in a circular pipe is known, then Equation E.4.1.1 reduces to

$$\forall = \int_0^R u_z 2\pi r dr \tag{E.4.1.2}$$

The trapezoidal rule can be used to estimate the value of the integral

$$\forall \cong \frac{h}{2}\left[f(x_0)+2\sum_{i=2}^{n-1} f(x_i)+f(x_n)\right]$$

Using $f(x)=u_z 2\pi r$, $h=\Delta r=\dfrac{1}{12}$, and n = 7, we have

$$\forall \cong \frac{\Delta r}{2}\left[\left(u_{z0}2\pi r_0\right)+2\sum_{i=2}^{6}\left(u_{z_i}2\pi r_i\right)+\left(u_{zR}2\pi R\right)\right]$$

or

$$\forall \cong \frac{2\pi\left(\frac{1}{12}\right)}{2}\Big[10(0)+2\left\{9.722\left(\tfrac{1}{12}\right)+8.889\left(\tfrac{1}{6}\right)+7.500\left(\tfrac{1}{4}\right)\right.$$

$$\left.+5.556\left(\tfrac{1}{3}\right)+3.056\left(\tfrac{5}{6}\right)\right\}+0\left(\tfrac{1}{2}\right)\Big]$$

or

$$\forall \cong 3.818\,\text{ft}^3/\text{s}$$

The average velocity in the pipe is estimated as

$$\bar{V}=\frac{\int_0^R u_z 2\pi r\,dr}{\pi R^2}=\frac{3.818}{\pi\left(1/2\right)^2}$$

or

$$\bar{V}=1.1259$$

4.1.2 Simpson's Integration Formula

This is the second **Newton–Cotes** integration formula and second-order or quadratic approximation using a parabola with three consecutive points. Approximation of the integral is obtained by connecting successive groups of three points by parabolas and summing the areas under the parabolas to approximate the area under the curve. Consider an area under two strips with three data points $\left(x_{i-1},f\left(x_{i-1}\right)\right),\left(x_i,f\left(x_i\right)\right),$and $\left(x_{i+1},f\left(x_{i+1}\right)\right)$, as shown in Figure 4.5.

The integration is approximated by

$$I=\int_{x_{i-1}}^{x_{i+1}} f(x)\mathrm{d}x \cong \int_{i-1}^{i+1} f_2(x)\mathrm{d}x \tag{4.13}$$

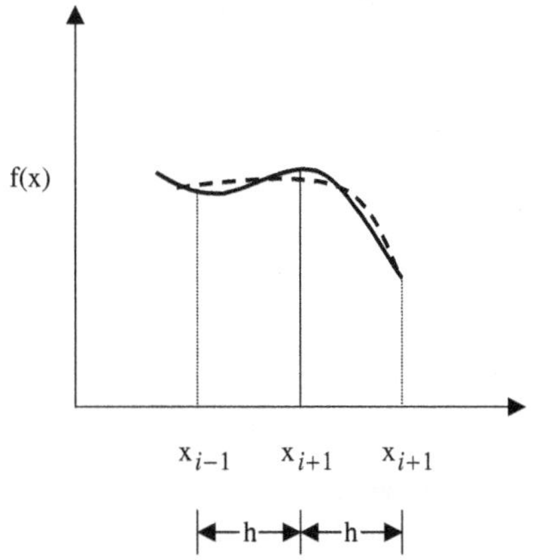

FIGURE 4.5 Simpson's 1/3rd rule using a three-point quadratic function.

where the second-order quadratic function $f_2(x) = a_0 + a_1 x + a_2 x^2$ is assumed to pass through the three points. The coefficients a_0, a_1, and a_2 are determined by substituting the co-ordinate of the three points as

$$a_0 = f(x_i), \quad a_1 = \frac{f(x_{i+1}) - f(x_{i-1})}{2h}, \quad \text{and} \quad a_2 = \frac{f(x_{i-1}) - 2f(x_i) + f(x_{i+1})}{2h^2} \tag{4.14}$$

The integral given by Equation 4.13 or the area under the function can now be evaluated by substituting the function as

$$I \cong \int_{x_{i-1}}^{x_{i+1}} f_2(x) = \int_{x_{i-1}}^{x_{i+1}} \left(a_0 + a_1 x + a_2 x^2\right) \mathrm{d}x$$

$$\cong a_0 \, x\Big|_{-h}^{h} + \frac{1}{2} a_1 \, x^2\Big|_{-h}^{h} + \frac{1}{3} a_2 \, x^3\Big|_{-h}^{h}$$

$$\cong 2a_0 h + \frac{2}{3} a_2 h^3$$

Now, by substituting the expressions of a_0, a_1, and a_2, we obtain the formula for the approximation of the integral or the area under the curve that passes through the three points as

$$I \cong \frac{h}{3}\left(f(x_{i-1}) + 4f(x_i) + f(x_{i+1})\right) \tag{4.15}$$

This formula is referred to as the **Simpson's 1/3rd rule** and it is based on the estimation of the area of two segments or three points. For the estimation of the integral, Equation 4.1, in the range (a, b), we need to divide the regions into multiple strips. For n even number of strips of width $h = (b - a)/n$, Equation 4.15 is applied many times in a group of two strips, and the general formula by Simpson's 1/3rd rule is given as

$$I \cong \frac{h}{3}\left[f(x_0) + 4\sum_{i=1,3,5} f(x_i) + 2\sum_{i=2,4,6} f(x_i) + f(x_n) \right] \tag{4.16}$$

Note that the *Simpson's 1/3rd rule* formula is applicable when the area is divided into n even number of strips of width h and $n + 1$ odd number of data points. In order to overcome the situation that involves odd number of strips and even number of data points, the *Simpson's 3/8th rule* is developed. This is a second-order approximation obtained by considering a third-order polynomial and fitting it with four consecutive data points.

Consider an area under three strips with four data points $(x_{i-1}, f(x_{i-1}))$, $(x_i, f(x_i))$, $(x_{i+1}, f(x_{i+1}))$ and $(x_{i+2}, f(x_{i+2}))$ as shown in Figure 4.6. The integration is approximated by

$$I = \int_{x_{i-1}}^{x_{i+2}} f(x)\mathrm{d}x \cong \int_{x_{i-1}}^{x_{i+2}} f_3(x)\mathrm{d}x \tag{4.17}$$

where the second-order quadratic function $f_3(x) = a_0 + a_1 x + a_2 x^2 + a_3 x^3$ is assumed to pass through the four points. The coefficients a_0, a_1, a_2, and a_3 are determined by substituting the coordinate of the three points into the function $f_3(x) = a_0 + a_1 x + a_2 x^2 + a_3 x^3$ and solving the system of equations for

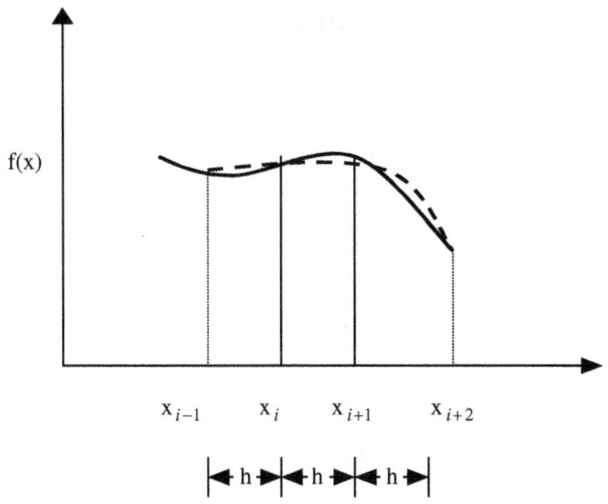

FIGURE 4.6 Simpson's 3/8th rule using a four-point quadratic function.

$$a_0 = f(x_i), \qquad a_1 = \frac{-f(x_{i+2}) + 6f(x_{i+1}) - 3f(x_i) - 2f(x_{i-1})}{6h}$$

$$a_2 = \frac{f(x_{i-1}) - 2f(x_i) + f(x_{i+1})}{2h^2}, \quad \text{and} \quad a_3 = \frac{f(x_{i+2}) - f(x_{i+1}) - 3f(x_i) - f(x_{i-1})}{6h^3}$$

The integral given by Equation 4.17 or the area under the function can now be evaluated by substituting the function as

$$I \cong \int_{x_{i-1}}^{x_{i+2}} f_3(x) = \int_{-h}^{2h} (a_0 + a_1 x + a_2 x^2 + a_3 x^3) dx$$

$$\cong a_0\, x\Big|_{-h}^{2h} + \frac{1}{2} a_1\, x^2\Big|_{-h}^{2h} + \frac{1}{3} a_2\, x^3\Big|_{-h}^{2h} + \frac{1}{4} a_3\, x^4\Big|_{-h}^{2h}$$

$$\cong 3a_0 h + \frac{1}{2} a_1 h^3 + 3a_2 h^3 + \frac{15}{3} a_3 h^4$$

Now, by substituting the expressions of a_0, a_1, a_2, and a_3, we obtain the formula for the approximation of the integral or the area under the curve that passes through the four points as

$$I \cong \frac{3h}{8}\left(f(x_{i-1}) + 3f(x_i) + 3f(x_{i+1}) + f(x_{i+2})\right) \tag{4.18}$$

It can be noted that that the Simpson's 3/8th formula, Equation 4.18, is used along with three strips or four points. For the estimation of the integral given by Equation 4.1 in the range (a, b), we need to divide the regions into n multiple strips with width $h = (b - a)/n$, where Equation 4.18 is applied many times in a group of three strips. The general formula by *Simpson's 3/8th rule* is derived as

$$I \cong \frac{3h}{8}\left[f(x_0) + 3\sum_{i=1,4,7} \{f(x_i) + f(x_{i+1})\} + 2\sum_{i=3,6,9} f(x_i) + f(x_n) \right] \tag{4.19}$$

It can be shown by using the Taylor series expansion that the truncation error associated with the multi-strip Simpson's 1/3rd rule is

$$E_t = -\frac{1}{180}(b-a)f^{\text{iv}}(\xi)_{\text{av}}\, h^4 \tag{4.20}$$

Since the point ξ is difficult to estimate and $f^{\text{iv}}(\xi)_{\text{av}}$ may not be known, Equation 4.20 is written in an alternate form as

$$E_t \cong ch^4 \tag{4.21a}$$

or

$$E_t \cong O\left(h^4\right) \tag{4.21b}$$

This indicates that truncation error in multistrip Simpson's 1/3rd rule is proportional to h^4, the error decreases with a decrease in step size at the fourth power.

Example 4.2: Simpson's 1/3rd Rule

Repeat Problem 4.1 using Simpson's 1/3rd rule.

Solution

We start with Equation E.4.1.1 for the volume flow rate in terms of the axial velocity distribution as

$$\forall = \int_0^R u_z 2\pi r \mathrm{d}r \tag{E.4.2.1}$$

Applying Simpson's 1/3rd rule to the data given in the table, we obtain the following estimate of the flow rate:

$$I \cong \frac{h}{3}\left[f(x_0) + 4\sum_{i=1,3,5} f(x_i) + 2\sum_{i=2,4,6} f(x_i) + f(x_n)\right]$$

Using $f(x) = u_z 2\pi r,\ h = \Delta r = \frac{1}{12}$ and $n = 7$, we have

$$\forall \cong \frac{\Delta r}{3}\left[(u_{z0}2\pi r_0) + 4\sum_{i=1,3,5}^{6}(u_{zi}2\pi r_i) + 2\sum_{i=2,4}^{6}(u_{zi}2\pi r_i) + (u_{z6}2\pi R)\right]$$

$$\forall \cong \frac{\Delta r}{3}\Big[(u_{z0}2\pi r_0) + 4\{(u_{z1}2\pi r_1) + (u_{z3}2\pi r_3) + (u_{z5}2\pi r_5)\}$$
$$+2\{(u_{z2}2\pi r_2) + (u_{z4}2\pi r_4)\} + (u_{z6}2\pi r_6)\Big]$$

or

$$\forall \cong \frac{\pi}{18}\Big[10(0) + 4\{9.722(\tfrac{1}{12}) + 7.500(\tfrac{1}{4}) + 3.056(\tfrac{5}{12})\}$$
$$+2\{8.889(\tfrac{1}{6}) + 5.556(\tfrac{1}{3}) + \} + 0(\tfrac{1}{2})\Big]$$

or

$$\forall \cong 3.927\,\text{ft}^3/\text{s}$$

The average velocity in the pipe is estimated as

$$\bar{V} = \frac{\int_0^R u_z 2\pi r dr}{\pi R^2} = \frac{3.927}{\pi(1)^2}$$

or

$$\bar{V} = 1.1506\ \text{ft/s}$$

4.1.3 Summary of Newton–Cotes Integration Formulas

A summary of some of the common Newton–Cotes integration formulas is given in Table 4.1.

Simpson's 1/3rd and 3/8th rules as well as the five-point and six-point formulas have the same order of error. Even segments/odd points formulas such as the 1/3rd rule and Boole's rule are usually the methods of preference. Higher-order formulas (greater than four points) are rarely used. Simpson's rule is sufficient for most applications. Accuracy can be improved by multiple applications schemes (**such as the Romberg integration algorithm**) rather than using higher-order formulas. If the function is known, and higher accuracy is needed, methods such as the **Romberg integration algorithm** or **Gauss quadrature** are used.

One of the important aspects of using **Newton–Cotes** integration formulas is that estimation of the integral increases with the decrease in step size or increase in the number of strips. One needs to progressively decrease the step size, h, or increase the number strips, n, until the desired level of accuracy is achieved. Approximate percent relative error can be computed at each stage

TABLE 4.1
Newton–Cotes Integration Formulas

Strips (n)	Points ($n+1$)	Name	Integration Formula	Truncation Error
1	2	Trapezoidal rule	$I \cong \frac{h}{2} f(x_i) + f(x_{i+1})$	$-\frac{1}{12}(b-a) f''(\xi)_{av}\, h^2$
2	3	Simpson's 1/3 rule	$I \cong \frac{h}{3}(f_{i-1} + 4f_i + f_{i+1})$	$-\frac{1}{180}(b-a) f^{iv}(\xi)_{av}\, h^4$
3	4	Simpson's 3/8 rule	$I \cong \frac{3h}{8}(f_{i-1} + 3f_i + 3f_{i+1} + f_{i+2})$	$-\frac{3}{240}(b-a) f^{iv}(\xi)_{av}\, h^4$
4	5	Boole's rule	$I \cong \frac{2h}{45}(7f_{i-2} + 32f_{i-1} + 12f_i + 32f_{i+1} + 7f_{i+2})$	$-\frac{2}{945}(b-a) f^{vi}(\xi)_{av}\, h^6$
5	6		$I \cong \frac{5h}{288}(19f_{i-2} + 75f_{i-1} + 50f_i + 50f_{i+1} + 75f_{i+2} + 19f_{i+3})$	$-\frac{55}{2096}(b-a) f^{vi}(\xi)_{av}\, h^6$

as numerical integration is carried out progressively with decrease in step size, h, as discussed in Chapter 2 and defined as follows:

$$\varepsilon_a = \frac{I_{h_{i+1}} - I_{h_i}}{I_{h,i}} \tag{4.22}$$

where

I_{h_i}= integration estimate based on step size, h_i, at iteration step i

$I_{h_{i+1}}$= integration estimate based on step size, h_{i+1}, at iteration step $i + 1$

$$\varepsilon_a \leq \varepsilon_s \tag{4.23}$$

where ε_s is the assigned tolerance limit.

4.2 ROMBERG INTEGRATION

The Romberg integration formula is the computational algorithm based on the Richardson's extrapolation, which combines two first-order numerical integration estimates and obtains a higher-order more accurate numerical estimate. The first-order estimates are obtained using the trapezoidal rule or Simpson's rule, or any other Newton–Cotes integration formula given in Table 4.1. Let us consider two approximate solutions I_{h1} and I_{h2} based on step sizes h_1 and h_2, respectively, and using the trapezoidal rule. The truncation error associated with these two first-order estimates can be written based on Equation 4.12 as

$$E_T \cong c_1 h^2 \quad or \quad o\left(h^2\right) \tag{4.24a}$$

or

$$E_{\mathrm{T}} = c_1 h^2 + o\left(h^4\right) \tag{4.24b}$$

Based on this error estimate, we can express the improved integration estimate as

$$I = I_{1,1} = I_{h1} + c_1 h_1^2 \tag{4.25a}$$

and

$$I = I_{2,1} = I_{h2} + c_1 h_2^2 \tag{4.25b}$$

Solving Equations 4.25a and 4.25b for the constant, c_1, we have

$$c_1 = \frac{I_{h2} - I_{h1}}{\left(h_1/h_2\right)^2 - 1} \tag{4.26}$$

Substituting Equation 4.26 into Equation 4.25b, we obtain a second-order improved estimate as

$$I_{1,2} = I_{h2} + \frac{I_{h2} - I_{h1}}{\left(h_1/h_2\right)^2 - 1} \tag{4.27}$$

If we consider increasing the number of strips such that $(h_1/h_2) = 2$, then Equation 4.27 reduces to

$$I_{1,2} = I_{h2} + \frac{I_{h2} - I_{h1}}{4 - 1} \tag{4.28a}$$

or

$$I_{1,2} = \frac{4^1 I_{h2} - I_{h1}}{4^1 - 1} \tag{4.28b}$$

It can be noted that this second-order estimate is obtained by eliminating the error term c_1h^2 or $o(h^2)$. With the elimination of this error term, the truncation error is expressed as

$$E_T = c_2h^4 + o(h^6) \tag{4.29}$$

In order to eliminate the error term c_2h^4 or $o(h^4)$ and obtain a third-order estimate, we need to have another second-order estimate, $I_{2,2}$. By considering two first-order estimates I_{h2} and I_{h3}, based on step sizes h_2 and h_3, respectively, and applying a similar procedure, we can obtain another second-order estimate as

$$I_{2,2} = I_{h3} + \frac{I_{h3} - I_{h2}}{(h_2/h_3)^2 - 1} \tag{4.30}$$

By considering $(h_2/h_3) = 2$, we can express Equation 4.30 as

$$I_{2,2} = \frac{4^1 I_{h3} - I_{h2}}{4^1 - 1} \tag{4.31}$$

Now, from the final two improved integrations given by Equations 4.28 and 4.31, we can compute a new improved integral value by eliminating the error term c_2h^4. This process of removing error terms is continued until the desired accuracy is obtained. The integration estimate at any accuracy level can be expressed by a general formula given as

$$I_{i,j} \cong \frac{4^{j-1} I_{i+1,j-1} - I_{i,j-1}}{4^{j-1} - 1} \tag{4.32}$$

where j is the index for the present order of integration and $j - 1$ represents the previous lower order of integration. The indices i and $i + 1$ represent the less and more accurate integration estimates. So, the integration $I_{i,j}$ represents a new improved integration estimation at the jth order of integration approximation, whereas $I_{i+1,j-1}$ and $I_{i,j-1}$ represent the more accurate and less accurate integration estimates at the $(j-1)$th order of integration approximations, respectively. A pseudo code for the Romberg integration formula is given in Table 4.2.

4.3 GAUSS QUADRATURE

One of the important characteristics of the Newton–Cotes integration formulas is that the integral estimate is based on evenly-spaced function values or the values are known at fixed interior points and fixed weights. The number of points on each strip is selected to define a nth degree polynomial that can be integrated exactly. However, the positions of the interior points and weights are not

TABLE 4.2
Pseudo Code for Romberg Integration

Input a, b:	integration limits
Input: ε_s:	pre-specified tolerance limit
$n = 1$:	number of strips
call trapz (n, a, b, int):	use trapezoidal rule for first-order estimate
$I_{1,1} = \text{int}$	
$\varepsilon_a = 1.05\varepsilon_s$	
i = 0	
dowhile ($\varepsilon_a > \varepsilon_s$)	
i = i + 1	
$n = 2^i$	
call trapz (n, a, b, int)	
$I_{i+1,1} = \text{int}$	
dofor k = 2 to i + 1	
j = 2 + i − k	
$I_{j,k} = \dfrac{4^{k-1} I_{j+1,k-1} - I_{j,k-1}}{4^{k-1} - 1}$	
enddo	
$\varepsilon_a = \left\| \dfrac{I_{1,i+1} - I_{1,i}}{I_{1,i+1}} \right\|$	
enddo	

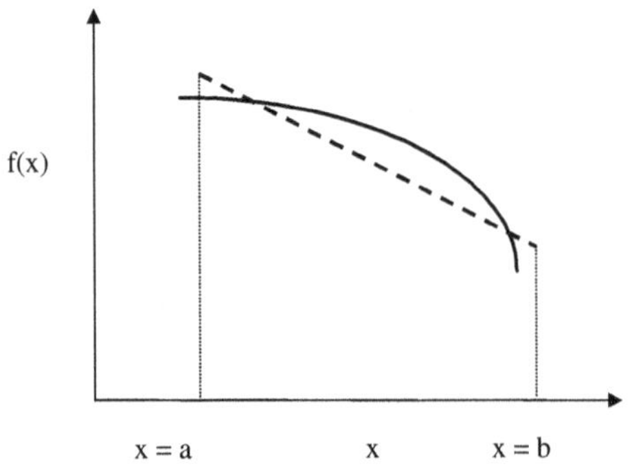

FIGURE 4.7 Integral estimate based on the area under a straight line passing through two points other than the integration limits.

optimized to minimize the truncation error. Accuracy is improved by considering an increased number of strips or using a higher-order polynomial. Also, since the formula is based on using the function values at the limit points, it will lead to inaccurate results for selecting a strip by joining any two points, such as shown in Figure 4.7.

It can be seen that such an approximation will involve both a positive error and a negative error. We could, however, select these points and straight line in such a way that these errors will cancel each other and give an improved estimate. **Gauss quadrature** is one such class of techniques that implement such a strategy. One gauss quadrature that has become extremely popular in engineering applications is the **Gauss–Legendre formula**, which is derived using the method of undetermined coefficients. The development of this formula is done in two steps. First, the integral is transformed from Equation 4.1 into a general form as

$$I = \int_{-1}^{1} g(x')\mathrm{d}x' \tag{4.33}$$

It can be noted that the integral limits have been changed from (a, b) to $(-1, 1)$. This is done to make the formulation simple and general. However, we need to make a change of variable to translate other limits of integration into this general form. This is achieved by substituting the following two equations

$$x = \frac{b+a}{2} + \frac{b-a}{2}x' \tag{4.34a}$$

and

$$\mathrm{d}x = \frac{b-a}{2}\mathrm{d}x' \tag{4.34b}$$

into the integral given by Equation 4.1.

In the second step, the integral is approximated by a quadrature formula that is expressed in a general form as

$$I = \int_{-1}^{1} g(x')\mathrm{d}x' \cong \sum_{i=0}^{n-1} w_i g(x') \tag{4.35a}$$

where w are the unknown co-efficients called **weighting factors**; x' are the unknown points, called **function arguments**, and n is the number of points. The major objective is to determine unknown weighting factors, w, and function arguments, x'.

$$I \cong w_0 g(x_0') + w_1 g(x_1') + w_2 g(x_2') + \cdots + w_{n-1} g(x_{n-1}') \tag{4.35b}$$

Let us now consider the development of a two-point **Gauss–Legendre formula** and present a summary of other commonly used higher-order **Gauss–Legendre formulas** in the following section.

4.3.1 Two-Point Gauss–Legendre Formula

In the two-point formula, the objective is to approximate the integral as

$$I \cong w_0 g(x_0') + w_1 g(x_1') \tag{4.36}$$

where the $\boldsymbol{w}$ are the unknown coefficients that need to be determined. The x' are the selected points that also must be determined. Thus, we have four unknowns and we require four conditions to determine them exactly. These four conditions are established by selecting four expressions for the function that will yield exact results. Four simple equations that represent these cases are that it fits the integral of a constant function $g(x')=1$, a straight line $g(x')=x'$, a parabola $g(x')=x'^2$, and a cubic function $g(x')=x'^3$. In order to satisfy these conditions, we need to solve equations

$$I \cong w_0 g(x_0') + w_1 g(x_1') = \int_{-1}^{1} 1\,\mathrm{d}x' = 2\frac{n!}{r!(n-r)!} \tag{4.37a}$$

$$I \cong w_0 g(x_0') + w_1 g(x_1') = \int_{-1}^{1} x\,\mathrm{d}x' = 0 \tag{4.37b}$$

$$I \cong w_0 g(x_0') + w_1 g(x_1') = \int_{-1}^{1} x'^2\,\mathrm{d}x' = \frac{2}{3} \tag{4.37c}$$

$$I \cong w_0 g(x_0') + w_1 g(x_1') = \int_{-1}^{1} x'^3\,\mathrm{d}x' = 0 \tag{4.37d}$$

Solving these equations simultaneously, we get

$$w_0 = 1 \tag{4.38a}$$

$$w_1 = 1 \tag{4.38b}$$

$$x_0' = -\frac{1}{\sqrt{3}} = -0.577350269 \tag{4.38c}$$

$$x_1' = \frac{1}{\sqrt{3}} = 0.577350269 \tag{4.38d}$$

4.3.2 Higher-Point Gauss–Legendre Formulas

Higher-point formulas are developed based on the equation

$$I \cong w_0 g(x_0') + w_1 g(x_1') + w_2 g(x_2') + \cdots + w_{n-1} g(x_{n-1}') \tag{4.39}$$

The weighting factors and the function arguments are derived using a similar procedure, and the values are summarized in Table 4.3.

One limitation of the Gauss–Legendre formulas is that it requires function evaluations at non-uniformly spaced points within the integration limits. For problems where the function is known and requires repetitive evaluations of the integrals, such formulas are usually the method of choice for efficient evaluation of the integrals. However, these formulas are not suitable for problems that involve discrete values of the variables at selected points such as velocity or temperature values at nodal points given by the finite difference–control volume or finite element methods. In such cases Newton–Cotes formulas are preferred.

TABLE 4.3
Function Arguments and Weight Factors in Gauss–Legendre Quadrature Formula

Number of Points (n)	Weighting Factors (w_j)	Function Points (x'_j)
2	$w_0 = 1.000000000$	$x_0 = -0.577350269$
	$w_1 = 1.000000000$	$x_1 = 0.577350269$
3	$w_0 = 0.555555556$	$x_0 = -0.774596669$
	$w_1 = 0.888888889$	$x_1 = 0.0$
	$w_2 = 0.555555556$	$x_2 = 0.774596669$
4	$w_0 = 0.347854845$	$x_0 = -0.861136312$
	$w_1 = 0.652145155$	$x_1 = -0.339981044$
	$w_2 = 0.652145155$	$x_2 = 0.339981044$
	$w_3 = 0.347854845$	$x_3 = -0.861136312$
5	$w_0 = 0.236926885$	$x_0 = -0.906179846$
	$w_1 = 0.478628670$	$x_1 = -0.538469310$
	$w_2 = 0.568888889$	$x_2 = 0.0$
	$w_3 = 0.478628670$	$x_3 = 0.538469310$
	$w_4 = 0.236926885$	$x_4 = 0.906179846$
6	$w_0 = 0.171324492$	$x_0 = -0.932469514$
	$w_1 = 0.360761573$	$x_1 = -0.661209386$
	$w_2 = 0.467913935$	$x_2 = -0.238619186$
	$w_3 = 0.467913935$	$x_3 = 0.238619186$
	$w_4 = 0.360761573$	$x_4 = 0.661209386$
	$w_5 = 0.171324492$	$x_5 = 0.932469514$

Example 4.4: Gauss–Legendre Quadrature

Use three-point, four-point, and six-point Gauss–Legendre formulas to evaluate the integral

$$I = \int_{-1}^{1} [N]^{\mathrm{T}} [N] \mathrm{d}x \tag{E.4.4.1}$$

where

$$[N]^{\mathrm{T}} = \begin{Bmatrix} N_i \\ N_j \\ N_k \end{Bmatrix} = \begin{Bmatrix} \frac{x}{2}(x-1) \\ -(x^2-1) \\ \frac{x}{2}(x+1) \end{Bmatrix} \tag{E.4.4.2}$$

Solution

Note that the integral represents a matrix whose each element involves an integral given as

$$I = \int_{-1}^{1} \begin{bmatrix} N_i \\ N_j \\ N_k \end{bmatrix} \begin{bmatrix} N_i & N_j & N_k \end{bmatrix} \tag{E.4.4.3}$$

or

$$I=\int_{-1}^{1}\begin{bmatrix} N_i^2 & N_iN_j & N_iN_k \\ N_jN_i & N_j^2 & N_jN_k \\ N_kN_i & N_kN_j & N_k^2 \end{bmatrix}\mathrm{d}x \quad \text{(E.4.4.4)}$$

Substituting the expressions for shape functions, we get

$$I=\int_{-1}^{1}\begin{bmatrix} \frac{x^2}{4}(x-1)^2 & -\frac{x}{2}(x-1)(x^2-1) & \frac{x^2}{4}(x^2-1) \\ -\frac{x}{2}(x-1)(x^2-1) & (x^2-1)^2 & -\frac{x}{2}(x+1)(x^2-1) \\ \frac{x^2}{4}(x^2-1) & -\frac{x}{2}(x+1)(x^2-1) & \frac{x^2}{4}(x+1)^2 \end{bmatrix}\mathrm{d}x \quad \text{(E.4.4.5)}$$

Note that the matrix is symmetric and we just need to estimate the elements in the upper triangular part. Let us consider evaluation of the first integral element as

$$I_{11}=\int_{-1}^{1}N_i^2\,\mathrm{d}x=\int_{-1}^{1}\left[\frac{x}{2}(x-1)\right]^2\mathrm{d}x \quad \text{(E.4.4.6)}$$

or

$$I_{11}=\int_{-1}^{1}\left(\frac{x^4}{4}-\frac{2x^3}{4}+\frac{x^2}{4}\right)\mathrm{d}x$$

or

$$I_{11}=\int_{-1}^{1}f(x)\mathrm{d}x$$

where

$$f(x)=\frac{x^4}{4}-\frac{2x^3}{4}+\frac{x^2}{4} \quad \text{(E.4.4.7)}$$

Three-Point Gauss–Legendre Formula

$$I_{11}\cong w_0f(x_0)+w_1f(x_1)+w_2f(x_2) \quad \text{(E.4.4.8)}$$

where

$$w_0 = 0.555555556, \quad w_1 = 0.888888889, \quad w_2 = 0.555555556$$
$$x_0 = -0.774596669, \quad x_1 = 0.0, \quad x_0 = 0.774596669$$

The function is evaluated at the function points as

$$\begin{aligned} f(x_0) &= \frac{(-0.774596669)^4}{4} - \frac{(-0.774596669)^3}{2} + \frac{(0.774596669)^2}{4} \\ &= 0.089999999 + 0.232379 + 0.149999999 \\ &= 0.472378998 \\ f(x_1) &= 0.0 \\ f(x_2) &= 0.089999999 - 0.232379 + 0.149999999 \\ &= 0.007620998 \end{aligned}$$

Substituting the function values, we have the first integral element estimated from Equation E.4.4.8 as

$$\begin{aligned} I_{11} &\cong w_0\, f(x_0) + w_1\, f(x_1) + w_2 f(x_2) \\ &= 0.555555556 \times 0.472378998 + 0.888888889 \times 0 \\ &\quad + 0.555555556 \times 0.007620998 \\ &= 0.262432776 + 0 + 0.004233888 \\ &= 0.266\,666\,664 \end{aligned}$$

The rest of the integral elements are computed in a similar manner and the integral based on the three-point formula is given by

$$I \cong \begin{bmatrix} 0.2666666668 & -1.20000001 & -0.0666666667 \\ -1.20000001 & 6.400000002 & 0.1333333335 \\ -0.0666666667 & 0.1333333335 & 0.2666666668 \end{bmatrix}$$

Four-Point Gauss–Legendre Formula

$$I_{11} \cong w_0 f(x_0) + w_1 f(x_1) + w_2 f(x_2) + w_3 f(x_3) \qquad \text{(E.4.4.9)}$$

where

$$w_0 = 0.347854845, \quad w_1 = 0.652145155,$$

$$w_2 = 0.652145155, \quad w_3 = 0.347854845$$

$$x_0 = -0.861136312, \quad x_1 = -0.339981044,$$

$$x_2 = 0.339981044, \quad x_3 = 0.861136312$$

$$f(x_0) = \frac{(-0.861136312)^4}{4} - \frac{(-0.861136312)^3}{2} + \frac{(-0.861136312)^2}{4}$$

$$= 0.137476231 + 0.31929029 + 0.185388937$$

$$= 0.642155458$$

$$f(x_1) = \frac{(-0.339981044)^4}{4} - \frac{(-339981044)^3}{2} + \frac{(-0.339981044)}{4}$$

$$= 0.003340095 + 0.019648713 + 0.028896777$$

$$= 0.051885585$$

$$f(x_2) = \frac{(0.339981044)^4}{4} - \frac{(339981044)^3}{2} + \frac{(0.339\,981\,044)}{4}$$

$$= 0.003340095 - 0.019648713 + 0.028896777$$

$$= 0.012588159$$

$$f(x_3) = \frac{(0.861136312)^4}{4} - \frac{(0.861136312)^3}{2} + \frac{(0.861136312)^2}{4}$$

$$= 0.137476231 - 0.31929029 + 0.185388937$$

$$= 0.003574878$$

Substituting the function values, we have the first integral element estimated from Equation E.4.4.9 as

$$I_{11} \cong w_0 f(x_0) + w_1 f(x_1) + w_2 f(x_2) + w_3 f(x_3)$$

$$= 0.223376887 + 0.03836932 + 0.007996768 + 0.0022331338$$

$$= 0.267541925$$

The rest of the integral elements are computed in a similar manner and the integral based on four-point formula is given by

$$I \cong \begin{bmatrix} 0.2666666670 & -1.20000001 & -0.0666666666 \\ -1.20000001 & 6.400000004 & 0.1333333335 \\ -0.0666666666 & 0.1333333333 & 0.2666666670 \end{bmatrix}$$

Six-Point Gauss–Legendre Formula

$$I_{11} \cong w_0 f(x_0) + w_1 f(x_1) + w_2 f(x_2) + w_3 f(x_3) + w_4 f(x_4) + w_5 f(x_5) \quad \text{(E.4.4.10)}$$

where

$$w_0 = 0.171324492, \quad w_1 = 0.360761573, \quad w_2 = 0.467913935$$

$$w_3 = 0.467913935, \quad w_4 = 0.360761573, \quad w_5 = 0.171324492$$

$$x_0 = -0.932469514, \quad x_1 = -0.661209386, \quad x_2 = -0.238619186$$

$$x_3 = 0.238619186, \quad x_4 = 0.661209386, \quad x_5 = 0.932469514$$

$$f(x_0) = \frac{(-0.932469514)^4}{4} - \frac{(-0.932469514)^3}{2} + \frac{(-0.932469514)^2}{4}$$

$$= 0.189007299 + 0.405390838 + 0.217374848$$

$$= 0.811772985$$

$$f(x_1) = \frac{(-0.661209386)^4}{4} - \frac{(-0.661209386)^3}{2} + \frac{(-0.661209386)^2}{4}$$

$$= 0.4778549 + 0.144539661 + 0.109299463$$

$$= 0.301624614$$

$$f(x_2) = \frac{(-0.238619186)^4}{4} - \frac{(-0.238619186)^3}{2} + \frac{(-0.238619186)^2}{4}$$

$$= 0.000810515 + 0.006793382 + 0.014234778$$

$$= 0.0021838675$$

$$f(x_3) = \frac{(0.238619186)^4}{4} - \frac{(0.238619186)^3}{2} + \frac{(0.238619186)^2}{4}$$

$$= 0.000810515 - 0.006793382 + 0.014234778$$

$$= 0.008257911$$

$$f(x_4) = \frac{(0.661209386)^4}{4} - \frac{(0.661209386)^3}{2} + \frac{(0.661209386)^2}{4}$$

$$= 0.4778549 - 0.144539661 + 0.109299463$$

$$= 0.012545292$$

$$f(x_5) = \frac{(0.932469514)^4}{4} - \frac{(0.932469514)^3}{2} + \frac{(0.932469514)^2}{4}$$

$$= 0.189007299 - 0.405390838 + 0.217374848$$

$$= 0.00091309$$

Substituting the function values, we have the first integral element estimated from Equation E.4.4.10 as

$$I_{11} \cong w_0 f(x_0) + w_1 f(x_1) + w_2 f(x_2) + w_3 f(x_3) + w_4 f(x_4) + w_5 f(x_5)$$

$$= 0.139076590 + 0.10881457 + 0.01021862 + 0.003861184$$

$$+ 0.004525859 + 0.000169835$$

$$= 0.266666662$$

The rest of the integral elements are computed in a similar manner and the integral based on the six-point formula is given by

$$I \cong \begin{bmatrix} 0.2666666664 & -1.199999997 & -0.0666666666 \\ -1.199999997 & 6.399999992 & 0.1333333311 \\ -0.0666666666 & 0.1333333311 & 0.2666666664 \end{bmatrix}$$

4.4 MULTIDIMENSIONAL NUMERICAL INTEGRATION

Let us consider multidimensional integrals of the following forms:

$$I = \int_c^d \int_a^b f(x,y)\mathrm{d}y\mathrm{d}x \quad \text{for two dimensions} \tag{4.40a}$$

and

$$I = \int_e^f \int_c^d \int_a^b f(x,y,z)\mathrm{d}x\mathrm{d}y\mathrm{d}z \quad \text{(for three dimensions)} \tag{4.40b}$$

Multidimensional integrals can be evaluated numerically using the methods discussed for one-dimensional case, i.e., by **Newton–Cotes formulas** or by **Gauss–Legendre formula**. The method is first applied in the first dimension while keeping each value of the second variable held constant. This will give an array of integrated values. Then the method is applied again for the second dimension using the array of integrated values. The procedure is described as follows:

Integration Using Newton–Cotes Formulas

$$I = \int_c^d \int_a^b f(x,y)\mathrm{d}y\mathrm{d}x = \int_c^d \left[\int_a^b f(x,y)\mathrm{d}y \right] \mathrm{d}x \tag{4.41}$$

The inner integral for y-dimension is evaluated using one-dimensional integration formula for different x-values. This gives the array of integrated values as

$$I(x) = \int_a^b f(x,y)\mathrm{d}y \tag{4.42}$$

One-dimensional integration is then applied for the x-direction using the array of integrated values as

$$I = \int_a^b I(x)\mathrm{d}x \tag{4.43}$$

Let us consider the double integral in the general form as

$$I = \int_{-1}^{1} \int_{-1}^{1} g(x',y')\mathrm{d}y'\mathrm{d}x' \tag{4.44}$$

Again, this two-dimensional integral is solved using the one-dimensional formula for each coordinate system. For example, we can use an n-term Gauss–Legendre quadrature in the y-direction to obtain

$$I = \int_{-1}^{1} \left(\sum_{j=1}^{n} w_j g\left(x', y_j'\right) \right) dx' \tag{4.45}$$

This is followed by the use of another n-term Gauss–Legendre quadrature in the x-direction that gives

$$I \cong \sum_{i=1}^{n} \sum_{j=1}^{n} w_i w_j g\left(x_i', y_j'\right) \tag{4.46}$$

The weight factors w_i and w_j, and the corresponding Gauss points x_i and x_j, are the same as those given in Table 4.3, and n is the number of Gauss points.

In a similar manner, we can evaluate a triple integral by repeatedly applying the one-dimensional Gauss–Legendre quadrature as

$$I \cong \int_{-1}^{1}\int_{-1}^{1}\int_{-1}^{1} g(x', y', z')dz'dy'dx' \cong \sum_{i=1}^{n}\sum_{j=1}^{n}\sum_{k=1}^{n} w_i w_j w_k g\left(x_i', y_j', z_k'\right) \tag{4.47}$$

where w_i, w_j, and w_k are weight factors, and x_i, x_j, and x_k are corresponding Gauss points as given in Table 4.3.

PROBLEMS

4.1 Local Nusselt numbers for developing flow in a circular tube are computed using a numerical scheme, and results are shown in the table below.

Location (x)	Local Nusselt Number (Nu_D)
1.0	13.2
2.0	11.1
3.0	8.4
4.0	6.7
5.0	7.3
6.0	6.8
7.0	5.9
8.0	5.0
9.0	4.3
10.0	3.9

Estimate the average Nusselt number using the trapezoidal rule.

4.2 Use the trapezoidal rule to evaluate the integral

$$I = \int_{1}^{3} \left(\frac{x+2}{x}\right)^2 dx$$

with eight segments.

a. Develop a computer code **Trapz** using the quadrature formula for the trapezoidal rule.
b. Use the code and progressively increase the number of segments as $n = 2, 4, 8$, and 16 to calculate the integral. Estimate approximate percent relative error for each integration estimate.
c. Use first-order estimates obtained in part (b) to estimate the higher-order estimates using the Romberg integration algorithm.

4.4 Power intensity of a high energy laser beam is given by

$$I_x = I_0 e^{-x^2/R_0^2}$$

where

$$I_0 = 2.30 \times 10^8 \,\mathrm{W/m^2}$$

$$R_0 = 0.001\mathrm{m}$$

Determine the average power of the laser beam using Simpsons 1/3rd rule.

4.4 Consider the integral

$$I = \int_0^8 \left(12 + 2x - 6x^2 + 5x^3\right) dx$$

a. Evaluate the integral with **Simpson's 1/3rd rule** and using eight segments.
b. Develop a computer code **simp** using the algorithm for Simpson's rule.
c. Use the code and progressively increase the number of segments as $n = 2, 4, 8$, and 16 to calculate the integral. Estimate approximate percent relative error for each integration estimate.
d. Use results obtained in part (c) and obtain improved estimates of the integral using the **Romberg integration formula**. Show approximate percent relative error in each step.

4.5 Develop a computer code to implement the Romberg integration formula to integrate the integral

$$I = \int_0^{\pi/2} \cos x \, dx$$

with $\varepsilon_s = 0.000001$ and a trapezoidal rule for the first-order estimates.

4.6 Use three-point *Gauss–Legendre* quadrature formula to evaluate the integral given in Problem 4.2.

4.7 Evaluate the following integral using the four-point Gauss–Legendre quadrature formula

$$I = \int_{-1}^{1} N_1 N_3$$

where

$$N_1 = \frac{x}{2}(x-1) \quad \text{and} \quad N_2 = \frac{x}{2}(x+1)$$

4.8 Write a computer code to implement three-, four-, six-, and eight-point Gauss–Legendre quadrature formula and evaluate the integral

$$I = \int_{-1}^{1} 0.1x\left[N\right]^{\mathrm{T}}\left[N\right]\mathrm{d}x$$

where

$$\left[N\right]^{\mathrm{T}} = \begin{bmatrix} N_1 \\ N_2 \\ N_3 \end{bmatrix} = \begin{bmatrix} \frac{x}{2}(x-1) \\ -(x^2-1) \\ \frac{x}{2}(x+1) \end{bmatrix}$$

Present your result including percent relative error in a summary table.

4.9 Develop a computer code based on the Romberg integration algorithm using the subroutine *trapz* to evaluate

$$\int_{0}^{2} \frac{e^x}{1+x^4}\mathrm{d}x$$

to an accuracy of 0.1%. Present your results in each step including the percent relative errors.

4.10 Develop a computer code using two through six-point Gauss–Legendre formulas to solve

$$\int_{-2}^{2} \frac{3}{1+3x^3}\mathrm{d}x$$

Present your result in a table.

a. Evaluate the following integral using four- and six-point Gauss–Legendre quadrature

$$I = \int_{-1}^{1} q(x)\left[N\right]^{\mathrm{T}}\left[N\right]\mathrm{d}x$$

where

$$\left[N\right]^{\mathrm{T}} = \begin{Bmatrix} N_i \\ N_j \\ N_k \end{Bmatrix} = \begin{Bmatrix} \frac{x}{2}(x-1) \\ -(x^2-1) \\ \frac{x}{2}(x+1) \end{Bmatrix}$$

$$q(x) = \mathrm{e}^{-0.2x}$$

4.11 Use Simpson's algorithm and computer program to evaluate the average temperature distribution in a square slab. The temperature distribution at selected nodal points is given below:

300	430	545	625	710
220	280	410	515	605
125	192	254	335	440
100	157	184	241	307

Note that you must use quadrature formula for two dimensional integrals.

4.12 Develop a computer code to evaluate the double integral using subroutine **simp**.

$$\int_{-4}^{4}\int_{0}^{8}\left(x^3 - 4y^2 + 3xy^3\right)\mathrm{d}x\,\mathrm{d}y$$

Part II

Finite Difference – Control Volume Method

5 Basic Steps in Finite Difference–Control Volume Method

The finite difference method has been in use for a long time. The method is popular due to its simplicity in the discretization procedure and the relative ease of implementation into a computer code. In this chapter, we will present the basic steps in the formulation and application of the finite difference method, using a general steady-state equation containing diffusion and source terms only. Also, we will restrict our formulation procedure to one-dimensional problems. To comprehend the basic concepts and steps clearly, several simple classical problems will be considered. Detailed formulation and application to multidimensional steady-state problems will be given in Chapter 6. Treatment of the unsteady state or storage term and convection terms will be considered in the subsequent Chapters 7 and 8.

5.1 INTRODUCTION AND BASIC STEPS IN FINITE DIFFERENCE METHOD

The mathematical formulation or model of a physical process is developed based on using fundamental laws of science and engineering, subject to some assumptions related to the process, and with the application of relevant mathematical methods. Such formulation results in mathematical statements, often differential equations, relating quantities of interest in the understanding and/or design of physical processes. The complexity in the mathematical model and in the geometry of the problem very often precludes the use of analytical solution methods. In such cases, we rely on numerical solution methods such as the finite difference and finite element methods. In a numerical simulation, we use an approximate method to evaluate the mathematical model and estimate the characteristics of the process using computers.

In the **finite difference method**, the governing equations are approximated by a pointwise discretization scheme where derivatives are replaced by difference equations that involve the values of the solution at the nodal points. In this method, the discretization process involves first dividing the solution region into a network of grid or mesh of intersecting lines, which are drawn parallel to the coordinate axes. The discrete intersecting points of these gridlines are called the grid or nodal points. The number or the distances between the gridlines affect the number of the resulting grid points or grid size, and hence affect the accuracy of the numerical solution. The accuracy improves with the increase of grid points or decrease of the grid size. The time coordinate is also discretized in a similar manner and it is assumed that solution progresses in a sequence of time steps. Once the grid is generated, the governing equations and boundary conditions are then transformed into discretization equations at each nodal point to obtain a set of algebraic equations that involve the unknown values at the grid points. Finally, this set of algebraic equations is solved for the unknown nodal values using linear solvers as discussed in Chapter 3.

The basic steps in obtaining a numerical solution using finite difference are categorized as follows.

1. State the mathematical statement of the problem in terms of governing equations, boundary conditions, and initial conditions.

2. Discretize the solution domain into a network of discrete nodal points. The unknown values are sought only at those discrete points rather than obtaining a continuous solution in the domain.
3. Obtain discretization equations for all nodal points by approximating the governing differential equations and boundary conditions. This discretization procedure leads to a set of algebraic equations involving the unknown values at the nodal points.
4. Use an appropriate solver algorithm to solve the system of algebraic equations involving the unknown values at the nodal points.
5. Postprocessing of the data to evaluate secondary quantities.

We will outline the basic principles of the finite difference method in the following sections.

5.2 DISCRETIZATION OF THE DOMAIN

In this discretization method, the two-dimensional calculation domain is divided into equal or non-equal small regions with increments of Δx and Δy in the x- and y-directions, respectively, as shown in Figure 5.1. The values of the dependent variables $\phi(x, y)$ are calculated at a finite number of discrete points in the solution domain as shown in the figure. These discrete points are referred to as the grid points or nodal points.

Each nodal point is designated by a numbering scheme i and j, where i and j locations indicate the x and y increments defined by $x = i\Delta x$ and $y = j\Delta y$, respectively. So, the continuous function behavior of the dependent variables $\phi(x, y)$ contained in the analytical solution of the governing equations is replaced by the discrete values $\phi(x_i, y_j)$ at these grid points. The discrete value of the function ϕ at a location (x_i, y_j) is designated by $\phi_{i,j}$. The discrete values at selected grid points are established by a system of algebraic equations, which we will refer to as discretization equations or finite difference equations. When the number of grid points is small, the resulting discretization or numerical solution is approximate, which may deviate considerably from the analytical solution of the governing equations. The accuracy of the numerical solution increases as we increase the number of grid points.

This designation procedure can be extended to three-dimensional and unsteady-state problems with the discrete values represented as $\phi_{i,j,k}^{l} = \phi\left(x_i, y_j, z_k, t_l\right)$. The index k represents z space locations with increments of Δz, and l represents the time level with time increments of Δt. We will describe the systematic approach for obtaining these discretization equations for interior as well as boundary nodes in the following sections.

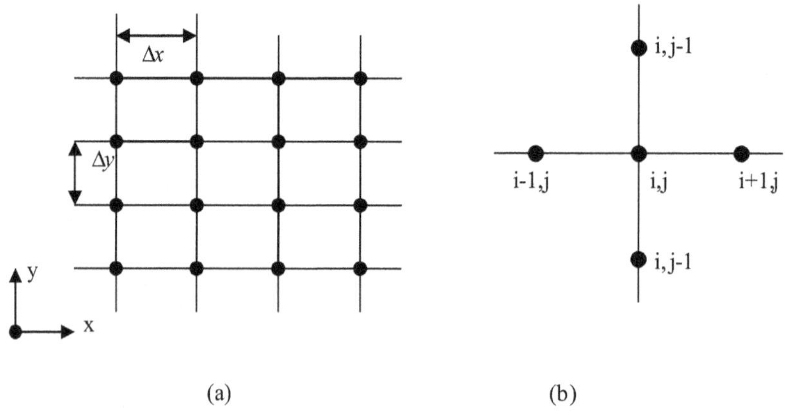

FIGURE 5.1 Nodal network for finite difference method.

5.3 DISCRETIZATION OF THE MATHEMATICAL MODEL

The discretization equation or finite difference equations can be derived by using several approaches, the most popular ones being the **Taylor series method** and the **control volume method**. The governing equations for fluid flows, heat, and mass transfer involve first- and second-order partial derivatives of the independent variables such as velocities, pressure, concentration, and temperature. In the Taylor series method, each derivative is approximated with the help of a difference formula that is derived from the truncated Taylor series expansion as discussed in the following section, whereas in a control volume method, the governing equation is discretized by integrating it over an appropriately chosen control volume. In the following sections, we present a summary of these discretization methods.

5.3.1 The Taylor Series Method

As we have already mentioned, in the Taylor series method, each space and time derivatives are approximated by appropriate finite difference approximation formulas, which are derived by using the Taylor series expansions. So, this method involves two steps. First, appropriate numerical **finite difference formulas** are selected and then the discretization equation is obtained by substituting these formulas in the governing equation. Let us briefly discuss the availability of such finite difference formulas followed by the derivation of discretization equations in the following section.

5.3.1.1 Numerical Differentiation – Finite Difference Formulas

Numerical differentiation of a governing differential equation is carried out by approximating the derivatives with the finite difference formulas, which are derived utilizing the truncated Taylor series expansions. A function $\phi(x, y)$ at the point $(i+1, j)$ can be expressed in terms of the function value and its derivatives at a neighboring point (i, j) using forward Taylor series expansion

$$\phi_{i+1,j} = \phi_{i,j} + \phi'_{i,j}\Delta x + \frac{1}{2!}\phi''_{i,j}(\Delta x)^2 + \frac{1}{3!}\phi'''_{i,j}\ (\Delta x)^3 + \cdots + \frac{1}{n!}\phi^{(n)}_{i,j}(\Delta x)^n + R_n \tag{5.1}$$

where $\Delta x = x_{i+1} - x_i$ is the step size in the x-direction, and R_n is the remainder term that is included to account for all the remaining terms from $n+1$ to infinity. The remainder term is basically a representation of the truncation error, and it is expressed as

$$R_n = \frac{1}{(n+1)!}\phi_i^{(n+1)}(\Delta x)^{n+1} \tag{5.2}$$

Similarly, a function $\varphi(x, y)$ at the point $(i-1, j)$ can be expressed in terms of the function value and its derivatives at a neighboring point (i, j) using the backward Taylor series expansion

$$\phi_{i-1,j} = \phi_{i,j} - \phi'_{i,j}\ \Delta x + \frac{1}{2!}\phi''_{i,j}(\Delta x)^2 - \frac{1}{3!}\phi'''_{i,j}\ (\Delta x)^3 + \cdots + \frac{1}{n!}\phi^{(n)}_{i,j}(\Delta x)^n + R_n \tag{5.3}$$

Equations 5.2 and 5.3 can be used to derive many different finite difference approximation formulas. Let us discuss here the derivation of some basic approximation formulas for the first derivatives.

Forward Difference Approximation This approximation formula is derived by neglecting all the terms of $o(\Delta x)^2$ and higher in the forward Taylor series expansion given by Equation 5.1, and solving for the first derivative term as

$$\phi'_{i,j} = \frac{\phi_{i+1,j} - \phi_{i,j}}{\Delta x} + \frac{R_1}{\Delta x} \tag{5.4}$$

where the second term represents the truncation error associated with the approximation of the first derivative, and is expressed as

$$\frac{R_1}{\Delta x} = \frac{\phi''}{2!}\Delta x \tag{5.5a}$$

or

$$\frac{R_1}{\Delta x} = o(\Delta x) \tag{5.5b}$$

Using Equation 5.5, we can rewrite the approximation, Equation 5.4, as

$$\phi'_{i,j} = \frac{\phi_{i+1,j} - \phi_{i,j}}{\Delta x} + o(\Delta x) \tag{5.6a}$$

or

$$\phi'_{i,j} \cong \frac{\phi_{i+1,j} - \phi_{i,j}}{\Delta x} \tag{5.6b}$$

This approximation is called the **forward difference approximation** because it estimates the derivative at the point (i, j) using the values at the point (i, j) and a forward point $(i+1, j)$ (Figure 5.1a), and it has an error of $o(\Delta x)$. Physically this approximation means that the derivative at or the tangent at the point x_i, which represents the slope or the tangent at x_i, is assumed to be equal to the slope of the straight-line joining points x_i and x_{i+1} as shown in Figure 5.2a.

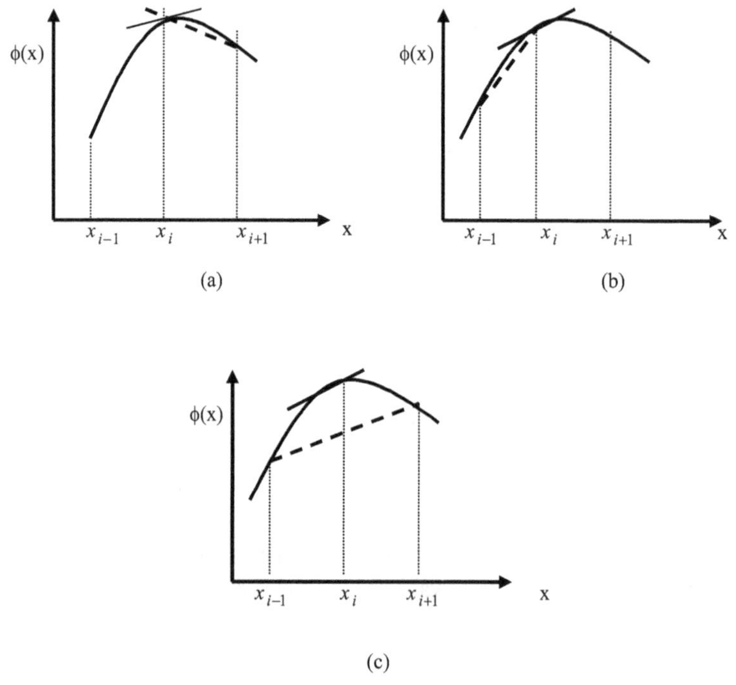

FIGURE 5.2 Graphical representation of the difference approximations: (a) forward difference, (b) backward difference, and (c) central difference.

Backward Difference Approximation Similarly, the backward difference approximation formula is derived by neglecting all the terms of $o(\Delta x)^2$ and higher in the backward Taylor series expansion given by Equation 5.3, and solving for the first derivative term as

$$\phi'_{i,j} = \frac{\phi_{i,j} - \phi_{i-1,j}}{\Delta x} + o(\Delta x) \tag{5.7a}$$

or

$$\phi'_{i,j} \cong \frac{\phi_{i,j} - \phi_{i-1,j}}{\Delta x} \tag{5.7b}$$

This approximation is called the **backward difference approximation** because it estimates the derivative at the point (i, j) using the values at the point (i, j) and a backward point $(i-1, j)$ (Figure 5.1b), and it has an error of $o(\Delta x)$. Physically this approximation means that the derivative at x_i is assumed to be equal to the slope of the straight-line joining points x_{i-1} and x_i as shown in Figure 5.2b.

Central Difference Approximation In order to derive this approximation formula, Equation 5.3 is subtracted from Equation 5.1 to obtain

$$\phi_{i+1,j} - \phi_{i-1,j} = 2\phi'_{i,j}\Delta x + \frac{2\phi'''_{i,j}}{3!}(\Delta x)^3 + \cdots + R_n \tag{5.8}$$

Now, neglecting the terms of $o(\Delta x)^3$ and higher, we have

$$\phi_{i+1,j} - \phi_{i-1,j} = 2\phi'_{i,j}\Delta x + R_2 \tag{5.9}$$

Solving for the first derivative, we get the approximation as

$$\phi'_i = \frac{\phi_{i+1} - \phi_{i-1}}{2\Delta x} + \frac{R_2}{2\Delta x} \tag{5.10a}$$

or

$$\phi'_i = \frac{\phi_{i+1} - \phi_{i-1}}{2\Delta x} + 0\left(\Delta x^2\right) \tag{5.10b}$$

or

$$\phi'_i \cong \frac{\phi_{i+1} - \phi_{i-1}}{2\Delta x} \tag{5.10c}$$

This approximation is called the **central difference approximation** as it expresses the derivative at the point (i, j) in terms of a forward point $(i+1, j)$ and a backward point $(i-1, j)$ (Figure 5.1c). Basically, the derivative at x_i is assumed to be equal to the slope of the straight-line joining points x_{i-1} and x_{i+1}. It can be noted that this central difference approximation of the first derivative has an error of the order of $o(\Delta x)^2$ as compared to the forward and backward differences that have an error of $o(\Delta x)$. The level of accuracy depends on the number of terms retained in the Taylor series expansion during the derivation of these formulas. By including higher-order terms of the Taylor series we can develop more accurate approximations of the first derivative. In a similar manner, we can also develop approximations for second- and higher-order derivatives. Also, the above equations can be derived for derivatives in other space and time coordinates in a similar manner. A summary of these formulas is given in Table 5.1.

TABLE 5.1
Two-, Three-, and Five-Point Finite Difference Formulas for First and Second Derivatives

Derivatives	Number of Points	Difference Formula	Order of Error
		First Derivative	
Forward Difference			
$\left.\frac{d\phi}{dx}\right)_i$	2	$\phi_i' = \frac{\phi_{i+1}-\phi_i}{\Delta x}$	$O(\Delta x)$
$\left.\frac{d\phi}{dx}\right)_i$	3	$\phi_i' = \frac{-\phi_{i+2}+4\phi_{i+1}-3\phi_i}{2\Delta x}$	$O(\Delta x^2)$
Backward Difference			
$\left.\frac{d\phi}{dx}\right)_i$	2	$\phi_i' = \frac{\phi_i-\phi_{i-1}}{\Delta x}$	$O(\Delta x)$
$\left.\frac{d\phi}{dx}\right)_i$	3	$\phi_i' = \frac{3\phi_i-4\phi_{i-1}+\phi_{i-2}}{2\Delta x}$	$O(\Delta x^2)$
Central Difference			
$\left.\frac{\partial\phi}{\partial x}\right)_i$	2	$\phi_i' = \frac{\phi_{i+1}-\phi_{i-1}}{2\Delta x}$	$O(\Delta x^2)$
		Second Derivative	
Forward Difference			
$\frac{d^2\phi}{dx^2}$	3	$\phi_i'' = \frac{\phi_{i+2}-2\phi_{i+1}+\phi_i}{\Delta x^2}$	$O(\Delta x)$
$\frac{d^2\phi}{dx^2}$	5	$\phi_i'' = \frac{-\phi_{i+3}+4\phi_{i+2}-5\phi_{i+1}-2\phi_i}{\Delta x^2}$	$O(\Delta x^2)$
Backward Difference			
$\frac{d^2\phi}{dx^2}$	3	$\phi_i'' = \frac{\phi_i-2\phi_{i-1}+\phi_{i-2}}{\Delta x^2}$	$O(\Delta x)$
$\frac{d^2\phi}{dx^2}$	5	$\phi_i'' = \frac{2\phi_i-5\phi_{i-1}+4\phi_{i-2}-\phi_{i-3}}{\Delta x^2}$	$O(\Delta x^2)$
Central Difference			
$\frac{d^2\phi}{dx^2}$	3	$\phi_i'' = \frac{\phi_{i+1}-2\phi_i+\phi_{i-1}}{\Delta x^2}$	$O(\Delta x^2)$
$\frac{d^2\phi}{dx^2}$	5	$\phi_i'' = \frac{-\phi_{i+2}+16\phi_{i+1}-30\phi_i+16\phi_{i-1}-\phi_{i-2}}{12\Delta x^2}$	$O(\Delta x^4)$

The list of derivatives given in Table 5.1 uses values at only two, three, or five grid points. We generally try to use two- or three-point formulas to represent partial derivatives in the governing equations of fluid flow and heat transfer. Finite difference approximations for derivatives involving more than three grid points are used when higher accuracy is required by reducing the truncation error.

Discretization Equations Even though this chapter primarily covers one-dimensional problems, we will show the derivation of the discretization equation considering a two-dimensional equation. We will describe the procedure for deriving the discretization equation using a two-dimensional general governing equation of the form

$$\frac{\partial}{\partial x}\left(\Gamma_x \frac{\partial \phi}{\partial x}\right) + \frac{\partial}{\partial y}\left(\Gamma_y \frac{\partial \phi}{\partial y}\right) + S = 0 \tag{5.11a}$$

and

$$\frac{\partial^2 \phi}{\partial x^2} + \frac{\partial^2 \phi}{\partial y^2} + \frac{5}{\Gamma} = 0 \tag{5.11b}$$

for isotropic materials with

$$\Gamma_x = \Gamma_y = \Gamma$$

This equation contains diffusion terms and a source term. Such a mathematical equation represents a mathematical statement of many different physical processes including heat conduction, mass diffusion, and fully developed flows in channels. To derive the discretization equation for any interior nodes (i, j) in the solution domain, we replace the derivatives by a suitable finite difference formula given in Table 5.1. For example, the second derivative at the node (i, j) can first be approximated by using a two-point central difference formula for the first derivatives

$$\left.\frac{\partial}{\partial x}\left(\Gamma_x \frac{\partial \phi}{\partial x}\right)\right|_{i,j} \approx \frac{\left.\Gamma_x \frac{\partial \phi}{\partial x}\right)_{i+\frac{1}{2},j} - \left.\Gamma_x \frac{\partial \phi}{\partial x}\right)_{i-\frac{1}{2},j}}{\Delta x} \tag{5.12a}$$

$$\left.\frac{\partial}{\partial y}\left(\Gamma_y \frac{\partial \phi}{\partial y}\right)\right|_{i,j} \approx \frac{\left.\Gamma_y \frac{\partial \phi}{\partial y}\right)_{i,j+\frac{1}{2}} - \left.\Gamma_y \frac{\partial \phi}{\partial y}\right)_{i,j-\frac{1}{2}}}{\Delta y} \tag{5.12b}$$

Next, the first derivatives located at

$$i+\tfrac{1}{2},j, \quad i-\tfrac{1}{2},j, \quad i,j+\tfrac{1}{2} \quad \text{and} \quad i,j-\tfrac{1}{2}$$

are replaced by two-point forward difference formulas

$$\left.\Gamma_x \frac{\partial \phi}{dx}\right)_{i+\frac{1}{2},j} \cong \Gamma_x \frac{\phi_{i+1,i} - \phi_{i,j}}{\Delta x} \tag{5.13a}$$

$$\left.\Gamma_x \frac{\partial \phi}{dx}\right)_{i-\frac{1}{2},j} \cong \Gamma_x \frac{\phi_{i,i} - \phi_{i-1,j}}{\Delta x} \tag{5.13b}$$

$$\left.\Gamma_y \frac{\partial \phi}{\partial y}\right)_{i,j+\frac{1}{2}} \cong \Gamma_y \frac{\phi_{i,j+1} - \phi_{i,j}}{\Delta y} \tag{5.13c}$$

$$\left.\Gamma_y \frac{\partial \phi}{\partial y}\right)_{i,j-\frac{1}{2}} \cong \Gamma_y \frac{\phi_{i,j} - \phi_{i,j-1}}{\Delta y} \tag{5.13d}$$

where Γ_x and Γ_y are assumed to be constant.

Substituting Equations 5.12 and 5.13 into Equation 5.11a, we have

$$\frac{\Gamma_x\left[\frac{\phi_{i+1,j}-\phi_{i,j}}{\Delta x}\right]-\Gamma_x\left[\frac{\phi_{i,j}-\phi_{i-1,j}}{\Delta x}\right]}{\Delta x}+\frac{\Gamma_y\left[\frac{\phi_{i,j+1}-\phi_{i,j}}{\Delta y}\right]-\Gamma_y\left[\frac{\phi_{i,j}-\phi_{i,j-1}}{\Delta y}\right]}{\Delta y}+S=0$$

$$\frac{\Gamma_x(\phi_{i+1,j}-2\phi_{i,j}+\phi_{i-1,j})}{\Delta x^2}+\frac{\Gamma_y(\phi_{i,j+1}-2\phi_{i,j}+\phi_{i,j-1})}{\Delta y^2}+S=0 \tag{5.14a}$$

or

$$\frac{\Gamma_x}{\Delta x^2}\left(\phi_{i+1,j}+\phi_{i-1,j}\right)+\frac{\Gamma_y}{(\Delta y)^2}\left(\phi_{i,j+1}+\phi_{i,j-1}\right)-\left(\frac{2\Gamma_x}{\Delta x^2}+\frac{2\Gamma_y}{(\Delta y)^2}\right)\phi_{i,j}+S=0 \tag{5.14b}$$

For isotropic materials with $\Gamma_x=\Gamma_y=\Gamma$, we get

$$\left(\phi_{i+1,j}+\phi_{i-1,j}\right)+\left(\frac{\Delta x}{\Delta y}\right)^2\left(\phi_{i,j+1}+\phi_{i,j-1}\right)-2\left[1+\left(\frac{\Delta x}{\Delta y}\right)^2\right]\phi_{i,j}+\frac{S}{\Gamma}(\Delta x)^2 \tag{5.15}$$

Note that we can also derive the discretization Equation 5.15 directly by replacing the second derivative terms in Equation 5.11b by the central difference formula given in Table 5.1 as

$$\left.\frac{\partial^2\phi}{\partial x^2}\right)_{i,j}\approx\frac{\phi_{i+1,j}-2\phi_{i,j}+\phi_{i-1,j}}{(\Delta x)^2} \tag{5.16a}$$

$$\left.\frac{\partial^2\phi}{\partial y^2}\right)_{i,j}=\frac{\phi_{i,j+1}-2\phi_{i,j}+\phi_{i,j-1}}{(\Delta y)^2} \tag{5.16b}$$

Substituting Equation 5.16 into Equation 5.11b, we get

$$\frac{\phi_{i+1,j}-2\phi_{i,j}+\phi_{i-1,j}}{(\Delta x)^2}+\frac{\phi_{i,j+1}-2\phi_{i,j}+\phi_{i,j-1}}{(\Delta y)^2}+\frac{S}{\Gamma}=0$$

$$\phi_{i+1,j}+\phi_{i-1,j}+\left(\frac{\Delta x}{\Delta y}\right)^2\left[\phi_{i,j+1}+\phi_{i,j-1}\right]-2\phi_{i,j}\left[1+\left(\frac{\Delta x}{\Delta y}\right)^2\right]+\frac{S}{\Gamma}(\Delta x)^2=0 \tag{5.17}$$

If we choose $\Delta x=\Delta y$, then Equation 5.17 reduces to

$$4\phi_{i,j}=\phi_{i+1,j}+\phi_{i-1,j}+\phi_{i,j+1}+\phi_{i,j-1}+\frac{S}{\Gamma}(\Delta x)^2 \tag{5.18}$$

Equations 5.17 and 5.18 represent finite difference approximations of the Poisson equation for any interior nodal points. This equation can be reduced to some limiting case problems as follows.

Two-dimensional steady-state problems without any source term

$$4\phi_{i,j}=\phi_{i+1,j}+\phi_{i-1,j}+\phi_{i,j+1}+\phi_{i,j-1} \tag{5.19}$$

One-dimensional steady state with the source term

$$2\phi_i=\phi_{i+1}+\phi_{i-1}+\frac{S}{\Gamma}(\Delta x)^2 \tag{5.20}$$

5.3.2 Control Volume Method

Another approach to derive the finite difference equation is by the control volume method, in which the solution domain is discretized into several nonoverlapping control volumes. Integration is then carried out over a control volume surrounding the grid point (i, j). The integral formulation will satisfy the overall balance of an extensive property such as energy in a heat-transfer problem. For each grid point, we consider a control volume bounded by the dashed lines located at

$$\left(i-\tfrac{1}{2},j\right),\ \left(i+\tfrac{1}{2},j\right),\ \left(i,j+\tfrac{1}{2}\right)\ \text{and}\ \left(i,j-\tfrac{1}{2}\right) \tag{5.21}$$

as shown in Figure 5.3.

Integrating Equation 5.11 over the control volume, we get

$$\int_{i,j-\frac{1}{2}}^{i,j+\frac{1}{2}}\int_{i-\frac{1}{2},j}^{i+\frac{1}{2},j}\frac{\partial}{\partial x}\left(\Gamma_x\frac{\partial\phi}{\partial x}\right)dx\,dy+\int_{i,j-\frac{1}{2}}^{i,j+\frac{1}{2}}\int_{i-\frac{1}{2},j}^{i+\frac{1}{2},j}\frac{\partial}{\partial y}\left(\Gamma_y\frac{\partial\phi}{\partial y}\right)dx\,dy$$
$$+\int_{i,j-\frac{1}{2}}^{i,j+\frac{1}{2}}\int_{i-\frac{1}{2},j}^{i+\frac{1}{2},j}S\,dx\,dy=0 \tag{5.22}$$

$$\left(\Gamma_x\frac{\partial\phi}{\partial x}\right)_{i+\frac{1}{2},j}\Delta y-\left(\Gamma_x\frac{\partial\phi}{\partial x}\right)_{i-\frac{1}{2},j}\Delta y+\left(\Gamma_y\frac{\partial\phi}{\partial y}\right)_{i,j+\frac{1}{2}}\Delta x-\left(\Gamma_y\frac{\partial\phi}{\partial y}\right)_{i,j-\frac{1}{2}}$$
$$\times\Delta x+\int_{i,j-\frac{1}{2}}^{i,j+\frac{1}{2}}\int_{i-\frac{1}{2},j}^{i+\frac{1}{2}}S\,dx\,dy=0 \tag{5.23}$$

In the next step the derivative terms are approximated with a profile assumption using an interpolation formula. Two common profile assumptions are a stepwise constant profile and a piece-wise linear profile. Figure 5.4 shows such profiles for a variable ϕ in the x-direction.

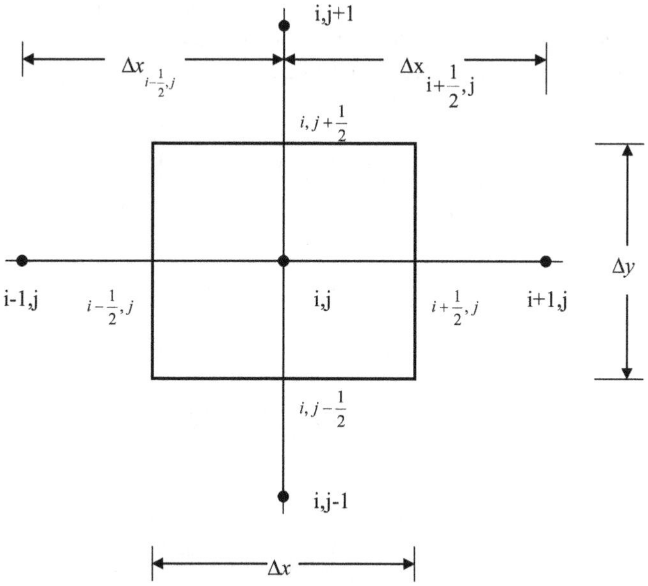

FIGURE 5.3 A typical control volume for an interior node.

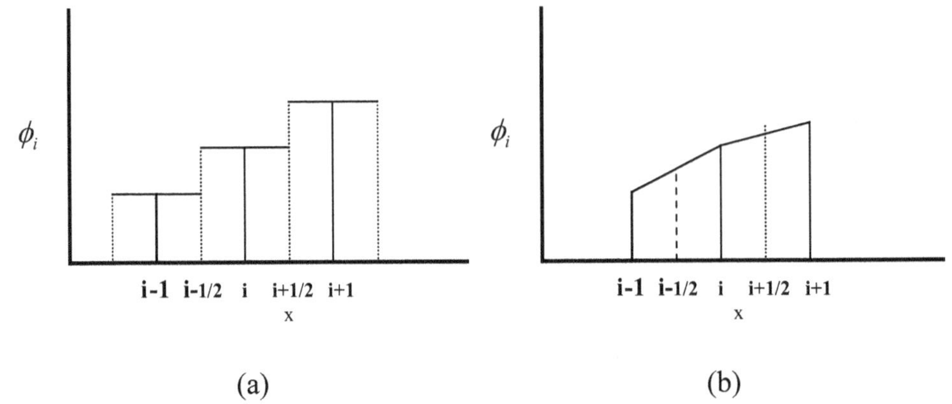

FIGURE 5.4 Profile assumptions in a control volume: (a) stepwise constant profile; (b) piece-wise linear profile.

A step-wise constant profile, shown in Figure 5.4a, can be used for approximating material transport or thermo-physical properties such as Γ and also for a dependent variable ϕ. For example, the temperature, T, heat generation, and thermal conductivity, k, at a node i can be assumed to be a constant value over the entire control volume that surrounds the node i. A piece-wise linear profile, shown in Figure 5.4b, is suitable for approximating a flux quantity that is continuous across the control volume surface and involves a first derivative. For example, the heat transfer rate or mass transfer rate functions involve first derivatives that must be continuous across the control volume surfaces.

Considering a stepwise constant profile for the source term we can express the source term integral as

$$\int_{i,j-\frac{1}{2}}^{i,j+\frac{1}{2}} \int_{i-\frac{1}{2},j}^{i+\frac{1}{2},j} S \, dx \, dy = \bar{S} \Delta x \, \Delta y \tag{5.24}$$

where $\bar{S}$ is the mean or average heat generation rate in the control volume.

Also, considering a piece-wise constant profile for the transport properties Γ and a piece-wise linear profile to approximate the first derivative terms, we can express the flux quantities across the control volume surfaces in Equation 5.23 as

$$\left.\Gamma_x \frac{\partial \phi}{\partial x}\right)_{i+\frac{1}{2},j} \approx \frac{\Gamma_{xi+\frac{1}{2}}\left(\phi_{i+1,j} - \phi_{i,j}\right)}{\Delta x_{i+\frac{1}{2}}} \tag{5.25a}$$

$$\left.\Gamma_x \frac{\partial \phi}{\partial x}\right)_{i-\frac{1}{2},j} \approx \frac{\Gamma_{xi-\frac{1}{2}}\left(\phi_{i,j} - \phi_{i-1,j}\right)}{\Delta x_{i-\frac{1}{2}}} \tag{5.25b}$$

$$\left.\Gamma_y \frac{\partial \phi}{\partial y}\right)_{i,j+\frac{1}{2}} = \frac{\Gamma_{yj+\frac{1}{2}}\left(\phi_{i,j+1} - \phi_{i,j}\right)}{\Delta y_{j+\frac{1}{2}}} \tag{5.25c}$$

$$\left.\Gamma_y \frac{\partial \phi}{\partial y}\right)_{i,j-\frac{1}{2}} = \frac{\Gamma_{yj-\frac{1}{2}}\left(\phi_{i,j} - \phi_{i,j-1}\right)}{\Delta y_{j-\frac{1}{2}}} \tag{5.25d}$$

Substituting these approximations (Equations 5.24 and 5.25) into Equation 5.23, we get

$$\frac{\Gamma_{xi+\frac{1}{2}}\left(\phi_{i+1,j}-\phi_{i,j}\right)}{\Delta x_{i+\frac{1}{2}}}\Delta y-\frac{\Gamma_{xi-\frac{1}{2}}\left(\phi_{i,j}-\phi_{i-1,j}\right)}{\Delta x_{i-\frac{1}{2}}}\Delta y+\frac{\Gamma_{yj+\frac{1}{2}}\left(\phi_{i,j+1}-\phi_{i,j}\right)}{\Delta y_{j+\frac{1}{2}}}\Delta x$$
$$-\frac{\Gamma_{yj-\frac{1}{2}}\left(\phi_{i,j}-\phi_{i,j-1}\right)}{\Delta y_{j-\frac{1}{2}}}\Delta x+\overline{S}\,\Delta x\,\Delta y=0 \tag{5.26}$$

Rearranging, we get the finite difference equation as

$$\frac{\Gamma_{xi+\frac{1}{2}}}{\Delta x_{i+\frac{1}{2}}}\Delta y\phi_{i+1}+\frac{\Gamma_{xi-\frac{1}{2}}}{\Delta x_{i-\frac{1}{2}}}\Delta y\phi_{i-1}+\frac{\Gamma_{yj+\frac{1}{2}}}{\Delta y_{j+\frac{1}{2}}}\Delta x\phi_{j+1}+\frac{\Gamma_{yj-\frac{1}{2}}}{\Delta y_{j-\frac{1}{2}}}\Delta x\phi_{j-1}$$
$$-\left(\frac{\Gamma_{xi+\frac{1}{2}}}{\Delta x_{i+\frac{1}{2}}}\Delta y+\frac{\Gamma_{xi-\frac{1}{2}}}{\Delta x_{i-\frac{1}{2}}}\Delta y+\frac{\Gamma_{yj+\frac{1}{2}}}{\Delta y_{j+\frac{1}{2}}}\Delta x+\frac{\Gamma_{yj-\frac{1}{2}}}{\Delta y_{j-\frac{1}{2}}}\Delta x\right)\phi_{i,j}+\overline{S}\Delta x\Delta y=0 \tag{5.27a}$$

Equation 5.27a can be written in a compact form as

$$a_{i,j}\phi_{i,j}=a_{i+1,j}\phi_{i+1,j}+a_{i-1,j}\phi_{i-1,j}+a_{i,j+1}\phi_{i,j+1}+a_{i,j-1}\phi_{i,j-1}+d \tag{5.27b}$$

where

$$a_{i,j}=a_{i+1,j}+a_{i-1,j}+a_{i,j+1}+a_{i,j-1} \tag{5.28a}$$

$$a_{i+1,j}=\frac{\Gamma_{xi+\frac{1}{2}}}{\Delta x_{i+\frac{1}{2}}}\Delta y \tag{5.28b}$$

$$a_{i-1,j}=\frac{\Gamma_{xi-\frac{1}{2}}}{\Delta x_{i-\frac{1}{2}}}\Delta y \tag{5.28c}$$

$$a_{i,j+1}=\frac{\Gamma_{yj+\frac{1}{2}}}{\Delta y_{j+\frac{1}{2}}}\Delta x \tag{5.28d}$$

$$a_{i,j-1}=\frac{\Gamma_{yj-\frac{1}{2}}}{\Delta y_{j-\frac{1}{2}}}\Delta x \tag{5.28e}$$

$$d=\overline{S}\Delta x\Delta y \tag{5.28f}$$

For $\Gamma_x=\Gamma_y=\Gamma$ and for uniform grid size distributions in x- and y-directions, i.e., $\Delta x_{i+\frac{1}{2}}=\Delta x_{i-\frac{1}{2}}=\Delta x$ and $\Delta y_{j+\frac{1}{2}}=\Delta y_{j-\frac{1}{2}}=\Delta y$, Equation 5.27b reduces to

$$a_{i,j}\phi_{i,j}=a_{i+1,j}\phi_{i+1,j}+a_{i-1,j}\phi_{i-1,j}+a_{i,j+1}\phi_{i,j+1}+a_{i,j-1}\phi_{i,j-1}+d \tag{5.29}$$

where the coefficients are given as

$$a_{i,j}=a_{i+1,j}+a_{i-1,j}+a_{i,j+1}+a_{i,j-1}=\frac{2\Delta y}{\Delta x}\left(1+\frac{\Delta x^2}{\Delta y^2}\right) \tag{5.30a}$$

$$a_{i+1,j}=\frac{1}{\Delta x}\Delta y \tag{5.30b}$$

$$a_{i-1,j} = \frac{1}{\Delta x}\Delta y \tag{5.30c}$$

$$a_{i,j+1} = \frac{1}{\Delta y}\Delta x \tag{5.30d}$$

$$a_{i,j-1} = \frac{1}{\Delta y}\Delta x \tag{5.30e}$$

$$d = \frac{\bar{S}\,\Delta x \Delta y}{\Gamma} \tag{5.30f}$$

For $\Delta x = \Delta y$, Equation 5.30 reduces to

$$a_{i,j} = 4$$

$$a_{i+1,j} = a_{i-1,j} = a_{i,j+1} = a_{i,j-1} = 1 \tag{5.31}$$

$$d = \frac{\bar{s}\Delta x^2}{\Gamma}$$

and discretization Equation 5.29 for interior nodes becomes

$$4\phi_{i,j} = \phi_{i+1,j} + \phi_{i-1,j} + \phi_{i,j+1} + \phi_{i,j-1} + \frac{\bar{S}\Delta x^2}{\Gamma} \tag{5.32}$$

It can be mentioned here that the assumption of the piecewise linear profile is the same as that made in the derivation of two-point approximate formulas for derivatives using Taylor's series. However, the primary advantage in control volume formulation is that we can choose many other profile assumptions within the control volume. Also, we are free to make different profile assumptions for different derivative terms in the governing equations and for different quantities such as material properties, Γ, and source term, S.

5.4 ONE-DIMENSIONAL STEADY-STATE DIFFUSION

For a one-dimensional steady-state diffusion with uniform source, the appropriate governing equation is

$$\frac{\mathrm{d}}{\mathrm{d}x}\left(\Gamma_x \frac{\mathrm{d}\phi}{\mathrm{d}x}\right) + \bar{S} = 0 \tag{5.33}$$

The discretization equation for interior nodes can be written from Equation 5.29 as

$$a_i\phi_i = a_{i+1}\phi_{i+1} + a_{i-1}\phi_{i-1} + d \tag{5.34}$$

where

$$a_i = a_{i+1} + a_{i-1} \tag{5.35a}$$

$$a_{i+1} = \frac{\Gamma_{xi+\frac{1}{2}}}{\Delta x_{i+\frac{1}{2}}} \tag{5.35b}$$

$$a_{i-1} = \frac{\Gamma_{xi-\frac{1}{2}}}{\Delta x_{i-\frac{1}{2}}} \tag{5.35c}$$

$$d = \bar{S}\Delta x \tag{5.35d}$$

Example 5.1: One-Dimensional Steady-State Conduction

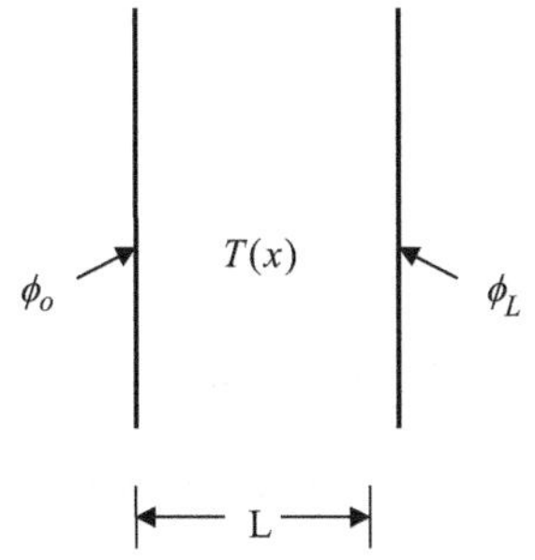

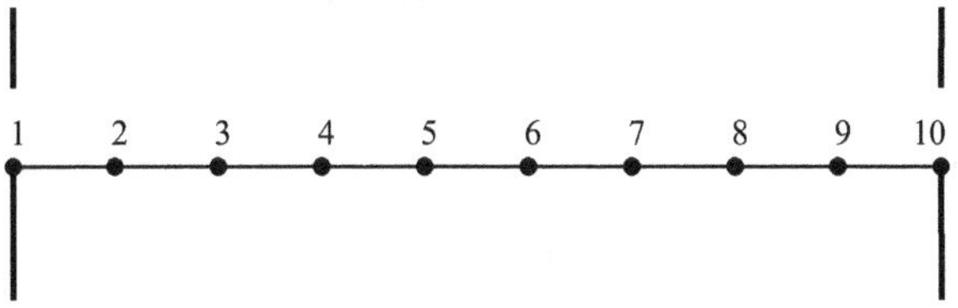

Consider a one-dimensional steady-state conduction without heat generation in a plane slab, shown in the figure below. The boundary surfaces at $x = 0$ and $x = L$ are maintained at constant temperatures T_0 and T_L, respectively. Determine the temperature distribution in the slab and heat transfer rates at the surfaces. Use $k = 120\,\text{W/cm·C}$, $L = 10$ cm, $T_0 = 100°\text{C}$, $T_L = 200°\text{C}$.

The mathematical statement of the problem is

Governing Equation:

$$\frac{d}{dx}\left(k\frac{dT}{dx}\right) = 0 \tag{E.5.1.1}$$

Boundary Condition:

$$1.\quad x = 0, \qquad T = T_0 \tag{E.5.1.2}$$

$$2.\quad x = L, \qquad T = T_L \tag{E.5.1.3}$$

Let us obtain the numerical solution with the given data.

Solution

In the **first step**, let us discretize the domain using nine uniform divisions as shown in the figure.

With this grid size distribution, we have $\Delta x_{i+\frac{1}{2}} = \Delta x_{i-\frac{1}{2}} = 1.111\,\text{cm}$ and number of grid points $N = 10$ with grids $i = 1$ and $i = 10$ located at $x = 0$ and $x = 10$. In the **second step**, we select an appropriate discretization equation. As the boundary temperature is known, we set

$$T_1 = T_0 = 100°\text{C} \quad \text{and} \quad T_{10} = T_L = 200°\text{C}$$

The rest of the nodes are interior nodes and we can apply the discretization Equation 5.34

$$a_i T_i = a_{i+1} T_{i+1} + a_{i-1} T_{i-1} + d \tag{E.5.1.4}$$

where

$$a_i = \frac{2k}{\Delta x} = 216$$

$$a_{i+1} = a_{i-1} = \frac{k}{\Delta x} = 108$$

$$d = \dot{q}\Delta x = 0$$

Substituting these coefficients into Equation E.5.1.4, we have

$$216T_i = 108T_{i+1} + 108T_{i-1} \tag{E.5.1.5a}$$

Applying this equation succeeding to all interior nodes 2, 3, 4, 5, 6, 7, 8 and 9, we get

$$\textit{Node 2} \quad 216\mathrm{T}_2 = 108T_3 + 108T_1$$

$$216T_2 = 108T_3 + 108 \times 100 \tag{E.5.1.5b}$$

$$\textit{Node 3} \quad 216T_3 = 108T_4 + 108T_2$$

$$-108T_2 + 216T_3 - 108T_4 = 0 \tag{E.5.1.5c}$$

$$\textit{Node 4} \quad 216T_4 = 108T_5 + 108T_3$$

$$-108T_3 + 216T_4 - 108T_5 = 0 \tag{E.5.1.5d}$$

$$\textit{Node 5} \quad 216T_5 = 108T_6 + 108T_4$$

$$-108T_4 + 216T_s - 108T_6 = 0 \tag{E.5.1.5e}$$

$$\textit{Node 6} \quad 216T_6 = 108T_7 + 108T_5$$

$$-108T_5 + 216T_6 - 108T_7 = 0 \tag{E.5.1.5f}$$

$$\textit{Node 7} \quad 216T_7 = 108T_8 + 108T_6$$

$$-108T_6 + 216T_7 - 108T_8 = 0 \tag{E.5.1.5g}$$

$$\textit{Node 8} \quad 216T_8 = 108T_9 + 108T_7$$

$$-108T_7 + 216T_8 - 108T_9 = 0 \tag{E.5.1.5h}$$

$$\textit{Node 9} \quad 216T_9 = 108T_{10} + 108T_8$$

$$-108T_8 + 216T_9 = 108 \times 200 \tag{E.5.1.5i}$$

The system of algebraic Equations E.5.1.5a–g can be written in matrix form as

$$\begin{bmatrix} 2 & -1 & 0 & 0 & 0 & 0 & 0 & 0 \\ -1 & 2 & -1 & 0 & 0 & 0 & 0 & 0 \\ 0 & -1 & 2 & -1 & 0 & 0 & 0 & 0 \\ 0 & 0 & -1 & 2 & -1 & 0 & 0 & 0 \\ 0 & 0 & 0 & -1 & 2 & -1 & 0 & 0 \\ 0 & 0 & 0 & 0 & -1 & 2 & -1 & 0 \\ 0 & 0 & 0 & 0 & 0 & -1 & 2 & -1 \\ 0 & 0 & 0 & 0 & 0 & 0 & -1 & 2 \end{bmatrix} \begin{Bmatrix} T_2 \\ T_3 \\ T_4 \\ T_4 \\ T_6 \\ T_7 \\ T_8 \\ T_9 \end{Bmatrix} = \begin{Bmatrix} 100 \\ 0 \\ 0 \\ 0 \\ 0 \\ 0 \\ 0 \\ 200 \end{Bmatrix}$$

In the **final step**, we choose an appropriate solver for this system of equations. Using the Gaussian elimination program, we get the numerical solution as

$$T_2 = 111.11°\text{C} \qquad T_3 = 122.22°\text{C}$$

$$T_4 = 133.33°\text{C} \qquad T_5 = 144.44°\text{C}$$

$$T_6 = 155.55°\text{C} \qquad T_7 = 166.66°\text{C}$$

$$T_8 = 177.77°\text{C} \qquad T_9 = 188.88°\text{C}$$

The heat transfer rate at $x = 0$ is estimated as

$$q''_{x=0} = k\frac{T_1 - T_2}{\Delta x} = 120\frac{100 - 111.11}{1.111}$$

$$q''_{x=0} = -1199.89\ \text{W/cm}^2$$

The heat transfer rate at $x = L$ is estimated as

$$q''_{x=L} = k\frac{T_9 - T_{10}}{\Delta x} = 120 \times \frac{188.88 - 200}{1.111}$$

$$q''_{x=L} = -1200.96\ \text{W/cm}^2$$

The negative sign indicates that the heat transfer rate is in the negative x-direction.

Example 5.2: Fully Developed Flow in a Channel

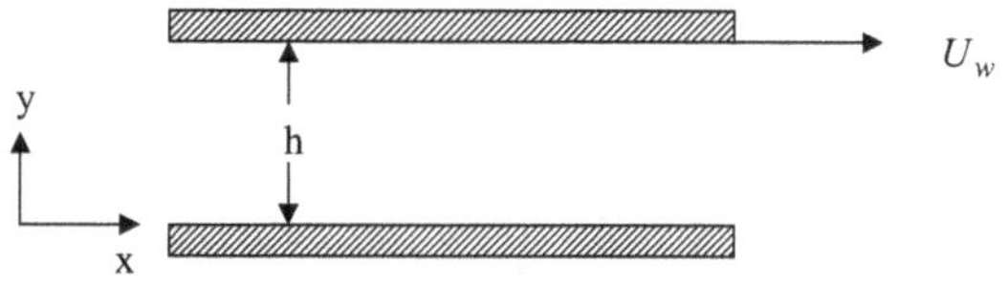

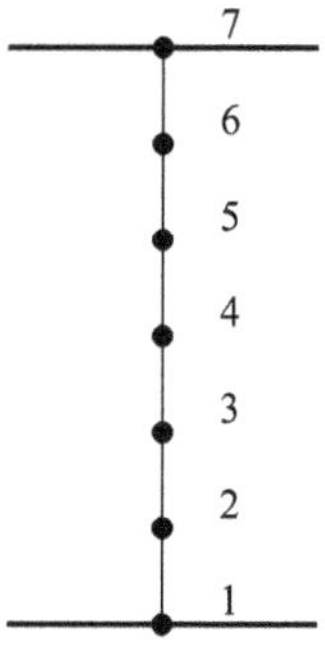

Consider the fully developed laminar steady flow of Newtonian viscous fluid through a two-dimensional channel formed by two infinite parallel plates as shown below. The lower plate is stationary and the upper plate is moving in the x-direction with a constant speed, U_w, in the positive x-direction.

The mathematical statement of the problem can be derived from mass and momentum equations of fluid flow and it is given as follows.

Governing Equation:

$$\frac{\mathrm{d}}{\mathrm{d}y}\left(\mu\frac{\mathrm{d}u}{\mathrm{d}y}\right)=\frac{\mathrm{d}p}{\mathrm{d}x} \tag{E.5.2.1}$$

Boundary Conditions:

$$1.\quad y=0,\qquad u=0 \tag{E.5.2.2}$$

$$2.\quad y=h,\qquad u=U_w \tag{E.5.2.3}$$

Determine (a) the velocity distribution, (b) volume flow rate, and (c) the shear stress at the plate and use the following data for the calculation: h = 6.00 mm, μ = 0.02 kg/m s, $\partial P/\partial x=\bar{S}=-500\text{Pa/m}$, and U_w = 10 mm/s.

Solution:

In the **first step**, let us discretize the domain using six uniform divisions as shown in the figure.

With this grid size distribution, we have $\Delta y_{i+\frac{1}{2}}=\Delta y_{i-\frac{1}{2}}=\Delta y=0.001\text{m}$ and number of grid points N = 7 with grids i = 1 and i = 7 located at y = 0 and y = 6 mm, respectively. In the **second step**, we select an appropriate discretization equation. As the boundary velocities are known, we set

$$u_1=0=0 \quad \text{and} \quad u_7=U_w=10\text{ mm/s}$$

The rest of the nodes are interior nodes and we select the discretization Equation 5.34 for a one-dimensional problem as

$$2u_{i,j}=u_{i+1}+u_{i-1}+\frac{\bar{S}\Delta y^2}{\mu} \tag{E.5.2.4}$$

where

$$S=-\frac{\bar{S}\Delta y^2}{\mu}=-\frac{-500\text{ N/m}^2/\text{m}(0.001)^2\text{m}^2}{0.02\left(\text{kg}\cdot\text{m/s}^2\right)}=0.025\text{ m/s} \tag{E.5.2.5}$$

Substituting these coefficients into the equation, we have

$$2u_i=u_{i+1}+u_{i-1}+S \tag{E.5.2.6}$$

Applying this equation succeeding to all interior nodes 2, 3, 4, 5, 6, and 7, we get

$$\textit{Node 2}\quad 2u_2=u_3+u_1+S$$

$$2u_2-u_3=S+u_1$$

$$\textit{Node 3}\quad 2u_3=u_4+u_2+S$$

$$-u_2+2u_3-u_4=S$$

$$\textit{Node 4}\quad 2u_4=u_5+u_3+S$$

$$-u_3+2u_4-u_5=S$$

$$\textit{Node 5}\quad 2u_5=u_6+u_4+S$$

$$-u_4+2u_5-u_6=S$$

$$\textit{Node 6}\quad 2u_6=u_7+u_5$$

$$-u_5+2u_6=u_7+S$$

Assembly of all nodal equations forms the system of equations as

$$\begin{bmatrix} 2 & -1 & 0 & 0 & 0 \\ -1 & 2 & -1 & 0 & 0 \\ 0 & -1 & 2 & -1 & 0 \\ 0 & 0 & -1 & 2 & -1 \\ 0 & 0 & 0 & -1 & 2 \end{bmatrix} \begin{Bmatrix} u_2 \\ u_3 \\ u_4 \\ u_5 \\ u_6 \end{Bmatrix} = \begin{Bmatrix} S+u_1 \\ S \\ S \\ S \\ S+u_7 \end{Bmatrix} \tag{E.5.2.7}$$

Substituting $u_1 = 0$, $u_7 = U_w = 0.01$ m/s and $S = 0.025$ m/s, we get

$$\begin{bmatrix} 2 & -1 & 0 & 0 & 0 \\ -1 & 2 & -1 & 0 & 0 \\ 0 & -1 & 2 & -1 & 0 \\ 0 & 0 & -1 & 2 & -1 \\ 0 & 0 & 0 & -1 & 2 \end{bmatrix} \begin{Bmatrix} u_2 \\ u_3 \\ u_4 \\ u_5 \\ u_6 \end{Bmatrix} = \begin{Bmatrix} 0.025 \\ 0.025 \\ 0.025 \\ 0.025 \\ 0.035 \end{Bmatrix} \tag{E.5.2.8}$$

In the **final step**, we choose an appropriate solver for this system of equations. The coefficient is tridiagonal and so we can use the tridiagonal matrix algorithm to get the numerical solution

$$\begin{Bmatrix} u_2 \\ u_3 \\ u_4 \\ u_5 \\ u_6 \end{Bmatrix} = \begin{Bmatrix} 0.064166 \\ 0.103333 \\ 0.117500 \\ 0.106666 \\ 0.070833 \end{Bmatrix}$$

5.5 VARIABLE SOURCE TERM

So far in our derivation of discretization equations we have treated the source term S as a constant. We now consider the source term to be a function of the dependent variable, i.e., $S(\varphi)$. To include this dependence in our formulation, let us assume that for the grid shown in Figure 5.5 φ_i can be assumed to be the representative dependent variable of the control volume surrounding the grid point i.

If we like to solve the set of discretization equations by the techniques of linear algebraic equations, then it is essential to express $S(\varphi)$ by a linear relation. The procedure for linearization will be discussed in a later section. Here we assume a linear variation of average heat source $\bar{S}$ with φ_i as

$$\bar{S} = S_0 + S_1\phi_i \tag{5.36}$$

where S_0 and S_1 are constants in the linear expression.

Substituting Equation 5.36 into Equation 5.23, we get the discretization equation and the coefficients as

$$a_{i,j}\phi_{i,j} = a_{i+1,j}\phi_{i+1,j} + a_{i-1,j}\phi_{i-1,j} + a_{i,j+1}\phi_{i,j+1} + a_{i,j-1}\phi_{i,j-1} + d \tag{5.37}$$

where

$$a_{i+1,j} = \frac{\Gamma_{i+\frac{1}{2}}}{\Delta x_{i+\frac{1}{2}}}\Delta y \tag{5.38a}$$

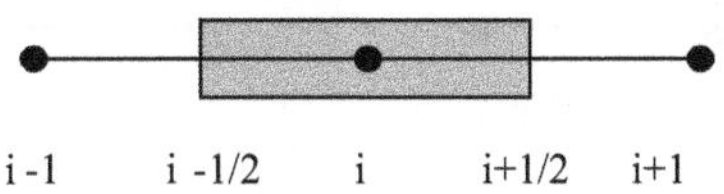

FIGURE 5.5 Control volume for an internal node.

$$a_{i-1,j} = \frac{\Gamma_{i-\frac{1}{2}}}{\Delta x_{i-\frac{1}{2}}} \Delta y \tag{5.38b}$$

$$a_{i,j+1} = \frac{\Gamma_{j+\frac{1}{2}}}{\Delta y_{j+\frac{1}{2}}} \Delta x \tag{5.38c}$$

$$a_{i,j-1} = \frac{\Gamma_{j-\frac{1}{2}}}{\Delta y_{j-\frac{1}{2}}} \Delta x \tag{5.38d}$$

$$a_{i,j} = a_{i+1,j} + a_{i-1,j} + a_{i,j+1} + a_{i,j-1} - S_1 \Delta x \Delta y \tag{5.38e}$$

$$d = S_0 \Delta x \Delta y \tag{5.38f}$$

When the source form does not depend on the temperature, S_1 becomes zero and $S_0 = \overline{S}$, and Equation 5.38 reduces to Equation 5.28.

5.6 BOUNDARY CONDITIONS

When the boundary condition is of the **first kind**, i.e., of **constant surface value**, we can simply assign the known value to the boundary nodes and solve the system of equations as was done in the previous example. However, when the boundary conditions are of the **second** or **third kind**, we need to use appropriate discretization equations for boundary nodes, which are obtained by applying overall balance of conservation quantities such as mass, heat, and momentum, or applying control volume formulation over a half control volume surrounding the boundary nodes. Let us show the derivation of such boundary equations for a one-dimensional problem for which the boundary control volume is a half control volume surrounding boundary node i as shown in Figure 5.6.

Let us first consider the boundary nodes with constant surface heat flux in a one-dimensional problem. The governing equation is

$$\frac{d}{dx}\left(\Gamma \frac{d\phi}{dx}\right) + S = 0 \tag{5.39}$$

Boundary Condition:

$$\text{at } x = 0, \quad \left(-\Gamma \frac{\partial \phi}{\partial x} = f''_{ls}\right) \tag{5.40}$$

where f''_{ls} is constant surface flux at the left boundary surface. Integrating Equation 5.39 over the half control volume

$$\int_i^{i+\frac{1}{2}} \frac{d}{dx}\left(\Gamma \frac{d\phi}{dx}\right) dx + \int_i^{i+\frac{1}{2}} S dx = 0 \tag{5.41a}$$

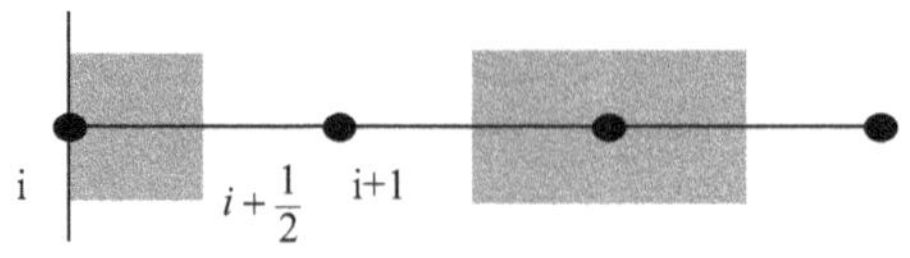

FIGURE 5.6 Half control volume for boundary nodes.

or

$$\left[\left(\Gamma\frac{d\phi}{dx}\right)_{i+\frac{1}{2}}-\left(\Gamma\frac{d\phi}{dx}\right)_{i}\right]+\bar{S}\frac{\Delta x_{i+\frac{1}{2}}}{2} \quad (5.41b)$$

Let us assume the piecewise linear profile to approximate the temperature gradient at the face of the control volume as

$$\left(\frac{d\phi}{dx}\right)_{i+\frac{1}{2}}=\frac{(\phi_{i+1}-\phi_i)}{\Delta x_{i+\frac{1}{2}}} \quad (5.42)$$

The source term $\bar{S}$ is linearized with a stepwise profile for the dependent variable as

$$\bar{S}=S_0+S_1\phi_i \quad (5.43)$$

The gradient at the boundary node is described by the constant flux boundary condition as

$$\left(-\Gamma\frac{d\phi}{dx}\right)_{i}=f''_{ls} \quad (5.44)$$

Substituting Equations 5.42–5.44 into Equation 5.41

$$\frac{\Gamma_{i+\frac{1}{2}}(\phi_{i+1}-\phi_i)}{\Delta x_{i+\frac{1}{2}}}+f''_{ls}+S_0\frac{\Delta x_{i+\frac{1}{2}}}{2}+S_1\phi_i\frac{\Delta x_{i+\frac{1}{2}}}{2}=0 \quad (5.45)$$

Rearranging, we have the discretization equation for the boundary node as

$$\left(\frac{\Gamma_{i+\frac{1}{2}}}{\Delta x_{i+\frac{1}{2}}}-S_1\frac{\Delta x_{i+\frac{1}{2}}}{2}\right)\phi_i=\frac{\Gamma_{i+\frac{1}{2}}\phi_{i+1}}{\Delta x_{i+\frac{1}{2}}}+S_0\frac{\Delta x_{i+\frac{1}{2}}}{2}+f''_{ls} \quad (5.46a)$$

or

$$a_i\phi_i=a_{i+1}\phi_{i+1}+d \quad (5.46b)$$

where

$$a_i=\left(\frac{\Gamma_{i+\frac{1}{2}}}{\Delta x_{i+\frac{1}{2}}}-S_1\frac{\Delta x_{i+\frac{1}{2}}}{2}\right) \quad (5.47a)$$

$$a_{i+1}=\frac{\Gamma_{i+\frac{1}{2}}}{\Delta x_{i+\frac{1}{2}}} \quad (5.47b)$$

$$d=S_0\frac{\Delta x_{i+\frac{1}{2}}}{2}+f''_{ls} \quad (5.47c)$$

We can see that boundary flux term f''_{ls} has been lumped together with the source term d.

Special Cases

1. For **zero surface flux or at a symmetric boundary** $f''_{ls} = 0$, so the discretization equation for boundary nodes becomes

$$a_i\phi_i = a_{i+1}\phi_{i+1} + d \tag{5.48}$$

where

$$a_i = \frac{\Gamma_{i+\frac{1}{2}}}{\Delta x_{i+\frac{1}{2}}} - S_1\frac{\Delta x_{i+\frac{1}{2}}}{2} \tag{5.49a}$$

$$a_{i+1} = \frac{\Gamma_{i+\frac{1}{2}}}{\Delta x_{i+\frac{1}{2}}} \tag{5.49b}$$

$$d = S_0\frac{\Delta x_{i+\frac{1}{2}}}{2} \tag{5.49c}$$

2. Another possible boundary condition is the boundary condition of the third kind or the mixed boundary conditions given as

$$-\Gamma\frac{d\phi}{dx} = h(\phi - \phi_\infty) \tag{5.50a}$$

or

$$f''_{ls} = h(\phi_\infty - \phi) \tag{5.51}$$

In heat transfer problems, such boundary conditions usually arise from convective boundary conditions such as

$$k\frac{dT}{dx} = h_c(\mathrm{T}_\infty - T_i) \tag{5.52}$$

or

$$q''_{ls} = h_c(T_\infty - T_i) \tag{5.53}$$

Now, if we substitute Equation 5.51 into Equation 5.48, we get the discretization equation for nonzero heat flux as

$$a_i\phi_i = a_{i+1}\phi_{i+1} + d \tag{5.54}$$

where

$$a_i = \frac{\Gamma_{i+\frac{1}{2}}}{\Delta x_{i+\frac{1}{2}}} - S_1\frac{\Delta x_{i+\frac{1}{2}}}{2} + h \tag{5.55a}$$

$$a_{i+1} = \frac{\Gamma_{i+\frac{1}{2}}}{\Delta x_{i+\frac{1}{2}}} \tag{5.55b}$$

$$d = s_0 \frac{\Delta x_{i+\frac{1}{2}}}{2} + h\phi_\infty \tag{5.55c}$$

After we solve the grid-point values, the flux at the boundary can be checked by performing an overall integral balance as

$$f_{\text{is}}'' = \frac{\Gamma_{i+\frac{1}{2}}\left(\phi_i - \phi_{i+1}\right)}{\Delta x_{i+\frac{1}{2}}} - \left(S_0 + S_1\phi_i\right)\frac{\Delta x_{i+\frac{1}{2}}}{2} \tag{5.56}$$

Similarly, we can derive the discretization equations for other boundary nodes and boundary conditions. A summary of these discretization equations is given in Table 5.2.

We can also use discretization, Equations 5.42 and 5.43, for problems when boundary heat flux is a nonlinear function of φ_i. In such cases, we express the nonlinear function in the form of Equation 5.43 through linearization techniques.

Example 5.3: One-Dimension Steady-State Fin

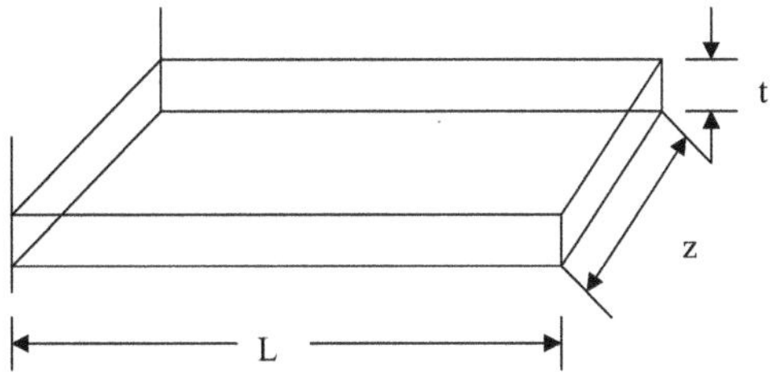

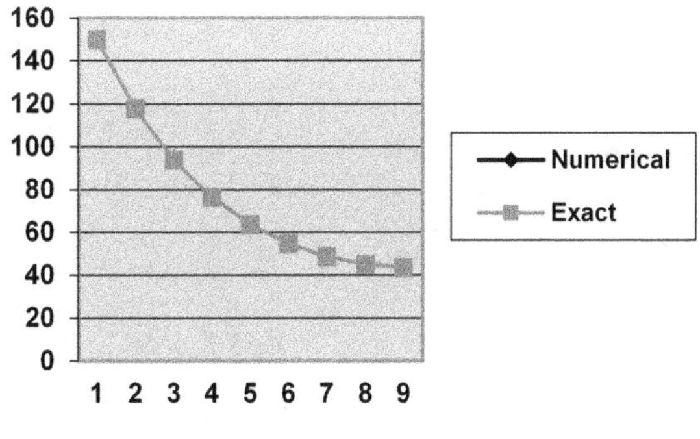

Consider a straight fin of uniform cross-sectional area A, length L = 2.0 cm, and a thickness t = 1.4 mm. The fin thermal conductivity is $k = 60\ \text{W/m}\cdot\text{C}$ and it is exposed to a convection environment at $T_\infty = 20°\text{C}$ and $h = 500\ \text{W/m}^2\cdot\text{C}$. The base of the fin is at a constant temperature T_0 = 150°C and the tip of the fin is assumed to be convective.

The mathematical statement of the problem is

Governing Equation:

$$\frac{\mathrm{d}}{\mathrm{d}x}\left(k\frac{\mathrm{d}T}{\mathrm{d}x}\right) + \frac{hP}{A}\left(T_\infty - T\right) = 0 \tag{E.5.3.1}$$

TABLE 5.2
Summary of Finite Difference Formulas for Interior and Boundary Nodes in One-Dimensional Steady-State Problems

Nodes	Finite Difference Formula
Interior Nodes	
a. Variable transport property and non-uniform grid 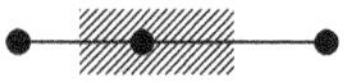	$a_i\phi_i = a_{i+1}\phi_{i+1} + a_{i-1}\phi_{i-1} + d$ $a_{i+1} = \Gamma_{i+\frac{1}{2}}/\Delta x_{i+\frac{1}{2}}$, $a_{i-1} = \Gamma_{i-\frac{1}{2}}/\Delta x_{i-\frac{1}{2}}$ $a_i = a_{i+1} + a_{i-1}$, $d = \bar{S}\Delta x$
b. Constant transport property and uniform grid in x and y 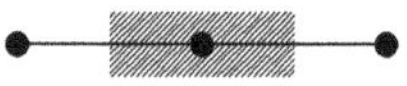	$a_i\phi_i = a_{i+1}\phi_{i+1} + a_{i-1}\phi_{i-1} + d$ $a_{i+1} = a_{i-1} = \Gamma/\Delta x$ $a_i = a_{i+1} + a_{i-1}$ $d = \bar{S}\Delta x$
c. Variable transport property, variable source ($S = S_0 + S_1\,T_i$), and nonuniform grid 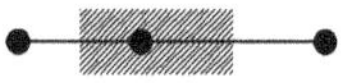	$a_i\phi_i = a_{i+1}\phi_{i+1} + a_{i-1}\phi_{i-1} + d$ $a_{i+1} = \Gamma_{i+\frac{1}{2}}/\Delta x_{i+\frac{1}{2}}$, $a_w = \Gamma_{i-\frac{1}{2}}/\Delta x_{i-\frac{1}{2}}$ $a_i = a_{i+1} + a_{i-1} - S_1\Delta x$ $d = S_0\Delta x$
Boundary Nodes	
d. Generalized form 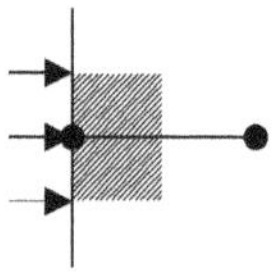	$a_i\phi_i = a_{i+1}\phi_{i+1} + d$ $a_{i+1} = \Gamma_{i+\frac{1}{2}}/\Delta x_{i+\frac{1}{2}}$, $a_i = \left(\Gamma_{i+\frac{1}{2}}/\Delta x_{i+\frac{1}{2}}\right) - S_1\left(\Delta x_{i+\frac{1}{2}}/2\right)$ $d = S_0\left(\Delta x_{i+\frac{1}{2}}/2\right) + f''_{ls}$
e. Zero flux or symmetric boundary: $f''_{ls} = 0$ 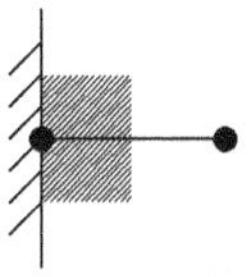	$a_i\phi_i = a_{i+1}\phi_{i+1} + d$ $a_{i+1} = \Gamma_{i+\frac{1}{2}}/\Delta x_{i+\frac{1}{2}}$, $a_i = \left(\Gamma_{i+\frac{1}{2}}/\Delta x_{i+\frac{1}{2}}\right) - S_1\left(\Delta x_{i+\frac{1}{2}}/2\right)$ $d = S_0\left(\Delta x_{i+\frac{1}{2}}/2\right)$
f. Variable surface flux: $f''_{ls} = f_0 + f_1\phi_i$ 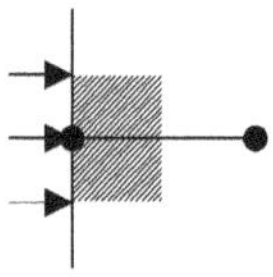	$a_i = a_{i+1}\phi_{i+1} + d$ $a_{i+1} = \Gamma_{i+\frac{1}{2}}/\Delta x_{i+\frac{1}{2}}$, $a_i = \left(\Gamma_{i+\frac{1}{2}}/\Delta x_{i+\frac{1}{2}}\right) - S_1\left(\Delta x_{i+\frac{1}{2}}/2\right) - f_1$, $d = S_0\left(\Delta x_{i+\frac{1}{2}}/2\right) + f_0$
g. Mixed boundary condition: a special case of (c) 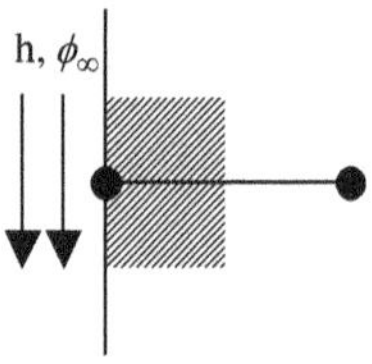	$a_i\phi_i = a_{i+1}\phi_{i+1} + d$ $a_{i+1} = \Gamma_{i+\frac{1}{2}}/\Delta x_{i+\frac{1}{2}}$ $a_i = \left(\Gamma_{i+\frac{1}{2}}/\Delta x_{i+\frac{1}{2}}\right) - S_1\left(\Delta x_{i+\frac{1}{2}}/2\right) + h$ $d = S_0\left(\Delta x_{i+\frac{1}{2}}/2\right) + hT_\infty$

Boundary Conditions:

$$1.\quad x=0,\qquad T=T_0 \tag{E.5.3.2a}$$

$$2.\quad x=L,\qquad -k\frac{dT}{dx}=h(T-T_\infty) \tag{E.5.3.2b}$$

Determine the temperature distribution in the fin and fin heat loss using one-dimensional heat conduction analysis.

Solution

In this problem, the fin is rectangular with uniform cross-section. We can estimate the cross-sectional area and the perimeter as

$$P=2(z+t)=2\left(1.0\times10^{-2}+1.4\times10^{-3}\right)=0.0228\text{ m}$$

$$A=zt=\left(1.0\times10^{-2}\right)\left(1.4\times10^{-3}\right)=1.4\times10^{-5}\text{ m}^2$$

The source term $(hP/A)(T_\infty-T)$ can be linearized as

$$S_0=\frac{hP}{A}T_\infty=\frac{(500)\left(2.28\times10^{-2}\right)}{\left(1.4\times10^{-5}\right)}\times(20)=16\,285\,714.29$$

$$S_1=-\frac{hP}{A}=-\frac{(500)\left(2.28\times10^{-2}\right)}{\left(1.4\times10^{-5}\right)}=-814\,285.7145$$

For constant conductivity

$$k_{i+\frac{1}{2}}=k_{i-\frac{1}{2}}=k=60\text{ W/m}\cdot\text{C}$$

Now the solution of the problem by the finite difference method is given as follows.

In the **first step**, we discretize the domain using a uniform grid size and number of division $N_{\text{div}}=8$ as shown below. This leads to

$$\text{Number of grid points, } N=9$$

$$\text{Grid size, } \Delta x_{i+\frac{1}{2}}=\Delta x_{i-\frac{1}{2}}=\Delta x=\frac{L}{N_{\text{div}}}=\frac{0.002}{8}=2.5\times10^{-3}\text{m}$$

In the **second step**, we select the following discretization equation for the node.

For **node 1**, we use the known constant temperature, i.e.,

$$T_1=T_0=200 \tag{E.5.3.3}$$

For the **interior nodes**, we use Equation (f) from Table 5.2.

$$a_iT_i=a_{i+1}T_{i+1}+a_{i-1}T_{i-1}+d \tag{E.5.3.4}$$

where

$$a_{i+1}=\frac{k_{i+\frac{1}{2}}}{\Delta x_{i+\frac{1}{2}}},\quad a_{i-1}=\frac{k_{i-\frac{1}{2}}}{\Delta x_{i-\frac{1}{2}}}$$

$$a_i=a_{i+1}+a_{i-1}-s_1\Delta x$$

$$d=s_0\Delta x$$

Substituting the input variables, we get the coefficients as

$$a_{i+1} = \frac{k_{i+\frac{1}{2}}}{\Delta x_{i+\frac{1}{2}}} = \frac{60}{2.5\times 10^{-3}} = 24\,000$$

$$a_{i-1} = \frac{k_{i-\frac{1}{2}}}{\Delta x_{i-\frac{1}{2}}} = \frac{60}{2.5\times 10^{-3}} = 24\,000$$

$$a_i = a_{i+1} + a_{i-1} - S_1\Delta x = 24\,000 + 24\,000$$

$$+\left(814\,285.7145\times 2.5\times 10^{-3}\right) = 50\,035.71429$$

$$d = S_0\Delta x = \left(16\,285\,714.29\times 2.5\times 10^{-3}\right) = 40\,714.28573$$

Applying this discretization equation in turn to all interior nodes, we have

$$\textit{Node 2}\quad 50\,035.71429T_2 = 24\,000T_1 + 24000T_3 + 40\,714.28573$$

$$\textit{Node 3}\quad 50\,035.71429T_3 = 24\,000T_2 + 24000T_4 + 40\,714.28573$$

$$\textit{Node 4}\quad 50\,035.71429T_4 = 24\,000T_3 + 24000T_5 + 40\,714.28573$$

$$\textit{Node 5}\quad 50\,035.71429T_5 = 24\,000T_4 + 24000T_6 + 40\,714.28573$$

$$\textit{Node 6}\quad 50\,035.71429T_6 = 24\,000T_5 + 24000T_7 + 40\,714.28573$$

$$\textit{Node 7}\quad 50\,035.71429T_7 = 24\,000T_6 + 24000T_8 + 40\,714.28573$$

$$\textit{Node 8}\quad 50\,035.71429T_8 = 24\,000T_7 + 24000T_9 + 40\,714.28573$$

Node 9 is a convective boundary node on the right-side boundary, and we use Equation 5.54 as the discretization equation, which is given as

$$a_iT_i = a_{i-1}T_{i-1} + d \tag{E.5.3.5}$$

where the coefficients are given as

$$hT_\infty = 500\times 20, \qquad h = 500$$

$$a_{i-1} = \frac{k_{i-1}}{\Delta x_{i-1}} = \frac{k}{\Delta x} = 2400$$

$$a_i = \frac{k_{i-1}}{\Delta x_{i-1}} - S_1\frac{\Delta x_{i-\frac{1}{2}}}{2} + h = 24\,000 + (814\,285.7145)\left(1.25\times 10^{-3}\right)$$

$$+500 = 25\,517.85714$$

$$d = S_0\frac{\Delta x_{i-1}}{2} + h\,T_\infty = (16.285714.29)\left(1.25\times 10^{-3}\right) + 10\,000 = 30\,357.14286$$

Substituting the coefficients, we get the discretization equation for the boundary node as

$$\textit{Node 9}\quad 25\,517.85714T_9 = 24\,000T_8 + 30\,357.14286$$

Now by assembling all nodal equations, we obtain the system of algebraic equation as

$$
\begin{bmatrix}
1 & 0 & 0 & 0 & 0 & 0 & 0 & 0 & 0 \\
-24\,000 & 5\,003\,571\,429 & -24\,000 & 0 & 0 & 0 & 0 & 0 & 0 \\
0 & -24\,000 & 5\,003\,571\,429 & -24\,000 & 0 & 0 & 0 & 0 & 0 \\
0 & 0 & -24\,000 & 5\,003\,571\,429 & -24\,000 & 0 & 0 & 0 & 0 \\
0 & 0 & 0 & -24\,000 & 5\,003\,571\,429 & -24\,000 & 0 & 0 & 0 \\
0 & 0 & 0 & 0 & -24\,000 & 5\,003\,571\,429 & -24\,000 & 0 & 0 \\
0 & 0 & 0 & 0 & 0 & -24\,000 & 5\,003\,571\,429 & -24\,000 & 0 \\
0 & 0 & 0 & 0 & 0 & 0 & -24\,000 & 5\,003\,571\,429 & -24\,000 \\
0 & 0 & 0 & 0 & 0 & 0 & 0 & -24\,000 & 2\,551\,785\,714
\end{bmatrix}
\times
\begin{Bmatrix} T_1 \\ T_2 \\ T_3 \\ T_4 \\ T_5 \\ T_6 \\ T_7 \\ T_8 \\ T_9 \end{Bmatrix}
=
\begin{Bmatrix} 150 \\ 40\,714:28573 \\ 40\,714:28573 \\ 40\,714:28573 \\ 40\,714:28573 \\ 40\,714:28573 \\ 40\,714:28573 \\ 40\,714:28573 \\ 30\,357:14286 \end{Bmatrix}
\quad \text{(E.5.3.6)}
$$

In the **third step**, we use the **tridiagonal matrix algorithm** given in Figure 3.1 to solve the system of equations. The solution given by this algorithm is given in the table below.

Node	a_i	b_i	c_i	d_i	P_i	Q_i	T_i
1	1	0	0	150	0	150	150
2	50 035.71429	24 000	24 000	40 714.28573	0.47966	72.76231	117.88670
3	50 035.71429	24 000	24 000	40 714.28573	0.62299	46.38707	94.07579
4	50 035.71429	24 000	24 000	40 714.28573	0.68407	32.89264	76.54814
5	50 035.71429	24 000	24 000	40 714.28573	0.71390	24.69322	63.81729
6	50 035.71429	24 000	24 000	40 714.28573	0.72944	19.24958	54.80330
7	50 035.71429	24 000	24 000	40 714.28573	0.73780	15.45400	48.74711
8	50 035.71429	24 000	24 000	40 714.28573	0.74238	12.73208	45.11671
9	25 517.85714	0	24 000	30 357.14286	0	43.62272	43.62272

In the **fourth step**, we estimate the fin heat loss and check the overall balance of heat.

Fin Heat Loss

Heat loss per unit area is estimated as

$$q_f = q_i\big|_{x=0} = \frac{k_{i+\frac{1}{2}}(T_i - T_{i+1})}{\Delta x_{i+\frac{1}{2}}} - (S_o + S_1 T_i)\frac{\Delta x_{i+\frac{1}{2}}}{2}$$

$$= \frac{60(150 - 117.89)}{2.5\times10^{-3}} - \left(16\,285\,714.29 - 814\,285.71456150\right)\times 1.25\times10^{-3}$$

$$= 902\,961.4286\ \mathrm{W/m^2}$$

Total Heat Loss

$$Q_f = q_f \times A = 902\,061.4286 \times 1.4\times10^{-5} = 12.64146\ \mathrm{W}$$

Check for Overall Integral Heat Balance Heat transfer rate at the base = heat generation or heat transferred by convection

$$Q_f = A\int_0^L (S_0 + S_1 T_i)\mathrm{d}x = A\sum (S_0 + S_1 T_i)\Delta x$$

Substituting the temperature values on the right-hand side of the equation, we get

$$\begin{aligned}
\sum (S_o + S_1 T_i)\Delta x &= \left[16\,285\,714.29 - 8.7145(150)\right]\times 1.25\times10^{-3} \\
&+ \left[16\,285\,714.29 - 8.7145(117.89)\right]\times 2.5\times10^{-3} \\
&+ \left[16\,285\,714.29 - 8.7145(94.08)\right]\times 2.5\times10^{-3} \\
&+ \left[16\,285\,714.29 - 8.7145(76.55)\right]\times 2.5\times10^{-3} \\
&+ \left[16\,285\,714.29 - 8.7145(63.82)\right]\times 2.5\times10^{-3} \\
&+ \left[16\,285\,714.29 - 8.7145(54.80)\right]\times 2.5\times10^{-3} \\
&+ \left[16\,285\,714.29 - 8.7145(48.74)\right]\times 2.5\times10^{-3} \\
&+ \left[16\,285\,714.29 - 8.7145(45.12)\right]\times 2.5\times10^{-3} \\
&+ \left[16\,285\,714.29 - 8.7145(43.62)\right]\times 2.5\times10^{-3}
\end{aligned}$$

$$\sum (S_o + S_1 T_i)\Delta x = -891\,256.0717\ \mathrm{W/m^2}$$

The total heat convection heat loss is

$$A\sum(S_o + S_1T_i)\Delta x = -891\,256.0717 \times 1.4 \times 10^{-3} = -12.477585$$

Comparison with the Exact Solution The analytical solutions for temperature distribution and fin heat loss are given as

$$\text{Temperature distribution:} \quad \frac{T - T_\infty}{T_0 - T_\infty} = \frac{\cosh m(L-x) + (h/mk)\sinh m(L-x)}{\cosh mL + (h/mk)\sinh mL}$$

$$\text{Fin heat loss:} \quad \sqrt{hPkA}(T_0 - T_\infty)\frac{\sinh mL + (h/mk)\cosh mL}{\cosh mL + (h/mk)\sinh mL}$$

where $m^2 = hP/kA$.

The percent error for temperature can be estimated as

$$\varepsilon_T = \left|\frac{(T_{\text{num}} - T_{\text{exact}})}{(T_0 - T_\infty)}\right| \times 100\%$$

The percent error for heat flux can be estimated as

$$\varepsilon_q = \left|\frac{q_{f,\text{num}} - q_{f,\text{exact}}}{q_{f,\text{exact}}}\right| \times 100\%$$

A comparison of the computational solution with the exact solution is presented in the table and figure below.

Node	T_i (Exact)	T_i (Numerical)	Error (ε_T)
1	150	150	0
2	112.78	117.89	0.0008462
3	93.91	94.08	0.0013077
4	76.35	76.55	0.001385
5	63.61	63.82	0.0016154
6	54.60	54.80	0.0015385
7	48.54	48.74	0.0015385
8	44.91	45.12	0.0016154
9	43.42	43.62	0.0015385

5.7 GRID SIZE DISTRIBUTION

In the preceding section we have used uniform grid size distributions, which is convenient for computation. However, nonuniform grid size distribution may become more effective in achieving desired accuracy for some problems that involve regions of steep variation of dependent variables. For such problems, we can use finer grid size distribution in regions when there is a very sharp variation of the dependent variables, and coarser grid size distribution in regions where there is very gradual variation. Let us consider the temperature variation in the straight fin as shown in Figure 5.7. One possible grid size distribution that was used in the previous example problem is the uniform grid size distribution as shown in Figure 5.8a.

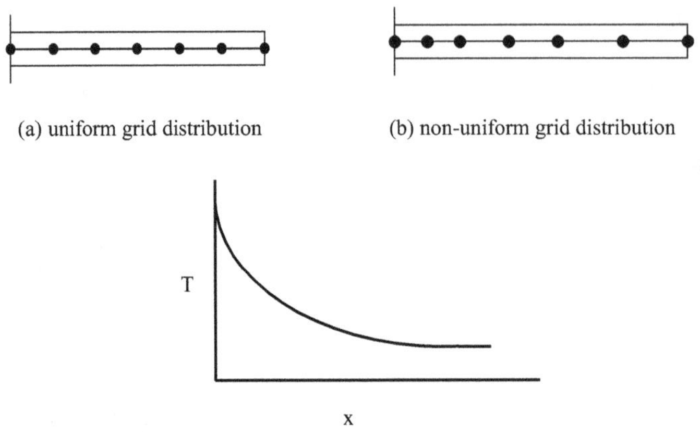

FIGURE 5.7 Typical temperature distribution in a fin.

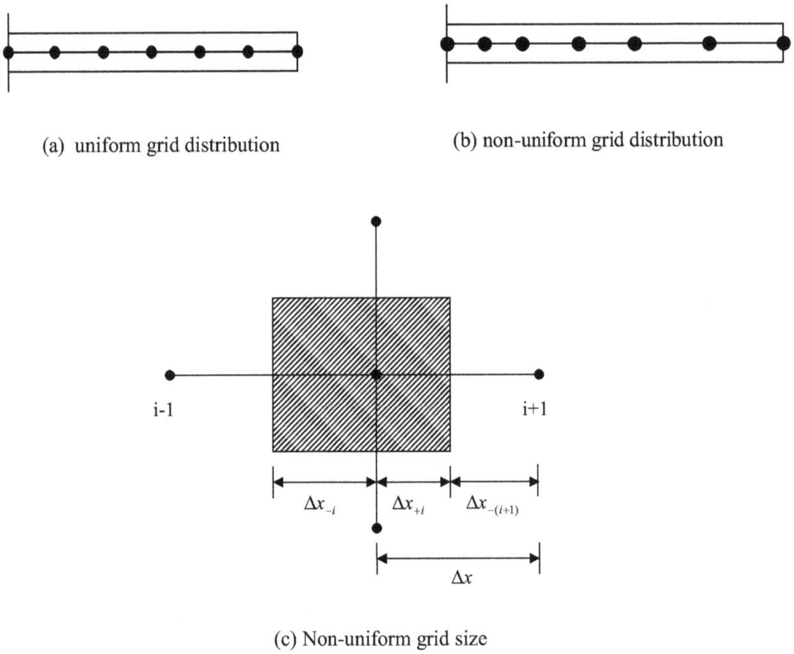

FIGURE 5.8 Grid size distributions: (a) uniform grid, (b) nonuniform grid, and (c) nonuniform grid size.

However, a more effective grid size would be the nonuniform grid size as shown in Figure 5.8b. For this problem, we could make this decision because of our prior knowledge of the problem and availability of the analytical solution. In most problems, we must conduct some preliminary numerical experimentation to gain some physical knowledge before making such a decision. Several parametric studies are to be made before deciding the number of grids and the degree of nonuniformity of the grid. Enough grid divisions and nonuniformity are to be used to capture the sharp variation near the boundaries and/or at locations near the source and sink. Some initial calculation with various degrees of nonuniformity will show the extent of regions of sharp velocity, temperature, and concentration gradients. Enough grid divisions are then used to capture the sharp variations in these regions.

5.8 NONUNIFORM TRANSPORT PROPERTY

Nonuniform transport property can result due to the presence of nonhomogeneous materials such as composite materials, and even in homogeneous materials with strong variations of transport property with dependent variable. In order to consider this nonuniform transport property, we have to estimate the transport properties $\Gamma_{i+\frac{1}{2}}$ and $\Gamma_{i-\frac{1}{2}}$ at the interfaces $i+\frac{1}{2}$ and $i-\frac{1}{2}$ of two adjacent control volumes as shown in Figure 5.9.

As was discussed in Section 5.3.2, a stepwise profile, i.e., a constant value, is usually assumed for the transport and thermophysical properties in a control volume surrounding a node, i. One common procedure for obtaining interface transport property, $\Gamma_{i+\frac{1}{2}}$, is to assume a linear variation of Γ between nodes i and $i+1$, and this leads to the expression

$$\Gamma_{i+\frac{1}{2}} = \frac{\Delta x_{+i}}{\Delta x_{i+\frac{1}{2}}}\Gamma_i + \left(1 - \frac{\Delta x_{-(i+1)}}{\Delta x_{i+\frac{1}{2}}}\right)\Gamma_{i+1} \tag{5.57}$$

For the case of a uniform grid size distribution, interface transport property would be given by the **arithmetic mean** of Γ_i and Γ_{i+1} as

$$\Gamma_{i+\frac{1}{2}} = \left(\frac{\Gamma_i + \Gamma_{i+1}}{2}\right) \tag{5.58}$$

This approach is quite suitable for a homogeneous material with strong temperature dependent transport properties. However, these equations may lead to erroneous results for composite materials that involve abrupt changes in transport properties between two adjacent materials. In such cases an alternate equation is derived based on evaluating the surface flux using a one-dimensional steady-state analysis as

$$f_s'' = \Gamma_i \frac{\phi_i - \phi_{i+\frac{1}{2}}}{\Delta x_{+i}} \tag{5.59a}$$

$$f_s'' = \Gamma_{i+1} \frac{\phi_{i+\frac{1}{2}} - \phi_{i+1}}{\Delta x_{-(i+1)}} \tag{5.59b}$$

$$f_s'' = \Gamma_{i+\frac{1}{2}} \frac{\phi_i - \phi_{i+1}}{\Delta x_{i+\frac{1}{2}}} \tag{5.59c}$$

Eliminating $\phi_{i+\frac{1}{2}}$ from Equations 5.59a and 5.59b, and equating with Equation 5.59c, we can derive the interface transport property as

$$\Gamma_{i+\frac{1}{2}} = \frac{1}{\dfrac{\Delta x_{+i}/\Delta x_{i+\frac{1}{2}}}{\Gamma_i} + \dfrac{\Delta x_{-(i+1)}/\Delta x_{i+\frac{1}{2}}}{\Gamma_{i+1}}} \tag{5.60}$$

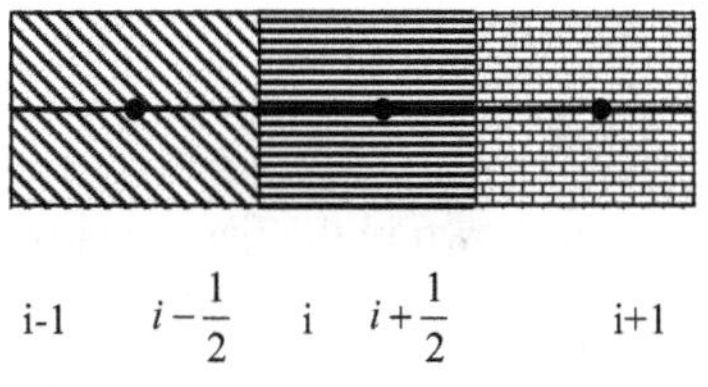

FIGURE 5.9 Grid distribution and interfaces in a composite material.

Again, for a uniform grid size distribution, Equation 5.60 reduces to a simplified form, which estimates the interface transport property as a **harmonic mean** of Γ_i and Γ_{i+1} as

$$\Gamma_{i+\frac{1}{2}} = \frac{2\Gamma_i\Gamma_{i+1}}{\Gamma_i + \Gamma_{i+1}} \tag{5.61}$$

Equations 5.65 and 5.57 are desirable for maintaining the continuity of surface flux at the interface of two dissimilar materials such as in composite materials without requiring an excessively fine grid size distribution. These equations are also desirable to ensure a zero-surface flux at the interface of two materials, one of which has zero transport property ($\Gamma = 0$) such as in an impermeable material or in an insulating material.

Example 5.4: One-Dimensional Steady-State Conduction in Composite Wall

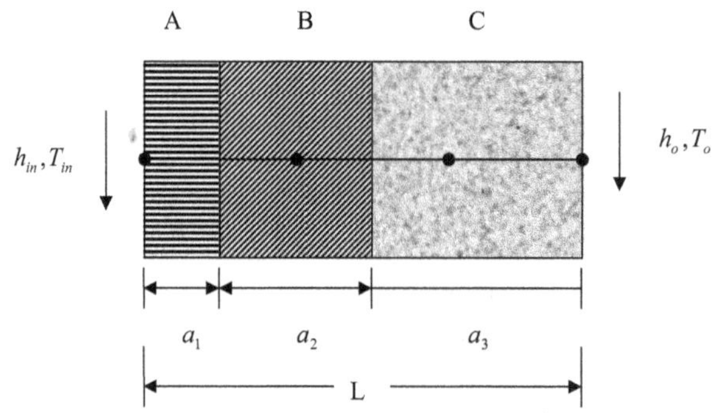

Consider steady-state heat conduction in a composite wall as shown below.

The mathematical statement of the problem is given by

Governing Equation:

$$\frac{d}{dx}\left(k(x)\frac{dT_n}{dx}\right) = 0 \tag{E.5.4.1}$$

where T_n is the temperature distribution in the nth layer.

Boundary Conditions:

$$1. \quad x = 0, \quad k\frac{dT}{dx} = h_i(T - T_i) \tag{E.5.4.2}$$

$$2. \quad x = L, \quad -k\frac{dT}{dx} = h_0(T - T_0) \tag{E.5.4.3}$$

Calculate the nodal temperature values within the composite wall and evaluate the heat flow per unit width through each material based on the following data: a_1 = 5 cm, a_2 = 10 cm, a_3 = 15 cm, k_A = 2.0 W/m°C, k_B = 30 W/m °C, k_C = 0.1 W/m°C, h_i = 20 W/cm²°C, h_0 = 20 W/cm²°C, $T_{\infty,i}$ = 100°C and $T_{\infty,0}$ = 20°C.

Solution

Let us consider a uniform grid system shown in the figure below. The step size is given as Δx = 1.0 cm.

For the **interior nodes**, we use Equation 5.34

$$a_i T_i = a_{i+1}T_{i+1} + a_{i-1}T_{i-1} + b \tag{E.5.4.4}$$

where

$$a_{i+1} = \frac{k_{i+\frac{1}{2}}}{\Delta x_{i+\frac{1}{2}}}, \qquad a_{i-1} = \frac{k_{i-\frac{1}{2}}}{\Delta x_{i-\frac{1}{2}}}$$

$$a_i = a_{i+1} + a_{i-1}$$

$$d = \bar{S}_o \Delta x$$

Nodes 2 and 3 have control volume surfaces that are the interface of two dissimilar materials. The thermal conductivity at the interface of a uniform grid is given as a **harmonic mean** Equation 5.61 as

$$k_{i+\frac{1}{2}} = \frac{2k_i k_{i+1}}{k_i + k_{i+1}} \tag{E.5.4.5}$$

Substituting material conductivities, the conductivities at the interfaces are computed as

$$k_{2-\frac{1}{2}} = \frac{2k_A k_B}{k_A + k_{B+1}} = \frac{2 \times 2.0 \times 30.0}{2.0 + 30.0} = 3.75$$

and

$$k_{2+\frac{1}{2}} = k_{3-\frac{1}{2}} = \frac{2k_B k_C}{k_B + k_C} = \frac{2 \times 30.0 \times 0.1}{30.0 + 0.1} = 0.1993$$

$$k_{3+\frac{1}{2}} = k_{4-\frac{1}{2}} = k_C = 0.1$$

For **node 2**, the discretization equation becomes

$$a_2 T_2 = a_3 T_3 + a_1 T_1 + d$$

where

$$a_3 = \frac{k_{2+\frac{1}{2}}}{\Delta x} = \frac{0.1993}{0.1} = 1.933, \qquad a_1 = \frac{k_{2-\frac{1}{2}}}{\Delta x} = \frac{3.75}{0.1} = 37.5$$

$$a_2 = a_3 + a_1 = 1.933 + 37.5 = 39.433, \qquad d = 0 \tag{E.5.4.6a}$$

$$39.433T_2 = 1.933T_3 + 37.5T_1$$

For **node 3**, the discretization equation becomes

$$a_3 T_3 = a_2 T_2 + a_4 T_4 + d$$

where

$$a_4 = \frac{k_{3+\frac{1}{2}}}{\Delta x} = \frac{0.1}{0.1} = 1.0, \qquad a_2 = \frac{k_{3-\frac{1}{2}}}{\Delta x} = \frac{0.1993}{0.1} = 1.933$$

$$a_3 = a_4 + a_2 = 1.0 + 1.933 = 2.933, \qquad d = 0$$

With the substitution of the coefficients, the equation becomes

$$2.933T_3 = 1.933T_2 + 1.0T_4 \tag{E.5.4.6b}$$

For boundary **node 1**, we use the convective boundary equation from Table 5.2 with $S_0 = 0$ and $S_1 = 0$ as

$$a_1T_1 = a_2T_2 + d$$

where

$$k_{1+\frac{1}{2}} = k_{2-\frac{1}{2}} = \frac{2k_Ak_B}{k_A + k_{B+1}} = 3.75$$

$$a_1 = \frac{k_{1+\frac{1}{2}}}{\Delta x_{1+\frac{1}{2}}} + h_i = \frac{3.75}{0.1} + 20 = 57.5$$

$$a_2 = \frac{k_{1+\frac{1}{2}}}{\Delta x_{1+\frac{1}{2}}} = \frac{3.75}{0.1} = 37.5$$

$$d = h_iT_{\infty,1} = 20.0 \times 100 = 2000$$

and

$$k_{1+\frac{1}{2}} = k_{2-\frac{1}{2}} = \frac{2k_Ak_B}{k_A + k_{B+1}} = 3.75$$

With the substitution of the coefficients, the equation becomes

$$57.5T_1 = 37.5T_2 + 2000 \tag{E.5.4.6c}$$

Similarly, for the boundary **node 4**, we use the convective boundary equation

$$a_4T_4 = a_3T_3 + d$$

where

$$a_3 = \frac{k_{4-\frac{1}{2}}}{\Delta x_{4-\frac{1}{2}}} = \frac{0.1}{0.1} = 1.0$$

$$a_4 = \frac{k_{4-\frac{1}{2}}}{\Delta x_{4-\frac{1}{2}}} + h_{\infty,0} = \frac{0.1}{0.1} + 50 = 51$$

$$d = h_0T_{\infty,0} = 50.0 \times 20.0 = 1000$$

With the substitution of the coefficients, we have the equation

$$51T_4 = T_3 + 1000 \tag{E.5.4.6d}$$

Assembling all nodal equations, we have the system of equations

$$\begin{bmatrix} 57.5 & -37.5 & 0 & 0 \\ -37.5 & 39.433 & -1.933 & 0 \\ 0 & -1.933 & 2.933 & -1.0 \\ 0 & 0 & -1.0 & 51 \end{bmatrix} \begin{Bmatrix} T_1 \\ T_2 \\ T_3 \\ T_4 \end{Bmatrix} = \begin{Bmatrix} 2000 \\ 0 \\ 0 \\ 1000 \end{Bmatrix} \tag{E.5.4.7}$$

Solving the system of equations, we get

$$\begin{Bmatrix} T_1 \\ T_2 \\ T_3 \\ T_4 \end{Bmatrix} = \begin{Bmatrix} 97.5217^\circ\text{C} \\ 96.1999^\circ\text{C} \\ 70.5577^\circ\text{C} \\ 20.9913^\circ\text{C} \end{Bmatrix} \tag{E.5.4.8}$$

Let us now estimate heat transfer rates at two surfaces as follows.

- The heat transfer rate at the left surface is estimated as

$$q_L = h_i\left(T_{\infty,i} - T_1\right) = 20.0(100 - 97.5217) = 49.566\ \text{W/m}^2$$

- The heat transfer rate at the right surface is estimated as

$$q_R'' = h_0\left(T_4 - T_{\infty,0}\right) = 50.0(20.9913 - 20.0) = 49.565\ \text{W/m}^2$$

As expected, the overall energy balance under steady state is satisfied even with the coarse grid considered.

5.9 NONLINEARITY

So far, we have considered the governing equations as linear, and so discretization of these equations leads to a system of linear algebraic equations, which is solved by a linear solver. Governing differential equations could become nonlinear due to the presence of the convective terms and/or due to a variable material or fluid properties or a variable source term. Nonlinearity due to the inclusion of convective terms will be addressed in subsequent chapters on convection problems.

Nonlinearity due to the dependence of the transport and thermo-physical properties or dependence of the source term or film transfer coefficient on the dependent variable such as $\Gamma(\varphi)$, $S(\varphi)$, or $h(T)$ are usually tackled by an iterative process. The iteration process starts with a guess for all nodal values of dependent variables. Using these guess values, the coefficients a and d of the discretization equations that include the properties and source term are estimated. This is followed by solving the resulting linear system of algebraic equations by the usual linear equation solver. The process is repeated with the latest iterated nodal values of dependent variables until convergence is reached.

5.10 LINEARIZATION OF A VARIABLE SOURCE TERM

In Section 5.5, we have mentioned that to solve the set discretization equations by the solution techniques of linear algebraic equations, it is essential to express a variable source $S(\varphi)$ by a linear relation. We have assumed a linear variation of an average source $\overline{S}$ with φ_i as $\overline{S} = S_0 + S_1\phi_i$ in deriving discretization equations 5.37 and 5.38 for the two-dimensional steady-state problem with source term. In many situations, the selections of S_0 and S_1 are quite straightforward. However, it is the situations where source functions are such that we need to decide whether we would simply evaluate the source function based on the previous iterated value of the dependent variable or try to linearize the function through rearrangement or using expansion. In this section, let us describe one of the ways of linearizing a source term as given by Patankar (1991). In this approach, the source function $\overline{S}(\phi)$ is expressed using a truncated Taylor series in the following manner:

$$\overline{S}(\phi) = \overline{S}(\phi^m) + \left(\frac{\partial \overline{S}}{\partial \phi}\right)^m (\phi_i - \phi_i^m) \tag{5.62}$$

where the superscript m represents the previous iteration number, and ϕ_i^m represents the value of the dependent variable at a previous iteration. Equation 5.62 now is expressed in linear form as

$$\overline{S}(\phi) = S_0 + S_1\phi_i \tag{5.63}$$

where

$$S_0 = \overline{S}(\phi^m) - \left(\frac{\partial \overline{S}}{\partial \phi}\right)^m \phi_i^m \tag{5.64a}$$

$$S_1 = \left(\frac{\partial \overline{S}}{\partial \phi}\right)^m \tag{5.64b}$$

The selection of S_0 and S_1 through Equations 5.64 is only done when it gives a negative value of S_1. If this is not achievable, then we simply select $S_0 = \overline{S}(\phi)$ and $S_1 = 0$.

PROBLEMS

5.1 Consider one-dimensional steady-state conduction without heat generation in a plane wall shown in the figure below. The boundary surfaces at $x = 0$ and $x = L$ are exposed to indoor and outdoor air environment, respectively. Determine the temperature distribution in the slab and heat transfer rates at the surfaces using six equally spaced nodes. Use $k = 20\,\text{W/cm}\cdot\text{C}$, $L = 10\,\text{cm}$, $T_i = 25°\text{C}$, $T_0 = 40°\text{C}$, $h_i = 50\,\text{W/m}^2\cdot\text{C}$, and $h_0 = 100\,\text{W/m}^2\cdot\text{C}$. Solve the system of equations using **TDMA** solver.

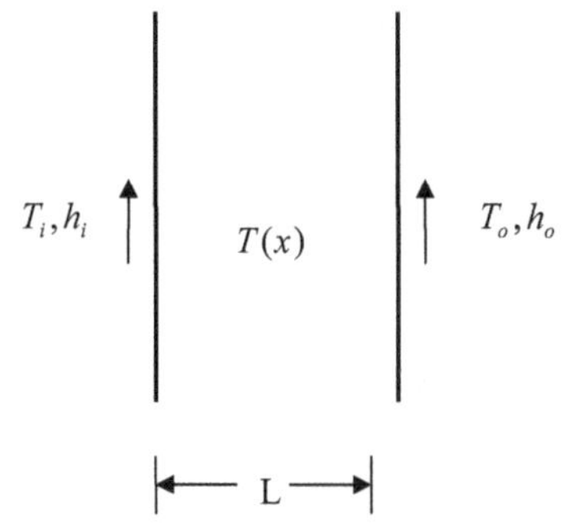

5.2 Consider one-dimensional steadystate heat conduction without heat generation in a plane wall with surface at $x = 0$ subjected to incident radiation q_0'' as shown in the figure below. The boundary surface at $x = L$ is exposed to convective fluid environment with $T_0 = 40°\text{C}$, $h_0 = 100\,\text{W/m}^2\cdot\text{C}$. Thermal conductivity of the material is as $\text{k} = 20\,\text{W/cm}\cdot\text{C}$. Use finite difference/control volume scheme to derive the system of equations for temperatures using number of nodal points as six. Solve the system of equations for temperature distribution in the wall and heat transfer rates at the surfaces using Gauss-elimination solver. Use $L = 10$ cm and $q_0'' = 10^3\ \text{W/m}^2$.

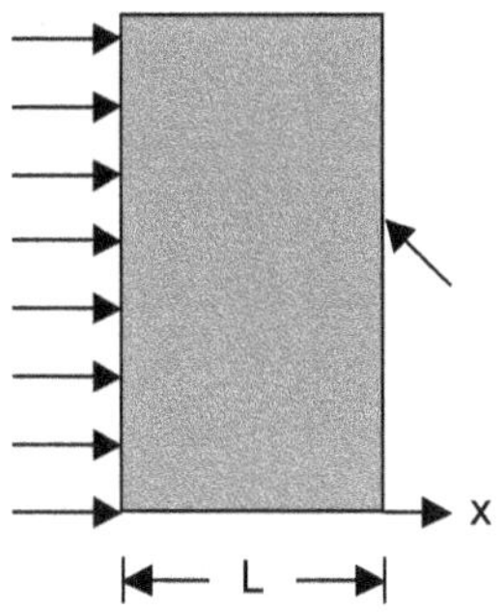

5.3 An exterior wall of a building is a composite wall made of four layers as shown below:

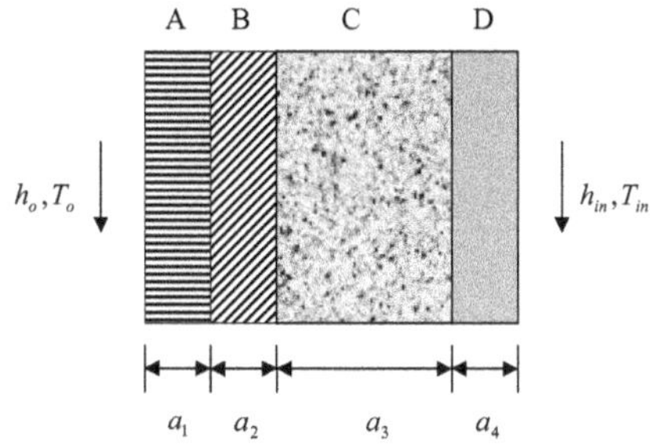

Consider the layers as homogeneous with following data: $a_1 = 2$ cm, $a_2 = 1$ cm, $a_3 = 5$ cm, $a_4 = 1$ cm, $k_A = 0.72$ W/m K, $k_B = 0.12$ W/m K w/cm C, $k_C = 0.026$ W/m w/cm C, and $k_D = 0.22$ W/m K. Calculate the nodal temperature values within the composite wall and evaluate the heat flow per unit width based assuming a winter design indoor and outdoor conditions as: $T_{\text{in}} = 25$°C, $h_{\text{in}} = 10$ w/cm^2 C and $T_{\text{out}} = -25$°C, $h_o = 50$ w/cm^2 C. Also, compare these numerical solutions to the analytical solution given by

Temperature Distribution:

$$T_n = T_{\text{in}} - \frac{(T_{\text{in}} - T_o)\left[h_{\text{in}}\left(\frac{x - x_{n-1}}{k_{n-1}} + \sum_{m=1}^{n-1} \frac{a_m}{k_m}\right) + 1\right]}{h_{\text{in}} \sum_{m=1}^{N} \frac{a_m}{k_m} + \frac{h_{\text{in}}}{h_o} + 1}$$

Heat Flux:

$$q'' = \frac{T_{\text{in}} - T_o}{\frac{1}{h_{\text{in}}} + \sum_{m=1}^{N} \frac{a_m}{k_m} + \frac{1}{h_o}}$$

5.4 The exposed surface ($x = 0$) of a plane wall of thermal conductivity k is subjected to thermal radiation that causes volumetric heat generation in the media to vary as

$$\dot{q}(x) = \dot{q}_0\left(1 - \frac{x}{L}\right)$$

where $\dot{q}_0\left(\text{W/m}^3\right)$ is a constant. The boundary at $x = L$ is convectively cooled, while the exposed surface is maintained at a constant temperature T_0.

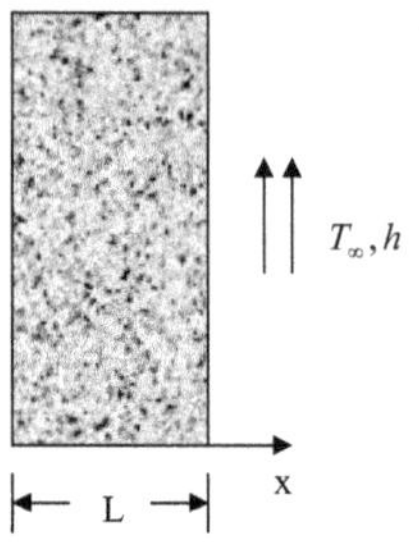

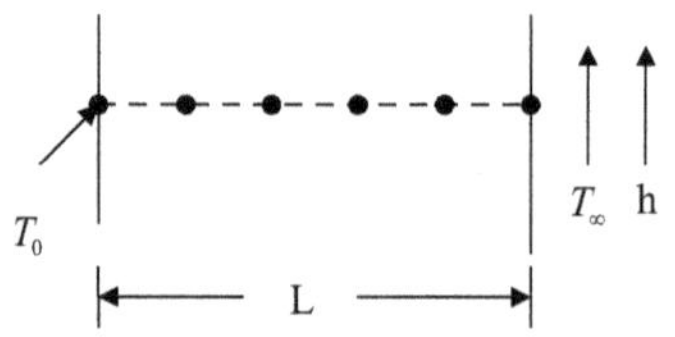

The mathematical statement of the problem is given as

Governing Equation:

$$\frac{\partial}{\partial x}\left(k\frac{\partial T}{\partial x}\right)+\dot{q}=0$$

Boundary Conditions:

$$1.\ x=0,\quad T(x,0)=T_0,\quad 2.\ x=L,\quad -k\frac{\partial T}{\partial x}=h(T-T_\infty)$$

Consider grid size distribution shown with six equally spaced grid points and determine the temperature distribution in the slab. Use following data for computation: $\dot{q}=1.0\times10^5\,\dot{q}_0\,\mathrm{W/m^3}, L=12\,\mathrm{cm}, T_0=30°\mathrm{C}, k=20\ \ \mathrm{W/m°C}\,h=50\,\mathrm{W/m^2\cdot C}, T_\infty=25°\mathrm{C}.$

Solve the system of equations using Gauss-elimination solver.

5.5 Semitransparent wall is irradiated at the surface at $x = 0$ by incident radiation such that such the absorbed radiation in the media results in an internal heat generation given as

$$\dot{q}=q_o''ae^{-ax}$$

where a is the coefficient absorption of the material and q_0'' is the incident radiation flux. The wall at $x = 0$ is also exposed to a convective environment of h_0 and T_0, and the surface at $x = L$ is maintained at constant temperature T_L. Derive discretization equations for interior nodes and for the convective boundary node for the nonuniform grid system shown. Derive the system of equations for the solution of nodal temperatures and heat transfer rates at inner and outer surfaces.

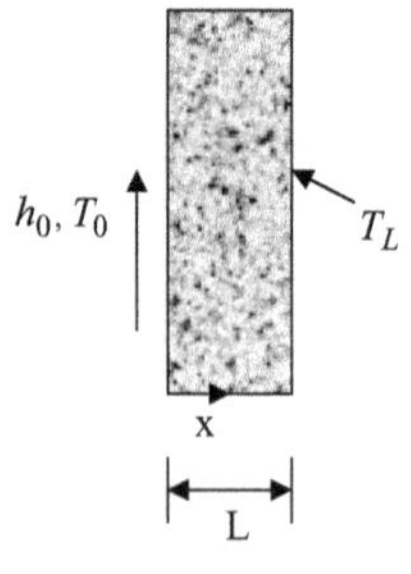

Solve the system of equations by (a) Gauss-elimination solver and (b) Gauss-Seidel solver.

5.6 Consider one-dimensional steady-state analysis for the cylindrical fin with a diameter of 10 mm and material thermal conductivity of 30 W/m · K. Surrounding fluid temperature and convection heat transfer coefficients are 30°C and 40 $W/m^2 \cdot K$. Obtain the system of equations for the grid system shown. Solve the equations for the temperature distribution

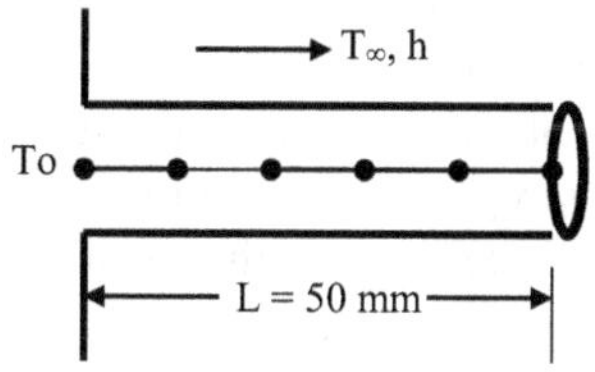

and fin heat loss. Use fin base temperature at $x = 0$ as $T_0 = 150°C$. Make an energy balance to check the accuracy.

5.7 An aluminum pin fin having a diameter D = 1 cm and length L = 6 cm is exposed to surrounding fluid with temperature, $T_0 = 150°C$, and convection heat transfer coefficient, $h = 8.0\,w/m^2 \cdot C$. The fin base temperature is $T_0 = 100°C$. Determine the temperature distribution and fin heat loss using finite difference method and using the grid system shown. Assume thermal conductivity of the material as $k = 200\,W/m°C$. Show overall energy balance.

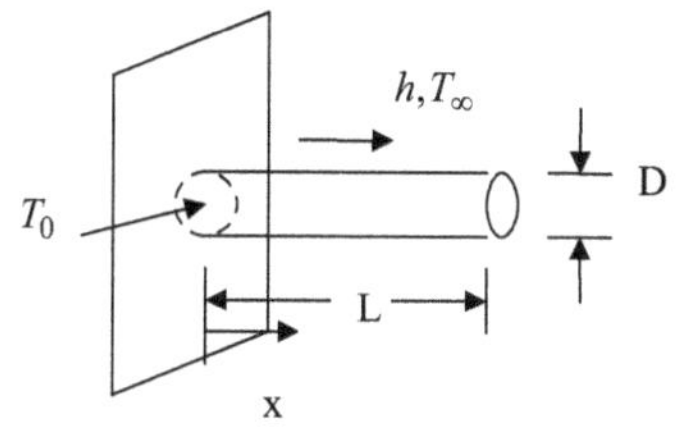

Solve the system of equations using TDMA solver.

5.8 Consider the fully developed laminar steady flow of Newtonian viscous fluid through a two-dimensional channel formed by two infinite parallel plates as shown below. The lower plate is moving in the negative x-direction with a constant speed U_{w1} and the upper plate is moving in the positive x-direction with a constant speed, U_{w2}, in the positive x-direction.

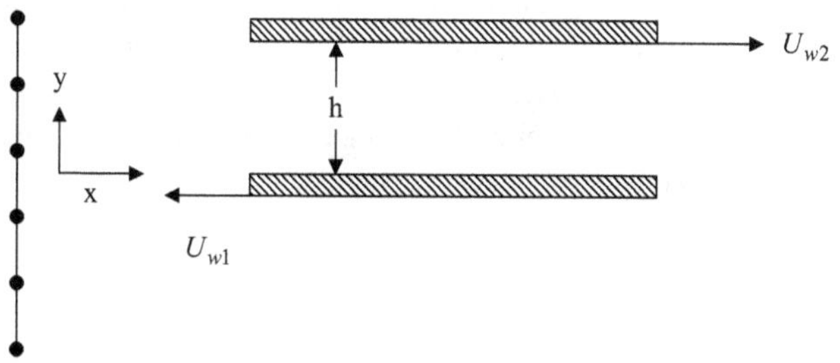

Determine (a) the velocity distribution, (b) volume flow rate, and (c) the shear stress at the plate and use the following data for the calculation: $h = 3.00$ mm,

$$\mu = 0.02\,\text{kg/m}\cdot\text{s}, \quad \frac{\partial P}{\partial x} = \bar{S} = -1000\,\text{Pa/m}, \quad U_{w1} = 10\,\text{mm/s} \quad \text{and} \quad U_{w2} = 20\,\text{mm/s}.$$

Determine the shear stresses at the bottom and top surfaces.

5.9 Consider one-dimensional steady-state diffusion of hydrogen in an anode electrode made of porous diffusion layer with integrated catalyst. Gas flow channel maintains a convective mass transfer with constant hydrogen concentration of C_∞ at the left surface of the electrode. The right surface of the electrode is assumed as impermeable to hydrogen. Assume constant rate of hydrogen consumption, $\dot{m}H_2$, due to the catalytic reaction throughout porous layer and determine the hydrogen concentration profile in the electrode.

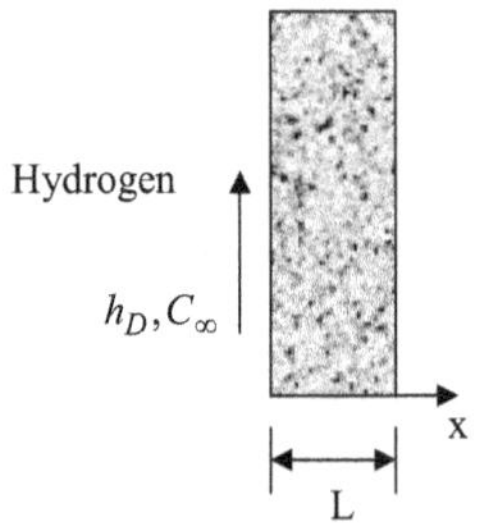

The mathematical statement of the problem is given as

Governing Equation:

$$\frac{d}{dx}\left(D_{H_2}\frac{dC}{dx}\right) - \dot{m}H_2 = 0$$

Boundary Conditions:

1. $x = 0,\ D_{H_2} = \dfrac{dC}{dx} = h_D(C - C_\infty)$
2. $x = L,\ \dfrac{dC}{dx} = 0$ due to impermeable right surface

Use a nonuniform grid size distribution shown to derive the system of equations for the nodal hydrogen concentration.

5.10 Consider one-dimensional steady-state diffusion of hydrogen in an anode electrode made of porous diffusion layer with a thin catalyst layer placed at the right side as shown. The catalyst layer provides a heterogeneous chemical reaction for the consumption of hydrogen at a rate m''_{H_2}, defined as the consumption per unit area of the catalyst. Assume the reaction rate as given by a **first-order reaction**, i.e., $m''_{H_2} = -k''_1 C(0)$, where k''_1 is the reaction rate constant. Gas flow channel maintains a constant hydrogen concentration of C_0 at the left surface of the electrode.

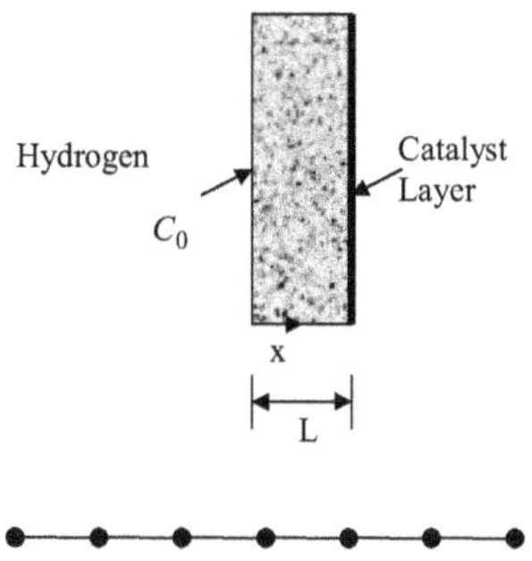

The mathematical statement of the problem is given as

Governing Equation:

$$\frac{d}{dx}\left(D\frac{dC}{dx}\right) = 0$$

Boundary Conditions:

1. $x = 0, C(0) = C_0$
2. $x = L, -D\frac{dC}{dx}\Big|_{x=L} = -k_1''C(L)$

Use a uniform grid size distribution shown to derive the system of equations for the nodal hydrogen concentration.

5.11 Develop a computer code for the solution one-dimensional diffuse equation in a plane slab of thick ness L that generates the system of equations for unknown nodal values using discretization equations for interior nodes and that for the boundary nodes at the left and right surface. Solve the system of equations using Gauss-elimination solver.

a. Use the code for the solution of the problem given in Problem 5.4.
b. Repeat the problem by progressively increasing the number of nodes as $N_x = 8$, $N_x = 10$, and $N_x = 12$. Present approximate percent relative error for the estimation of heat transfer rate at the left surface at $x = 0$ with increasing number of nodes.

6 Finite Difference–Control Volume Method

Multidimensional Problems

The objective of this chapter is to consider the application of the finite difference method to multi-dimensional problems. Basics for the finite difference method were presented in Chapter 5 dealing with one-dimensional steady-state problems. Even though the introduction of the discretization procedure was given using a two-dimensional steady-state equation, Chapter 5 was primarily involved with applications to one-dimensional problems. In this chapter, we present detailed formulation for obtaining discretization equations for two-dimensional boundary nodes on plane surfaces and corners. The objective is to build enough knowledge to formulate discretization equations for nodes involving three-dimensional problems and fore complex physical phenomena.

Figure 6.1 shows examples of heat transfer in one-dimensional, two-dimensional, and three-dimensional problems:

6.1 TWO-DIMENSIONAL STEADY-STATE PROBLEMS

To demonstrate the formulation of discretization equations for two-dimensional problems, let us consider a control volume in a two-dimensional grid system as shown in Figure 6.2 below:

For two-dimensional steady-state problems, the appropriate governing equation is

$$\frac{\partial}{\partial x}\left(\Gamma\frac{\partial\phi}{\partial x}\right)+\frac{\partial}{\partial y}\left(\Gamma\frac{\partial\phi}{\partial y}\right)+S=0 \tag{6.1}$$

The discretization equation for interior nodes was derived in Chapter 5 as

$$a_{i,j}\phi_{i,j}=a_{i+1,j}\phi_{i+1,j}+a_{i-1,j}\phi_{i-1,j}+a_{i,j+1}\phi_{i,j+1}+a_{i,j-1}\phi_{i,j-1}+d \tag{6.2a}$$

where

$$a_{i+1,j}=\frac{k_{i+\frac{1}{2},j}}{\Delta x_{i+\frac{1}{2},j}},\quad a_{i-1,j}=\frac{k_{i-\frac{1}{2},j}}{\Delta x_{i-\frac{1}{2},j}},\quad a_{i,j+1}=\frac{k_{i,j+\frac{1}{2}}}{\Delta y_{i,j+\frac{1}{2}}},\quad a_{i,j-1}=\frac{k_{i,j-\frac{1}{2}}}{\Delta x_{i,j}-\frac{1}{2}}$$

$$d=S_o\Delta x\Delta y \tag{6.2b}$$

$$a_{i,j}=a_{i+1,j}+a_{i-1,j}+a_{i,j+1}+a_{i,j-1}-S_1\Delta x\Delta y$$

Equation 6.2 can be further simplified for many different cases. A summary of finite difference formulas for such cases is given in Table 6.1.

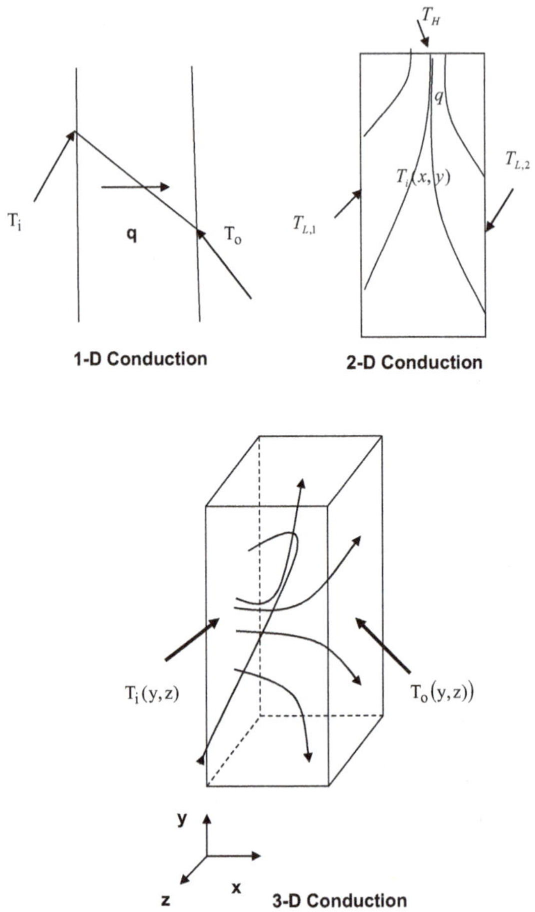

FIGURE 6.1 Examples of multidimensional problems.

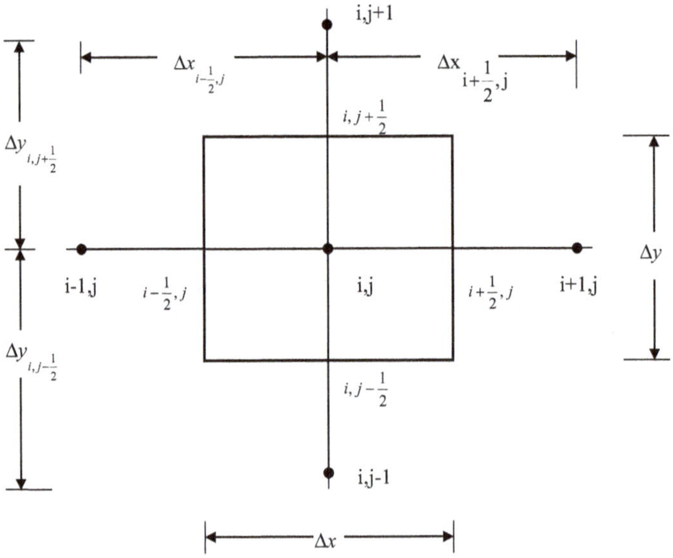

FIGURE 6.2 Control volume for two-dimensional problems.

TABLE 6.1

A Summary of Finite Difference Formulas for Interior Nodes in Two-dimensional Steady State Problems

Physical Situation | **Finite-Difference Equation**

a. Non-uniform grid size

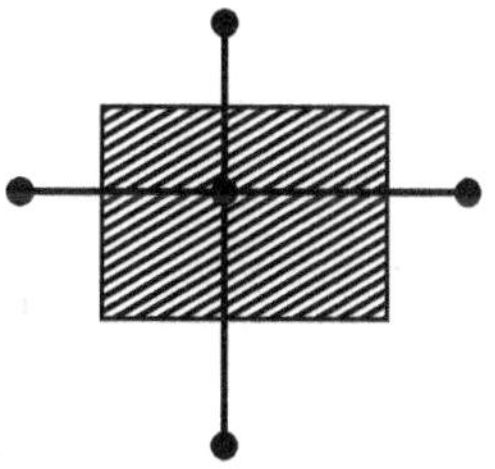

$$a_{i,j}\phi_{i,j} = a_{i+1,j}\phi_{i+1,j} + a_{i-1,j}\phi_{i-1,j} + a_{i,j+1}\phi_{i,j+1} + a_{i,j-1}\phi_{i,j-1} + b$$

$$a_{i+1,j} = \frac{\Gamma_{i+\frac{1}{2}}}{\Delta x_{1+\frac{1}{2}}}\Delta y, \qquad a_{i-1,j} = \frac{\Gamma_{1-\frac{1}{2},j}}{\Delta x_{i-\frac{1}{2},j}}\Delta y$$

$$a_{i,j+1} = \frac{\Gamma_{i,j+\frac{1}{2}}}{\Delta y_{i+\frac{1}{2}}}\Delta x, \qquad a_{i,j-1} = \frac{\Gamma_{i,j-\frac{1}{2}}}{\Delta y_{i,j-\frac{1}{2}}}\Delta x$$

$$a_p = a_E + a_w + a_N + a_s$$

b. For constant transport property and uniform grid size distribution in x and y

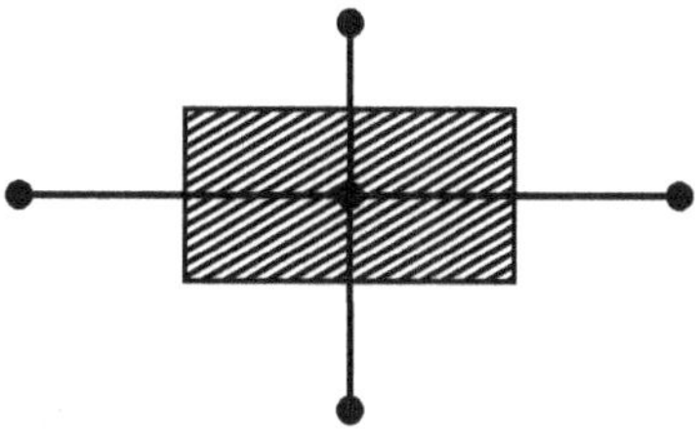

$$a_{i,j}\phi_{i,j} = a_{i+1,j}\phi_{i+1,j} + a_{i-1,j}\phi_{i-1,j} + a_{i,j+1}\phi_{i,j+1} + a_{i,j-1}\phi_{i,j-1} + b$$

$$a_{i+1,j} = a_{i-1,j} = \frac{\Delta y}{\Delta x}$$

$$a_{i,j+1} = a_{i,j-1} = \frac{\Delta x}{\Delta y}$$

$$a_{i,j} = a_{i+1,j} + a_{i-1,j} + a_{i,j=1} + a_{i,j-1} = 2\frac{\Delta y}{\Delta x}\left(1 + \frac{\Delta x^2}{\Delta y^2}\right)$$

$$b = \bar{S}\Delta x\Delta y / \Gamma$$

c. Constant transport property and $\Delta x = \Delta y$

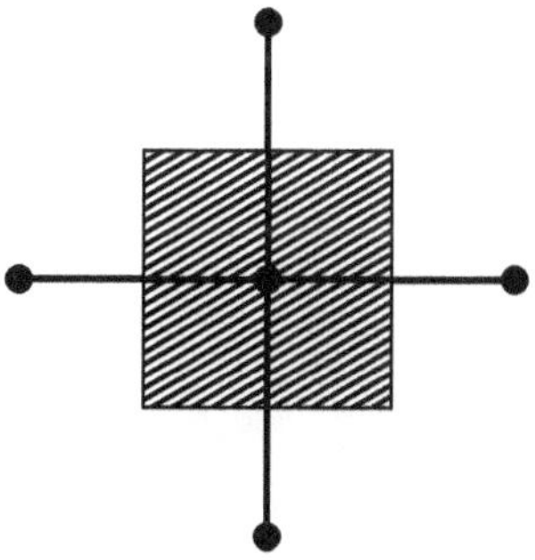

$$a_{i,j} = a_{i+1,j}\phi_{i+1,j} + a_{i-1,j}\phi_{i-1,j} + a_{i,j+1}\phi_{i,j+1} + a_{i,j-1}\phi_{i,j-1} + b$$

$$a_{i+1,j} = a_{i-1,j} = a_{i,j+1} = a_{i,j-1} = 1, \quad a_{i,j} = 4$$

$$b = \bar{S}\Delta x^2 / \Gamma$$

d. Constant transport property, $\Delta x = \Delta y$, and no source

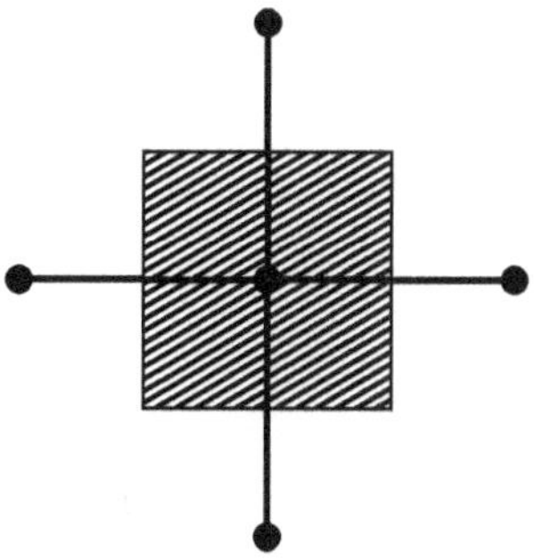

$$4\phi_{i,j} = \phi_{i+1,j} + \phi_{i-1,j} + \phi_{i,j+1} + \phi_{i,j-1}$$

Example 6.1: Two-Dimensional Steady-State Conduction

Consider a 2 m by 2 m square slab with the top surface maintained at 400°C and all other surfaces maintained at 20°C. Apply the finite difference method to calculate the two-dimensional temperature distribution using a grid with $\Delta x = \Delta y = 0.5$ m as shown in the figure below. Considering the thermal conductivity of the slab material as 40 W/m·K, estimate the heat transfer rates at all surfaces and show the overall energy balance.

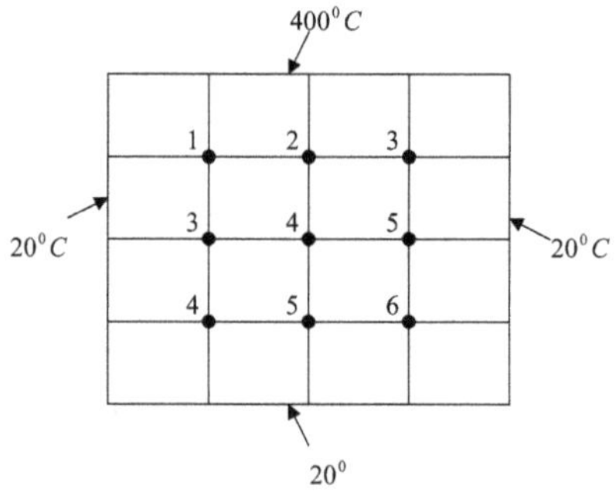

Solution

In the prescribed grid, the boundary nodes can be assigned with the prescribed surface temperatures. There are nine interior nodes for which the appropriate finite difference equation is derived based on Equation 6.2 in the following manner:

$$a_{i,j}T_{i,j} = a_{i+1,j}T_{i+1,j} + a_{i-1,j}T_{i-1,j} + a_{i,j+1}\,T_{i,j-1} + d$$

where

$$a_{i+1,j} = \frac{k_{i+\frac{1}{2},j}}{\Delta x_{i+\frac{1}{2},j}}, \quad a_{i-1,j} = \frac{k_{i-\frac{1}{2},j}}{\Delta x_{i-\frac{1}{2},j}}, \quad a_{i,j+1} = \frac{k_{i,j+\frac{1}{2}}}{\Delta y_{i,j+\frac{1}{2}}}, \quad a_{i,j-1} = \frac{k_{i,j-\frac{1}{2}}}{\Delta x_{i,j} - \frac{1}{2}}$$

$$d = S_o \Delta x \Delta y$$

$$a_{i,j} = a_{i+1,j} + a_{i-1,j} + a_{i,j+1} + a_{i,j-1} - S_1 \Delta x \Delta y$$

Considering the uniform grid size distributions and constant thermal conductivity, coefficients are given as

$$a_{i+1,j} = \frac{k}{\Delta x}, \; a_{i-1,j} = \frac{k}{\Delta x}, \; a_{i,j+1} = \frac{k}{\Delta y}, \; a_{i,j-1} = \frac{k}{\Delta y}$$

Substituting the coefficients values; setting $S_0 = 0$, $S_1 = 0$ and d = 0 for no volume heat generation in the slab, we can derive the discretization equations for the interior nodes as

$$\frac{4k}{\Delta x}T_{i,j} = \frac{k}{\Delta x}\,T_{i+1,j} + \frac{k}{\Delta x}\,T_{i-1,j} + \frac{k}{\Delta y}T_{i,j+1} + \frac{k}{\Delta y}T_{i,j-1}$$

Simplifying,

$$T_{i+1,j} + T_{i-1,j} + \frac{\Delta x}{\Delta y}\left(T_{i,j+1} + T_{i,j-1}\right) - 4T_{i,j} = 0$$

Further setting $\Delta x = \Delta y$ for equal grid size distributions in both x- and y-directions, we obtain the discretization equations for all interior nodes for this problem as $T_{i+1,j} + T_{i-1,j} + T_{i,j+1} + T_{i,j-1} - 4T_{i,j} = 0$

Using this equation, the discretization equations for the nine interior nodes are obtained as follows:

$$\begin{aligned}
&\textit{Node 1} \quad T_2 + T_4 + 20 + 400 - 4T_1 = 0\\
&\textit{Node 2} \quad T_1 + T_5 + T_3 + 400 - 4T_2 = 0\\
&\textit{Node 3} \quad T_2 + T_6 + 20 + 400 - 4T_3 = 0\\
&\textit{Node 4} \quad 20 + T_1 + T_5 + T_7 - 4T_4 = 0\\
&\textit{Node 5} \quad T_2 + T_4 + T_8 + T_6 - 4T_5 = 0\\
&\textit{Node 6} \quad T_3 + T_5 + T_9 + 20 - 4T_6 = 0\\
&\textit{Node 7} \quad T_4 + 20 + 20 + T_8 - 4T_7 = 0\\
&\textit{Node 8} \quad T_5 + T_7 + 20 + T_9 - 4T_8 = 0\\
&\textit{Node 9} \quad T_6 + T_8 + 20 + 20 - 4T_9 = 0
\end{aligned}$$

Rearranging these equations, we get the linear system of equations as

$$\begin{aligned}
&-4T_1 + T_2 + 0 + T_4 + 0 + 0 + 0 + 0 + 0 = -420\\
&T_1 - 4T_2 + T_3 + 0 + T_5 + 0 + 0 + 0 + 0 = -400\\
&0 + T_2 - 4T_3 + 0 + 0 + T_6 + 0 + 0 + 0 = -420\\
&T_1 + 0 + 0 - 4T_4 + T_5 + 0 + T_7 + 0 + 0 = -20\\
&0 + T_2 + 0 + T_4 - 4T_5 + T_6 + 0 + T_8 + 0 = 0\\
&0 + 0 + T_3 + 0 + T_5 - 4T_6 + 0 + 0 + T_9 = -20\\
&0 + 0 + 0 + T_4 + 0 + 0 - 4T_7 + T_8 + 0 = -40\\
&0 + 0 + 0 + 0 + T_5 + 0 + T_7 - 4T_8 + T_9 = -20\\
&0 + 0 + 0 + 0 + 0 + T_6 + 0 + T_8 - 4T_9 = -40
\end{aligned}$$

The system of equations can be expressed in matrix notation as

$$\begin{bmatrix}
-4 & 1 & 0 & 1 & 0 & 0 & 0 & 0 & 0\\
1 & -4 & 1 & 0 & 1 & 0 & 0 & 0 & 0\\
0 & 1 & -4 & 0 & 0 & 1 & 0 & 0 & 0\\
1 & 0 & 0 & -4 & 1 & 0 & 1 & 0 & 0\\
0 & 1 & 0 & 1 & -4 & 1 & 0 & 1 & 0\\
0 & 0 & 1 & 0 & 1 & -4 & 0 & 0 & 1\\
0 & 0 & 0 & 1 & 0 & 0 & -4 & 0 & 0\\
0 & 0 & 0 & 0 & 1 & 0 & 1 & -4 & 1\\
0 & 0 & 0 & 0 & 0 & 1 & 0 & 1 & -4
\end{bmatrix}
\begin{Bmatrix} T_1\\ T_2\\ T_3\\ T_4\\ T_5\\ T_6\\ T_7\\ T_8\\ T_9 \end{Bmatrix}
=
\begin{Bmatrix} -420\\ -400\\ -420\\ -20\\ 0\\ -20\\ -40\\ -20\\ -40 \end{Bmatrix}$$

Using the Gauss elimination routine, the solution to this system of equations is obtained as

$$\begin{Bmatrix} T_1 \\ T_2 \\ T_3 \\ T_4 \\ T_5 \\ T_6 \\ T_7 \\ T_8 \\ T_9 \end{Bmatrix} = \begin{Bmatrix} 182.86°\text{C} \\ 220.16°\text{C} \\ 182.86°\text{C} \\ 91.25°\text{C} \\ 115.00°\text{C} \\ 91.25°\text{C} \\ 47.14°\text{C} \\ 57.32°\text{C} \\ 47.14°\text{C} \end{Bmatrix}$$

The heat transfer rate from the top surface is computed from the expression

$$q_T = \sum kA_y \frac{\Delta T}{\Delta y} = k\Delta x \cdot 1 \frac{(400 - T_1)}{\Delta y} + k\Delta x \cdot 1 \frac{(400 - T_2)}{\Delta y} + k\Delta x \cdot 1 \frac{(400 - T_3)}{\Delta y}$$

$$= 40.0 \times 0.5 \times \frac{(400 - 182.86)}{0.5} + 40.0 \times 0.5 \times \frac{(400 - 220.16)}{0.5} + 40.0 \times 0.5 \times \frac{(400 - 182.86)}{0.5}$$

$$= 24\,564.8\,\text{W}$$

The heat transfer rate from the left surface is computed from the expression

$$q_L = \sum kA_x \frac{\Delta T}{\Delta x} = k\Delta y \cdot 1 \frac{(20 - T_1)}{\Delta x} + k\Delta y \cdot 1 \frac{(20 - T_4)}{\Delta x} + k\Delta y \cdot 1 \frac{(20 - T_7)}{\Delta x}$$

$$= 40.0 \times 0.5 \times \frac{(20 - 182.86)}{0.5} + 40.0 \times 0.5 \times \frac{(20 - 91.25)}{0.5} + 40.0 \times 0.5 \times \frac{(20 - 47.14)}{0.5}$$

$$= -10\,450\,\text{W}$$

The heat transfer rate from the right surface is computed from the expression

$$q_R = \sum kA_x \frac{\Delta T}{\Delta x} = k\Delta y \cdot 1 \frac{(20 - T_3)}{\Delta x} + k\Delta y \cdot 1 \frac{(20 - T_6)}{\Delta x} + k\Delta y \cdot 1 \frac{(20 - T_9)}{\Delta x}$$

$$= 40.0 \times 0.5 \times \frac{(20 - 182.86)}{0.5} + 40.0 \times 0.5 \times \frac{(20 - 91.25)}{0.5} + 40.0 \times 0.5 \times \frac{(20 - 47.14)}{0.5}$$

$$= -10 \cdot 450\,\text{W}$$

The heat transfer rate from the bottom surface is computed from the expression

$$q_B = \sum kA_y \frac{\Delta T}{\Delta y} = k\Delta x \cdot 1 \frac{(20 - T_7)}{\Delta y} + k\Delta x \cdot 1 \frac{(20 - T_8)}{\Delta y} + k\Delta x \cdot 1 \frac{(20 - T_9)}{\Delta y}$$

$$= 40.0 \times 0.5 \times \frac{(20 - 47.14)}{0.5} + 40.0 \times 0.5 \times \frac{(20 - 57.32)}{0.5} + 40.0 \times 0.5 \times \frac{(20 - 47.14)}{0.5}$$

$$= -3664\,\text{W}$$

It can be noted that the heat transfer rates at the left and right surfaces are identical. This is due to the symmetric nature of the boundary conditions and geometry of the problem. The negative heat transfer rates indicate that heat is lost from these surfaces. The overall energy balance for the two-dimensional steady-state problem states that rate of energy inflow is balanced by the rate of energy outflow, i.e.,

$$E_{\text{in}} = E_{\text{out}}$$

In this problem energy inflow is given by the heat transfer rate at the top surface

$$E_{\text{in}} = q_T = 24\,564.8\,\text{W}$$

The energy outflow is given by the summation of all heat transfer rates at the left, right, and bottom surfaces

$$E_{\text{out}} = q_L + q_R + q_B = -10\,450 - 10\,450 - 3664$$

$$E_{\text{out}} = -24\,564.0\,\text{W}$$

The percent relative error in the overall energy balance is given as

$$\varepsilon_{\text{ae}} = \frac{E_{\text{in}} - E_{\text{out}}}{E_{\text{in}}} = \frac{24\,564.8 - 24\,564.0}{24\,564.8} \times 100$$

$$\varepsilon_{\text{ae}} = 0.003\%$$

It can be noted that there is an excellent agreement in the overall energy balance as expected, and this indicates the accuracy in the finite difference formulation and the solution of the system of equations. However, it should be remembered here that the even though the computed temperatures at the nodal points satisfy the overall energy balance, they do not necessarily give us the correct temperature distribution. This approximate temperature distribution can be improved further by performing a *grid refinement* study by reducing the grid size or using increased number of nodal points. Such grid refinement study will lead to a converged and more accurate solution. Another way of checking the accuracy of the approximate temperature distribution is by comparing this solution with that given by an exact solution for a simpler or limiting case problem for which an exact solution exists.

6.2 BOUNDARY CONDITIONS

When the boundary condition is of the **first kind**, i.e., **of constant surface value**, we can simply assign the known value of temperature to the boundary nodes and solve the system of equations as was done in the previous example. However, for boundary conditions of the **second** or **third kind**, we need to use discretization equations derived for those boundary nodes. Figure 6.3 shows typical two-dimensional boundary nodes, such as **boundary surface node, exterior corner node**, and **interior corner node**.

Figure 6.4 shows control volumes for boundary nodes on all four sides of a two-dimensional coordinate system. It can be noted that only one-half of a control volume surrounds the boundary node i, as marked by the shading in the figures.

Let us show the derivation of one such boundary node based on the control volume formulation. Consider the grid point located on the **right-hand side boundary** as shown in Figure 6.4a. Integrating the governing Equation 6.1 over this half control volume, we get

$$\int_{i,j-\frac{1}{2}}^{i,j+\frac{1}{2}} \int_{i-\frac{1}{2},j}^{i,j} \frac{\partial}{\partial x}\left(\Gamma \frac{\partial \phi}{\partial x}\right) dxdy + \int_{i,j-\frac{1}{2}}^{i,j+\frac{1}{2}} \int_{i-\frac{1}{2},j}^{i,j} \frac{\partial}{\partial y}\left(\Gamma \frac{\partial \phi}{\partial y}\right) dxdy + \int_{i,j-\frac{1}{2}}^{i,j+\frac{1}{2}} \int_{i-\frac{1}{2},j}^{i,j} S\, dxdy$$

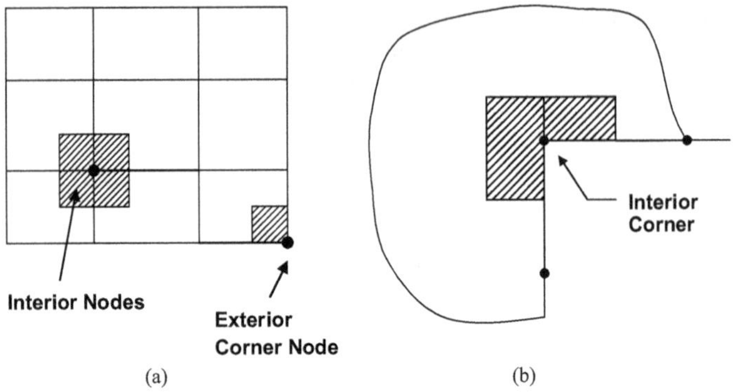

FIGURE 6.3 Control volumes for boundary nodes in two-dimensional problems.

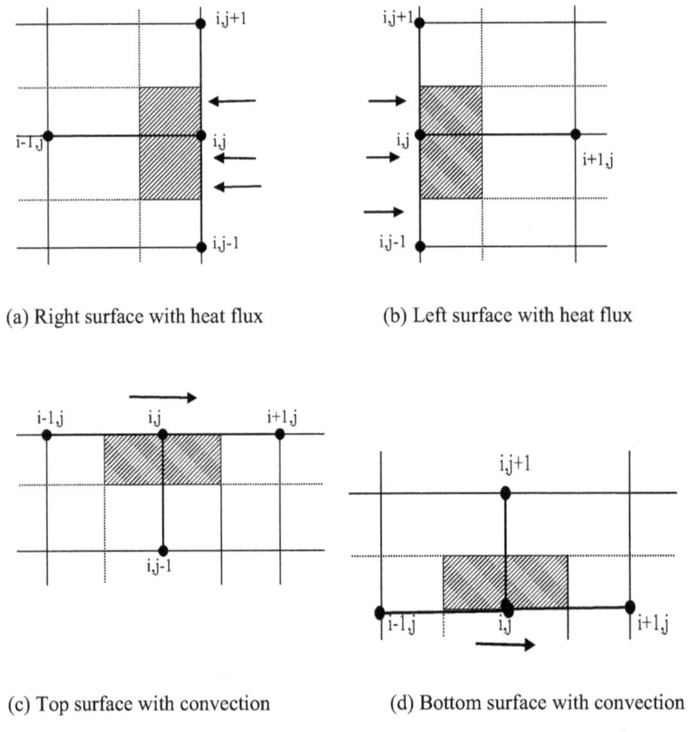

FIGURE 6.4 Control volumes for boundary nodes on two-dimensional plane surfaces with different types of boundary conditions.

or

$$\left[\left(\Gamma\frac{\partial\phi}{\partial x}\right)_{i,j}-\left(\Gamma\frac{\partial\phi}{\partial x}\right)_{i-\frac{1}{2},j}\right]\Delta y+\left[\left(\Gamma\frac{\partial\phi}{\partial y}\right)_{i,j+\frac{1}{2}}-\left(\Gamma\frac{\partial\phi}{\partial y}\right)_{i,j-\frac{1}{2}}\right]\frac{\Delta x}{2}+\Delta y\left(\int_{i-\frac{1}{2},j}^{i,j} S\Delta x\right) \tag{6.3}$$

The first term in the equation is described by the constant surface flux condition as

$$\left(\Gamma\frac{\partial\phi}{\partial x}\right)_{i,j}=f''_{\text{rs}} \tag{6.4}$$

We assume piecewise-linear profiles to approximate gradients at the faces of the control volume as follows:

$$\left(\Gamma\frac{\partial\phi}{\partial y}\right)_{i,j+\frac{1}{2}} = \Gamma_{i,j+\frac{1}{2}}\frac{\phi_{i,j+1}-\phi_{i,j}}{\Delta y} \tag{6.5a}$$

$$\left(\Gamma\frac{\partial\phi}{\partial y}\right)_{i,j-\frac{1}{2}} = \Gamma_{i,j-\frac{1}{2}}\frac{\phi_{i,j}-\phi_{i,j-1}}{\Delta y} \tag{6.5b}$$

$$\left(\Gamma\frac{\partial\phi}{\partial x}\right)_{i-\frac{1}{2},j} = \Gamma_{i-\frac{1}{2},j}\frac{\phi_{i,j}-\phi_{i-1,j}}{\Delta x} \tag{6.5c}$$

A stepwise constant profile for the dependent variable is assumed in the linearized form of the source term

$$S = S_0 + S_1\phi_i \tag{6.6}$$

Substituting Equations 6.4–6.6 into Equation 6.3, we get

$$\begin{aligned}&\left(\Gamma_{i-\frac{1}{2},j}\frac{\Delta y}{\Delta x}+\frac{\Gamma_{i,j+\frac{1}{2}}}{2}\frac{\Delta x}{\Delta y}+\frac{\Gamma_{i,j-\frac{1}{2}}}{2}\frac{\Delta x}{\Delta y}-\frac{S_1}{2}\Delta x\Delta y\right)\phi_{i,j}\\&=\Gamma_{i-\frac{1}{2},j}\frac{\Delta y}{\Delta x}\phi_{i-1,j}+\frac{\Gamma_{i,j+\frac{1}{2}}}{2}\frac{\Delta x}{\Delta y}\phi_{i,j+1}+\frac{\Gamma_{i,i-\frac{1}{2}}}{2}\frac{\Delta x}{\Delta y}\phi_{i,j-1}+\frac{S_0}{2}\Delta x\Delta y+f''_{rs}\Delta y\end{aligned} \tag{6.7}$$

Simplifying, we get

$$a_{i,j}\phi_{i,j} = a_{i-1,j}\phi_{i-1,j} + a_{i,j+1}\phi_{i,j+1} + a_{i,j-1}\phi_{i,j-1} + d \tag{6.8}$$

where

$$a_{i-1,j} = \Gamma_{i-\frac{1}{2},j}\frac{\Delta y}{\Delta x} \tag{6.9a}$$

$$a_{i,j+1} = \Gamma_{i,j+\frac{1}{2}}\frac{\Delta x}{2\Delta y} \tag{6.9b}$$

$$a_{i,j-1} = \Gamma_{i,j-\frac{1}{2}}\frac{\Delta x}{2\Delta y} \tag{6.9c}$$

$$a_{i,j} = a_{i-1,j} + a_{i,j+1} + a_{i,j-1} - \frac{S_1}{2}\Delta x\Delta y \tag{6.9d}$$

$$d = \frac{S_0}{2}\Delta x\Delta y + f''_{rs}\Delta y \tag{6.9e}$$

Zero Surface Flux

For the case of a **zero surface flux** or **along a line of symmetry**, the appropriate boundary condition is defined as

$$\Gamma\frac{\partial \phi}{\partial x} = 0 \quad \text{or} \quad f''_{rs} = 0 \tag{6.10}$$

We can obtain the appropriate finite difference for this case by setting $f''_{rs} = 0$ in Equation 6.10.

Similarly, we can derive the discretization equations for other boundary nodes as shown in. A summary of these equations is given in Table 6.2.

TABLE 6.2
A Summary of Finite Difference Equations for Boundary Nodes with Constant Surface Flux

Physical Situations — **Finite Difference Formula**

a. Grid point located at the right-hand side boundary

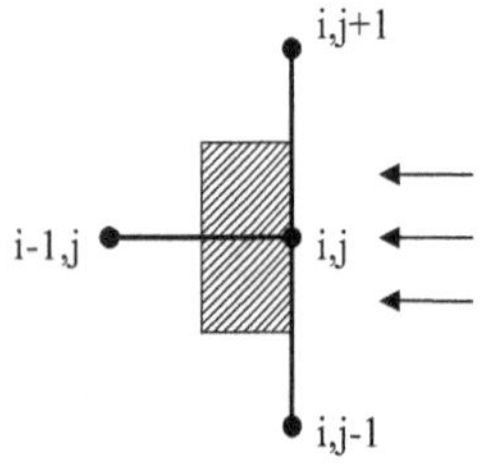

$$a_{i,j}\phi_{i,j} = a_{i-1,j}\phi_w + a_{i,j+1}\phi_{i,j+1} + a_{i,j-1}\phi_{i,j-1s} + d$$

$$a_{i-1,j} = \Gamma_{i-\frac{1}{2},j}\frac{\Delta y}{\Delta x}, \quad a_{i,j+1} = \frac{\Gamma_{i,j+\frac{1}{2}}}{2}\frac{\Delta x}{\Delta y}, \quad a_{i,j-1} = \frac{\Gamma_{i,j-\frac{1}{2}}}{2}\frac{\Delta x}{\Delta y}$$

$$a_{i,j} = a_{i-1,j} + a_{i,j+1} + a_{i,j-1} - \frac{S_1}{2}\Delta x\Delta y$$

$$d = \frac{S_0}{2}\Delta x \ \Delta y + f_{rs}^{||}\Delta y$$

b. Grid point located at the left-hand side boundary

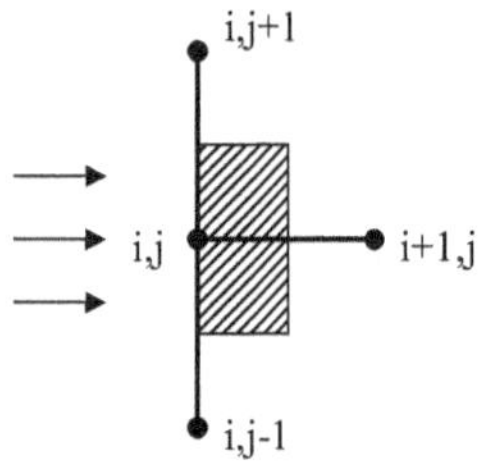

$$a_{i,j}\phi_{i,j} = a_{i+1,j}\phi_w + a_{i,j+1}\phi_{i,j+1} + a_{i,j-1}\phi_{i,j-1s} + d$$

$$a_{i+1,j} = \Gamma_{i+\frac{1}{2},j}\frac{\Delta y}{\Delta x}, \quad a_{i,j+1} = \frac{\Gamma_{i,j+\frac{1}{2}}}{2}\frac{\Delta x}{\Delta y}, \quad a_{i,j-1} = \frac{\Gamma_{i,j-\frac{1}{2}}}{2}\frac{\Delta x}{\Delta y}$$

$$a_{i,j} = a_{i-1,j} + a_{i,j+1} + a_{i,j-1} - \frac{S_1}{2}\Delta x\Delta y$$

$$d = \frac{S_0}{2}\Delta x \ \Delta y + f_{ls}^{||}\Delta y$$

c. Grid point located at the top boundary

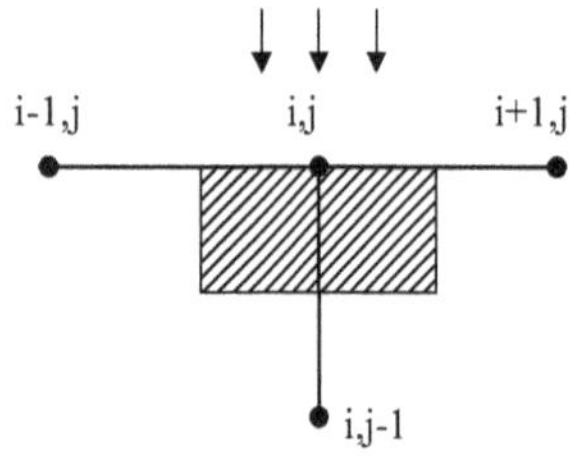

$$a_{i,j}\phi_{i,j} = a_{i+1,j}\phi_{i+1,j} + a_{i-1,j}\phi_{-1,j} + a_{i,j-1}\phi_{i,j-1s} + d$$

$$a_{i+1,j} = \frac{\Gamma_{i+\frac{1}{2},j}}{2}\frac{\Delta y}{\Delta x}, \quad a_{i-1,j} = \frac{\Gamma_{i-\frac{1}{2},j}}{2}\frac{\Delta y}{\Delta x}, a_{i,j-1} = \Gamma_{i,j-\frac{1}{2}}\frac{\Delta x}{\Delta y}$$

$$a_{i,j} = a_{i+1,j} + a_{i-1,j+1} + a_{i,j-1} - \frac{S_1}{2}\Delta x\Delta y$$

$$d = \frac{S_0}{2}\Delta x \ \Delta y + f_{ts}^{||}\Delta y$$

d. Grid point located at bottom boundary

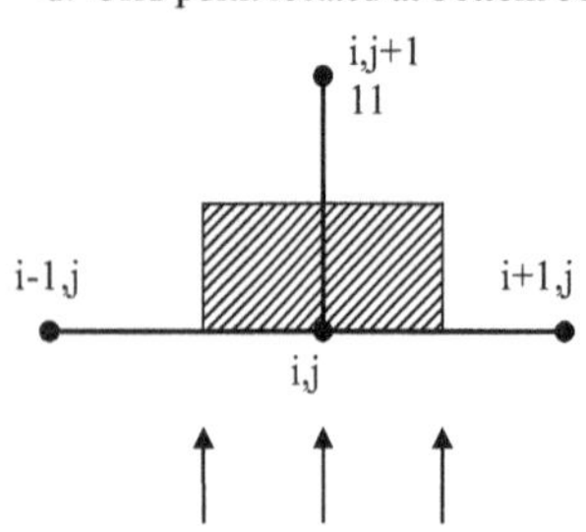

$$a_{i,j}\phi_{i,j} = a_{i+1,j}\phi_{i+1,j} + a_{i-1,j}\phi_{-1,j} + a_{i,j-1}\phi_{i,j-1s} + d$$

$$a_{i+1,j} = \frac{\Gamma_{i+\frac{1}{2},j}}{2}\frac{\Delta y}{\Delta x}, \quad a_{i-1,j} = \frac{\Gamma_{i-\frac{1}{2},j}}{2}\frac{\Delta y}{\Delta x}, a_{i,j+1} = \Gamma_{i,j+\frac{1}{2}}\frac{\Delta x}{\Delta y}$$

$$a_{i,j} = a_{i+1,j} + a_{i-1,j+1} + a_{i,j-1} - \frac{S_1}{2}\Delta x\Delta y$$

$$d = \frac{S_0}{2}\Delta x \ \Delta y + f_{bs}^{||}\Delta y$$

Mixed Boundary Condition Another possible boundary condition is the boundary condition of the **third kind** or **the mixed boundary condition** defined as

$$-\Gamma\frac{\partial\phi}{\partial x}=h(\phi-\phi_\infty) \tag{6.11a}$$

or

$$f_{\text{rs}}''=h(\phi_\infty-\phi) \tag{6.11b}$$

Substituting Equation 6.11b into Equation 6.7, we get

$$\left(\Gamma_{i-\frac{1}{2},j}\frac{\Delta y}{\Delta x}+\frac{\Gamma_{i,j+\frac{1}{2}}}{2}\frac{\Delta x}{\Delta y}+\frac{\Gamma_{i,j-\frac{1}{2}}}{2}\frac{\Delta x}{\Delta y}-\frac{S_1}{2}\Delta x\Delta y\right)\phi_{i,j}$$

$$=\Gamma_{i-\frac{1}{2},j}\frac{\Delta y}{\Delta x}\phi_{i-1,j}+\frac{\Gamma_{i,j+\frac{1}{2}}}{2}\frac{\Delta x}{\Delta y}\phi_{i,j+1}+\frac{\Gamma_{i,j-\frac{1}{2}}}{2}\frac{\Delta x}{\Delta y}\phi_{i,j-1}+\frac{S_0}{2}\Delta x\Delta y+h(\phi_\infty-\phi_{i,j})\Delta y$$

Rearranging

$$\left(\Gamma_{i-\frac{1}{2},j}\frac{\Delta y}{\Delta x}+\frac{\Gamma_{i,j+\frac{1}{2}}}{2}\frac{\Delta x}{\Delta y}+\frac{\Gamma_{i,j-\frac{1}{2}}}{2}\frac{\Delta x}{\Delta y}-\frac{S_1}{2}\Delta x\Delta y+h\Delta y\right)\phi_{i,j}$$

$$=\Gamma_{i-\frac{1}{2},j}\frac{\Delta y}{\Delta x}\phi_{i-1,j}+\frac{\Gamma_{i,j+\frac{1}{2}}}{2}\frac{\Delta x}{\Delta y}\phi_{i,j+1}+\frac{\Gamma_{i,j-\frac{1}{2}}}{2}\frac{\Delta x}{\Delta y}\phi_{i,j-1}+\frac{S_0}{2}\Delta x\Delta y+h\phi_\infty\Delta y$$

Simplifying, we get

$$a_{i,j}\phi_{i,j}=a_{i-1,j}\phi_{i-1,j}+a_{i,j+1}\phi_{i,j+1}+a_{i,j-1}\phi_{i,j-1}+d \tag{6.12}$$

where

$$a_{i-1,j}=\Gamma_{i-\frac{1}{2},j}\frac{\Delta y}{\Delta x} \tag{6.13a}$$

$$a_{i,j+1}=\Gamma_{i,j+\frac{1}{2}}\frac{\Delta x}{2\Delta y} \tag{6.13b}$$

$$a_{i,j-1}=\Gamma_{i,j-\frac{1}{2}}\frac{\Delta x}{2\Delta y} \tag{6.13c}$$

$$a_{i,j}=a_{i-1,j}+a_{i,j+1}+a_{i,j-1}-\frac{S_1}{2}\Delta x\Delta y+h\Delta y \tag{6.13d}$$

$$d=\frac{S_0}{2}\Delta x\Delta y+h\phi_\infty\Delta y \tag{6.13e}$$

For uniform grid size distributions, $\Delta x=\Delta y$, and constant transport property, Γ, Equations 6.12 and 6.13 are simplified to

$$\left(\frac{h\Delta x}{\Gamma}+2-\frac{S_1}{2\Gamma}\Delta x^2\right)\phi_{i,j}=\frac{1}{2}\left(2\phi_{i-1,j}+\phi_{i,j+1}+\phi_{i,j-1}\right)+\frac{S_0}{2\Gamma}\Delta x^2+\frac{h\Delta x}{\Gamma}\phi_\infty \tag{6.14}$$

For the case with no source term, we can set $S_1 = 0$ and $S_0 = 0$, and Equation 6.14 transforms into

$$\left(\frac{h\Delta x}{\Gamma}+2\right)\phi_{i,j}=\frac{1}{2}\left(2\phi_{i-1,j}+\phi_{i,j+1}+\phi_{i,j-1}\right)+\frac{h\Delta x}{\Gamma}\phi \tag{6.15}$$

Similarly, we can derive the discretization equations for other boundary nodes as shown in Figure 6.2. A summary of these equations is given in Table 6.3.

TABLE 6.3
A Summary of Finite Difference Equations for Boundary Nodes with Mixed Condition

Physical Situations	Finite Difference Formula
a. Grid point located at the right-hand side boundary 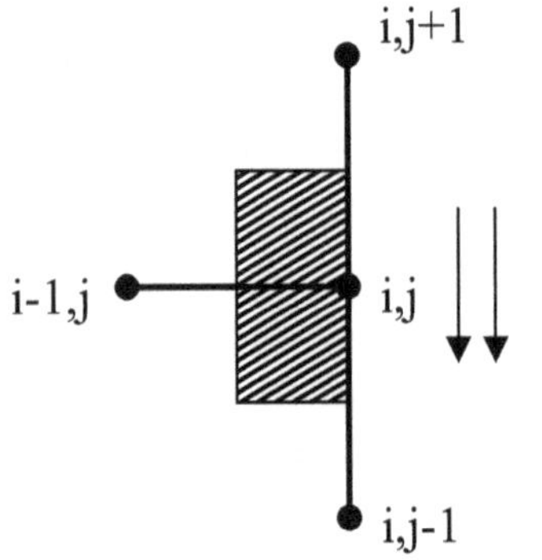	$\left(\frac{h\Delta x}{\Gamma}+2\right)\phi_{i,j}=\frac{1}{2}\left(2\phi_{i-1,j}+\phi_{i,j+1}+\phi_{i,j-1}\right)+\frac{h\Delta x}{\Gamma}\phi_\infty$
b. Grid point located at the left-hand side boundary 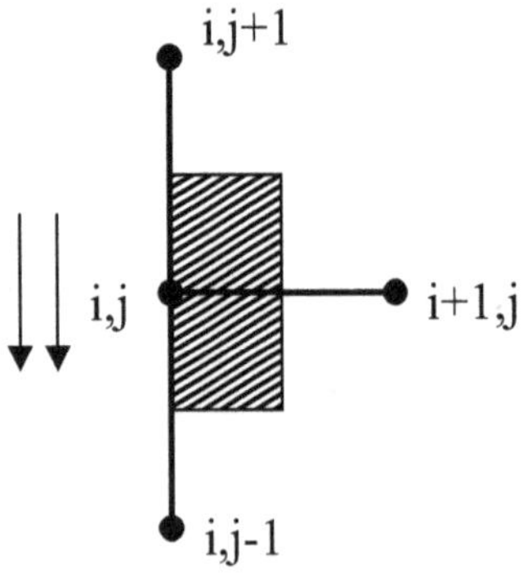	$\left(\frac{h\Delta x}{\Gamma}+2\right)\phi_{i,j}=\frac{1}{2}\left(2\phi_{i+1,j}+\phi_{i,j+1}+\phi_{i,j-1}\right)+\frac{h\Delta x}{\Gamma}\phi_\infty$
c. Grid point located at the top boundary 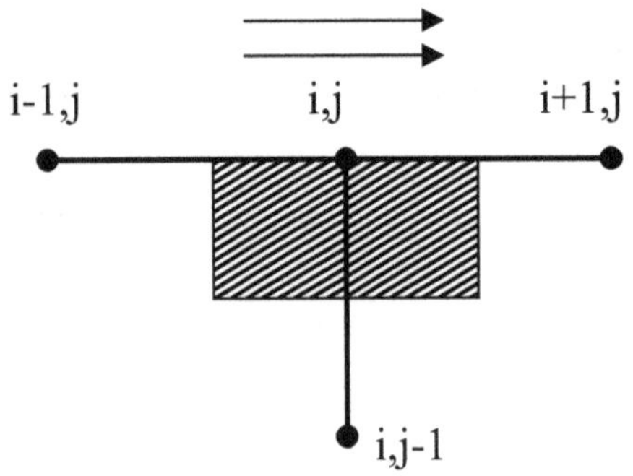	$\left(\frac{h\Delta x}{\Gamma}+2\right)\phi_{i,j}=\frac{1}{2}\left(2\phi_{i,j-1}+\phi_{i+1,j}+\phi_{i-1,j}\right)+\frac{h\Delta x}{\Gamma}\phi_\infty$

(Continued)

TABLE 6.3 (*Continued*)
A Summary of Finite Difference Equations for Boundary Nodes with Mixed Condition

Physical Situations	Finite Difference Formula
d. Grid point located at bottom boundary 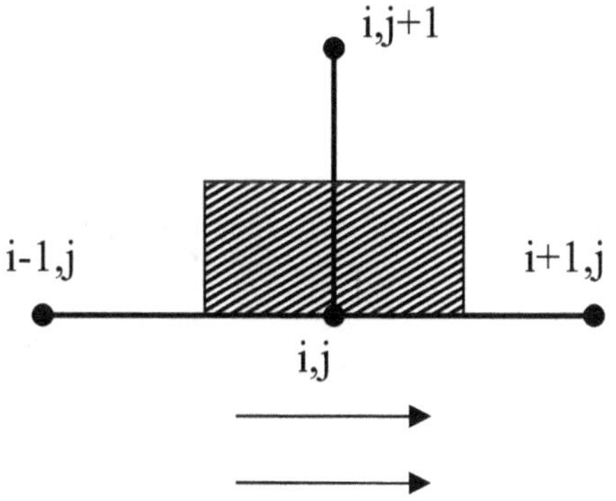	$\left(\frac{h\Delta x}{\Gamma}+2\right)\phi_{i,j}=\frac{1}{2}\left(2\phi_{i,j+1}+\phi_{i+1,j}+\phi_{i-1,j}\right)+\frac{h\Delta x}{\Gamma}\phi_\infty$
e. Convective boundary node as above with uniform heat generation	$\left(\frac{h\Delta x}{k}+2\right)\phi_{i,j}=\frac{1}{2}\left(2\phi_{i,j+1}+\phi_{i+1,j}+\phi_{i-1,j}\right)+\frac{h\Delta x}{\Gamma}\phi_\infty+\frac{\bar{S}(\Delta x)^2}{k}$

6.2.1 Corner Boundary Nodes

The procedure outlined for boundary nodes on a plane boundary surface can be used to derive the discretization equations for corner boundary nodes. Let us show the derivation of one such equation for the northeast exterior corner node shown in Figure 6.5 below.

Integrating the governing equation (Equation 6.1) over the quarter control volume surrounding the node (i, j), we get

$$\int_{i,j-\frac{1}{2}}^{i,j}\int_{i-\frac{1}{2},j}^{i,j}\frac{\partial}{\partial x}\left(\Gamma\frac{\partial\phi}{\partial x}\right)dxdy+\int_{i,j-\frac{1}{2}}^{i,j}\int_{i-\frac{1}{2},j}^{i,j}\frac{\partial}{\partial y}\left(\Gamma\frac{\partial\phi}{\partial y}\right)dxdy+\int_{i,j-\frac{1}{2}}^{i,j}\int_{i-\frac{1}{2},j}^{i,j}Sdxdy \tag{6.16}$$

or

$$\left[\left(\Gamma\frac{\partial\phi}{\partial x}\right)_{i,j}-\left(\Gamma\frac{\partial\phi}{\partial x}\right)_{i-\frac{1}{2},j}\right]\frac{\Delta y}{2}+\left[\left(\Gamma\frac{\partial\phi}{\partial y}\right)_{i,j}-\left(\Gamma\frac{\partial\phi}{\partial y}\right)_{i,j-\frac{1}{2}}\right]\frac{\Delta x}{2}+\frac{\Delta y}{2}\left(\int_{i-\frac{1}{2},j}^{i,j}S\Delta x\right) \tag{6.17}$$

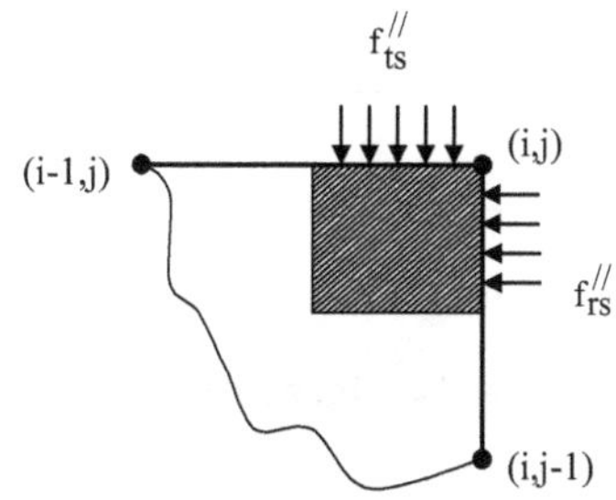

FIGURE 6.5 Control volume for an exterior corner node.

The first and the third terms in the equation are described by the constant surface flux condition as

$$\left(\Gamma \frac{\partial \phi}{\partial x}\right)_{i,j} = f''_{\rm rs} \tag{6.18a}$$

and

$$\left(\Gamma \frac{\partial \phi}{\partial y}\right)_{i,j} = f''_{\rm ts} \tag{6.18b}$$

We assume piecewise-linear profiles to approximate gradients at the faces of the control volume as

$$\left(\Gamma \frac{\partial \phi}{\partial y}\right)_{i,j-\frac{1}{2}} = \Gamma_{i,j-\frac{1}{2}} \frac{\phi_{i,j} - \phi_{i,j-1}}{\Delta y} \tag{6.19a}$$

$$\left(\Gamma \frac{\partial \phi}{\partial x}\right)_{i-\frac{1}{2},j} = \Gamma_{i-\frac{1}{2},j} \frac{\phi_{i,j} - \phi_{i-1,j}}{\Delta x} \tag{6.19b}$$

A stepwise constant profile for the dependent variable is assumed in the linearized form of the source term

$$S = S_0 + S_1 \phi_i \tag{6.20}$$

Substituting Equations 6.18 to 6.20 into Equation 6.17, we get

$$\left(\Gamma_{i-\frac{1}{2},j} \frac{\Delta y}{2\Delta x} + \Gamma_{i,j-\frac{1}{2}} \frac{\Delta x}{2\Delta y} - S_1 \frac{\Delta x \Delta y}{4}\right)\phi_{i,j} = \Gamma_{i-\frac{1}{2},j} \frac{\Delta y}{2\Delta x}\phi_{i-1,j} + \frac{\Gamma_{i,i-\frac{1}{2}}}{2}\frac{\Delta x}{\Delta y}\phi_{i,j-1} + \frac{S_0}{4}\Delta x \Delta y + f''_{rs}\frac{\Delta y}{2} + f''_{ts}\frac{\Delta x}{2} \tag{6.21}$$

Simplifying, we get

$$a_{i,j}\phi_{i,j} = a_{i-1,j}\phi_{i-1,j} + a_{i,j-1}\phi_{i,j-1} + d \tag{6.22}$$

where

$$a_{i-1,j} = \Gamma_{i-\frac{1}{2},j} \frac{\Delta y}{2\Delta x} \tag{6.23a}$$

$$a_{i,j-1} = \Gamma_{i,j-\frac{1}{2}} \frac{\Delta x}{2\Delta y} \tag{6.23b}$$

$$a_{i,j} = a_{i-1,j} + a_{i,j-1} - \frac{S_1}{4}\Delta x \Delta y \tag{6.23c}$$

$$d = \frac{S_0}{4}\Delta x \Delta y + f''_{\rm rs}\frac{\Delta y}{2} + f''_{\rm ts}\frac{\Delta x}{2} \tag{6.23d}$$

Zero Surface Flux For the case of a **zero surface flux** or **along a line of symmetry**, the appropriate boundary condition is defined as

$$\Gamma \frac{\partial \phi}{\partial x} = 0 \quad \text{or} \quad f''_{\rm rs} = 0 \tag{6.24a}$$

$$\Gamma\frac{\partial\phi}{\partial y}=0 \quad or \quad f''_{ts}=0 \tag{6.24b}$$

We can obtain the appropriate finite difference equation for this case by setting $f''_{rs}=0$, $f''_{ts}=0$ in Equation 6.23.

Mixed Boundary Condition Another possible boundary condition is the boundary condition of the **third kind** or **the mixed boundary condition** defined as

$$-\Gamma\frac{\partial\phi}{\partial x}=h(\phi-\phi_\infty) \tag{6.25a}$$

or

$$f''_{rs}=h(\phi_\infty-\phi) \tag{6.25b}$$

and

$$-\Gamma\frac{\partial\phi}{\partial y}=h(\phi-\phi_\infty) \tag{6.26a}$$

or

$$f''_{ts}=h(\phi_\infty-\phi) \tag{6.26b}$$

Substituting Equations 6.25 and 6.26 into Equation 6.21, we get

$$\left(\Gamma_{i-\frac{1}{2},j}\frac{\Delta y}{2\Delta x}+\frac{\Gamma_{i,i-\frac{1}{2}}}{2}\frac{\Delta x}{\Delta y}-\frac{S_1}{4}\Delta x\Delta y\right)\phi_{i,j}=\Gamma_{i-\frac{1}{2},j}\frac{\Delta y}{2\Delta x}\phi_{i-1,j}+\frac{\Gamma_{i,i-\frac{1}{2}}}{2}\frac{\Delta x}{\Delta y}\phi_{i,j-1}$$
$$+\frac{S_0}{4}\Delta x\Delta y+h_{rs}\left(\phi_\infty-\phi_{i,j}\right)\Delta y+h_{ts}\left(\phi_\infty-\phi_{i,j}\right)\Delta x \tag{6.27}$$

Rearranging

$$\left(\Gamma_{i-\frac{1}{2},j}\frac{\Delta y}{2\Delta x}+\frac{\Gamma_{i,i-\frac{1}{2}}}{2}\frac{\Delta x}{\Delta y}-\frac{S_1}{4}\Delta x\Delta y+h_{rs}\frac{\Delta y}{2}+h_{ts}\frac{\Delta x}{2}\right)\phi_{i,j}$$
$$=\Gamma_{i-\frac{1}{2},j}\frac{\Delta y}{2\Delta x}\phi_{i-1,j}+\frac{\Gamma_{i,i-\frac{1}{2}}}{2}\frac{\Delta x}{\Delta y}\phi_{i,j-1}+\frac{S_0}{4}\Delta x\Delta y+h''_{rs}\phi_\infty\frac{\Delta y}{2}+h''_{ts}\phi_\infty\frac{\Delta x}{2} \tag{6.28}$$

Simplifying, we get

$$a_{i,j}\phi_{i,j}=a_{i-1,j}\phi_{i-1,j}+a_{i,j-1}\phi_{i,j-1}+d \tag{6.29}$$

where

$$a_{i-1,j}=\Gamma_{i-\frac{1}{2},j}\frac{\Delta y}{2\Delta x} \tag{6.30a}$$

$$a_{i,j-1}=\Gamma_{i,j-\frac{1}{2}}\frac{\Delta x}{2\Delta y} \tag{6.30b}$$

$$a_{i,j} = a_{i-1,j} + a_{i,j-1} - \frac{S_1}{4}\Delta x \Delta y + h_{rs}\frac{\Delta y}{2} + h_{ts}\frac{\Delta x}{2} \quad (6.30c)$$

$$d = \frac{S_0}{4}\Delta x \Delta y + h_{rs}\frac{\Delta y}{2}\phi_\infty + h_{ts}\frac{\Delta x}{2}\phi_\infty \quad (6.30d)$$

For uniform grid size distributions, $\Delta x = \Delta y$, and using constant transport property Γ and constant coefficients $h''_{rs} = h''_{ts} = h$, Equation 6.30 is simplified to

$$a_{i-1,j} = \frac{\Gamma}{2} \quad (6.31a)$$

$$a_{i,j-1} = \frac{\Gamma}{2} \quad (6.31b)$$

$$a_{i,j} = a_{i-1,j} + a_{i,j-1} - \frac{S_1}{4}\Delta x^2 + h\Delta x \quad (6.31c)$$

$$d = \frac{S_0}{4}\Delta x^2 + h\phi_\infty \Delta x \quad (6.31d)$$

For the case with no source term, we can set $S_1 = 0$ and $S_0 = 0$ to get the simplified form

$$2\left(\frac{h\Delta x}{\Gamma} + 1\right)\phi_{i,j} = \phi_{i-1,j} + \phi_{i,j-1} + \frac{2h\Delta x}{\Gamma}\phi_\infty \quad (6.32)$$

Similarly, we can derive the discretization equations for other corner boundary nodes. A summary of these equations is given in Table 6.4.

TABLE 6.4
A Summary of Finite Difference Equations for Corner Boundary Nodes with Mixed Condition

Physical Situations	Finite Difference Formula
Exterior Corner Nodes	
a. **Northeast Corner**	
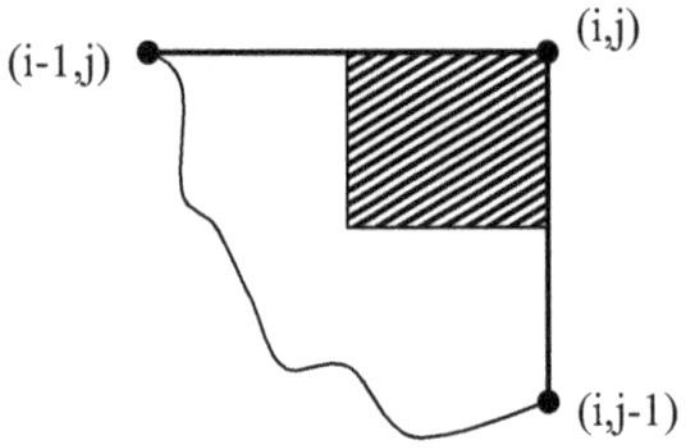	$2\left(\frac{h\Delta x}{k} + 1\right)T_{i,j} = \left(T_{i-1,j} + T_{i,j-1}\right) + 2\frac{h\Delta x}{k}T_\infty$

(*Continued*)

TABLE 6.4 (*Continued*)

A Summary of Finite Difference Equations for Corner Boundary Nodes with Mixed Condition

Physical Situations	Finite Difference Formula
b. **Southeast Corner** 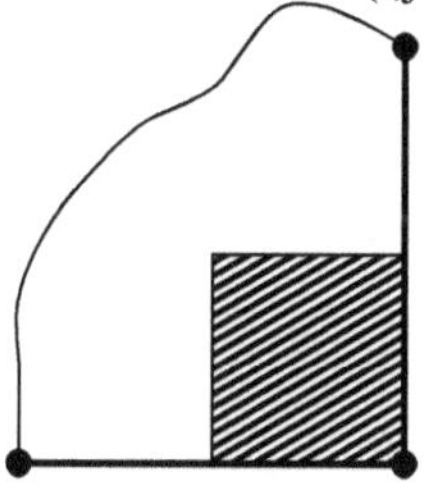	$2\left(\frac{h\Delta x}{k}+1\right)T_{i,j}=\left(T_{i,j+1}+T_{i-1,j}\right)+2\frac{h\Delta x}{k}T_\infty$
c. **Northwest Corner** 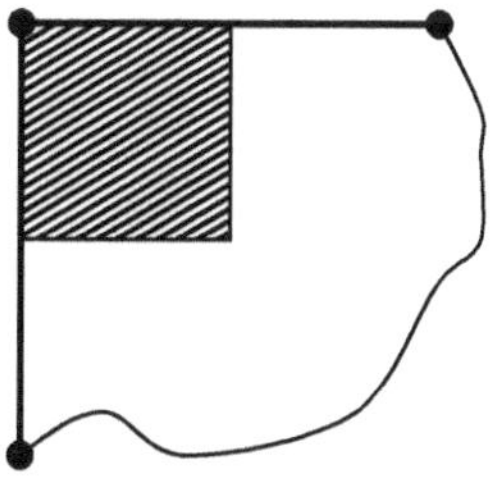	$2\left(\frac{h\Delta x}{k}+1\right)T_{i,j}=\left(T_{i+1,j}+T_{i,j-1}\right)+2\frac{h\Delta x}{k}T_\infty$
d. **Southwest Corner** 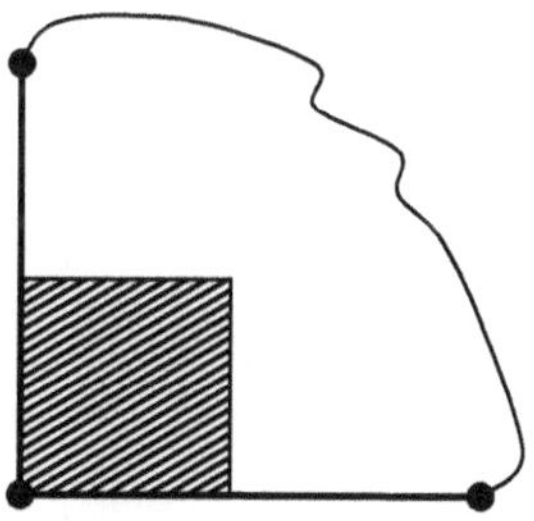	$2\left(\frac{h\Delta x}{k}+1\right)T_{i,j}=\left(T_{i,j+1}+T_{i+1,j}\right)+2\frac{h\Delta x}{k}T_\infty$
Interior Corner Nodes	
e. **Northwest Corner** 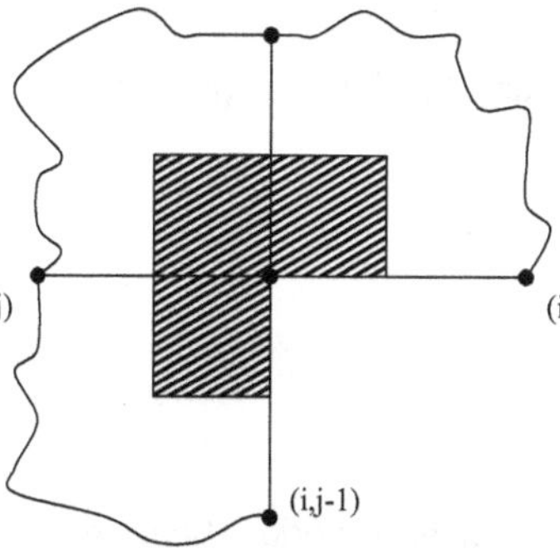	$2\left(3+\frac{h\Delta x}{k}+1\right)T_{i,j}=T_{i+1,j}+2T_{i-1,j}+2T_{i,j+1}+T_{i,j-1}+2\frac{h\Delta x}{k}T_\infty$

Example 6.2: Two-Dimensional Steady-State Heat Conduction with Mixed Boundary Conditions

Consider a 1.5 m by 1.5 m square slab with the top and the left surfaces maintained at a high temperature, $T_H = 600°C$, and a low temperature, $T_C = 20°C$, respectively. The right and the bottom surfaces are subjected to convection conditions with $h_c = 30\,W/m^2\,°C$ and $T_\infty = 25°C$. The material conductivity is $k = 30$ W/m °C. Apply the finite difference method to calculate the two-dimensional temperature distribution in the slab based on the grid distribution shown in the figure. Estimate the heat transfer rates at all surfaces and show the overall energy balance.

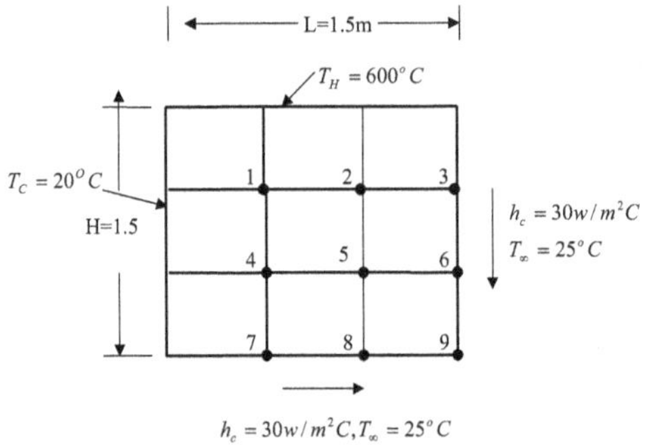

The mathematical statement of the problem is as follows.

Governing Equation

$$\frac{\partial}{\partial x}\left(k\frac{\partial T}{\partial x}\right) + \frac{\partial}{\partial y}\left(k\frac{\partial T}{\partial y}\right) = 0$$

Boundary Conditions

1. $x = 0, \quad T(0,y) = T_C$

2. $x = L, \quad -k\left.\frac{\partial T}{\partial x}\right|_{x=L} = h_c\left(T|_{x=L} - T_\infty\right)$

3. $y = 0, \quad k\left.\frac{\partial T}{\partial y}\right|_{y=0} = h\left(T|_{y=0} - T_\infty\right)$

4. $y = H, \quad T(x, H) = T_H$

Solution

For the prescribed grid size distribution with uniform grid sizes and number of division $N_{div} = 3$, we can estimate the following parameters as

$$\Delta x = \frac{L}{N_{xdiv}} = \frac{1.5\text{m}}{3} = 0.5\text{m}, \quad \Delta y = \frac{H}{N_{ydiv}} = \frac{1.5\text{m}}{3} = 0.5\text{m}$$

$$\frac{h_c \Delta x}{k} = \frac{30.0 \times 0.5}{30.0} = 0.5$$

In the next step, we select appropriate discretization equations for each node. There are four interior nodes 1, 2, 4, and 5 for which the appropriate finite difference equation is given by

$$a_{i,j} = a_{i+1,j}T_{i+1,j} + a_{i-1,j}T_{i-1,j} + a_{i,j+1}T_{i,j+1} + a_{i,j-1}T_{i,j-1} + d$$

$$a_{i+1,j} = a_{i-1,j} = a_{i,j+1} = a_{i,j-1} = 1, \quad a_{i,j} = 4$$

$$d = \overline{S}\Delta x^2/k$$

With no heat source term, $\overline{S} = 0$, and $d = 0$, the equation is expressed in a simplified form as

$$4T_{i,j} = T_{i+1,j} + T_{i-1,j} + T_{i,j+1} + T_{i,j-1}$$

Using this equation, the finite difference equations for the interior nodes 1, 2, 4, and 5 are obtained as

$$\textit{Node 1} \quad 4T_1 = T_2 + T_\mathrm{C} + T_\mathrm{H} + T_4$$

$$\textit{Node 2} \quad 4T_2 = T_3 + T_1 + T_\mathrm{H} + T_5$$

$$\textit{Node 4} \quad 4T_4 = T_5 + T_\mathrm{C} + T_1 + T_7$$

$$\textit{Node 5} \quad 4T_5 = T_6 + T_4 + T_2 + T_8$$

Nodes 3 and 6 are convective boundary nodes on right side surface, and we use the appropriate discretization equation which is given as

$$a_{i,j}T_{i,j} = a_{i-1,j}T_{i-1,j} + a_{i,j+1}T_{i,j+1} + a_{i,j-1}T_{i,j-1} + d$$

$$a_{i-1,j} = k_{i-\frac{1}{2},j}\frac{\Delta y}{\Delta x}, \quad a_{i,j+1} = \frac{k_{i,j+\frac{1}{2}}}{2}\frac{\Delta x}{\Delta y}, \quad a_{i,j-1} = \frac{k_{i,j-\frac{1}{2}}}{2}\frac{\Delta x}{\Delta y}$$

$$a_{i,j} = a_{i-1,j} + a_{i,j+1} + a_{i,j-1} - \frac{S_1}{2}\Delta x\Delta y + h\Delta y$$

$$d = \frac{S_0}{2}\Delta x\Delta y + hT_\infty\Delta y$$

Again, for uniform grid $\Delta x = \Delta y$, constant conductivity, $k_{i-\frac{1}{2},j} = k_{i+\frac{1}{2},j} = k_{i,j+\frac{1}{2}} = k_{i,j-\frac{1}{2}} = k$, and no heat source, $S_0 = S_1 = 0$, the equation reduces to

$$\left(\frac{h\Delta x}{k} + 2\right)T_{i,j} = \frac{1}{2}\left(2T_{i-1,j} + T_{i,j+1} + T_{i,j-1}\right) + \frac{h\Delta x}{k}T_\infty$$

Applying this equation to nodes 3 and 5, we get

$$\textit{Node 3}\,(0.5+2)T_3 = \frac{1}{2}(2T_2 + T_H + T_6) + 0.5T_\infty$$

$$5.0T_3 = 2T_2 + T_H + T_6 + T_\infty$$

$$\textit{Node 6}\,(0.5+2)T_6 = \frac{1}{2}(2T_5 + T_3 + T_9) + 0.5T_\infty$$

$$5.0T_6 = 2T_5 + T_3 + T_9 + T_\infty$$

Nodes 7 and 8 are convective boundary nodes on the bottom surface, and we select the discretization as

$$a_{i,j}T_{i,j} = a_{i+1,j}T_{i+1,j} + a_{i-1,j}T_{i-1,j} + a_{i,j-1}T_{i,j-1} + d$$

$$a_{i+1,j} = \frac{\Gamma_{i+\frac{1}{2},j}}{2}\frac{\Delta y}{\Delta x}, \quad a_{i-1,j} = \frac{\Gamma_{i-\frac{1}{2},j}}{2}\frac{\Delta y}{\Delta x}, a_{i,j+1} = \Gamma_{i,j+\frac{1}{2}}\frac{\Delta x}{\Delta y}$$

$$a_{i,j} = a_{i+1,j} + a_{i-1,j+1} + a_{i,j-1} - \frac{S_1}{2}\Delta x\,\Delta y + h\,\Delta x$$

$$d = \frac{S_0}{2}\Delta x\,\Delta y + hT_\infty\,\Delta x$$

This equation is simplified in a similar manner for a uniform grid, constant conductivity, and no heat source of the form

$$\left(h\frac{\Delta x}{k}+2\right)T_{i,j}=\frac{1}{2}\left(2T_{i,j+1}+T_{i+1,j}+T_{i-1,j}\right)+\frac{h\Delta x}{k}T_\infty$$

Applying this equation to nodes 7 and 8, we get

$$\textit{Node 7}\quad (0.5+2)T_7=\frac{1}{2}(2T_4+T_8+T_c)+0.5T_\infty$$

$$5.0T_7=2T_4+T_8+T_c+T_\infty$$

$$\textit{Node 8}\quad (0.5+2)T_8=\frac{1}{2}(2T_s+T_9+T_7)+0.5T_\infty$$

$$5.0T_8=2T_s+T_9+T_7+T_\infty$$

Node 9 is an exterior corner node, and the discretization equation is selected for a southeast corner node as

$$2\left(\frac{h\Delta x}{k}+1\right)T_{i,j}=\left(T_{i,j+1}+T_{i-1,i}\right)+2\frac{h\Delta x}{k}T_\infty$$

Applying this equation to corner node 9, we get

$$\textit{Node 9}\quad 2(0.5+1)T_9=T_6+T_8+T_\infty$$

$$3T_9=T_6+T_8+T_\infty$$

Rearranging and assembling all nodal equations, we get the following system of equations:

$$-4T_1+T_2+0+T_4+0+0+0+0+0=-(T_C+T_H)$$

$$T_1-4T_2+T_3+0+T_5+0+0+0+0=-T_H$$

$$0+2T_2-5T_3+0+0+T_6+0+0+0=-(T_H+T_\infty)$$

$$T_1+0+0-4T_4+T_5+0+T_7+0+0=-T_C$$

$$0+T_2+0+T_4-4T_5+T_6+0+T_8+0=0$$

$$0+0+T_3+0+2T_5-5T_6+0+0+T_9=-T_\infty$$

$$0+0+0+2T_4+0+0-5T_7+T_8+0=-(T_C+T_\infty)$$

$$0+0+0+0+2T_5+0+T_7-5T_8+T_9=-T_\infty$$

$$0+0+0+0+0+T_6+0+T_8-3T_9=-T_\infty$$

Now, substituting $T_H = 600°C$, $T_C = 20°C$, and $T_\infty = 25°C$, and writing this in matrix form, we get the system of equations as

$$\begin{bmatrix} -4 & 1 & 0 & 1 & 0 & 0 & 0 & 0 & 0 \\ 1 & -4 & 1 & 0 & 1 & 0 & 0 & 0 & 0 \\ 0 & 2 & -5 & 0 & 0 & 1 & 0 & 0 & 0 \\ 1 & 0 & 0 & -4 & 1 & 0 & 1 & 0 & 0 \\ 0 & 1 & 0 & 1 & -4 & 1 & 0 & 1 & 0 \\ 0 & 0 & 1 & 0 & 2 & -5 & 0 & 0 & 1 \\ 0 & 0 & 0 & 2 & 0 & 0 & -5 & 1 & 0 \\ 0 & 0 & 0 & 0 & 2 & 0 & 1 & -5 & 1 \\ 0 & 0 & 0 & 0 & 0 & 1 & 0 & 1 & -3 \end{bmatrix} \begin{Bmatrix} T_1 \\ T_2 \\ T_3 \\ T_4 \\ T_5 \\ T_6 \\ T_7 \\ T_8 \\ T_9 \end{Bmatrix} = \begin{Bmatrix} -620 \\ -600 \\ -625 \\ -20 \\ 0 \\ -25 \\ -45 \\ -25 \\ -25 \end{Bmatrix}$$

Using the Gauss elimination routine, the solution to this system of equations is obtained as

$$\begin{Bmatrix} T_1 \\ T_2 \\ T_3 \\ T_4 \\ T_5 \\ T_6 \\ T_7 \\ T_8 \\ T_9 \end{Bmatrix} = \begin{Bmatrix} 276.1977°C \\ 340.1860°C \\ 293.1860°C \\ 144.6047°C \\ 191.3605°C \\ 160.5581°C \\ 90.8605°C \\ 120.0930°C \\ 101.8837°C \end{Bmatrix}$$

The heat transfer rate from the top surface is computed based on the *Fourier* law as

$$q_T = \sum kA_y \frac{\Delta T}{\Delta y} = k\Delta x \cdot 1 \frac{(600 - T_1)}{\Delta y} + k\Delta x \cdot 1 \frac{(600 - T_2)}{\Delta y} + k\frac{\Delta x}{2} \cdot 1 \frac{(600 - T_3)}{\Delta y}$$

$$= 30.0 \times 0.5 \times \frac{(600 - 276.1977)}{0.5} + 30.0 \times 0.5 \times \frac{(600 - 340.1860)}{0.5} + 30.0 \times \frac{0.5}{2} \times \frac{(600 - 293.1860)}{0.5}$$

$$= 9714.069 + 7794.42 + 4602.2$$

$$= 22\,110.699\,\text{W}$$

The heat transfer rate from the left surface is computed based on the *Fourier* law as

$$q_L = \sum kA_x \frac{\Delta T}{\Delta x} = k\Delta y \cdot 1 \frac{(T_1 - 20)}{\Delta x} + k\Delta y \cdot 1 \frac{(T_4 - 20)}{\Delta x} + k\Delta y \cdot 1 \frac{(T_7 - 20)}{\Delta x}$$

$$= 30.0 \times 0.5 \times \frac{(276.1977 - 20)}{0.5} + 30.0 \times 0.5 \times \frac{(144.6047 - 20)}{0.5} + 30.0 \times \frac{0.5}{2} \times \frac{(90.8605 - 20)}{0.5}$$

$$= 7685.931 + 3738.141 + 1062.9075$$

$$= 12\,486.9795\,\text{W}$$

The heat transfer rate from the right surface is computed based on *Newton's* law of cooling as

$$q_B = \Sigma h\,\Delta y\left(T_{i,j} - T_\infty\right) = h\Delta y \cdot 1(T_3 - T_\infty) + h\Delta y \cdot 1(T_6 - T_\infty) + h\frac{\Delta y}{2} \cdot 1(T_9 - T_\infty)$$

$$= 30.0 \times 0.5 \times (293.1860 - 25) + 30.0 \times 0.5 \times (160.5581 - 25) + 30.0 \times \frac{0.5}{2} \times (101.8837 - 25)$$

$$= 4022.79 + 2033.3715 + 576.62775$$

$$= 6632.78925\,\text{W}$$

The heat transfer rate from the bottom surface is computed based on *Newton's* law of cooling as

$$q_R = \Sigma h\,\Delta x\left(T_{i,j} - T_\infty\right) = h\Delta x \cdot 1(T_7 - T_\infty) + h\Delta x \cdot 1(T_8 - T_\infty) + h\frac{\Delta x}{2} \cdot 1(T_9 - T_\infty)$$

$$= 30.0 \times 0.5 \times (90.8605 - 25) + 30.0 \times 0.5 \times (120.0930 - 25) + 30.0 \times \frac{0.5}{2} \times (101.8837 - 25)$$

$$= 987.9075 + 1426.395 + 576.62775$$

$$= 2990.93025\,\text{W}$$

The overall energy balance for the two-dimensional steady-state problems states that rate of energy inflow is balanced by the rate of energy outflow, i.e.,

$$E_{\text{in}} = E_{\text{out}}$$

In this problem, energy inflow is given by the heat transfer rate at the top surface

$$E_{\text{in}} = q_T = 22\,110.699\,\text{W}$$

The energy outflow is the sum of heat transfer rates at the left, right, and bottom surfaces, i.e.,

$$E_{\text{out}} = q_L + q_R + q_B = 12\,486.9795 + 6632.78925 + 2990.93025\,\text{W}$$

$$E_{\text{out}} = 22\,110.699\,\text{W}$$

Example 6.3: Two-Dimensional Steady-State Conduction in a Slab Irradiated by a High Energy Laser Beam at the Surface

Use the finite difference method to solve a two-dimensional steady-state conduction in a rectangular aluminum (k = 200 W/m °C) slab subjected to a constant surface heat flux irradiated by a high-energy laser beam at the top surface. For simplicity, assume the heat flux distribution to be a constant average value, $I_0 = 2\times10^8\,\text{W/m}^2$, acting over a section of the surface equal to the beam diameter, d = 4 mm as shown in the figure. The remaining portion of the top surface is subjected to convection with h_c = 100 W/m² °C. All other surfaces are assumed to be maintained at constant temperature of T_∞ = 25°C.

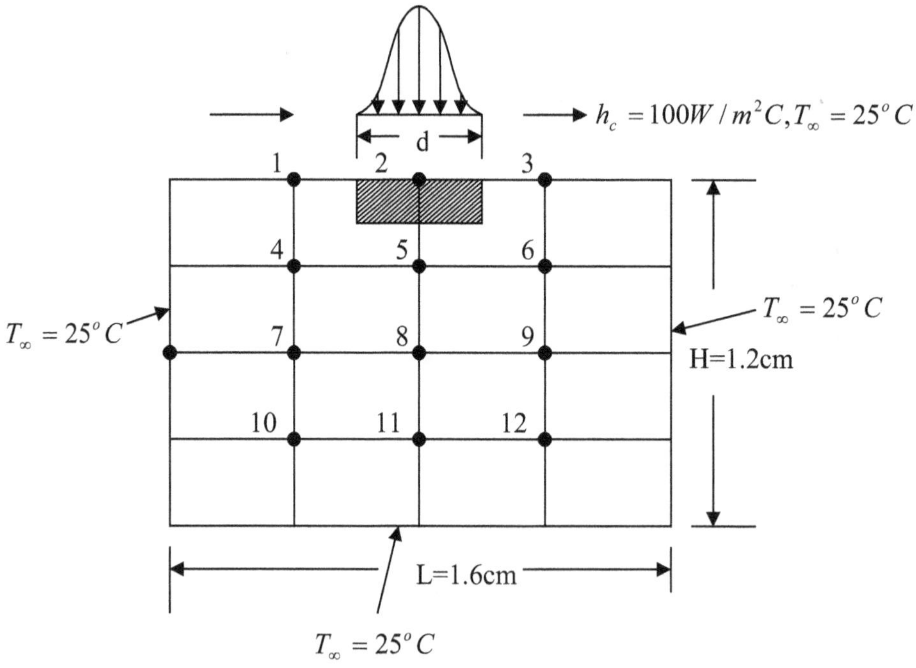

The mathematical statement of the problem is given as:

Governing Heat Equation

$$\frac{\partial}{\partial x}\left(k_x \frac{\partial T}{\partial x}\right)+\frac{\partial}{\partial y}\left(k_y \frac{\partial T}{\partial y}\right)=0$$

Boundary Conditions

1. at the left surface, $T\left(-\frac{L}{2}, y\right)=T_\infty$
2. at the right surface, $T\left(\frac{L}{2}, y\right)=T_\infty$
3. at the bottom surface, $T(x, H)=T_\infty$
4. at the top surface, the boundary condition is specified in two regions

Region 1

$$-\frac{L}{2}<x<-\frac{d}{2} \quad \text{and} \quad \frac{d}{2}<x<\frac{L}{2} \qquad k\left.\frac{\partial T}{\partial y}\right|_{y=0}=h\left(T|_{y=0}-T_\infty\right)$$

Region 2

$$-\frac{d}{2}<\frac{d}{2} \; -k\frac{\partial T}{\partial x}=q_0''$$

Use the finite difference method to calculate the two-dimensional temperature distribution in the slab based on grid distribution shown in the figure. Estimate the heat transfer rates at all surfaces and check the overall energy balance.

Solution

For the prescribed 4 × 4 grid size distribution with $N_{xdiv}=N_{ydiv}=4$, we can estimate the parameters as

$$\Delta x=\frac{L}{N_{xdiv}}=\frac{1.6\text{cm}}{4}=0.4\text{cm}, \quad \Delta y=\frac{L}{N_{ydiv}}=\frac{1.2\text{cm}}{4}=0.3\text{cm}$$

$$\frac{\Delta x}{\Delta y}=\frac{0.4}{0.3}=1.333, \quad \frac{\Delta y}{\Delta x}=\frac{0.3}{0.4}=0.75, \quad \frac{h_c\Delta x}{k}=\frac{100\times 0.004}{200}=0.002$$

In the next step, we select appropriate discretization equations for each node. There are six interior nodes 4, 5, 6, 7, 8, 9, 10, 11, and 12 for which the appropriate finite difference equation is given by the equation (b) in Table 6.1.

$$a_{i,j}\phi_{i,j}=a_{i+1,j}\phi_{i+1,j}+a_{i-1,j}\phi_{i-1,j}+a_{i,j+1}\phi_{i,j+1}+a_{i,j-1}\phi_{i,j-1}+d$$

$$a_{i+1,j}=a_{i-1,j}=\frac{\Delta y}{\Delta x}=0.75$$

$$a_{i,j+1}=a_{i,j-1}=\frac{\Delta x}{\Delta y}=1.333$$

$$a_{i,j}=a_{i+1,j}+a_{i-1,j}+a_{i,j=1}+a_{i,j-1}=2\frac{\Delta y}{\Delta x}\left(1+\frac{\Delta x^2}{\Delta y^2}\right)$$

$$=2\times 0.75\times\left(1+(1.333)^2\right)=4.165$$

$$d=\bar{S}\,\Delta x\Delta y/k=0$$

With no heat source term, $\bar{S}=0$, and $d=0$, and the equation is expressed in a simplified form as

$$4.165T_{i,j}=0.75T_{i+1,j}+0.75T_{i-1,j}+1.333T_{i,j+1}+1.333T_{i,j-1}$$

Using this equation, the finite difference equation for the interior nodes 4–12 is written as

$$\textit{Node 4} \quad 4.165T_4 = 0.75T_5 + 0.75T_0 + 1.333T_1 + 1.333T_7$$

$$\textit{Node 5} \quad 4.165T_5 = 0.75T_6 + 0.75T_4 + 1.333T_2 + 1.333T_8$$

$$\textit{Node 6} \quad 4.165T_6 = 0.75T_0 + 0.75T_5 + 1.333T_3 + 1.333T_9$$

$$\textit{Node 7} \quad 4.165T_7 = 0.75T_8 + 0.75T_0 + 1.333T_4 + 1.333T_{10}$$

$$\textit{Node 8} \quad 4.165T_8 = 0.75T_9 + 0.75T_7 + 1.333T_5 + 1.333T_{11}$$

$$\textit{Node 9} \quad 4.165T_9 = 0.75T_0 + 0.75T_8 + 1.333T_6 + 1.333T_{12}$$

$$\textit{Node 10} \quad 4.165T_{10} = 0.75T_{11} + 0.75T_0 + 1.333T_7 + 1.333T_0$$

$$\textit{Node 11} \quad 4.165T_{11} = 0.75T_{12} + 0.75T_{10} + 1.333T_8 + 1.333T_0$$

$$\textit{Node 12} \quad 4.165T_{12} = 0.75T_0 + 0.75T_{11} + 1.333T_9 + 1.333T_0$$

Nodes 1 and 3 are convective boundary nodes on the top surface, and we use Equations 6.12 and 6.13 as the appropriate discretization equation, which is given as

$$a_{i,j}T_{i,j} = a_{i-1,j}T_{i-1,j} + a_{i+1,j}T_{i+1,j} + a_{i,j-1}T_{i,j-1} + d$$

$$a_{i,j-1} = k_{i,j-\frac{1}{2}}\frac{\Delta x}{\Delta y}, \quad a_{i+1,j} = \frac{k_{i+\frac{1}{2},j}}{2}\frac{\Delta y}{\Delta x}, \quad a_{i-1,j} = \frac{k_{i-\frac{1}{2},j}}{2}\frac{\Delta y}{\Delta x}$$

$$a_{i,j} = a_{i-1,j} + a_{i+1,j} + a_{i,j-1} - \frac{S_1}{2}\Delta x\,\Delta y + h\,\Delta x$$

$$d = \frac{S_0}{2}\Delta x\,\Delta y + hT_\infty\,\Delta x$$

Again, for constant conductivity, $k_{i-\frac{1}{2},j} = k_{i+\frac{1}{2},j} = k_{i,j+\frac{1}{2}} = k_{i,j-\frac{1}{2}} = k$, and no heat source, $S_0 = S_1 = 0$, the equation reduces to

$$\left(\frac{h_c\Delta x}{k} + 2.833\right)T_{i,j} = \frac{1}{2}\left(2.666T_{i,j-1} + 0.75T_{i+1,j} + 0.75T_{i-1,j}\right) + \frac{h_c\Delta x}{k}T_\infty$$

or

$$(0.002 + 2.833)T_{i,j} = \frac{1}{2}\left(2.666T_{i,j-1} + 0.75T_{i+1,j} + 0.75T_{i-1,j}\right) + 0.002T_\infty$$

or

$$5.67T_{i,j} = \left(2.666T_{i,j-1} + 0.75T_{i+1,j} + 0.75T_{i-1,j}\right) + 0.004T_\infty$$

Applying this equation to nodes 1 and 3, we get

$$\textit{Node 1} \quad 5.67T_1 = (2.666T_4 + 0.75T_2 + 0.75T_0) + 0.004T_\infty$$

or

$$5.67T_1 = 2.666T_4 + 0.75T_2 + 0.75T_0 + 0.754T_\infty$$

$$\textit{Node 3} \quad 5.67T_3 = (2.666T_6 + 0.75T_0 + 0.75T_2) + 0.004T_\infty$$

or

$$5.67T_3 = 2.666T_6 + 0.75T_2 + 0.75T_0 + 0.004T_\infty$$

Node 2 is a boundary node on the top surface with constant surface heat flux, and we select the discretization equation (c) from Table 6.2 as

$$a_{i,j}T_{i,j} = a_{i+1,j}T_{i+1,j} + a_{i-1,j}T_{i-1,j} + a_{i,j-1}T_{i,j-1} + d$$

$$a_{i+1,j} = \frac{k_{i+\frac{1}{2},j}}{2}\frac{\Delta y}{\Delta x}, \quad a_{i-1,j} = \frac{k_{i-\frac{1}{2},j}}{2}\frac{\Delta y}{\Delta x}, \quad a_{i,j-1} = k_{i,j-\frac{1}{2}}\frac{\Delta x}{\Delta y}$$

$$a_{i,j} = a_{i+1,j} + a_{i-1,j+1} + a_{i,j-1} - \frac{S_1}{2}\Delta x\,\Delta y$$

$$d = \frac{S_0}{2}\Delta x\,\Delta y + f''_{\text{ts}}\,\Delta x$$

This equation is simplified in a similar manner for constant conductivity and no heat source to the form

$$\left(\frac{\Delta x}{\Delta y} + \frac{\Delta y}{\Delta x}\right)T_{i,j} = \frac{1}{2}\left(2\frac{\Delta x}{\Delta y}T_{i,j-1} + \frac{\Delta y}{\Delta x}T_{i+1,j} + \frac{\Delta y}{\Delta x}T_{i-1,j}\right) + \frac{I_0\Delta x}{k}$$

Substituting the numerical values, we get

$$(1.333 + 0.75)T_{i,j} = \frac{1}{2}\left(2.666T_{i,j-1} + 0.75T_{i+1,j} + 0.75T_{i-1,j}\right) + \frac{I_0 \times 0.004}{200}$$

or

$$4.166T_{i,j} = 2.666T_{i,j-1} + 0.75T_{i+1,j} + 0.75T_{i-1,j} + 4\times 10^{-5}I_0$$

Applying this equation to node 2, we get

$$\textit{Node 2} \quad 4.166T_2 = 2.666T_5 + 0.75T_3 + 0.75T_1 + 4.0\times 10^{-5}I_0$$

Rearranging and assembling all nodal equations, we get the following system of equations.

$$5.67T_1 - 0.75T_2 + 0T_3 - 2.666T_4 + 0T_5 + 0T_6 + 0T_7 + 0T_8 + 0T_9 + 0T_{10} + 0T_{11} + 0T_{12} = 0.75T_0 + 0.754T_\infty$$

$$-0.75T_1 + 4.166T_2 - 0.75T_3 + 0T_4 - 2.666T_5 + 0T_6 + 0T_7 + 0T_8 + 0T_9 + 0T_{10} + 0T_{11} + 0T_{12} = 4.0\times 10^{-5}I_0$$

$$0T_1 - 0.75T_2 + 5.67T_3 + 0T_4 + 0T_5 - 2.666T_6 + 0T_7 + 0T_8 + 0T_9 + 0T_{10} + 0T_{11} + 0T_{12} = 0.75T_0 + 0.004T_\infty$$

$$-1.333T_1 + 0T_2 + 0T_3 + 4.165T_4 - 0.75T_5 + 0T_6 - 1.333T_7 + 0T_8 + 0T_9 + 0T_{10} + 0T_{11} + 0T_{12} = 0.75T_0$$

$$0T_1 - 1.333T_2 + 0T_3 - 0.75T_4 + 4.165T_5 - 0.75T_6 + 0T_7 - 1.333T_8 + 0T_9 + 0T_{10} + 0T_{11} + 0T_{12} = 0$$

$$0T_1 + 0T_2 - 1.333T_3 + 0T_4 - 0.75T_5 + 4.165T_6 + 0T_7 + 0T_8 - 1.333T_9 + 0T_{10} + 0T_{11} + 0T_{12} = 0.75T_0$$

$$0T_1 + 0T_2 + 0T_3 - 1.333T_4 + 0T_5 + 0T_6 + 4.165T_7 - 0.75T_8 + 0T_9 - 1.333T_{10} + 0T_{11} + 0T_{12} = 0.75T_0$$

$$0T_1 + 0T_2 + 0T_3 + 0T_4 - 1.333T_5 + 0T_6 - 0.75T_7 + 4.165T_8 - 0.75T_9 + 0T_{10} - 1.333T_{11} + 0T_{12} = 0$$

$$0T_1 + 0T_2 + 0T_3 + 0T_4 + 0T_5 - 1.333T_6 + 0T_7 - 0.75T_8 + 4.165T_9 + 0T_{10} + 0T_{11} - 1.333T_{12} = 0.75T_0$$

$$0T_1 + 0T_2 + 0T_3 + 0T_4 + 0T_5 + 0T_6 - 1.333T_7 + 0T_8 + 0T_9 + 4.165T_{10} - 0.75T_{11} + 0T_{12} = 0.75T_0 + 1.333T_0$$

$$0T_1 + 0T_2 + 0T_3 + 0T_4 + 0T_5 + 0T_6 + 0T_7 - 1.333T_8 + 0T_9 - 0.75T_{10} + 4.165T_{11} - 0.75T_{12} = 1.333T_0$$

$$0T_1 + 0T_2 + 0T_3 + 0T_4 + 0T_5 + 0T_6 + 0T_7 + 0T_8 - 1.333T_9 + 0T_{10} - 0.75T_{11} + 4.165T_{12} = 0.75T_0 + 1.333T_0$$

Now, substituting $I_0 = 2\times10^8 \text{W/m}^2$, $T_0 = 25°\text{C}$, and $T_\infty = 25°\text{C}$, and writing this in matrix form, we get the system of equations as

$$
\begin{bmatrix}
5.67 & -0.75 & 0 & 2.666 & 0 & 0 & 0 & 0 & 0 & 0 & 0 & 0 \\
-0.75 & 4.166 & 0.75 & 0 & -2.666 & 0 & 0 & 0 & 0 & 0 & 0 & 0 \\
0 & -0.75 & 5.67 & 0 & 0 & -2.666 & 0 & 0 & 0 & 0 & 0 & 0 \\
-1.333 & 0 & 0 & 4.165 & -0.75 & 0 & -1.333 & 0 & 0 & 0 & 0 & 0 \\
0 & -1.333 & 0 & -0.75 & 4.165 & -0.75 & 0 & -1.333 & 0 & 0 & 0 & 0 \\
0 & 0 & -1.333 & 0 & -0.75 & 4.165 & 0 & 0 & -1.333 & 0 & 0 & 0 \\
0 & 0 & 0 & -1.333 & 0 & 0 & 4.165 & -0.75 & 0 & -1.333 & 0 & 0 \\
0 & 0 & 0 & 0 & -1.333 & 0 & -0.75 & 4.165 & -0.75 & 0 & -1.333 & 0 \\
0 & 0 & 0 & 0 & 0 & -1.333 & 0 & -0.75 & 4.165 & 0 & 0 & -1.333 \\
0 & 0 & 0 & 0 & 0 & 0 & -1.333 & 0 & 0 & 4.165 & -0.75 & 0 \\
0 & 0 & 0 & 0 & 0 & 0 & 0 & -1.333 & 0 & -0.75 & 4.165 & -0.75 \\
0 & 0 & 0 & 0 & 0 & 0 & 0 & 0 & -1.333 & 0 & -0.75 & 4.165
\end{bmatrix}
$$

$$
\left\{\begin{bmatrix} T_1 \\ T_2 \\ T_3 \\ T_4 \\ T_5 \\ T_6 \\ T_7 \\ T_8 \\ T_9 \\ T_{10} \\ T_{11} \\ T_{12} \end{bmatrix}\right\} \left\{\begin{bmatrix} 18.85 \\ 8000 \\ 18.85 \\ 18.75 \\ 0 \\ 18.75 \\ 18.75 \\ 0 \\ 18.75 \\ 52.075 \\ 33.325 \\ 52.075 \end{bmatrix}\right\}
$$

Using the Gauss elimination routine, the solution to this system of equations is obtained as

$$
\left\{\begin{bmatrix} T_1 \\ T_2 \\ T_3 \\ T_4 \\ T_5 \\ T_6 \\ T_7 \\ T_8 \\ T_9 \\ T_{10} \\ T_{11} \\ T_{12} \end{bmatrix}\right\} = \left\{\begin{bmatrix} 100.533°\text{C} \\ 401.784°\text{C} \\ 100.533°\text{C} \\ 93.712°\text{C} \\ 196.186°\text{C} \\ 93.712°\text{C} \\ 67.825°\text{C} \\ 105.754°\text{C} \\ 67.825°\text{C} \\ 44.640°\text{C} \\ 57.924°\text{C} \\ 44.640°\text{C} \end{bmatrix}\right\}
$$

Note that the solution shows symmetric temperature distribution in the workpiece. This is true because the geometry as well as the boundary conditions are symmetric.

Example 6.4: Cooling of Electronic Chips

Consider the problem of cooling an electronic chip that is mounted on a substrate. The electric power generation in the chip is $\dot{q} = 1.6\times10^8\,\text{W/m}^3$ and the size of the chip is 1 cm×0.5 cm. The top surface is assumed to be convectively cooled with $h_C = 50\,\text{W/m}^2\,°\text{C}$ and $T_\infty = 20°\text{C}$. The left, right, and bottom surfaces are assumed to be at constant temperature $T_0 = 25°\text{C}$. Assume the conductivity of the chip and substrate materials as $k = 15\,\text{W/m·K}$.

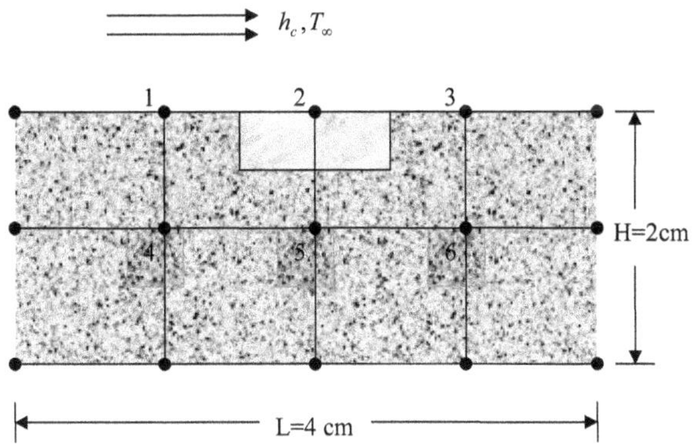

(a) Derive the finite difference equation for node 2; (b) select the finite difference equations for the rest of the nodes 1, 3, 4, 5, and 6; (c) assemble the equations to form the system of equations; and (d) obtain the solution for the temperature distribution.

Solution

The governing heat equation is given as

$$\frac{\partial}{\partial x}\left(k_x\frac{\partial T}{\partial x}\right)+\frac{\partial}{\partial y}\left(k_y\frac{\partial T}{\partial y}\right)=0$$

For the selected uniform grid system, we have

$$\Delta x=\frac{L}{N_{\text{div}x}}=\frac{4}{4}=1\text{cm}=0.01\text{m},\qquad \Delta y=\frac{H}{N_{\text{div}y}}=\frac{2}{2}=1\text{cm}=0.01\text{m}$$

$$\frac{h\Delta x}{k}=\frac{50\times0.01}{15}=0.0333$$

The following discretization equations are selected from Table 6.4.

Node 2 Convective boundary nodes with uniform heat generation

$$\left(\frac{h\Delta x}{k}+2\right)T_{i,j}=\frac{1}{2}\left(2T_{i,j-1}+T_{i+1,j}+T_{i-1,j}\right)+\frac{h\Delta x}{\Gamma}T_\infty+\frac{q\left(\dot{\Delta x}\right)^2}{k}$$

or

$$2\left(\frac{h\Delta x}{k}+2\right)T_2=\left(2T_5+T_3+T_1\right)+\frac{2h\Delta x}{\Gamma}T_\infty+\frac{q\left(\dot{\Delta x}\right)^2}{k}$$

$$2(0.0333+2)T_2=\left(2T_5+T_3+T_1\right)+2\times0.0333\times20+\frac{1.6\times10^8(0.01)^2}{15} \tag{E.6.4.1}$$

$$T_1-4.0666T_2+T_3+2T_5=-1.332-\frac{2\times8\times10^7(0.01)^2}{15}$$

Nodes 1 and 3 Convective boundary nodes

$$\left(\frac{h\Delta x}{k}+2\right)T_{i,j}=\frac{1}{2}\left(2T_{i,j-1}+T_{i+1,j}+T_{i-1,j}\right)+\frac{h\Delta x}{k}T_{\infty}$$

Node 1 $2(0.0333+2)T_1=(2T_4+T_2+25)+2\times 0.0333\times 20$
or

$$-4.0666T_1+T_2+2T_4=-26.332 \tag{E.6.4.2}$$

Node 3 $2(0.0333+2)T_3=(2T_6+25+T_2)+2\times 0.0333\times 20$
or

$$T_2-4.0666T_3+2T_6=-26.332 \tag{E.6.4.3}$$

Nodes 4, 5, and 6 Interior nodes

$$4T_{i,j}=T_{i+1,j}+T_{i-1,j}+T_{i,j+1}+T_{i,j-1}$$

Node 4 $4T_4=T_5+25+T_1+25$
or

$$T_1-4T_4+T_5=-50 \tag{E.6.4.4}$$

Node 5 $4T_5=T_6+T_4+T_2+25$
or

$$T_2+T_4-4T_5+T_6=-25 \tag{E.6.4.5}$$

Node 6 $4T_6=25+T_5+T_3+25$
or

$$T_3+T_5-4T_6=-50 \tag{E.6.4.6}$$

Assembly of all nodal equations (E.6.4.1–E.6.4.6) leads to the system of equations as

$$\begin{bmatrix} -4.0666 & 1 & 0 & 2 & 0 & 0 \\ 0 & -4.0666 & 1 & 0 & 2 & 0 \\ 0 & 1 & -4.0666 & 0 & 0 & 2 \\ 1 & 0 & 0 & -4.0 & 1 & 0 \\ 0 & 1 & 0 & 1 & -4.0 & 1 \\ 0 & 0 & 1 & 0 & 1 & -4 \end{bmatrix}\begin{Bmatrix} T_1 \\ T_2 \\ T_3 \\ T_4 \\ T_5 \\ T_6 \end{Bmatrix}=\begin{Bmatrix} -26.332 \\ -1067.992 \\ -26.332 \\ -50.0 \\ -25.0 \\ -50.0 \end{Bmatrix}$$

Solution of the system of equations gives the temperature distribution as

$$\begin{Bmatrix} T_1 \\ T_2 \\ T_3 \\ T_4 \\ T_5 \\ T_6 \end{Bmatrix}=\begin{Bmatrix} 135.6302 \\ 363.6288 \\ 135.6302 \\ 80.7964 \\ 137.5554 \\ 80.7964 \end{Bmatrix}$$

Example 6.5: Fully Developed Flow in a Rectangular Channel with Top Surface Moving

Consider a fully developed flow in a rectangular channel with top surface moving with an axial velocity U_0 and the pressure drop in the axial direction assumed to be $dP/dz = S$. Determine the axial-component velocity distribution.

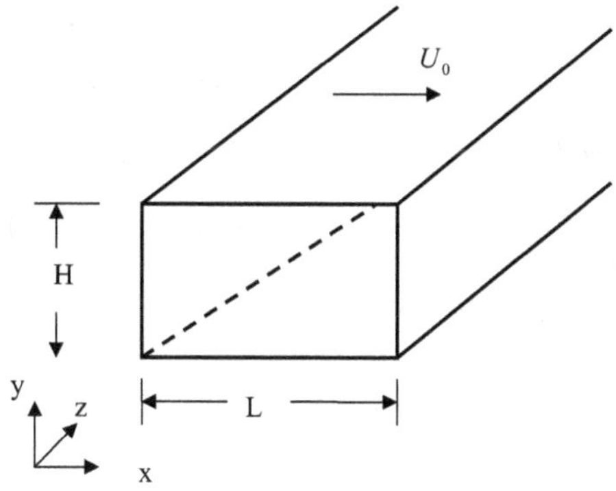

(a)

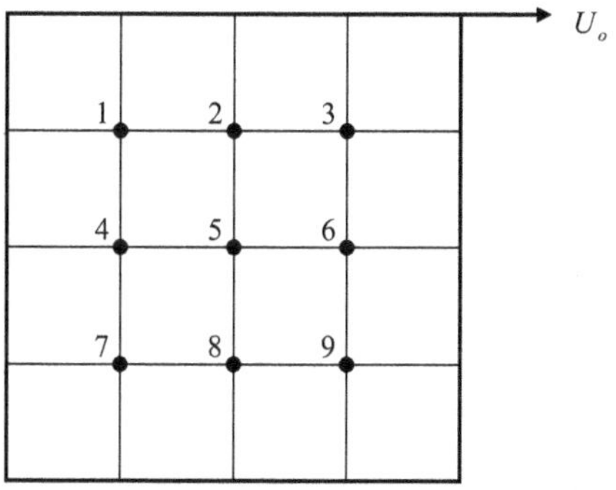

(b)

A mathematical statement of the problem can be derived from the continuity and momentum equations for two-dimensional steady incompressible flow equations given as:

Governing Equation

$$\frac{\partial^2 w}{\partial x^2} + \frac{\partial^2 w}{\partial y^2} = \frac{1}{\mu}\frac{dp}{dz} \quad \text{(E.6.5.1)}$$

Boundary Conditions

1. $y = 0,\ w = 0$
2. $y = H,\ w = w_0$
3. $x = 0,\ w = 0$
4. $x = L,\ w = 0$

The finite difference equation reduces to

$$w_{i+1,j} + w_{i-1,j} + w_{i,j+1} + w_{i,j-1} - 4w_{i,j} + S\Delta x^2 = 0 \quad \text{(E.6.5.2)}$$

where

$$S = -\frac{1}{\mu}\frac{dP}{dz} \tag{E.6.5.3}$$

Applying this equation successively to all nodes, we get

$$\begin{aligned}
&\textit{Node 1} \quad w_2 + w_L + w_0 + w_4 - 4w_1 + S\Delta x^2 = 0 \\
&\textit{Node 2} \quad w_3 + w_1 + w_0 + w_5 - 4w_2 + S\Delta x^2 = 0 \\
&\textit{Node 3} \quad w_R + w_2 + w_0 + w_6 - 4w_3 + S\Delta x^2 = 0 \\
&\textit{Node 4} \quad w_5 + w_L + w_1 + w_7 - 4w_4 + S\Delta x^2 = 0 \\
&\textit{Node 5} \quad w_6 + w_4 + w_2 + w_8 - 4w_5 + S\Delta x^2 = 0 \\
&\textit{Node 6} \quad w_R + w_5 + w_3 + w_9 - 4w_6 + S\Delta x^2 = 0 \\
&\textit{Node 7} \quad w_8 + w_L + w_4 + w_B - 4w_7 + S\Delta x^2 = 0 \\
&\textit{Node 8} \quad w_9 + w_7 + w_5 + w_B - 4w_8 + S\Delta x^2 = 0 \\
&\textit{Node 9} \quad w_R + w_8 + w_6 + w_B - 4w_9 + S\Delta x^2 = 0
\end{aligned}$$

Rearranging and writing in matrix form, we get the system of equations as

$$\begin{bmatrix}
-4 & 1 & 0 & 1 & 0 & 0 & 0 & 0 & 0 \\
1 & -4 & 1 & 0 & 1 & 0 & 0 & 0 & 0 \\
0 & 1 & -4 & 0 & 0 & 1 & 0 & 0 & 0 \\
1 & 0 & 0 & -4 & 1 & 0 & 1 & 0 & 0 \\
0 & 1 & 0 & 1 & -4 & 1 & 0 & 1 & 0 \\
0 & 0 & 1 & 0 & 1 & -4 & 0 & 0 & 1 \\
0 & 0 & 0 & 1 & 0 & 0 & -4 & 1 & 0 \\
0 & 0 & 0 & 0 & 1 & 0 & 1 & -4 & 1 \\
0 & 0 & 0 & 0 & 0 & 1 & 0 & 1 & -4
\end{bmatrix}
\begin{Bmatrix} w_1 \\ w_2 \\ w_3 \\ w_4 \\ w_5 \\ w_6 \\ w_7 \\ w_8 \\ w_9 \end{Bmatrix}
=
\begin{Bmatrix} -\left(w_0 + S\Delta x^2\right) \\ -\left(w_0 + S\Delta x^2\right) \\ -\left(w_0 + S\Delta x^2\right) \\ -S\Delta x^2 \\ -S\Delta x^2 \\ -S\Delta x^2 \\ -S\Delta x^2 \\ -S\Delta x^2 \\ -S\Delta x^2 \end{Bmatrix} \tag{E.6.5.4}$$

6.3 IRREGULAR GEOMETRIES

So far, we have considered geometries involving only regular boundaries, which match well with the grid system. However, there are many physical problems that involve complex boundaries including slant and curved surfaces. As was discussed in Chapter 5, a nonuniform grid size distribution is essential for problems that have regions of steep variation in dependent variables, and thus finer grids are used to capture the necessary resolution. Such nonuniform grids are also necessary near an irregular surface. For example, a uniform grid size distribution is used at the interior regions, away from the curved surface, as shown in Figure 6.3. However, the spacing became nonuniform for spaces between the boundary nodes and the neighboring interior nodes due to the irregular shape of the boundary surface. Another option could be using a nonuniform grid with finer grids near the irregular surface and coarser ones away from the boundary.

The procedure for developing the finite difference discretization equations for such a nonuniform grid near an irregular surface is, however, the same as those discussed for regular geometries. Let us show the development of one such finite difference equation for an interior node near a curved surface as well as nodes on such boundaries as shown in Figure 6.6. As can be noted, a uniform grid size distribution with grid sizes Δx and Δy is used in the x- and y-directions for most of the regions. The grid sizes near the curved boundary are defined arbitrarily as $\Delta x' = \alpha \Delta x$ between the interior node (i, j) and boundary node b_1, $\Delta y' = \beta \Delta y$ between the interior node (i, j) and the boundary node b_2, and $\Delta y'' = \gamma \Delta y$ between boundary nodes b_2 and b_3 as shown in Figure 6.7.

The control volume method can be applied to develop the finite difference equation for interior nodes near, as well as on, the irregular surfaces by considering overlapping control volumes that surround a node and extend halfway to its neighboring nodes.

Interior Nodes Near a Curved Surface In order to derive the discretization equation for the interior node (i, j) close to the boundary, we consider a control volume bounded by the dashed lines located at $-\Delta x'/2, -\Delta x/2, -\Delta y'/2$ and $\Delta y/2$ as shown in Figure 6.4. Integrate the two-dimensional steady-state governing Equation 6.1 over the control volume surrounding the grid point (i, j) as

$$\int_{-\Delta x'/2}^{\Delta x/2} \int_{-\Delta y/2}^{\Delta y'/2} \frac{\partial}{\partial x}\left(\Gamma_x \frac{\partial \phi}{\partial x}\right) dx\,dy + \int_{-\Delta x'/2}^{\Delta x/2} \int_{-\Delta y/2}^{\Delta y'/2} \frac{\partial}{\partial y}\left(\Gamma_y \frac{\partial \phi}{\partial y}\right) dx\,dy + \int_{-\Delta x'/2}^{\Delta x/2} \int_{-\Delta y/2}^{\Delta y'/2} S\,dx\,dy = 0 \quad (6.33)$$

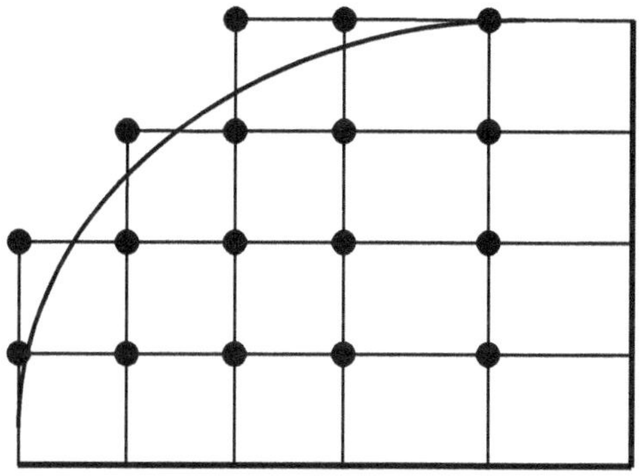

FIGURE 6.6 Grid distribution in a region with an irregular boundary surface.

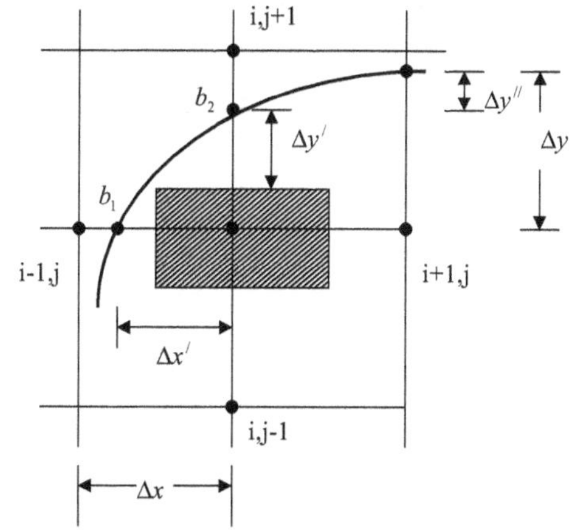

FIGURE 6.7 Nodes near and on a curved boundary.

Now, the source term is approximated with a stepwise profile and integrated over x and y. The diffusion terms in x and y are integrated over x and y, respectively, to reduce the equation to

$$\left[\int_{-\Delta x'/2}^{\Delta x/2} \frac{\partial}{\partial x}\left(\Gamma_x \frac{\partial \phi}{\partial x}\right) \mathrm{d}x \left(y\Big|_{-\Delta y/2}^{\Delta y'/2}\right)\right] + \left[\int_{-\Delta y/2}^{\Delta y'/2} \frac{\partial}{\partial y}\left(\Gamma_y \frac{\partial \phi}{\partial y}\right) \mathrm{d}y \left(x\Big|_{-\Delta x'/2}^{\Delta x/2}\right)\right] + \bar{S}\left(y\Big|_{-\Delta y/2}^{\Delta y'/2}\right)\left(x\Big|_{-\Delta x'/2}^{\Delta x/2}\right) = 0$$

Applying the limit values, we get

$$\left[\int_{-\Delta x'/2}^{\Delta x/2} \frac{\partial}{\partial x}\left(\Gamma_x \frac{\partial \phi}{\partial x}\right) \mathrm{d}x \left(\frac{\Delta y'}{2} + \frac{\Delta y}{2}\right)\right] + \left[\int_{-\Delta y/2}^{\Delta y'/2} \frac{\partial}{\partial y}\left(\Gamma_y \frac{\partial \phi}{\partial y}\right) \mathrm{d}y \left(\frac{\Delta x}{2} + \frac{\Delta x'}{2}\right)\right]$$
$$+ \bar{S}\left(\frac{\Delta y'}{2} + \frac{\Delta y}{2}\right)\left(\frac{\Delta x}{2} + \frac{\Delta x'}{2}\right) = 0$$

Substituting $\Delta x' = \alpha \Delta x$ and $\Delta y' = \beta \Delta y$, we get

$$\left[\int_{-\Delta x'/2}^{\Delta x/2} \frac{\partial}{\partial x}\left(\Gamma_x \frac{\partial \phi}{\partial x}\right) dx \left(\frac{\Delta y}{2}(\beta + 1)\right)\right] + \left[\int_{-\Delta y/2}^{\Delta y'/2} \frac{\partial}{\partial y}\left(\Gamma_y \frac{\partial \phi}{\partial y}\right) dy \left(\frac{\Delta x}{2}(1 + \alpha)\right)\right]$$
$$+ \bar{S}\left(\frac{\Delta y}{2}(1 + \beta)\right)\left(\frac{\Delta x}{2}(1 + \alpha)\right) = 0$$

Now, integrating over x, we get

$$\left[\Gamma_x \frac{\partial \phi}{\partial x}\Big|_{\Delta x/2} - \Gamma_x \frac{\partial \phi}{\partial x}\Big|_{-\Delta x'/2}\right]\frac{\Delta y}{2}(1 + \beta) + \left[\Gamma_y \frac{\partial \phi}{\partial y}\Big|_{\Delta y'/2} - \Gamma_y \frac{\partial \phi}{\partial y}\Big|_{-\Delta y/2}\right]\frac{\Delta x}{2}(1 + \alpha)$$
$$+ \bar{S}(1 + \alpha)(1 + \beta)\frac{\Delta x \, \Delta y}{4} = 0 \tag{6.34}$$

Also, considering a stepwise constant profile for the transport properties Γ and a piece-wise linear profile to approximate the first derivative terms, we can express the flux quantities across the control volume surfaces in Equation 6.23 as

$$\Gamma_x \frac{\partial \phi}{\partial x}\Big|_{-\Delta x'/2} = \Gamma_x \frac{\phi_{i,j} - \phi_{b1}}{\Delta x'} = \Gamma_x \frac{\phi_{i,j} - \phi_{b1}}{\alpha \Delta x} \tag{6.35a}$$

$$\Gamma_x \frac{\partial \phi}{\partial x}\Big|_{\Delta x/2} = \Gamma_x \frac{\phi_{i+1,j} - \phi_{i,j}}{\Delta x} \tag{6.35b}$$

$$\Gamma_y \frac{\partial \phi}{\partial y}\Big|_{-\Delta y/2} = \Gamma_y \frac{\phi_{i,j} - \phi_{i,j-1}}{\Delta y} \tag{6.35c}$$

$$\Gamma_y \frac{\partial \phi}{\partial y}\Big|_{-\Delta y'/2} = \Gamma_y \frac{\phi_{b2} - \phi_{i,j}}{\Delta y'} = \Gamma_y \frac{\phi_{b2} - \phi_{i,j}}{\beta \Delta y} \tag{6.35d}$$

Substituting Equation 6.35 into Equation 6.34, we get

$$\left[\Gamma_x \frac{\phi_{i+1,j}-\phi_{i,j}}{\Delta x} - \Gamma_x \frac{\phi_{i,j}-\phi_{b1}}{\alpha\Delta x}\right]\frac{\Delta y}{2}(1+\beta) + \left[\Gamma_y \frac{\phi_{b2}-\phi_{i,j}}{\beta\Delta y} - \Gamma_y \frac{\phi_{i,j}-\phi_{i,j-1}}{\Delta y}\right]\frac{\Delta x}{2}(1+\alpha)$$
$$+\bar{S}(1+\alpha)(1+\beta)\frac{\Delta x\Delta y}{4} = 0$$

Rearranging

$$\left[\Gamma_x \frac{\Delta y}{2\Delta x}(1+\beta)\left(\frac{1}{\alpha}+1\right) + \Gamma_y \frac{\Delta x}{2\Delta y}(1+\alpha)\left(1+\frac{1}{\beta}\right)\right]\phi_{i,j} - \Gamma_x \frac{\Delta y}{2\Delta x}\frac{1+\beta}{\alpha}\phi_{b1} - \Gamma_x \frac{\Delta y}{2\Delta x}(1+\beta)\phi_{i+1,j}$$
$$-\Gamma_y \frac{\Delta x}{2\Delta y}(1+\alpha)\phi_{i,j-1} - \Gamma_y \frac{\Delta x}{2\Delta y}\frac{1+\alpha}{\beta}\phi_{b2} - \bar{S}(1+\alpha)(1+\beta)\frac{\Delta x\Delta y}{4} = 0 \tag{6.36}$$

We can write this in a compact form as

$$a_{i,j}\phi_{i,j} = a_{i+1,j}\phi_{i+1,j} + a_{i,j-1}\phi_{i,j-1} + d \tag{6.37}$$

where

$$a_{i+1,j} = \Gamma_x \frac{\Delta y}{2\Delta x}(1+\beta)$$

$$a_{i,j-1} = \Gamma_y \frac{\Delta x}{2\Delta y}(1+\alpha)$$

$$a_{i,j} = a_{i+1,j}\left(1+\frac{1}{\alpha}\right) + a_{i,j-1}\left(1+\frac{1}{\beta}\right)$$

$$d = \Gamma_x \frac{\Delta y}{2\Delta x}\frac{1+\beta}{\alpha}\phi_{b1} + \Gamma_y \frac{\Delta x}{2\Delta y}\frac{1+\alpha}{\beta}\phi_{b2} + \bar{S}(1+\alpha)(1+\beta)\frac{\Delta x\Delta y}{4}$$

For constant transport property $\Gamma_x = \Gamma_y$ and for uniform grid size in x and y directions, i.e., $\Delta x = \Delta y$, Equation 6.37 reduces to

$$a_{i,j}\phi_{i,j} = a_{i+1,j}\phi_{i+1,j} + a_{i,j-1}\phi_{i,j-1} + d \tag{6.38}$$

where

$$a_{i+1,j} = \frac{1}{2}(1+\beta)$$

$$a_{i,j-1} = \frac{1}{2}(1+\alpha)$$

$$a_{i,j} = a_{i+1,j}\left(1+\frac{1}{\alpha}\right) + a_{i,j-1}\left(1+\frac{1}{\beta}\right)$$

$$d = \frac{1+\beta}{2\alpha}T_{b1} + \frac{1+\alpha}{2\beta}T_{b2} + \bar{S}(1+\alpha)(1+\beta)\frac{\Delta x^2}{4\Gamma}$$

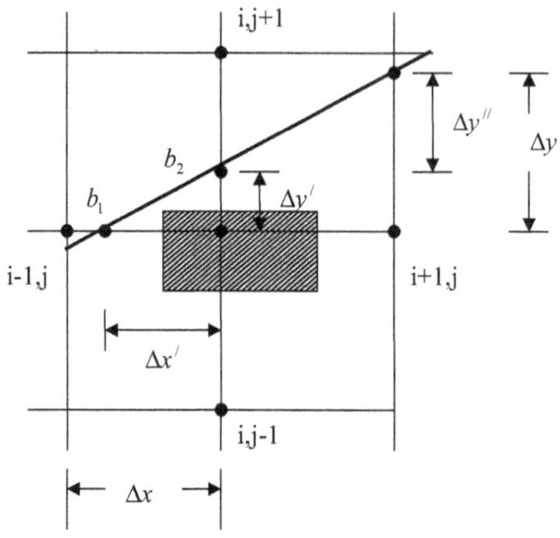

FIGURE 6.8 Interior nodes near a slanted boundary surface.

Boundary Nodes on a Curved Surface A similar procedure can be applied to derive finite difference discretization equations for other boundary nodes with **mixed boundary** or **Newman's** condition. By considering a half control volume surrounding the boundary node, b_2, we can derive the finite difference equation as

$$\left(\Gamma\frac{\beta\Delta y}{\sqrt{\alpha^2\Delta x^2+\beta^2\Delta y^2}}+\Gamma\frac{\beta\Delta y}{\sqrt{\gamma^2\Delta y^2+\Delta x^2}}+\Gamma\frac{\Delta x}{\Delta y}\frac{\alpha+1}{\beta}+h_c\sqrt{\gamma^2\Delta y^2+\Delta x^2}+h_c\sqrt{\alpha^2\Delta x^2+\beta^2\Delta y^2}\right)\phi_{b2}$$

$$=\Gamma\frac{\beta\Delta y}{\sqrt{\alpha^2\Delta x^2+\beta^2\Delta y^2}}\phi_{b1}+\Gamma\frac{\beta\Delta y}{\sqrt{\gamma^2\Delta y^2+\Delta x^2}}\phi_{b3}+\Gamma\frac{\Delta x}{\Delta y}\frac{\alpha+1}{\beta}T_{i,j}$$

$$+h_c\left(\sqrt{\gamma^2\Delta y^2+\Delta x^2}+\sqrt{\alpha^2\Delta x^2+\beta^2\Delta y^2}\right)\phi_\infty+\frac{1}{2}\bar{S}(\alpha+1)\beta\Delta x\Delta y \tag{6.39}$$

Boundary Nodes with No Flux Conditions The finite difference equations for nodes on a boundary surface with no flux conditions, such as an insulated surface or an impermeable surface, can be obtained by setting $h_c = 0$ in Equation 6.39.

Slanted Surface In the case of a slanted surface (Figure 6.8), we can deduce the finite difference equations for interior nodes and boundary nodes from Equations 6.38 and 6.39, respectively, by setting $\alpha=\frac{1}{2}, \beta=\frac{1}{2}$ and $\gamma=\frac{3}{2}$.

6.4 THREE-DIMENSIONAL STEADY-STATE PROBLEMS

So far, we have considered only two-dimensional problems, assuming that the dimension in the z-direction is significantly larger than those in the x- and y-directions, and thus net flux quantity in the z-direction is negligible. Let us now consider three-dimensional problems in which dependent variables are a function of three space variables. A discretization procedure outlined for the two-dimensional problem can be extended and applied in a similar manner to the three-dimensional problem. The three-dimensional calculation domain is divided into equal or nonequal small regions with increments of Δx, Δy, and Δz in x-, y-, and z-directions, respectively, as shown in Figure 6.9. The values of the dependent variables ϕ (x, y, z) are calculated at a finite number of discrete points in the solution domain as shown in the figure.

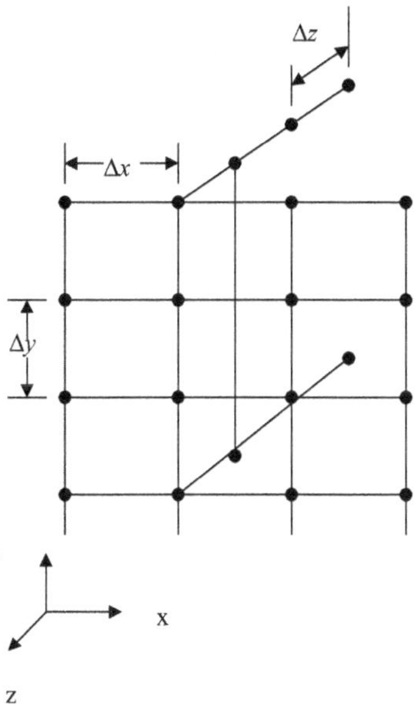

FIGURE 6.9 Nodal network for a three-dimensional region.

Each nodal point is designated by an additional index k, which indicates z increments defined $z = k\Delta z$. The discrete value of the function ϕ at a location (x_i, y_j, z_k) is designated as $\phi_{i,j,k}$.

For three-dimensional steady-state problems, the appropriate governing equation is

$$\frac{\partial}{\partial x}\left(\Gamma\frac{\partial\phi}{\partial x}\right)+\frac{\partial}{\partial y}\left(\Gamma\frac{\partial\phi}{\partial y}\right)+\frac{\partial}{\partial z}\left(\Gamma\frac{\partial\phi}{\partial z}\right)+S=0 \tag{6.40}$$

Let us again consider control volume method to obtain the discretization equation for the interior nodes. For each grid point we consider a control volume bounded by the dashed lines localized at

$$\left(i-\frac{1}{2},j,k\right),\left(i+\frac{1}{2},j,k\right),\left(i,j+\frac{1}{2},k\right),\left(i,j-\frac{1}{2},k\right),\left(i,j,k+\frac{1}{2}\right) \text{ and } \left(i,j,k-\frac{1}{2}\right)$$

as shown in Figure 6.10. Integrate the governing equation over a three-dimensional control volume surrounding the grid point (i, j, k) as shown.

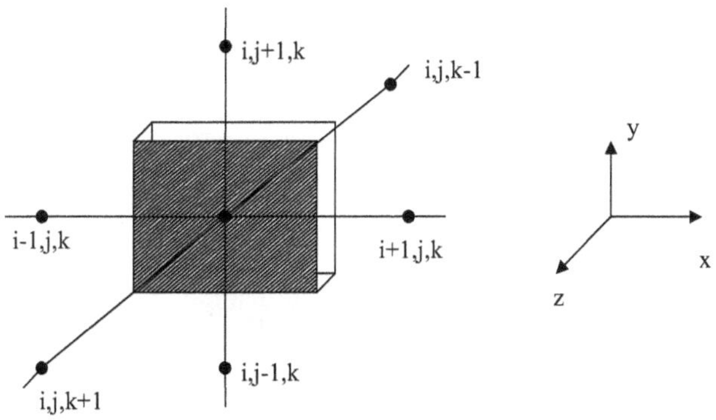

FIGURE 6.10 Details of grid points in a three-dimensional grid system.

Integrating Equation 6.40 over the control volume, we get

$$\int_{i,j-\frac{1}{2},k}^{i,j+\frac{1}{2},k}\int_{i-\frac{1}{2},j,k}^{i+\frac{1}{2},j,k}\int_{i,j,k-\frac{1}{2}}^{i,j,k+\frac{1}{2}}\frac{\partial}{\partial x}\left(\Gamma_x\frac{\partial\phi}{\partial x}\right)dx\,dy\,dz+\int_{i,j-\frac{1}{2}}^{i,j+\frac{1}{2}}\int_{i-\frac{1}{2},j}^{i+\frac{1}{2},j}\int_{i,j,k-\frac{1}{2}}^{i,j,k+\frac{1}{2}}\frac{\partial}{\partial y}\left(\Gamma_y\frac{\partial\phi}{\partial y}\right)\mathrm{d}x\mathrm{d}y\mathrm{d}z$$

$$+\int_{i,j-\frac{1}{2},k}^{i,j+\frac{1}{2},k}\int_{i-\frac{1}{2},j,k}^{i+\frac{1}{2},j,k}\int_{i,j,k-\frac{1}{2}}^{i,j,k+\frac{1}{2}}\frac{\partial}{\partial z}\left(\Gamma_z\frac{\partial\phi}{\partial z}\right)\mathrm{d}x\mathrm{d}y\mathrm{d}z+\int_{i,j-\frac{1}{2},k}^{i,j+\frac{1}{2},k}\int_{i-\frac{1}{2},j,k}^{i+\frac{1}{2},j,k}\int_{i,j,k-\frac{1}{2}}^{i,j,k+\frac{1}{2}}S\mathrm{d}x\mathrm{d}y\mathrm{d}z=0 \tag{6.41}$$

$$\left(\Gamma_x\frac{\partial\phi}{\partial x}\right)_{i+\frac{1}{2},j,k}\Delta y\Delta z-\left(\Gamma_x\frac{\partial\phi}{\partial x}\right)_{i-\frac{1}{2},j,k}\Delta y\Delta z+\left(\Gamma_y\frac{\partial\phi}{\partial y}\right)_{i,j+\frac{1}{2},k}\Delta x\Delta z-\left(\Gamma_y\frac{\partial\phi}{\partial y}\right)_{i,j-\frac{1}{2},k}\Delta x\Delta z$$

$$+\left(\Gamma_z\frac{\partial\phi}{\partial z}\right)_{i,j,k+\frac{1}{2}}\Delta x\Delta y-\left(\Gamma_z\frac{\partial\phi}{\partial z}\right)_{i,j,k-\frac{1}{2}}\Delta x\Delta y+\int_{i,j-\frac{1}{2}}^{i,j+\frac{1}{2}}\int_{i-\frac{1}{2},j}^{i+\frac{1}{2}}S\mathrm{d}x\,\mathrm{d}y=0 \tag{6.42}$$

Considering a stepwise constant profile for the source term we can express the source term integral as

$$\int_{i,j-\frac{1}{2},k}^{i,j+\frac{1}{2},k}\int_{i-\frac{1}{2},j,k}^{i+\frac{1}{2},j,k}\int_{i,j,k-\frac{1}{2}}^{i,j,k+\frac{1}{2}}S\mathrm{d}x\mathrm{d}y=\overline{S}\Delta x\Delta y\Delta z \tag{6.43}$$

where $\overline{S}$ is the mean or average heat generation rate in the control volume, and it can be linearized as discussed before.

Also, considering a stepwise constant profile for the transport properties Γ and a piece-wise linear profile to approximate the first derivative terms, we can express the flux quantities across the control volume surfaces in Equation 6.42 as

$$\left.\Gamma_x\frac{\partial\phi}{\partial x}\right)_{i+\frac{1}{2},j,k}\approx\frac{\Gamma_{x\,i+\frac{1}{2}}\left(\phi_{i+1,j,k}-\phi_{i,j,k}\right)}{\Delta x_{i+\frac{1}{2}}} \tag{6.44a}$$

$$\left.\Gamma_x\frac{\partial\phi}{\partial x}\right)_{i-\frac{1}{2},j,k}\approx\frac{\Gamma_{x\,i-\frac{1}{2}}\left(\phi_{i,j,k}-\phi_{i-1,j,k}\right)}{\Delta x_{i-\frac{1}{2}}} \tag{6.44b}$$

$$\left.\Gamma_y\frac{\partial\phi}{\partial y}\right)_{i,j+\frac{1}{2},k}=\frac{\Gamma_{y\,j+\frac{1}{2}}\left(\phi_{i,j+1,k}-\phi_{i,j,k}\right)}{\Delta y_{j+\frac{1}{2}}} \tag{6.44c}$$

$$\left.\Gamma_y\frac{\partial\phi}{\partial y}\right)_{i,j-\frac{1}{2},k}=\frac{\Gamma_{y\,j-\frac{1}{2}}\left(\phi_{i,j,k}-\phi_{i,j-1,k}\right)}{\Delta y_{j-\frac{1}{2}}} \tag{6.44d}$$

$$\left.\Gamma_z\frac{\partial\phi}{\partial z}\right)_{i,j,k+\frac{1}{2}}=\frac{\Gamma_{z\,k+\frac{1}{2}}\left(\phi_{i,j,k+1}-\phi_{i,j,k}\right)}{\Delta z_{k+\frac{1}{2}}} \tag{6.44e}$$

$$\Gamma_z \frac{\partial \phi}{\partial z}\bigg)_{i,j,k-\frac{1}{2}} = \frac{\Gamma_{z k-\frac{1}{2}}\left(\phi_{i,j,k} - \phi_{i,j,k-1}\right)}{\Delta z_{k-\frac{1}{2}}} \tag{6.44f}$$

Substituting Equations 6.43 and 6.44 into Equation 6.42 and simplifying, we get

$$a_{i,j,k}\phi_{i,j,k} = a_{i+1,j,k}\phi_{i+1,j,k} + a_{i-1,j,k}\phi_{i-1,j,k} + a_{i,j+1,k}\phi_{i,j+1,k} + a_{i,j-1,k}\phi_{i,j-1,k}$$
$$+ a_{i,j,k+1}\phi_{i,j,k+1} + a_{i,j,k-1}\phi_{i,j,k-1} + d \tag{6.45}$$

where

$$a_{i+1,j,k} = \frac{\Gamma_{i+\frac{1}{2},j,k}}{\Delta x_{i+\frac{1}{2},j,k}}\Delta y\,\Delta z \quad a_{i-1,j,k} = \frac{\Gamma_{i-\frac{1}{2},j,k}}{\Delta x_{i-\frac{1}{2},j,k}}\Delta y\,\Delta z \quad a_{i,j+1,k} = \frac{\Gamma_{i,j+\frac{1}{2},k}}{\Delta y_{i,j+\frac{1}{2},k}}\Delta x\,\Delta z$$

$$a_{i,j-1,k} = \frac{\Gamma_{i,j-\frac{1}{2},k}}{\Delta y_{i,j-\frac{1}{2},k}}\Delta x\,\Delta z \quad a_{i,j,k+1} = \frac{\Gamma_{i,j,k+\frac{1}{2}}}{\Delta z_{i,j,k+\frac{1}{2}}}\Delta x\,\Delta y \quad a_{i,j,k-1} = \frac{\Gamma_{i,j,k-\frac{1}{2}}}{\Delta y_{i,j,k-\frac{1}{2}}}\Delta x\Delta y$$

$$a_{i,j,k} = a_{i+1,j,k} + a_{i-1,j,k} + a_{i,j+1,k} + a_{i,j-1,k} + a_{i,j,k+1} + a_{i,j,k-1} - S_1\,\Delta x\,\Delta y\,\Delta z$$

$$d = S_0\,\Delta x\,\Delta y\,\Delta z$$

Equation 6.45 can be simplified further for many different cases as was shown for the two-dimensional case.

Boundary Nodes The finite difference equations for boundary nodes on x- and y-planes are the same as those derived for the two-dimensional case and presented in Tables 6.2–6.4. Equations for the boundary nodes on z-planes, i.e., the front and the back planes, can be deduced from those derived for x- or y-planes.

6.5 SOLUTION TECHNIQUES AND COMPUTER IMPLEMENTATION

It can be noted from Examples 6.1–6.5 worked out in the previous section that the system of equations resulting from a two-dimensional diffusion equation is no longer tridiagonal, as it was in the case of a one-dimensional diffusion problem. The reason for this is the fact that, while the nodal discretization equation can be rearranged so that three of the terms are at or adjacent to the main diagonal, the remaining two terms are displaced. Hence, unlike the unidirectional diffusion problem, the resulting system of equations for the two-dimensional heat diffusion cannot be solved by Tridiagonal Matrix Algorithm (TDMA). The resulting system of equations is in fact very sparse, containing many zeros. We can certainly use a direct solver such as the Gauss elimination or *LU* decomposition method discussed in Chapter 3. However, with more refined grid size distributions such direct methods are subject to increased computer memory limitation and become computationally expansive. As we have discussed in Chapter 3, that iterative point-by-point methods such as Gauss–Seidel are the method of choice for an increased number of nodes and for such a sparse system. Additional discussion about the solution techniques for multidimensional problems is given in Chapter 7 for multidimensional unsteady-state problems. In the following section, we present the solution algorithm or pseudo code for solving a two-dimensional diffusion equation based on the Gauss–Seidel algorithm.

6.5.1 Solution Algorithm Based on the Gauss–Seidel Method

This is the simplest to implement of all iterative methods as discussed in Chapter 3. It generally involves the following steps.

1. Before assembling the discretization equations for all nodes, it is desirable to reorder each equation such that the resulting system of equations is diagonally dominant, a sufficient condition for the convergence of the Gauss–Seidel method.
2. Write each equation in the system in explicit form for the nodal value associated with the diagonal element as

$$\phi_i^{k+1} = \frac{c_i - \sum_{j=1}^{i-1} a_{ij}\phi_j^{k+1} - \sum_{j=i+1}^{n} a_{ij}\phi_j^{k}}{a_{ii}}, \quad \text{for } i = 1,2,\ldots,n \tag{6.46}$$

   where the superscript index k represents the previous iteration number and $k+1$ represents the present iteration.
3. An initial set of values for ϕ_i is assumed as ϕ_i^0.
4. New improved iterative values ϕ_i^1 for $k = 1$ are calculated by substituting initial guess value ϕ_i^0 at $k = 0$ and any recent calculated value ϕ_i^1 at $k = 1$ into the right-hand side of Equation 6.46. This concludes the first iteration step.
5. Step 3 is repeated for all new iterated values of ϕ_i^{k+1} using the most recent values of ϕ.
6. The iteration process is continued until a prescribed convergence criterion such as

$$\left|\frac{\phi_i^{k+1} - \phi_i^k}{\phi_i^{k+1}}\right| \le \varepsilon_s \tag{6.47a}$$

   or

$$\frac{\left\|\phi_i^{k+1} - \phi_i^k\right\|_e}{\left\|\phi_i^{k+1}\right\|_e} \le \varepsilon_s \tag{6.47b}$$

   is satisfied.

A pseudo code for solving a two-dimensional diffusion equation by the Gauss–Seidel method is given in Table 6.5.

6.5.2 Solution by Combination of TDMA and Gauss–Seidel Method (Line-by-Line Method)

This procedure is also referred to as the **line-by-line method** as suggested by Patankar (1980). The **line-by-line method** is primarily a combination of a TDMA algorithm for a one-dimensional problem and a Gauss–Seidel iteration method. The basic approach involves either sweeping in the x direction while determining all nodal values along a vertical line using TDMA or sweeping in the y direction while determining all nodal values along a horizontal line using TDMA. The values of all nodal points ($\phi_{i,j}^{n+1}$, $j = 1, \ldots, N$) along a vertical line (solid line in Figure 6.11) are determined in the present iteration ($k + 1$) as a one-dimensional problem by using the TDMA algorithm while assuming known values for the nodal points along two adjacent vertical lines, shown as dotted lines at x_{i-1} and x_{i+1} in Figure 6.11. The nodal values in the x_{i-1} line are assumed to be known in the present iteration ($k + 1$), while the nodal values in the x_{i+1} line are known based on the previous iteration (k).

This is repeated for all vertical lines in the x-direction, and one iteration is said to be complete when all vertical lines are scanned. In the next iteration, one can choose to sweep in the y-direction and estimate all nodal values along all horizontal lines.

TABLE 6.5
Pseudo Code for Steady State Solution of Two-Dimensional Diffusion Equation Using Finite Difference Scheme and Gauss-Seidel Method

Input dimensions, properties, select number of divisions, maximum number of iteration parameters

```
    Input dimension L and H
    Input transport property Γ

   Input number of divisions NDIVX and NDIVY
   Number of node in x-direction NX = NDIVX+1
   Number of node in y-direction NY = NDIVY+1
   Input maximum number of iteration, itmax
Set initial guess values
   dofor I = 1, NX
     dofor J = 1, NY
          φ(I,J) = φ_I
     enddo
   enddo
   iflag = 1
   iter = 0
   dowhile (it < itmax and iflag = 1)
      iter = iter+1
Set new values to old values
          dofor I = 2, NX-1
           dofor J=2, NY-1
           φo(I,J) = φ(I,J)
          Enddo
Calculate new iterated nodal values
          dofor I = 1, NX
              sum1=0.0
              do for j = 1, (i-1)
                sum1 = sum1 + A(I,J) * φ(I,J)
              end do
              sum2 = 0.0
              do for j = (i+1), NY
                sum2 = sum2 + A(I,J) * φo(I,J)
              end do
              φ(I,J) = [C(I) - sum1 - sum2]/A(I,I)
          enddo
Estimate approximate percent relative error and check for convergence
          dofor I = 1, NX
            dofor J=1, NY
                  ε_a = ABS( (φ(I,J) - φo(I,J)) / φ(I,J) )
                  if (ε_a>ε_s) then
                     iflag = 1
                  else
                     iflag = 0
                  endif
               enddo
            enddo
          enddo
```

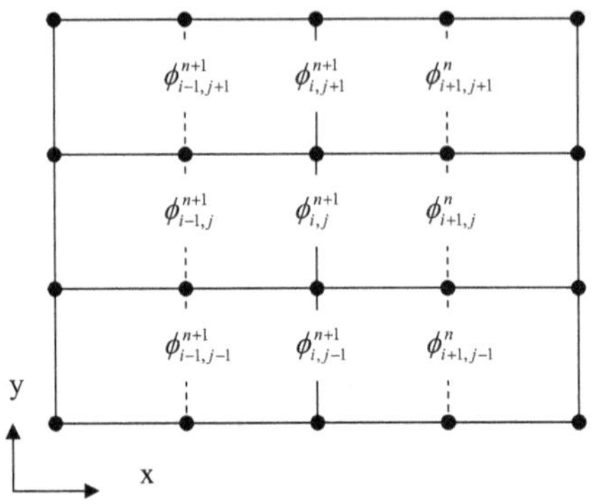

FIGURE 6.11 Computational grid in the line-by-line method with sweeping in the x-direction.

For example, to sweep all vertical lines in the x-direction, the discretization equation for the two-dimensional diffusion equation (5.27) can be rearranged in the following manner:

$$a_{i,j}\phi^{k+1}_{i,j} = a_{i,j+1}\phi^{k+1}_{i,j+1} + a_{i,j-1}\phi^{k+1}_{i,j-1} + a_{i-1,j}\phi^{k+1}_{i-1,j} + a_{i+1,j}\phi^{k}_{i+1,j} + d \quad (6.48)$$

Notice that the term $a_{i-1,j}\phi^{k+1}_{i-1,j}$ on right-hand side is known as it has already been computed at the present iteration (k+1) as we seep from the left and the term $a_{i+1,j}\phi^{k}_{i+1,j}$ is known based on its value at the previous iteration (k). To apply the TDMA algorithm at any y line, we can first rewrite the discretization equation in the form of a tridiagonal matrix format as

$$A_j\phi^{k+1}_{i,j} = B_j\phi^{k+1}_{i,j+1} + C_j\phi^{k+1}_{i,j-1} + D_j \quad (6.49)$$

where

$$A_j = a_{i,j},\ B_j = a_{i,j+1},\ C_j = a_{i,j-1}$$

$$D_j = a_{i-1,j}\phi^{k+1}_{i-1,j} + a_{i+1,j}\phi^{k}_{i+1,j} + d$$

Equation 6.49 can now be solved using the TDMA algorithm. Let us demonstrate the development of a pseudo code for solving a two-dimensional diffusion problem by considering two different cases involving different types of boundary conditions in a rectangular slab.

Case 6.1: With Boundary Condition of the First Kind

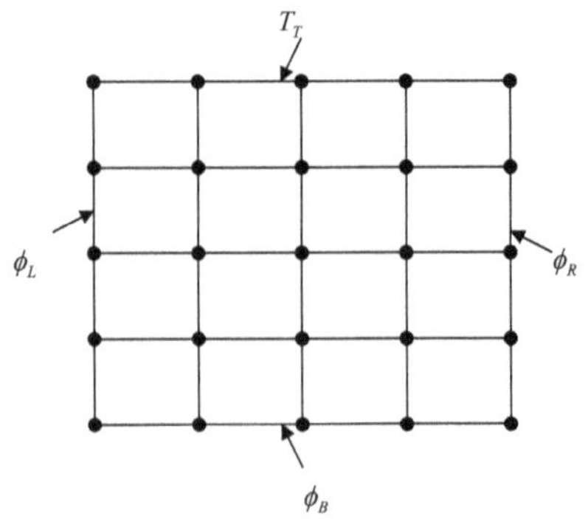

See Table 6.6.

TABLE 6.6
Pseudo-Code for Line-by-Line Method with Sweeping in *x*-Direction and with Boundary Condition of the First Kind at the Top and Bottom Surfaces

```
Dofor I = 2, (NX – 1)
  A(1) = 0.0
  A(NX)=0.0
  B(1) = 0.0
  B(NX)=0.0
  Dofor J=2, (NY – 1)

    A(J) = a_{i,j}
    B(J) = a_{i,j+1}
    C(J) = a_{i,j-1}

    D(J) = a_{i+1,j}φ_{i+1,j} + a_{i-1,j}φ_{i-1,j} + b
  Enddo
  Call tdma (NX, A, B, C, D,φ)
Enddo
```

Case 6.2: With Boundary Condition of the First Kind at the Top and No Flux at the Bottom

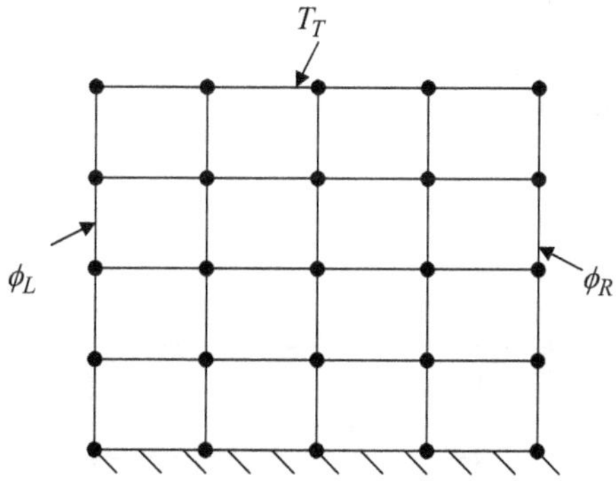

See Table 6.7.

We can see from the pseudo codes that the boundary condition information is brought in to calculate interior nodes values. This results in an improved convergence rate in a line-by-line method compared to the Gauss–Seidel method. Alternating the sweeping directions essentially improves the convergence rate. However, selection of the preferred sweeping direction may be dictated, in many problems, by the geometry and magnitude of the boundary conditions, as well as magnitudes of the coefficient values. Alternating the sweeping directions essentially improves the convergence rate.

TABLE 6.7
Pseudo-Code for Line-by-Line Method with Sweeping in *x*-Direction and with Boundary Condition of the First Kind at the Top Surface and No Flux at the Bottom Surface

Dofor I = 2, $(NX-1)$
  A(1) = 0.0
  A(NX**)=0.0**
  B(1) = 0.0
  B(NX**)=0.0**
  Dofor J=2, $(NY-1)$

   A(J) = $a_{i,j}$
   B(J) = $a_{i,j+1}$
   C(J) = $a_{i,j-1}$

   D(J) = $a_{i+1,j}\phi_{i+1,j} + a_{i-1,j}\phi_{i-1,j} + b$
  Enddo
  Call tdma (NX, A, B, C, D,φ**)**
Enddo

PROBLEMS

6.1 Consider a 1.5 m by 1.5 m square slab with the top and left surfaces maintained at high temperature, $T_H = 600°C$. The left, right, and bottom surfaces are subjected to a convection condition with $h_c = 30\,W/m^2°C$ and $T_\infty = 25°C$. The material conductivity is $k = 30$ W/m °C. Apply the finite difference method to calculate the two-dimensional temperature distribution in the slab based on the grid distribution shown in the figure. Estimate the heat transfer rates at all surfaces and check for the overall energy balance.

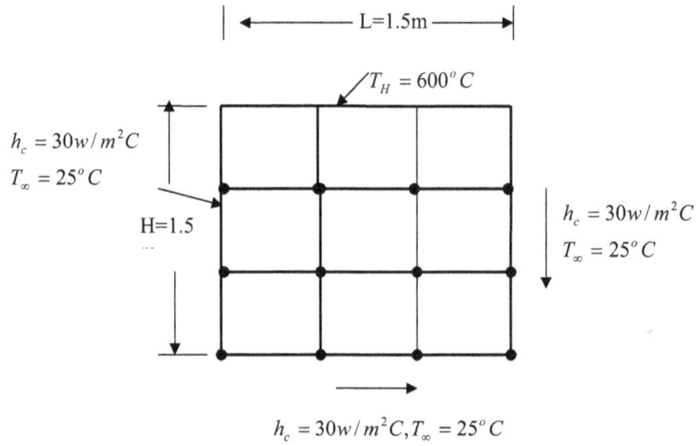

The mathematical statement of the problem is as follows.

Governing Equation

$$\frac{\partial^2 T}{\partial x^2} + \frac{\partial^2 T}{\partial x^2} = 0$$

Boundary Conditions

1. $x=0, \quad k\left.\frac{\partial T}{\partial x}\right|_{x=0} = h_c\left(\left.T\right|_{x=0} - T_\infty\right)$

2. $x=L, \quad -k\left.\frac{\partial T}{\partial x}\right|_{x=L} = h_c\left(\left.T\right|_{x=L} - T_\infty\right)$

3. $y=0, \quad k\left.\frac{\partial T}{\partial y}\right|_{y=0} = h\left(\left.T\right|_{y=0} - T_\infty\right)$

4. $y=H, \quad T(x,H) = T_{\mathrm{H}}$

Solve the system of equations by using (a) the Gauss-elimination method and (b) the Gauss–Seidel method.

6.2 Develop a computer code for solving a two-dimensional diffusion equation in a rectangular slab using the pseudo code given in Table 6.5 based on using the Gauss–Seidel algorithm. Use the computer code to solve Problem 6.1 with refining the grid as 3×3, 4×4, 6×6, and 8×8. Check the convergence of the solution by comparing the temperature distribution at the mid-section with the progressively refined grids.

6.3 Consider a 0.75 m by 1.0 m porous rectangular slab with the top and bottom surfaces maintained at a constant moisture concentration of $C_0 = 0.002\,\mathrm{kg/m^3}$. The left surface is exposed to moving air with a mass transfer coefficient of $h_D = 0.02\,\mathrm{m/s}$ and a moisture concentration of $C_\infty = 0.02\,\mathrm{kg/m^3}$. The right surface is attached to an impermeable aluminum foil. The material is assumed to be homogeneous with diffusion coefficient $D_0 = 4.0\times10^{-5}\,\mathrm{m^2/s}$.

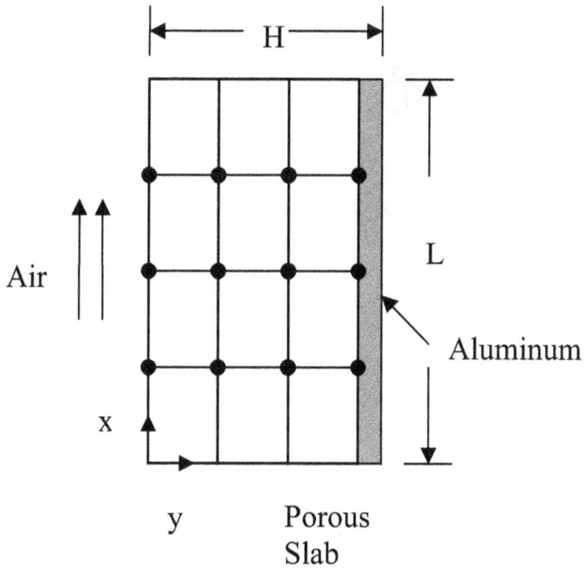

The mathematical statement of the problem is as follows.

Governing Equation

$$\frac{\partial}{\partial x}\left(D_x\frac{\partial C}{\partial x}\right)+\frac{\partial}{\partial y}\left(D_y\frac{\partial C}{\partial y}\right)=0$$

Boundary Conditions

1. $x=0, \quad -D_0\left.\frac{\partial C}{\partial x}\right|_{x=0}=h_D(C_\infty-C|_x=0)$

2. $x=H, \quad \left.\frac{\partial C}{\partial x}\right|_{x=H}=0$

3. $y=0, \quad C(x,0)=C_0$

4. $y=L, \quad C(x,L)=C_0$

Apply a finite difference method to calculate the two-dimensional moisture concentration distribution in the slab based on the grid distribution shown in the figure. Estimate the moisture transfer rates at all surfaces and show the overall moisture balance.

6.4 Consider temperature rise in a metal cutting tool. Heat is generated in primary and secondary shear zones of a metallic workpiece and metal chips during metal cutting processes. This heat is transmitted by conduction into the cutting tools. To study the thermal effects on the cutting tool, it is necessary to evaluate the temperature distribution in the tools.

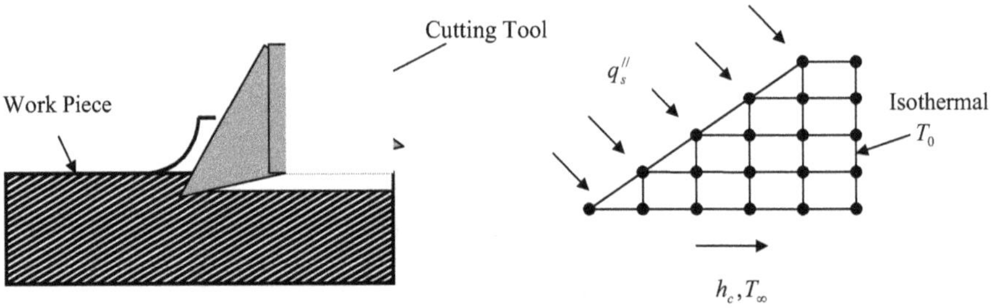

A simplified heating condition is considered in this problem by considering a constant surface heat flux at the top rake surface and the bottom surface exposed to forced convective cooling. Further interior regions, i.e., the right surface, are assumed to be isothermal or at a constant temperature T_0. Consider the grid distribution, write finite difference equations for each node, and obtain the system of equations for the unknown nodal values.

6.5 Consider the problem of cooling an electronic chip that is mounted on a substrate. The electric power generation in the chip is $q = 80$ W/cm^3 and the size of the chip is 10 mm × 10 mm×5 mm. The top and bottom surfaces are assumed to be convectively cooled with $h_c = 50$ W/m^2°C and $T_\infty = 20$°C. The two-side surfaces are assumed to be at a constant temperature $T_0 = 25$°C. The conductivities of the substrate and chip materials are $K_s = 4$ W/m °C and $K_c = 60$ W/m °C, respectively.

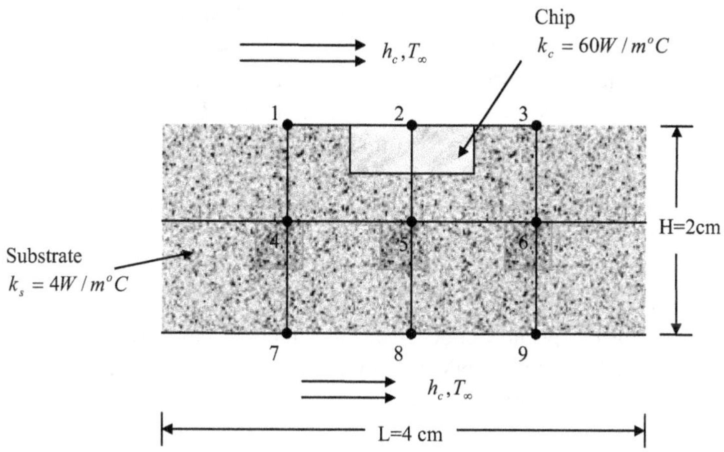

Use the finite difference method/control volume to calculate the two-dimensional temperature distribution in the chip and the substrate using the grid distribution shown in the figure. Check the overall energy balance.

6.6 Use the finite difference method to solve the two-dimensional steady-state conduction in a rectangular carbon steel ($k = 200$ W/m °C) slab subjected to a constant surface heat flux irradiated by a high-energy laser beam at the top surface. For simplicity, assume the heat flux distribution to be a constant average value, $q_0'' = 2 \times 10^8$ W/m^2, acting over a section of the surface. The top surface is also subjected to forced convection with $h_c = 500$ W/m^2°C. The left surface is assumed to be adiabatic as a line of symmetry. All other surfaces are assumed to be subjected to free convection with $T_\infty = 25$°C and $h = 40$ W/m^2°C. Derive the finite difference equations for all nodes and represent the system in matrix form. Assume $\Delta x = \Delta y = 1.0\ cm$.

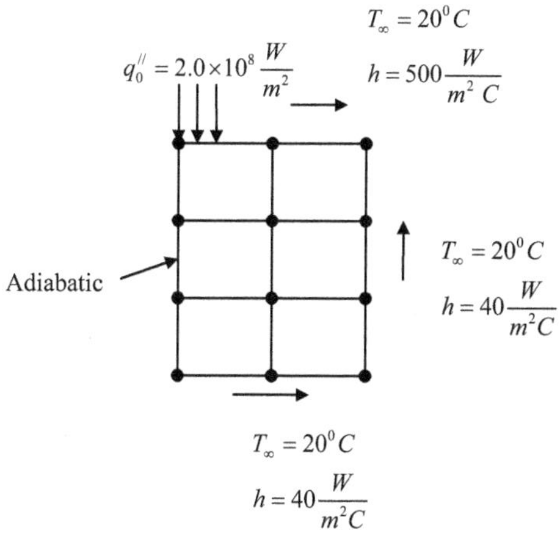

a. Derive the finite difference equations for all nodes and represent the system in matrix form.
b. Solve the system of equations using the Gauss elimination solver and check for the overall energy balance.

6.7 Solve the velocity distribution for the example Problem 6.5 using the computer code based on the Gauss–Seidel method in Problem 6.1. Estimate the average velocity and volume flow rate using two-dimensional quadrature formulas presented in Chapter 4. Use the following data points for the computation. $H = L = 8.00\,\text{mm}$, $\mu = 0.02$ kg/m/s, $\partial P/\partial x = \overline{S} = -750\text{Pa/m}$, and $U_0 = 10$ mm/s.

Repeat the computation by further refining the grid to 3×3, 4×4, 6×6, and 8×8. Check the convergence by observing velocity distribution at the mid-section along the y-direction and by observing percent relative error in average velocity or volume flow rate. Use the assigned tolerance limit, $\varepsilon_s = 0.001$.

6.8 Develop a computer code for solving a two-dimensional diffusion equation in a rectangular slab using the pseudo code given in Table 6.7, based on using the line-by-line algorithm. Use the computer code to solve the problem presented in Example 6.1 by progressively refining the grid as 4×4, 6×6, and 8×8. Check the convergence of the solution by comparing temperature distribution at the mid-section with progressively refined grids. Use the assigned tolerance limit, $\varepsilon_s = 0.001$.

7 Finite Difference–Control Volume Method

Unsteady State Diffusion Equation

In this chapter, we will introduce the basic treatment of an unsteady-state term in the finite difference method using the one-dimensional steady-state problem. Different temporal approximations that are needed to discretize time derivative are introduced. The procedure will be extended to two- and three-dimensional problems.

In previous chapters on steady-state problems, we have demonstrated the use of both Taylor series approach and control volume approach in deriving the finite difference equations. In our treatment of the unsteady term, we also employ both these approaches to demonstrate the discretization procedure.

7.1 TIME DISCRETIZATION PROCEDURE

Let us demonstrate the time discretization procedure by considering the governing equation as the one-dimensional diffusion equation with one diffusion term in the x-direction and a storage term or unsteady-state term

$$C\frac{\partial \phi}{\partial t} = \frac{\partial}{\partial x}\left(\Gamma_x \frac{\partial \phi}{\partial x}\right) \tag{7.1}$$

Also, we will demonstrate the procedure using the control volume approach and considering an interior node shown in Figure 7.1.

As time is a one-way coordinate, we obtain the solution by marching in time, i.e., for a given ϕ value at time step t, we find ϕ at $t + \Delta t$. Integrating Equation 7.1 over the control volume (Figure 7.1) from $i - \frac{1}{2}$ to $i + \frac{1}{2}$ and over the time interval t to $t + \Delta t$, we get

$$\int_t^{t+\Delta t}\int_{i-\frac{1}{2}}^{i+\frac{1}{2}} C\frac{\partial \phi}{\partial t}\mathrm{d}x\mathrm{d}t = \int_t^{t+\Delta t}\int_{i-\frac{1}{2}}^{i+\frac{1}{2}} \frac{\partial}{\partial x}\left(\Gamma_x \frac{\partial \phi}{\partial x}\right)\mathrm{d}x\mathrm{d}t \tag{7.2}$$

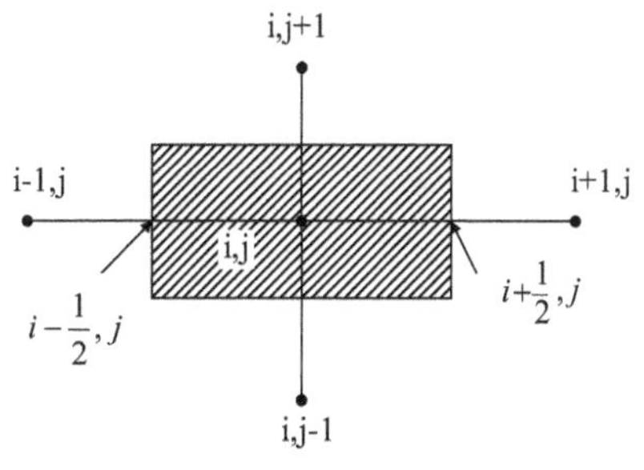

FIGURE 7.1 Control volume for an interior node.

The left-hand side of the integral is evaluated by assuming that the value of ϕ at the node i, i.e., ϕ_i, prevails over the entire control volume surrounding the node

$$C\int_{i-\frac{1}{2}}^{i+\frac{1}{2}}\int_{t}^{t+\Delta t}\frac{\partial\phi}{\partial t}\mathrm{d}t\mathrm{d}x = C\Delta x_i\int_{t}^{t+\Delta t}\frac{\partial\phi_i}{\partial t}\mathrm{d}t = C\Delta x_i\,\phi_i\Big|_t^{t+\Delta t} = C(\phi_i^{l+1}-\phi_i^l)\Delta x_i \tag{7.3}$$

where

$$\phi_i^{l+1} = \text{new value at time } t+\Delta t$$

and

$$\phi_i^l = \text{old value at time } t$$

The right-hand side integral of Equation 7.2 is evaluated based on the discretization procedure outlined in Chapter 5 for the diffusion term as

$$\int_{t}^{t+\Delta t}\int_{i-\frac{1}{2}}^{i+\frac{1}{2}}\frac{\partial}{\partial x}\left(\Gamma_x\frac{\partial\phi}{\partial x}\right) = \int_{t}^{t+\Delta t}\left[\left(\Gamma_x\frac{\partial\phi}{\partial x}\right)_{i+\frac{1}{2}} - \left(\Gamma_x\frac{\partial\phi}{\partial x}\right)_{i-\frac{1}{2}}\right]\mathrm{d}t \tag{7.4}$$

Approximating the first-order derivative terms by piece-wise linear profile, we get

$$\int_{t}^{t+\Delta t}\int_{i-\frac{1}{2}}^{i+\frac{1}{2}}\frac{\partial}{\partial x}\left(\Gamma_x\frac{\partial\phi}{\partial x}\right)\mathrm{d}x\mathrm{d}t = \int_{t}^{t+\Delta t}\left[\frac{\Gamma_{i+\frac{1}{2}}(\phi_{i+1}-\phi_i)}{(\Delta x)_{i+\frac{1}{2}}} - \frac{\Gamma_{i-\frac{1}{2}}(\phi_i-\phi_{i-1})}{(\Delta x)_{i-\frac{1}{2}}}\right]\mathrm{d}t \tag{7.5}$$

In the next step, we need a profile assumption for the variation of ϕ over the time step t to $t+\Delta t$. Let us assume a general profile such as

$$\phi_i = \left[f\phi_i^{l+1} + (1-f)\phi_i^l\right] \tag{7.6}$$

where f is called the time approximation weighting factor.

Using this general temporal profile assumption, we can write each one of the integral terms in Equation 7.3 as

$$\int_{t}^{t+\Delta t}\phi_i\,\mathrm{d}t = \left[f\phi_i^{l+1} + (1-f)\phi_i^l\right]\Delta t \tag{7.7a}$$

$$\int_{t}^{t+\Delta t}\phi_{i+1}\,\mathrm{d}t = \left[f\phi_{i+1}^{l+1} + (1-f)\phi_{i+1}^l\right]\Delta t \tag{7.7b}$$

$$\int_{t}^{t+\Delta t}\phi_{i-1}\,\mathrm{d}t = \left[f\phi_{i-1}^{l+1} + (1-f)\phi_{i-1}^l\right]\Delta t \tag{7.7c}$$

Substituting Equation 7.7 into Equation 7.5, we get

$$\int_t^{t+\Delta t}\int_{i-\frac{1}{2}}^{i+\frac{1}{2}} \frac{\partial}{\partial x}\left(\Gamma_x \frac{\partial \phi}{\partial x}\right) \mathrm{d}x\mathrm{d}t = f\left[\frac{\Gamma_{i+\frac{1}{2}}\left(\phi_{i+1}^{l+1}-\phi_i^{l+1}\right)}{(\Delta x)_{i+\frac{1}{2}}} - \frac{\Gamma_{i-\frac{1}{2}}\left(\phi_i^{l+1}-\phi_{i-1}^{l+1}\right)}{(\Delta x)_{i-\frac{1}{2}}}\right]\Delta t$$
$$+(l-f)\left[\frac{\Gamma_{i+\frac{1}{2}}\left(\phi_{i+1}^{l}-\phi_i^{l}\right)}{(\Delta x)_{i+\frac{1}{2}}} - \frac{\Gamma_{i-\frac{1}{2}}\left(\phi_i^{l}-\phi_{i-1}^{l}\right)}{(\Delta x)_{i-\frac{1}{2}}}\right]\Delta t \quad (7.8)$$

Substituting Equations 7.3 and 7.8 into Equation 7.2, we have the discretization equation for an interior node as

$$C\left(\phi_i^{l+1}-\phi_i^l\right)\Delta x_i = f\left[\frac{\Gamma_{i+\frac{1}{2}}\left(\phi_{i+1}^{l+1}-\phi_i^{l+1}\right)}{(\Delta x)_{i+\frac{1}{2}}} - \frac{\Gamma_{i-\frac{1}{2}}\left(\phi_i^{l+1}-\phi_{i-1}^{l+1}\right)}{(\Delta x)_{i-\frac{1}{2}}}\right]\Delta t$$
$$+(1-f)\left[\frac{\Gamma_{i+\frac{1}{2}}\left(\phi_{i+1}^{l}-\phi_i^{l}\right)}{(\Delta x)_{i+\frac{1}{2}}} - \frac{\Gamma_{i-\frac{1}{2}}\left(\phi_i^{l}-\phi_{i-1}^{l}\right)}{(\Delta x)_{i-\frac{1}{2}}}\right]\Delta t \quad (7.9)$$

Rearranging and regrouping terms, we have

$$a_i\phi_i^{l+1} = a_{i+1}\left[f\phi_{i+1}^{l+1}+(1-f)\phi_{i+1}^l\right] + a_{i-1}\left[f\phi_{i-1}^{l+1}+(1-f)\phi_{i-1}^l\right]$$
$$+\left[a_i^l-(1-f)a_{i+1}-(1-f)a_{i-1}\right]\phi_i^l \quad (7.10a)$$

where

$$a_i = fa_{i+1}+fa_{i-1}+a_i^l \quad (7.10b)$$

$$a_{i+1} = \frac{\Gamma_{i+\frac{1}{2}}}{(\Delta x)_{i+\frac{1}{2}}} \quad (7.10c)$$

$$a_{i-1} = \frac{\Gamma_{i-\frac{1}{2}}}{(\Delta x)_{i-\frac{1}{2}}} \quad (7.10d)$$

$$a_i^l = \frac{C\Delta x}{\Delta t} \quad (7.10e)$$

The selection of the time approximation weighting factor leads to different time approximation schemes, which are named as the **explicit scheme** with **$f = 0$, the Crank–Nicolson scheme** with $f = 0.5$, and **fully implicit scheme** with $f = 1$. A brief description along with the physical meaning of these schemes is given below.

7.2 EXPLICIT SCHEME

7.2.1 Discretization Equation by Control Volume Approach

An explicit form of the discretization equation is obtained substituting $f = 0$ into Equation 7.10 as

$$a_i\phi_i^{l+1} = a_{i+1}\phi_{i+1}^l + a_{i-1}\phi_{i-1}^l + \left[a_i^l - a_{i+1} - a_{i-1}\right]\phi_i^l \quad (7.11a)$$

where

$$a_i = a_i^l \tag{7.11b}$$

$$a_{i+1} = \frac{\Gamma_{i+\frac{1}{2}}}{(\Delta x)_{i+\frac{1}{2}}} \tag{7.11c}$$

$$a_{i-1} = \frac{\Gamma_{i-\frac{1}{2}}}{(\Delta x)_{i-\frac{1}{2}}} \tag{7.11d}$$

$$a_i^l = \frac{C\Delta x_i}{\Delta t} \tag{7.11e}$$

Physically, in the **explicit scheme**, it is assumed that the old value of ϕ_i^l prevails throughout the entire time step except at time $t+\Delta t$, as depicted in Figure 7.2, which show temporal variation of ϕ with time.

Substituting the coefficient values in Equation 7.11, we get

$$\frac{C\Delta x_i}{\Delta t}\phi_i^{l+1} = \frac{\Gamma_{i+\frac{1}{2}}}{\Delta x_{i+\frac{1}{2}}}\phi_{i+1}^l + \frac{\Gamma_{i-\frac{1}{2}}}{\Delta x_{i-\frac{1}{2}}}\phi_{i-1}^l + \left(\frac{C\Delta x_i}{\Delta t} - \frac{\Gamma_{i+\frac{1}{2}}}{\Delta x_{i+\frac{1}{2}}} - \frac{\Gamma_{i-\frac{1}{2}}}{\Delta x_{i-\frac{1}{2}}}\right)\phi_i^l \tag{7.12}$$

For a special case of constant transport property Γ and uniform grid size distribution, i.e., $\Delta x_{i+\frac{1}{2}} = \Delta x_{i-\frac{1}{2}} = \Delta x_i = \Delta x$, we get

$$\phi_i^{l+1} = \frac{\Gamma\Delta t}{C\Delta x^2}\left(\phi_{i+1}^l + \phi_{i-1}^l\right) + \left(1 - \frac{2\Gamma\Delta t}{C\Delta x^2}\right)\phi_i^l \tag{7.13}$$

Similarly, for a two-dimensional problem, we get

$$\phi_{i,j}^{l+1} = \frac{\Gamma\Delta t}{C(\Delta x)^2}\left[\phi_{i+1,j}^l + \phi_{i-1,j}^l + \phi_{i,j+1}^l + \phi_{i,j-1}^l\right] + \left[1 - \frac{4\Gamma\Delta t}{C(\Delta x)^2}\right]\phi_{i,j}^l \tag{7.14}$$

The finite difference equations presented here are called the **explicit form**, because variable ϕ_i^{l+1} at time step $(l+1)$ can be determined directly from the knowledge of the variables ϕ_{i+1}^l, ϕ_i^l, and ϕ_{i-1}^l at the previous time step l.

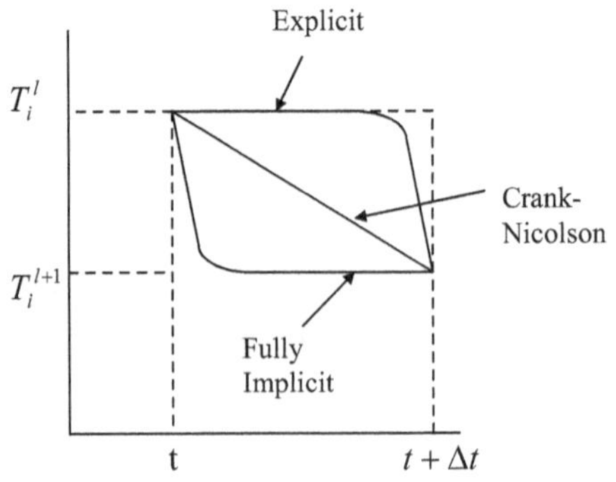

FIGURE 7.2 Time approximation schemes.

Equation (7.14) can be extended to **three-dimension problem**, we get

$$\phi_{i,j}^{l+1} = \frac{\Gamma\Delta t}{C(\Delta t)^2}\left(\phi_{i+1,j,k}^{l} + \phi_{i-1,j,k}^{l} + \phi_{i,j+1,k}^{l} + \phi_{i,j-1,k}^{l} + \phi_{i,j,k+1}^{l} + \phi_{i,j,k-1}^{l}\right) + \left[1 - \frac{6\Gamma\Delta t}{C(\Delta x)^2}\right]\Gamma_{i,j,k}^{l} \quad (7.15)$$

Note that Equations 7.13–7.15 are derived assuming uniform grid-size distributions in x-, y-, and z-directions, respectively. The procedure outlined in Section 7.1 can be utilized to derive discretization equations for nodes in problems that require nonuniform grid-size distributions.

7.2.2 Finite Difference Equation by Taylor Series Expansion

As was demonstrated before, we can derive the discretization equation directly by applying finite difference approximations to the derivatives in the unsteady-state diffusion equation (7.1). Let us demonstrate this by considering the one-dimensional, time-dependent diffusion equation of the form

$$C\frac{\partial \phi}{\partial t} = \Gamma\frac{\partial^2 \phi}{\partial x^2} \quad (7.16)$$

Using the second-order central-difference formula for the second derivative diffusion term and the forward difference formula for the first-order time derivative, we get

$$C\frac{\phi_i^{l+1} - \phi_i^l}{\Delta t} = \Gamma\frac{\phi_{i+1}^l - 2\phi_i^l + \phi_{i-1}^l}{(\Delta x)^2} \quad (7.17)$$

with error of the $0(\Delta t) + 0(\Delta x)^2$.

Rearranging, we obtain

$$\phi_i^{l+1} = \frac{\Gamma\Delta t}{C(\Delta x)^2}\left[\phi_{i+1}^l + \phi_{i-1}^l\right] + \left[1 - \frac{2\Gamma\Delta t}{C(\Delta x)^2}\right]\phi_i^l \quad (7.18)$$

Similarly, for the two-dimensional unsteady diffusion equation of the form

$$C\frac{\partial \phi}{\partial t} = \Gamma\left(\frac{\partial^2 \phi}{\partial x^2} + \frac{\partial^2 \phi}{\partial y^2}\right) \quad (7.19)$$

we obtain

$$C\frac{\phi_{i,j}^{l+1} - \phi_{i,j}^l}{\Delta t} = \Gamma\frac{\phi_{i+1,j}^l - 2\phi_{i,j}^l + \phi_{i-1,j}^l}{(\Delta x)^2} + \Gamma\frac{\phi_{i,j+1}^l - 2\phi_{i,j}^l + \phi_{i,j-1}^l}{\Delta y^2} \quad (7.20)$$

Rearranging we obtain

$$\phi_{i,j}^{l+1} = \frac{\Gamma\Delta t}{C}\left[\left(\phi_{i+1,j}^l + \phi_{i-1,j}^l\right) + \left(\frac{\Delta x}{\Delta y}\right)^2\left(\phi_{i,j+1}^l + \phi_{i,j-1}^l\right)\right] + \left[1 - \frac{4\Gamma\Delta t}{C(\Delta x)^2}\right]\phi_{i,j}^l \quad (7.21a)$$

For uniform grid size distribution, i.e., $\Delta x = \Delta y$

$$\phi_{i,j}^{l+1} = \frac{\Gamma\Delta t}{C(\Delta x)^2}\left[\phi_{i+1,j}^l + \phi_{i-1,j}^l + \phi_{i,j+1}^l + \phi_{i,j-1}^l\right] + \left[1 - \frac{4\Gamma\Delta t}{C(\Delta x)^2}\right]\phi_{i,j}^l \quad (7.21b)$$

Similarly, for the three-dimensional unsteady diffusion equation of the form

$$C\frac{\partial \phi}{\partial t}=\Gamma\left(\frac{\partial^2 \phi}{\partial x^2}+\frac{\partial^2 \phi}{\partial y^2}+\frac{\partial^2 \phi}{\partial z^2}\right) \tag{7.22}$$

We obtain

$$C\frac{\phi^{l+1}_{i,j,k}-\phi^{l}_{i,j,k}}{\Delta t}=\Gamma\frac{\phi^{l}_{i+1,j,k}-2\phi^{l}_{i,j,k}+\phi^{l}_{i-1,j,k}}{(\Delta x)^2}+\Gamma\frac{\phi^{l}_{i,j+1,k}-2\phi^{l}_{i,j,k}+\phi^{l}_{i,j-1,k}}{(\Delta y)^2}+\Gamma\frac{\phi^{l}_{i,j,k+1}-2\phi^{l}_{i,j,k}+\phi^{l}_{i,j,k-1}}{(\Delta x)^2} \tag{7.23}$$

Rearranging we obtain

$$\phi^{l+1}_{i,j,k}=\frac{\Gamma\Delta t}{C}\left[\left(\phi^{l}_{i+1,j,k}+\phi^{l}_{i-1,j,k}\right)+\left(\frac{\Delta x}{\Delta y}\right)^2\left(\phi^{l}_{i,j+1,k}+\phi^{l}_{i,j-1,k}\right)+\left(\frac{\Delta x}{\Delta z}\right)^2\left(\phi^{l}_{i,j+1,k}+\phi^{l}_{i,j-1,k}\right)\right] \tag{7.24a}$$

and for uniform grid size distributions, i.e., $\Delta x = \Delta y = \Delta y$

$$\phi^{l+1}_{i,j,k}=\frac{\Gamma\Delta t}{C}\left[\phi^{l}_{i+1,j,k}+\phi^{l}_{i-1,j,k}+\phi^{l}_{i,j+1,k}+\phi^{l}_{i,j-1,k}+\phi^{l}_{i,j+1,k}+\phi^{l}_{i,j-1,k}\right]+\left[1-\frac{6\Gamma\Delta t}{C(\Delta x)^2}\right]\phi^{l}_{i,j,k} \tag{7.24b}$$

In order to demonstrate the derivation of the discretization equation for problems with uniform volume heat generation, let us consider the with following governing heat equation for the three-dimensional heat diffusion equation as an example

$$C\frac{\partial \phi}{\partial t}=\Gamma\left(\frac{\partial^2 \phi}{\partial x^2}+\frac{\partial^2 \phi}{\partial y^2}+\frac{\partial^2 \phi}{\partial z^2}\right)+\dot{Q}''' \tag{7.25}$$

The appropriate discretization equation for the interior nodes can be derived as

$$\begin{aligned}C\frac{\phi^{l+1}_{i,j,k}-\phi^{l}_{i,j,k}}{\Delta t}&=\Gamma\frac{\phi^{l}_{i+1,j,k}-2\phi^{l}_{i,j,k}+\phi^{l}_{i-1,j,k}}{(\Delta x)^2}+\Gamma\frac{\phi^{l}_{i,j+1,k}-2\phi^{l}_{i,j,k}+\phi^{l}_{i,j-1,k}}{(\Delta y)^2}\\&\quad+\Gamma\frac{\phi^{l}_{i,j,k+1}-2\phi^{l}_{i,j,k}+\phi^{l}_{i,j,k-1}}{(\Delta x)^2}+\dot{Q}'''\end{aligned} \tag{7.26}$$

Rearranging we obtain

$$\begin{aligned}\phi^{l+1}_{i,j,k}&=\frac{\Gamma\Delta t}{C}\left[\left(\phi^{l}_{i+1,j,k}+\phi^{l}_{i-1,j,k}\right)+\left(\frac{\Delta x}{\Delta y}\right)^2\left(\phi^{l}_{i,j+1,k}+\phi^{l}_{i,j-1,k}\right)+\left(\frac{\Delta x}{\Delta z}\right)^2\left(\phi^{l}_{i,j+1,k}+\phi^{l}_{i,j-1,k}\right)\right]\\&\quad+\left[1-\frac{6\Gamma\Delta t}{C(\Delta x)^2}\right]\phi^{l}_{i,j,k}+\frac{\dot{Q}'''\,\Delta t}{C}\end{aligned} \tag{7.27}$$

The explicit finite difference scheme discussed here is also referred to as the **FTCS (forward in time central in space) scheme**. This explicit form of finite-difference representation provides a relatively straightforward procedure for computer implementation. However, the explicit scheme suffers serious limitation from a **stability** point of view.

7.2.3 Stability Consideration

An important issue that needs to be examined before using finite difference equations based on an explicit scheme is that of stability. Although it is possible for any fluctuations resulting from round-off errors in numerical solutions to dampen out and reach convergence, this is not always guaranteed to occur. This error may grow at any stage of the computation and cause termination of the program. A finite difference scheme is assumed to be stable when the errors do not grow in the successive time steps. It may be indicated that the round-off error is the one that is relevant to the stability of the numerical solution. There are several methods for stability analysis that examine the behavior of the round-off errors. The most common one is the **von Neumann** stability analysis method, also known as **Fourier stability analysis method** as applied to the linear partial differential equations. This stability analysis method is introduced and described in the following references: Nicolson (1947); Charney et al. (1950); Anderson (1994)]. Readers are recommended to refer to these literatures for more understanding and application of stability analysis method to different application problems. A brief description of von Newman analysis method is described here while considering one-dimensional heat diffusion equation

Fourier or von Neumann Stability Analysis and Criterion The von Neumann method is based on the prediction of the growth or decay of the errors identified as the difference between the exact solution and the numerical solution. A limitation of this method is that it yields necessary and sufficient conditions for the stability of mathematical models that can be described as linear, initial value problems with constant coefficients. Otherwise, this method provides only necessary conditions for stability. Even though the von Newman method pertains to interior nodes of the domain, it can provide useful approximate information on the effect of the boundary conditions on the stability solution if it is applied to the boundaries of the solution domain.

The method involves solving an error equation using the Fourier series and the method of separation of variables. The method yields the **von Newman** stability criterion, which needs to be satisfied for the stability of the solution. Let us briefly outline the analysis procedure considering one-dimensional diffusion equation in a parabolic differential equation form as

Let us consider the one-dimensional diffusion equation in a parabolic differential equation form as

$$C\frac{\partial \phi}{\partial t} = \frac{\partial}{\partial x}\left(\Gamma \frac{\partial \phi}{\partial x}\right) \tag{7.28a}$$

Considering constant values of Γ and C, and setting $\alpha_0 = \frac{\Gamma}{C}$, we get

$$\frac{\partial \phi}{\partial t} = \alpha_0 \frac{\partial^2 \phi}{\partial x^2} \tag{7.28b}$$

Introducing explicit scheme, we can transform the differential equation to the following discretization equation:

$$\frac{\phi_i^{l+1} - \phi_i^l}{\Delta t} = \alpha_0 \frac{\phi_{i+1}^l - 2\phi_i^l + \phi_{i-1}^l}{(\Delta x)^2} \tag{7.29}$$

Rearranging,

$$\phi_i^{l+1} = \phi_i^l + \frac{\alpha_0 \Delta t}{(\Delta x)^2}\left(\phi_{i+1}^l - 2\phi_i^l + \phi_{i-1}^l\right) \tag{7.30}$$

The numerical solution, $T_{i,\text{num}}$, to the discretization equation (7.30) is the approximation to the exact solution, $T_{i,\text{exact}}$, at the time step, and the difference is characterized as the round-off error

$$\varepsilon_i^l = \phi_{i,\text{exact}}^l - \phi_{i,\text{num}}^l \tag{7.31}$$

Substituting $\phi_{i,\text{num}}^l == \phi_{i,\text{exact}}^l + \varepsilon_i^l$ into Equation 7.30, we get

$$\frac{\phi_{i,\text{exact}}^{l+1} + \varepsilon_i^{l+1} - \phi_{i,\text{exact}}^l - \varepsilon_i^{l+1}}{\Delta t} = \alpha_0 \frac{\phi_{i+1}^l + \varepsilon_{i+1}^l - 2\phi_i^l - \varepsilon_i^l + \phi_{i-1}^l + \varepsilon_{i-1}^l}{(\Delta x)^2} \tag{7.32a}$$

Rearranging,

$$\frac{\phi_i^{l+1} - \phi_i^l}{\Delta t} + \frac{\varepsilon_i^{l+1} - \varepsilon_i^l}{\Delta t} = \alpha_0 \frac{\phi_{i+1}^l - 2\ \phi_i^l + \phi_{i-1}^l}{(\Delta \mathrm{x})^2} + \alpha_0 \frac{\varepsilon_{i+1}^l - 2\varepsilon_i^l + \varepsilon_{i-1}^l}{(\Delta x)^2} \tag{7.32b}$$

Since exact solution satisfies the differential equation (7.28b), we can also obtain the error equation satisfying the same discretization

$$\frac{\varepsilon_i^{l+1} - \varepsilon_i^l}{\Delta t} = \alpha_0 \frac{\varepsilon_{i+1}^l - 2\varepsilon_i^l + \varepsilon_{i-1}^l}{(\Delta x)^2} \tag{7.33a}$$

or

$$\varepsilon_i^{l+1} = \varepsilon_i^l + \frac{\alpha_0 \Delta t}{(\Delta x)^2}\left(\varepsilon_{i+1}^l - 2\varepsilon_i^l + \varepsilon_{i-1}^l\right) \tag{7.33b}$$

Let introduce the following parameter:

$$\gamma_0 = \frac{\alpha_0\ \Delta t}{(\Delta x)^2} = \frac{\Gamma\ \Delta t}{C\ (\Delta x)^2} \tag{7.34}$$

and write the error difference equation as

$$\varepsilon_i^{l+1} = \varepsilon_i^l + \gamma_0\left(\varepsilon_{i+1}^l - 2\varepsilon_i^l + \varepsilon_{i-1}^l\right) \tag{7.35}$$

Based on the principles of linear differential equations with can express spatial variation of error distribution in series form as follows:

$$\varepsilon_m(x,t) = \sum_m A_m(t) e^{iw_m x} \tag{7.36a}$$

where the **amplitude** $A_m\ (t)$ of the error is a function of time, t, and ω_m is the **wavenumber** given as

$$\omega_m = \frac{m\pi}{L}, \quad m = 0,\ 1,\ 2,\ldots M \tag{7.36b}$$

$M = \dfrac{L}{\Delta x}$ = number of intervals for the spatial step size Δx over spatial dimension of L.

For the linear discretization equation under consideration, superimposition solution can be applied and we can just consider the error distribution of a single term of the series function as

$$\varepsilon_m\ (x,t) = A_m(t)(t) e^{iw_m x} \tag{7.37}$$

Following the Fourier series solution method (Churchill and Brown, 1978), we can seek a solution considering superimposition principles for linear difference equation. We consider solution of a form that consists of two separate components: one for the for time function and other for the spatial function in the following form:

$$e_m(x,t) = \mathrm{e}^{at}\mathrm{e}^{iw_m x} \tag{7.38}$$

Substituting this solution form into the error difference Equation 7.35, we get

$$\varepsilon_i^{l+1}\mathrm{e}^{iw_m x} = \mathrm{e}^{at}\mathrm{e}^{iw_m x} + \frac{\alpha_0 \Delta t}{(\Delta x)^2}\left(\mathrm{e}^{at}\mathrm{e}^{iw_m (x+\Delta x)} - 2\mathrm{e}^{at}\mathrm{e}^{iw_m (x)} + \mathrm{e}^{at}\mathrm{e}^{iw_m (x-\Delta x)}\right) \tag{7.39}$$

This can be simplified to the following form:

$$\mathrm{e}^{a\Delta t} = 1 - 4\gamma_0 \sin^2\frac{\theta}{2} \tag{7.40}$$

where $\theta = w_m \Delta x$

We can see that if $\mathrm{e}^{a\Delta t} \leq 1$, then the growth of the error will be suppressed, and we can state the necessary and sufficient condition or condition for the stability of error growth as

$$1 - 4\gamma_0 \sin^2\frac{\theta}{2} \leq 1 \tag{7.41a}$$

or

$$4\gamma_0 \sin^2\frac{\theta}{2} \leq 2 \tag{7.41b}$$

where $\gamma_0 = \dfrac{\alpha \Delta t}{C(\Delta x)^2}$ is defined as the **stability parameter**.

For **one-dimensional heat diffusion problems**, the stability parameter is referred to as the **Fourier number** f_0 and stability condition written in the following form:

$$\gamma_0 = f_0 = \frac{\alpha \Delta t}{(\Delta x)^2} \leq \frac{1}{2} \tag{7.42}$$

For **two-dimensional problems**, this criterion is derived as

$$\frac{\Gamma \Delta t}{C\Delta x^2} + \frac{\Gamma \Delta t}{C\Delta y^2} < \frac{1}{2} \tag{7.43a}$$

For uniform grid size distribution, i.e., $\Delta x = \Delta y$, the criterion becomes

$$\frac{\Gamma \Delta t}{C\Delta x^2} < \frac{1}{4} \tag{7.43b}$$

As we have defined, the parameter $\gamma_0 = \Gamma \Delta t / C\Delta x^2$ is the stability parameter. The stability criterion tells us that for given values of Γ, C, and selected grid size, Δx, there is a maximum permissible time step Δt, which cannot be exceeded for stability of the solution. In other words, if the time step Δt exceeds the limit imposed by the criterion, then the numerical computations become unstable resulting from the amplification of errors.

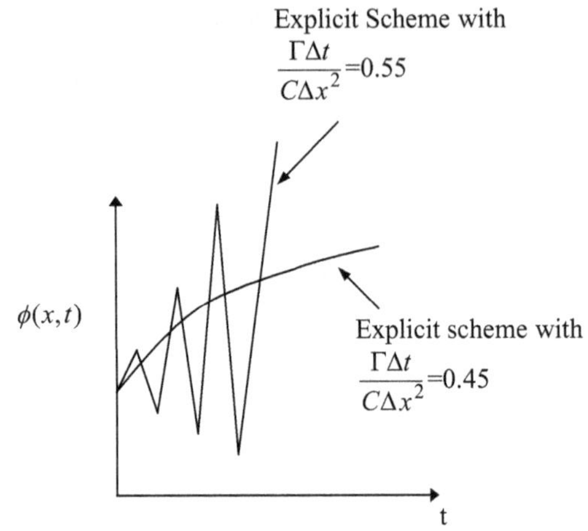

FIGURE 7.3 Effect of stability parameter in finite difference solution using the explicit scheme.

Figure 7.3 illustrates typical numerical results with different stability parameters.

Similarly, the stability conditions for **hyperbolic differential** can be derived using the von-Neumann stability method. Let us consider a linear first-order wave form of the equation involving first-order advection spatial derivative and transient terms in the following general form:

$$\frac{\partial \phi}{\partial t} + c\frac{\partial \phi}{\partial x} \tag{7.44}$$

where c is the constant wave speed with which the wave travels in spatial direction x.

Let us introduce a finite difference approximation of the first-order time derivative and first-order space derivative using the following approximation:

For the first-order time derivative, we can use first-order forward-difference scheme as

$$\frac{\partial \phi}{\partial t} = \frac{\phi_i^{l+1} - \dfrac{\phi_{i+1}^l + \phi_{i-1}^l}{2}}{\Delta t} \tag{7.45}$$

where the function value at time step t is taken as the average of the upstream and downstream values at time step t.

For the space derivative, we can us first-order central difference scheme

$$\frac{\partial \phi}{\partial x} = \frac{\phi_{i+1}^l - \phi_{i-1}^l}{2\Delta x} \tag{7.46}$$

Substituting difference approximation equations (7.43 and 7.44) into hyperbolic wave equation 7.42

$$\frac{\phi_i^{l+1} - \dfrac{\phi_{i+1}^l + \phi_{i-1}^l}{2}}{\Delta t} = c\,\frac{\phi_{i+1}^l - \phi_{i-1}^l}{2\Delta x} \tag{7.47a}$$

Rearranging,

$$\phi_i^{l+1} = \frac{\phi_{i+1}^l + \phi_{i-1}^l}{2} - \frac{C\Delta t}{\Delta x}\frac{\phi_{i+1}^l - \phi_{i-1}^l}{2} \tag{7.47b}$$

Introducing the constant $\gamma_c = \dfrac{c\Delta t}{\Delta x}$, we get the difference equation as

$$\phi_i^{l+1} = \frac{\phi_{i+1}^l + \phi_{i-1}^l}{2} - \gamma_c \frac{\phi_{i+1}^l - \phi_{i-1}^l}{2} \tag{7.48}$$

and the error difference equation

$$\varepsilon_i^{l+1} = \frac{\varepsilon_{i+1}^l + \varepsilon_{i-1}^l}{2} - \gamma_c \frac{\varepsilon_{i+1}^l - \varepsilon_{i-1}^l}{2} \tag{7.49}$$

Following the von Neumann stability method and use of superimposition principle for a linear difference equation, we can assume solution for the error as

$$\varepsilon_m(x,t) = e^{at} e^{iw_m x} \tag{7.50}$$

Substitution of the error solution form into error difference equation (7.47) leads to the stability criterion for hyperbolic wave form of equation as

$$1 - i\gamma_c \sin\frac{\theta}{2} \le 1 \quad \text{Where} \quad \theta = w_m \Delta x \tag{7.51a}$$

and taking the squares on both sides of Equation 7.49a, we get the stability criterion as

$$\gamma_c \le 1 \qquad \text{(7.51b) Or}$$

$$\frac{C\Delta t}{\Delta x} \le 1 \tag{7.51c}$$

where $\gamma_c = \dfrac{C\Delta t}{\Delta x}$ is defined as the stability parameter, **Courant-Friedrichs-Lewy (CFL) number,** and Equation 7.51 is referred to as the **CFL stability condition.** This stability condition is often used in computational fluid dynamic and heat transfer studies.

Again, it may be noted here that the above criterion provides the necessary and sufficient stability condition for linear problems with constant coefficients. For more complex problems with nonlinearities and variable coefficients, it usually provides only the necessary (but not always sufficient) stability condition. For more detailed discussion on stability analysis, refer to Jaluria and Torrance (1986), Smith (1978), Fletcher (1991), and Poulikakos (1994).

A pseudo code for solving one-dimensional transient problems based on an explicit finite difference scheme is given in Table 7.1.

Example 7.1: Explicit Scheme with Constant Surface Temperatures

Consider a one-dimensional unsteady-state heat conduction in a plane slab of thickness L = 3.6 mm as shown in the figure. Initially the slab is at a temperature T_I = 0°C. The right side of the plate is suddenly brought to a temperature T_L = 200°C. The left side is maintained at T_0 = 0°C. The material's thermo-physical properties are c_p = 760 J/kg °C, $k = 50\,\text{W/m°C}$, and $\alpha = 2\times10^{-5}\,\text{m}^2/\text{s}$.

TABLE 7.1
Pseudo Code for an Explicit Finite Difference Scheme in a Plane Slab with Constant Surface Values

Input dimensions, properties, select number of divisions, maximum time, and stability parameters

Input dimension L_x

Input thermo-physical properties Γ and C

Select number of division in x-direction NDIVX

Select stability parameter γ_0

Number of node NX = NDIVX + 1

Calculate step size DX = $L_x/_{\text{NDIVX}}$

Calculate time step from stability parameter

$Dt = \gamma_0\, C(\text{DX})^2/\Gamma$

Initialize time

time = 0.0

Set initial condition

dofor $I = 1$, NX

$\phi(I) = \phi_I$

Enddo

Set boundary conditions

$\phi(1) = \phi_L$

$\phi(\text{NX}) = \phi_R$

dowhile (time<tmax)

time = time+Dt

Calculate nodal values at interior nodes

dofor $I = 2$, NX − 1

$\phi(I) = \gamma_0 * \left(\phi(I+1) + \phi(I-1)\right) + \left(1 - 2 * \gamma_0\right) * \phi(I)$

enddo

Estimate flux quantity at right surface

$f_x'' = \Gamma\left(\phi(NX) - \phi(NX-1)/DX\right)$

Enddo

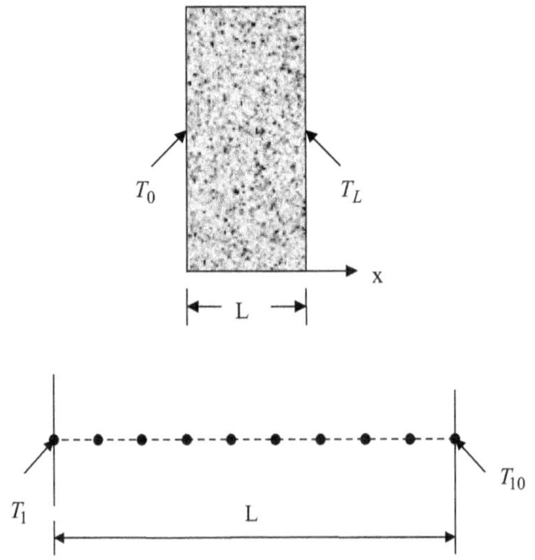

Governing Equation

$$\rho c \frac{\partial T}{\partial t} = \frac{\partial}{\partial x}\left(k \frac{\partial T}{\partial x}\right)$$

For constant properties, this equation reduces to

$$\frac{\partial T}{\partial t} = \alpha \frac{\partial^2 T}{\partial x^2}$$

Boundary Conditions

1. $x = 0, T(0, t) = T_L$
2. $x = L, T(L, t) = T_R$

Initial Condition

$$t = 0, T(x,0) = T_I$$

Solution

For number of nodes $n = 10$, the spatial step size is

$$\Delta x = \frac{3.6\times 10^{-3}}{9} = 4\times 10^{-4} \text{ m}$$

Noting that for the heat transfer problems $\Gamma = k$ and $C = \rho c$, we can write

$$\frac{\Gamma \Delta t}{C\Delta x^2} = \frac{\alpha \Delta t}{\Delta x^2} = f_0$$

For satisfying the stability criterion, let us assume $\left(\Gamma\Delta t/C\Delta x^2\right) = \left(\alpha\Delta t/\Delta x^2\right) = 0.4$, and this gives a time step estimate of

$$\Delta t = \frac{0.4\times \Delta x^2}{2\times 10^{-5}} = 0.00395 \text{ s}$$

We select the explicit discretization equation (7.13) for interior nodes as

$$T_i^{l+1} = f_0\left(T_{i+1}^l + T_{i-1}^l\right) + (1-2f_0)T_i^l$$

Applying this equation to all interior nodes, we get

$$\textit{Node 2} \quad T_2^{l+1} = f_0\left(T_3^l + T_1^l\right) + (1-2f_0)T_2^l$$

$$\textit{Node 3} \quad T_3^{l+1} = f_0\left(T_4^l + T_2^l\right) + (1-2f_0)T_3^l$$

$$\textit{Node 4} \quad T_4^{l+1} = f_0\left(T_5^l + T_3^l\right) + (1-2f_0)T_4^l$$

$$\textit{Node 5} \quad T_5^{l+1} = f_0\left(T_6^l + T_4^l\right) + (1-2f_0)T_5^l$$

$$\textit{Node 6} \quad T_6^{l+1} = f_0\left(T_7^l + T_5^l\right) + (1-2f_0)T_6^l$$

$$\textit{Node 7} \quad T_7^{l+1} = f_0\left(T_8^l + T_6^l\right) + (1-2f_0)T_7^l$$

$$\textit{Node 8} \quad T_8^{l+1} = f_0\left(T_9^l + T_7^l\right) + (1-2f_0)T_8^l$$

$$\textit{Node 9} \quad T_9^{l+1} = f_0\left(T_{10}^l + T_8^l\right) + (1-2f_0)T_9^l$$

Since the nodal equations are explicit, the system of equations can be solved at the present time with all temperature values on the right-hand side of the equations known at the previous time. The heat transfer rate at the right surface is then estimated as

$$q''_L = k\frac{T(N)-T(N-1)}{\Delta X}$$

A computer code based on the pseudo code given in Table 7.1 is used to solve the problem. Results for transient temperature distribution at selected time intervals are summarized in the table and figure below.

	Time					
Node	**0.01 s**	**0.03 s**	**0.05 s**	**0.07 s**	**0.09 s**	**0.1**
T_1	0	0	0	0	0	0
T_2	0	0	0.4876	3.8687	8.1739	11.8623
T_3	0	0	2.4662	10.4488	18.8527	25.6856
T_4	0	0	8.099	21.4315	33.2186	42.3091
T_5	0	2.048	20.2003	38.1834	52.0149	62.2003
T_6	0	12.4928	41.1502	61.362	75.4229	85.4112
T_7	0	12.8	71.7258	90.6736	103.0272	111.5885
T_8	0	55.04	110.5805	124.8823	133.8867	140.0408
T_9	80	123.776	154.6249	162.0634	166.6964	169.8489
T_{10}	200	200	200	200	200	200

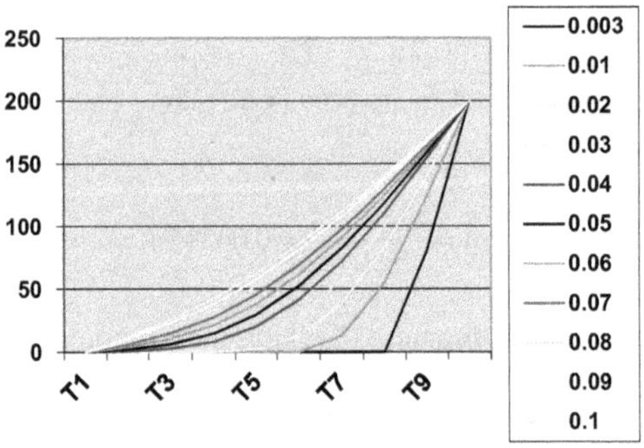

The heat transfer rates at the right surface are also estimated at selected time intervals and presented in the table and figure below:

Time	q''_L
0	13 500000
0.01	8 575200
0.02	6 765777
0.03	5 763110
0.04	5 104696
0.05	4 630217
0.06	4 267871
0.07	3 980437
0.08	3 746659
0.09	3 553408
0.1	3 391994

Time (s)

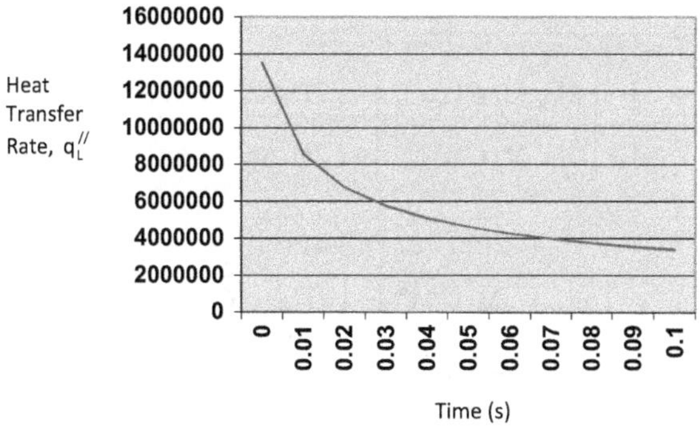

7.2.4 Other Explicit Schemes

The explicit scheme discussed so far is called a **FTCS** scheme, and it uses a first-order-accurate forward-difference approximation for the time derivative. There are many other explicit schemes that are proposed to improve the accuracy and stability using higher-order finite difference formulas. A few of them are discussed here.

Richardson Scheme   In this scheme, a second order-accurate central-difference approximation for the time derivative is used to improve the accuracy. The discretized equation is

$$C\frac{\phi_{i,j}^{l+1} - \phi_{i,j}^{l-1}}{2\Delta t} = \Gamma\frac{\phi_{i+1,j}^{l} - 2\phi_{i,j}^{l} + \phi_{i-1,j}^{l}}{(\Delta x)^2} + \Gamma\frac{\phi_{i,j+1}^{l} - 2\phi_{i,j}^{l} + \phi_{i,j-1}^{l}}{(\Delta y)^2} \qquad (7.52\text{a})$$

Rearranging,

$$\phi_{i,j}^{l+1} = \phi_{i,j}^{l-1} + \frac{2\Gamma\Delta t}{C(\Delta x)^2}\left[\phi_{i+1,j}^{l} + \phi_{i-1,j}^{l}\right] + \frac{2\Gamma\Delta t}{C(\Delta y)^2}\left[\phi_{i,j+1}^{l} + \phi_{i,j-1}^{l}\right] - \left[\frac{2\Gamma\Delta t}{C(\Delta x)^2} + \frac{2\Gamma\Delta t}{C(\Delta y)^2}\right]\phi_{i,j}^{l} \quad (7.52\text{b})$$

For uniform grid size distributions, i.e., $\Delta x = \Delta y$, we have

$$\phi_{i,j}^{l+1} = \phi_{i,j}^{l-1} + \frac{2\Gamma\Delta t}{C(\Delta x)^2}\left[\phi_{i+1,j}^{l} + \phi_{i-1,j}^{l} + \phi_{i,j+1}^{l} + \phi_{i,j-1}^{l} - 2\phi_{i,j}^{l}\right] \qquad (7.52\text{c})$$

The equation is second-order accurate of the order of $O(\Delta x^2) + O(\Delta t^2)$. However, a stability analysis indicates that it is unconditionally unstable. Therefore, it is usually not used for heat and fluid flow problems.

DuFort–Frankel Scheme   In this scheme, $\phi_{i,j}^{l}$ in the above Richardson's equation is replaced by $\frac{1}{2}\left(\phi_{i,j}^{l-1} + \phi_{i,j}^{l+1}\right)$ to avoid the stability problem. The resulting scheme is

$$C\frac{\phi_{i,j}^{l+1} - \phi_{i,j}^{l-1}}{2\Delta t} = \Gamma_x\frac{\phi_{i+1,j}^{l} - \left(\phi_{i,j}^{l-1} + \phi_{i,j}^{l+1}\right) + \phi_{i-1,j}^{l}}{(\Delta x)^2} + \Gamma_y\frac{\phi_{i,j+1}^{l} - \left(\phi_{i,j}^{l-1} + \phi_{i,j}^{l+1}\right) + \phi_{i,j-1}^{l}}{(\Delta y)^2}$$

or

$$\left(1 + \frac{2\Gamma_x\Delta t}{C(\Delta x)^2} + \frac{2\Gamma_y\Delta t}{C(\Delta y)^2}\right)\phi_i^{l+1} = \frac{2\Gamma_x\Delta t}{C(\Delta x)^2}\left(\phi_{i+1,j}^{l} + \phi_{i-1,j}^{l}\right) + \frac{2\Gamma_y\Delta t}{C(\Delta y)^2}\left(\phi_{i,j+1}^{l} + \phi_{i-1,j-1}^{l}\right)$$

$$+ \left(1 - \frac{2\Gamma_x\Delta t}{C(\Delta x)^2} - \frac{2\Gamma_y\Delta t}{C(\Delta y)^2}\right)\phi_{i,j}^{l-1}$$

Simplifying

$$\phi_{i,j}^{l+1} = \frac{1-\left[2\Gamma_x\Delta t/C(\Delta x)^2\right]-\left[2\Gamma_y\Delta t/C(\Delta y)^2\right]}{1+\left[2\Gamma_x\Delta t/C(\Delta x)^2\right]+\left[2\Gamma_y\Delta t/C(\Delta y)^2\right]}\phi_{i,j}^{l-1}$$

$$+\frac{2\left[2\Gamma_x\Delta t\,/\,C(\Delta x)^2\right]}{1+\left[2\Gamma_x\Delta t/C\Delta x^2\right]+\left[2\Gamma_y\Delta t/C\Delta y^2\right]}\left(\phi_{i+1,j}^{l}+\phi_{i-1,j}^{l}\right)$$

$$+\frac{2\left[2\Gamma_y\Delta t/C(\Delta y)^2\right]}{1+\left[2\Gamma_x\Delta t/C\Delta x^2\right]+\left[2\Gamma_y\Delta t/C\Delta y^2\right]}\left(\phi_{i,j+1}^{l}+\phi_{i,j-1}^{l}\right) \tag{7.53}$$

For uniform grid size distribution and for an isotropic media, Equation 7.53 reduces to

$$\phi_{i,j}^{l+1} = \frac{1-\left[4\Gamma\Delta t/C(\Delta x)^2\right]}{1+\left[4\Gamma\Delta t/C(\Delta x)^2\right]}\phi_{i,j}^{l-1}+\frac{\left[2\Gamma\Delta t/C(\Delta x)^2\right]}{1+\left[4\Gamma\Delta t/C\left(\Delta x^2\right)\right]}\left(\phi_{i+1,j}^{l}+\phi_{i-1,j}^{l}\right)$$

$$+\frac{\left[2\Gamma\Delta t/C\Delta x^2\right]}{1+\left[4\Gamma\Delta t/C(\Delta x)^2\right]}\left(\phi_{i,j+1}^{l}+\phi_{i,j-1}^{l}\right) \tag{7.54a}$$

For **one-dimensional problems**, Equation 7.53 reduces to

$$\phi_{i}^{l+1} = \frac{1-\left[2\Gamma\Delta t/C\Delta x^2\right]}{1+\left[2\Gamma\Delta t/C(\Delta x)^2\right]}\phi_{i}^{l-1}+\frac{\left[2\Gamma\Delta t/C\Delta x^2\right]}{1+\left[2\Gamma\Delta t/C\Delta x^2\right]}\left(\phi_{i+1,j}^{l}+\phi_{i-1,j}^{l}\right) \tag{7.54b}$$

It can be shown by application of stability analysis that the DuFort–Frankel scheme is unconditionally stable. It can also be noted that the DuFort–Frankel scheme utilizes three levels in time: $l-1$, l, and $l+1$, and hence brings in additional complexity in computations.

7.2.5 Boundary Conditions

The discretization equations given in the previous section are applicable for determining the temperature of interior nodes of a solid as a function of space and time. These equations can be solved along with the boundary conditions of constant surface temperature or first kind. For other types of boundary conditions, we need to use appropriate discretization equations for boundary nodes, which are derived based on assuming a half control volume surrounding the boundary nodes.

Let us briefly discuss the derivation of one such equation for the grid point located at the **right-hand side boundary** shown below (Figure 7.4).

Let us demonstrate derivation of the boundary node equation by considering **the two-dimensional unsteady-state diffusion equation** given as

$$\frac{\partial}{\partial x}\left(\Gamma_x\frac{\partial\phi}{\partial x}\right)+\frac{\partial}{\partial y}\left(\Gamma_y\frac{\partial\phi}{\partial y}\right)=C\,\frac{\partial\phi}{\partial t} \tag{7.55}$$

To proceed, we integrate Equation 7.55 over the half control volume and over the time interval t to $t+\Delta t$:

$$\int_{i-\frac{1}{2}}^{i}\int_{j-\frac{1}{2}}^{j+\frac{1}{2}}\int_{t}^{t+\Delta t}\frac{\partial}{\partial x}\left(\Gamma_x\frac{\partial\phi}{\partial x}\right)\mathrm{d}x\mathrm{d}y\mathrm{d}t+\int_{i-\frac{1}{2}}^{i}\int_{j-\frac{1}{2}}^{j+\frac{1}{2}}\int_{t}^{t+\Delta t}\frac{\partial}{\partial y}\left(\Gamma_y\frac{\partial\phi}{\partial y}\right)\mathrm{d}x\mathrm{d}y\mathrm{d}t=C\int_{i-\frac{1}{2}}^{i}\int_{j-\frac{1}{2}}^{j+\frac{1}{2}}\int_{t}^{t+\Delta t}\frac{\partial\phi}{\partial t}\mathrm{d}x\mathrm{d}y\mathrm{d}t \tag{7.56}$$

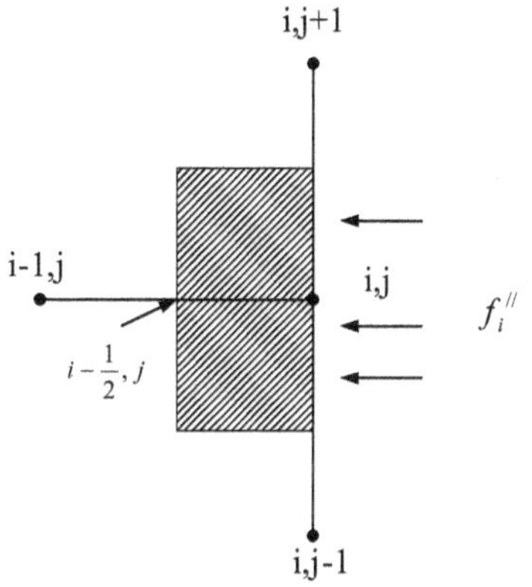

FIGURE 7.4 Control volume for a boundary node.

Integrating,

$$\Gamma_x \frac{\partial \phi}{\partial x}\bigg)_i \Delta y \Delta t - \Gamma_y \frac{\partial \Gamma}{\partial x}\bigg)_{i-\frac{1}{2}} \Delta y \Delta t + \Gamma_y \frac{\partial \phi}{\partial y}\bigg)_{j+\frac{1}{2}} \frac{\Delta x \Delta t}{2} - \Gamma_y \frac{\partial \phi}{\partial y}\bigg)_{j-\frac{1}{2}} \frac{\Delta x \Delta t}{2} = C \frac{\Delta x \Delta y}{2} \int_t^{t+\Delta t} \frac{\partial \phi}{\partial t} \mathrm{d}t \tag{7.57}$$

The first term in the equation is described by the heat flux boundary conditions as

$$-\Gamma_x \frac{\partial \phi}{\partial x}\bigg)_i = -f_i'' \tag{7.58a}$$

or

$$\Gamma_x \frac{\partial \phi}{\partial x}\bigg)_i = +f_i'' \tag{7.58b}$$

Next, we assume a piecewise-linear profile or central difference formula to approximate temperature gradients at points $j+\frac{1}{2}$, $j-\frac{1}{2}$, and $i-\frac{1}{2}$, and forward difference formula for first-order time derivative

$$\Gamma_x \frac{\partial \phi}{\partial x}\bigg)_{i-\frac{1}{2}} = \Gamma_{xi-\frac{1}{2}} \frac{\phi_{i,j} - \phi_{i-1,j}}{\Delta x}, \quad \Gamma_y \frac{\partial \phi}{\partial y}\bigg)_{j+\frac{1}{2}} = \Gamma_{yj+\frac{1}{2}} \frac{\phi_{i,j+1} - \phi_{i,j}}{\Delta y} \tag{7.59a}$$

$$\Gamma_y \frac{\partial \phi}{\partial y}\bigg)_{j-\frac{1}{2}} = \Gamma_{yj-\frac{1}{2}} \frac{(\phi_{i,j} - \phi_{i,j-1})}{\Delta y}, \quad \frac{\partial \phi}{\partial t} = \frac{\phi_{i,j}^{l+1} - \phi_{i,j}^{l}}{\Delta t} \tag{7.59b}$$

Substituting these approximations, we get

$$f_i'\Delta y \Delta t - \frac{\Gamma_{i-\frac{1}{2}}\left(\phi_{i,j}^l - \phi_{i-1,j}^l\right)}{\Delta x} \Delta y \Delta t - \frac{\Gamma_{j+-\frac{1}{2}}\left(\phi_{i,j+1}^l - \phi_{i,j}^l\right)}{\Delta y} \frac{\Delta x}{2} \Delta t - \frac{\Gamma_{j-\frac{1}{2}}\left(\phi_{i,j}^l - \phi_{i,j-1}^l\right)}{\Delta y} \frac{\Delta x}{2} \Delta t = C \frac{\Delta x}{2} \Delta y \left(\phi_{i,j}^{l+1} - \phi_{i,j}^l\right) \tag{7.60a}$$

Rearranging

$$\frac{C\Delta x\Delta y}{2\Delta t}\phi_{i,j}^{l+1} = \frac{\Gamma_{i-\frac{1}{2}}\Delta y}{\Delta x}\phi_{i-1}^{l} + \frac{\Gamma_{j+\frac{1}{2}}\Delta x}{2\Delta y}\phi_{i,j+1}^{l} + \frac{\Gamma_{j-\frac{1}{2}}\Delta x}{2\Delta y}\phi_{i,j-1}^{l}$$

$$+\left(\frac{C\Delta x\Delta y}{2\Delta t} - \frac{\Gamma_{i-\frac{1}{2}}\Delta y}{\Delta x} - \frac{\Gamma_{j+\frac{1}{2}}\Delta x}{2\Delta y} - \frac{\Gamma_{j-\frac{1}{2}}\Delta x}{2\Delta y}\right)\phi_{i,j}^{l} + f''\Delta y_i \tag{7.60b}$$

Simplifying

$$a_i\phi_{i,j}^{l+1} = a_{i-1}\phi_{i-1,j}^{l} + a_{j+1}\phi_{i,j+1}^{l} + a_{j-1}\phi_{i,j-1}^{l} + \left(a_i^l - a_{i-1} - a_{j+1} - a_{j-1}\right)\phi_{i,j}^{l} + d \tag{7.61a}$$

where

$$a_i = \frac{C\Delta x\Delta y}{2\Delta t},\ a_{i-1} = \frac{\Gamma_{i-\frac{1}{2}}\Delta y}{\Delta x},\ a_{j+1} = \frac{\Gamma_{j+\frac{1}{2}}\Delta x}{2\Delta y}, a_{j-1} = \frac{\Gamma_{j-\frac{1}{2}}\Delta x}{2\Delta y}$$

$$a_i^l = \frac{C\Delta x\Delta y}{2\Delta t},\quad d = f_i'' \tag{7.61b}$$

For constant transport property, Γ, and for uniform grid size distributions, i.e., $\Delta x = \Delta y$, we can write the above equation in an alternative form as

$$\phi_{i,j}^{i+1} = \frac{\Gamma\Delta t}{C\Delta x^2}\left[2\phi_{i-1,j}^{l} + \phi_{i,j+1}^{l} + \phi_{i,j-1}^{l}\right] + \left[1 - 4\frac{\Gamma\Delta t}{C\Delta x^2}\right]\phi_i^l + \frac{2\Delta t}{C\Delta x}f_i'' \tag{7.62}$$

For the **one-dimensional case**, the equation reduces to

$$\phi_i^{l+1} = \frac{\Gamma\Delta t}{C\Delta x^2}\left[2\phi_{i-1}^{l}\right] + \left[1 - \frac{2\Gamma\Delta t}{C\Delta x^2}\right]\phi_i^l + \frac{2\Delta t}{C\Delta x}f_i'' \tag{7.63}$$

Let us now consider two special cases.

1. **Adiabatic boundary condition or a symmetric boundary (Figure 7.5)**
   For an adiabatic surface, the surface flux is zero, i.e., $f_i'' = 0$.

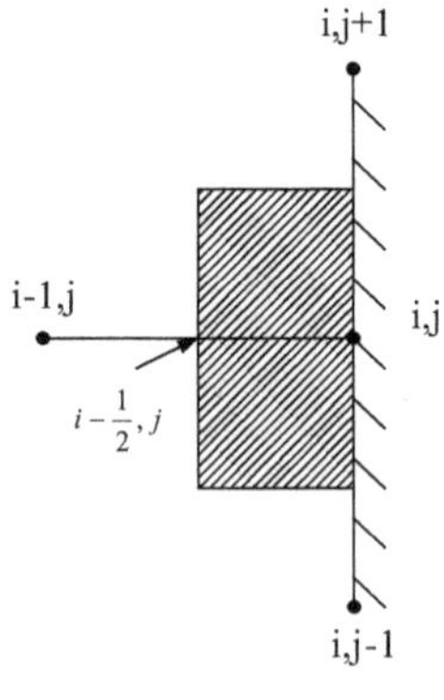

FIGURE 7.5 Adiabatic boundary node.

Substituting Equations 7.35 and 7.36, we get

$$\phi_{i,j}^{l+1} = \frac{\Gamma\Delta t}{C\Delta x^2}\left[2\phi_{i-1,j}^{l} + \phi^{l}_{i,j+1} + \phi_{i,j-1}^{l}\right] + \left[1 - \frac{4\Gamma\Delta t}{C\Delta x^2}\right]\phi_{i,j}^{l} \quad (7.64a)$$

for two-dimensional problems

and

$$\phi_i^{l+1} = \frac{\Gamma\Delta t}{C\Delta x^2}\left[2\phi_{i-1}^{l}\right] + \left[1 - \frac{2\Gamma\Delta t}{C\Delta x^2}\right]\phi_i^{l} \quad (7.64b)$$

for one-dimensional problems

2. **Convective boundary (Figure 7.6)**

   For a convective boundary condition,

$$-\Gamma\frac{\partial\phi}{\partial x} = h(\phi - \phi_\infty) \quad (7.65a)$$

or

$$f_i'' = h(\phi_{i,j} - \phi_\infty) \quad (7.65b)$$

Substituting Equation 7.65 into Equation 7.35, we get

$$\phi_{i,j}^{l+1} = \frac{\Gamma\Delta t}{C\Delta x^2}\left[2\phi_{i-1,j}^{l} + \phi_{i,j+1}^{l} + \phi_{i,j-1}^{l} + \frac{2h\Delta x}{\Gamma}\phi_\infty\right] + \left[1 - \frac{4\Gamma\Delta t}{C\Delta x^2} - \frac{2h\Delta t}{C\Delta x}\right]\phi_{i,j}^{l} \quad (7.66)$$

A summary of some explicit discretization equations for boundary nodes is given in Table 7.2.

Boundary Conditions for Three-Dimensional Problems

Discretization equations for boundary nodes in **three-dimensional diffusion problems** can also be derived following the procedure outlined. Let us consider the following three-dimensional transient diffusion equation and a three-dimensional boundary control volume on a negative z-plane:

$$\frac{\partial}{\partial x}\left(\Gamma_x\frac{\partial\phi}{\partial x}\right) + \frac{\partial}{\partial y}\left(\Gamma_y\frac{\partial\phi}{\partial y}\right) + \frac{\partial}{\partial z}\left(\Gamma_z\frac{\partial\phi}{\partial y}\right) = C\frac{\partial\phi}{\partial t} \quad (7.67)$$

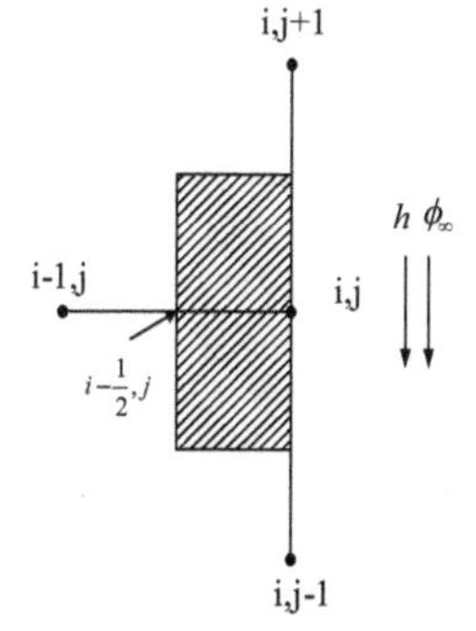

FIGURE 7.6 Convective boundary node.

TABLE 7.2
A Summary of Explicit Discretization Equations with Uniform Grid Size Distributions

Physical Situation	Discretization Equation	Stability Requirement
a. Interior node 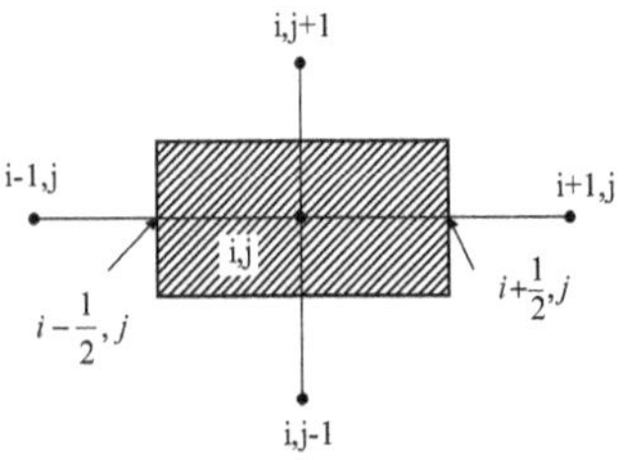	$T_{i,j}^{l+1} = \frac{\Gamma\Delta t}{C\Delta x^2}\left[\phi^l_{i+1,j}+\phi^l_{i-1,j}+\phi^l_{i,j+1}+\phi^l_{i,j-1}\right] + \left[1-\frac{4\Gamma\Delta t}{C\Delta x^2}\right]\phi^l_{i,j}$	$\frac{\Gamma\Delta t}{\Delta x^2}\le\frac{1}{4}$
b. Convective boundary 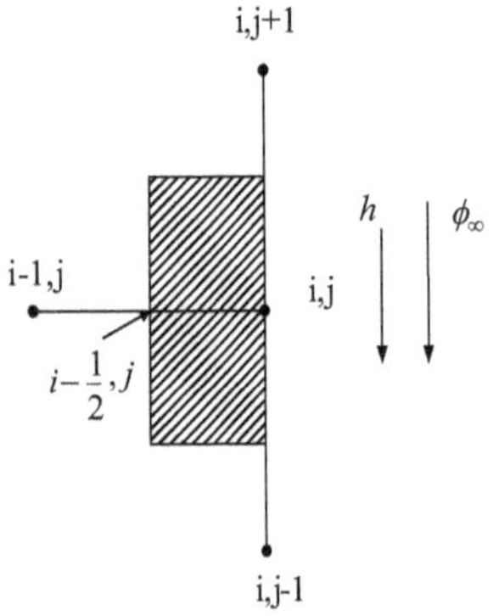	$\phi^{l+1}_{i,j} = \frac{\Gamma\Delta t}{C\Delta x^2}\left[2\phi^l_{i-1,j}+\phi^l_{i,j+1}+\phi^l_{i,j-1}+2h\Delta x/\Gamma\phi_\infty\right] + \left[1-\frac{4\Gamma\Delta t}{C\Delta x^2}-\frac{2h\Delta t}{C\Delta x}\right]\phi^l_{i,j}$	$\frac{\Gamma\Delta t}{C\Delta x^2}\left(2+\frac{h\Delta x}{\Gamma}\right)\le\frac{1}{2}$
c. Exterior corner with convection boundary 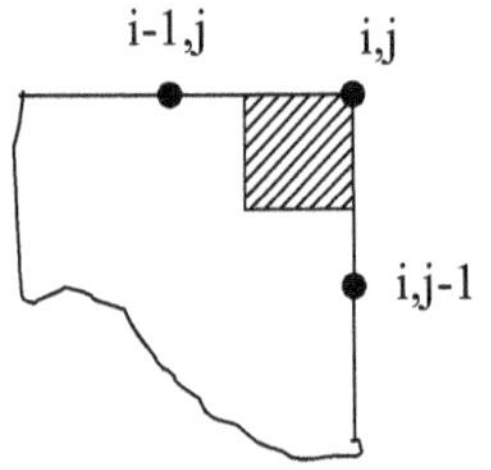	$\phi^{l+1}_{i,j}=\frac{2\Gamma\Delta t}{C\Delta x^2}\left[\phi^l_{i-1,j}+\phi^l_{i,j-1}+\frac{2h\Delta x}{\Gamma}T_\infty\right] + \left[1-\frac{4\Gamma\Delta t}{C\Delta x^2}-\frac{4\Gamma\Delta t}{C\Delta x^2}\frac{h\Delta x}{\Gamma}\right]\phi^l_{i,j}$	$\frac{\Gamma\Delta t}{C\Delta x^2}\left(1+\frac{h\Delta x}{\Gamma}\right)\le\frac{1}{4}$
d. Interior corner with convection boundary 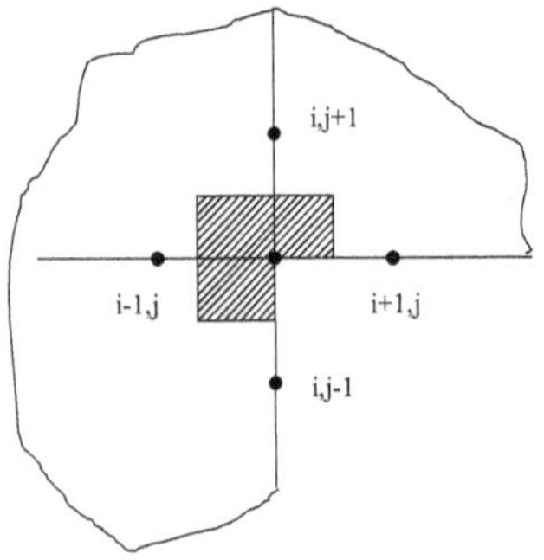	$\phi^{l+1}_{i,j}=\frac{\frac{2}{3}\Gamma\Delta t}{C\Delta x^2}\left[2\phi^l_{i,j+1}+\phi^l_{i+1,j}+2\phi^l_{i-1,j}+\phi^l_{i,j-1}+2\frac{h\Delta x}{K}T_\infty\right] + \left[1-4\frac{\Gamma\Delta t}{C\Delta x^2}-\frac{4}{3}\frac{\Gamma\Delta t}{C\Delta x^2}\frac{h\Delta x}{\Gamma}\right]\phi^l_{i,j}$	$\frac{\Gamma\Delta t}{C\Delta x^2}\left(3+\frac{h\Delta x}{\Gamma}\right)\le\frac{3}{4}$

(Continued)

TABLE 7.2 (*Continued*)
A Summary of Explicit Discretization Equations with Uniform Grid Size Distributions

Physical Situation	Discretization Equation	Stability Requirement
e. Adiabatic boundary 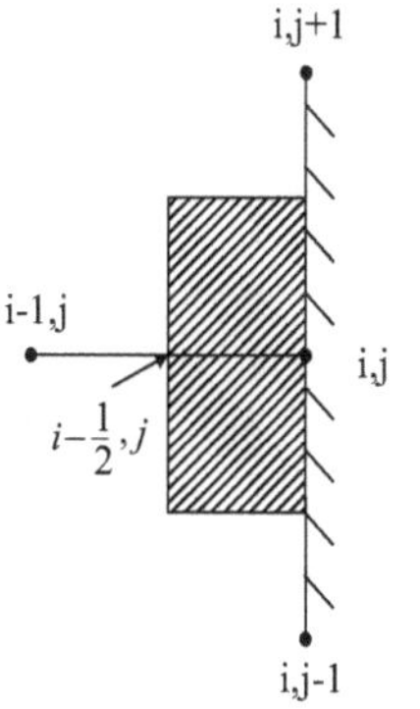	$\phi_{i,j}^{l+1} = \frac{\Gamma\Delta t}{C\Delta x^2}\left[2\phi_{i-1,j}^{l} + \phi_{i,j+1}^{l} + \phi_{i,j-1}^{l}\right] + \left[1 - 4\frac{\Gamma\Delta t}{C\Delta x^2}\right]\phi_{i,j}^{l}$	$\frac{\Gamma\Delta t}{C\Delta x^2} \leq \frac{1}{4}$
f. Constant surface heat flux 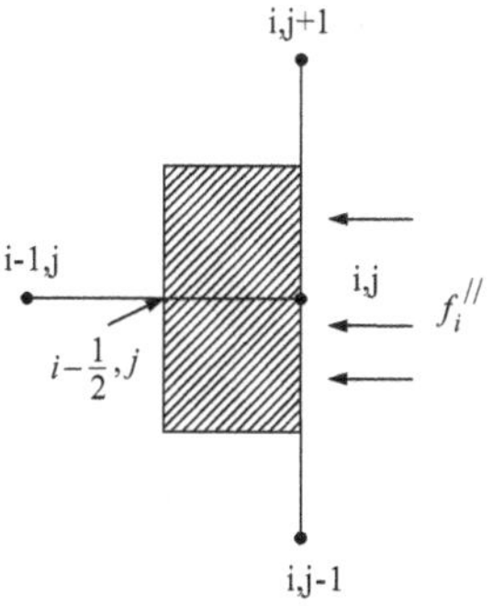	$\phi_{i,j}^{l+1} = \frac{\Gamma\Delta t}{C\Delta x^2}\left[2\phi_{i-1,j}^{l} + \phi_{i,j+1}^{l} + \phi_{i,j-1}^{l}\right]_i^{\prime\prime} + \left[1 - 4\frac{\Gamma\Delta t}{C\Delta x^2}\right]\phi_i^{l} - \frac{2\Delta t}{C\Delta x}f_i^{\prime\prime}$	

To proceed, we integrate the equation over the half control volume and over the time interval t to $t+\Delta t$:

$$+\int_{i-\frac{1}{2}}^{i+\frac{1}{2}}\int_{j-\frac{1}{2}}^{j+\frac{1}{2}}\int_{k-\frac{1}{2}}^{k}\int_{t}^{t+\Delta t}\frac{\partial}{\partial z}\left(\Gamma_z\frac{\partial\phi}{\partial z}\right)\mathrm{d}x\mathrm{d}y\mathrm{d}z\mathrm{d}t = \int_{i-\frac{1}{2}}^{i+\frac{1}{2}}\int_{j-\frac{1}{2}}^{j+\frac{1}{2}}\int_{k-\frac{1}{2}}^{k}\int_{t}^{t+\Delta t}C\frac{\partial\phi}{\partial t}\,\mathrm{d}x\mathrm{d}y\mathrm{d}z\mathrm{d}t \tag{7.68}$$

Integrating and substituting,

$$\Gamma_x\left.\frac{\partial\phi}{\partial x}\right)_{i+\frac{1}{2}}\frac{\Delta y\Delta z\Delta t}{2} - \Gamma_x\left.\frac{\partial\phi}{\partial x}\right)_{i-\frac{1}{2}}\frac{\Delta x\Delta z\Delta t}{2} + \Gamma_y\left.\frac{\partial\phi}{\partial y}\right)_{j+\frac{1}{2}}\frac{\Delta x\Delta z\Delta t}{2} - \Gamma_y\left.\frac{\partial\phi}{\partial y}\right)_{i-\frac{1}{2}}\frac{\Delta x\Delta z\Delta t}{2} \tag{7.69}$$

The heat flux terms on the boundary node k can be replaced with heat flux boundary condition as

$$\Gamma_z\frac{\partial\phi}{\partial z} = f_k^{\prime\prime} \tag{7.70}$$

Substituting this boundary condition and approximating the rest of the heat flux derivatives by first-order piecewise linear profiles or central difference formula to approximate temperature gradients at points $i+\frac{1}{2}, i-\frac{1}{2}, j+\frac{1}{2}, j-\frac{1}{2}$, and $k-\frac{1}{2}$, and forward difference formula for first-order time derivative, we get

$$\frac{\Gamma_{i+\frac{1}{2}}\left(\phi_{i+1,j,k}^{l}-\phi_{i,j,k}^{l}\right)}{\Delta x}\frac{\Delta y\Delta z\Delta t}{2}-\frac{\Gamma_{i-\frac{1}{2}}\left(\phi_{i,j,k}^{l}-\phi_{i-1,j,k}^{l}\right)}{\Delta x}\frac{\Delta y\Delta z\Delta t}{2}+\frac{\Gamma_{j+\frac{1}{2}}\left(\phi_{i,j+1,k}^{l}-\phi_{i,j,k}^{l}\right)}{\Delta y}\frac{\Delta x\Delta z\Delta t}{2}$$

$$-\frac{\Gamma_{j-\frac{1}{2}}\left(\phi_{i,j,k}^{l}-\phi_{i,j-1,k}^{l}\right)}{\Delta y}\frac{\Delta x\Delta z\Delta t}{2}-f_k''\,\Delta x\Delta y\Delta t-\frac{\Gamma_{k-\frac{1}{2}}\left(\phi_{i,j,k}^{l}-\phi_{i,j,k-1}^{l}\right)}{\Delta z}\Delta x\Delta y\Delta t$$

$$=C\frac{\Delta x\Delta y\Delta z}{2}\left(\frac{\phi_{i,j,k}^{l+1}-\phi_{i,j,k}^{l}}{\Delta t}\right) \tag{7.71}$$

Rearranging,

$$\frac{C\Delta x\Delta y\Delta z}{2\Delta t}\phi_{i,j,k}^{l+1}=\frac{\Gamma_{i+\frac{1}{2}}\Delta y\Delta z}{2\Delta x}\phi_{i+1}^{l}+\frac{\Gamma_{i-\frac{1}{2}}\Delta y\Delta z}{2\Delta x}\phi_{i-1}^{l}+\frac{\Gamma_{j+\frac{1}{2}}\Delta x\Delta z}{2\Delta y}\phi_{i,j+1}^{l}+\frac{\Gamma_{j-\frac{1}{2}}\Delta x\Delta z}{2\Delta y}\phi_{i,j-1}^{l}$$

$$+\frac{\Gamma_{k-\frac{1}{2}}\Delta x\Delta y\Delta}{\Delta z}\phi_{k-1}^{l}+\left(\frac{C\Delta x\Delta y\Delta z}{2\Delta t}-\frac{\Gamma_{i-\frac{1}{2}}\Delta y\Delta z}{2\Delta x}-\frac{\Gamma_{i-\frac{1}{2}}\Delta y\Delta z}{2\Delta x}\right.$$

$$\left.-\frac{\Gamma_{j+\frac{1}{2}}\Delta x\Delta z}{2\Delta y}-\frac{\Gamma_{j-\frac{1}{2}}\Delta x\Delta z}{2\Delta y}-\frac{\Gamma_{k-\frac{1}{2}}\Delta x\Delta y}{\Delta z}\right)\phi_{i,j,k}^{l}-f_k''\Delta x\Delta y \tag{7.72}$$

Simplifying

$$a_i\phi_{i,j,k}^{l+1}=a_{i+1}\phi_{i+1,j,k}^{l}+a_{i-1}\phi_{i-1,j,k}^{l}+a_{j+1}\phi_{i,j+1,k}^{l}+a_{j-1}\phi_{i,j-1,k}^{l}+a_{k-1}\phi_{i,j,k-1}^{l}$$

$$+\left(a_i^l-a_{i+1}-a_{i-1}-a_{j+1}-a_{j-1}-a_{k-1}\right)\phi_{i,j,k}^{l}+d \tag{7.73a}$$

where

$$a_i=\frac{C\Delta x\Delta y\Delta z}{2\Delta t},\ a_{i+1}\frac{\Gamma_{i+\frac{1}{2}}\Delta y\Delta z}{2\Delta x},\ a_{i-1}\frac{\Gamma_{i-\frac{1}{2}}\Delta y\Delta z}{2\Delta x},a_{j+1}\frac{\Gamma_{j+\frac{1}{2}}\Delta x\Delta z}{2\Delta y},\ a_{j-1}\frac{\Gamma_{j-\frac{1}{2}}\Delta x\Delta z}{2\Delta y},\ a_{k-1}\frac{\Gamma_{k-\frac{1}{2}}\Delta x\Delta y}{\Delta z}$$

$$a_i^l=\frac{C\Delta x\Delta y\Delta z}{2\Delta t},\ d=-f_k''\Delta x\Delta y \tag{7.73b}$$

For constant transport property, Γ, and for uniform grid size distributions, i.e., $\Delta x=\Delta y=\Delta z$, we can write the above equation in an alternative form as

$$\phi_{i,j,k}^{l+1}=\frac{\Gamma\Delta t}{C\Delta x^2}\left[2\phi_{i,j,k-1}^{l}+\phi_{i+1,j,k}^{l}+\phi_{i-1,j,k}^{l}+\phi_{i,j+1,k}^{l}+\phi_{i,j-1,k}^{l}\right]+\left[1-\frac{6\Gamma\Delta t}{C\Delta x^2}\right]\phi_{i,j,k}^{l}-\frac{2\Delta t}{C\Delta x}f_k'' \tag{7.74}$$

Let us now consider two special cases.

1. Adiabatic boundary condition or a symmetric boundary.

   For an adiabatic surface, the surface flux is zero, i.e., $f_k''=0$.

   Substituting, we get the discretization equation for the convective boundary node on negative z-plane

$$\phi_{i,j,k}^{l+1} = \frac{\Gamma \Delta t}{C\Delta x^2}\left[2\phi_{i,j,k-1}^{l} + \phi_{i+1,j,k}^{l} + \phi_{i-1,j,k}^{l} + \phi_{i,j+1,k}^{l} + \phi_{i,j-1,k}^{l}\right] + \left[1 - \frac{6\Gamma\Delta t}{C\Delta x^2}\right]\phi_{i,j,k}^{l} \tag{7.75}$$

2. Convective boundary

   For a convective boundary condition,

$$-\Gamma\frac{\partial \phi}{\partial x} = h(\phi - \phi_\infty) \tag{7.76a}$$

or

$$f_k'' = h\left(\phi_{i,j,k} - \phi_\infty\right) \tag{7.76b}$$

Substituting Equation 7.76b into Equation 7.74, we get

$$\phi_{i,j,k}^{l+1} = \frac{\Gamma \Delta t}{C\Delta x^2}\left[2\phi_{i,j,k-1}^{l} + \phi_{i+1,j,k}^{l} + \phi_{i-1,j,k}^{l} + \phi_{i,j+1,k}^{l} + \phi_{i,j-1,k}^{l} + \frac{2h\Delta x}{\Gamma}\phi_\infty\right] + \left[1 - \frac{6\Gamma\Delta t}{C\Delta x^2} - \frac{2h\Delta t}{C\Delta x}\right]\phi_{i,j,k}^{l} \tag{7.77}$$

The explicit method discussed here is simple computationally, but due to its stability requirement, it may require a very small time increment Δt and hence requires a large number of time steps for problems that need to be analyzed over a longer period of time. In such situations, other finite difference schemes in which the time increment is not restricted by the stability criterion are used. Two such widely used schemes are the **fully implicit scheme** and the **Crank–Nicholson scheme**. These are discussed in the subsequent section.

A pseudo code for a transient solution in a plane slab with a convective boundary condition on the right side and using the explicit finite difference scheme is given in Table 7.3.

Example 7.2: Explicit Scheme with Convective Boundary

Consider a one-dimensional unsteady-state conduction in a plane slab of thickness L = 4 mm. Initially, the slab is at a temperature T_I = 20°C. The right side of the plate is suddenly brought to a convective environment with temperature T_∞ = 100°C and h = 80 $80\,\mathrm{W/m^2\,°C}$. The other side is maintained at 20°C. The material's thermo-physical properties are c_p =760 J/kg °C, $k = 50\,\mathrm{W/m\,°C}$, and $\alpha = 2\times10^{-5}\,\mathrm{m^2/s}$.

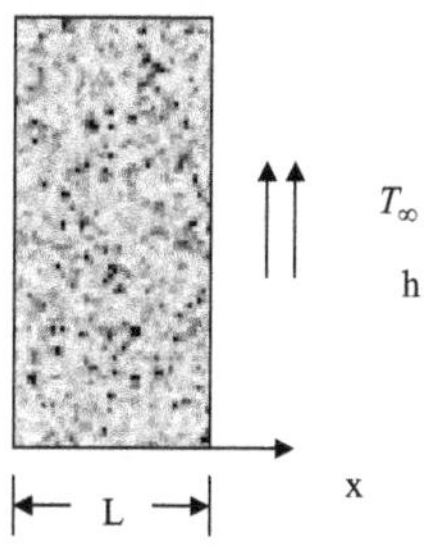

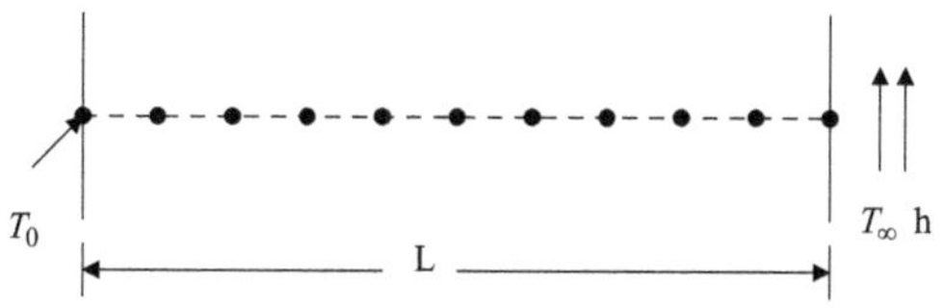

TABLE 7.3
Pseudo Code for a Transient Solution in a Plane Slab with a Convective Boundary on One Side and Using the Explicit Finite Difference Scheme

Input dimensions, properties, select number of divisions, maximum time, and stability parameters

```
   Input dimension L_x
   Input thermo-physical properties Γ and C
   Input convective boundary parameters h and φ_∞
   Select number of division in x-direction NDIVX
   Select stability parameter γ_0
   Number of node NX = NDIVX + 1
   Calculate step size DX = L_x/NDIVX
Calculate time step using stability parameter
   Dt = γ_0 C (DX)^2 / Γ
Initialize time
   time = 0.0
Set initial condition
   dofor i = 1, NX
      φ(I) = φ_I
   enddo
Set boundary conditions
   φ(1) = φ_L
dowhile (time < tmax)
   time = time + Dt
Calculate nodal values at interior nodes
   dofor I = 2, NX – 1
      φ(I) = γ_0 * (φ(I + 1) + φ(I − 1)) + (1 − 2 * γ_0) * φ(I)
   enddo
Calculate nodal values at the right boundary nodes
      φ(NX)) = γ_0 * (2φ(NX − 1) + (2hDX/Γ)φ_∞) + (1 − 2 * γ_0 − 2 * γ_0 * (hDX/Γ)) * φ(NX)
Estimate flux quantity at right surface
   f_x'' = h * (φ(NX) − φ_∞)
Enddo
```

Governing Equation:

$$\rho c \frac{\partial T}{\partial t} = \frac{\partial}{\partial x}\left(k \frac{\partial T}{\partial x}\right)$$

For constant properties, this equation reduces to

$$\frac{\partial T}{\partial t} = \alpha \frac{\partial^2 T}{\partial x^2}$$

Boundary Conditions:

1. $x = 0, \quad T(x,0) = T_L$

2. $x = L, \quad -k\frac{\partial T}{\partial x} = h(T - T_\infty)$

Initial Condition:

$$t = 0,\ T\,(x,0) = T_I$$

Consider the grid size distribution shown with ten grid points and examine the effect of the time step or the stability parameter. Determine the transient temperature distribution in the slab.

Solution

For number of nodes $n = 11$, spatial step size is

$$\Delta x = \frac{L}{n-1} = \frac{4\times10^{-3}}{10} = 4\times10^{-4}\,\text{m}$$

Noting that for heat transfer problems $\Gamma = k$ and $C = \rho c$, we can write

$$\frac{\Gamma\Delta t}{C\Delta x^2} = \frac{\alpha\Delta t}{\Delta x^2} = f_0$$

and we select the discretization equation for the nodes as follows.

Interior Nodes 1, 2, 3, 4, 5, 6, 7, 8, 9

$$T_i^{l+1} = f_0\left(T_{i+1}^l + T_{i-1}^l\right) + \left(1-2f_0\right)T_i^l$$

$$\textit{Node 1}\quad T_1^{l+1} = f_0\left(T_2^l + T_0^l\right) + \left(1-2f_0\right)T_1^l$$

$$\textit{Node 2}\quad T_2^{l+1} = f_0\left(T_3^l + T_1^l\right) + \left(1-2f_0\right)T_2^l$$

$$\textit{Node 3}\quad T_3^{l+1} = f_0\left(T_4^l + T_2^l\right) + \left(1-2f_0\right)T_3^l$$

$$\textit{Node 4}\quad T_4^{l+1} = f_0\left(T_5^l + T_3^l\right) + \left(1-2f_0\right)T_4^l$$

$$\textit{Node 5}\quad T_5^{l+1} = f_0\left(T_6^l + T_4^l\right) + \left(1-2f_0\right)T_5^l$$

$$\textit{Node 6}\quad T_6^{l+1} = f_0\left(T_7^l + T_5^l\right) + \left(1-2f_0\right)T_6^l$$

$$\textit{Node 7}\quad T_7^{l+1} = f_0\left(T_8^l + T_6^l\right) + \left(1-2f_0\right)T_7^l$$

$$\textit{Node 8}\quad T_8^{l+1} = f_0\left(T_9^l + T_7^l\right) + \left(1-2f_0\right)T_8^l$$

$$\textit{Node 9}\quad T_9^{l+1} = f_0\left(T_{10}^l + T_8^l\right) + \left(1-2f_0\right)T_9^l$$

Right-Hand Convective Node 10

$$T_i^{l+1} = f_0\left(2T_{i-1}^l + \frac{2h\Delta x}{k}T_\infty\right) + \left(1-2f_0-2f_0\frac{h\Delta x}{k}\right)T_i^l$$

$$\textit{Node 10}\quad T_{10}^{l+1} = f_0\left(2T_9^l + \frac{2h\Delta x}{k}T_\infty\right) + \left(1-2f_0-2f_0\frac{h\Delta x}{k}\right)T_{10}^l$$

The stability condition for interior nodes is given by Equation 7.22 as

$$f_0 = \frac{\alpha\Delta t}{(\Delta x)^2} < 0.5$$

The time step is estimated based on the value of f_0 used

$$\Delta t < \frac{(\Delta x)^2}{2\alpha} = 0.004\text{s}$$

A pseudo-code for this solution methodology is shown in Table 7.3.

In order to show the effect of the time step, the heat transfer rate at the right surface can be evaluated with increase in time for different Fourier numbers and results could be plotted for visual observation of the convergence and transient temperature distribution in the slab.

7.3 IMPLICIT SCHEME

7.3.1 Discretization Equation by Control Volume Approach

The implicit form of the finite difference equation for the governing differential equation (7.1) can be derived directly from generalized formulation given before using the control volume approach. By considering the **time approximation weighting factor** $f = 1$, the discretization equation (7.9) becomes

$$a_i\phi_i^{l+1} = a_{i+1}\phi_{i+1}^{l+1} + a_{i-1}\phi_{i-1}^{l+1} + a_i^l\phi_i^l \tag{7.78}$$

where

$$a_i = a_{i+1} + a_{i-1} + a_i^l \tag{7.79a}$$

$$a_{i+1} = \frac{\Gamma_{i+\frac{1}{2}}}{(\Delta x)_{i+\frac{1}{2}}} \tag{7.79b}$$

$$a_{i-1} = \frac{\Gamma_{i-1\frac{1}{2}}}{(\Delta x)_{i-\frac{1}{2}}} \tag{7.79c}$$

$$a_i^l = \frac{C\Delta x_i}{\Delta t} \tag{7.79d}$$

Physically, the fully **implicit scheme** assumes that, at time t, ϕ changes from ϕ_i^l to ϕ_i^{l+1} and then remains constant at this value over the entire time step as depicted in Figure 7.2.

Substituting the coefficient values in Equation 7.78, we get

$$\left(\frac{\Gamma_{i+\frac{1}{2}}}{\Delta x_{i+\frac{1}{2}}} + \frac{\Gamma_{i-\frac{1}{2}}}{\Delta x_{i-\frac{1}{2}}} + \frac{C\Delta x}{\Delta t}\right)\phi_i^{l+1} = \frac{\Gamma_{i+\frac{1}{2}}}{\Delta x_{i+\frac{1}{2}}}\phi_{i+1}^{l+1} + \frac{\Gamma_{i-\frac{1}{2}}}{\Delta x_{i-\frac{1}{2}}}\phi_{i-1}^{l+1} + \frac{C\Delta x}{\Delta t}\phi_i^l \tag{7.80}$$

For the special case of constant transport property Γ and uniform grid size distribution, i.e., $\Delta x_{i+\frac{1}{2}} = \Delta x_{i-\frac{1}{2}} = \Delta x$, we get

$$\left(\frac{2\Gamma\Delta t}{C\Delta x^2} + 1\right)\phi_i^{l+1} = \frac{\Gamma\Delta t}{C\Delta x^2}\left(\phi_{i+1}^{l+1} + \phi_{i-1}^{l+1}\right) + \phi_i^l \tag{7.81a}$$

Substituting $\gamma_0 = \Gamma\Delta t / C\Delta x^2$

$$(2\gamma_o + 1)\phi_i^{l+1} = \gamma_o\left(\phi_{i+1}^{l+1} + \phi_{i-1}^{l+1}\right) + \phi_i^l \tag{7.81b}$$

This is called an implicit form of the finite difference representation, because to determine the unknown dependent variable ϕ at the $(l+1)$th time step, a system of simultaneous algebraic equations has to be solved. Based on **von Neumann stability analysis**, it can be shown that the fully implicit scheme is **unconditionally stable**. The main advantage of the implicit scheme is that it is stable for all time steps Δt.

The outlined procedure can easily be extended to two-dimensional and three-dimensional cases as follows.

Two-dimensional Fully Implicit Discretization Equation For the *two-dimensional diffusion equation with a source term,*

$$C\frac{\partial \phi}{\partial t} = \frac{\partial}{\partial x}\left(\Gamma\frac{\partial \phi}{\partial x}\right) + \frac{\partial}{\partial y}\left(\Gamma\frac{\partial \phi}{\partial y}\right) + S \tag{7.82}$$

the fully implicit form of the finite difference equations is

$$a_{i,j}\phi_{i,j}^{l+1} = a_{i+1}\phi_{i+1,j}^{l+1} + a_{i-1}\phi_{i-1,j}^{l+1} + a_{j+1}\phi_{i,j+1}^{l+1} + a_{j-1}\phi_{i,j-1}^{l+1} + d \tag{7.83}$$

where

$$a_{i+1} = \frac{\Gamma_{i+\frac{1}{2}}\Delta y}{\Delta x_{i+\frac{1}{2}}},\quad a_{i-1} = \frac{\Gamma_{i-\frac{1}{2}}\Delta y}{\Delta x_{i-\frac{1}{2}}},\quad a_{j+1} = \frac{\Gamma_{j+\frac{1}{2}}\Delta x}{\Delta y_{j+\frac{1}{2}}},\quad a_{j-1} = \frac{\Gamma_{j-\frac{1}{2}}\Delta x}{\Delta y_{j-\frac{1}{2}}}$$

$$a_i = a_{i+1} + a_{i-1} + a_{j+1} + a_{j-1} + a_i^l - S_1\Delta x\Delta y \tag{7.84}$$

$$a_i^l = \frac{C\Delta x\Delta y}{\Delta t},\quad d = S_0\Delta x\Delta y + a_i^l\phi_i^l$$

For the special case of a constant transport property Γ, uniform grid size distribution and uniform source, i.e.,

$$\Gamma_{i+\frac{1}{2}} = \Gamma_{i-\frac{1}{2}} = \Gamma_{j+\frac{1}{2}} = \Gamma_{j-\frac{1}{2}} = \Gamma,\quad \Delta x_{i+\frac{1}{2}} = \Delta x_{i-\frac{1}{2}} = \Delta y_{j+\frac{1}{2}} = \Delta y_{j-\frac{1}{2}} = \Delta x$$

$$S_0 = \bar{S},\quad S_1 = 0$$

the coefficients of Equation 7.84 reduce to

$$a_{i+1} = a_{i-1} = a_{j+1} = a_{j-1} = \Gamma$$

$$a_i^l = \frac{C\Delta x^2}{\Delta t},\quad b = a_i^l\phi_i^l + S\Delta x^2$$

$$a_i = 4\Gamma + \frac{C\Delta x^2}{\Delta t} \tag{7.85}$$

$$a_{i+1} = a_{i-1} = a_{j+1} = a_{j-1} = \Gamma,\quad a_i = 4\Gamma + \frac{C\Delta x^2}{\Delta t}$$

$$a_i^l = \frac{C\Delta x^2}{\Delta t},\quad d = a_i^l\phi_i^l + \bar{S}\Delta x^2$$

Substituting the coefficients, we get the simplified form

$$\left(4\Gamma + \frac{C\Delta x^2}{\Delta t}\right)\phi_{i,j}^{l+1} = \Gamma\left[\phi_{i+1,j}^{l+1} + \phi_{i-1,j}^{l+1} + \phi_{i,j+1}^{l+1} + \phi_{i,j-1}^{l+1}\right] + \frac{C\Delta x^2}{\Delta t}\phi_i^l + \bar{S}\Delta x^2 \tag{7.86a}$$

or

$$\left[4 + \frac{C\Delta x^2}{\Gamma\Delta t}\right]\phi_{i,j}^{l+1} = \left[\phi_{i+1,j}^{l+1} + \phi_{i-1,j}^{l+1} + \phi_{i,j+1}^{l+1} + \phi_{i,j-1}^{l+1}\right] + \frac{C\Delta x^2}{\Gamma\Delta t}\phi_i^l + \frac{\bar{S}\Delta x^2}{\Gamma} \tag{7.86b}$$

Substituting

$$\gamma_0 = \frac{\Gamma\Delta t}{C\Delta x^2} \tag{7.87}$$

We have

$$\left[1 + 4\gamma_0\right]\phi_{i,j}^{l+1} - \gamma_0\left[\phi_{i+1,j}^{l+1} + \phi_{i-1,j}^{l+1} + \phi_{i,j+1}^{l+1} + \phi_{i,j-1}^{l+1}\right] = \phi_{i,j}^l + \frac{\bar{S}\Delta t}{C} \tag{7.88}$$

Three-Dimensional Fully Implicit Finite Difference Equation Consider the three-dimensional governing equation as

$$C\frac{\partial\phi}{\partial t} = \frac{\partial}{\partial x}\left(\Gamma\frac{\partial\phi}{\partial x}\right) + \frac{\partial}{\partial y}\left(\Gamma\frac{\partial\phi}{\partial y}\right) + \frac{\partial}{\partial z}\left(\Gamma\frac{\partial\phi}{\partial z}\right) + S \tag{7.89}$$

with a three-dimensional grid shown in Figure 7.7 as

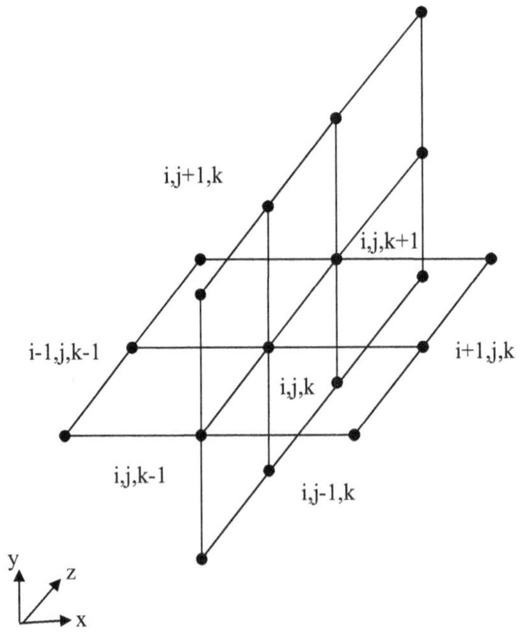

FIGURE 7.7 Three-dimensional grid.

Integration over the control volume gives

$$C\int_{i-\frac{1}{2}}^{i+\frac{1}{2}}\int_{j-\frac{1}{2}}^{j+\frac{1}{2}}\int_{k-\frac{1}{2}}^{k+\frac{1}{2}}\int_{t}^{t+\Delta t}\frac{\partial\phi}{\partial t}\mathrm{d}x\mathrm{d}y\mathrm{d}z\mathrm{d}t = \int_{t}^{\Delta t}\int_{i-\frac{1}{2}}^{i+\frac{1}{2}}\int_{j-\frac{1}{2}}^{j+\frac{1}{2}}\int_{k-\frac{1}{2}}^{k+\frac{1}{2}}\frac{\partial\phi}{\partial x}\mathrm{d}x\mathrm{d}y\mathrm{d}z\mathrm{d}t$$

$$= \int_{t}^{t+\Delta t}\int_{i-\frac{1}{2}}^{i+\frac{1}{2}}\int_{j-\frac{1}{2}}^{j+\frac{1}{2}}\int_{k-\frac{1}{2}}^{k+\frac{1}{2}}\frac{\partial}{\partial y}\left(\Gamma\frac{\partial\phi}{\partial y}\right)\mathrm{d}x\mathrm{d}y\mathrm{d}z\mathrm{d}t$$

$$+ \int_{t}^{t+\Delta t}\int_{i-\frac{1}{2}}^{i+\frac{1}{2}}\int_{j-\frac{1}{2}}^{j+\frac{1}{2}}\int_{k-\frac{1}{2}}^{k+\frac{1}{2}}\frac{\partial}{\partial z}\left(\Gamma\frac{\partial\phi}{\partial z}\right)\mathrm{d}x\mathrm{d}y\mathrm{d}z\mathrm{d}t$$

$$+ \int_{t}^{t+\Delta t}\int_{i-\frac{1}{2}}^{i+\frac{1}{2}}\int_{j-\frac{1}{2}}^{j+\frac{1}{2}}\int_{k-\frac{1}{2}}^{k+\frac{1}{2}}S\mathrm{d}x\mathrm{d}y\mathrm{d}z\mathrm{d}t \qquad (7.90)$$

The fully implicit form of the finite difference equation

$$a_i\phi_{i,j,k} = a_{i+1}\phi_{i+1,j,k} + a_{i-1}\phi_{i-1,j,k} + a_{j+1}\phi_{i,j+1,k} + a_{j-1}\phi_{i,j-1,k}$$

$$+ a_{k+1}\phi_{i,j,k+1} + a_{k-1}\phi_{i,j,k-1} + d \qquad (7.91a)$$

where

$$a_{i+1} = \frac{\Gamma_{i+\frac{1}{2}}\Delta y\Delta z}{\Delta x_{i+\frac{1}{2}}}, \quad a_{i-1} = \frac{\Gamma_{i-\frac{1}{2}}\Delta y\Delta z}{\Delta x_{i-\frac{1}{2}}} \qquad (7.91b)$$

$$a_{j+1} = \frac{\Gamma_{j+\frac{1}{2}}\Delta z\Delta x}{\Delta z_{j+\frac{1}{2}}}, \quad a_{j-1} = \frac{\Gamma_{j-\frac{1}{2}}\Delta z\Delta x}{\Delta z_{j-\frac{1}{2}}} \qquad (7.91c)$$

$$a_{k+1} = \frac{\Gamma_{k+\frac{1}{2}}\Delta x\Delta y}{\Delta z_{k+\frac{1}{2}}}, \quad a_{k-1} = \frac{\Gamma_{k-\frac{1}{2}}\Delta x\Delta y}{\Delta z_{k-\frac{1}{2}}} \qquad (7.91d)$$

$$a_i^l = \frac{C\ \Delta x\Delta y\ \Delta z}{\Delta t}, \quad d = S_0\ \Delta x\Delta y\ \Delta z + a_i^l\ \phi_{i,j,k}^l \qquad (7.91e)$$

$$a_i = a_{i+1} + a_{i-1} + a_{j+1} + a_{j-1} + a_{k+1} + a_{k-1} + a_i^l - S_1\Delta x\Delta y\Delta z \qquad (7.91f)$$

7.3.2 Finite Difference Equation by Taylor Series Expansion

Let us consider the two-dimensional transient diffusion equation given as

$$C\frac{\partial\phi}{\partial t} = \Gamma\left[\frac{\partial^2\phi}{\partial x^2} + \frac{\partial^2\phi}{\partial y^2}\right] \qquad (7.92)$$

Using the central-difference formula for the second derivative at the $(l+1)$th time step as

$$\left.\frac{\partial^2\phi}{\partial x^2}\right)_{i,j} = \frac{\phi_{i+1,j}^{l+1} - 2\phi_{i,j}^{l+1} + \phi_{i-1,j}^{l+1}}{\Delta x^2} \qquad (7.93a)$$

$$\left.\frac{\partial^2\phi}{\partial y^2}\right)_{l,i} = \frac{\phi_{i,j+1}^{l+1} - 2\phi_{i,j}^{l+1} + \phi_{i,j-1}^{l+1}}{\Delta y^2} \tag{7.93b}$$

and backward-difference formula at the $(l+1)$th time step for the first-order time derivative from Table 5.1 as

$$\left.\frac{\partial\phi}{\partial t}\right|_{i,j} = \frac{\phi_{i,j}^{l+1} - \phi_{i,j}^{l}}{\Delta t} \tag{7.93c}$$

Substituting Equation 7.93 into Equation 7.92, we get

$$\begin{aligned}\phi_{i,j}^{l+1} = {} & \frac{\Gamma \Delta t / C\Delta x^2}{1+2\left[\Gamma \Delta t / C\Delta x^2\right]+2\left[\Gamma \Delta t / C\Delta x^2\right]}\left[\phi_{i+1,j}^{l+1}+\phi_{i+1,j}^{l+1}\right] \\ & + \frac{\Gamma \Delta t / C\Delta x^2}{1+2\left[\Gamma \Delta t / C\Delta x^2\right]+2\left[\Gamma \Delta t / C\Delta x^2\right]}\left[\phi_{i,j+1}^{l+1}+\phi_{i,j-1}^{l+1}\right] \\ & + \frac{1}{1+2\left[\Gamma \Delta t / C\Delta x^2\right]+2\left[\Gamma \Delta t / C\Delta x^2\right]}\phi_{i,j}^{l}\end{aligned} \tag{7.94}$$

The above equation can be simplified to the following special cases.

For Uniform Grid Size Distribution, i.e., $\Delta x = \Delta y$

$$\phi_{i,j}^{l+1} = \frac{\Gamma\Delta t/C\Delta x^2}{1+4\Gamma\Delta t/C\Delta x^2}\left[\phi_{i+1,j}^{l+1}+\phi_{i-1,j}^{l+1}\right]+\frac{\Gamma\Delta t/C\Delta x^2}{1+4\Gamma\Delta t/C\Delta x^2}\left[\phi_{l,j+1}^{l+1}+\phi_{i,j-1}^{l+1}\right]+\frac{1}{1+4\Gamma\Delta t/C\Delta x^2}\phi_{l,i}^{l}. \tag{7.95a}$$

or

$$\left[1+4\gamma_0\right]\phi_{i,j}^{l+1} - \gamma_o\left[\phi_{i+1,j}^{l+1}+\phi_{i-l,j}^{l+1}+\phi_{i,j+1}^{l+1}+\phi_{i,j-1}^{l+1}\right] = \phi_{i,j}^{l} \tag{7.95b}$$

For One-Dimensional Unsteady-State Case

$$\phi_i^{l+1} = \frac{\Gamma\Delta t/C\Delta x^2}{1+2\Gamma\Delta t/C\Delta x^2}\left[\phi_{i+1}^{l+1}+\phi_{i-1}^{l+1}\right]+\frac{1}{1+2\Gamma\Delta t/C\Delta x^2}\phi_i^{l} \tag{7.96a}$$

or

$$\left[1+2\gamma_o\right]\phi_i^{l+1} - \gamma_o\left[\phi_{i+1}^{l+1}+\phi_{i-l}^{l+1}\right] = \phi_i^{l} \tag{7.96b}$$

Three-Dimensional Problems

Let us consider a general three-dimension transient diffusion equation a constant source term given as

$$C\frac{\partial\phi}{\partial t} = \Gamma_x\frac{\partial^2\phi}{\partial x^2}+\Gamma_y\frac{\partial^2\phi}{\partial y^2}+\Gamma_z\frac{\partial^2\phi}{\partial z^2}+S \tag{7.97}$$

Substituting central difference finite difference approximation for the second derivative at the $(l+1)$th time step and fist-order forward difference approximation for first-order time derivate, we get

$$C\frac{\phi_{i,j,k}^{l+1}-\phi_{i,j,k}^{l}}{\Delta t}=\Gamma_x\frac{\phi_{i+1,j,k}^{l+1}-2\,\phi_{i,j,k}^{l+1}+\phi_{i-1,j,k}^{l+1}}{\Delta x^2}+\Gamma_y\frac{\phi_{i,j+1,k}^{l+1}-2\,\phi_{i,j,k}^{l+1}+\phi_{i,1-1,j,k}^{l+1}}{\Delta y^2}$$
$$+\Gamma_x\frac{\phi_{i,j,k+1}^{l+1}-2\,\phi_{i,j,k}^{l+1}+\phi_{i,j,k-1}^{l+1}}{\Delta x^2}+S \tag{7.98}$$

Rearranging

$$\left[\frac{C}{\Delta t}+\frac{2\Gamma_x}{\Delta x^2}+\frac{2\Gamma_y}{\Delta y^2}+\frac{2\Gamma_z}{\Delta y^2}\right]\phi_{i,j,k}^{l+1}=\frac{\Gamma_x}{\Delta \mathrm{x}^2}\left[\phi_{i+1,j,k}^{l+1}+\phi_{i-1,j,k}^{l+1}\right]+\frac{\Gamma_y}{\Delta \mathrm{y}^2}\left[\phi_{i,j+1,k}^{l+1}+\phi_{i,j-1,k}^{l+1}\right]$$
$$+\frac{\Gamma_z}{\Delta \mathrm{z}^2}\left[\phi_{i,jk+1}^{l+1}+\phi_{i,j,k-1}^{l+1}\right]+\frac{C}{\Delta t}\,\phi_{i,j,k}^{l}+S \tag{7.99}$$

Divide both side by $C/_{\Delta t}$, we get

$$\left[1+\frac{2\Gamma_x\Delta t}{C\Delta x^2}+\frac{2\Gamma_y\Delta t}{C\Delta y^2}+\frac{2\Gamma_z\Delta t}{C\Delta y^2}\right]\phi_{i,j,k}^{l+1}=\frac{\Gamma_x\Delta t}{C\Delta \mathrm{x}^2}\left[\phi_{i+1,j,k}^{l+1}+\phi_{i-1,j,k}^{l+1}\right]+\frac{\Gamma_y\Delta t}{C\Delta \mathrm{y}^2}\left[\phi_{i,j+1,k}^{l+1}+\phi_{i,j-1,k}^{l+1}\right]$$
$$+\frac{\Gamma_z\Delta t}{C\Delta \mathrm{z}^2}\left[\phi_{i,jk+1}^{l+1}+\phi_{i,j,k-1}^{l+1}\right]+\phi_{i,j,k}^{l}+S\frac{\Delta t}{C} \tag{7.100}$$

Simplifying

$$\phi_{i,j,k}^{l+1}=\frac{\dfrac{\Gamma_X\Delta t}{C\Delta x^2}}{\left[1+\dfrac{2\Gamma_X\Delta t}{C\Delta x^2}+\dfrac{2\Gamma_y\Delta t}{C\Delta y^2}+\dfrac{2\Gamma_z\Delta t}{C\Delta y^2}\right]}\left[\phi_{i,j,k}^{l+1}+\phi_{i-1,j,k}^{l+1}\right]$$
$$+\frac{\dfrac{\Gamma_y\Delta t}{C\Delta y^2}}{\left[1+\dfrac{2\Gamma_x\Delta t}{C\Delta x^2}+\dfrac{2\Gamma_y\Delta t}{C\Delta y^2}+\dfrac{2\Gamma_z\Delta t}{C\Delta y^2}\right]}\left[\phi_{i,j+1,k}^{l+1}+\phi_{i,j-1,k}^{l+1}\right]$$
$$+\frac{\dfrac{\Gamma_z\Delta t}{C\Delta y^2}}{\left[1+\dfrac{2\Gamma_x\Delta t}{C\Delta x^2}+\dfrac{2\Gamma_y\Delta t}{C\Delta y^2}+\dfrac{2\Gamma_z\Delta t}{C\Delta y^2}\right]}\left[\phi_{i,j,k+1}^{l+1}+f_{i,j,k-1}^{l+1}\right]$$
$$+\frac{1}{\left[1+\dfrac{2\Gamma_x\Delta t}{C\Delta x^2}+\dfrac{2\Gamma_y\Delta t}{C\Delta y^2}+\dfrac{2\Gamma_z\Delta t}{C\Delta y^2}\right]}\phi_{i,j,k}^{l}+\frac{\Delta t/C}{\left[1+\dfrac{2\Gamma_x\Delta t}{C\Delta x^2}+\dfrac{2\Gamma_y\Delta t}{C\Delta y^2}+\dfrac{2\Gamma_z\Delta t}{C\Delta y^2}\right]}S \tag{7.101}$$

For $\Gamma_x=\Gamma_y=\Gamma_z=\Gamma$ and uniform grid size distribution in all dimensions,

$$\Delta x==\Delta y=\Delta z$$

$$\phi_{i,j,k}^{l+1} = \frac{\frac{\Gamma_x \Delta t}{C\Delta x^2}}{\left[1+\frac{6\Gamma\Delta t}{C\Delta x^2}\right]}\left[\phi_{i+1,j,k}^{l+1}+\phi_{i-1,j,k}^{l+1}\right]+\frac{\frac{\Gamma_y \Delta t}{C\Delta y^2}}{\left[1+\frac{6\Gamma\Delta t}{C\Delta x^2}\right]}\left[\phi_{i,j+1,k}^{l+1}+\phi_{i,j-1,k}^{l+1}\right]$$

$$+\frac{\frac{\Gamma_z \Delta t}{C\Delta z^2}}{\left[1+\frac{6\Gamma\Delta t}{C\Delta x^2}\right]}\left[\phi_{i,j,k+1}^{l+1}+\phi_{i,j,k-1}^{l+1}\right]+\frac{1}{\left[\left[1+\frac{6\Gamma\Delta t}{C\Delta x^2}\right]\right]}\phi_{i,j,k}^{l}+\frac{\Delta t/C}{\left[1+\frac{6\Gamma\Delta t}{C\Delta x^2}\right]}S \tag{7.102}$$

7.3.3 A General Formulation of Fully Implicit Scheme for One-Dimensional Problems

Let us consider a general one-dimensional problem with the grid system shown in the figure and boundary conditions of the first kind at the left- and right-hand side of the solution domain (Figure 7.8).

Assuming that the boundary conditions are of constant value (i.e., boundary temperatures ϕ_1^{l+1} and ϕ_m^{l+1} are known), the following system of equations can be constructed using Equation 7.96b.

$$\begin{bmatrix} 1+2\gamma_0 & -\gamma_0 & & & \\ -\gamma_0 & 1+2\gamma_0 & -\gamma_0 & & \\ & -\gamma_0 & 1+2\gamma_0 & -\gamma_0 & \\ \vdots & \vdots & \vdots & \vdots & \vdots \\ & & & -\gamma_0 & 1+2\gamma_0 \end{bmatrix} \begin{bmatrix} \phi_2^{l+1} \\ \phi_3^{l+1} \\ \vdots \\ \phi_{m-2}^{l+1} \\ \phi_{m-1}^{l+1} \end{bmatrix} = \begin{Bmatrix} \phi_2^{l}+\gamma_o\phi_1^{l+1} \\ \phi_3^{l} \\ \vdots \\ \phi_{m-2}^{l} \\ \phi_{m-1}^{l}+\gamma_o\phi_m^{l+1} \end{Bmatrix} \tag{7.103}$$

It can be noted that the coefficient matrix is tridiagonal and all its elements are known. So, the unknown solution vector

$$\begin{Bmatrix} \phi_2^{l+1} \\ \phi_3^{l+1} \\ \vdots \\ \phi_{m-1}^{l+1} \end{Bmatrix}$$

can be solved by using a direct or iterative solution method.

Since the matrix is tridiagonal, the TDMA algorithm is quite suitable as a direct solver. Another convenient way to solve the system with increase in time steps is to evaluate the inverse of the matrix A^{-1} and solve the system as

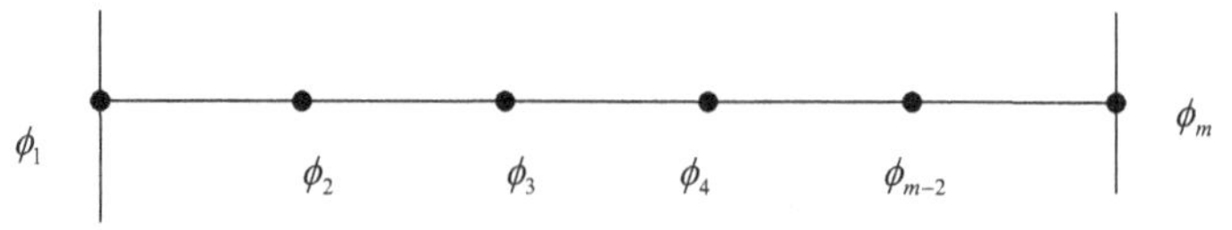

FIGURE 7.8 One-dimensional grid.

$$\begin{bmatrix} \phi_2^{l+1} \\ \phi_3^{l+1} \\ \vdots \\ \phi_{m-2}^{l+1} \\ \phi_{m-1}^{l+1} \end{bmatrix} = \begin{bmatrix} 1+2\gamma_0 & -\gamma_0 & & & \\ -\gamma_0 & 1+2\gamma_0 & -\gamma_0 & & \\ & -\gamma_0 & 1+2\gamma_0 & -\gamma_0 & \\ \vdots & \vdots & \vdots & \vdots & \vdots \\ & & & -\gamma_0 & 1+2\gamma_0 \end{bmatrix}^{-1} \begin{Bmatrix} \phi_2^l + \gamma_o \phi_1^{l+1} \\ \phi_3^l \\ \vdots \\ \phi_{m-2}^l \\ \phi_{m-1}^l + \gamma_o \phi_m^{l+1} \end{Bmatrix} \quad (7.104)$$

The above equation can be used to determine the solution vector with marching in time.

Example 7.3: One-Dimensional Unsteady State with Fully Implicit Scheme

Let us consider a one-dimensional unsteady-state heat conduction in a plane slab of thickness $L = 10\,\text{cm}$. Initially the slab is a uniform temperature at $T_I = 0°\text{C}$. At time $t>0$, the left and right sides of the slab are simultaneously brought to a temperature of 100°C and 50°C, respectively, as shown in the figure.

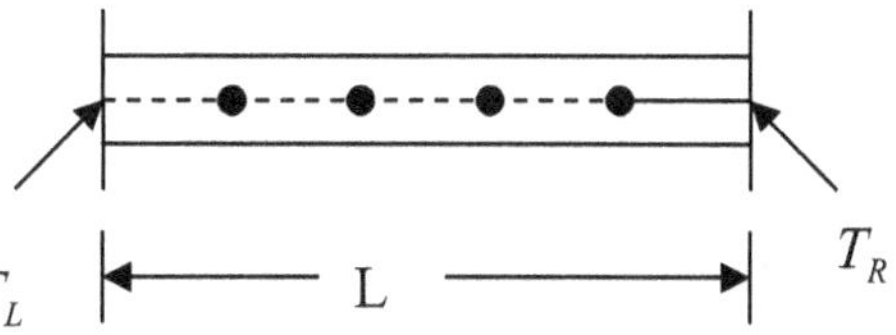

Governing Equation:

From one-dimensional transient heat diffusion equation,

$$\rho c_p \frac{\partial \phi}{\partial t} = \frac{\partial}{\partial x}\left(k \frac{\partial \phi}{\partial x}\right)$$

Considering constant thermal conductivity and introducing thermal diffusivity $\alpha = \dfrac{k}{\rho c_p}$,

we have

$$\frac{\partial T}{\partial t} = \alpha \frac{\partial^2 T}{\partial x^2}$$

Boundary Conditions:

1. $x = 0, \quad T = T_L = 100°\text{C}$
2. $x = L, \quad T = T_R = 50°\text{C}$

Initial Condition:

$$t = 0, \quad T(x) = T_I = 0$$

The rod is aluminum with thermal diffusivity at $\alpha = 0.835\text{cm}^2/\text{s}$. Use a fully implicit scheme to obtain the system of equations for the grid shown in the figure and a time step $\Delta t = 0.1\,\text{s}$. Obtain temperature distribution in the rod at $t = 0.1$ s and $t = 0.2$ s.

Solution

For the selected grid system, the spatial step

$$\Delta x = \frac{L}{N\text{div}_x} = \frac{10}{5} = 2\text{cm}$$

and the dimensionless time $\gamma_0 = f_0$ = Fourier number is

$$f_0 = \frac{\alpha \Delta t}{\Delta x^2} = \frac{0.835(0.1)}{(2)^2} = 0.04175$$

The appropriate discretization equation for all interior nodes is the fully implicit Equation 7.59, which for the heat diffusion equation becomes

$$[1+2f_o]T_i^{l+1} - f_0\left[T_{i+1}^{l+1} + T_{i-1}^{l+1}\right] = T_i^l$$

Applying this equation successively to all nodes, we get

Node 1

$$[1+2f_0]T_1^{l+1} - f_0\left[T_2^{l+1} + T_L^{l+1}\right] = T_1^l$$

$$[1+2(0.04175)]T_1^1 - 0.04175\left[T_2^1 + T_L^{l+1}\right] = T_1^0$$

$$[1+2(0.04175)]T_1^1 - 0.04175\left[T_2^1 + 100\right] = 0$$

$$1.0835T_1^1 - 0.04175T_2^1 = 4.175$$

Node 2

$$[1+2f_0]T_2^{l+1} - f_0\left[T_3^{l+1} + T_1^{l+1}\right] = T_2^l$$

$$[1+2(0.04175)]T_2^1 - 0.04175\left[T_3^{l+1} + T_1^{l+1}\right] = T_2^0$$

$$-0.04175T_1^1 - 1.0835T_2^1 - 0.04175T_3^1 = 0$$

Node 3

$$[1+2f_0]T_3^{l+1} - f_0\left[T_4^{l+1} + T_2^{l+1}\right] = T_3^l$$

$$[1+2(0.04175)]T_3^1 - 0.04175\left[T_4^{l+1} + T_2^{l+1}\right] = T_3^0$$

$$-0.041\,75T_2^1 - 1.0835T_3^1 - 0.04175T_4^1 = 0$$

Node 4

$$[1+2f_0]T_4^{l+1} - f_0\left[T_R^{l+1} + T_3^{l+1}\right] = T_4^l$$

$$[1+2(0.04175)]T_4^1 - 0.04175\left[T_R^{l+1} + T_3^{l+1}\right] = T_4^0$$

$$-0.04175T_3^1 - 1.0835T_4^1 = 2.0875$$

Assembling all nodal equations, we get the system of equations in matrix form as

$$\begin{bmatrix} 1.0835 & -0.04175 & 0 & 0 \\ -0.04175 & 1.0835 & -0.04175 & 0 \\ 0 & -0.04175 & 1.0835 & -0.04175 \\ 0 & 0 & -0.04175 & 1.0835 \end{bmatrix} \begin{Bmatrix} T_1^1 \\ T_2^1 \\ T_3^1 \\ T_4^1 \end{Bmatrix} = \begin{Bmatrix} T_1^0 + 4.175 \\ T_2^0 \\ T_3^0 \\ T_4^0 + 2.0875 \end{Bmatrix}$$

Setting $T_1^0 = T_2^0 = T_3^0 = T_4^0 = 0$ as initial condition and solving the system of equations, we have the temperature distribution at time $t = 0.1$ s as

$$\begin{Bmatrix} T_1^1 \\ T_2^1 \\ T_3^1 \\ T_4^1 \end{Bmatrix} = \begin{Bmatrix} 3.85910°\text{C} \\ 0.15179°\text{C} \\ 0.08021°\text{C} \\ 1.92972°\text{C} \end{Bmatrix}$$

Similarly, at time $t = 2\Delta t$, the system of equations for nodal temperatures is obtained as

$$\begin{bmatrix} 1.0835 & -0.04175 & 0 & 0 \\ -0.04175 & 1.0835 & -0.04175 & 0 \\ 0 & -0.04175 & 1.0835 & -0.04175 \\ 0 & 0 & -0.04175 & 1.0835 \end{bmatrix} \begin{Bmatrix} T_1^2 \\ T_2^2 \\ T_3^2 \\ T_4^2 \end{Bmatrix} = \begin{Bmatrix} T_1^1 + 4.175 \\ T_2^1 \\ T_3^1 \\ T_4^1 + 2.0875 \end{Bmatrix}$$

or

$$\begin{bmatrix} 1.0835 & -0.04175 & 0 & 0 \\ -0.04175 & 1.0835 & -0.04175 & 0 \\ 0 & -0.04175 & 1.0835 & -0.04175 \\ 0 & 0 & -0.04175 & 1.0835 \end{bmatrix} \begin{Bmatrix} T_1^2 \\ T_2^2 \\ T_3^2 \\ T_4^2 \end{Bmatrix} = \begin{Bmatrix} 3.85910 + 4.175 \\ 0.15179 \\ 0.08021 \\ 1.92972 + 2.0875 \end{Bmatrix}$$

or

$$\begin{bmatrix} 1.0835 & -0.04175 & 0 & 0 \\ -0.04175 & 1.0835 & -0.04175 & 0 \\ 0 & -0.04175 & 1.0835 & -0.04175 \\ 0 & 0 & -0.04175 & 1.0835 \end{bmatrix} \begin{Bmatrix} T_1^2 \\ T_2^2 \\ T_3^2 \\ T_4^2 \end{Bmatrix} = \begin{Bmatrix} 8.0341 \\ 0.15179 \\ 0.08021 \\ 4.01722 \end{Bmatrix}$$

The solution of the system gives temperature distribution at time $t = 0.2$ s as

$$\begin{Bmatrix} T_1^2 \\ T_2^2 \\ T_3^2 \\ T_4^2 \end{Bmatrix} = \begin{Bmatrix} 7.43173°\text{C} \\ 0.43547°\text{C} \\ 0.23402°\text{C} \\ 3.71665°\text{C} \end{Bmatrix}$$

7.3.4 A General Formulation of Fully Implicit Scheme for Two-Dimensional Problems

Let us consider a general two-dimensional problem with the grid system and boundary conditions as shown in Figure 7.9.

The appropriate discretization equation for the interior node is given by Equation 7.95b as

$$\left[1 + 4\gamma_0\right]\phi_{i,j}^{l+1} - \gamma_o\left[\phi_{i+1,j}^{l+1} + \phi_{i-l,j}^{l+1} + \phi_{i,j+1}^{l+1} + \phi_{i,j-1}^{l+1}\right] = \phi_{i,j}^{l} \tag{7.105}$$

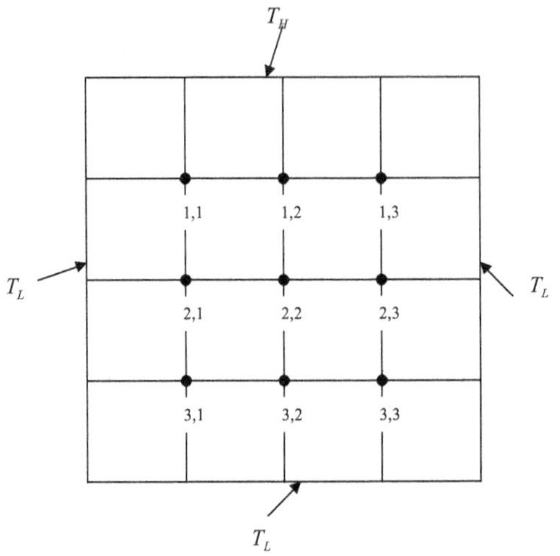

FIGURE 7.9 Two-dimensional grid.

Let us apply this equation to all interior nodes as follows:

$$Node\,(1,1)\quad [1+4\gamma_0]T_{1,1}^{l+1} - \gamma_o\left[T_{1,2}^{l+1} + T_{\mathrm{L}}^{l+1} + T_{\mathrm{H}}^{l+1} + T_{2,1}^{l+1}\right] = T_{1,1}^{l}$$

$$Node\,(1,2)\quad [1+4\gamma_0]T_{1,2}^{l+1} - \gamma_o\left[T_{1,3}^{l+1} + T_{1,1}^{l+1} + T_{\mathrm{H}}^{l+1} + T_{2,2}^{l+1}\right] = T_{1,2}^{l}$$

$$Node\,(1,3)\quad [1+4\gamma_0]T_{1,3}^{l+1} - \gamma_o\left[T_{\mathrm{L}}^{l+1} + T_{1,2}^{l+1} + T_{\mathrm{H}}^{l+1} + T_{2,3}^{l+1}\right] = T_{1,3}^{l}$$

$$Node\,(2,1)\quad [1+4\gamma_0]T_{2,1}^{l+1} - \gamma_o\left[T_{2,2}^{l+1} + T_{\mathrm{L}}^{l+1} + T_{1,1}^{l+1} + T_{3,1}^{l+1}\right] = T_{2,1}^{l}$$

$$Node\,(2,2)\quad [1+4\gamma_0]T_{2,2}^{l+1} - \gamma_o\left[T_{2,3}^{l+1} + T_{2,1}^{l+1} + T_{1,2}^{l+1} + T_{3,2}^{l+1}\right] = T_{2,2}^{l}$$

$$Node\,(2,3)\quad [1+4\gamma_0]T_{2,3}^{l+1} - \gamma_o\left[T_{\mathrm{L}}^{l+1} + T_{2,2}^{l+1} + T_{1,3}^{l+1} + T_{3,3}^{l+1}\right] = T_{2,3}^{l}$$

$$Node\,(3,1)\quad [1+4\gamma_0]T_{3,1}^{l+1} - \gamma_o\left[T_{3,2}^{l+1} + T_{\mathrm{L}}^{l+1} + T_{2,1}^{l+1} + T_{\mathrm{L}}^{l+1}\right] = T_{3,1}^{l}$$

$$Node\,(3,2)\quad [1+4\gamma_0]T_{3,2}^{l+1} - \gamma_o\left[T_{3,3}^{l+1} + T_{3,1}^{l+1} + T_{2,2}^{l+1} + T_{\mathrm{L}}^{l+1}\right] = T_{3,2}^{l}$$

$$Node\,(3,3)\quad [1+4\gamma_0]T_{3,3}^{l+1} - \gamma_o\left[T_{\mathrm{L}}^{l+1} + T_{3,2}^{l+1} + T_{2,3}^{l+1} + T_{\mathrm{L}}^{l+1}\right] = T_{3,3}^{l}$$

Writing the system of equations in matrix form, we have

$$\begin{bmatrix} 1+4\gamma_0 & -\gamma_0 & 0 & -\gamma_0 & 0 & 0 & 0 & 0 & 0 \\ -\gamma_0 & 1+4\gamma_0 & -\gamma_0 & 0 & -\gamma_0 & 0 & 0 & 0 & 0 \\ 0 & -\gamma_0 & 1+4\gamma_0 & 0 & 0 & -\gamma_0 & 0 & 0 & 0 \\ -\gamma_0 & 0 & 0 & 1+4\gamma_0 & -\gamma_0 & 0 & -\gamma_0 & 0 & 0 \\ 0 & -\gamma_0 & 0 & -\gamma_0 & 1+4\gamma_0 & -\gamma_0 & 0 & -\gamma_0 & 0 \\ 0 & 0 & -\gamma_0 & 0 & -\gamma_0 & 1+4\gamma_0 & 0 & 0 & -\gamma_0 \\ 0 & 0 & 0 & -\gamma_0 & 0 & 0 & 1+4\gamma_0 & -\gamma_0 & 0 \\ 0 & 0 & 0 & 0 & -\gamma_0 & 0 & -\gamma_0 & 1+4\gamma_0 & -\gamma_0 \\ 0 & 0 & 0 & 0 & 0 & -\gamma_0 & 0 & -\gamma_0 & 1+4\gamma_0 \end{bmatrix} \begin{Bmatrix} T_{1,1} \\ T_{1,2} \\ T_{1,3} \\ T_{2,2} \\ T_{2,3} \\ T_{3,3} \\ T_{3,1} \\ T_{3,2} \\ T_{3,3} \end{Bmatrix}$$

$$= \begin{Bmatrix} T_{1,1}+\gamma_0(T_L+T_H) \\ T_{1,2}+\gamma_0 T_L \\ T_{1,3}+\gamma_0(T_L+T_H) \\ T_{2,1}+\gamma_0 T_L \\ T_{2,2} \\ T_{2,3}+\gamma_0 T_L \\ T_{3,1}+2\gamma_0 T_L \\ T_{3,2}+\gamma_0 T_L \\ T_{3,3}+\gamma_0 T \end{Bmatrix} \tag{7.106}$$

7.3.5 Solution Methods for a Two-Dimensional Implicit Scheme

It can be noted that Equation 7.63 resulting from a two-dimensional diffusion equation is no longer tridiagonal, and hence, unlike the unidirectional transient diffusion problem, the discretization equations for the two-dimensional transient heat conduction cannot be solved by TDMA. The reason for this is the fact that while Equation 7.58 can be rearranged so that three of the terms are at or adjacent to the main diagonal, the remaining two terms are displaced. We can certainly use a direct solver such as the Gauss elimination or *LU* decomposition method or iterative point-by-point methods such as Gauss–Seidel or Successive overrelaxation (SOR) methods. However, there are two other alternate methods, **line-by-line method** and **block tridiagonal method**, that are effectively used for some class of problems.

Application of Gauss–Seidel Method (Point-by-Point Method) This is the simplest of all iterative methods to implement, as discussed in Chapters 2 and 6. It stores the most recent calculated values of nodal temperatures. An iterative process at each time step is executed before proceeding to the next time step. For example, for the two-dimensional diffusion problems, the temperature at each grid point in a body is obtained from Equation 7.95b by writing it in an explicit form as

$$\phi_{i,j}^{n+1} = \frac{\gamma_0\left(\phi_{i+1,j}^{n+1} + \phi_{i-1,j}^{n+1}\right) + \gamma_0\left(\phi_{i,j+1}^{n+1} + \phi_{i,j-1}^{n+1}\right) + \phi_{i,j}^{n}}{1+4\gamma_0} \tag{7.107}$$

where $\phi_{i+l,j}^{n+1}, \phi_{i-l,j}^{n+1}, \phi_{i,j+1}^{n+1}, \phi_{i,j-1}^{n+1}$ are temperatures at four grid points that are currently stored in the memory. The iterative scheme is continued until all nodal values converge with assigned tolerance at each time step before increasing. This procedure can also be extended to the SOR algorithm.

Line-By-Line Method The **line-by-line method** is primarily a combination of a TDMA algorithm for a one-dimensional problem and a Gauss–Seidel iteration method. Instead of estimating one nodal value at a time, as in a Gauss–Seidel method, all nodal values in a vertical line or in a horizontal line are computed directly by using a TDMA algorithm and sweeping from left to right or from bottom to top. For example, in sweeping all vertical lines in an x-direction, the values of all nodal points ($\phi_{i,j}^{n+1}, j = 1, \ldots, N$) along a vertical line or an x line (x_i) are determined in a present iteration ($n+1$) by using a TDMA algorithm while assuming the nodal values along two adjacent vertical or x lines (x_{i-1} and x_{i+1}) are known, as shown in Figure 7.10. The nodal values in the x_{i-1} line are assumed to be known in the present iteration ($n+1$), while the nodal values in the x_{i+1} line are obtained in the previous iteration (n).

This is repeated for all vertical lines in an x-direction, and one iteration is said to be completed when all vertical lines are scanned. In the next iteration, we can choose to sweep in a y-direction and estimate all nodal values along all horizontal lines. Alternating the sweeping directions essentially

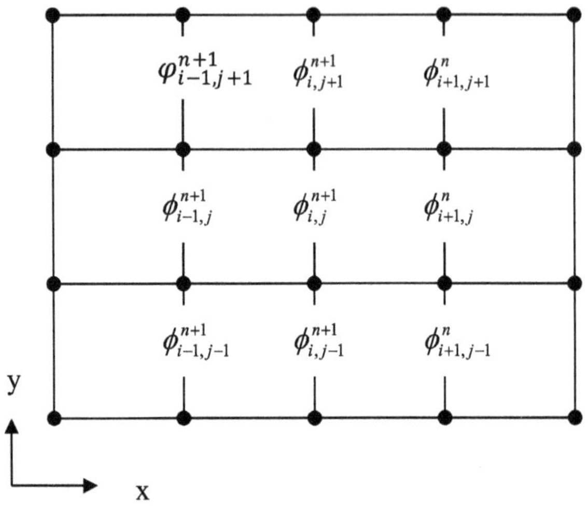

FIGURE 7.10 Line-by-line method with sweep in the *x*-direction.

improves the convergence rate. Once convergence is reached at the present given time, solution procedure is repeated for the next subsequent time steps.

Block Tridiagonal Matrix Algorithm The second approach, known as the **block tridiagonal matrix algorithm** as discussed in Chapter 2, is also applicable to the system of Equation 7.58 by writing it in a $N^2 \times N^2$ coefficient matrix in a block tridiagonal form as

$$\gamma_0 \begin{bmatrix} T_N & -I_N & & & & & & & \\ -I_N & T_N & & & & & & & \\ & & T_N & & & & & & \\ & & & T_N & & & & & \\ & & & & T_N & & & & \\ & & & & & T_N & & & \\ & & & & & & T_N & & \\ & & & & & & & T_N & -I_N \\ & & & & & & & -I_N & T_N \end{bmatrix} \begin{Bmatrix} \\ \\ \\ \end{Bmatrix} = \begin{Bmatrix} \\ \\ \\ \end{Bmatrix} \quad (7.108)$$

where

$$T_N = \begin{bmatrix} \gamma_0^{-1}+4 & -1 & & & & & & & \\ -1 & \gamma_0^{-1}+4 & -1 & & & & & & \\ & -1 & \gamma_0^{-1}+4 & -1 & & & & & \\ & & -1 & \gamma_0^{-1}+4 & -1 & & & & \\ & & & -1 & \gamma_0^{-1}+4 & -1 & & & \\ & & & & -1 & \gamma_0^{-1}+4 & -1 & & \\ & & & & & -1 & \gamma_0^{-1}+4 & -1 & \\ & & & & & & -1 & \gamma_0^{-1}+4 & -1 \\ & & & & & & & -1 & \gamma_0^{-1}+4 \end{bmatrix}$$

and I_N is an $N \times N$ identity matrix.

7.3.6 Boundary Conditions for Implicit Scheme

Implicit discretization equations for boundary nodes can be derived using the procedure outlined in the case of an explicit scheme.

Let us briefly discuss the derivation of one such equation for the grid point located at the **right-hand side boundary** shown below.

Let us demonstrate derivation of the boundary node equation by considering the two-dimensional unsteady-state diffusion equation given as

$$\frac{\partial}{\partial x}\left(\Gamma_x \frac{\partial \phi}{\partial x}\right)+\frac{\partial}{\partial x}\left(\Gamma_y \frac{\partial \phi}{\partial y}\right)=C\frac{\partial \phi}{\partial t} \tag{7.109}$$

To proceed, we integrate the equation over the half control volume and over the time interval t to $t+\Delta t$ (Figure 7.11):

$$\int_{i-\frac{1}{2}}^{i}\int_{j-\frac{1}{2}}^{j+\frac{1}{2}}\int_{t}^{t+\Delta t}\frac{\partial}{\partial x}\left(\Gamma_x \frac{\partial \phi}{\partial x}\right)\mathrm{d}x\mathrm{d}y\mathrm{d}t+\int_{i-\frac{1}{2}}^{i}\int_{j-\frac{1}{2}}^{j+\frac{1}{2}}\int_{t}^{t+\Delta t}\frac{\partial}{\partial y}\left(\Gamma_y \frac{\partial \phi}{\partial y}\right)\mathrm{d}x\mathrm{d}y\mathrm{d}t=C\int_{i-\frac{1}{2}}^{i}\int_{j-\frac{1}{2}}^{j+\frac{1}{2}}\int_{t}^{t+\Delta t}\frac{\partial \phi}{\partial t}\mathrm{d}x\mathrm{d}y\mathrm{d}t \tag{7.110}$$

$$\left.\Gamma_x \frac{\partial \phi}{\partial x}\right)_i \Delta y\Delta t-\left.\Gamma_y \frac{\partial \Gamma}{\partial x}\right)_{i-\frac{1}{2}} \Delta y\Delta t+\left.\Gamma_y \frac{\partial \phi}{\partial y}\right)_{j+\frac{1}{2}}\frac{\Delta x\Delta t}{2}-\left.\Gamma_y \frac{\partial \phi}{\partial y}\right)_{j-\frac{1}{2}}\frac{\Delta x\Delta t}{2}=C\frac{\Delta x\Delta y}{2}\int_{t}^{t+\Delta t}\frac{\partial \phi}{\partial t}\mathrm{d}t \tag{7.111}$$

The first term in the equation is described by the heat flux boundary conditions as

$$-\left.\Gamma_x \frac{\partial \phi}{\partial x}\right)_i=-f_i'' \tag{7.112a}$$

or

$$\left.\Gamma_x \frac{\partial \phi}{\partial x}\right)_i=f_i'' \tag{7.112b}$$

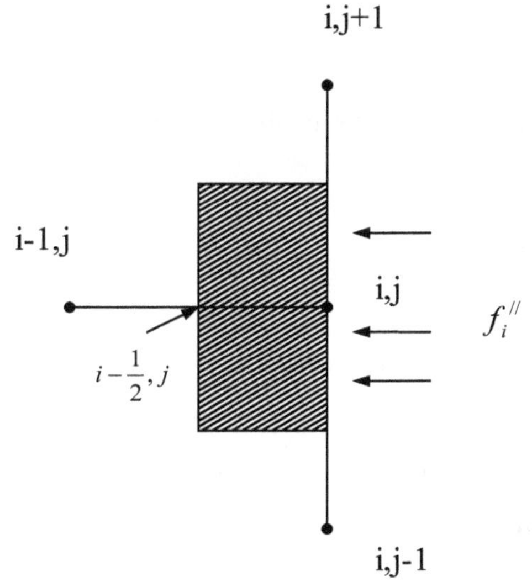

FIGURE 7.11 Control volume for a boundary node.

Next, we assume a piecewise-linear profile or a central difference formula to approximate temperature gradients at points $j+\frac{1}{2}$, $j-\frac{1}{2}$, and $i-\frac{1}{2}$, and a forward difference formula for the first-order time derivative

$$\Gamma_x \frac{\partial \phi}{\partial x}\bigg)_{i-\frac{1}{2}} = \Gamma_{x_{i-\frac{1}{2}}} \frac{\phi_{i,j}^{l+1} - \phi_{i-1,j}^{l+1}}{\Delta x}, \quad \Gamma_y \frac{\partial \phi}{\partial y}\bigg)_{j+\frac{1}{2}} = \Gamma_{y_{j+\frac{1}{2}}} \frac{\phi_{i,j+1}^{l+1} - \phi_{i,j}^{l+1}}{\Delta y} \tag{7.113a}$$

$$\Gamma_y \frac{\partial \phi}{\partial y}\bigg)_{j-\frac{1}{2}} = \Gamma_{y_{j-\frac{1}{2}}} \frac{\left(\phi_{i,j}^{l+1} - \phi_{i,j-1}^{l+1}\right)}{\Delta y}, \quad \frac{\partial \phi}{\partial t} = \frac{\phi_{i,j}^{l+1} - \phi_{i,j}^{l}}{\Delta t} \tag{7.113b}$$

Substituting these approximations, we get

$$+ f_i'' \Delta y \Delta t - \frac{\Gamma_{i-\frac{1}{2}}\left(\phi_{i,j}^{l+1} - \phi_{i-1,j}^{l+1}\right)}{\Delta x} \Delta y \Delta t + \frac{\Gamma_{j+\frac{1}{2}}\left(\phi_{i,j+1}^{l+1} - \phi_{i,j}^{l+1}\right)}{\Delta y} \frac{\Delta x}{2} \Delta t - \frac{\Gamma_{j-\frac{1}{2}}\left(\phi_{i,j}^{l+1} - \phi_{i,j-1}^{l+1}\right)}{\Delta y} \frac{\Delta x}{2} \Delta t$$

$$= C \frac{\Delta x}{2} \Delta y \left(\phi_{i,j}^{l+1} - \phi_{i,j}^{l}\right) \tag{7.114}$$

$$\frac{\Gamma_{i-\frac{1}{2}} \Delta y}{\Delta x} \phi_{i-1,j}^{l+1} + \frac{\Gamma_{j+\frac{1}{2}} \Delta x}{2\Delta y} \phi_{i,j+1}^{l+1} + \frac{\Gamma_{j-\frac{1}{2}} \Delta x}{2\Delta y} \phi_{i,j-1}^{l+1} + \left(-\frac{C\Delta x \Delta y}{2\Delta t} - \frac{\Gamma_{i-\frac{1}{2}} \Delta y}{\Delta x} - \frac{\Gamma_{i,j+\frac{1}{2}} \Delta x}{2\Delta y} - \frac{\Gamma_{j-\frac{1}{2}} \Delta x}{2\Delta y}\right) \phi_{i,j}^{l+1}$$

$$= -\frac{C\Delta x \Delta y}{2\Delta t} \phi_{i,j}^{l} + f_i'' \Delta y$$

Simplifying

$$a_{i-1}\phi_{i-1,j}^{l+1} + a_{j+1}\phi_{i,j+1}^{l+1} + a_{j-1}\phi_{i,j-1}^{l+1} - \left(a_i^l + a_{i-1} + a_{j+1} + a_{j-1}\right)\phi_{i,j}^{l+1} = a_i^l \phi_{i,j}^l + d \tag{7.115a}$$

where

$$a_i = a_i^l + a_{i-1} + a_{j+1} + a_{j-1}, \quad a_{i-1} = \frac{\Gamma_{i-\frac{1}{2}} \Delta y}{\Delta x}, \quad a_{j+1} = \frac{\Gamma_{j+\frac{1}{2}} \Delta x}{2\Delta y}, \quad a_{j-1} = \frac{\Gamma_{j-\frac{1}{2}} \Delta x}{2\Delta y}$$

$$a_i^l = -\frac{C\Delta x \Delta y}{2\Delta t}, \quad d = f_i'' \Delta y \tag{7.115b}$$

For constant transport property, Γ, and for uniform grid size distributions, i.e., $\Delta x = \Delta y$, we can write the above equation in an alternate form as

$$\phi_{i,j}^{l+1} = \frac{\Gamma \Delta t}{C\Delta x^2}\left[2\phi_{i-1,j}^{l+1} + \phi_{i,j+1}^{l+1} + \phi_{i,j-1}^{l+1}\right] + \left[1 - 4\frac{\Gamma \Delta t}{C\Delta x^2}\right]\phi_i^l + \frac{2\Delta t}{C\Delta x} f_i'' \tag{7.116}$$

For a **one-dimensional case**, the equation reduces to

$$\phi_i^{l+1} = \frac{\Gamma \Delta t}{C\Delta x^2}\left[2\phi_{i-1}^{l+1}\right] + \left[1 - \frac{2\Gamma \Delta t}{C\Delta x^2}\right]\phi_i^l + \frac{2\Delta t}{C\Delta x} f_i'' \tag{7.117}$$

Let us now consider special cases.

1. **Adiabatic boundary condition or a symmetric boundary**
   For an adiabatic surface, the surface flux is zero, i.e., $f_i'' = 0$

Equations 7.116 and 7.117 reduce to

$$\phi_{i,j}^{l+1} = \frac{\Gamma \Delta t}{C\Delta x^2}\left[2\phi_{i-1,j}^{l+1} + \phi_{i,j+1}^{l+1} + \phi_{i,j-1}^{l+1}\right] + \left[1 - \frac{4\Gamma\Delta t}{C\Delta x^2}\right]\phi_{i,j}^{l} \tag{7.118}$$

for two-dimensional problems
and

$$\phi_{i}^{l+1} = \frac{\Gamma \Delta t}{C\Delta x^2}\left[2\phi_{i-1}^{l+1}\right] + \left[1 - \frac{2\Gamma\Delta t}{C\Delta x^2}\right]\phi_{i}^{l} \tag{7.119}$$

for one-dimensional problems

2. **Convective boundary (Figure 7.12)**
   For convective boundary condition,

$$-\Gamma\frac{\partial\Gamma}{\partial x} = h(\phi - \phi_\infty) \tag{7.120a}$$

or

$$f_i'' = h(\phi_{i,j} - \phi_\infty) \tag{7.120b}$$

Substituting Equation 7.120b into Equation 7.116, we get

$$\phi_{i,j}^{l+1} = \frac{\Gamma \Delta t}{C\Delta x^2}\left[2\phi_{i-1,j}^{l+1} + \phi_{i,j+1}^{l+1} + \phi_{i,j-1}^{l+1} + \frac{2h\Delta x}{\Gamma}\phi_\infty\right] + \left[1 - \frac{4\Gamma\Delta t}{C\Delta x^2} - \frac{2h\Delta t}{C\Delta x}\right]\phi_{i,j}^{l} \tag{7.121}$$

A summary of these implicit discretization formulas is given in Table 7.4.

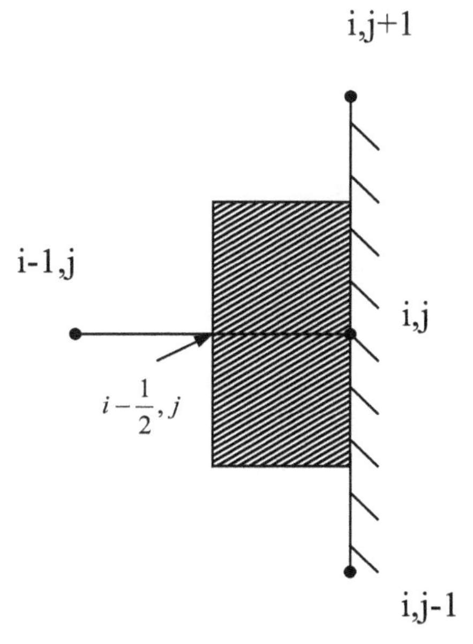

FIGURE 7.12 Convective boundary node.

TABLE 7.4
A Summary of Implicit Discretization Equations with $\Delta x = \Delta y$

Physical Situation	Discretization Equation
a. Interior node 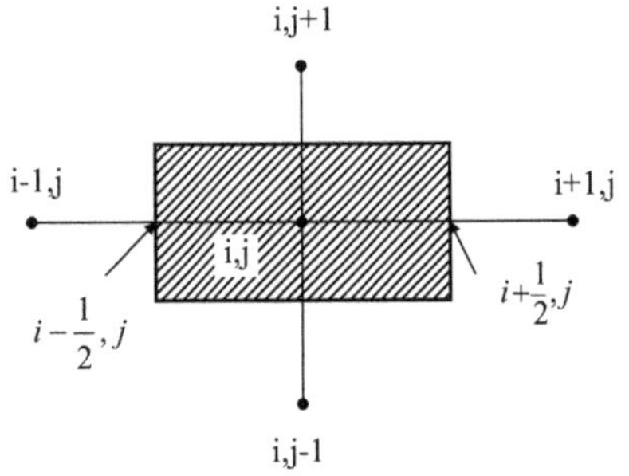	$\left[1+4\gamma_o\right]\phi_{i,j}^{l+1} - \gamma_o\left[\phi_{i+1,j}^{l+1}+\phi_{i-1,j}^{l+1}+\phi_{i,j+1}^{l+1}+\phi_{i,j-1}^{l+1}\right] = \phi_{i,j}^{l}$
b. Convective boundary node 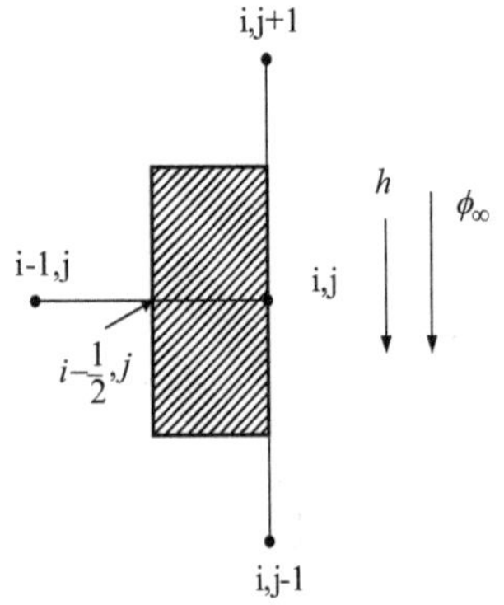	$\left[1+2\gamma_0\left(2+\frac{h\Delta x}{\Gamma}\right)\right]\phi_{i,j}^{l+1} - \gamma_0\left[2\phi_{i-1,j}^{l+1}+\phi_{i,j-1}^{l+1}+\phi_{i,j+1}^{l+1}+2\frac{h\Delta x}{\Gamma}\phi_\infty\right] = \phi_{i,j}^{l}$
c. Exterior corner with convection boundary 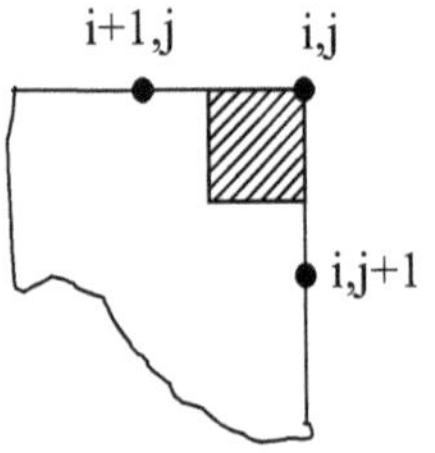	$\left[1+4\gamma_0\left(1+\frac{h\Delta x}{\Gamma}\right)\right]\phi_{i,j}^{l+1} - 2\gamma_0\left[\phi_{i-1,j}^{l+1}+\phi_{i,j-1}^{l+1}+2\frac{h\Delta x}{\Gamma}\phi_\infty\right] = \phi_{i,j}^{l}$
d. Interior corner with convection boundary 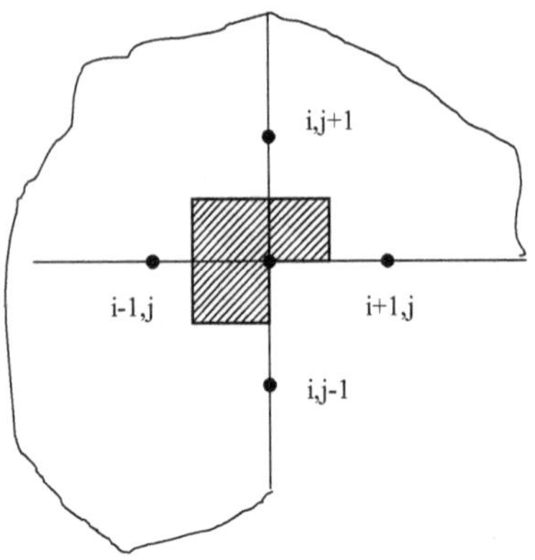	$\left[1+4\gamma_o\left(1+\frac{h\Delta x}{3\Gamma}\right)\right]\phi_{i,j}^{l+1} - 2\left(\frac{\gamma_0}{3}\right)\left[2\phi_{i-1,j}^{l+1}+\phi_{i,j-1}^{l+1}+2\phi_{i,j+1}^{l+1}+\phi_{i+1,j}^{l+1}+2\frac{h\Delta x}{\Gamma}\phi_\infty\right] = \phi_{i,j}^{l}$

(*Continued*)

TABLE 7.4 (*Continued*)

A Summary of Implicit Discretization Equations with $\Delta x = \Delta y$

Physical Situation	Discretization Equation
e. Adiabatic boundary 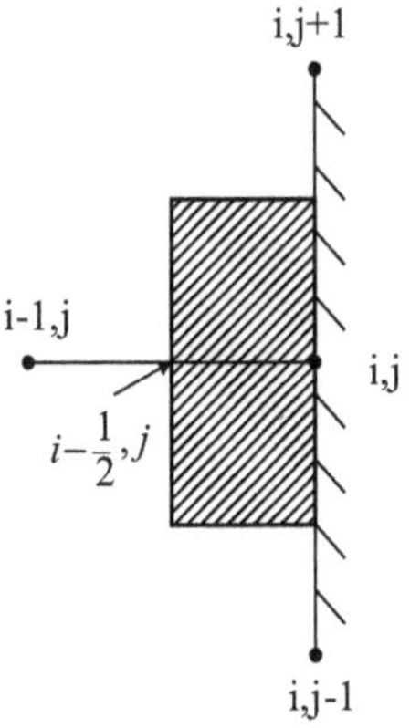	$\left[1+4\gamma_o\right]\phi_{i,j}^{l+1}-\frac{2\gamma_o}{3}\left[2\phi_{i-1,j}^{l+1}+\phi_{i,j+1}^{l+1}+\phi_{i,j-1}^{l+1}\right]=\phi_{i,j}^{l}$
f. Constant surface heat flux 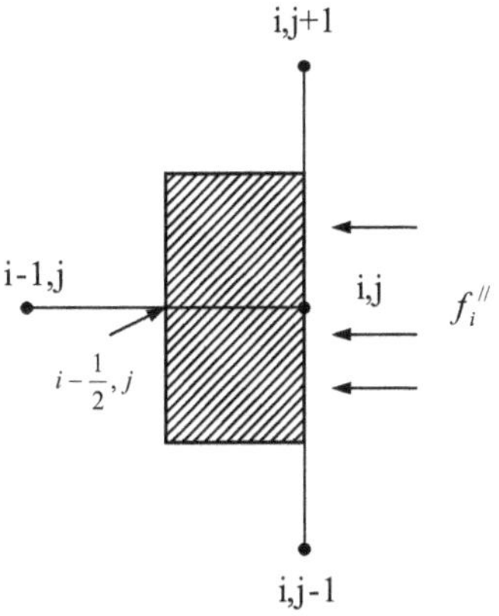	$\phi_{i,j}^{l+1}-\frac{\Gamma\Delta t}{C\Delta x^2}\left[2\phi_{i-1,j}^{l+1}+\phi_{i,j+1}^{l+1}+\phi_{i,j-1}^{l+1}\right]=\left[1-4\frac{\Gamma\Delta t}{C\Delta x^2}\right]\phi_i^l+\frac{2\Delta t}{C\Delta x}f_i''$

Note: $\gamma_0 = \Gamma\Delta t/C\Delta x^2$.

Example 7.4: Two-Dimensional Unsteady-State Problem with Fully Implicit Scheme

A fin shown below is initially at a uniform temperature $T_I = 300°\text{C}$ and then suddenly exposed to the convection environment at a temperature of $T_\infty = 20°\text{C}$ and $h = 40\ \text{W/m}^2\,°\text{C}$. The base of the fin is maintained at a temperature of $T_0 = 300°\text{C}$. Consider two-dimensional unsteady-state conduction in the fin and determine temperature as a function of time using an 8×2 grid. The material thermo-physical properties are $\rho = 7800\ \text{kg/m}^3$, $c = 800\ \text{J/kg}\,°\text{C}$, and $k = 10\ \text{W/m}\,°\text{C}$. The length and thickness of the fin are $L = 8\ \text{cm}$ and $H = 2\ \text{cm}$, respectively.

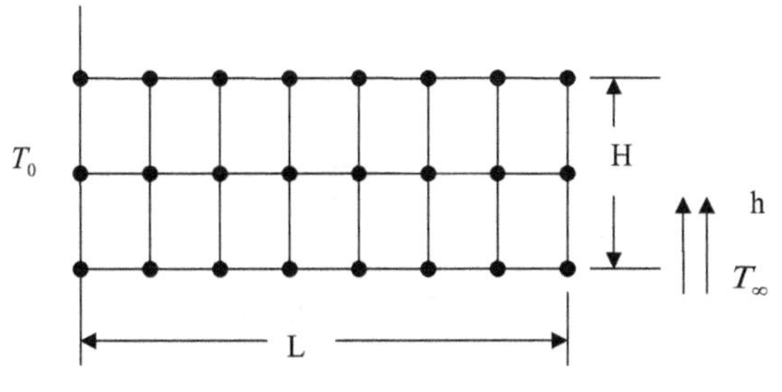

Solution

Governing Equation:

$$\frac{\partial}{\partial x}\left(k_x \frac{\partial \phi}{\partial x}\right)+\frac{\partial}{\partial y}\left(k_y \frac{\partial \phi}{\partial y}\right)=\rho c_p \frac{\partial \phi}{\partial t}$$

For constant thermal conductivity, $k_x = k_y = k$, we have

$$\frac{\partial^2 T}{\partial x^2}+\frac{\partial^2 T}{\partial y^2}=\frac{1}{\alpha}\frac{\partial T}{\partial t}$$

where $\alpha = \dfrac{k}{\rho c}$

Initial Condition $t = 0,\ T(x, y, 0) = T_0 = 300^\circ\mathrm{C}$

Boundary Conditions

1. $x = 0,\quad T(0, y, t) = T_0 = 300°\mathrm{C}$
2. $x = L$ and on boundary surface, $-k(\partial T/\partial n) = h(T - T_\infty)$

The grid sizes in the *x*- and *y*-directions are

$$\Delta x = \frac{L}{NDIV_x} = \frac{8}{8} = 1\ \mathrm{cm} = 0.01\,\mathrm{m},\quad \Delta y = \frac{L}{NDIV_y} = \frac{2}{2} = 1\,\mathrm{cm} = 0.01\,\mathrm{m}.$$

Discretization equations for the nodes are selected as follows:

Interior Nodes: 9, 10, 11, 12, 13, 14, 15

$$(1+4f_0)T_{i,j}^{l+1} - f_0\left(T_{i+1,j}^{l+1}+T_{i-1,j}^{l+1}+T_{i,j+1}^{l+1}+T_{i,j-1}^{l+1}\right)=T_{i,j}^{l}$$

where $f_0 = \dfrac{\alpha \Delta t}{\Delta x^2}$

Node – 9

$$(1+4f_0)T_9^{l+1} - f_0\left(T_{10,}^{l+1}+T_0+T_1^{l+1}+T_{17}^{l+1}\right)=T_9^{l}$$

Node – 15

$$(1+4f_0)T_{15}^{l+1} - f_0\left(T_{16,}^{l+1}+T_{14}+T_7^{l+1}+T_{23}^{l+1}\right)=T_{15}^{l}$$

Convective nodes on top surface: 1, 2, 3, 4, 5, 6, 7 (Figure 7.13)

$$\left[1+2f_0(2+B_i\right]T_{i,j}^{l+1} - f_0\left[2T_{i,j-1}^{l+1}+T_{i-1,j}^{l+1}+T_{i+1,j}^{l+1}+2B_iT_\infty\right]=T_{i,j}^{l}$$

Node 1

$$\left[1+2f_0(2+B_i\right]T_1^{l+1} - f_0\left[2T_9^{l+1}+T_0+T_2^{l+1}+2B_iT_\infty\right]=T_1^{l}$$

⋮

⋮

Node 7

$$\left[1+2f_0(2+B_i\right]T_7^{l+1} - f_0\left[2T_{15}^{l+1}+T_6+T_8^{l+1}+2B_iT_\infty\right]=T_7^{l}$$

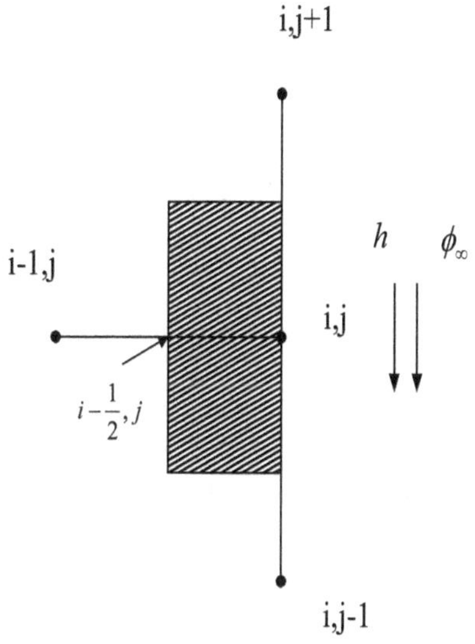

FIGURE 7.13 Convective boundary node.

Convective nodes on bottom surface: 17, 18, 19, 20, 21, 22, 23

$$\left[1+2f_0(2+B_i\right]T_{i,j}^{l+1}-f_0\left[2T_{i,j+1}^{l+1}+T_{i-1,j}^{l+1}+T_{i+1,j}^{l+1}+2B_iT_\infty\right]=T_{i,j}^{l}$$

Node 17

$$\left[1+2f_0(2+B_i\right]T_{17}^{l+1}-f_0\left[2T_{9}^{l+1}+T_0+T_{18}^{l+1}+2B_iT_\infty\right]=T_{17}^{l}$$

⋮

⋮

Node 23

$$\left[1+2f_0(2+B_i\right]T_{23}^{l+1}-f_0\left[2T_{15}^{l+1}+T_{22}^{l+1}+T_{24}^{l+1}+2B_iT_\infty\right]=T_{23}^{l}$$

Convective node right surface: 16

$$\left[1+2f_0(2+B_i\right]T_{i,j}^{l+1}-f_0\left[2T_{i-1,j}^{l+1}+T_{i,j-1}^{l+1}+T_{i,j+1}^{l+1}+2B_iT_\infty\right]=T_{i,j}^{l}$$

Node 16

$$\left[1+2f_0(2+B_i\right]T_{16}^{l+1}-f_0\left[2T_{15}^{l+1}+T_{24}^{l+1}+T_{8}^{l+1}+2B_iT_\infty\right]=T_{16}^{l}$$

North-east corner with convective node: 8

$$\left[1+4f_0(1+B_i\right]T_{i,j}^{l+1}-2f_0\left[2T_{i-1,j}^{l+1}+T_{i,j-1}^{l+1}+2B_iT_\infty\right]=T_{i,j}^{l}$$

Node 8

$$\left[1+4f_0(1+B_i\right]T_{8}^{l+1}-2f_0\left[2T_{7}^{l+1}+T_{16}^{l+1}+2B_iT_\infty\right]=T_{8}^{l}$$

South-east corner with convective node: 24

$$\left[1+4f_0(1+B_i\right]T_{i,j}^{l+1}-2f_0\left[2T_{i-1,j}^{l+1}+T_{i,j+1}^{l+1}+2B_iT_\infty\right]=T_{i,j}^{l}$$

Node 24

$$\left[1+4f_0(1+B_i\right]T_{24}^{l+1}-2f_0\left[2T_{23}^{l+1}+T_{16}^{l+1}+2B_iT_\infty\right]=T_{24}^{l}$$

Note that for convective nodes, we have $f_0=\dfrac{\alpha\Delta t}{\Delta x^2}$ and $Bi=\dfrac{h\Delta x}{k}$

Node	Temperature
	Top Surface
1	261.73332029
2	233.95998474
3	213.05863823
4	197.13343750
5	184.78988480
6	175.09333805
7	167.27218777
8	160.95470105
	Mid-Section
9	265.63788848
10	238.09027107
11	216.90011772
12	200.66013908
13	188.09017058
14	178.17752625
15	170.21842068
16	163.77223699
	Bottom Surface
17	261.70592749
18	233.91145214
19	212.99992949
20	197.05738540
21	184.72851089
22	175.01658512
23	167.21506168
24	160.90214644

A pseudo code implementing these discretization equations and solving the system of equation using the Gauss–Seidel method is given in Table 7.5. A solution to the problem is obtained using a computer code based on the pseudo code and results for temperature distribution at time $t = 10\Delta t$ are given below.

As expected, the top and bottom surface temperature distributions are almost identical. The discrepancy is due to the round-off error.

TABLE 7.5
Pseudo Code for Transient Solution in a Rectangular Slab Using Implicit Finite Difference Scheme and Gauss–Seidel Method

```
Input dimensions, properties, select number of divisions, maximum time, and stability parameters
  Input dimension L and H
  Input thermo-physical properties Γ and C
  Input parameters γ0
  Select number of divisions NDIVX and NDIVY
  Number of node NX = NDIVX+1
  Number of node NY = NDIVY+1
  Time step Dt
Initialize time
  time = 0.0
Set initial condition
  dofor I = 1, NX
    dofor J = 1, NY
      φ(I,J) = φI
    enddo
  enddo
dowhile (time<tmax)
  time = time+Dt
    iflag = 1
    iter = 0
  dowhile (it<itmax and iflag = 1)
    iter = iter+1
Set new values to old values
      dofor I = 2, NX−1
        dofor J = 2, NY−1
          φo(I,J) = φ(I,J)
      enddo
Calculate nodal values at interior nodes
    dofor I = 2, NX−1
      dofor J=2, NY−1
        φ(I,J) = (γ0 (φ(I+1,J)+φ(I−1,j)+φ(I,J+1)+φ(I,J−1))+φo (I,J))/(1+4γ0)
    enddo
Similarly calculate nodal values at other boundary nodes or assign boundary values
Estimate approximate percent relative error
    dofor I = 1, NX
      dofor J = 1, NY
        εa = ABS(φ(I,J)−φo(I,J)/φ(I,J))
        If (εa > εs) then
          iflag = 1
        else
          iflag = 0
        endif
```

(Continued)

TABLE 7.5 (*Continued*)
Pseudo Code for Transient Solution in a Rectangular Slab Using Implicit Finite Difference Scheme and Gauss–Seidel Method

```
        enddo
      enddo
    enddo
Enddo
```

7.4 CRANK–NICOLSON SCHEME

Crank and Nicolson suggested a modified implicit method. In this scheme, the space derivatives are averaged between the $(l+1)$th time step and lth time step. A forward-difference formula is used to approximate the time derivative. The resulting discretization equation for two-dimensional unsteady case with constant thermo-physical properties and without heat generation is

$$C\frac{\phi_{i,j}^{l+1}-\phi_{i,j}^{l}}{\Delta t}=\Gamma\left[\frac{1}{2}\frac{\phi_{i+1,j}^{l+1}-2\phi_{i,j}^{l+1}+\phi_{i-1,j}^{l+1}}{\Delta x^2}+\frac{1}{2}\frac{\phi_{i+1,j}^{l}-2\phi_{i,j}^{l}+\phi_{i-1,j}^{l}}{\Delta x^2}\right]$$
$$+\Gamma\left[\frac{1}{2}\frac{\phi_{i,j+1}^{l+1}-2\phi_{i,j}^{l+1}+\phi_{i,j-1}^{l+1}}{\Delta y^2}+\frac{1}{2}\frac{\phi_{i,j+1}^{l}-2\phi_{i,j}^{l}+\phi_{i,j-1}^{l}}{\Delta y^2}\right] \quad (7.122)$$

Solving for $\phi_{i,j}^{l+1}$, we obtain

$$\phi_{i,j}^{l+1}=\frac{1-\left[\Gamma\Delta t/C\Delta x^2\right]-\left[\Gamma\Delta t/C\Delta y^2\right]}{1+\left[\Gamma\Delta t/C\Delta x^2+\Gamma\Delta t/C\Delta y^2\right]}\phi_{i,j}^{l}+\frac{\left[\alpha\Delta t/2\Delta x^2\right]}{1+\left[\alpha\Delta t/\Delta x^2\right]+\left[\alpha\Delta t/\Delta y^2\right]}\left[\phi_{i-1,j}^{l+1}+\phi_{i+1,j}^{l+1}+\phi_{i-1,j}^{l}+\phi_{i+1,j}^{l}\right]$$
$$+\frac{\left[\Gamma\Delta t/2C\Delta x^2\right]}{1+\left[\Gamma\Delta t/C\Delta x^2\right]+\left[\Gamma\Delta t/C\Delta y^2\right]}\left[\phi_{i,j+1}^{l+1}+\phi_{i,j-1}^{l+1}+\phi_{i,j+1}^{l}+\phi_{i,j-1}^{l}\right] \quad (7.123)$$

It can be mentioned here that the discretization equation 7.80 can also be obtained by setting the weighting factor $f = 0.5$ in general form of discretization equation (7.9) obtained by using the control volume method.

The truncation error is of the order of $O(\Delta x^2 + \Delta t^2)$. The advantage of this method is that, for given values of space and time steps Δx and Δt, the resulting solution involves less truncation error due to Δt than the explicit and implicit schemes. A von Neumann stability analysis shows that the *Crank–Nicolson scheme* is *unconditionally stable*. However, the Crank–Nicolson scheme involves additional computation. If there are N internal grid points over the region, then this method involves the solution of N simultaneous algebraic equations for each time step.

7.4.1 Solution Methods for Crank–Nicolson Method

Regarding the solution of the system of algebraic equations, the methodologies utilized in the fully implicit scheme can be used in this case as well. For example, in the case of one-dimensional heat conduction, the above equation can be rearranged as

$$-\frac{1}{2}\frac{\Gamma\Delta t}{C\Delta x^2}\phi_{i-1}^{l+1}+\left[1+\frac{\Gamma\Delta t}{C\Delta x^2}\right]\phi_i^{l+1}-\frac{1}{2}\frac{\Gamma\Delta t}{C\Delta x^2}\phi_{i+1}^{l+1}=\frac{1}{2}\frac{\Gamma\Delta t}{C\Delta x^2}\phi_{i-1}^{l}+\left[1-\frac{\Gamma\Delta t}{C\Delta x^2}\right]\phi_i^{l}+\frac{1}{2}\frac{\Gamma\Delta t}{C\Delta x^2}\phi_{i+1}^{l} \quad (7.124)$$

For two-dimensional problems, the resulting system of discretization equations is not tridiagonal.

7.5 SPLITTING METHODS

The difficulties encountered in solving two-dimensional diffusion equations by a conventional implicit scheme led to the development of splitting algorithms such as the **alternating-direction implicit (ADI) method** and the **alternating direction explicit (ADE) method**. In the ADI method, marching of the solution in a time step takes place in two half steps with the implicit scheme being used alternatively between the x- and y-directions. This leads to the formation of a tridiagonal system at each time step and the system can be solved by a direct solver such as TDMA.

7.5.1 ADI METHOD

In the ADI method given by Peaceman and Rachford (1955), the two-step procedure for the solution of Equation 7.55 is given as follows.

Step 1 In this step, the equation is solved at intermediate time step $\Delta t/2$ with an implicit scheme used only for the derivative in the x-direction and an explicit scheme for the derivative in y-direction

$$C\frac{\phi_{i,j}^{l+\frac{1}{2}} - \phi_{i,j}^{l}}{\Delta t/2} = \Gamma\frac{\phi_{i+1,j}^{l+\frac{1}{2}} - 2\phi_{i,j}^{l+\frac{1}{2}} + \phi_{i-1,j}^{l+\frac{1}{2}}}{\Delta x^2} + \Gamma\frac{\phi_{i,j+1}^{l} - 2\phi_{i,j}^{l} + \phi_{i,j-1}^{l}}{\Delta y^2} \tag{7.125}$$

Rearranging,

$$\left(2+\frac{2C\Delta x^2}{\Gamma\Delta t}\right)\phi_{i,j}^{l+\frac{1}{2}} - \phi_{i+1,j}^{i+\frac{1}{2}} - \phi_{i-1,j}^{i+\frac{1}{2}} = \left(\frac{2C\Delta x^2}{\Gamma\Delta t} - 2\frac{\Delta x^2}{\Delta y^2}\right)\phi_{i,j}^{l} + \frac{\Delta x^2}{\Delta y^2}\left(\phi_{i,j+1}^{l} + \phi_{i,j-1}^{l}\right) \tag{7.126}$$

Equation 7.126 represents a tridiagonal system for a fixed j location and as approximate solution is obtained at this intermediate time step for nodes $i = 1, 2, \ldots, N$.

Step 2 The solution is now advanced to time step $l+1$ with $\Delta t/2$ as the time step and using an implicit scheme only for the derivative in the y-direction and an explicit scheme for the derivative in the x-direction as follows:

$$C\frac{\phi_{i,j}^{l+1} - \phi_{i,j}^{l+\frac{1}{2}}}{\Delta t/2} = \Gamma\frac{\phi_{i+1,j}^{l+\frac{1}{2}} - 2\phi_{i,j}^{l+\frac{1}{2}} + \phi_{i-1,j}^{l+\frac{1}{2}}}{\Delta x^2} + \Gamma\frac{\phi_{i,j+1}^{l+1} - 2\phi_{i,j}^{l+1} + \phi_{i,j-1}^{l+1}}{\Delta y^2} \tag{7.127}$$

Rearranging,

$$\left(2+\frac{2C\Delta x^2}{\Gamma\Delta t}\right)\phi_{i,j}^{l+1} - \phi_{i,j+1}^{i+1} - \phi_{i,j-1}^{i+1} = \left(\frac{2C\Delta x^2}{\Gamma\Delta t} - 2\frac{\Delta x^2}{\Delta y^2}\right)\phi_{i,j}^{l+\frac{1}{2}} + \frac{\Delta x^2}{\Delta y^2}\left(\phi_{i+1,j}^{l+\frac{1}{2}} + \phi_{i-1,j}^{l+\frac{1}{2}}\right) \tag{7.128}$$

Equation 7.128 represents a tridiagonal system for a fixed i location and a solution is obtained at the time step for all nodes $j = 1, 2, \ldots, N$.

The solution is marched forward by alternating between the rows and columns with the implicit scheme being used only in one direction and the explicit scheme in the other direction. The ADI method is second-order accurate with an error of the order $O\left[(\Delta t)^2, (\Delta x)^2, (\Delta y)^2\right]$. A stability analysis leads to the conclusion that the ADI scheme is unconditionally stable. The method can also be employed along with the Crank–Nicolson scheme with two intermediate time steps and for three-dimensional problems.

7.5.2 ADE Method

In the ADE method, the two-step procedure for the solution of Equation 7.55 is used alternately for x- and y-directions, and the resulting system is solved in an explicit manner. For example, the two-step marching procedure for the x-direction is used as follows.

Step 1 In this step, the equation is solved at an intermediate time step $\Delta t/2$ with the solution obtained explicitly by marching from left to right (i.e., $i = 1, 2, \ldots, N$). The scheme is shown as

$$C\frac{\phi_{i,j}^{l+\frac{1}{2}} - \phi_{i,j}^{l}}{\Delta t/2} = \Gamma\frac{\phi_{i+l,j}^{l} - \phi_{i,j}^{l} - \phi_{i,j}^{l+\frac{1}{2}} + \phi_{i-1,j}^{l+\frac{1}{2}}}{\Delta x^2} + \Gamma\frac{\phi_{i,j+1}^{l} - 2\phi_{i,j}^{l} + \phi_{l,j-1}^{l}}{\Delta y^2} \tag{7.129}$$

Rearranging,

$$\left(1+\frac{2C\Delta x^2}{\Gamma\Delta t}\right)\phi_{i,j}^{l+\frac{1}{2}} = \phi_{i-1,j}^{i+\frac{1}{2}} + \phi_{i+1,j}^{i} + \left(\frac{2C\Delta x^2}{\Gamma\Delta t} - 2\frac{\Delta x^2}{\Delta y^2} - 1\right)\phi_{i,j}^{l} + \frac{\Delta x^2}{\Delta y^2}\left(\phi_{i,j+1}^{l} + \phi_{i,j-1}^{l}\right) \tag{7.130}$$

Note that in marching from left to right ($i = 1, 2, \ldots, N$), all items including $\phi_{i-1,j}^{i+\frac{1}{2}}$ on the right-hand side are known.

Step 2 In this step, the equation is solved at the time step Δt with solution obtained explicitly by marching from right to left, i.e., $i = N, (N-1) \ldots 2, 1$

$$C\frac{\phi_{i,j}^{l+1} - \phi_{i,j}^{l+\frac{1}{2}}}{\Delta t/2} = \Gamma\frac{\phi_{i+l,j}^{l+1} - \phi_{i,j}^{l+1} - \phi_{i,j}^{l+\frac{1}{2}} + \phi_{i-1,j}^{l+\frac{1}{2}}}{\Delta x^2} + \Gamma\frac{\phi_{i,j+1}^{l+\frac{1}{2}} - 2\phi_{i,j}^{l+\frac{1}{2}} + \phi_{l,j-1}^{l+\frac{1}{2}}}{\Delta y^2} \tag{7.131}$$

Rearranging,

$$\left(1+\frac{2C\Delta x^2}{\Gamma\Delta t}\right)\phi_{i,j}^{l+1} = \phi_{i+1,j}^{l+1} + \phi_{i-1,j}^{i+\frac{1}{2}} + \left(\frac{2C\Delta x^2}{\Gamma\Delta t} - 2\frac{\Delta x^2}{\Delta y^2} - 1\right)\phi_{i,j}^{l+\frac{1}{2}} + \frac{\Delta x^2}{\Delta y^2}\left(\phi_{i,j+1}^{l+\frac{1}{2}} + \phi_{i,j-1}^{l+\frac{1}{2}}\right) \tag{7.132}$$

Note that in marching from left to right ($i = 1, 2, \ldots, N$), all items including $\phi_{i+1,j}^{i+1}$ on the right-hand side are known. The two-step procedure is repeated for the y-direction by alternating the marching direction from top to bottom and then from bottom to top.

For a one-dimensional problem, Equations 7.130 and 7.132 reduce to

$$\left(1+\frac{2C\Delta x^2}{\Gamma\Delta t}\right)\phi_{i}^{l+\frac{1}{2}} = \phi_{i-1}^{i+\frac{1}{2}} + \phi_{i+1}^{i} + \left(\frac{2C\Delta x^2}{\Gamma\Delta t} - 1\right)\phi_{i}^{l} \tag{7.133}$$

and

$$\left(1+\frac{2C\Delta x^2}{\Gamma\Delta t}\right)\phi_{i}^{l+1} = \phi_{i+1}^{l+1} + \phi_{i-1}^{i+\frac{1}{2}} + \left(\frac{2C\Delta x^2}{\Gamma\Delta t} - 1\right)\phi_{i}^{l+\frac{1}{2}} \tag{7.134}$$

PROBLEMS

7.1 Consider a one-dimensional unsteady-state conduction in a plane slab [$k = 50\ \mathrm{W/m\,°C}$, $\alpha = 2\times10^{-5}\ \mathrm{m^2/s}$] of thickness $L = 4\,\mathrm{mm}$. Initially, the slab is at 20°C. The right side of the plate is suddenly brought to a temperature of 200°C. The left side is maintained at 20°C.

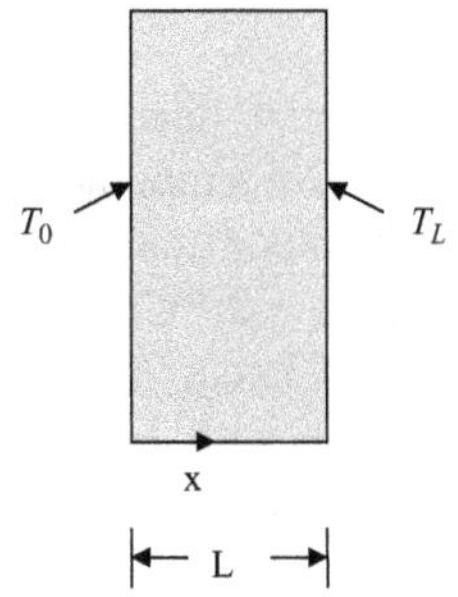

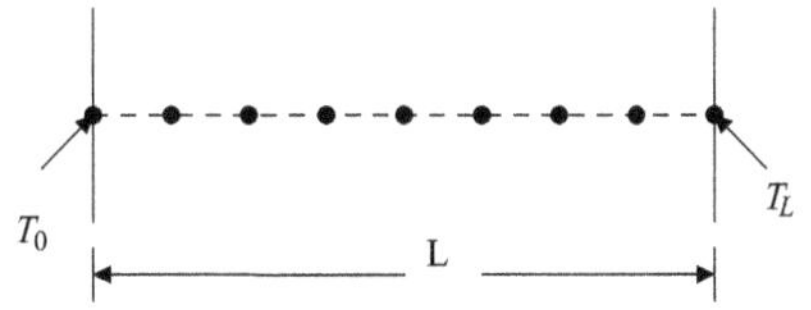

Solve the problem using an explicit finite difference scheme to study the transient temperature distribution and rate of heat transfer from the slab. Use the number of nodes as 10 and the stability parameters as 0.4 and 0.55. Carry out the computation for four consecutive time steps and plot heat flux at the right surface as a function of time with each stability parameter.

7.2 Consider a one-dimensional unsteady-state conduction in a plane slab [$k = 50\ \mathrm{W/m\,°C}$, $\alpha = 2\times10^{-5}\ \mathrm{m^2/s}$] of thickness $L = 4\,\mathrm{mm}$. Initially, the slab is at a temperature of 20°C. The right side of the plate is suddenly brought to a convective environment with a temperature of $T_\infty = 200°\mathrm{C}$ and $h = 80\ \mathrm{W/m^2\,°C}$. The left side is maintained at 20°C. Consider the grid size distribution shown with ten grid points and determine the transient temperature distribution in the slab.

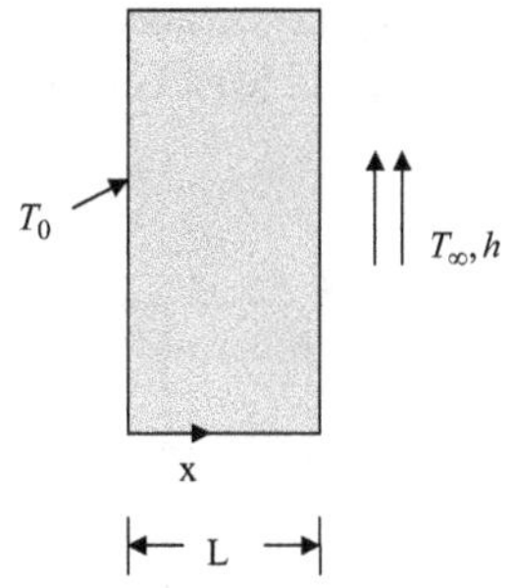

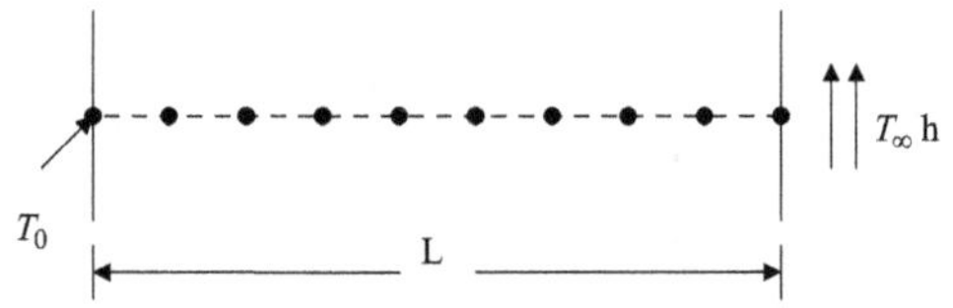

Develop a computer program using an explicit finite difference scheme and the pseudo code presented in Table 7.3 to study the transient temperature distribution and rate of heat transfer from the slab.

a. Examine the effect of stability parameter, $\gamma_0=0.4$, 0.45, 0.5, 0.55, and 0.6. Plot the heat flux at $x=L$ as a function of time for different values of stability parameter.
b. Select the appropriate stability parameter and determine the transient temperature distribution with increase in time. Present temperature distributions at ten different time steps up to 5 s.

7.3 Consider a one-dimensional unsteady-state heat conduction in a plane slab of thickness $L = 10\,\text{cm}$. Initially, the slab is at a uniform temperature of $T_I = 20°\text{C}$. At time $t>0$, the left side of the slab is subjected to a uniform constant surface heat flux $q_s'' = 2.0\times 10^4\ \text{W/cm}^2$ and the right surface is maintained at $T_R = 20°\text{C}$.

The slab is carbon steel with $\alpha = 0.835\ \text{cm}^2/\text{s}$. Use the fully implicit finite difference scheme to obtain the system of equations with the number of nodes as six and a time step of $\Delta t = 0.2$ s. Show computations for two time steps.

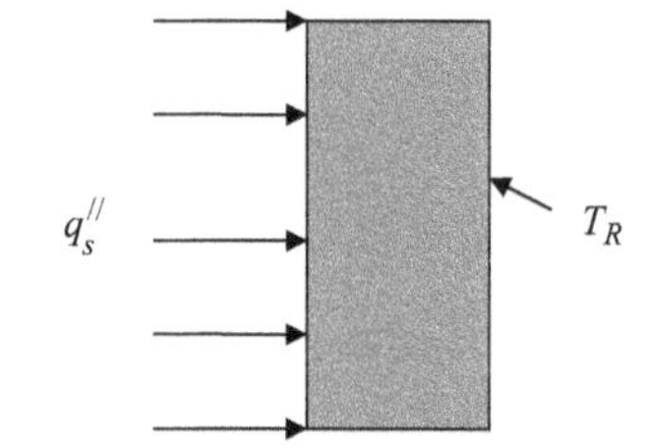

7.4 Develop a computer program using a fully implicit finite difference scheme and TDMA solver to study the transient temperature distribution and rate of heat transfer for Problem 7.3. Present results for Problem 7.3 in terms of temperature distribution with the number of nodes as 10 and a time step of $\Delta t = 0.2$ s up to 5 s.

7.5 Redo Problem 7.4 based on developing a computer code that uses a matrix inversion procedure with a Gauss–Jordan solver.

7.6 Let us consider a general two-dimensional unsteady-state problem with the grid system and boundary conditions as shown in the figure.

Derive the system of equations using a fully implicit scheme and the following data: $W=8\,\text{cm}$, $H=8\,\text{cm}$, $T_I = 25°\text{C}$, $T_H = 400°\text{C}$, $T_L = 20°\text{C}$, and $f_0 = 0.05$. Solve the system for two time steps with $\Delta t=0.2$ s and using (a) Gauss elimination, (b) a matrix inversion with Gauss–Jordan, (c) LU decomposition, and (d) a Gauss–Seidel method.

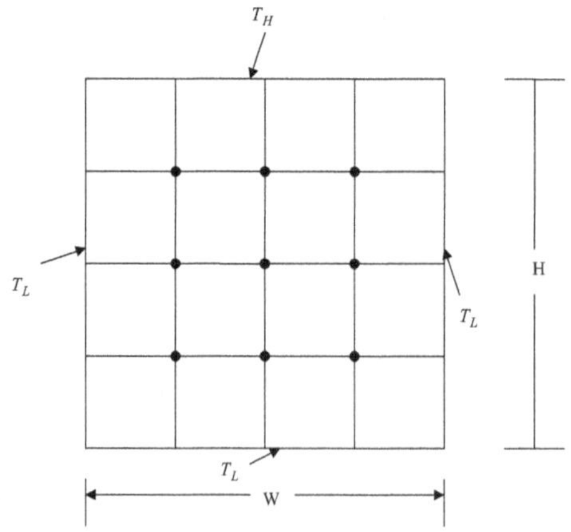

7.7 A two-dimensional $8\text{ cm} \times 8\text{ cm}$ square slab, initially at a uniform temperature of 300°C, is suddenly exposed to a convection environment at a temperature of 20°C and $h = 40\ \text{W/m}^2\,°\text{C}$ all around.

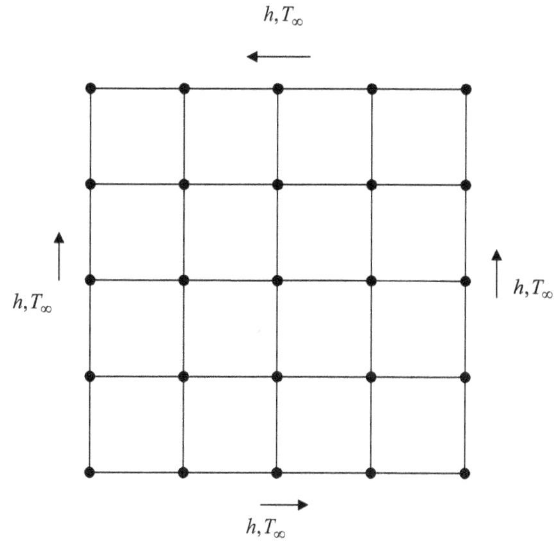

Consider a two-dimensional unsteady-state conduction in the slab and determine temperature distribution and rate of heat loss as a function of time.

a. Use the 4×4 grid and derive the system of equations for grid temperatures using a fully implicit scheme.

b. Select an appropriate Δt and calculate the grid point temperature for five time increments. Material thermo-physical properties are $\rho = 7900\ \text{kg/m}^3$, $c = 477$ J/kg K, and $k = 14.9$ W/m K.

7.8 Repeat Problem 7.7 using the Crank–Nicolson scheme.

7.9 Develop a computer program using the fully implicit scheme and Gauss–Seidel method. Redo Problem 7.7 with refined grid size distributions as 4×4, 6×6, and 8×8 using this code.

a. Present temperature distribution along with percent relative error for the vertical center line with refined grid size distributions at the fifth time step.

b. Show the temperature distribution for the vertical center line with increase in time for the grid of 8×8.

7.10 Develop a computer program using the fully implicit scheme and line-by-line method. Repeat Problem 7.9 using this code.

7.11 The rectangular slab shown below is initially uniform in temperature at 300°C and then suddenly exposed to the convection environment at a temperature of 20°C and $h = 40\ \text{W/m}^2$ K all around except the left surface, which is maintained at $T_0 = 300$°C.

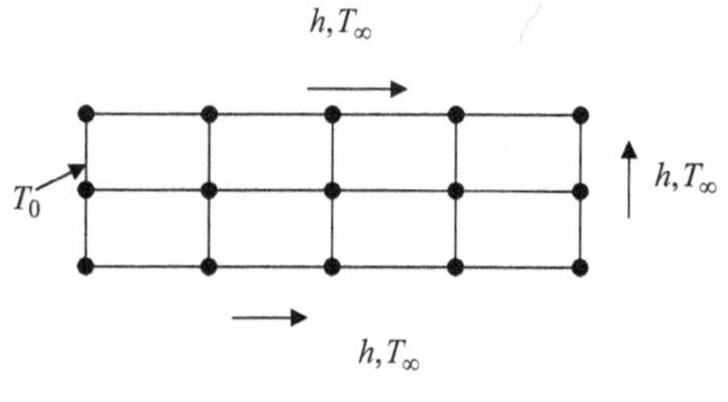

Consider a two-dimensional unsteady-state conduction in the slab and determine temperature distribution and total heat loss as a function of time. Material thermo-physical properties are $\rho = 2700\ \text{kg/m}^3$, $c = 900$ J/kg K, and $k = 230$ W/m K. The length and height of the slab are 4 cm and 2 cm, respectively. Use the Gauss elimination solver.

7.12 Consider the diffusion of moisture from an airstream in a two-dimensional porous adsorbing felt material lined on the side of the air flow channel as shown. The other three sides of the felt are lined with thin aluminum foil, thus making it impermeable to mass diffusion. Initial moisture concentration in the felt is $C_I = 0.001\ \text{kg}\,\text{H}_2\text{O/m}^3$. The left surface of the felt is exposed to moving air in the channel with mass transfer coefficient $h_D = 0.02$ m/s and moisture concentration $C_\infty = 0.01\ \text{kg}\,\text{H}_2\text{O/m}^3$. The material is assumed to be homogeneous with a constant moisture adsorption rate of $\dot{m}_{ad} = 0.00004\ \text{kg}\,\text{H}_2\text{O/kg}\,\text{s}$.

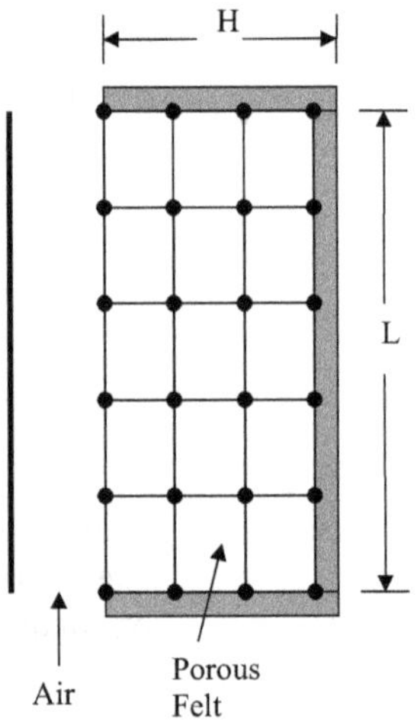

The mathematical statement of the problem is as follows.

Governing Equation:

$$\frac{\partial C}{\partial t} = \frac{\partial}{\partial x}\left(D\frac{\partial C}{\partial x}\right) + \frac{\partial}{\partial y}\left(D\frac{\partial C}{\partial y}\right) + \frac{(1-\varepsilon)}{\varepsilon}\rho\dot{m}_{\text{ad}}$$

Boundary Condition:

1. $y = 0, \quad -D\left.\frac{\partial C}{\partial x}\right|_{y=0} = h_D\left(C_\infty - C|_{y=0}\right)$

2. $y = H, \quad \left.\frac{\partial C}{\partial x}\right|_{x=H} = 0$

3. $x = 0, \quad \left.\frac{\partial C}{\partial x}\right|_{x=0} = 0$

4. $x = L, \quad \left.\frac{\partial C}{\partial x}\right|_{x=L} = 0$

where ε = porosity and ρ_b = material bulk density.

a. Apply a finite difference–control volume method using an explicit scheme and derive the system of equations to calculate the two-dimensional transient moisture concentration distribution in the slab based on the grid distribution shown in the figure. Assume diffusion coefficient as $D = 4.0 \times 10^{-6}\ \text{m}^2/\text{s}$ and use the following data for computation: $H = 4$ mm, $L = 0.1$ m, $\rho_b = 450\ \text{kg}/\text{m}^3$, $\varepsilon = 0.5$.

b. Solve the system of equations using the Gauss–Seidel method and present results along the four vertical lines as a function of time.

7.13 Use a finite difference-control volume method to solve a two-dimensional unsteady-state conduction in a rectangular carbon steel ($k = 200$ W/m°C) slab subjected to a constant surface heat flux irradiated by a continuous high-energy laser beam at the top surface. Assume the heat flux distribution to be a constant average value, $q_0'' = 2 \times 10^8\ \text{W}/\text{m}^2$, acting over a section of the surface. The top surface is also subjected to forced convection with $h_c = 500\ \text{W}/\text{m}^2\,°\text{C}$. The left surface is assumed to be adiabatic as a line of symmetry. All other surfaces are assumed to be subjected to free convection with $T_\infty = 25°\text{C}$ and $h = 40$ W/m°C. Consider the height as 3-cm and length as 2-cm.

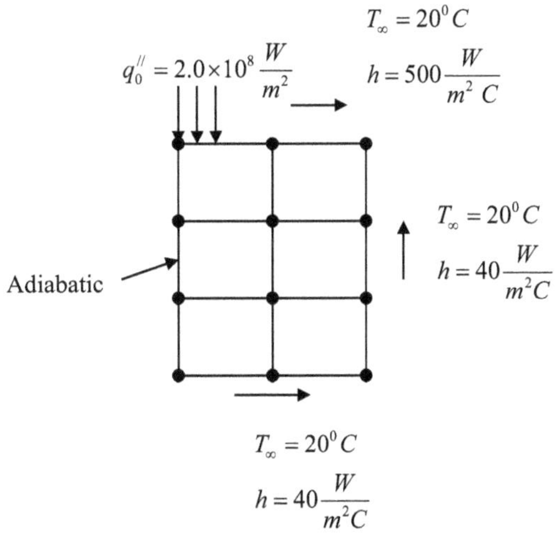

a. Derive the finite difference equations for all nodes using a fully implicit scheme and represent the system in matrix form.

b. Solve the system of equations using the Gauss elimination solver and present results for temperature distribution as a function of time.

7.14 A three-dimensional block $(0.4\ \text{m} \times 0.4\ \text{m} \times 0.4\ \text{m})$ is heated in a furnace to a uniform temperature of $T_I = 600°\text{C}$ and then suddenly submerged in cooling fluid with convection environment $h = 40\ \text{W}/\text{m}^2\,\text{K}$ and $T_\infty = 20°\text{C}$. Develop a computer program using a fully implicit scheme and line-by-line method to determine the temperature distribution in the slab with increase in time. Use TDMA with alternate sweeping on the x–y plane and marching scheme in the z-direction. Use thermo-physical properties of the material as $\rho = 7854\ \text{kg}/\text{m}^3$, $c = 434$ J/kg K, and $k = 60$ W/m K.

a. Determine temperature distribution using a grid of $10 \times 10 \times 10$ and assigned tolerance limit of $\varepsilon_s = 0.001$, and plot the temperature distribution at the mid-section along the z-direction with increase in time.

b. Repeat step (a) with an assigned tolerance limit of $\varepsilon_s = 0.00001$.

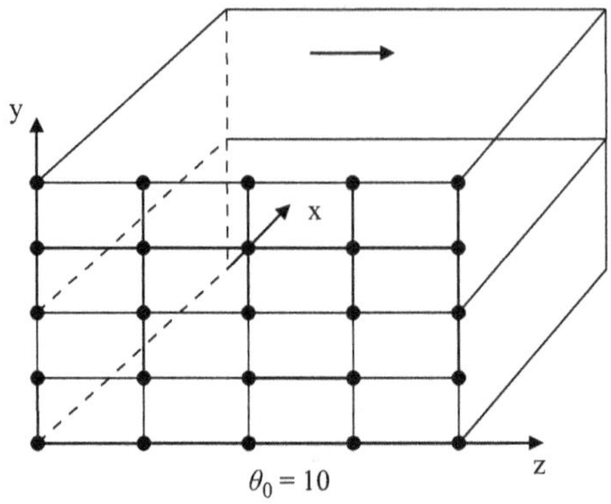

8 Finite Difference–Control Volume Method

Convection Problems

Convection transport mechanism is primarily influenced by the associated flow field, and hence, the solution of convection problems involves solution of conservation of mass and momentum equations for the flow field, followed by the solution of energy or mass concentration equations. The governing equations for convection problems differ from the diffusion equations discussed in previous chapters with the inclusion of additional terms such as the convective and pressure derivative terms. Major difficulties in numerical solution of convection problems arise from the discretization of the convective and pressure derivative terms in momentum equations. In this chapter, we will begin with the discussion of discretization procedures for convective and pressure derivative terms. This will be followed by the discussion of solution algorithms for solving coupled systems of partial differential equations for velocity field, pressure, and temperature and/or mass concentration fields.

8.1 SPATIAL DISCRETIZATION USING CONTROL VOLUME METHOD

Let us consider a one-dimensional steady-state problem involving convection and diffusion terms as

$$\frac{d}{dx}(\rho u\phi) = \frac{d}{dx}\left(\Gamma_x \frac{d\phi}{dx}\right) \tag{8.1}$$

In the control volume method, we integrate the equation over a control volume shown in Figure 8.1.

$$\int_{i-\frac{1}{2}}^{i+\frac{1}{2}} \frac{d}{dx}(\rho u\phi)dx = \int_{i-\frac{1}{2}}^{i+\frac{1}{2}} \frac{d}{dx}\left(\Gamma_x \frac{d\phi}{dx}\right)dx \tag{8.2a}$$

or

$$(\rho u\phi)\Big|_{i+\frac{1}{2}} - (\rho u\phi)\Big|_{i-\frac{1}{2}} = \left(\Gamma_x \frac{d\phi}{dx}\right)_{i+\frac{1}{2}} - \left(\Gamma_x \frac{d\phi}{dx}\right)_{i-\frac{1}{2}} \tag{8.2b}$$

In the next step, quantities such as $\rho u, \phi, \Gamma_x$ and $d\phi/dx$ are approximated with a profile assumption using interpolation formulas. We have mentioned before that the two common profile assumptions are (a) stepwise constant profile and (b) piece-wise linear profile, as depicted in Figure 5.4.

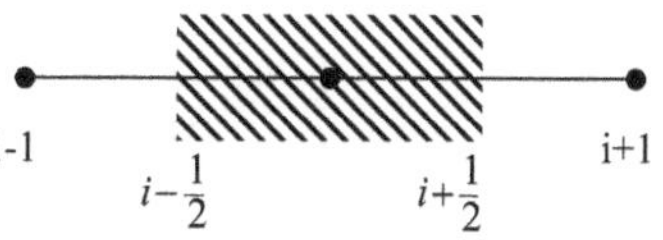

FIGURE 8.1 One-dimensional control volume with grid.

Assuming a step-wise constant profile for the convection quantity, ρu, and transport property, Γ_x, and a piece-wise linear profile for the first derivative terms, we get

$$(\rho u)_{i+\frac{1}{2}}\phi_{i+\frac{1}{2}}-(\rho u)_{i-\frac{1}{2}}\phi_{i-\frac{1}{2}}=\Gamma_{xi+\frac{1}{2}}\frac{\phi_{i+1}-\phi_i}{\Delta x_{i+\frac{1}{2}}}-\Gamma_{xi-\frac{1}{2}}\frac{\phi_i-\phi_{i-1}}{\Delta x_{i-\frac{1}{2}}} \tag{8.3}$$

A profile assumption for the dependent variable, φ ($\phi_{i-\frac{1}{2}}$ and $\phi_{i+\frac{1}{2}}$ at left and right interface of the control volume), in the convection term leads to the so-called different discretization schemes for the convection term. Some of these schemes are (a) **central difference scheme**, (b) **upwind scheme**, (c) **exponential scheme**, (d) **hybrid scheme**, and (e) **power law scheme**. A brief discussion of these basic schemes is now given.

8.1.1 Central Difference Scheme

In this scheme, the dependent variable φ in the convection term is approximated by a piece-wise linear profile, such as the **central difference formula**, as

$$\phi_{i+\frac{1}{2}}=\frac{1}{2}(\phi_{i+1}+\phi_i) \tag{8.4a}$$

and

$$\phi_{i-\frac{1}{2}}=\frac{1}{2}(\phi_i+\phi_{i-1}) \tag{8.4b}$$

Substituting Equations 8.4a and 8.4b into Equation 8.3, we get

$$\frac{1}{2}(\rho u)_{i+\frac{1}{2}}(\phi_{i+1}+\phi_i)-\tfrac{1}{2}(\rho u)_{i-\frac{1}{2}}(\phi_i+\phi_{i-1})=\frac{\Gamma_{xi+\frac{1}{2}}(\phi_{i+1}-\phi_i)}{\Delta x_{i+\frac{1}{2}}}-\frac{\Gamma_{xi-\frac{1}{2}}(\phi_i-\phi_{i-1})}{\Delta x_{i-\frac{1}{2}}}$$

Rearranging and grouping terms, we get the discretization equation

$$a_i\phi_i=a_{i+1}\phi_{i+1}+a_{i-1}\phi_{i-1} \tag{8.5a}$$

where

$$a_{i+1}=\frac{\Gamma_{xi+\frac{1}{2}}}{\Delta x_{i+\frac{1}{2}}}-\frac{1}{2}(\rho u)_{i+\frac{1}{2}}=\frac{\Gamma_{xi+\frac{1}{2}}}{\Delta x_{i+\frac{1}{2}}}\left(1-\frac{P_{i+\frac{1}{2}}}{2}\right) \tag{8.5b}$$

$$a_{i-1}=\frac{\Gamma_{xi-\frac{1}{2}}}{\Delta x_{i-\frac{1}{2}}}+\tfrac{1}{2}=\frac{1}{2}(\rho u)_{i-\frac{1}{2}}=\frac{\Gamma_{xi-\frac{1}{2}}}{\Delta x_{i-\frac{1}{2}}}\left(1+\frac{P_{i-\frac{1}{2}}}{2}\right) \tag{8.5c}$$

$$a_i=a_{i+1}+a_{i-1}+\left\{(\rho u)_{i+\frac{1}{2}}-(\rho u)_{i-\frac{1}{2}}\right\} \tag{8.5d}$$

$$P=\frac{(\rho u)}{\Gamma/\Delta x} \tag{8.5e}$$

Application of this three-node **central difference scheme** to all internal nodes leads to a tridiagonal system of linear algebraic equations. An inspection of the coefficient terms a in this scheme reveals that the condition for the stable and converging solution of the linear system, i.e., $a_i \geq \left(|a_{i+1}| + |a_{i-1}|\right)$, may not always be satisfied because the net mass velocity term $\left\{(\rho u)_{i+\frac{1}{2}} - (\rho u)_{i-\frac{1}{2}}\right\}$ may become negative during computations. Under this situation, this scheme becomes computationally unstable. A central difference scheme works well for problems where diffusion dominates over convection, i.e., for low values of $P = \rho u/\left(\Gamma/\Delta x\right)$ or for low control volume Reynolds number, $\mathrm{Re}_{\Delta x} = \left((\rho u \Delta x)/\mu\right)$.

8.1.2 Upwind Scheme

In the upwind scheme, the dependent variable at the interface of a control volume is assumed to be equal to the upstream or upwind side nodal values for positive values of the mass velocity at the interface and equal to the downstream-side nodal value for negative values of mass velocity at the interface. This is expressed as

$$\phi_{i+\frac{1}{2}} = \phi_i \quad \text{for } (\rho u)_{i+\frac{1}{2}} > 0 \tag{8.6a}$$

and

$$\phi_{i+\frac{1}{2}} = \phi_{i+1} \quad \text{for } (\rho u)_{i+\frac{1}{2}} < 0 \tag{8.6b}$$

Similarly

$$\phi_{i-\frac{1}{2}} = \phi_{i-1} \quad \text{for } (\rho u)_{i-\frac{1}{2}} > 0 \tag{8.6c}$$

and

$$\phi_{i-\frac{1}{2}} = \phi_i \quad \text{for } (\rho u)_{i-\frac{1}{2}} < 0 \tag{8.6d}$$

Using a compact notation given by Patankar (1980), we expressed the upwind scheme as

$$(\rho u)_{i+\frac{1}{2}} \phi_{i+\frac{1}{2}} = \phi_i \left\langle \left((\rho u)_{i+\frac{1}{2}}\right), 0 \right\rangle - \phi_{i+1} \left\langle \left(-(\rho u)_{i+\frac{1}{2}}\right), 0 \right\rangle \tag{8.7a}$$

and

$$(\rho u)_{i-\frac{1}{2}} \phi_{i-\frac{1}{2}} = \phi_{i-1} \left\langle \left((\rho u)_{i-\frac{1}{2}}\right), 0 \right\rangle - \phi_i \left\langle \left(-(\rho u)_{i-\frac{1}{2}}\right), 0 \right\rangle \tag{8.7b}$$

where the **operator** $\langle a, b \rangle$ represents the maximum of a and b.

Substituting the upwind scheme given by Equation 8.7 into Equation 8.3, we get

$$\phi_i \left\langle \left((\rho u)_{i+\frac{1}{2}}\right), 0 \right\rangle - \phi_{i+1} \left\langle \left((-\rho u)_{i+\frac{1}{2}}\right), 0 \right\rangle - \phi_{i-1} \left\langle \left((\rho u)_{i-\frac{1}{2}}\right), 0 \right\rangle - \phi_i \left\langle \left(-(\rho u)_{i-\frac{1}{2}}\right), 0 \right\rangle$$
$$= \frac{\Gamma_{xi+\frac{1}{2}} (\phi_{i+1} - \phi_i)}{\Delta x_{i+\frac{1}{2}}} - \frac{\Gamma_{xi-\frac{1}{2}} (\phi_i - \phi_{i-1})}{\Delta x_{i-\frac{1}{2}}} \tag{8.8}$$

Simplifying and rearranging, we get

$$a_i \phi_i = a_{i+1} \phi_{i+1} + a_{i-1} \phi_{i-1} \tag{8.9}$$

where

$$a_{i+1} = \frac{\Gamma_{xi+\frac{1}{2}}}{\Delta x_{i+\frac{1}{2}}} + \left\langle -(\rho u)_{i+\frac{1}{2}}, 0 \right\rangle$$

$$a_{i-1} = \frac{\Gamma_{xi-\frac{1}{2}}}{\Delta x_{i-\frac{1}{2}}} + \left\langle (\rho u)_{i-\frac{1}{2}}, 0 \right\rangle$$

$$a_i = a_{i+1} + a_{i-1} + (\rho u)_{i+\frac{1}{2}} - (\rho u)_{i-\frac{1}{2}}$$

It can be noted from these expressions for the coefficient a that they do not become negative during computation, thus ensuring stability.

8.1.3 Exponential Scheme

In the exponential scheme, the convection and diffusion terms are combined and expressed as a single flux term as

$$j = \rho u \phi - \Gamma \frac{\mathrm{d}\phi}{\mathrm{d}x} \tag{8.10}$$

Substituting Equation 8.10 into Equation 8.1, we get

$$\frac{\mathrm{d}J}{\mathrm{d}x} = 0 \tag{8.11}$$

Integrating Equation 8.11 over the control volume, we get

$$J_{i+\frac{1}{2}} - J_{i-\frac{1}{2}} = 0 \tag{8.12}$$

Considering the analytical solution for the one-dimensional problem given by Equation 8.1 in a plane slab with boundary conditions of the first kind, the combined J flux is approximated as

$$J_{i+\frac{1}{2}} = (\rho u)_{i+\frac{1}{2}} \left(\phi_i + \frac{\phi_i - \phi_{i+1}}{\exp\left(P_{i+\frac{1}{2}}\right) - 1} \right) \tag{8.13a}$$

where

$$P_{i+\frac{1}{2}} = \frac{(\rho u)_{i+\frac{1}{2}}}{\Gamma_{i+\frac{1}{2}} / \Delta x_{i+\frac{1}{2}}} \tag{8.13b}$$

and

$$J_{i-\frac{1}{2}} = (\rho u)_{i-\frac{1}{2}} \left(\phi_{i-1} + \frac{\phi_{i-1} - \phi_i}{\exp\left(P_{i-\frac{1}{2}}\right) - 1} \right) \tag{8.14a}$$

where

$$P_{i-\frac{1}{2}} = \frac{(\rho u)_{i-\frac{1}{2}}}{\Gamma_{i-\frac{1}{2}} / \Delta x_{i-\frac{1}{2}}} \tag{8.14b}$$

Substituting Equations 8.13 and 8.14 into Equation 8.12, we get

$$(\rho u)_{i+\frac{1}{2}}\left(\phi_i + \frac{\phi_i - \phi_{i+1}}{\exp\left(P_{i+\frac{1}{2}}\right) - 1}\right) - (\rho u)_{i-\frac{1}{2}}\left(\phi_{i-1} + \frac{\phi_{i-1} - \phi_i}{\exp\left(P_{i-\frac{1}{2}}\right) - 1}\right) = 0 \tag{8.15a}$$

Rearranging

$$a_i\phi_i = a_{i+1}\phi_{i+1} + a_{i-1}\phi_{i-1} \tag{8.15b}$$

where

$$a_{i+1} = \frac{(\rho u)_{i+\frac{1}{2}}}{\exp\left(P_{i+\frac{1}{2}}\right) - 1} = \frac{\Gamma_{x_{i+\frac{1}{2}}}}{\Delta x_{i+\frac{1}{2}}} \frac{P_{i+\frac{1}{2}}}{\exp\left(P_{i+\frac{1}{2}}\right) - 1}$$

$$a_{i-1} = \frac{(\rho u)_{i-\frac{1}{2}} \exp\left(P_{i-\frac{1}{2}}\right)}{\exp\left(P_{i-\frac{1}{2}}\right) - 1} = \frac{\Gamma_{x_{i-\frac{1}{2}}}}{\Delta x_{i-\frac{1}{2}}} \frac{P_{i-\frac{1}{2}} \exp\left(P_{i-\frac{1}{2}}\right)}{\exp\left(P_{i-\frac{1}{2}}\right) - 1}$$

$$a_i = a_{i+1} + a_{i-1} + \left\{(\rho u)_{i+\frac{1}{2}} - (\rho u)_{i-\frac{1}{2}}\right\}$$

The main disadvantage of this scheme is that it involves exponential functions, which are computationally more expensive and time consuming, particularly for multidimensional problems involving a large system of equations.

8.1.4 Hybrid Scheme

The hybrid scheme is a combination of the central difference scheme and the upwind scheme. The exponential form of the analytical solution is approximated by three straight lines. It is identical with the central difference scheme in the mid-range for $-2 \le P_{i+\frac{1}{2}} < 2$, and outside this range, it reduces to the upwind scheme with zero diffusion as

$$\text{for } P_{i+\frac{1}{2}} < -2, \quad a_{i+1} = -\frac{\Gamma_{x_{i+\frac{1}{2}}}}{\Delta x_{i+\frac{1}{2}}}\left(P_{i+\frac{1}{2}}\right) \tag{8.16a}$$

$$\text{for } -2 \le P_{i+\frac{1}{2}} < 2, \quad a_{i+1} = \frac{\Gamma_{x_{i+\frac{1}{2}}}}{\Delta x_{i+\frac{1}{2}}}\left(1 - \frac{P_{i+\frac{1}{2}}}{2}\right) \tag{8.16b}$$

$$\text{for } P_{i+\frac{1}{2}} > 2, \quad a_{i+1} = 0 \tag{8.16c}$$

In compact notation, Equation 8.16 is written as

$$a_{i+1} = \frac{\Gamma_{x_{i+\frac{1}{2}}}}{\Delta x_{i+\frac{1}{2}}}\left\langle -P_{i+\frac{1}{2}}, 1-\frac{P_{i+\frac{1}{2}}}{2}, 0\right\rangle$$

$$= \frac{\Gamma_{x_{i+\frac{1}{2}}}}{\Delta x_{i+\frac{1}{2}}}\left\langle 1-\frac{P_{i+\frac{1}{2}}}{2}, 0\right\rangle + \left\langle 0, -(\rho u)_{i+\frac{1}{2}}\right\rangle \quad (8.17a)$$

Similarly

$$a_{i-1} = \frac{\Gamma_{x_{i-\frac{1}{2}}}}{\Delta x_{i-\frac{1}{2}}}\left\langle P_{i-\frac{1}{2}}, 1+\frac{P_{i-\frac{1}{2}}}{2}, 0\right\rangle$$

$$= \frac{\Gamma_{x_{i+\frac{1}{2}}}}{\Delta x_{i+\frac{1}{2}}}\left\langle 1+\frac{P_{i-\frac{1}{2}}}{2}, 0\right\rangle + \left\langle 0, (\rho u)_{i-\frac{1}{2}}\right\rangle \quad (8.17b)$$

Based on this approximation, the discretization scheme for the convection-diffusion equation given by Equation 8.1 can be derived as

$$a_i\phi_i = a_{i+1}\phi_{i+1} + a_{i-1}\phi_{i-1} \quad (8.18)$$

where

$$a_{i+1} = \frac{\Gamma_{x_{i+\frac{1}{2}}}}{\Delta x_{i+\frac{1}{2}}}\left\langle 1-\frac{P_{i+\frac{1}{2}}}{2}, 0\right\rangle + \left\langle 0, -(\rho u)_{i+\frac{1}{2}}\right\rangle$$

$$a_{i-1} = \frac{\Gamma_{x_{i-\frac{1}{2}}}}{\Delta x_{i-\frac{1}{2}}}\left\langle 1+\frac{P_{i-\frac{1}{2}}}{2}, 0\right\rangle + \left\langle 0, (\rho u)_{i-\frac{1}{2}}\right\rangle$$

$$a_i = a_{i+1} + a_{i-1} + \left\{(\rho u)_{i+\frac{1}{2}} - (\rho u)_{i-\frac{1}{2}}\right\}$$

8.1.5 Power Law Scheme

In the power law scheme, the exponential function is approximated by straight lines and power law profiles, which are computationally less expensive in the following way:

$$\text{for } P_{i+\frac{1}{2}} < -10, \quad a_{i+1} = -\frac{\Gamma_{x_{i+\frac{1}{2}}}}{\Delta x_{i+\frac{1}{2}}}\left(P_{i+\frac{1}{2}}\right) \quad (8.19a)$$

$$\text{for } -10 \le P_{i+\frac{1}{2}} < 0, \quad a_{i+1} = \frac{\Gamma_{x_{i+\frac{1}{2}}}}{\Delta x_{i+\frac{1}{2}}}\left\{\left(1+0.1P_{i+\frac{1}{2}}\right)^5 - P_{i+\frac{1}{2}}\right\} \quad (8.19b)$$

$$\text{for } 0 \le P_{i+\frac{1}{2}} < 10, \quad a_{i+1} = \frac{\Gamma_{x_{i+\frac{1}{2}}}}{\Delta x_{i+\frac{1}{2}}}\left(1-0.1P_{i+\frac{1}{2}}\right)^5 \quad (8.19c)$$

$$\text{for } P_{i+\frac{1}{2}} > 10, \quad a_{i+1} = 0 \quad (8.19d)$$

Equation 8.19 can also be written in the compact form as

$$a_{i+1} = \frac{\Gamma_{x_{i+\frac{1}{2}}}}{\Delta x_{i+\frac{1}{2}}} \left\langle 0, \left(1 - 0.1\left|P_{i+\frac{1}{2}}\right|\right)^5 \right\rangle + \frac{\Gamma_{x_{i+\frac{1}{2}}}}{\Delta x_{i+\frac{1}{2}}} \left\langle 0, -P_{i+\frac{1}{2}} \right\rangle$$

or

$$a_{i+1} = \frac{\Gamma_{x_{i+\frac{1}{2}}}}{\Delta x_{i+\frac{1}{2}}} \left\langle 0, \left(1 - 0.1\left|P_{i+\frac{1}{2}}\right|\right)^5 \right\rangle + \left\langle 0, -(\rho u)_{i+\frac{1}{2}} \right\rangle \tag{8.20a}$$

Similarly

$$a_{i-1} = \frac{\Gamma_{x_{i-\frac{1}{2}}}}{\Delta x_{i-\frac{1}{2}}} \left\langle 0, \left(1 - 0.1\left|P_{i-\frac{1}{2}}\right|\right)^5 \right\rangle + \frac{\Gamma_{x_{i-\frac{1}{2}}}}{\Delta x_{i-\frac{1}{2}}} \left\langle 0, P_{i-\frac{1}{2}} \right\rangle$$

or

$$a_{i-1} = \frac{\Gamma_{x_{i-\frac{1}{2}}}}{\Delta x_{i-\frac{1}{2}}} \left\langle 0, \left(1 - 0.1\left|P_{i-\frac{1}{2}}\right|\right)^5 \right\rangle + \left\langle 0, (\rho u)_{i-\frac{1}{2}} \right\rangle \tag{8.20b}$$

The discretized form of Equation 8.1 based on the power law scheme is given as

$$a_i\phi_i = a_{i+1}\phi_{i+1} + a_{i-1}\phi_{i-1} \tag{8.21}$$

where

$$a_{i+1} = \frac{\Gamma_{x_{i+\frac{1}{2}}}}{\Delta x_{i+\frac{1}{2}}} \left\langle 0, \left(1 - 0.1\left|P_{i+\frac{1}{2}}\right|\right)^5 \right\rangle + \left\langle 0, -(\rho u)_{i+\frac{1}{2}} \right\rangle$$

$$a_{i-1} = \frac{\Gamma_{x_{i-\frac{1}{2}}}}{\Delta x_{i-\frac{1}{2}}} \left\langle 0, \left(1 - 0.1\left|P_{i-\frac{1}{2}}\right|\right)^5 \right\rangle + \left\langle 0, (\rho u)_{i-\frac{1}{2}} \right\rangle$$

$$a_i = a_{i+1} + a_{i-1} + \left\{ (\rho u)_{i+\frac{1}{2}} - (\rho u)_{i-\frac{1}{2}} \right\}$$

8.1.6 Generalized Convection–Diffusion Scheme

A generalized form of all approximation schemes for the convection–diffusion terms is given by Patankar (1980) as

$$a_i\phi_i = a_{i+1}\phi_{i+1} + a_{i-1}\phi_{i-1} \tag{8.22}$$

where

$$a_{i+1} = \frac{\Gamma_{x_{i+\frac{1}{2}}}}{\Delta x_{i+\frac{1}{2}}} A\left|P_{i+\frac{1}{2}}\right| + \left\langle 0, -(\rho u)_{i+\frac{1}{2}} \right\rangle$$

$$a_{i-1} = \frac{\Gamma_{x_{i-\frac{1}{2}}}}{\Delta x_{i-\frac{1}{2}}} A\left|P_{i-\frac{1}{2}}\right| + \left\langle 0, (\rho u)_{i-\frac{1}{2}} \right\rangle$$

TABLE 8.1
Expression for the Function $A|P|$ in Different Discretization Schemes for Convection–Diffusion

Approximation Scheme	Function $A\|P\|$
Central difference	$1-\frac{\|P\|}{2}$
Upwind	1
Hybrid	$\left\langle 0, 1-\frac{\|P\|}{2} \right\rangle$
Exponential	$\left\langle \frac{\|P\|}{\exp(\|P\|)}-1 \right\rangle$
Power	$\left\langle 0, (1-0.1\|P\|)^5 \right\rangle$

$$a_i = a_{i+1} + a_{i-1} + \left\{ (\rho u)_{i+\frac{1}{2}} - (\rho u)_{i-\frac{1}{2}} \right\}$$

It can be seen that the only difference among difference approximation schemes is the expression for the function $A|P|$, which is listed in Table 8.1.

8.1.7 Higher-Order Discretization Schemes for Convective Terms

Although, the first-order upwind scheme is simple and ensures stability, it also results in an unwanted diffusion error in space and increases the numerical error. While the numerical error can be controlled to some extent using grid refinements, diffusion error becomes more predominant in multidimensional and more complex problems.

To reduce such numerical errors, higher order approximation schemes are proposed. Higher accurate solution can be produced while using larger grid spacing than used for first-order schemes. Few example of such higher order schemes are **second-order upwind scheme**, **third-order QUICK (Quadratic Upwind Interpolation for Convective Kinematics Scheme** (Leonard, 1979)), and Third-order MUSCL (Monotonic Upstream-Centered Scheme for Conservation Laws) (van Leer, 1979).

8.1.7.1 Second-Order Upwind Scheme

This is an improvement to the first-order scheme that retains higher-order terms in the Taylor series and hence reduces the truncation error due to approximation or discretization of the convective first-order derivative term. Basic idea is to use the discretization equation for the first-order convective derivative involving three grid values and bring in additional flow information from adjacent upstream grid points such as the nodes at $(i-2)$ and $(i+2)$ as demonstrated in Figure 8.2. For simplicity, let us consider here a uniform grid size distribution in summarizing **the second-order upwind scheme**:

For positive flow with $\left(\rho u_{i-\frac{1}{2}}\right) > 0$ and $\left(\rho u_{i+\frac{1}{2}}\right) > 0$

$$\phi_{i-\frac{1}{2}} = \frac{3}{2}\phi_{i-1} - \frac{1}{2}\phi_{i-2}$$
$$\phi_{i+\frac{1}{2}} = \frac{3}{2}\phi_{i} - \frac{1}{2}\phi_{i-1} \tag{8.23a}$$

and

For negative flow $\left(\rho u_{i-\frac{1}{2}}\right) < 0$ and $\left(\rho u_{i+\frac{1}{2}}\right) < 0$

$$\phi_{i-\frac{1}{2}} = \frac{3}{2}\phi_i - \frac{1}{2}\phi_{i+1}$$
$$\phi_{i+\frac{1}{2}} = \frac{3}{2}\phi_{i+1} - \frac{1}{2}\phi_{i+2} \tag{8.23b}$$

8.1.7.2 Third-Order QUICK Scheme

Quadratic Upwind Interpolation for Convective Kinematics (QUICK) scheme was proposed by Leonard (1979). In this scheme for discretizing the convective term, the face values are approximated using a quadratic interpolation function passing through two upstream node-points and one node value at the downstream depending on the flow directions. A quadratic interpolation function fitting through the two adjacent nodes and one next upstream node is used in one of the flow directions as demonstrated in Figure 8.2.

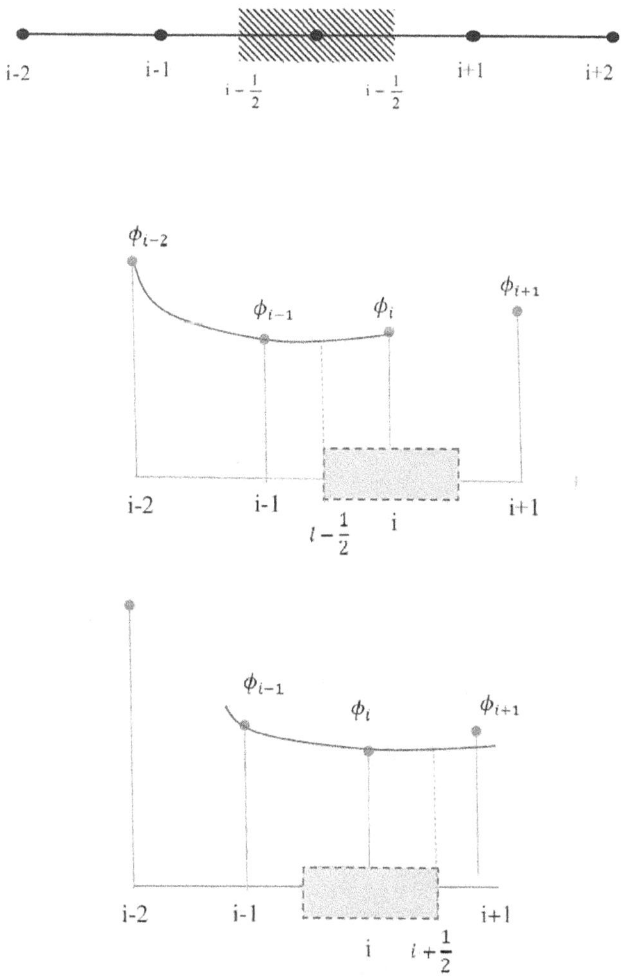

FIGURE 8.2 One-dimensional control volume with grids for deriving QUICK scheme.

Let us demonstrate the scheme consider for **positive flows** (if $\left(\rho u_{i-\frac{1}{2}}\right) > 0$ and $\left(\rho u_{i+\frac{1}{2}}\right) > 0$) at the faces of the control volume and for the cell face values of each dependent variable in a one-dimensional convection-diffusion problem.

For the control volume face located at $i+\frac{1}{2}$, we can calculate the value $\phi_{1+\frac{1}{2}}$ using a quadratic interpolation function passing through the node at $i-1$, i, and $i+1$ as

$$\phi_{i+\frac{1}{2}} = -\frac{1}{8}\phi_{i-1} + \frac{6}{8}\phi_i + \frac{3}{8}\phi_{i+1} \tag{8.24a}$$

for $\left(\rho u_{i-\frac{1}{2}}\right) > 0$ and $\left(\rho u_{i+\frac{1}{2}}\right) > 0$

For the control volume face located at $i-\frac{1}{2}$, we can calculate the value $\phi_{1-\frac{1}{2}}$ using a quadratic interpolation function passing through the node at $i-2$, $i-1$, and i as shown in the figure below:

$$\phi_{i-\frac{1}{2}} = -\frac{1}{8}\phi_{i-2} + \frac{6}{8}\phi_{i-1} + \frac{3}{8}\phi_i$$

$$\text{for}\left(\rho u_{i-\frac{1}{2}}\right) > 0 \quad \text{and} \quad \left(\rho u_{i+\frac{1}{2}}\right) > 0 \tag{8.24b}$$

Third-Order QUICK Scheme is summarized below:

For positive flows at the faces of the control volume

For $u_{i-\frac{1}{2}} > 0$ and $u_{i+\frac{1}{2}} > 0$

$$\phi_{i-\frac{1}{2}} = -\frac{1}{8}\phi_{i-2} + \frac{6}{8}\phi_{i-1} + \frac{3}{8}\phi_i \tag{8.24a}$$

$$\phi_{i+\frac{1}{2}} = -\frac{1}{8}\phi_{i-1} + \frac{6}{8}\phi_i + \frac{3}{8}\phi_{i+1} \tag{8.24b}$$

For $u_{i-\frac{1}{2}} < 0$ and $u_{i+\frac{1}{2}} < 0$

$$\phi_{i-\frac{1}{2}} = -\frac{1}{8}\phi_{i+1} + \frac{6}{8}\phi_i + \frac{3}{8}\phi_{i-1} \tag{8.24c}$$

$$\phi_{i+\frac{1}{2}} = -\frac{1}{8}\phi_{i+2} + \frac{6}{8}\phi_i + \frac{3}{8}\phi_i \tag{8.24d}$$

8.1.7.3 Derivation of Discretization Equation Using QUICK Scheme

Let us show the derivation of the discretization for an internal node assuming a one-dimensional convection-diffusion problem assuming a one-dimensional steady-state problem **involving convection and diffusion terms** as below

$$\frac{d}{dx}(\rho u\phi) = \frac{d}{dx}\left(\Gamma_x \frac{d\phi}{dx}\right) \tag{8.25}$$

Integrating of the control volume, we get (from Equation 8.26)

$$\left.(\rho u\phi)\right|_{i+\frac{1}{2}}-\left.(\rho u\phi)\right|_{i-\frac{1}{2}}=\left(\Gamma_x\frac{\mathrm{d}\phi}{\mathrm{d}x}\right)_{i+\frac{1}{2}}-\left(\Gamma_x\frac{\mathrm{d}\phi}{\mathrm{d}x}\right)_{i-\frac{1}{2}} \tag{8.26}$$

As demonstrated before the quantities such as ρu, ϕ, Γ_x and $\frac{\mathrm{d}\phi}{\mathrm{d}x}$ are approximated with a profile assumption using interpolation formulas. Assuming, a *stepwise constant profile* for the convection quantity, ρu, and transport property, Γ_x, and a piece-wise linear profile for the first derivative terms, we get

$$(\rho u)_{i+\frac{1}{2}}\phi_{i+\frac{1}{2}}-(\rho u)_{i-\frac{1}{2}}\phi_{i-\frac{1}{2}}=\Gamma_{xi+\frac{1}{2}}\frac{\phi_{i+1}-\phi_i}{\Delta x_{i+\frac{1}{2}}}-\Gamma_{xi-\frac{1}{2}}\frac{\phi_i-\phi_{i-1}}{\Delta x_{i-\frac{1}{2}}} \tag{8.27}$$

Need profile approximation

Let us now approximate the **dependent variable** ϕ in the convection term by the QUICK scheme using Equations 8.24a and 8.24b for the **case of a positive flow** in faces of the control volume as

$$\begin{aligned}&(\rho u)_{i+\frac{1}{2}}\left(-\frac{1}{8}\phi_{i-1}+\frac{6}{8}\phi_i+\frac{3}{8}\phi_{i+1}\right)-(\rho u)_{i-\frac{1}{2}}\left(-\frac{1}{8}\phi_{i-2}+\frac{6}{8}\phi_{i-1}+\frac{3}{8}\phi_i\right)\\&=\Gamma_{i+\frac{1}{2}}\frac{(\phi_{i+1}-\phi_i)}{\Delta x_{i+\frac{1}{2}}}-\Gamma_{i-\frac{1}{2}}\frac{(\phi_i-\phi_i-1)}{\Delta x_{i-\frac{1}{2}}}\end{aligned} \tag{8.28}$$

Rearranging and grouping terms, we get the following discretization equation:

$$\begin{aligned}&(\rho u)_{i+\frac{1}{2}}\left(-\frac{1}{8}\phi_{i-1}+\frac{6}{8}\phi_i+\frac{3}{8}\phi_{i+1}\right)-(\rho u)_{i-\frac{1}{2}}\left(-\frac{1}{8}\phi_{i-2}+\frac{6}{8}\phi_{i-1}+\frac{3}{8}\phi_i\right)\\&=\Gamma_{i+\frac{1}{2}}\frac{(\phi_{i+1}-\phi_i)}{\Delta x_{i+\frac{1}{2}}}-\Gamma_{i-\frac{1}{2}}\frac{(\phi_i-\phi_{i-1})}{\Delta x_{i-\frac{1}{2}}}\end{aligned} \tag{8.29a}$$

Rearranging,

$$\begin{aligned}&\left(\frac{6}{8}(\rho u)_{i+\frac{1}{2}}-\frac{3}{8}(\rho u)_{i-\frac{1}{2}}+\frac{\Gamma_{i+\frac{1}{2}}}{\Delta x_{i+\frac{1}{2}}}+\frac{\Gamma_{i-\frac{1}{2}}}{\Delta x_{i-\frac{1}{2}}}\right)\phi_i\\&=\left(\frac{\Gamma_{i+\frac{1}{2}}}{\Delta x_{i+\frac{1}{2}}}-\frac{3}{8}(\rho u)_{i+\frac{1}{2}}\right)\phi_{i+1}+\left(\frac{\Gamma_{i-\frac{1}{2}}}{\Delta x_{i-\frac{1}{2}}}+\frac{1}{8}(\rho u)_{i+\frac{1}{2}}+\frac{6}{8}(\rho u)_{i-\frac{1}{2}}\right)\phi_{i-1}\\&+\left(-\frac{1}{8}(\rho u)_{i-\frac{1}{2}}\right)\phi_{i-2}\end{aligned} \tag{8.29b}$$

Writing the discretization equation in compact form as

$$a_i\phi_i=a_{i+1}\phi_{i+1}+a_{i-1}\phi_{i-1}+a_{i-2}\phi_{i-2} \tag{8.30}$$

where

$$a_{i+1} = \frac{\Gamma_{i+\frac{1}{2}}}{\Delta x_{i+\frac{1}{2}}} - \frac{3}{8}(\rho u)i + \frac{1}{2} \tag{8.31a}$$

$$a_{i-1} = \frac{\Gamma_{i-\frac{1}{2}}}{\Delta x_{i-\frac{1}{2}}} + \frac{1}{8}(\rho u)_{i+\frac{1}{2}} + \frac{6}{8}(\rho u)_{i-\frac{1}{2}} \tag{8.31b}$$

$$a_{i-2} = -\frac{1}{8}(\rho u)_{i-\frac{1}{2}} \tag{8.31c}$$

$$a_i = \frac{6}{8}(\rho u)_{i+\frac{1}{2}} - \frac{3}{8}(\rho u)_{i-\frac{1}{2}} + \frac{\Gamma_{i+\frac{1}{2}}}{\Delta x_{i+\frac{1}{2}}} + \frac{\Gamma_{i-\frac{1}{2}}}{\Delta x_{i-\frac{1}{2}}} \tag{8.31d}$$

Rearranging

$$a_i = a_{i+1} + a_{i-1} + a_{i-2} + \left\{(\rho u)_{i+\frac{1}{2}} - (\rho u)_{i-\frac{1}{2}}\right\} \tag{8.31e}$$

In a similar manner, we can derive the discretization equation for the interior nodes for the **case of negative flow directions** through the faces of the control volume using following QUICK scheme equations (8.24c and 8.24d) and substituting into Equation 8.27.

For $u_{i-\frac{1}{2}} < 0$ and $u_{i+\frac{1}{2}} < 0$

$$\phi_{i-\frac{1}{2}} = \frac{1}{8}\phi_{i+1} + \frac{6}{8}\phi_i + \frac{3}{8}\phi_{i-1} \tag{8.24c}$$

$$\phi_{i+\frac{1}{2}} = -\frac{1}{8}\phi_{i+2} + \frac{6}{8}\phi_{i+1} + \frac{3}{8}\phi_i \tag{8.24d}$$

Substituting this set into Equation 8.27, we get

$$(\rho u)_{i+\frac{1}{2}}\left(-\frac{1}{8}\phi_{i+2} + \frac{6}{8}\phi_{i+1} + \frac{3}{8}\phi_i\right) - (\rho u)_{i-\frac{1}{2}}\left(-\frac{1}{8}\phi_{i+1} + \frac{6}{8}\phi_i + \frac{3}{8}\phi_{i-1}\right)$$

$$= \Gamma_{i+\frac{1}{2}}\frac{(\phi_{i+1} - \phi_i)}{\Delta x_{i+\frac{1}{2}}} - \Gamma_{i-\frac{1}{2}}\frac{(\phi_i - \phi_{i-1})}{\Delta x_{i-\frac{1}{2}}} \tag{8.32}$$

Rearranging and grouping terms, we get the following discretization equation:

$$\left(\frac{6}{8}(\rho u)_{i+\frac{1}{2}} - \frac{3}{8}(\rho u)_{i-\frac{1}{2}} + \frac{\Gamma_{i+\frac{1}{2}}}{\Delta x_{i+\frac{1}{2}}} + \frac{\Gamma_{i-\frac{1}{2}}}{\Delta x_{i-\frac{1}{2}}}\right)\phi_i$$

$$= \left(\frac{\Gamma_{i+\frac{1}{2}}}{\Delta x_{i+\frac{1}{2}}} - \frac{3}{8}(\rho u)_{i+\frac{1}{2}} - \frac{1}{8}(\rho u)_{i-\frac{1}{2}}\right)\phi_{i+1} + \left(\frac{\Gamma_{i-\frac{1}{2}}}{\Delta x_{i-\frac{1}{2}}} + \frac{6}{8}(\rho u)_{i-\frac{1}{2}}\right)\phi_{i-1} \tag{8.33}$$

$$+ \left(\frac{1}{8}(\rho u)_{i+\frac{1}{2}}\right)\phi_{i+2}$$

Writing the discretization equation in compact form as

$$a_i\phi_i = a_{i+1}\phi_{i+1} + a_{i-1}\phi_{i-1} + a_{i+2}\phi_{i+2} \tag{8.34}$$

where

$$a_{i+1} = \frac{\Gamma_{i+\frac{1}{2}}}{\Delta x_{i+\frac{1}{2}}} - \frac{3}{8}(\rho u)_{i+\frac{1}{2}} - \frac{1}{8}(\rho u)_{i-\frac{1}{2}} \tag{8.35a}$$

$$a_{i-1} = \frac{\Gamma_{i-\frac{1}{2}}}{\Delta x_{i-\frac{1}{2}}} + \frac{6}{8}(\rho u)_{i-\frac{1}{2}} \tag{8.35b}$$

$$a_{i+2} = \frac{1}{8}(\rho u)_{i+\frac{1}{2}} \tag{8.35c}$$

$$a_i = \frac{6}{8}(\rho u)_{i+\frac{1}{2}} - \frac{3}{8}(\rho u)_{i-\frac{1}{2}} + \frac{\Gamma_{i+\frac{1}{2}}}{\Delta x_{i+\frac{1}{2}}} + \frac{\Gamma_{i-\frac{1}{2}}}{\Delta x_{i-\frac{1}{2}}}$$

Rearranging

$$a_i = a_{i+1} + a_{i-1} + a_{i+2} + \left\{(\rho u)_{i+\frac{1}{2}} - (\rho u)_{i-\frac{1}{2}}\right\}$$

Mass Velocity term

8.1.7.4 MUSCL Scheme: Monotonic Upstream-Centered Scheme for Conservation Laws

The **MUSCL scheme** for convection term was proposed by van Leer (1979) as a third-order discretization scheme for the convective terms in a finite volume method. The scheme not only potentially provides higher spatial accuracy but also shown to provide stable solution problems involving shocks and discontinuities or steep gradients. Another major advantage of this scheme is its applicability to wide varieties of mesh: structured, unstructured, or any arbitrary mesh. This scheme has been implemented in many codes including the commercial code such as ANSYS-Fluent. For more discussions on MUSCL scheme and its usage, refer van Leer [1979] and Sohn [2005].

8.2 DISCRETIZATION OF A GENERAL TRANSPORT EQUATION

Let us now consider an unsteady-state problem consisting of the continuity or conservation of a mass equation and a transport equation, which represents a general form for the momentum, energy, and other transport equations.

8.2.1 One-Dimensional Unsteady-State Problems

Let us now consider a one-dimensional unsteady-state problem with mass conservation and transport equation including a linearized source term that is described by the following governing differential equation.

Mass Conservation:

$$\frac{\partial \rho}{\partial t}+\frac{\partial(\rho u)}{\partial x}=0 \tag{8.36a}$$

Transport Equations:

$$\frac{\partial(\rho\phi)}{\partial t}+\frac{\partial}{\partial x}(\rho u\phi)=\frac{\partial}{\partial x}\left(\Gamma_x\frac{\partial\phi}{\partial x}\right)+(S_0+S_1\phi) \tag{8.36b}$$

With the representation of the convection and the diffusion terms in the form of a total flux given by Equation 8.10, then Equation 8.36b reduces to

$$\frac{\partial(\rho\phi)}{\partial t}+\frac{\partial J_x}{\partial x}=(S_0+S_1\phi) \tag{8.37}$$

Integrating the transport equation (Equation 8.37) over the control volume (Figure 8.1) from $i-\frac{1}{2}$ to $i+\frac{1}{2}$ and over the time interval t to $t+\Delta t$, we get

$$\int_{i-\frac{1}{2}}^{i+\frac{1}{2}}\int_{t}^{t+\Delta t}\frac{\partial(\rho\phi)}{\partial t}\mathrm{d}x\mathrm{d}t+\int_{i-\frac{1}{2}}^{i+\frac{1}{2}}\int_{t}^{t+\Delta t}\frac{\partial J_x}{\partial x}\mathrm{d}x\mathrm{d}t=\int_{t}^{t+\Delta t}\int_{i-\frac{1}{2}}^{i+\frac{1}{2}}(S_0+S_1\phi)\mathrm{d}x\mathrm{d}t$$

Using the implicit time integration scheme

$$\left(\rho_i\phi_i\right)\Big|_t^{t+\Delta t}\Delta x+J_x^{m+1}\Big|_{i-\frac{1}{2}}^{i+\frac{1}{2}}\Delta t=S_0\Delta x\Delta t+S_1\phi_i^{m+1}\Delta x\Delta t \tag{8.38}$$

or

$$\frac{\rho_i^{m+1}\phi_i^{m+1}-\rho_i^m\phi_i^m}{\Delta t}\Delta x+J_{x_{i+\frac{1}{2}}}^{m+1}-J_{x_{i-\frac{1}{2}}}^{m+1}=S_0\Delta x+S_1\phi_i^{m+1}\Delta x \tag{8.39}$$

where the superscript $m+1$ represents the new values at time $t+\Delta t$ and m represents the old value at time t.

In a similar manner, we integrate the mass conservation equation (8.36a) over the control volume as

$$\int_{i-\frac{1}{2}}^{i+\frac{1}{2}}\int_{t}^{t+\Delta t}\frac{\partial(\rho)}{\partial t}\mathrm{d}x\mathrm{d}t+\int_{i-\frac{1}{2}}^{i+\frac{1}{2}}\int_{t}^{t+\Delta t}\frac{\partial(\rho u)}{\partial x}\mathrm{d}x\mathrm{d}t=0$$

or

$$\frac{\rho_i^{m+1}-\rho_i^m}{\Delta t}\Delta x+(\rho u)_{i+\frac{1}{2}}^{m+1}-(\rho u)_{i-\frac{1}{2}}^{m+1}=0 \tag{8.40}$$

Multiplying Equation 8.40 by ϕ_i^{m+1} and subtracting from Equation 8.39, we get

$$\left(\phi_i^{m+1}-\phi_i^m\right)\frac{\rho_i^m\Delta x}{\Delta t}+\left(J_{x_{i+\frac{1}{2}}}^{m+1}-(\rho u)_{i+\frac{1}{2}}^{m+1}\phi_i^{m+1}\right)-\left(J_{x_{i-\frac{1}{2}}}^{m+1}-(\rho u)_{i-\frac{1}{2}}^{m+1}\phi_i^{m+1}\right)$$

$$=S_0\Delta x+S_1\phi_i^{m+1}\Delta x \tag{8.41}$$

Let us now use the approximations (Patankar, 1980)

$$\left(J_{x_{i+\frac{1}{2}}}^{m+1}-(\rho u)_{i+\frac{1}{2}}^{m+1}\phi_i^{m+1}\right)=a_{i+1}\left(\phi_i^{m+1}-\phi_{i+1}^{m+1}\right) \tag{8.42a}$$

and

$$\left(J_{x_{i-\frac{1}{2}}}^{m+1}-(\rho u)_{i-\frac{1}{2}}^{m+1}\phi_i^{m+1}\right)=a_{i-1}\left(\phi_{i-1}^{m+1}-\phi_i^{m+1}\right) \tag{8.42b}$$

where a_{i+1} and a_{i-1} are given by Equation 8.39 as

$$a_{i+1}=\frac{\Gamma_{x_{i+\frac{1}{2}}}}{\Delta x_{i+\frac{1}{2}}}A\left|P_{i+\frac{1}{2}}\right|+\left\langle 0,-(\rho u)_{i+\frac{1}{2}}\right\rangle \tag{8.39b}$$

$$a_{i-1}=\frac{\Gamma_{x_{i-\frac{1}{2}}}}{\Delta x_{i-\frac{1}{2}}}A\left|P_{i-\frac{1}{2}}\right|+\left\langle 0,(\rho u)_{i-\frac{1}{2}}\right\rangle \tag{8.39c}$$

Substituting Equation 8.43 into Equation 8.41, we get

$$a_i\phi_i^{m+1}=a_{i+1}\phi_{i+1}^{m+1}+a_{i-1}^{m+1}\phi_{i-1}^{m+1}+d \tag{8.43}$$

where

$$a_i=a_{i+1}+a_{i-1}+a_i^m-S_1\Delta x$$

$$a_{i+1}=\frac{\Gamma_{x_{i+\frac{1}{2}}}}{\Delta x_{i+\frac{1}{2}}}A\left|P_{i+\frac{1}{2}}\right|+\left\langle 0,-(\rho u)_{i+\frac{1}{2}}\right\rangle$$

$$a_{i-1}=\frac{\Gamma_{x_{i-\frac{1}{2}}}}{\Delta x_{i-\frac{1}{2}}}A\left|P_{i-\frac{1}{2}}\right|+\left\langle 0,(\rho u)_{i-\frac{1}{2}}\right\rangle$$

$$a_i^m=\frac{\rho_i^m\Delta x}{\Delta t}$$

$$d=S_0\Delta x+a_i^m\phi_i^m$$

Equation 8.43 represents an implicit form of the discretization equation for a general transport equation. This equation can now be applied to the *x*, *y*, and *z* components of the momentum equation, and to heat and mass transport equations.

8.2.2 Two-Dimensional Unsteady-State Problem

A two-dimensional unsteady-state problem with mass conservation and a general form of transport equation is given as

Mass Conservation:

$$\frac{\partial \rho}{\partial t}+\frac{\partial(\rho u)}{\partial x}+\frac{\partial(\rho v)}{\partial y}=0 \tag{8.44a}$$

Transport Equations:

$$\frac{\partial(\rho\phi)}{\partial t}+\frac{\partial}{\partial x}(\rho u\phi)+\frac{\partial}{\partial y}(\rho v\phi)=\frac{\partial}{\partial x}\left(\Gamma_x\frac{\partial\phi}{\partial x}\right)+\frac{\partial}{\partial y}\left(\Gamma_y\frac{\partial\phi}{\partial y}\right)+(S_0+S_1\phi) \tag{8.44b}$$

The implicit discretization of these equations is obtained by integrating over the two-dimensional control volume, shown in Figure 8.3, and expressed as

$$a_i\phi_i^{m+1}=a_{i+1}\phi_{i+1}^{m+1}+a_{i-1}^{m+1}\phi_{i-1}^{m+1}+a_{j+1}\phi_{j+1}^{m+1}+a_{j-1}^{m+1}\phi_{j-1}^{m+1}+d \tag{8.45}$$

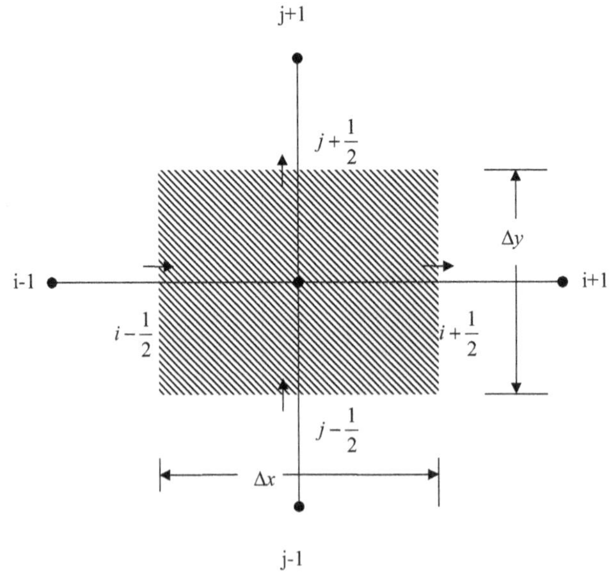

FIGURE 8.3 Two-dimensional control volume.

where

$$a_i = a_{i+1} + a_{i-1} + a_{j+1} + a_{j-1} + a_i^m - S_1 \Delta x \Delta y$$

$$a_{i+1} = \frac{\Gamma_{x_{i+\frac{1}{2}}}}{\Delta x_{i+\frac{1}{2}}} A\left|P_{i+\frac{1}{2}}\right| + \left\langle 0, -(\rho u)_{i+\frac{1}{2}} \right\rangle$$

$$a_{i-1} = \frac{\Gamma_{x_{i-\frac{1}{2}}}}{\Delta x_{i-\frac{1}{2}}} A\left|P_{i-\frac{1}{2}}\right| + \left\langle 0, (\rho u)_{i-\frac{1}{2}} \right\rangle$$

$$a_{j+1} = \frac{\Gamma_{y_{j+\frac{1}{2}}}}{\Delta y_{i+\frac{1}{2}}} A\left|P_{j+\frac{1}{2}}\right| + \left\langle 0, -(\rho v)_{j+\frac{1}{2}} \right\rangle$$

$$a_{j-1} = \frac{\Gamma_{y_{j-\frac{1}{2}}}}{\Delta y_{i-\frac{1}{2}}} A\left|P_{j-\frac{1}{2}}\right| + \left\langle 0, -(\rho v)_{j-\frac{1}{2}} \right\rangle$$

$$a_i^m = \frac{\rho_i^m \Delta x \Delta y}{\Delta t}$$

$$d = S_0 \Delta x \Delta y + a_i^m \phi_i^m$$

8.2.3 Three-Dimensional Unsteady-State Problem

A three-dimensional unsteady-state problem with mass conservation and a general form of transport equation is given as

Mass Conservation:

$$\frac{\partial \rho}{\partial t} + \frac{\partial(\rho u)}{\partial x} + \frac{\partial(\rho v)}{\partial y} + \frac{\partial(\rho w)}{\partial z} = 0 \tag{8.46a}$$

Transport Equations:

$$\frac{\partial(\rho\phi)}{\partial t} + \frac{\partial}{\partial x}(\rho u \phi) + \frac{\partial}{\partial y}(\rho v \phi) + \frac{\partial}{\partial z}(\rho w \phi)$$

$$= \frac{\partial}{\partial x}\left(\Gamma_x \frac{\partial \phi}{\partial x}\right) + \frac{\partial}{\partial y}\left(\Gamma_y \frac{\partial \phi}{\partial y}\right) + \frac{\partial}{\partial z}\left(\Gamma_z \frac{\partial \phi}{\partial z}\right) + (S_0 + S_1\phi) \tag{8.46b}$$

The implicit discretization of these equations leads to

$$a_i\phi_i^{m+1} = a_{i+1}\phi_{i+1}^{m+1} + a_{i-1}^{m+1}\phi_{i-1}^{m+1} + a_{j+1}\phi_{j+1}^{m+1} + a_{j-1}^{m+1}\phi_{j-1}^{m+1}$$

$$+ a_{k+1}\phi_{k+1}^{m+1} + a_{k-1}^{m+1}\phi_{k-1}^{m+1} + c \tag{8.47a}$$

or

$$a_i\phi_i^{m+1} = \sum a_{nb}\phi_{nb}^{m+1} + d \tag{8.47b}$$

where

$$a_i = a_{i+1} + a_{i-1} + a_{j+1} + a_{j-1} + a_{k+1} + a_{k-1} + a_i^m - S_1 \Delta x \Delta y \Delta z$$

$$a_{i+1} = \frac{\Gamma_{x_{i+\frac{1}{2}}}}{\Delta x_{i+\frac{1}{2}}} A \left| P_{i+\frac{1}{2}} \right| + \left\langle 0, -(\rho u)_{i+\frac{1}{2}} \right\rangle$$

$$a_{i-1} = \frac{\Gamma_{x_{i-\frac{1}{2}}}}{\Delta x_{i-\frac{1}{2}}} A \left| P_{i-\frac{1}{2}} \right| + \left\langle 0, (\rho u)_{i-\frac{1}{2}} \right\rangle$$

$$a_{j+1} = \frac{\Gamma_{y_{j+\frac{1}{2}}}}{\Delta y_{i+\frac{1}{2}}} A \left| P_{j+\frac{1}{2}} \right| + \left\langle 0, -(\rho v)_{j+\frac{1}{2}} \right\rangle$$

$$a_{j-1} = \frac{\Gamma_{y_{j-\frac{1}{2}}}}{\Delta y_{i-\frac{1}{2}}} A \left| P_{j-\frac{1}{2}} \right| + \left\langle 0, (\rho v)_{j-\frac{1}{2}} \right\rangle$$

$$a_{k+1} = \frac{\Gamma_{z_{k+\frac{1}{2}}}}{\Delta z_{k+\frac{1}{2}}} A \left| P_{k+\frac{1}{2}} \right| + \left\langle 0, -(\rho w)_{k+\frac{1}{2}} \right\rangle$$

$$a_{k-1} = \frac{\Gamma_{z_{k-\frac{1}{2}}}}{\Delta z_{k-\frac{1}{2}}} A \left| P_{k-\frac{1}{2}} \right| + \left\langle 0, (\rho w)_{k-\frac{1}{2}} \right\rangle$$

$$a_i^m = \frac{\rho_i^m \Delta x \Delta y \Delta z}{\Delta t}$$

$$d = S_0 \Delta x \Delta y \Delta z + a_i^m \phi_i^m$$

8.3 SOLUTION OF FLOW FIELD

Solution of a flow field is obtained in terms of the velocity field $\vec{V} = \hat{i}u + \hat{j}v + \hat{k}w$ and pressure field $p(x, y, z)$, which are determined by solving continuity or conservation of mass and momentum. As we have discussed in Chapter 1, the conservation of mass and momentum for Newtonian viscous fluid flow is given by the Navier–Stokes equation

Continuity:

$$\frac{\partial \rho}{\partial t} + \frac{\partial (\rho u)}{\partial x} + \frac{\partial (\rho v)}{\partial y} + \frac{\partial (\rho w)}{\partial z} = 0 \tag{8.48a}$$

***x*–momentum:**

$$\rho \left(\frac{\partial u}{\partial t} + u \frac{\partial u}{\partial x} + v \frac{\partial u}{\partial y} + w \frac{\partial u}{\partial z} \right) = \rho g_x - \frac{\partial p}{\partial x} + \mu \left(\frac{\partial^2 u}{\partial x^2} + \frac{\partial^2 u}{\partial y^2} + \frac{\partial^2 u}{\partial z^2} \right) \tag{8.48b}$$

***y*–momentum:**

$$\rho \left(\frac{\partial v}{\partial t} + u \frac{\partial v}{\partial x} + v \frac{\partial v}{\partial y} + w \frac{\partial v}{\partial z} \right) = \rho g_y - \frac{\partial p}{\partial y} + \mu \left(\frac{\partial^2 v}{\partial x^2} + \frac{\partial^2 v}{\partial y^2} + \frac{\partial^2 v}{\partial z^2} \right) \tag{8.48c}$$

z-**momentum:**

$$\rho\left(\frac{\partial w}{\partial t}+u\frac{\partial w}{\partial x}+v\frac{\partial w}{\partial y}+w\frac{\partial w}{\partial z}\right)=\rho g_z-\frac{\partial p}{\partial z}+\mu\left(\frac{\partial^2 w}{\partial x^2}+\frac{\partial^2 w}{\partial y^2}+\frac{\partial^2 w}{\partial z^2}\right) \tag{8.48d}$$

This set of coupled partial differential equations is nonlinear due to the presence of convective terms on the left-hand side of Equations 8.34b and 8.34c, and the equations are solved in an iterative manner. The major difficulty is that there is no equation that we can solve directly for the pressure. There are two major approaches for solving the flow field: (a) **the direct solution involving the primitive variable** and (2) **the stream function/vorticity-based method**.

In the direct solution method involving the primitive variables, the three momentum and continuity equations are solved directly for three velocity components *u*, *v*, and *w* with a preassumed pressure field. The iteration process is continued with an updated pressure field. The correct pressure field is established when the estimated velocity components satisfy the continuity equation (8.34a). In the subsequent sections, we will present the development of a pressure equation by using this indirect information of pressure in the continuity equation. This procedure of solving the flow field directly in terms of four primitive variables, three velocity components and pressure, poses considerable complexity and computational difficulties.

8.3.1 Stream Function–Vorticity-Based Method

The vorticity/stream function-based method is primarily used for two-dimensional problems. The set of equations for the two-dimensional incompressible fluid flow with constant viscosity is deduced from Equation 8.44 as

Continuity:

$$\frac{\partial u}{\partial x}+\frac{\partial v}{\partial y}=0 \tag{8.49a}$$

x-**momentum:**

$$\rho\left(\frac{\partial u}{\partial t}+u\frac{\partial u}{\partial x}+v\frac{\partial u}{\partial y}\right)=\rho g_x-\frac{\partial P}{\partial x}+\mu\left(\frac{\partial^2 u}{\partial x^2}+\frac{\partial^2 u}{\partial y^2}\right) \tag{8.49b}$$

y-**momentum:**

$$\rho\left(\frac{\partial v}{\partial t}+u\frac{\partial v}{\partial x}+v\frac{\partial v}{\partial y}\right)=\rho g_y-\frac{\partial P}{\partial y}+\mu\left(\frac{\partial^2 v}{\partial x^2}+\frac{\partial^2 v}{\partial y^2}\right) \tag{8.49c}$$

In order to transform the two momentum equations for *u* and *v* into a single equation, we introduce **vorticity** as the **curl** of velocity vector or $\zeta=\nabla\times\vec{V}$, which reduces to the *z*-component vorticity as

$$\zeta=\zeta_z=\frac{\partial u}{\partial y}-\frac{\partial v}{\partial x} \tag{8.50}$$

Now by taking a **curl** of the momentum equations (Schlichting, 1955; Yuan, 1967) and eliminating pressure based on the *z*-component vorticity, we have

$$\rho\left(\frac{\partial \zeta}{\partial t}+u\frac{\partial \zeta}{\partial x}+v\frac{\partial \zeta}{\partial y}\right)=\mu\left(\frac{\partial^2 \zeta}{\partial x^2}+\frac{\partial^2 \zeta}{\partial y^2}\right) \tag{8.51}$$

Equation 8.37 is referred to as the **vorticity transport equation**, which is like the general transport Equation 8.28 considered in an earlier section. The terms on the left-hand side of the equation represent the local and convective variation of the vorticity, and the terms on the right-hand side represent the diffusion dissipation of vorticity. It can be noted that the pressure no longer appears in the equation as normal forces do not have any effect on the angular rotation of the fluid. The vorticity transport Equation 8.37, along with the continuity Equation 8.35a, forms a set of two equations for two unknowns u and v without the need for a third equation for the pressure. The numerical approximation schemes for convective and diffusion terms, and the solution procedure described for the two-dimensional transport equation in Section 8.2.2 is also applicable for the solution of this set.

The number of equations can further be reduced by introducing the stream function, ψ, as

$$u = \frac{\partial \psi}{\partial y} \quad \text{and} \quad v = -\frac{\partial \psi}{\partial x} \tag{8.52}$$

This satisfies the continuity Equation 8.35a automatically and transforms the vorticity given by Equation 8.36 and vorticity transport Equation 8.37 into

$$\frac{\partial^2 \psi}{\partial x^2} + \frac{\partial^2 \psi}{\partial y^2} = \zeta \tag{8.53}$$

and

$$\rho \left\{ \frac{\partial (\nabla^2 \psi)}{\partial t} + \frac{\partial \psi}{\partial y} \frac{\partial (\nabla^2 \psi)}{\partial x} - \frac{\partial \psi}{\partial x} \frac{\partial (\nabla^2 \psi)}{\partial y} \right\} = \mu \Delta^4 \psi \tag{8.54}$$

In this form, the vorticity transport equation contains only one unknown variable, ψ, but has the form of a nonlinear fourth-order differential equation. Equation 8.40, referred to as the **biharmonic–stream function formulation**, can be solved directly by applying finite difference formulas for the derivatives. Such a formulation and solution are given in detail by Shih (1984).

A major advantage of this vorticity-based method is that there is no need to solve for the pressure. This is, however, also a disadvantage in some class of problems where it is essential to know the pressure distribution. Since the stream function can only be defined for two dimensions, the vorticity-based method is primarily restricted to two-dimensional problems. It will pose considerable difficulties if it were to be extended to three-dimensions where it involves solving a set of six differential equations: three for the three components of vorticity vector, $\varsigma = \hat{i}\varsigma_x + \hat{j}\varsigma_y + \hat{k}\varsigma_z$, and velocity potentials, φ. Numerical solution methods for the vorticity/stream function-based method are discussed by Jaluria and Torrance (1986) and Roaches (1982). As we have previously mentioned, for problems where estimation of pressure is essential and for three-dimensional problems, it is more appropriate to solve continuity and momentum equations directly for the primitive variables. In fact, in recent times, this approach has become increasingly popular. In the following sections, we will focus on the direct solution method involving the primitive variables, i.e., three velocity components u, v, w, and pressure, p.

8.3.2 Direct Solution with the Primitive Variables

We have previously mentioned that in the direct solution method involving the primitive variable, the three momentum equations are solved directly for three velocity components u, v, and w, with an estimated value of the pressure field. The iteration process is continued with an updated pressure field. The correct pressure field is established when estimated velocity components satisfy continuity Equation 8.31a. There was considerable effort in developing solution algorithms based on this basic criterion. Among the most popular algorithms are the semi-implicit time algorithm such as **SIMPLE** (Patankar and Spalding, 1972; Patankar, 1980) and its variations, and the explicit time algorithm such as Marker-and-cell **(MAC)** (Harlow and Welch, 1965; Hirt and Cook, 1972; Majumdar and Deb, 2003). We will describe these two algorithms in the following sections.

In deriving the discretization equations for the momentum equations, we note that the momentum Equations 8.34b to 8.34d are identical to the general form of the transport Equation 8.21b except for the pressure derivative terms. The discretization schemes for convective and diffusion terms described for the general transport equation in Section 8.2.2 are also applicable for deriving the discretization equation for the momentum equation. However, before we proceed to the solution of the flow field, we need to present the discretization scheme for the pressure derivative terms in the momentum equation as well as the derivation of a separate equation for the pressure variable.

Discretization of Pressure Derivative Let us consider the one-dimensional x-momentum equation by simply substituting $\phi = u$ and $\Gamma_x = \mu$ into Equation 8.26b, and including the pressure derivative term as

$$\frac{\partial(\rho u)}{\partial t} + \frac{\partial}{\partial x}(\rho u u) = -\frac{\partial p}{\partial x} + \frac{\partial}{\partial x}\left(\mu \frac{\partial u}{\partial x}\right) + (S_0 + S_1\phi) \tag{8.55}$$

Integration of the pressure gradient term over the control volume shown in Figure 8.1 gives

$$\int_t^{t+\Delta t}\int_{i-\frac{1}{2}}^{i+\frac{1}{2}} \frac{dP}{dx}dx = \left(P_{i+\frac{1}{2}}^{m+1} - P_{i-\frac{1}{2}}^{m+1}\right)\Delta t \tag{8.56}$$

Assuming a piece-wise linear profile for pressure at interfaces of the control volume, Equation 8.56 reduces to

$$\int_t^{t+\Delta t}\int_{i-\frac{1}{2}}^{i+\frac{1}{2}} \frac{dP}{dx}dx = \left(\frac{P_{i+1}^{m+1} + P_i^{m+1}}{2} - \frac{P_i^{m+1} + P_{i-1}^{m+1}}{2}\right)\Delta t = \tfrac{1}{2}\left(P_{i+1}^{m+1} - P_{i-1}^{m+1}\right)\Delta t \tag{8.57}$$

With the inclusion of this pressure derivative term in the momentum equation, velocity at the node i will be calculated based on Equation 8.43.

$$a_i u_i^{m+1} = a_{i+1}u_{i+1}^{m+1} + a_{i-1}u_{i-1}^{m+1} + d - \frac{1}{2}(P_{i+1} - P_{i-1}) \tag{8.58a}$$

or

$$a_i u_i^{m+1} = \sum a_{nb}u_{nb}^{m+1} + d - \tfrac{1}{2}(P_{i+1} - P_{i-1}) \tag{8.58b}$$

We can see that the velocity component u at a node i is calculated based on difference of pressure values at two alternate nodes $i-1$ and $i+1$, and does not involve the pressure values at the node i. Consequently, this velocity discretization equation may cause instability during the iteration process and may not converge or it may converge to unrealistic values. A detailed discussion of such a behavior is given by Patankar (1980). To avoid such unrealistic situations, a recommended practice is to use a **staggered grid**.

Staggered Grid In a staggered grid, the velocity components are calculated using a grid system that is staggered from the grid system that is being used for the estimation of pressure or any other scalar variables such as temperature or mass concentration. The nodal points of a staggered grid system, represented by broken lines and smaller bullets, lie at the faces of the control volumes for the regular grid as represented by solid lines and bigger bullets in Figure 8.4.

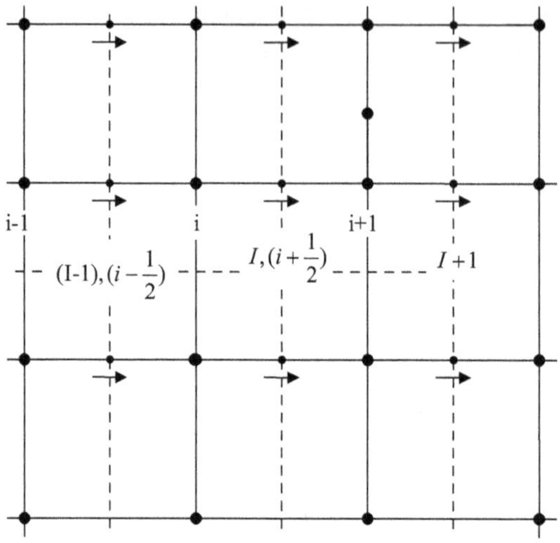

FIGURE 8.4 A staggered grid.

Also, the nodal points in the staggered grid are designated by i, $i+1$, and $i-1$ to differentiate from the regular grid (indicated by I, $I-1$, $I+1$) for the scalar quantities such as pressure. Using such a grid system and using Equation 8.56, the discretized equation for the x-component velocity is

$$a_i u_i^{m+1} = a_{i+1} u_{i+1}^{m+1} + a_{i-1}^{m+1} u_{i-1}^{m+1} + d - \left(P_{i+\frac{1}{2}}^{m+1} - P_{i-\frac{1}{2}}^{m+1} \right) \tag{8.59a}$$

or

$$a_i u_i^{m+1} = a_{i+1} u_{i+1}^{m+1} + a_{i-1}^{m+1} u_{i-1}^{m+1} + d - \left(P_I^{m+1} - P_{I-1}^{m+1} \right) \tag{8.59b}$$

Note that with this discretization equation, based on the staggered grid, the velocity at the faces of the regular grid is calculated based on the pressure difference between two adjacent nodal points of the regular grid. In a similar manner, this procedure is extended to derive y-component velocity and z-component velocity for two- and three-dimensional problems.

Discretization Equation for Velocity Component The discretization equations for the three velocity components u, v, and w momentum can now be written in terms of guessed pressure values, following the procedures outlined in Section 8.2.1 for the general form of the transport equation and in Section 8.3.2.2 for the pressure derivative term, and using a staggered grid system for each of the velocity components. For the u-component velocity, we can consider a control volume in a staggered grid system as shown in Figure 8.5 and discretize the x-momentum Equation 8.48b using Equations 8.46 and 8.59 as

$$a_i u_i^{m+1} = a_{i+1} u_{i+1}^{m+1} + a_{i-1} u_{i-1}^{m+1} + a_{j+1} u_{j+1}^{m+1} + a_{j-1} u_{j-1}^{m+1} + a_{k+1} u_{k+1}^{m+1} + a_{k-1} u_{k-1}^{m+1} + d + \left(p_{I-1} - p_I \right) \Delta y \Delta z \tag{8.60a}$$

or

$$a_i u_i^{m+1} = \sum a_{nb} u_{nb}^{m+1} + d + \left(p_{I-1} - p_I \right) \Delta y \Delta z \tag{8.60b}$$

where

$$a_i = a_{i+1} + a_{i-1} + a_{j+1} + a_{j-1} + a_{k+1} + a_{k-1} + a_i^m \tag{8.61a}$$

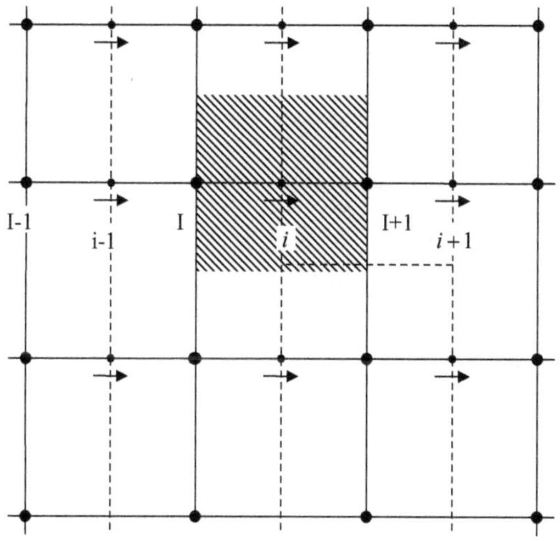

FIGURE 8.5 Control volume for u-component velocity.

$$a_{i+1} = \frac{\mu}{\Delta x_{i+\frac{1}{2}}} A\left|P_{i+\frac{1}{2}}\right| + \left\langle 0, -(\rho u)_{i+\frac{1}{2}} \right\rangle \tag{8.61b}$$

$$a_{i-1} = \frac{\mu}{\Delta x_{i-\frac{1}{2}}} A\left|P_{i-\frac{1}{2}}\right| + \left\langle 0, (\rho u)_{i-\frac{1}{2}} \right\rangle \tag{8.61c}$$

$$a_{j+1} = \frac{\mu}{\Delta y_{i+\frac{1}{2}}} A\left|P_{j+\frac{1}{2}}\right| + \left\langle 0, -(\rho v)_{j+\frac{1}{2}} \right\rangle \tag{8.61d}$$

$$a_{j-1} = \frac{\mu}{\Delta y_{i-\frac{1}{2}}} A\left|P_{j-\frac{1}{2}}\right| + \left\langle 0, (\rho v)_{j-\frac{1}{2}} \right\rangle \tag{8.61e}$$

$$a_{k+1} = \frac{\mu}{\Delta z_{k+\frac{1}{2}}} A\left|P_{k+\frac{1}{2}}\right| + \left\langle 0, -(\rho w)_{k+\frac{1}{2}} \right\rangle \tag{8.61f}$$

$$a_{k-1} = \frac{\mu}{\Delta z_{k-\frac{1}{2}}} A\left|P_{k-\frac{1}{2}}\right| + \left\langle 0, (\rho w)_{k-\frac{1}{2}} \right\rangle \tag{8.61g}$$

$$a_i^m = \frac{\rho \Delta x \Delta y \Delta z}{\Delta t} \tag{8.61h}$$

$$d = \rho g_x \Delta x \Delta y \Delta z + a_i^m u_i^m \tag{8.61i}$$

$$P = \frac{(\rho u)}{\mu / \Delta x} \tag{8.61j}$$

For the v-component velocity, we consider a control volume in a staggered grid system as shown in Figure 8.6 and discretize the y-momentum Equation 8.48c as

$$a_j v_j^{m+1} = a_{i+1} v_{i+1}^{m+1} + a_{i-1} v_{i-1}^{m+1} + a_{j+1} v_{j+1}^{m+1} + a_{j-1} v_{j-1}^{m+1} + a_{k+1} v_{k+1}^{m+1} + a_{k-1} v_{k-1}^{m+1} + d + \left(p_{J-1} - p_J\right) \Delta x \Delta z \tag{8.62a}$$

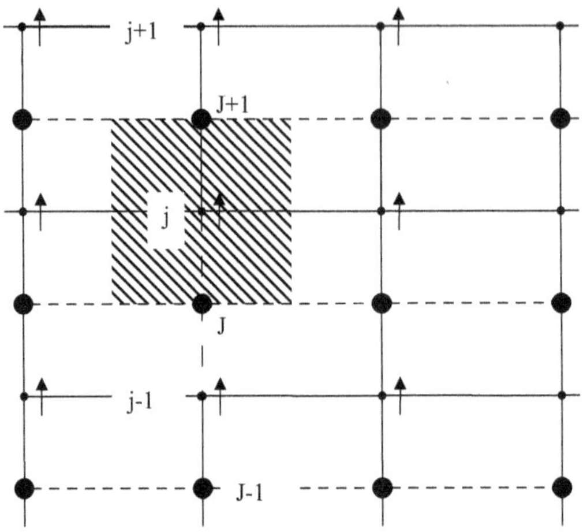

FIGURE 8.6 Control volume for v-component velocity.

or

$$a_j v_j^{m+1} = \sum a_{nb} v_{nb}^{m+1} + d + \left(p_{J-1} - p_J\right)\Delta x \Delta z \tag{8.62b}$$

where

$$a_j = a_{i+1} + a_{i-1} + a_{j+1} + a_{j-1} + a_{k+1} + a_{k-1} + a_j^m \tag{8.63a}$$

$$a_{i+1} = \frac{\mu}{\Delta x_{i+\frac{1}{2}}} A\left|P_{i+\frac{1}{2}}\right| + \left\langle 0, -(\rho u)_{i+\frac{1}{2}}\right\rangle \tag{8.63b}$$

$$a_{i-1} = \frac{\mu}{\Delta x_{i-\frac{1}{2}}} A\left|P_{i-\frac{1}{2}}\right| + \left\langle 0, (\rho u)_{i-\frac{1}{2}}\right\rangle \tag{8.63c}$$

$$a_{j+1} = \frac{\mu}{\Delta y_{i+\frac{1}{2}}} A\left|P_{j+\frac{1}{2}}\right| + \left\langle 0, -(\rho v)_{j+\frac{1}{2}}\right\rangle \tag{8.63d}$$

$$a_{j-1} = \frac{\mu}{\Delta y_{i-\frac{1}{2}}} A\left|P_{j-\frac{1}{2}}\right| + \left\langle 0, (\rho v)_{j-\frac{1}{2}}\right\rangle \tag{8.63e}$$

$$a_{k+1} = \frac{\mu}{\Delta z_{k+\frac{1}{2}}} A\left|P_{k+\frac{1}{2}}\right| + \left\langle 0, -(\rho w)_{k+\frac{1}{2}}\right\rangle \tag{8.63f}$$

$$a_{k-1} = \frac{\mu}{\Delta z_{k-\frac{1}{2}}} A\left|P_{k-\frac{1}{2}}\right| + \left\langle 0, (\rho w)_{k-\frac{1}{2}}\right\rangle \tag{8.63g}$$

$$a_j^m = \frac{\rho \Delta x \Delta y \Delta z}{\Delta t} \tag{8.63h}$$

$$d = \rho g_y \Delta x \Delta y \Delta z + a_j^m v_j^m \tag{8.63i}$$

$$P = \frac{(\rho v)}{\mu/\Delta y} \tag{8.63j}$$

In a similar manner, the discretization equation for the w-component velocity derived from z-momentum Equation 8.48d is written as

$$a_k w_k^{m+1} = a_{i+1} w_{i+1}^{m+1} + a_{i-1} w_{i-1}^{m+1} + a_{j+1} w_{j+1}^{m+1} + a_{j-1} w_{j-1}^{m+1} + a_{k+1} w_{k+1}^{m+1} + a_{k-1} w_{k-1}^{m+1}$$

$$+ d + \left(p_{K-1} - p_K \right) \Delta x \Delta y \tag{8.64a}$$

or

$$a_k w_k^{m+1} = \sum a_{nb} w_{nb}^{m+1} + d + \left(p_{K-1} - p_K \right) \Delta x \Delta y \tag{8.64b}$$

where

$$a_k = a_{i+1} + a_{i-1} + a_{j+1} + a_{j-1} + a_{k+1} + a_{k-1} + a_k^m \tag{8.65a}$$

$$a_{i+1} = \frac{\mu}{\Delta x_{i+\frac{1}{2}}} A\left|P_{i+\frac{1}{2}}\right| + \left\langle 0, -(\rho u)_{i+\frac{1}{2}} \right\rangle \tag{8.65b}$$

$$a_{i-1} = \frac{\mu}{\Delta x_{i-\frac{1}{2}}} A\left|P_{i-\frac{1}{2}}\right| + \left\langle 0, (\rho u)_{i-\frac{1}{2}} \right\rangle \tag{8.65c}$$

$$a_{j+1} = \frac{\mu}{\Delta y_{i+\frac{1}{2}}} A\left|P_{j+\frac{1}{2}}\right| + \left\langle 0, -(\rho v)_{j+\frac{1}{2}} \right\rangle \tag{8.65d}$$

$$a_{j-1} = \frac{\mu}{\Delta y_{i-\frac{1}{2}}} A\left|P_{j-\frac{1}{2}}\right| + \left\langle 0, (\rho v)_{j-\frac{1}{2}} \right\rangle \tag{8.65e}$$

$$a_{k+1} = \frac{\mu}{\Delta z_{k+\frac{1}{2}}} A\left|P_{k+\frac{1}{2}}\right| + \left\langle 0, -(\rho w)_{k+\frac{1}{2}} \right\rangle \tag{8.65f}$$

$$a_{k-1} = \frac{\mu}{\Delta z_{k-\frac{1}{2}}} A\left|P_{k-\frac{1}{2}}\right| + \left\langle 0, (\rho w)_{k-\frac{1}{2}} \right\rangle \tag{8.65g}$$

$$a_k^m = \frac{\rho \Delta x \Delta y \Delta z}{\Delta t} \tag{8.65h}$$

$$d = \rho g_z \Delta x \Delta y \Delta z + a_k^m w_k^m \tag{8.65i}$$

$$P = \frac{(\rho w)}{\mu/\Delta z} \tag{8.65j}$$

Equations 8.60, 8.62, and 8.64 are the discretization equations for estimated values u, v, and w based on the guess values of the pressure field. To differentiate these estimated values from the corrected values during the iteration process, let us designate these estimated values of velocity components and guess values of pressure with a superscript * and write the discretization equations as

$$a_i u_i^* = a_{i+1}u_{i+1}^* + a_{i-1}u_{i-1}^* + a_{j+1}u_{j+1}^* + a_{j-1}u_{j-1}^* + a_{k+1}u_{k+1}^* + a_{k-1}u_{k-1}^* + d + \left(p_{I-1}^* - p_I^*\right)\Delta y \Delta z \tag{8.66a}$$

$$a_j v_j^* = a_{i+1}v_{i+1}^* + a_{i-1}v_{i-1}^* + a_{j+1}v_{j+1}^* + a_{j-1}v_{j-1}^* + a_{k+1}v_{k+1}^* + a_{k-1}v_{k-1}^* + d + \left(p_{J-1}^* - p_J^*\right)\Delta x \Delta z \tag{8.66b}$$

$$a_k w_k^* = a_{i+1}w_{i+1}^* + a_{i-1}w_{i-1}^* + a_{j+1}w_{j+1}^* + a_{j-1}w_{j-1}^* + a_{k+1}w_{k+1}^* + a_{k-1}w_{k-1}^* + d + \left(p_{K-1}^* - p_K^*\right)\Delta x \Delta y \tag{8.66c}$$

The Velocity Correction Equations We have mentioned that the estimated starred velocity components given by Equation 8.52 must be updated using corrected pressure values given as

$$p = p^* + p' \tag{8.67}$$

where p' is the pressure correction, which will be derived in the following section using the continuity equation.

The iterative procedure for updating the velocity components is like the iterative refinement procedure outlined in Section 3.2.5, Chapter 3, for a linear system of algebraic equations. Let us designate the updated velocity components as

$$u = u^* + u' \tag{8.68a}$$

$$v = v^* + v' \tag{8.68b}$$

$$w = w^* + w' \tag{8.68c}$$

where u', v', and w' are the velocity corrections. Substituting Equations 8.68 into Equations 8.60, 8.62, and 8.64, and subtracting from Equation 8.66, we obtain the correction equations from the velocity components as

$$a_i u_i' = a_{i+1}u_{i+1}' + a_{i-1}u_{i-1}' + a_{j+1}u_{j+1}' + a_{j-1}u_{j-1}' + a_{k+1}u_{k+1}' + a_{k-1}u_{k+1}' + \left(p_{I-1}' - p_I'\right)\Delta y \Delta z \tag{8.69a}$$

$$a_j v_j' = a_{i+1}v_{i+1}' + a_{i-1}v_{i-1}' + a_{j+1}v_{j+1}' + a_{j-1}v_{j-1}' + a_{k+1}v_{k+1}' + a_{k-1}v_{k-1}' + \left(p_{J-1}' - p_J'\right)\Delta x \Delta z \tag{8.69b}$$

$$a_k w_k' = a_{i+1}w_{i+1}' + a_{i-1}w_{i-1}' + a_{j+1}w_{j+1}' + a_{j-1}w_{j-1}' + a_{k+1}w_{k+1}' + a_{k-1}w_{k-1}' + \left(p_{K-1}' - p_K'\right)\Delta x \Delta y \tag{8.69c}$$

The system of equations given by Equations 8.69a–c is similar to that given by Equation 8.66 with the exception of the body force term, and can be solved by the same linear solver discussed in Chapter 2. However, one of the major features of the **SIMPLE** algorithm, as discussed by Patankar (1980), is to simplify Equation 8.69 by dropping the velocity component of the surrounding nodes on the right-hand side of Equation 8.69, i.e., the term $\sum a_{nb}u_{nb}'$, and write the correction equations as

$$u_i' = \left(p_{I-1}' - p_I'\right)\frac{\Delta y \Delta z}{a_i} \tag{8.70a}$$

$$v_j' = \left(p_{J-1}' - p_J'\right)\frac{\Delta x \Delta z}{a_j} \tag{8.70b}$$

$$w_k' = \left(p_{K-1}' - p_K'\right)\frac{\Delta x \Delta z}{a_k} \tag{8.70c}$$

We can see that the use of Equation 8.70 instead of Equation 8.69 considerably reduces the computational difficulties but sacrifices the accuracy in velocity corrections to some extent. However, accuracy of the estimated velocity components is achieved through iterations and assigned tolerance limits.

Substituting the velocity correction values given by either Equation 8.69 or 8.70 into Equation 8.68, we get the expressions for corrected velocity components. If we use Equation 8.69 for velocity corrections, then the expressions for the **corrected velocity** components are given as

$$u_i = u_i^* + \left(p'_{I-1} - p'_I\right)\frac{\Delta y \Delta z}{a_i} \tag{8.71a}$$

$$v_j = v_j^* + \left(p'_{J-1} - p'_J\right)\frac{\Delta x \Delta z}{a_j} \tag{8.71b}$$

$$w_k = w_k^* + \left(p'_{K-1} - p'_K\right)\frac{\Delta x \Delta y}{a_k} \tag{8.71c}$$

The Pressure Correction Equation The pressure correction equation is derived by substituting the estimated velocity components into the continuity equation. Let us first integrate the continuity Equation 8.48a over the control volume in the regular grid system (Figure 8.7) for the pressure

$$\int_t^{t+\Delta t}\int_{i-1}^{i}\int_{j-1}^{j}\int_{k-1}^{k}\frac{\partial \rho}{\partial t}\mathrm{d}t\mathrm{d}x\mathrm{d}y\mathrm{d}z + \int_t^{t+\Delta t}\int_{i-1}^{i}\int_{j-1}^{j}\int_{k-1}^{k}\frac{\partial(\rho u)}{\partial x}\mathrm{d}t\mathrm{d}x\mathrm{d}y\mathrm{d}z + \int_t^{t+\Delta t}\int_{i-1}^{i}\int_{j-1}^{j}\int_{k-1}^{k}\frac{\partial(\rho v)}{\partial y}\mathrm{d}t\mathrm{d}x\mathrm{d}y\mathrm{d}z$$
$$+ \int_t^{t+\Delta t}\int_{i-1}^{i}\int_{j-1}^{j}\int_{k-1}^{k}\frac{\partial(\rho w)}{\partial z}\mathrm{d}t\mathrm{d}x\mathrm{d}y\mathrm{d}z = 0 \tag{8.72}$$

or

$$\rho_I\Big|_t^{t+\Delta t}\Delta x\Delta y\Delta z + \left\{(\rho u)_i - (\rho u)_{i-1}\right\}\Delta y\Delta z\Delta t + \left\{(\rho v)_j - (\rho v)_{j-1}\right\}\Delta x\Delta z\Delta t$$
$$+\left\{(\rho w)_k - (\rho w)_{k-1}\right\}\Delta x\Delta y\Delta t = 0 \tag{8.73}$$

Using the **implicit time integration scheme**, i.e., values of density and velocity components at time $t+\Delta t$ are assumed to prevail over the entire time step Δt, and substituting the expressions for the velocity field given by Equation 8.71, we get

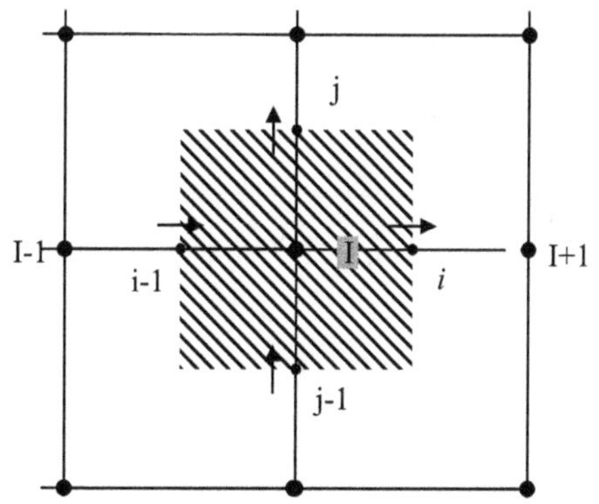

FIGURE 8.7 Control volume for pressure.

$$\frac{\left(\rho_I^{m+1}-\rho_I^m\right)\Delta x\Delta y\Delta z}{\Delta t}+\left[\left\{\left(\rho u^*\right)_i+\frac{(P_I'-P_{I+1}')\Delta y\Delta z}{a_i}\right\}-\left\{\left(\rho u^*\right)_{i-1}+\frac{(P_{I-1}'-P_I')\Delta y\Delta z}{a_{i-1}}\right\}\right]\Delta y\Delta z$$
$$+\left[\left\{\left(\rho v^*\right)_j+\frac{(P_J'-P_{J+1}')\Delta x\Delta z}{a_j}\right\}-\left\{\left(\rho v^*\right)_{j-1}+\frac{(P_{J-1}'-P_J')\Delta x\Delta z}{a_{j-1}}\right\}\right]\Delta x\Delta z$$
$$+\left[\left\{\left(\rho w^*\right)_k+\frac{(P_K'-P_{K+1}')\Delta x\Delta y}{a_k}\right\}-\left\{\left(\rho w^*\right)_{k-1}+\frac{(P_{K-1}'-P_K')\Delta x\Delta y}{a_{k-1}}\right\}\right]\Delta x\Delta y=0 \quad (8.74)$$

In evaluating the integrals, we assumed that the fluid is not strongly influenced by the pressure, and the equation is applicable to incompressible or weak compressible flows. Rearranging Equation 8.74, we get the **pressure correction equation** as

$$a_I p_I' = a_{I+1}p_{I+1}' + a_{I-1}p_{I-1}' + a_{J+1}p_{J+1}' + a_{J-1}p_{J-1}' + a_{K+1}p_{K+1}' + a_{K-1}p_{K-1}' + d \quad (8.75)$$

where

$$a_{I+1} = \rho_i\frac{\left(\Delta y\Delta z\right)^2}{a_i} \quad (8.76a)$$

$$a_{I-1} = \rho_{i-1}\frac{\left(\Delta y\Delta z\right)^2}{a_{i-1}} \quad (8.76b)$$

$$a_{J+1} = \rho_j\frac{\left(\Delta z\Delta x\right)^2}{a_j} \quad (8.76c)$$

$$a_{J-1} = \rho_{j-1}\frac{\left(\Delta z\Delta x\right)^2}{a_{j-1}} \quad (8.76d)$$

$$a_{K+1} = \rho_k\frac{\left(\Delta x\Delta y\right)^2}{a_k} \quad (8.76e)$$

$$a_{K-1} = \rho_{k-1}\frac{\left(\Delta x\Delta y\right)^2}{a_{k-1}} \quad (8.76f)$$

$$a_I = a_{I+1} + a_{I-1} + a_{J+1} + a_{J-1} + a_{K+1} + a_{K-1} \quad (8.76g)$$

$$d=\frac{\left(\rho_I^m-\rho_I^{m+1}\right)\Delta x\Delta y\Delta z}{\Delta t}+\left[\left(\rho u^*\right)_{i-1}-\left(\rho u^*\right)_i\right]\Delta y\Delta z+\left[\left(\rho v^*\right)_{j-1}-\left(\rho v^*\right)_j\right]\Delta z\Delta x$$
$$+\left[\left(\rho w^*\right)_{k-1}-\left(\rho w^*\right)_k\right]\Delta x\Delta y \quad (8.76h)$$

It can be noted that the source term d in the pressure correction Equation 8.75 is a residual, which is obtained while the estimated starred velocity components given by Equation 8.66 are substituted into the continuity equation. So, as this mass source term, d, approaches zero or a specified tolerance limit, a solution procedure leads to correct velocity and pressure fields.

Under-relaxation The iterative solution of velocity and pressure fields may encounter instability, particularly due to the simplification made in the derivation of velocity correction Equation 8.70. This necessitates the need for under-relaxation parameters for the estimated starred velocity components in the momentum equations and pressure with respect to the previous iterated values as

$$u = u^* + \alpha_V u' \tag{8.77a}$$

$$v = v^* + \alpha_V v' \tag{8.77b}$$

$$w = w^* + \alpha_V w' \tag{8.77c}$$

and

$$p = p^* + \alpha_p p' \tag{8.78}$$

where α_V and α_p are the under-relaxation parameters for velocity and pressure, respectively. The optimum values of these relaxation parameters are problem-dependent and are usually established through numerical experimentations.

SIMPLE Algorithm

The sequence of calculation to be performed using the equations derived for velocity, pressures, and their correction values constitutes an algorithm known as **semi-implicit method for pressure-linked equation (SIMPLE)** as proposed by Patankar and Spalding (Patankar and Spalding, 1972; Patankar, 1980). The algorithm is given as follows.

1. Guess the pressure field p^*
2. Solve discretization Equations 8.66 for u^*, v^*, and ι
3. Solve pressure correction Equations 8.75 and 8.76 for p'
4. Solve for velocity corrections u', v', w' using Equation 8.70 and using pressure correction values, P'
5. Calculate corrected velocity components, u_i, v_j, and w_k, using Equation 8.71 from their starred components and using pressure correction, P'.
6. Solve the discretization equation for other scalar quantities, such as turbulence quantities, temperature, and mass concentration if they influence the flow field though dependence, through fluid properties such as density, viscosity, and source terms. If a particular scalar quantity does not affect the velocity field, then that quantity is calculated after a converged velocity field is achieved.
7. Velocity and pressure filed values can be updated using under-relaxation values as per Equations 8.77 and 8.78, respectively.
8. Reset the corrected pressure p as the new guessed pressure p^*.
9. Repeat 2–7 until the solution converges within a specified tolerance limit.

The SIMPLE algorithm also shown in a flow chart below:

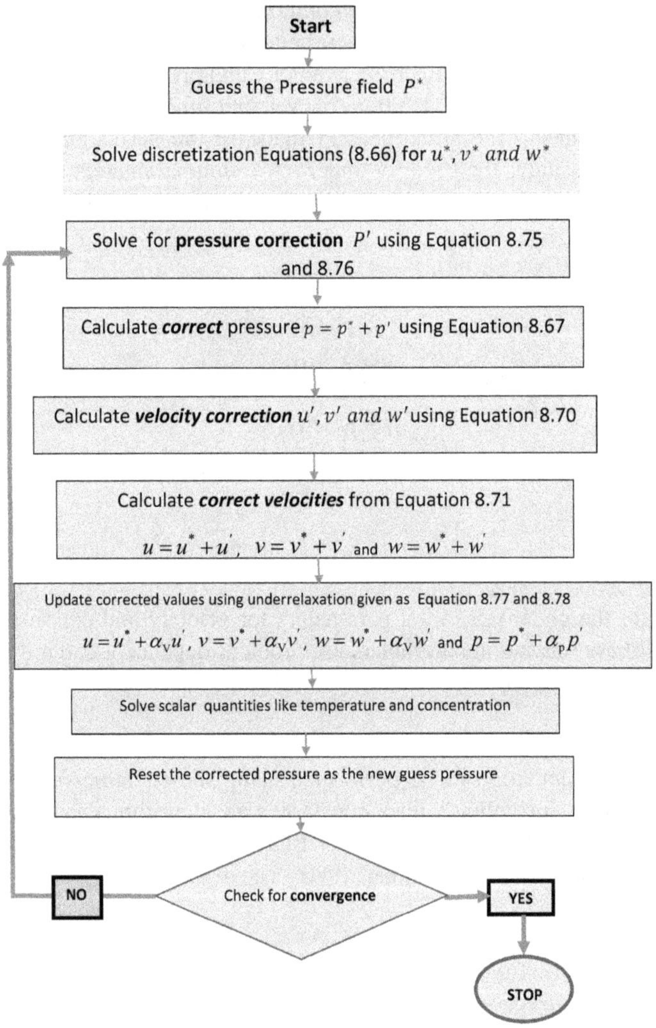
Start
Guess the Pressure field P^*
Solve discretization Equations (8.66) for u^*, v^* and w^*
Solve for pressure correction P' using Equation 8.75 and 8.76
Calculate correct pressure $p = p^* + p'$ using Equation 8.67
Calculate velocity correction u', v' and w' using Equation 8.70
Calculate correct velocities from Equation 8.71
$u = u^* + u'$, $v = v^* + v'$ and $w = w^* + w'$
Update corrected values using underrelaxation given as Equation 8.77 and 8.78
$u = u^* + \alpha_V u'$, $v = v^* + \alpha_V v'$, $w = w^* + \alpha_V w'$ and $p = p^* + \alpha_p p'$
Solve scalar quantities like temperature and concentration
Reset the corrected pressure as the new guess pressure
NO
Check for convergence
YES
STOP

Example 8.1

Consider a one-dimensional flow in a straight nozzle shown in the figure. The governing equations are given as

Continuity:

$$\frac{\partial(uA)}{\partial x} = 0$$

***x*-momentum:**

$$\frac{\partial(\rho uu)}{\partial x} = -\frac{\partial p}{\partial x}$$

where A is the cross-sectional area of the nozzle. Consider the data

$$\rho = 1.00\ \text{kg/m}^3,\quad A_1 = 5\ \text{m}^2,\quad A_2 = 4\ \text{m}^2,\quad A_3 = 3\ \text{m}^2$$

$$p_1 = 300\ \text{kPa},\quad p_4 = 100\ \text{kPa and } \alpha_v = 0.5$$

Show discretization equations for velocity and pressure and use the **SIMPLE** algorithm to compute velocity and pressure at the nodes for one iteration. Use initial guess for mass flow rate as $\dot{m} = \rho u A = 100$ kg/s.

Solution

For the initial guess value of the mass flow rate, we can compute the initial guess values for the u-component velocities at the staggered grid points as

$$u_1 = \frac{100}{5} = 20 \text{ m/s}, \quad u_2 = \frac{100}{4} = 25\text{m/s} \quad \text{and} \quad u_3 = \frac{100}{3} = 33.333 \text{ m/s}$$

The **SIMPLE** algorithm can be used with following steps:

1. Guess the pressure field p^* at regular grid points as

$$p_1^* = 300 \text{ kPa}, \quad p_2^* = 250 \text{ kPa}, \quad p_3^* = 200 \text{ kPa}, \quad p_4^* = 100 \text{ kPa}$$

2. Discretization equations for u^* are obtained from Equation 8.66 for a one-dimensional problem as

$$a_i u_i^* = a_{i+1}u_{i+1}^* + a_{i-1}u_{i-1}^* + d + \left(p_I^* - p_{I+1}^*\right)A_i$$

$$a_i = a_{i+1} + a_{i-1}$$

$$a_{i+1} = \frac{\mu}{\Delta x_{i+\frac{1}{2}}} A\left|P_{i+\frac{1}{2}}\right| + \left\langle 0, -(\rho u)_{i+\frac{1}{2}} \right\rangle$$

$$a_{i-1} = \frac{\mu}{\Delta x_{i-\frac{1}{2}}} A\left|P_{i-\frac{1}{2}}\right| + \left\langle 0, (\rho u)_{i-\frac{1}{2}} \right\rangle$$

$$P = \frac{(\rho u)}{\mu/\Delta x}, \quad d = 0$$

For steady-state and for zero-gravity force as in this problem, and with $\mu = 0$, the discretization equation for the u-component velocity reduces to

$$a_i u_i^* = a_{i+1}u_{i+1}^* + a_{i-1}u_{i-1}^* + \left(p_{I-1}^* - p_I^*\right)A_i$$

$$a_i = a_{i+1} + a_{i-1}$$

$$a_{i+1} = \left\langle 0, -(\rho u)_{i+\frac{1}{2}} \right\rangle$$

$$a_{i-1} = \left\langle 0, (\rho u)_{i-\frac{1}{2}} \right\rangle$$

Applying this equation successively to nodes 1–3 in a staggered system, we obtain the system of equations for the u component velocities as
Node 1

$$a_1 u_1^* = a_2 u_2^* + a_0 u_0^* + \left(p_1^* - p_2^*\right)A_1$$

$$a_2 = \left\langle 0, -(\rho u)_{i+\frac{1}{2}} \right\rangle = \left\langle 0, -(\rho u)_{1+\frac{1}{2}} \right\rangle = 0$$

$$a_0 = \left\langle 0, (\rho u)_{i-\frac{1}{2}} \right\rangle = \left\langle 0, (\rho u)_{\frac{1}{2}} \right\rangle = \frac{0 + u_1}{2} = \frac{20}{2} = 10$$

$$a_1 = a_2 + a_0 = 0 + 10 = 10$$

$$10u_1^* = 0u_2^* + 10 \times 0 + (300 - 250) \times 5$$

$$u_1^* = 25 \text{ m/s}$$

Node 2

$$a_2 u_2^* = a_3 u_3^* + a_1 u_1^* + \left(p_2^* - p_3^* \right) A_2$$

$$a_i = a_{i+1} + a_{i-1}$$

$$a_3 = \left\langle 0, -(\rho u)_{i+\frac{1}{2}} \right\rangle = \left\langle 0, -(\rho u)_{2+\frac{1}{2}} \right\rangle = 0$$

$$a_1 = \left\langle 0, (\rho u)_{1+\frac{1}{2}} \right\rangle = \frac{u_1 + u_2}{2} = \frac{20 + 25}{2} = 22.5$$

$$a_2 = a_3 + a_1 = 22.5$$

$$22.5u_2^* = 0u_3^* + 22.5 \times 25 + (250 - 200) \times 4$$

$$u_2^* = 33.888 \text{ m/s}$$

Node 3

$$a_3 u_3^* = a_2 u_2^* + a_4 u_4^* + \left(p_3^* - p_4^* \right) A_3$$

$$a_4 = \left\langle 0, -(\rho u)_{3+\frac{1}{2}} \right\rangle = 0$$

$$a_2 = \left\langle 0, (\rho u)_{3-\frac{1}{2}} \right\rangle = \frac{u_2 + u_3}{2} = \frac{25 + 33.333}{2} = 29.166$$

$$a_3 = 0 + 29.166 = 29.166$$

$$29.166u_3^* = 29.166 \times 33.888 + 0u_4^* + (200 - 100) \times 3$$

$$u_3^* = 44.174 \text{ m/s}$$

3. Solve for pressure correction Equation 8.75 for p'

$$a_I p_I' = a_{I+1} p_{I+1}' + a_{I-1} p_{I-1}' + d$$

where

$$a_{I+1} = \rho_{I+\frac{1}{2}} \frac{\left(A_{I+\frac{1}{2}} \right)^2}{a_{I+\frac{1}{2}}}$$

$$a_{I-1} = \rho_{I-\frac{1}{2}} \frac{\left(A_{I-\frac{1}{2}} \right)^2}{a_{I-\frac{1}{2}}}$$

$$a_I = a_{I+1} + a_{I-1}$$

$$d = \left[\left(\rho u^* \right)_{I-\frac{1}{2}} - \left(\rho u^* \right)_{I+\frac{1}{2}} \right] A_I$$

Applying this discretization equation to pressure grid points, we have
For Node 2

$$a_{p2}p_2' = a_{p3}p_3' + a_{p1}p_1' + \left[\left(\rho u_1^*\right)-\left(\rho u_2^*\right)\right]\frac{A_1 + A_2}{2}$$

where

$$a_{p3} = a_{pI+1} = \rho_{I+\frac{1}{2}}\frac{\left(A_{I+\frac{1}{2}}\right)^2}{a_{I+\frac{1}{2}}} = \frac{(A_2)^2}{u_2} = \frac{(4)^2}{25} = 0.64$$

$$a_{p1} = a_{pI-1} = \rho_{I-\frac{1}{2}}\frac{\left(A_{I-\frac{1}{2}}\right)^2}{a_{I-\frac{1}{2}}} = \frac{(A_1)^2}{u_1} = \frac{(5)^2}{20} = 1.25$$

$$a_{p2} = a_{p3} + a_{p1} = 0.64 + 1.25 = 1.89 \quad \text{(E.8.1.1)}$$

$$1.89p_2' = 0.64p_3' + 1.25p_1' \left[(1\times 25)-(1\times 33.88)\right]\frac{5+4}{2}$$

$$1.89p_2' = 0.64p_3' + 0 - 39.96$$

For Node 3

$$a_{p3}p_3' = a_{p4}p_4' + a_{p2}p_2' + d$$

where

$$a_{p4} = a_{pI+1} = \rho_{I+\frac{1}{2}}\frac{\left(A_{I+\frac{1}{2}}\right)^2}{a_{I+\frac{1}{2}}} = \frac{(A_3)^2}{u_3} = \frac{(3)^2}{33.333} = 0.270$$

$$a_{p2} = \rho_{I-\frac{1}{2}}\frac{\left(A_{I-\frac{1}{2}}\right)^2}{a_{I-\frac{1}{2}}} = \frac{(A_2)^2}{u_2} = \frac{(4)^2}{25} = 0.64$$

$$a_{p3} = a_{p2} + a_{p4} = 0.270 + 0.64 = 0.91 \quad \text{(E.8.1.2)}$$

$$a_{p3}p_3' = a_{p4}p_4' + a_{p2}p_2' + \left[\left(\rho u_2^*\right)-\left(\rho u_3^*\right)\right]\frac{A_2 + A_3}{2}$$

$$0.91p_2' = 0.270\times 0 + 0.64p_2' + \left[(1\times 33.888)-(1\times 44.174)\right]\frac{4+3}{2}$$

$$0.91p_3' = 0.64p_2' - 36.001$$

Solving Equations E.8.1.1 and E.8.1.2, we get

$$p_2' = -45.33290\ \text{kPa} \quad \text{and} \quad p_3' = -71.44397\ \text{kPa}$$

Calculate corrected pressure from Equation 8.75 with under-relaxation

$$p_2 = p_2^* + 0.5p_2' = 250 + 0.5(-45.332)$$

$$p_2 = 227.334\ \text{kPa}$$

and

$$p_3 = p_3^* + 0.5p_3' = 200 + 0.5(-71.44397)$$

$$p_3 = 164.278 \text{ kPa}$$

Calculate velocity components *u*, *v*, and *w* from their starred and correction components using Equation 8.71 or 8.77 with under-relaxation

$$u = u^* + \alpha_V u'$$

or

$$u_i = u_i^* + \alpha_V \left(p_I' - p_{I+1}'\right) \frac{\Delta y \Delta z}{a_i}$$

Node 1

$$u_1 = u_1^* + \alpha_V \left(p_1' - p_2'\right) \frac{A_1}{a_1}$$

$$= 25 + 0.5 \times \left(0 - (-45.3329)\right) \frac{5}{10}$$

$$u_1 = 36.3332 \text{ m/s}$$

Node 2

$$u_2 = u_2^* + \alpha_V \left(p_2' - p_3'\right) \frac{A_2}{a_2}$$

$$= 33.888 + 0.5 \times (-45.3329 + 71.44397) \frac{4}{22.5}$$

$$u_2 = 40.238 \text{ m/s}$$

Node 3

$$u_3 = u_3^* + \alpha_V \left(p_3' - p_4'\right) \frac{A_3}{a_3}$$

$$= 44.174 + 0.5(-71.44397 - 0) \frac{3}{29.166}$$

$$u_3 = 40.4999 \text{ m/s}$$

Reset the corrected pressure *p* as the new guessed pressure p^*

$$p_2^* = p_2 = 227.334 \text{ kPa}, \qquad p_3^* = p_3 = 164.278 \text{ kPa}$$

This concludes the first iteration.

Iteration 2 *u*-component velocities

Node 1

$$a_1 u_1^* = a_2 u_2^* + a_0 u_0^* + \left(p_1^* - p_2^*\right) A_1$$

$$a_2 = \left\langle 0, -(\rho u)_{i+\frac{1}{2}} \right\rangle = \left\langle 0, -(\rho u)_{1+\frac{1}{2}} \right\rangle = 0$$

$$a_0 = \left\langle 0, (\rho u)_{i-\frac{1}{2}} \right\rangle = \left\langle 0, (\rho u)_{\frac{1}{2}} \right\rangle = \frac{0+u_1}{2} = \frac{36.332}{2} = 18.166$$

$$a_1 = a_2 + a_0 = 0 + 18.166 = 18.166$$

$$18.166u_1^* = 0u_2^* + 18.166 \times 0 + (300 - 227.334) \times 5$$

$$u_1^* = 20.00 \text{ m/s}$$

Node 2

$$a_2 u_2^* = a_3 u_3^* + a_1 u_1^* + \left(p_2^* - p_3^* \right) A_2$$

$$a_i = a_{i+1} + a_{i-1}$$

$$a_3 = \left\langle 0, -(\rho u)_{i+\frac{1}{2}} \right\rangle = \left\langle 0, -(\rho u)_{2+\frac{1}{2}} \right\rangle = 0$$

$$a_1 = \left\langle 0, (\rho u)_{1.+\frac{1}{2}} \right\rangle = \frac{u_1 + u_2}{2} = \frac{36.3332 + 40.238}{2} = 38.2856$$

$$a_2 = a_3 + a_1 = 38.2856$$

$$38.2856u_2^* = 0u_3^* + 38.2856 \times 20.0 + (227.334 - 164.278) \times 4$$

$$= 765.712 + 252.224$$

$$u_2^* = 26.5879 \text{ m/s}$$

Node 3

$$a_3 u_3^* = a_2 u_2^* + a_4 u_4^* + \left(p_3^* - p_4^* \right) A_3$$

$$a_4 = \left\langle 0, -(\rho u)_{3.+\frac{1}{2}} \right\rangle = 0$$

$$a_2 = \left\langle 0, (\rho u)_{3-\frac{1}{2}} \right\rangle = \frac{u_2 + u_3}{2} = \frac{40.238 + 40.4999}{2} = 40.36895$$

$$a_3 = 0 + 40.3685 = 40.36895$$

$$40.36895u_3^* = 40.36895 \times 26.5879 + 0u_4^* + (164.278 - 100) \times 3$$

$$= 1073.3256 + 192.834$$

$$u_3^* = 31.3647 \text{ m/s}$$

Solve for pressure correction Equation 8.61 for p'

$$a_I p_I' = a_{I+1} p_{I+1}' + a_{I-1} p_{I-1}' + d$$

where

$$a_{I+1} = \rho_{I+\frac{1}{2}} \frac{\left(A_{I+\frac{1}{2}} \right)^2}{a_{I+\frac{1}{2}}}$$

$$a_{I-1} = \rho_{I-\frac{1}{2}} \frac{\left(A_{I-\frac{1}{2}}\right)^2}{a_{I-\frac{1}{2}}}$$

$$a_I = a_{I+1} + a_{I-1}$$

$$d = \left[\left(\rho u^*\right)_{I-\frac{1}{2}} - \left(\rho u^*\right)_{I+\frac{1}{2}}\right] A_I$$

Applying this discretization equation to pressure grid points, we have
For Node 2

$$a_{p2}p_2' = a_{p3}p_3' + a_{p1}p_1' + \left[\left(\rho u_1^*\right) - \left(\rho u_2^*\right)\right]\frac{A_1 + A_2}{2}$$

$$a_{p3} = a_{pI+1} = \rho_{I+\frac{1}{2}} \frac{\left(A_{I+\frac{1}{2}}\right)^2}{a_{I+\frac{1}{2}}} = \frac{(A_2)^2}{u_2} = \frac{(4)^2}{40.238} = 0.3976$$

$$a_{p1} = a_{pI-1} = \rho_{I-\frac{1}{2}} \frac{\left(A_{I-\frac{1}{2}}\right)^2}{a_{I-\frac{1}{2}}} = \frac{(A_1)^2}{u_1} = \frac{(5)^2}{36.3332} = 0.6880 \quad \text{(E.8.1.3)}$$

$$a_{p2} = a_{p3} + a_{p1} = 0.3976 + 0.6880 = 1.0856$$

$$1.0856p_2' = 0.3976p_3' + 0.6880p_1' + \left[(1\times 20) - (1\times 26.5879)\right]\frac{5+4}{2}$$

$$1.0856p_2' = 0.3976p_3' + 0 - 29.645$$

For Node 3

$$a_{p3}p_3' = a_{p4}p_4' + a_{p2}p_2' + d$$

where

$$a_{p4} = a_{pI+1} = \rho_{I+\frac{1}{2}} \frac{\left(A_{I+\frac{1}{2}}\right)^2}{a_{I+\frac{1}{2}}} = \frac{(A_3)^2}{u_3} = \frac{(3)^2}{40.238} = 0.22366$$

$$a_{p2} = \rho_{I-\frac{1}{2}} \frac{\left(A_{I-\frac{1}{2}}\right)^2}{a_{I-\frac{1}{2}}} = \frac{(A_2)^2}{u_2} = \frac{(4)^2}{40.238} = 0.3976$$

$$a_{p3} = a_{p2} + a_{p4} = 0.22366 + 0.3976 = 0.62126$$

$$a_{p3}p_3' = a_{p4}p_4' + a_{p2}p_2' + \left[\left(\rho u_2^*\right) - \left(\rho u_3^*\right)\right]\frac{A_2 + A_3}{2} \quad \text{(E.8.1.4)}$$

$$0.62126p_3' = 0.22366 \times 0 + 0.3976p_2' + \left[(1\times 26.5879) - (1\times 31.3647)\right]\frac{4+3}{2}$$

$$0.62126p_3' = 0.3976p_2' - 16.7188$$

Solving Equations E.8.1.3 and E.8.1.4, we get

$$P_2' = -48.539 \text{ kPa} \quad \text{and} \quad P_3' = -57.975 \text{ kPa}$$

Calculate corrected pressure from Equation 8.50 or 8.61 with under-relaxation.

$$p_2 = p_2^* + 0.5p_2' = 227.334 + 0.5(-48.539)$$

$$p_2 = 203.06 \text{ kPa}$$

and

$$p_3 = p_3^* + 0.5p_3' = 164.278 + 0.5(-57.975)$$

$$p_3 = 135.290 \text{ kPa}$$

Calculate velocity components *u*, *v*, and *w* from their starred and correction components using Equation 8.71 or 8.77 with under-relaxation

$$u = u^* + \alpha_V u'$$

or

$$u_i = u_i^* + \alpha_V \left(p_I' - p_{I+1}'\right)\frac{\Delta y \Delta z}{a_i}$$

Node 1

$$u_1 = u_1^* + \alpha_V \left(p_1' - p_2'\right)\frac{A_1}{a_1} = 18.166$$

$$= 20 + 0.5 \times \left(0 - (-48.539)\right)\frac{5}{18.166}$$

$$u_1 = 26.6799 \text{ m/s}$$

Node 2

$$u_2 = u_2^* + \alpha_V \left(p_2' - p_3'\right)\frac{A_2}{a_2} \qquad u_2^* = 26.5879 P_3' = -57.975$$

$$= 26.5879 + 0.5 \times (-48.539 + 57.975)\frac{4}{38.2856}$$

$$u_2 = 27.0808 \text{ m/s}$$

Node 3

$$u_3 = u_3^* + \alpha_V \left(p_3' - p_4'\right)\frac{A_3}{a_3} a_3 = 0 + 40.36895 = 40.36895$$

$$= 31.3647 + 0.5(-57.975 - 0)\frac{3}{40.36895}$$

$$u_3 = 29.210 \text{ m/s}$$

Reset the corrected pressure *p* as the new guessed pressure p^*

$$p_2^* = p_2 = 203.06 \text{ kPa}, \qquad p_3^* = p_3 = 135.290 \text{ kPa}$$

A summary of the two iteration results is given in the table below. The procedure can be repeated until convergence.

Iteration Number	u_1	u_2	u_3	p_2	p_3
0	20	25	33.333	250	200
1	36.3332	40.238	40.4999	227.334	164.278
2	26.6799	27.0808	29.210	203.06	135.290

SIMPLER Algorithm

The **SIMPLE** algorithm has been used very effectively for many fluid flow and heat transfer problems. However, it showed some difficulties in terms of instability, slow rate of convergence, and poor prediction of the pressure fields. This is primarily due to the simplification made in the derivation of velocity correction Equation 8.53 by dropping the velocity contributions from the neighboring nodes, i.e., $\sum a_{nb}u'_{nb}$ term, and using only the pressure correction equation to correct the velocity field and pressure field. Additional iterations are usually needed to obtain the correct pressure field even if the correct velocity field is achieved. To overcome these difficulties, some modification is made to the **SIMPLE** algorithm and a new algorithm named **SIMPLER** (**SIMPLE Revised**) is developed (Patankar, 1980, 1988b).

In this algorithm, the solution procedure starts with a guessed velocity field, and a separate pressure equation is derived for the calculation of the pressure field in addition to the pressure correction equation. Let us discuss here the development of this pressure equation and the basic steps of this algorithm.

The Pressure Equation The discretization equation for the velocity fields given by Equations 8.60b, 8.62b, and 8.64b is rewritten as

$$u_i = \frac{\sum a_{nb}u_{nb} + d}{a_i} + \left(p_{I-1} - p_I\right)\frac{\Delta y \Delta z}{a_i} \tag{8.79a}$$

$$v_j = \frac{\sum a_{nb}v_{nb} + d}{a_j} + \left(p_{J-1} - p_J\right)\frac{\Delta x \Delta z}{a_j} \tag{8.79b}$$

$$w_k = \frac{\sum a_{nb}w_{nb} + d}{a_k} + \left(p_{K-1} - p_K\right)\frac{\Delta x \Delta y}{a_k} \tag{8.79c}$$

The first term on the right-hand side of Equation 8.79 is defined as a new pseudo-velocity field in the following manner:

$$\hat{u}_i = \frac{\sum a_{nb}u_{nb} + d}{a_i} \tag{8.80a}$$

$$\hat{v}_j = \frac{\sum a_{nb}v_{nb} + d}{a_j} \tag{8.80b}$$

$$\hat{w}_k = \frac{\sum a_{nb}w_{nb} + d}{a_k} \tag{8.80c}$$

Substituting Equation 8.80 into Equation 8.79, we get

$$u_i = \hat{u}_i + \left(p_{I-1} - p_I\right)\frac{\Delta y \Delta z}{a_i} \tag{8.81a}$$

$$v_j = \hat{v}_j + \left(p_{J-1} - p_J\right)\frac{\Delta x \Delta z}{a_j} \tag{8.81b}$$

$$w_k = \hat{w}_k + \left(p_{K-1} - p_K\right)\frac{\Delta x \Delta y}{a_k} \tag{8.81c}$$

In the next step, the discretization equation for the pressure field can be derived in the same manner as for the pressure correction equation discussed earlier. The only difference is that the pressure equation is derived by substituting the estimated velocity components given by Equation 8.81 into the integral form of the continuity equation. To demonstrate this, let us start from integral form of the continuity equation given by Equation 8.63

$$\rho_I\Big|_t^{t+\Delta t}\Delta x\Delta y\Delta z + \left\{(\rho u)_i - (\rho u)_{i-1}\right\}\Delta y\Delta z\Delta t + \left\{(\rho v)_j - (\rho v)_{j-1}\right\}\Delta x\Delta z\Delta t$$
$$+\left\{(\rho w)_k - (\rho w)_{k-1}\right\}\Delta x\Delta y\Delta t \tag{8.63}$$

Using the **implicit time integration scheme** and substituting the expressions for the velocity field given by Equation 8.81, we get

$$\frac{\left(\rho_I - \rho_I^0\right)\Delta x\Delta y\Delta z}{\Delta t} + \left[\left\{(\rho\hat{u})_i + \frac{\rho_i(P_I - P_{I+1})\Delta y\Delta z}{a_i}\right\} - \left\{(\rho\hat{u})_{i-1} + \frac{\rho_{i-1}(P_{I-1} - P_I)\Delta y\Delta z}{a_{i-1}}\right\}\right]\Delta y\Delta z$$
$$+\left[\left\{(\rho\hat{v})_j + \frac{\rho_j(P_J - P_{J+1})\Delta x\Delta z}{a_j}\right\} - \left\{(\rho\hat{v})_{j-1} + \frac{\rho_{j-1}(P_{J-1} - P_J)\Delta x\Delta z}{a_{j-1}}\right\}\right]\Delta x\Delta z$$
$$+\left[\left\{(\rho\hat{w})_k + \frac{\rho_k(P_K - P_{K+1})\Delta x\Delta y}{a_k}\right\} - \left\{(\rho\hat{w})_{k-1} + \frac{\rho_{k-1}(P_{K-1} - P_K)\Delta x\Delta y}{a_{k-1}}\right\}\right]\Delta x\Delta y = 0 \tag{8.82}$$

Rearranging Equation 8.82, we get the **pressure equation** as

$$a_I p_I = a_{I+1}p_{I+1} + a_{I-1}p_{I-1} + a_{J+1}p_{J+1} + a_{J-1}p_{J-1} + a_{K+1}p_{K+1} + a_{K-1}p_{K-1} + d \tag{8.83}$$

where

$$a_{I+1} = \rho_i\frac{\left(\Delta y\Delta z\right)^2}{a_i} \tag{8.84a}$$

$$a_{I-1} = \rho_{i-1}\frac{\left(\Delta y\Delta z\right)^2}{a_{i-1}} \tag{8.84b}$$

$$a_{J+1} = \rho_j\frac{\left(\Delta z\Delta x\right)^2}{a_j} \tag{8.84c}$$

$$a_{J-1} = \rho_{j-1}\frac{(\Delta z\Delta x)^2}{a_{j-1}} \tag{8.84d}$$

$$a_{K+1} = \rho_k\frac{(\Delta x\Delta y)^2}{a_k} \tag{8.84e}$$

$$a_{K-1} = \rho_{k-1}\frac{(\Delta x\Delta y)^2}{a_{k-1}} \tag{8.84f}$$

$$a_I = a_{I+1} + a_{I-1} + a_{J+1} + a_{J-1} + a_{K+1} + a_{K-1} \tag{8.84g}$$

$$d = \frac{(\rho_I^0 - \rho_I)\Delta x\Delta y\Delta z}{\Delta t} + \left[(\rho\hat{u})_{i-1} - (\rho\hat{u})_i\right]\Delta y\Delta z + \left[(\rho\hat{v})_{j-1} - (\rho\hat{v})_j\right]\Delta z\Delta x + \left[(\rho\hat{w})_{k-1} - (\rho\hat{w})_k\right]\Delta x\Delta y \tag{8.84h}$$

It can be noted that the pressure equation given by Equations 8.83 and 8.84 is identical to that given for the pressure correction equations given by Equations 8.75 and 8.76, with the exception of the mass source term (d), which is calculated in terms of the velocity components $\hat{u}$, $\hat{v}$, and $\hat{w}$ instead of the starred velocity components. Also, as we approach a correct velocity field through iteration, the pressure field is also converged to a correct value.

The algorithm is given as follows.

1. Guess the velocity field u, v, and w.
2. Solve Equation 8.80 for $\hat{u}$, $\hat{v}$, and $\hat{v}$ using the guess velocity field for the surrounding nodes.
3. Solve the pressure Equations 8.83 and 8.84 for the pressure field, p.
4. Treat this pressure field as the pressure field p^* and solve discretization Equations 8.66 for u^*, v^*, and w^*.
5. Calculate the mass source term d given by Equation 8.76h and solve for pressure correction Equation 8.75 for p'.
6. Calculate correct velocity components u, v, and w from their starred components and pressure correction using Equation 8.71.
7. Solve the discretization equations for other scalar quantities such as turbulence quantities, temperature, and mass concentration.
8. Repeat steps 2–7 with the new velocity estimates until the solution converges within a specified tolerance limit.

The **SIMPLER** algorithm can achieve the correct velocity and pressure fields at the same time without requiring additional iterations for the pressure fields. However, it involves additional computational steps compared to that in the **SIMPLE** algorithm.

Example 8.2: Solution using SIMPLER

Consider a developing flow and heat transfer in a rectangular channel shown in the figure.

The flow enters the channel with uniform axial velocity w_0, pressure $\bar{p}_0$, and temperature T_0. The top surface is maintained at constant temperature T_0 and the other three surfaces are adiabatic.

The procedure for solving a three-dimensional developing fluid flow and heat transfer in a channel was given by Patankar and Spalding (1972).

The three-dimensional elliptic nature of the differential equations is transformed into parabolic equations by making the following assumptions: (a) assume that the flow is predominant in a positive flow direction (z direction) and there is no reverse flow in that direction; (b) assume negligible diffusion of heat and momentum in the flow direction (i.e., negligible shear stress and heat diffusion flux on the xy plane). Because of this parabolic nature of the problem, the equations can be solved with marching integration from the upstream side to the downstream side. With these assumptions, associated governing equations and boundary conditions for the parabolic three-dimensional developing flow and heat transfer are given as follows.

Continuity:

$$\frac{\partial u}{\partial x}+\frac{\partial v}{\partial y}+\frac{\partial w}{\partial z}=0 \tag{E.8.2.1a}$$

***x*-momentum:**

$$u\frac{\partial u}{\partial x}+v\frac{\partial u}{\partial y}+w\frac{\partial u}{\partial z}=-\frac{1}{\rho}\frac{\partial p}{\partial x}+\nu\left(\frac{\partial^2 u}{\partial x^2}+\frac{\partial^2 u}{\partial y^2}\right) \tag{E.8.2.1b}$$

***y*–momentum:**

$$u\frac{\partial v}{\partial x}+v\frac{\partial v}{\partial y}+w\frac{\partial v}{\partial z}=-\frac{1}{\rho}\frac{\partial p}{\partial y}+\nu\left(\frac{\partial^2 v}{\partial x^2}+\frac{\partial^2 v}{\partial y^2}\right) \tag{E.8.2.1c}$$

***z*-momentum:**

$$u\frac{\partial w}{\partial x}+v\frac{\partial w}{\partial y}+w\frac{\partial w}{\partial z}=-\frac{1}{\rho}\frac{\partial \bar{p}}{\partial z}+\nu\left(\frac{\partial^2 w}{\partial x^2}+\frac{\partial^2 w}{\partial y^2}\right) \tag{E.8.2.1d}$$

Energy

$$u\frac{\partial T}{\partial x}+v\frac{\partial T}{\partial y}+w\frac{\partial T}{\partial z}=\frac{k}{\rho c_p}\left(\frac{\partial^2 T}{\partial x^2}+\frac{\partial^2 T}{\partial y^2}\right) \tag{E.8.2.1e}$$

Another important modification is made by decoupling the axial and lateral pressure gradients terms. The pressure $\bar{p}$ in the z-momentum equation represents an area average pressure at any cross-section (xy plane). The pressure gradient $d\bar{p}/dz$ is assumed to be known or calculated first, and then, we proceed to solve x- and y-momentum equations and lateral pressure gradients $\partial p/\partial x$ and $\partial p/\partial y$.

Boundary Conditions

- **Inlet**
  - Uniform velocity and temperature at inlet, i.e., at $z = 0$, $w = w_0$, $T = T_0$.
- **Wall**
  - No-slip velocity on all walls, i.e., $u = 0$, $v = 0$, and $w = 0$ on all walls.
  - Constant surface temperature on top wall, i.e., at $y = 0$, $T = T_w$ and rest of the walls are adiabatic.

Assume dimensionless variables as

$$u=\frac{u}{w_0},\quad v=\frac{v}{w_0},\quad w=\frac{w}{w_0},\quad p=\frac{p}{\rho w_0^2},\quad \theta=\frac{T-T_w}{T_b-T_w} \tag{E.8.2.2}$$

$$x=\frac{x}{H},\quad y=\frac{y}{H},\quad z=\frac{z}{H}$$

Equations E.8.1–E.8.5 can be transformed into dimensionless equations as

$$\frac{\partial u}{\partial x}+\frac{\partial v}{\partial y}+\frac{\partial w}{\partial z}=0 \tag{E.8.2.3a}$$

$$u\frac{\partial u}{\partial x}+v\frac{\partial u}{\partial y}+w\frac{\partial u}{\partial z}=-\frac{\partial p}{\partial x}+\frac{1}{\mathrm{Re}_H}\left(\frac{\partial^2 u}{\partial x^2}+\frac{\partial^2 u}{\partial y^2}\right) \tag{E.8.2.3b}$$

$$u\frac{\partial v}{\partial x}+v\frac{\partial v}{\partial y}+w\frac{\partial v}{\partial z}=-\frac{\partial p}{\partial y}+\frac{1}{\mathrm{Re}_H}\left(\frac{\partial^2 v}{\partial x^2}+\frac{\partial^2 v}{\partial y^2}\right) \tag{E.8.2.3c}$$

$$u\frac{\partial w}{\partial x}+v\frac{\partial w}{\partial y}+w\frac{\partial w}{\partial z}=\frac{\partial p}{\partial z}+\frac{1}{\mathrm{Re}_H}\left(\frac{\partial^2 w}{\partial x^2}+\frac{\partial^2 w}{\partial y^2}\right) \tag{E.8.2.3d}$$

$$u\frac{\partial \theta}{\partial x}+v\frac{\partial \theta}{\partial y}+w\frac{\partial \theta}{\partial z}+w\theta\lambda=\frac{1}{\mathrm{Re}_H\,\mathrm{Pr}}\left(\frac{\partial^2 \theta}{\partial x^2}+\frac{\partial^2 \theta}{\partial y^2}\right) \tag{E.8.2.3e}$$

where $\mathrm{Re}_H=\dfrac{w_0 H}{v}, \mathrm{Pr}=\dfrac{y}{\alpha}$, and $\alpha=\dfrac{k}{\rho_{\mathrm{cp}}}$

Derive the discretization equations for *u*, *v*, *w*, and *p* and present the sequence of steps to be used for calculating velocity and temperature field using the **SIMPLER** algorithm.

Solution

Let us first consider the discretization of Equation E.8.2.3b by rearranging the equation as

$$\left(u\frac{\partial u}{\partial x}-\Gamma\frac{\partial^2 u}{\partial x^2}\right)+\left(v\frac{\partial u}{\partial y}-\Gamma\frac{\partial^2 u}{\partial y^2}\right)=-\frac{\partial \bar{P}}{\partial x}-w\frac{\partial u}{\partial z} \tag{E.8.2.4}$$

Equation E.8.2.4 can be written in a general form as

$$\frac{\partial J_x}{\partial x}+\frac{\partial J_y}{\partial y}=S \tag{E.8.2.5}$$

where

$$J_x=u^2-\Gamma\frac{\partial u}{\partial x},\quad J_y=vu-\Gamma\frac{\partial u}{\partial y}$$

$$S=-\frac{\mathrm{d}p}{\mathrm{d}x}-w\frac{\partial u}{\partial z},\quad \Gamma=\frac{1}{\mathrm{Re}_H}$$

Using Equation 8.60a we get the discretization equation for Equation E.8.2.3b as

$$ua_i u_i=ua_{i+1}u_{i+1}+ua_{i-1}u_{i-1}+ua_{j+1}u_{j+1}+ua_{j-1}u_{j-1}-\left(\frac{p_{I+1,J}-p_{I,J}}{\rho\Delta x}\right)\Delta y\Delta x-w_i^u\frac{u_i-u_i^u}{\Delta z}\Delta x\Delta y$$

or

$$\left(ua_i u_i+\frac{w_i^u}{\Delta z}\Delta x\Delta y\right)u_i=ua_{i+1}u_{i+1}+ua_{i-1}u_{i-1}+ua_{j+1}u_{j+1}+ua_{j-1}u_{j-1}-\left(\frac{p_{I+1,J}-p_{I,J}}{\rho}\right)\Delta y+w_i^u\frac{u_i^u}{\Delta z}\Delta x\Delta y \tag{E.8.2.6a}$$

or

$$ua_i u_i = \sum ua_{nb} u_{nb}^{m+1} + d \tag{E.8.2.6b}$$

where

$$ua_i = ua_{i+1} + ua_{i-1} + ua_{j+1} + ua_{j-1} + \frac{w_i^u}{\Delta z}\Delta x \Delta y$$

$$ua_{i+1} = \frac{\Gamma}{\Delta x} A\left|P_{i+\frac{1}{2}}\right| + \left\langle 0, -(\rho u)_{i+\frac{1}{2}}\right\rangle, \qquad ua_{i-1} = \frac{\Gamma}{\Delta x} A\left|P_{i-\frac{1}{2}}\right| + \left\langle 0, (\rho u)_{i-\frac{1}{2}}\right\rangle$$

$$ua_{j+1} = \frac{\Gamma}{\Delta y} A\left|P_{j+\frac{1}{2}}\right| + \left\langle 0, -(\rho v)_{j+\frac{1}{2}}\right\rangle, \qquad ua_{j-1} = \frac{\Gamma}{\Delta y_{i-\frac{1}{2}}} A\left|P_{j-\frac{1}{2}}\right| + \left\langle 0, (\rho v)_{j-\frac{1}{2}}\right\rangle$$

$$d = -\left(\frac{p_{I+1,J} - p_{,I,J}}{\rho}\right)\Delta y + w_i^u \frac{u_i^u}{\Delta z}\Delta x \Delta y$$

$$P_i = \frac{(\rho u)}{\Gamma/\Delta x}, \qquad P_j = \frac{(\rho v)}{\Gamma/\Delta y}$$

In the above discretization equation, values u_i^u and w_i^u with a superscript u indicate values at the upstream position as shown in the grid shown in the figure.

Similarly, Equation E.8.2.3c for the y-momentum equation can be discretized as

$$\left(u\frac{\partial v}{\partial x} - \Gamma\frac{\partial^2 v}{\partial x^2}\right) + \left(v\frac{\partial v}{\partial y} - \Gamma\frac{\partial^2 v}{\partial y^2}\right) = -\frac{1}{\rho}\frac{\partial p}{\partial y} - w\frac{\partial v}{\partial z}$$

Equation E.8.2.5 can be written in a general form as

$$\frac{\partial J_x}{\partial x} + \frac{\partial J_y}{\partial y} = S \tag{E.8.2.7}$$

where

$$J_x = uv - \Gamma\frac{\partial v}{\partial x}, \qquad J_y = v^2 - \Gamma\frac{\partial v}{\partial y}$$

$$S = -\frac{1}{\rho}\frac{\mathrm{d}p}{\mathrm{d}y} - w\frac{\partial v}{\partial z}, \qquad \Gamma = \frac{1}{\mathrm{Re}_H}$$

The discretization equation for Equation E.8.2.7 is written as

$$va_i v_i = va_{i+1}v_{i+1} + va_{i-1}v_{i-1} + va_{j+1}v_{j+1} + va_{j-1}va_{j-1} - \left(\frac{p_{I,J+1} - p_{I,J}}{\rho\Delta y}\right)\Delta y \Delta x - w_i^u \frac{v_i - v_i^u}{\Delta z}\Delta x \Delta y$$

or

$$\left(va_i + \frac{v_i^u}{\Delta z}\Delta x \Delta y\right)v_i = va_{i+1}v_{i+1} + va_{i-1}v_{i-1} + va_{j+1}v_{j+1} + va_{j-1}v_{j-1} - \left(\frac{p_{I,J+1} - p_{,I,J}}{\rho}\right)\Delta x - w_i^u \frac{v_i^u}{\Delta z}\Delta x \Delta y \tag{E.8.2.8a}$$

or

$$va_i v_i = \sum va_{\text{nb}} v_{\text{nb}} + d \tag{E.8.2.8b}$$

where

$$va_i = va_{i+1} + va_{i-1} + va_{j+1} + va_{j-1} + \frac{w_i^u}{\Delta z}\Delta x \Delta y$$

$$av_{i+1} = \frac{\Gamma}{\Delta x} A\left|P_{i+\frac{1}{2}}\right| + \left\langle 0, -(\rho u)_{i+\frac{1}{2}}\right\rangle, \quad av_{i-1} = \frac{v}{\Delta x_{i-\frac{1}{2}}} A\left|P_{i-\frac{1}{2}}\right| + \left\langle 0, -(\rho u)_{i-\frac{1}{2}}\right\rangle$$

$$av_{j+1} = \frac{\Gamma}{\Delta x} A\left|P_{i+\frac{1}{2}}\right| + \left\langle 0, -(\rho u)_{i+\frac{1}{2}}\right\rangle, \quad av_{i-1} = \frac{v}{\Delta x_{i-\frac{1}{2}}} A\left|P_{i-\frac{1}{2}}\right| + \left\langle 0, -(\rho u)_{i-\frac{1}{2}}\right\rangle$$

$$d = -\left(\frac{p_{I,J+1} - p_{,I,J}}{\rho}\right)\Delta x - w_i^u \frac{v_i^u}{\Delta z}\Delta x \Delta y$$

$$P_i = \frac{(\rho u)}{\frac{\Gamma}{\Delta x}}, \quad P_i = \frac{(\rho u)}{\frac{\Gamma}{\Delta x}}$$

Similarly, Equation E.8.2.3d for the z-momentum equation can be discretized as

$$\left(u\frac{\partial w}{\partial x} - \Gamma\frac{\partial^2 w}{\partial x^2}\right) + \left(v\frac{\partial w}{\partial y} - \Gamma\frac{\partial^2 w}{\partial y^2}\right) = -\frac{\partial \bar{p}}{\partial z} - w\frac{\partial w}{\partial z}$$

Equation E.8.2.5 can be written in a general form as

$$\frac{\partial J_x}{\partial x} + \frac{\partial J_y}{\partial y} = S \tag{E.8.2.9}$$

where

$$J_x = uw - \Gamma\frac{\partial w}{\partial x}, \quad J_y = vw - \Gamma\frac{\partial w}{\partial y}$$

$$S = -\Delta\bar{p} - w\frac{\partial w}{\partial z}, \quad \Gamma = \frac{1}{\text{Re}_H}$$

The discretization equation for Equation E.8.2.9 is written as

$$wa_i w_{i,j} = wa_{i+1} w_{i+1,j} + wa_{i-1} w_{i-1,j} + wa_{j+1} w_{i,j+1} + wa_{j-1} w_{i,j-1} - (\Delta\bar{p})\Delta y \Delta x - w_i^u \frac{w_i - w_i^u}{\Delta z}\Delta x \Delta y$$

or

$$\left(wa_{i,j} + \frac{w_{i,j}^u}{\Delta z}\Delta x \Delta y\right) w_{i,j} = wa_{i+1,j} w_{i+1,j} + wa_{i-1,j} w_{i-1,j} + wa_{i,j+1} w_{i,j+1} + wa_{i,j-1} w_{i,j-1} - \Delta\bar{p}\Delta x - w_i^u \frac{w_i^u}{\Delta z}\Delta x \Delta y \tag{E.8.2.10a}$$

or

$$wa_i w_{i,j} = \sum wa_{nb} w_{nb} + d \qquad \text{(E.8.2.10b)}$$

where

$$wa_i = wa_{i+1} + wa_{j-1} + wa_{j+1} + wa_{j-1} + \frac{w_{i,j}^u}{\Delta z}\Delta x \Delta y$$

$$wa_{i+1} = \frac{\Gamma}{\Delta x} A\left|P_{i+\frac{1}{2}}\right| + \left\langle 0, -(\rho u)_{i+\frac{1}{2}} \right\rangle$$

$$wa_{i-1} = \frac{\Gamma}{\Delta x} A\left|P_{i-\frac{1}{2}}\right| + \left\langle 0, (\rho u)_{i-\frac{1}{2}} \right\rangle$$

$$wa_{j+1} = \frac{\Gamma}{\Delta y} A\left|P_{j+\frac{1}{2}}\right| + \left\langle 0, -(\rho v)_{j+\frac{1}{2}} \right\rangle$$

$$wa_{j-1} = \frac{\Gamma}{\Delta y} A\left|P_{j-\frac{1}{2}}\right| + \left\langle 0, (\rho v)_{j-\frac{1}{2}} \right\rangle$$

$$d = -\left(\frac{\Delta \bar{p}}{\rho}\right)\Delta x \Delta y - w_i^u \frac{w_i^u}{\Delta z}\Delta x \Delta y$$

$$P_i = \frac{(\rho u)}{\Gamma/\Delta x}, \qquad P_j = \frac{(\rho v)}{\Gamma/\Delta y}$$

In a similar manner, the energy Equation E.8.2.3e is first transformed into the general form as

$$\left(u\frac{\partial \theta}{\partial x} - \Gamma_\theta \frac{\partial^2 \theta}{\partial x^2}\right) + \left(v\frac{\partial \theta}{\partial y} - \Gamma_\theta \frac{\partial^2 \theta}{\partial y^2}\right) = -w\theta\lambda - w\frac{\partial \theta}{\partial z}$$

or

$$\frac{\partial J_x}{\partial x} + \frac{\partial J_y}{\partial y} = S \qquad \text{(E.8.2.11)}$$

where

$$J_x = u\theta - \Gamma_\theta \frac{\partial \theta}{\partial x}, \qquad J_y = v\theta - \Gamma_\theta \frac{\partial \theta}{\partial y}$$

$$S = -w\theta\lambda - w\frac{\partial \theta}{\partial z}, \qquad \Gamma_\theta = \frac{1}{\mathrm{Re}_H \,\mathrm{Pr}}$$

Following Equation 8.33a, we get the discretization equation for the energy Equation E.8.2.11 as

$$Ta_{i,j}\theta_{i,j} = Ta_{i+1,j}\theta_{i+1,j} + Ta_{i-1,j}\theta_{i-1,j} + Ta_{i,j+1}\theta_{i,j+1} + Ta_{i,j-1}\theta_{i,j-1} - w_{i,j}\theta_{i,j}\lambda\Delta x \Delta y - w_{i,j}^u \frac{\theta_{i,j} - \theta_{i,j}^u}{\Delta z}\Delta x \Delta y$$

or

$$\left(Ta_i + \frac{w_{i,j}^u}{\Delta z}\Delta x \Delta y\right)\theta_{i,j} = Ta_{i+1}\theta_{i+1} + Ta_{i-1}\theta_{i-1} + Ta_{j+1}\theta_{j+1} + Ta_{j-1}\theta_{j-1} - w_{i,j}\theta_{i,j}\lambda - \frac{w_{i,j}^u \theta_{i,j}^u}{\Delta z}\Delta x \Delta y \qquad \text{(E.8.2.12a)}$$

or

$$Ta_{i,j}\theta_{i,j} = \sum Ta_{nb}\theta_{nb} + d \tag{E.8.2.12b}$$

where

$$Ta_{i,j} = Ta_{i+1,j} + Ta_{i-1,j} + Ta_{i,j+1} + Ta_{i,j-1} + \frac{w^u_{i,j}}{\Delta z}\Delta x \Delta y$$

$$Ta_{i+1,j} = \frac{\Gamma_\theta}{\Delta x} A\left|P_{i+\frac{1}{2}}\right| + \left\langle 0, -(\rho u)_{i+\frac{1}{2}}\right\rangle$$

$$Ta_{i-1,j} = \frac{\Gamma_\theta}{\Delta x} A\left|P_{i-\frac{1}{2}}\right| + \left\langle 0, (\rho u)_{i-\frac{1}{2}}\right\rangle$$

$$Ta_{i,j+1} = \frac{\Gamma_\theta}{\Delta y} A\left|P_{j+\frac{1}{2}}\right| + \left\langle 0, -(\rho v)_{j+\frac{1}{2}}\right\rangle$$

$$Ta_{i,j-1} = \frac{\Gamma_\theta}{\Delta y} A\left|P_{j-\frac{1}{2}}\right| + \left\langle 0, (\rho v)_{j-\frac{1}{2}}\right\rangle$$

$$d = -w_{i,j}\theta_{i,j}\lambda - \frac{w^u_{i,j}\theta^u_{i,j}}{\Delta z}\Delta x \Delta y$$

Using Equation 8.71, we get the pressure correction equation as

$$pa_I p'_I = pa_{I+1}p'_{I+1} + pa_{I-1}p'_{I-1} + pa_{J+1}p'_{J+1} + pa_{J-1}p'_{J-1} + d \tag{E.8.2.13}$$

where

$$pa_{I+1} = \rho_i \frac{(\Delta y \Delta z)^2}{a_i}$$

$$pa_{I-1} = \rho_{i-1} \frac{(\Delta y \Delta z)^2}{a_{i+1}}$$

$$pa_{J+1} = \rho_j \frac{(\Delta z \Delta x)^2}{a_j}$$

$$pa_{J-1} = \rho_{j-1} \frac{(\Delta z \Delta x)^2}{a_{j-1}} \tag{E.8.2.14}$$

$$a_I = a_{I+1} + a_{I-1} + a_{J+1} + a_{J-1}$$

$$d = \left[\left(\rho u^*\right)_{i-1} - \left(\rho u^*\right)_i\right]\Delta y \Delta z + \left[\left(\rho v^*\right)_{j-1} - \left(\rho v^*\right)_j\right]\Delta z \Delta x + \left[\left(\rho w^*\right)_{k-1} - \left(\rho w^*\right)_k\right]\Delta x \Delta y$$

The solution algorithm for this problem is given as follows.

1. Guess the velocity field *u*, *v*, and *w*, and pressure field on the *xy* plane and $\Delta\bar{p} = \partial\bar{p}/\partial z$.
2. Solve discretization Equations E.8.2.10 for w with the guess value of $\Delta\bar{p} = \partial\bar{p}/\partial z$ at new *z* location incremented.
3. Check mass flow rate based on w^* and update $\Delta\bar{p} = \partial\bar{p}/\partial z$.

4. Repeat steps 2 and 3 until convergence for w, $\Delta\bar{p} = \partial\bar{p}/\partial z$, and mass flow rate, and proceed to calculate velocity and pressure field on the xy-plane.
5. Solve Equations 8.66 for $\hat{u}$ and $\hat{v}$ guess velocity field for the surrounding nodes.
6. Solve the pressure equation 8.73 for the pressure field, p.
7. Solve Equations E.8.2.6 and E.8.2.8 for u^* and v^*, respectively.
8. Calculate the mass source term d given by Equation E.8.2.14 and solve the pressure correction Equation E.8.2.13 for p'.
9. Calculate corrected velocity components u and v from their starred components (u^* and v^*) and pressure correction (p') using Equation 8.71.
10. Repeat steps 5–9 with the new velocity estimates until solution converges within a specified tolerance limit.
11. Solve the discretization Equation E.8.2.12 for temperature.
12. Reset values of velocity components u, v, and w to upstream velocity components u^u, v^u, and w^u.
13. Repeat steps 2–12 for a new z-location using increment Δz.

Computer Implementation Let us briefly discuss a procedure that can be used for solving the system of equations for u, v, w, and p on the xy-plane (Figure 8.8).

Each set of equations could be solved either by using the Gauss–Seidel method or by a combination of the Tridiagonal Matrix Algorithm (TDMA) and Gauss–Seidel method. Let us show the solution algorithm using these two methods for the discretization Equation E.8.2.15 for the w-component equation given as

$$wa_{i,j}w_{i,j} = wa_{i+1,j}w_{i+1,j} + wa_{i-1,j}w_{i-1,j} + wa_{i,j+1}w_{i,j+1} + wa_{i,j-1}w_{i,j-1} - \Delta\bar{p}\Delta x - w_i^u \frac{w_i^u}{\Delta z}\Delta x\Delta y \tag{E.8.2.15}$$

Solution Algorithm Using the Gauss–Seidel Method In this method, the equation is written explicitly for the unknown

$$w_{i,j} = \left(wa_{i+1,j}w_{i+1,j} + wa_{i-1,j}w_{i-1,j} + wa_{i,j+1}w_{i,j+1} + wa_{i,j-1}w_{i,j-1} + c\right)/wa_{i,j} \tag{E.8.2.16}$$

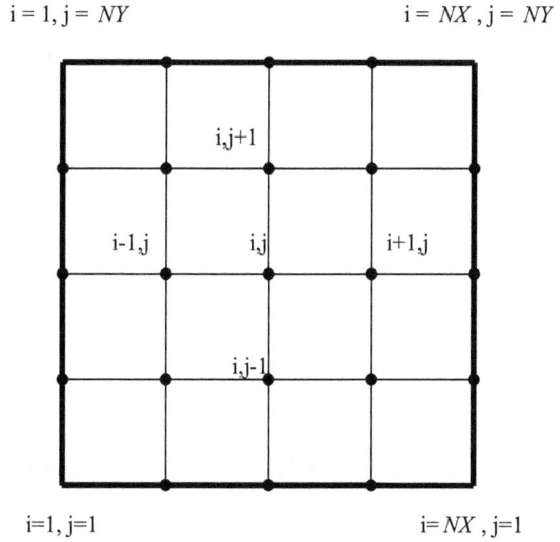

FIGURE 8.8 Two-dimensional grid system for computer implementation.

The equation is then solved iteratively based on the values of the neighboring nodes from the previous iteration. A pseudo code for sweeping all the nodes at any iteration step can be given as

dofor $I = 1, N_x$

dofor $J = 1, N_y$

$$w_{i,j} = \left(wa_{i+1,j}w_{i+1,j} + wa_{i-1,j}w_{i-1,j} + wa_{i,j+1}w_{i,j+1} + wa_{i,j-1}w_{i,j-1} + c\right)/wa_{i,j}$$

enddo

enddo

Solution by Combination of TDMA and the Gauss–Seidel Methods This procedure is also referred to as the **line-by-line method** as suggested by Patankar (1980). The system of equations is solved by applying the **TDMA** algorithm along an *x*-line or a *y*-line and based on the values of the neighboring lines from their previous iteration. This is repeated for all lines in one direction and subsequently repeated for the other direction. For example, we can first sweep for all *x*-lines and then all *y*-lines. To apply the TDMA algorithm at any *x*-line, we can first rewrite the discretization Equation E.8.2.15 in the form of a tridiagonal matrix format as

$$A_i w_{i,j} = B_i w_{i+1,j} + C_i w_{i-1,j} + D_i \tag{E.8.2.17}$$

where

$$A_i = wa_i, \quad B_i = wa_{i+1,j}, \quad C_i = wa_{i-1,j}$$

$$D_i = wa_{i,j+1}w_{i,j+1} + wa_{i,j-1}w_{i,j-1} - \Delta\bar{p}\Delta x - w_i^u \frac{w_i^u}{\Delta z}\Delta x \Delta y$$

Equation E.8.2.17 can now be solved using the TDMA algorithm. A pseudo code for this combination of procedure is outlined as

Dofor $J = 2,(\text{NY} - 1)$

$\text{A}(1) = 1.0$

$\text{B}(1) = 0.0$

$\text{C}(1) = 0.0$

$\text{D}(1) = 0.0$

$A(\text{NX}) = 1.0$

$B(\text{NX}) = 0.0$

$C(\text{NX}) = 0.0$

$D(\text{NX}) = 0.0$

Dofor $I = 2,(NX - 1)$

$A(I) = wa_i$

$B(I) = wa_{i+1}$

$C(I) = wa_{i-1}$

$$D(I) = wa_{i,j+1}w_{i,j+1} + wa_{i,j-1}w_{i,j-1} - \Delta\bar{p}\Delta x - w_i^u \frac{w_i^u}{\Delta z}\Delta x \Delta y$$

Enddo

Call tdma (NX,*A*,*B*,*C*,*D*,*w*)

Enddo

SIMPLEC Algorithm

The **SIMPLEC** algorithm is another variation of the **SIMPLE** algorithm, which was proposed by Doormaal and Raithby (1984) to improve the computational cost. The **SIMPLE** algorithm requires the use of optimum under-relaxation parameters for velocity and pressure to remove instability in the iterative solution and improve convergence rate. The **SIMPLEC** algorithm removes the need for the under-relaxation parameter α_p for pressure. To discuss the development of the **SIMPLEC** algorithm, let us write the x-component velocity correction Equation 8.69 in the compact form

$$a_i u_i' = \sum a_{\text{nb}} u_{\text{nb}}' + \left(p_{I-1}' - p_I'\right) \Delta y \Delta z \tag{8.85}$$

In order to simplify this equation, the approximation made in the **SIMPLE** algorithm is to drop the $\sum a_{\text{nb}} u_{\text{nb}}'$ term, which represents the correction contributions from the surrounding nodes. In the **SIMPLEC**, a "consistent" approximation is used by first subtracting a term $\sum a_{\text{nb}} u_i'$ from both sides of Equation 8.85 and that gives

$$\left(a_i - \sum a_{\text{nb}}\right) u_i' = \sum a_{\text{nb}} \left(u_{\text{nb}}' - u_i'\right) + \left(p_{I-1}' - p_I'\right) \Delta y \Delta z \tag{8.86}$$

In the next step, the term $\sum a_{\text{nb}} \left(u_{\text{nb}}' - u_i'\right)$ is neglected from the right-hand side of Equation 8.86, and a simplified form of the x-component velocity correction equation is written as

$$\left(a_i - \sum a_{\text{nb}}\right) u_i' = \left(p_{I-1}' - p_I'\right) \Delta y \Delta z$$

or

$$u_i' = \left(p_{I-1}' - p_I'\right) \frac{\Delta y \Delta z}{\left(a_i - \sum a_{\text{nb}}\right)} \tag{8.87a}$$

and similarly, for y- and z-component velocity correction equation

$$v_j' = \left(p_{J-1}' - p_J'\right) \frac{\Delta z \Delta x}{\left(a_j - \sum a_{\text{nb}}\right)} \tag{8.87b}$$

$$w_k' = \left(p_{K-1}' - p_K'\right) \frac{\Delta x \Delta y}{\left(a_k - \sum a_{\text{nb}}\right)} \tag{8.87c}$$

The expressions for the corrected velocity components then change to

$$u_i = u_i^* + \left(p_{I-1}' - p_I'\right) \frac{\Delta y \Delta z}{\left(a_i - \sum a_{\text{nb}}\right)} \tag{8.88a}$$

$$v_j = v_j^* + \left(p_{J-1}' - p_J'\right) \frac{\Delta x \Delta z}{\left(a_j - \sum a_{\text{nb}}\right)} \tag{8.88b}$$

$$w_k = w_k^* + \left(p_{K-1}' - p_K'\right) \frac{\Delta x \Delta y}{\left(a_k - \sum a_{\text{nb}}\right)} \tag{8.88c}$$

The rest of the steps is identical to those in **SIMPLE**. The **SIMPLEC** algorithm is listed below with major differences highlighted in italic.

1. Guess the pressure field p^*.
2. Solve discretization Equations 8.66 for u^*, v^*, and w^*.
3. Solve for pressure correction Equation 8.75 for p'.
4. Calculate the corrected pressure from Equation 8.67 without under-relaxation.
5. Calculate velocity components u, v, and w from their starred and correction components using Equation 8.88 and velocity correction by Equation 8.87.
6. Solve the discretization equations for other scalar quantities, such as turbulence quantities, temperature, and mass concentration if they influence the flow field through dependence through fluid properties such as density, viscosity, and source terms. If a particular scalar quantity does not affect the velocity field, then that quantity is calculated after a converged velocity field is achieved.
7. Reset the corrected pressure p as the new guessed pressure p^*.
8. Repeat steps 2–7 until the solution converges within a specified tolerance limit.

MAC Method The MAC method was proposed by Harlow and Welch (1965) for the finite-difference-based solution of incompressible Navier–Stokes equations. In this method, finite difference schemes are applied to both space and time derivatives directly. The marker-cell method also uses marker particles that are convected by the fluid to the location of the free surfaces. The computational regions are divided into an Eulerian mesh set of small rectangular cells having dimensions Δx, Δy, and Δz. With respect to this set of computational cells, the velocity components are located at cell faces and pressure and other scalar values at cell centers, as shown in Figure 8.9. The cells are numbered with indices i, j, and k, which count cell center positions in x, y, and z directions. The pressure at the center (i, j, k) of the cell is designated as $p_{i,j,k}$ while $u_{i+\frac{1}{2},j,k}$ is the u-component velocity at the center of the face between cells (i, j, k) and $(i+1, j, k)$, $v_{i,j+\frac{1}{2},k}$ is the v-component velocity at the center of the face between (i, j, k) and $(i, j+1, k)$, and $w_{i,j,k+\frac{1}{2}}$ is the w-component velocity at the center of the face between (i, j, k) and $(i, j, k+1)$. To designate the time cycles, superscript $m+1$ is used to designate time $t=(m+1)\Delta t$. When we drop the superscript for simplicity, it is assumed that its value is m and corresponds to the time $t = m\Delta t$, i.e., it refers to values at the previous time step.

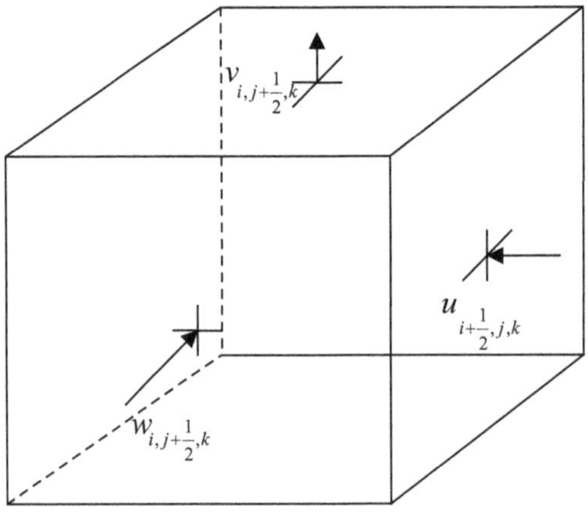

FIGURE 8.9 Computational cell with velocity components located at the faces.

The time-dependent solution is obtained by advancing the flow field variables through small time steps, Δt.

The Finite Difference Equations Let us show the derivation of finite difference equations for viscous incompressible Navier–Stokes equations in three-dimensional Cartesian coordinates given as

$$\frac{\partial u}{\partial x}+\frac{\partial v}{\partial y}+\frac{\partial v}{\partial y}=0 \tag{8.89a}$$

$$\frac{\partial u}{\partial t}+\frac{\partial\left(u^2\right)}{\partial x}+\frac{\partial(uv)}{\partial y}+\frac{\partial(uw)}{\partial z}=g_x-\frac{1}{\rho}\frac{\partial p}{\partial x}+\nu\left(\frac{\partial^2 u}{\partial x^2}+\frac{\partial^2 u}{\partial y^2}+\frac{\partial^2 u}{\partial z^2}\right) \tag{8.89b}$$

$$\frac{\partial v}{\partial t}+\frac{\partial(uv)}{\partial x}+\frac{\partial\left(v^2\right)}{\partial y}+\frac{\partial(vw)}{\partial z}=g_y-\frac{1}{\rho}\frac{\partial p}{\partial y}+\nu\left(\frac{\partial^2 v}{\partial x^2}+\frac{\partial^2 v}{\partial y^2}+\frac{\partial^2 v}{\partial z^2}\right) \tag{8.89c}$$

$$\frac{\partial w}{\partial t}+\frac{\partial(uw)}{\partial x}+\frac{\partial(vw)}{\partial y}+\frac{\partial\left(w^2\right)}{\partial z}=g_z-\frac{1}{\rho}\frac{\partial p}{\partial z}+\nu\left(\frac{\partial^2 w}{\partial x^2}+\frac{\partial^2 w}{\partial y^2}+\frac{\partial^2 w}{\partial z^2}\right) \tag{8.89d}$$

Substituting an explicit time approximation for the time derivatives, a forward difference scheme for the convective terms, and a central difference scheme for the diffusion terms in Equations 8.89b–d, we derive the discretization equations for u, v, and w as

$$\begin{aligned}
&\frac{u^{m+1}_{i+\frac{1}{2},j,k}-u_{i+\frac{1}{2},j.k}}{\Delta t}+\frac{\left(u_{i,j,k}\right)^2-\left(u_{i+1,j,k}\right)^2}{\Delta x}+\frac{\left(u_{i+\frac{1}{2},j+\frac{1}{2},k}v_{i+\frac{1}{2}\,j+\frac{1}{2},k}-u_{i+\frac{1}{2},j-\frac{1}{2},k}v_{i+\frac{1}{2},j-\frac{1}{2},k}\right)}{\Delta y}\\
&\quad+\frac{\left(u_{i+\frac{1}{2},j,k+\frac{1}{2}}w_{i+\frac{1}{2}\,j,k+\frac{1}{2}}-u_{i+\frac{1}{2},j,k-\frac{1}{2}}w_{i+\frac{1}{2},j,k-\frac{1}{2}}\right)}{\Delta z}=g_x-\frac{1}{\rho}\frac{p_{i,j,k}-p_{i+1,j,k}}{\Delta x}\\
&\quad+\nu\left(\frac{u_{i+\frac{3}{2},j,k}-2u_{i+\frac{1}{2},j,k}+u_{i-\frac{1}{2},j,k}}{\Delta x^2}+\frac{u_{i+\frac{1}{2},j+1.k}-2u_{i+\frac{1}{2},j,k}+u_{i+\frac{1}{2},j-1,k}}{\Delta y^2}+\frac{u_{i+\frac{1}{2},j,k+1}-2u_{i+\frac{1}{2},j,k}+u_{i+\frac{1}{2},j,k-1}}{\Delta z^2}\right)
\end{aligned} \tag{8.90a}$$

$$\begin{aligned}
&\frac{v^{m+1}_{i,j+\frac{1}{2},k}-v_{i,j+\frac{1}{2}.k}}{\Delta t}+\frac{\left(u_{i+\frac{1}{2},\,j+\frac{1}{2},k}v_{i+\frac{1}{2}\,j+\frac{1}{2},k}-u_{i-\frac{1}{2},\,j+\frac{1}{2},k}v_{i-\frac{1}{2},\,j-\frac{1}{2},k}\right)}{\Delta x}+\frac{\left(v_{i,j,k}\right)^2-\left(v_{i,j+1,k}\right)^2}{\Delta y}\\
&\quad+\frac{\left(w_{i,j+\frac{1}{2},k+\frac{1}{2}}v_{i,j+\frac{1}{2},k+\frac{1}{2}}-w_{i,j+\frac{1}{2},k-\frac{1}{2}}v_{i,j+\frac{1}{2},k-\frac{1}{2}}\right)}{\Delta z}=g_y-\frac{1}{\rho}\frac{p_{i,j,k}-p_{i,j+1,k}}{\Delta y}\\
&\quad+\nu\left(\frac{v_{i+1,j+\frac{1}{2},k}-2v_{i,j+\frac{1}{2},k}+v_{i-1,j+\frac{1}{2},k}}{\Delta x^2}+\frac{v_{i,j+\frac{3}{2},k}-2v_{i,j+\frac{1}{2},k}+v_{i,j-\frac{1}{2},k}}{\Delta y^2}+\frac{v_{i,j,k+1}-2v_{i,j+\frac{1}{2},k}+v_{i,j+\frac{1}{2},k-1}}{\Delta z^2}\right)
\end{aligned} \tag{8.90b}$$

$$\frac{w^{m+1}_{i,j,k+\frac{1}{2}} - w_{i,j,k+\frac{1}{2}}}{\Delta t} + \frac{\left(u_{i+\frac{1}{2},j,k+\frac{1}{2}} w_{i+\frac{1}{2},j,k+\frac{1}{2}} - u_{i-\frac{1}{2},j,k+\frac{1}{2}} w_{i-\frac{1}{2},j,k+\frac{1}{2}}\right)}{\Delta x}$$

$$+ \frac{\left(v_{i,j+\frac{1}{2},k+\frac{1}{2}} w_{i,j+\frac{1}{2},k+\frac{1}{2}} - v_{i,j-\frac{1}{2},k+\frac{1}{2}} w_{i,j-\frac{1}{2},k+\frac{1}{2}}\right)}{\Delta y} + \frac{\left(w_{i,j,k}\right)^2 - \left(w_{i,j,k+1}\right)^2}{\Delta z}$$

$$= g_z - \frac{1}{\rho}\frac{p_{i,j,k} - p_{i,j,k+1}}{\Delta z} + \nu\left(\frac{w_{i+1,j,k+\frac{1}{2}} - 2w_{i,j,k+\frac{1}{2}} + w_{i-1,j,k+\frac{1}{2}}}{\Delta x^2}\right.$$

$$\left. + \frac{w_{i,j+1,k+\frac{1}{2}} - 2w_{i,j,k+\frac{1}{2}} + w_{i,j-1,k+\frac{1}{2}}}{\Delta y^2} + \frac{w_{i,j,k+\frac{3}{2}} - 2w_{i,j,k+\frac{1}{2}} + w_{i,j,k-\frac{1}{2}}}{\Delta z^2}\right) \tag{8.90c}$$

We can see that some of the velocity values are not centered at the points shown on the cell. Such values are expressed as an average of the adjacent nodal values. For example, we substitute expressions

$$u_{i+\frac{1}{2},j+\frac{1}{2},k} = \frac{1}{2}\left(u_{i+\frac{1}{2},j,k} + u_{i+\frac{1}{2},j+1,k}\right) \tag{8.91a}$$

$$v_{i+\frac{1}{2},j+\frac{1}{2},k} = \frac{1}{2}\left(v_{i,j+\frac{1}{2},k} + v_{i+1,j+\frac{1}{2},k}\right) \tag{8.91b}$$

$$w_{i+\frac{1}{2},j,k+\frac{1}{2}} = \frac{1}{2}\left(w_{i,j,k+\frac{1}{2}} + w_{i+1,j,k+\frac{1}{2}}\right) \tag{8.91c}$$

$$u_{i,j,k} = \frac{1}{2}\left(u_{i-\frac{1}{2},j,k} + u_{i+\frac{1}{2},j,k}\right) \tag{8.91d}$$

$$v_{i,j,k} = \frac{1}{2}\left(v_{i,j-\frac{1}{2},k} + v_{i,j+\frac{1}{2},k}\right) \tag{8.91e}$$

$$w_{i,j,k} = \frac{1}{2}\left(w_{i,j,k-\frac{1}{2}} + w_{i,j,k+\frac{1}{2}}\right) \tag{8.91f}$$

Equation 8.90 along with Equation 8.91 represents an explicit finite difference approximation of the velocity field, and can be solved with the advancement of time using small time step, Δt. The velocity field given by this explicit calculation does not necessarily result in a correct velocity field but results in a nonzero mass deficit flux D in the mass conservation equation. So, the estimated velocity field must be corrected to ensure mass conservation in each cell. An iterative process is used for this purpose, in which the cell pressures are corrected to make all velocity field divergences become negligibly small, and so the mass deficit flux D becomes less than an assigned tolerance value ε_D. The mass deficit flux D is estimated from the mass conservation Equation 8.75a with the substitution of the estimated velocity field given by Equation 8.76.

$$D = \frac{\tilde{u}^{m+1}_{i+\frac{1}{2},j,k} - \tilde{u}^{m+1}_{i-\frac{1}{2},j,k}}{\Delta x} + \frac{\tilde{v}^{m+1}_{i,j+\frac{1}{2},k} - \tilde{v}^{m+1}_{i,j-\frac{1}{2},k}}{\Delta y} + \frac{\tilde{w}^{m+1}_{i,j,k+\frac{1}{2}} - \tilde{w}^{m+1}_{i,j,k-\frac{1}{2}}}{\Delta z} \tag{8.92}$$

Note that we have changed the notation of the velocity components $\tilde{u}$, $\tilde{v}$, and $\tilde{w}$ to designate them as the estimated values. If the magnitude of the mass deficit flux D is greater than the specified tolerance limit, then the pressure correction is estimated by

$$\Delta P' = \beta D \tag{8.93}$$

where β is given by

$$\beta = \frac{\beta_0}{2\Delta t\left[\dfrac{1}{\Delta x^2}+\dfrac{1}{\Delta y^2}+\dfrac{1}{\Delta z^2}\right]} \tag{8.94}$$

and constant β_0 is a relaxation parameter.

The corrected pressure in each cell is then estimated as

$$p_{i,j,k} = p_{i,j,k}+\Delta p' \tag{8.95}$$

and velocity components at the faces of the cell are corrected as

$$\begin{aligned}
u^{m+1}_{i+\frac{1}{2},j,k} &= \tilde{u}^{m+1}_{i+\frac{1}{2},j,k} + \frac{\Delta t}{\Delta x}\Delta p' \\
u^{m+1}_{i-\frac{1}{2},j,k} &= \tilde{u}^{m+1}_{i-\frac{1}{2},j,k} - \frac{\Delta t}{\Delta x}\Delta p' \\
v^{m+1}_{i,j+\frac{1}{2},k} &= \tilde{v}^{m+1}_{i,j+\frac{1}{2},k} + \frac{\Delta t}{\Delta y}\Delta p' \\
v^{m+1}_{i,j-\frac{1}{2},k} &= \tilde{v}^{m+1}_{i,j-\frac{1}{2},k} - \frac{\Delta t}{\Delta y}\Delta p' \\
w^{m+1}_{i,j,k+\frac{1}{2}} &= \tilde{w}^{m+1}_{i,j,k+\frac{1}{2}} + \frac{\Delta t}{\Delta z}\Delta p' \\
w^{m+1}_{i,j,k-\frac{1}{2}} &= \tilde{w}^{m+1}_{i,j,k-\frac{1}{2}} + \frac{\Delta t}{\Delta z}\Delta p'
\end{aligned} \tag{8.96}$$

This procedure is repeated for all cells to ensure that the mass deficit flux D becomes less than the tolerance value ε_D in all cells. After we reach convergence at a given time, we advance the time to the next time step.

The solution algorithm is given as follows.

1. Start with a guess velocity field. In addition, coordinates of a set of marker particles are assumed to be known and this shows which region is occupied by fluid and which is empty.
2. The corresponding pressure field is computed in such a way that the rate of change of velocity divergence or mass deficit term as given by the continuity equation vanishes in all cells.
3. The two components of acceleration are calculated. The products of these with time increment per cycle then give the changes in velocity to be added to the old values.
4. The marker particles are moved according to the velocity components in their vicinities.

PISO (Pressure Implicit with Splitting of Operator) Algorithm

PISO algorithm was proposed by Issa et al. (1986) with an intention to use noniterative solutions with larger time steps and reduce computational time. PISO is a pressure-velocity computation procedure that employs the concepts of splitting of operators like the ADI procedure outlined for solving discretized equations for differential equations. Even though PISO algorithm was originally developed for transient compressible and incompressible flows, it has been applied successfully for steady compressible and incompressible flows also. Some of the major benefits of PISO algorithm includes good temporal accuracy, larger time-steps can be employed while maintain stability, and faster solution than other iterative solution methods.

The PISO algorithm can be an extension to SIMPLE of SIMPLER algorithm with inclusion additional corrections steps. While SIMPLE algorithm includes **one predictor step** and **one corrector step** with an iteration loop using iterative refinements; the PISO algorithm uses **one predictor step** and **two correction steps**. The PISO algorithm is summarized below:

Predictor Step

1. Solve for **first velocity prediction** u^*, v^*, and w^* using discretization Equations 8.70, 8.72, and 8.74 at a new time step $(m+1)$. Noticed that at this predictor step, the velocity field is solved based on solving momentum discretization equation and assuming pressure field from old time-step (m) without satisfying the mass continuity equation.

$$a_i u_i^* = \sum a_{\text{nb}} u_{\text{nb}}^* + d + \left(P_I^m - P_I^m\right)\Delta y \Delta z \tag{8.74}$$

$$a_i v_i^* = \sum a_{\text{nb}} v_{\text{nb}}^* + d + \left(P_J^m - P_J^m\right)\Delta x \Delta z \tag{8.76}$$

$$a_k w_k^* = \sum a_{\text{nb}} w_{\text{nb}}^* + d + \left(P_K^m - P_K^m\right)\Delta x \Delta y \tag{8.78}$$

The coefficients in these equations are given by Equations 8.71, 8.73, and 8.75 and using velocity field values at old time step (m).

Correction Step #1

In this corrector step, new velocity field u^{**}, v^{**}, w^{**} and pressure field p^{**} are computed satisfying both the mass continuity and momentum equations.

1. Compute new pressure field value p^* by solving the **pressure discretization equation** 8.75 and 8.76 which are derived by satisfying mass continuity equation and momentum equation. Noticed that pressure discretization equation is used in an explicit manner with predicted velocity field values (u^*, v^*, and w^*).
2. Calculate corrected velocity components u_i^{**}, v_j^{**}, and w_k^{**} using discretization velocity field Equations 8.75 and 8.76 in an explicit form using surrounding node\values for u^*, v^*, and w^* and using pressure, P^{**}.

Correction Step #2

In this second corrector step, new velocity field u^{***}, v^{***}, and w^{***} and pressure field p^{***} are computed following the procedure outlined in steps 3–4, satisfying both the mass continuity and momentum equations.

1. Solve for **second pressure discretization equation**, p^{***}, using Equations 8.75 and 8.76 using u_i^{**}, v_j^{**}, and w_k^{**}.
2. Calculate the second corrected velocity field u^{***}, v^{***}, and w^{***} from the discretization equation (8.66), but in an explicit form using surrounding node values u_i^{**}, v_j^{**} and w_k^{**}, p^{***}.
3. Solve the discretization equation for other scalar quantities such as turbulence quantities, temperature, and mass concentration if they influence the flow field though dependence, through properties such as density, viscosity, and source terms. If a scalar quantity does not affect the velocity field, then that quantity is calculated after a converged velocity field is achieved.

The correction procedure can be continued, but it has been demonstrated that two correction steps are sufficient for achieving desired accuracy.

PROBLEMS

8.1 Consider a one-dimensional flow in a straight nozzle shown in the figure. The governing equations are given as

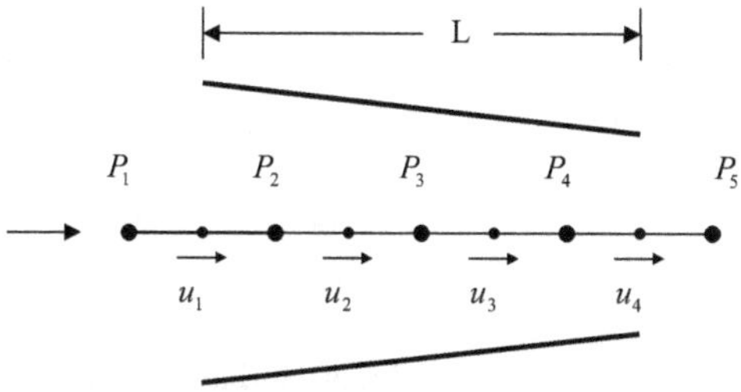

Continuity

$$\frac{\partial(uA)}{\partial x}=0$$

***x*-momentum**

$$\frac{\partial(\rho uu)}{\partial x}=-\frac{\partial p}{\partial x}$$

where A is the cross-sectional area of the nozzle. Consider the following data: $\rho=1.00\ \mathrm{kg/m^3}, A_1=5\ \mathrm{m^2}, A_2=4.333\ \mathrm{m^2}, A_3=3.666\ \mathrm{m^2}, A_4=3\ \mathrm{m^2}$, $P_1=300$ kPa, and $P_5=100$ kPa. Show discretization equations for velocity and pressure and use the SIMPLE algorithm to compute velocity and pressure at the nodes for two iterations. Use an initial guess for mass flow rate as $\dot{m}=\rho uA=100$ kg/s.

8.2 Consider a fully developed heat transfer in a duct of rectangular cross section as shown in the figure.

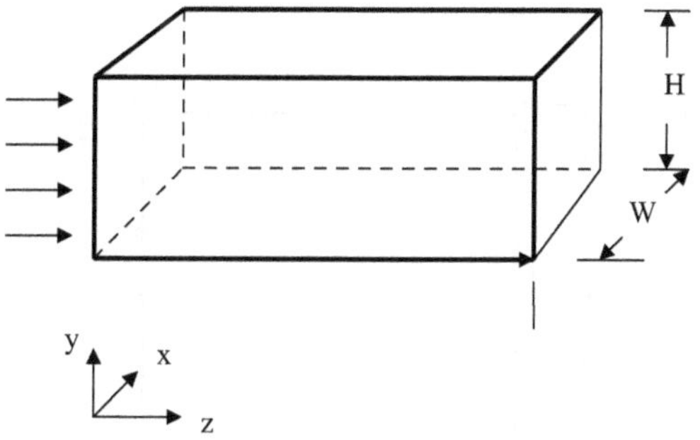

The top surface is maintained at a constant surface heat flux, q_w'', and the other three surfaces are adiabatic. The associated governing equations and boundary conditions for the fully developed flow and heat transfer with constant surface heat flux are given as follows.

Momentum

$$\nu\left(\frac{\partial^2 w}{\partial x^2}+\frac{\partial^2 w}{\partial y^2}\right)=-\frac{1}{\rho}\frac{\partial \bar{p}}{\partial z}$$

Energy

$$\frac{k}{\rho c_p}\left(\frac{\partial^2 T}{\partial x^2}+\frac{\partial^2 T}{\partial y^2}\right)=w\frac{\partial \overline{T}}{\partial z}$$

The pressure gradient $d\overline{p}/dz$ is assumed to be known. The temperature gradient $\partial \overline{T}/\partial x$ is constant and given based on an energy balance as

$$\frac{\partial \overline{T}}{\partial x}=\frac{q''_w}{\rho c_p \overline{w}}$$

where $\overline{w}$ and $\overline{T}$ are the average velocity and average temperature given as

$$\overline{w}=\frac{1}{WH}\int_A w\mathrm{d}A \quad \text{and} \quad \overline{T}=\frac{1}{WH\overline{w}}\int_A wT\mathrm{d}A$$

Boundary Conditions

No-slip velocity on all walls, i.e., $w = 0$ on all walls.
Constant surface heat flux on top wall, i.e., at $y = H$, $k\,\partial T/\partial x = q''_w$.
Rest of the walls are adiabatic, i.e., $\partial T/\partial n = 0$.

Derive the discretization equations for w and T. Present the sequence of steps to be used for calculating velocity, temperature field, and convection heat transfer coefficient.

8.3 Consider the example Problem 8.2 and derive all discretization equations, including the error equations for u, v, w, p, and T using the upwind scheme for the convective terms.

8.4 Consider free-convection motion and heat transfer in a two-dimensional rectangular chamber as shown. The free convection motion is created by maintaining the left wall at a high temperature T_H and right surface at a low temperature T_C. The top and bottom are assumed to be insulated. The governing equations for mass, momentum, and energy with Boussinesq assumption and for Newtonian fluid are given as

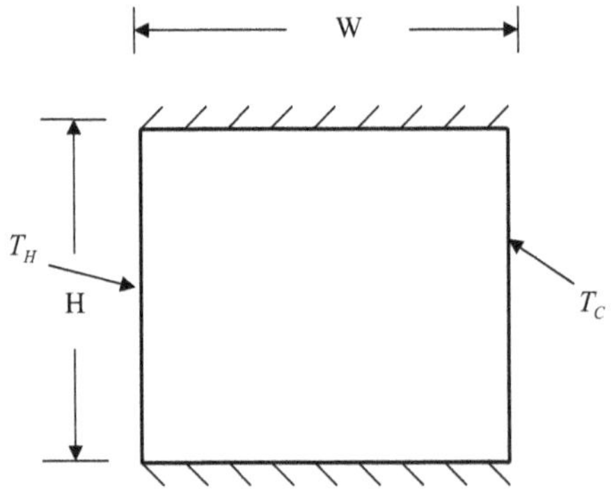

Continuity

$$\frac{\partial u}{\partial x}+\frac{\partial v}{\partial y}=0$$

***x*-momentum**

$$u\frac{\partial u}{\partial x}+v\frac{\partial u}{\partial y}=-\frac{1}{\rho}\frac{\partial p}{\partial x}+\nu\left(\frac{\partial^2 u}{\partial x^2}+\frac{\partial^2 u}{\partial y^2}\right)$$

y-momentum

$$u\frac{\partial v}{\partial x}+v\frac{\partial v}{\partial y}=-\frac{1}{\rho}\frac{\partial p}{\partial y}+v\left(\frac{\partial^2 v}{\partial x^2}+\frac{\partial^2 v}{\partial y^2}\right)-g\left[1-\beta(T-T_0)\right]$$

Energy

$$u\frac{\partial T}{\partial x}+v\frac{\partial T}{\partial y}=\alpha\left(\frac{\partial^2 T}{\partial x^2}+\frac{\partial^2 T}{\partial y^2}\right)$$

Boundary Conditions

Velocity No slip condition on all walls.

Temperature

1. at $x = 0,\ 0<y<H,\ T(0,y)=T_H$
2. at $x = W,\ 0<y<H,\ T(0,W)=T_C$
3. at $y = 0,\ 0<x<W,\ \left.\frac{\partial T}{\partial y}\right|_{y=0}=0$
4. at $y = H,\ 0<x<W,\ \left.\frac{\partial T}{\partial y}\right|_{y=H}=0$

a. Derive the discretization equations for u, v, p, and T using the upwind scheme for the convective term and central difference for the diffusion terms by considering a 5×5 grid. Show the system of equations for each variable.
b. Derive the correction equations for the u, v, p, and T.
c. Develop a computer program implementing the SIMPLE algorithm for velocity and pressure coupling and the line-by-line method with *tdma* solver for solving the system of equation for each variable.

8.5 Consider the developing laminar steady flow of Newtonian viscous fluid through a two-dimensional channel formed by two infinite parallel plates as shown below. The lower plate is stationary and insulated, and the upper plate is moving in the x-direction with a constant speed, U_w, in the positive x-direction and maintained at a constant temperature, T_W.

The mathematical statement of the problem is given as:

Continuity

$$\frac{\partial u}{\partial x}+\frac{\partial v}{\partial y}=0$$

Momentum

$$\frac{\partial(u^2)}{\partial x}+\frac{\partial(uv)}{\partial y}=-\frac{1}{\rho}\frac{\partial p}{\partial x}+v\frac{\partial^2 u}{\partial y^2}$$

Energy

$$\frac{\partial(uT)}{\partial x}+\frac{\partial(vT)}{\partial y}=\alpha\frac{\partial^2 T}{\partial y^2}$$

Boundary Conditions

Velocity

$$x = 0, \quad u = U_0, \quad v = 0$$

$$y = 0, \quad u = 0, \quad v = 0$$

$$y = h, \quad u = U_w, \quad v = 0$$

Temperature

$$x = 0, \qquad T = T_0$$

$$y = 0, \qquad \left.\frac{\partial T}{\partial y}\right|_{y=0} = 0$$

$$y = h, \qquad T = T_w$$

Derive the discretization equation for velocity, pressure, and temperature using the power law scheme for the convective term for use in SIMPLER algorithm.

8.6 Consider a fully developed laminar flow and a thermally developing flow and heat transfer, known as the thermally developing Hagen–Poiseuille flow, in a duct of rectangular cross-section as shown in the figure.

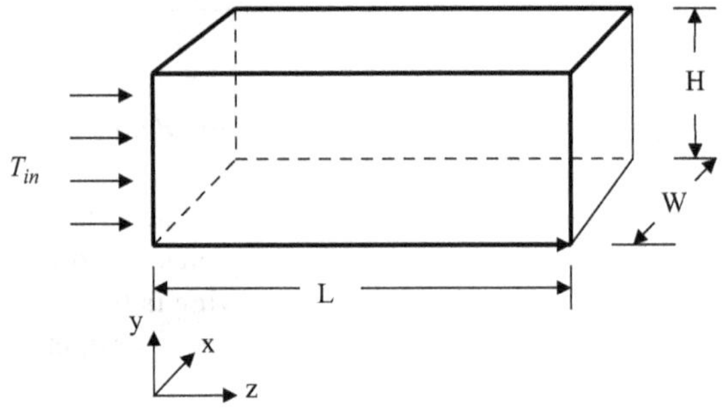

The top surface is maintained at a constant surface heat flux q_w'' and the other three surfaces are adiabatic. Assuming negligible axial conduction, the associated governing equations and boundary conditions for the fully developed flow and heat transfer with constant surface heat flux are given as follows.

Momentum

$$\nu\left(\frac{\partial^2 w}{\partial x^2}+\frac{\partial^2 w}{\partial y^2}\right) = -\frac{1}{\rho}\frac{\partial \bar{p}}{\partial z}$$

Energy

$$\alpha\left(\frac{\partial^2 T}{\partial x^2}+\frac{\partial^2 T}{\partial y^2}\right) = w\frac{\partial T}{\partial z}$$

The pressure gradient $d\bar{p}/dz$ is assumed to be known for a fully developed flow.

Boundary Conditions

No-slip velocity on all walls, i.e., $w = 0$ on all walls.

Constant surface heat flux on top wall, i.e., at $y = H$, $k\,\partial T/\partial x = q''_w$.

Rest of the walls are adiabatic, i.e., $\partial T/\partial n = 0$.

Derive the discretization equations for w and T. Present the sequence of steps to be used for calculating velocity and temperature field and convection heat transfer coefficient.

8.7 Consider diffusion of moisture from an air stream in a two-dimensional channel to the porous adsorbing material felt lined on the bottom of the air flow channel as shown. The felt is supported at the bottom by an aluminum plate, thus making it impermeable to mass diffusion. Air enters the channel at uniform velocity u_{in} and uniform concentration C_{in}. Initial concentration distribution in the felt is uniform at C_0. Assume constant rate of moisture adsorption, m''_{ad}.

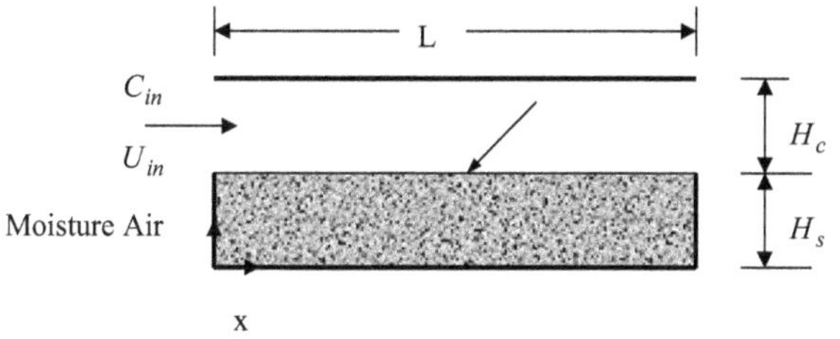

x

Governing Equation

Channel

$$\rho H_c U_{\text{in}} \frac{\partial C_a}{\partial x} = h_D \left(C_a - C|_{y=H} \right)$$

Solid

$$\frac{\partial C}{\partial t} = \frac{\partial}{\partial x}\left(D \frac{\partial C}{\partial x} \right) + \frac{\partial}{\partial y}\left(D \frac{\partial C}{\partial y} \right) + \frac{(1-\varepsilon)}{\varepsilon} \rho \dot{m}_{\text{ad}}$$

Boundary Conditions

1. $y = 0 \quad \left.\frac{\partial C}{\partial x}\right|_{x=0} = 0$
2. $y = H, \quad D\left.\frac{\partial C}{\partial x}\right|_{y=H_s} = h_D \left(C_a - C|_{y=H_s} \right)$
3. $x = 0, \quad \left.\frac{\partial C}{\partial x}\right|_{x=0} = 0$
4. $x = L, \quad \left.\frac{\partial C}{\partial x}\right|_{x=L} = 0$

where ε = porosity and ρ_b = material bulk density.Apply the finite difference–control volume method using an explicit scheme and derive the system of equations for the concentration distribution in the channel and the two-dimensional transient moisture concentration distribution in the slab based on assuming a 5×5 grid.

9 Additional Features in Computational Model and Mesh Generations

In this chapter, we will address some of the essential features and guidelines for developing computational fluid dynamics and heat transfer codes implementing discretization procedure for control volume/finite difference method discussed so far as well as in the finite element method to be discussed in the subsequent chapters. These important features include guidelines for mesh generation and quality; adapting meshing; MG methods; and initial, inlet, and boundary conditions.

9.1 BOUNDARY CONDITIONS

As we have discussed in Chapter 1 that types of boundary conditions are categorized mathematically as (a) **Dirichlet boundary conditions** or boundary conditions of the **first kind**; (b) **Neumann conditions** or boundary conditions of the **second kind**; and **mixed boundary conditions** or boundary conditions of the **third kind**. *Dirichlet conditions* are those where dependent variable values are given or assigned. Some examples of this type are assigning known value, ϕ=constant like the constant inlet velocity, pressure or temperature, and constant values at the boundary walls such as the constant boundary surface temperature or a moving wall velocity. *Neumann conditions* are those where normal gradients of dependent variables are assigned, $d\phi/dn$=constant. Examples of these include assigning a velocity gradient or a constant heat fluxes on the boundary surfaces. This also includes some special cases like **insulated boundary** or **adiabatic wall** or a **symmetric boundary**. Examples of mixed boundary conditions include convective or radiative heating or cooling of the boundary surface, and this can be derived based on an energy balances. Also, it can be noted here that for a given boundary, different types of boundary conditions can be applied for different dependent variables and at different sections of the same boundary.

Before assigning the boundary conditions, it is necessary to evaluate the geometry of the flow domains such as the external flow or the internals flow as depicted in Figure 9.1a and b, respectively

9.1.1 Inlet Conditions

The followings are some examples of the types of conditions that are assigned at the inlets depending on the type of physical problems and application types:

1. Velocity vector and scalar properties of the flow at inlet are defined at the velocity inlet boundaries. Inlet flow direction and magnitude of the velocity vector and other scalar properties of flow are assigned.
2. Uniform inlet velocity: Velocity at the inlet of a flow region is unidirectional and uniform across the cross-section.
3. Predetermined inlet velocity profile or tabular data.
4. A mass flow inlet is used for compressible flow, instead of velocity as in for incompressible.
5. Prescribed pressure inlet condition.
6. In addition to the velocity, turbulent flows require certain scalars turbulent quantities to define the turbulent parameters. Constant or predetermined profile of turbulence quantities needs to be assigned.

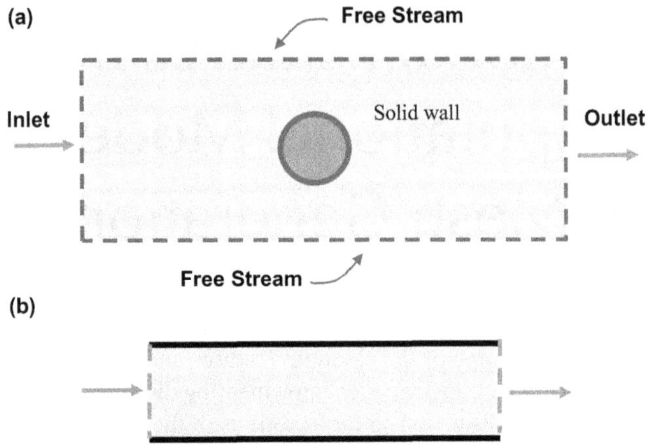

FIGURE 9.1 Boundary designations for flow domains. (a) External flow domain. (b) Internal flow domain.

7. Temperature inlet condition.
8. Special condition for considering any intake vents and fan.

Some of the examples are depicted in the figure below:

9.1.1.1 Restrictions on the Selection Inlet Location

The location of the inlet position to flow domain often is selected carefully to allow the flow to develop sufficiently by including additional inlet length section to achieve certain level of desired flow distributions before entering the region of interest. Care should be taken in placing a velocity inlet not too close to any solid flow obstructions such as tubes or cubes or not too close to the outer edge for flow in the presence of a backstep as shown in Figure 9.3. The placement of such a velocity inlet may impact the solutions in the wakes or recirculating regions downstream of the obstructions. For computational purposes, attention should be given in providing a sufficient entrance length to allow development and establish required flow condition upstream of the obstructions. Figure 9.3 shows some poor inlet positions over the backsteps.

As with any turbulent flow solutions, turbulent intensity at the inlet is necessary to be specified. This serves as the initial guess for the whole domain. Regions with higher shear stresses generate greater turbulence than the flow entering the domain. So, accurate estimation of inlet turbulence intensity is necessary based on experiments or based on established previous experiences. The range of turbulence intensity varies significantly depending on the application problems with intensity value ranging from below 1% to as high as 20%. Further discussion about inlet turbulent intensity is given in Chapter 10 on Turbulent Flow Modeling.

$$(\mathrm{Re}_{DH})^{1/8}$$

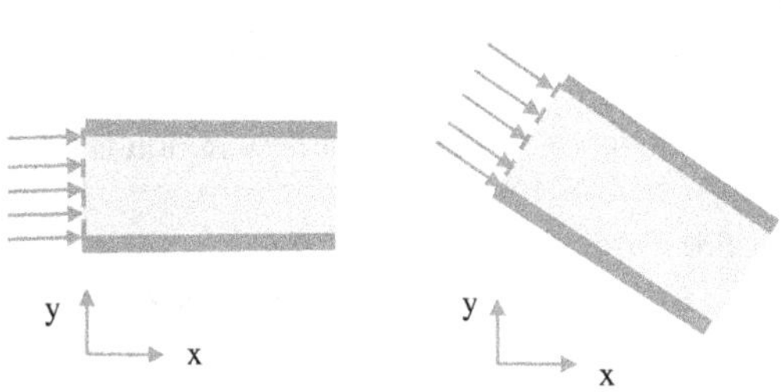

FIGURE 9.2 Assignment of boundary conditions at inlets.

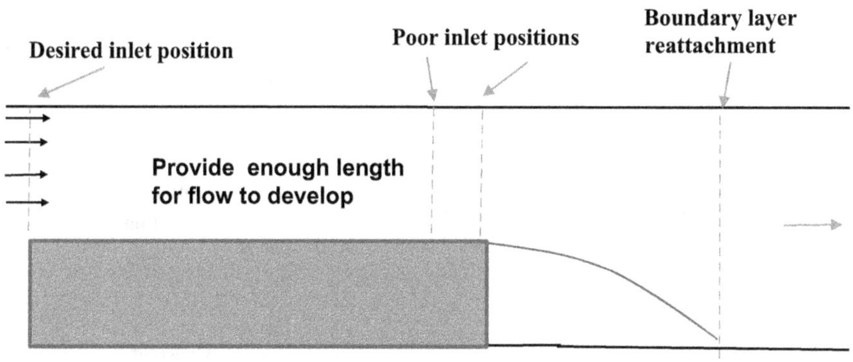

FIGURE 9.3 Requirement of selecting an inlet location.

9.1.2 Outlet Conditions

There is always some uncertainty in assigning an outlet boundary condition when the outlet flow conditions are not specifically known. In a computational analysis, one needs to perform few numerical experimentations to ascertain an appropriate outlet boundary condition. A list of several outlet boundary condition types is outlined below:

1. Target mass flow rate at inlet or mass flow is fixed from overall continuity. An overall mass balance correction is applied.
2. **Outflow boundary conditions** are applied to model flow at outlets where accurate information of the flow velocity and pressure conditions are not known beforehand. Outflow boundary conditions are often used when outlet flow approach close to a fully developed flow condition where the gradients of all dependent variables along the flow direction are taken to be zero. In order words, this implies that the convective derivative normal to the outlet boundary face is set to zero. For physical geometries or devices with shorter exit length, often additional computational exit length is used is emulate a correct flow exit condition.
3. Flow split outlets with assigned mass flow rates.
4. Prescribed pressure outlet condition.

Some of the examples are depicted in Figures 9.4 and 9.5 below:

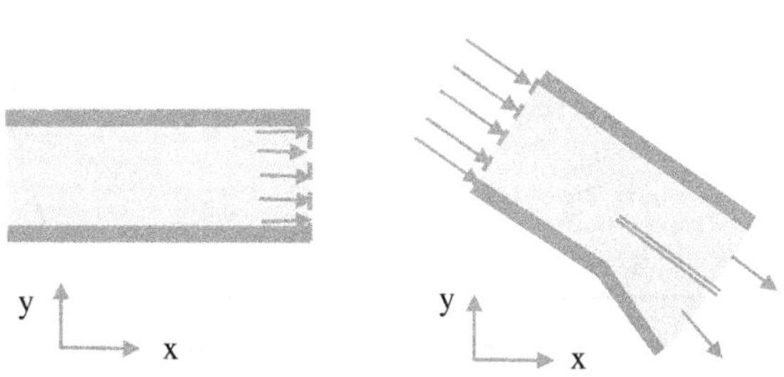

FIGURE 9.4 Assignment of boundary conditions.

FIGURE 9.5 Example of flow configures and conditions.

9.1.2.1 Restrictions of Assigning Fully Developed Outflow Conditions

There are some restrictions on the use of outflow boundary conditions in problems that involve adverse pressure gradient, flow separation and reattachment, and flow development. A computational domain consisting of three different flow regions in the downstream is considered and as shown in the figure below. As we can see that a wake region with recirculating flow regions is formed downstream for flows over a solid object. Because of the formation of adverse pressure gradient over the surface of the object, the flow boundary layer separates from the top sold surface and creates the wake and recirculating region. The recirculating extends over a length downstream of the object before the boundary layer reattaches as demonstrated in Figure 9.6 shows flow over backward-facing step geometry of height, *h*. The reattachment length varies depending on the shape and size of the object as well as on the flow conditions. This reattachment region is followed by a **developing flow region** and finally by a region of fully developed flow. A typical developing length of 3–8 h has been observed beyond the reattachment length and reported in many studies. As a guideline, an outlet boundary should be placed at a distance using an exit length of $x_L > 10$ h in the developed region to ensure that the outlet location do not fall within the recirculating regions and experience reverse flows as demonstrated using the axial-component velocity profiles at different sections within the recirculating zone. Exit length required to ensure fully developed outlet flow condition should be selected based on numerical experimentations.

Some of the other restrictions are (a) not recommended for compressible flows; (b) not recommended for use with pressure inlet condition; (c) do not use for situations where backflow may occur or in problems where flow condition downstream of the outlet plane may influence the flow domain.

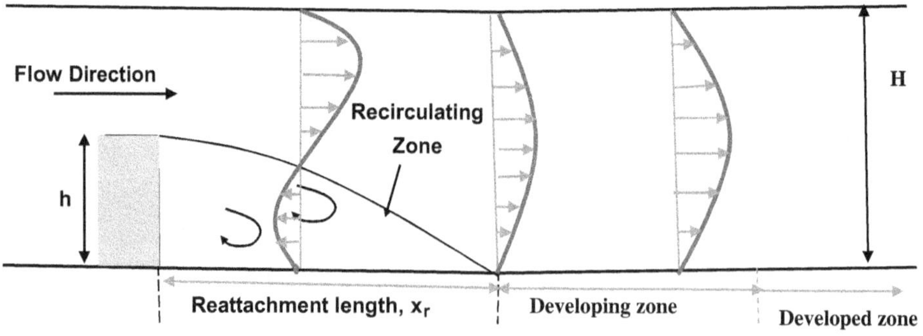

FIGURE 9.6 Reattachment length from flow separation point.

9.1.3 Wall Boundary Conditions

1. No slip velocity condition is generally applied to wall boundaries. A tangential component of velocity equals to wall velocity and the normal component of velocity set to zero. For example, for flow in channels formed between two parallel plates shown in Figure 9.7, the axial component of velocity equals to wall velocity, i.e., $u=0$ and normal component, $v=0$, at the stationary bottom boundary wall, and $u=uw$ and $v=0$ for the top wall which is moving at velocity of uw. Like translational velocity, rotating velocity can be assigned on the walls in the same manner.
2. Constant scalar value or flux is assigned on the wall surface; for example, a fixed temperature or an adiabatic wall or a constant surface heat flux or a convective mixed condition.
3. Wall material and thickness values need to be defined for conjugate heat transfer problems.
4. For momentum equations, a shear stress values can be specified. The wall roughness values need to be specified for turbulent flows.

9.1.4 Pressure Conditions at the Inlets and Outlets

Pressure boundary condition is applied for problems for which the mass flowrate or the velocity is unknown and for problems where backflow into the domain may occur. For pressure boundary

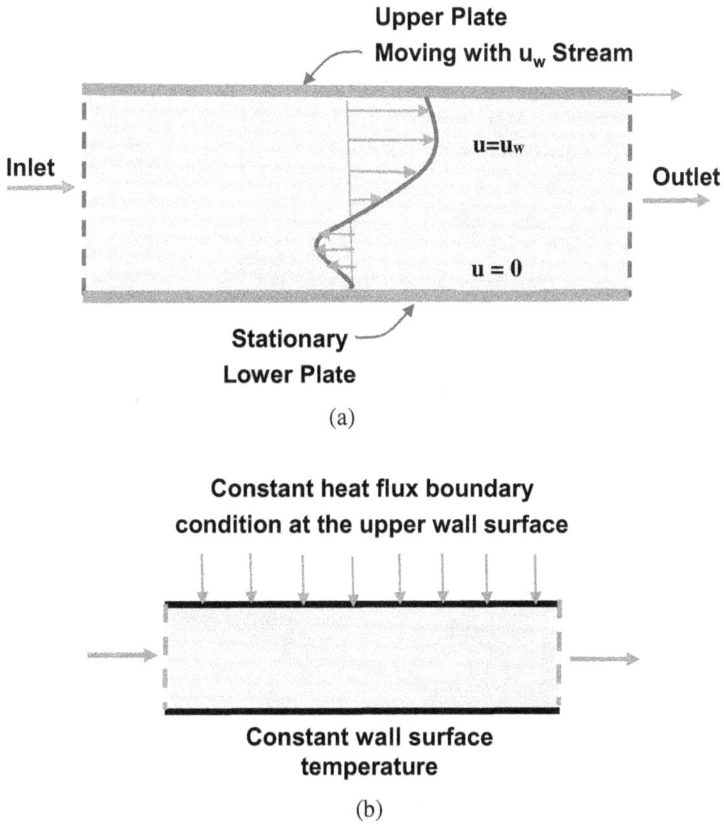

FIGURE 9.7 Wall boundary conditions. (a) No-slip velocity boundary condition. (b) Constant flux and constant scalar boundary.

conditions, the static gauge pressure is used while the operating pressure is specified separately. The total gauge pressure is defined for incompressible and compressible flows as follows:

$$P_{\text{total}} = P_{\text{static}} + \frac{1}{2}\rho V^2 \quad \text{For incompressible flows}$$

and

$$P_{\text{total}} = P_{\text{static}} + \left(1 + \frac{1}{2}(k-1)\right)M^2 \quad \text{for compressible flows}$$

where k=ratio of specific heats (c_p/c_v); M=Mach number ($M=V/c$); and c=speed of sound.

It is also essential to define the flow direction to avoid certain nonphysical solutions. While one needs to specify a temperature value for a nonisothermal incompressible flow, a total temperature value is required for a nonisothermal compressible flow defined in the following manner:

$$T_0 = T_{\text{static}}\left(1 + \frac{1}{2}(k-1)\right)M^2$$

9.1.5 Symmetric and Periodic Boundary Conditions

Symmetric and periodic boundary conditions are used when the both the flow and geometry exhibit symmetric and periodic properties, respectively, and allow solving a smaller subset of the entire flow geometry. This helps in reducing computational mesh, time, and efforts.

9.1.5.1 Symmetric Boundary Planes and Conditions

Figure 9.8 shows two such application problems. In Figure 9.8a, for liquid flow through rectangular channel, two planes of symmetry exist and this may allow us to consider only the one-quarter of the entire flow domain to reduce the computational effort.

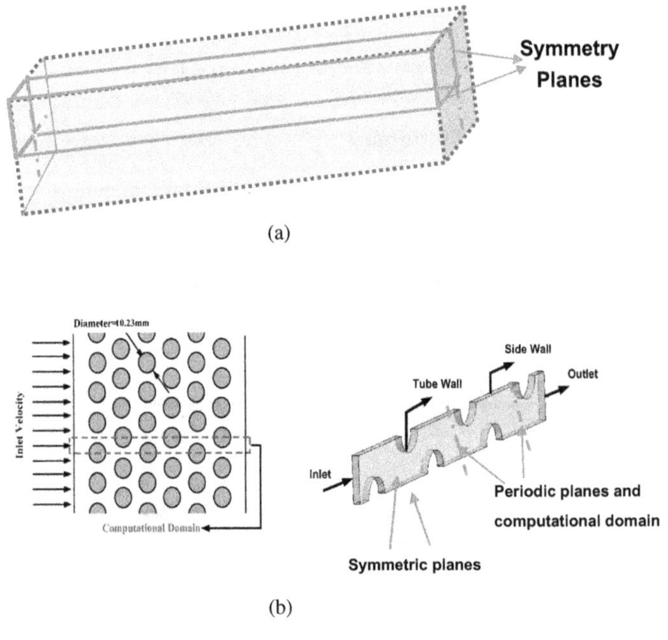

FIGURE 9.8 Examples showing symmetric and periodic planes flow over a bundle of tubes. (a) Symmetric planes in flow channel. (b) Symmetric and periodic planes for flow over a bundle of tubes.

Figure 9.8b shows external flow over a bank of tubes where symmetric planes (front, back top, and bottom planes) may be experience for cases with higher number rows or tubes along the flow direction in a bank for both staggered and in-line arrangements of tubes. Significant amount of computational time and effort could be saved by considering only a subregion bounded by the symmetric plans rather than solving it over the entire geometrical domain. Care must be taken to ensure that the errors due to edge effects or the presence of surrounding physical walls are minimal while selecting such symmetric planes and solution domains. The boundary conditions on symmetric planes are specified as (a) zero normal component of velocity at the symmetric planes and (b) zero normal gradient of all variables at the symmetric planes. Symmetric boundaries are also used to address slip walls in a viscous flow.

9.1.6 Periodic Boundary Planes and Boundary Conditions

Periodic flows and heat transfer seen in problems in which the physical geometry as well as flow and thermal conditions exhibit repeated or periodic nature. For example, as demonstrated in Figure 9.8b, because of the periodic nature of the tube arrays, the flow and heat transfer over a bank of tubes approaches periodic fully developed flow and heat transfer pattern after few tube rows in the flow directions. For such problems, considerable savings in computational effort can be achieved by considering a computational domain bounded by the pair of periodic planes. This is particularly applicable for a heat exchanger where major region of tube rows in the banks is far away from the inlet developing regions

Figure 9.9 shows another example showing periodic flow regions and planes for flow in a rotating fan-blade impeller.

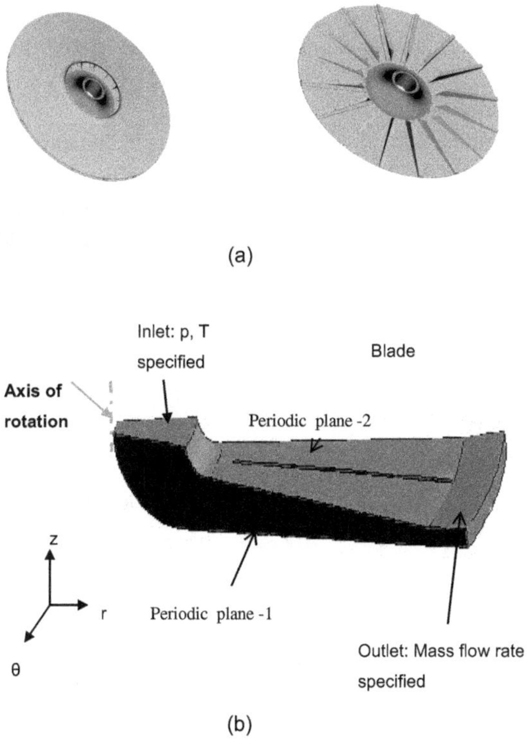

FIGURE 9.9 Example of periodic flow region and periodic boundary in a rotating fan impeller. (a) Rotating fan impeller. (b) A single blade passage of the centrifugal fan impeller with a pair of period planes.

In this example, a three-dimensional flow in a single blade passage of the centrifugal fan impeller is considered, considering cyclic nature of the flow and impeller geometry. The computational domain is depicted in Figure 9.9b. Periodic boundary planes extend from the inlet impeller eye section, flow passages on two sides of the impeller blade to the impeller exit.

The periodic boundary conditions refer to the pair of boundary planes where the flow repeats itself. The **period boundary conditions** are applied by considering that the flow leaving one periodic plane (periodic plane –2) is equal to the flow entering the other periodic plan (periodic plane –1) of the pair. There are two ways one can apply the periodic condition: (a) for transitionally periodic, a finite pressure drop, ΔP, per period or net mass flow is applied and (b) for rotational periodic ΔP can be applied around the axis of rotation.

9.2 MESH TYPES AND MESH GENERATION

The computational effort and quality of solution based on computational fluid dynamics and heat depend strongly on the quality mesh type and on mesh size distributions used. Hence it is especially important to have a good understanding of the type of meshes and good practices of mesh size distributions.

9.2.1 Mesh Types

Meshes can be classified based on dimensionality and configurations. Some examples of mesh types are described here.

9.2.1.1 Two-Dimensional Mesh

Two-dimensional meshes are used for studying two-dimensional problems and for surfaces of three-dimensional volumes. Some examples of the most popular two-dimensional meshes are quadrilateral, triangular, and polynomial shapes as demonstrated below (Figure 9.10).

9.2.1.2 Three-Dimensional Meshes

Three-dimensional meshes are used for three-dimensional problems and for volume meshing. Few examples are given below (Figure 9.11).

Hexahedral/quadrilateral meshes are more desirable than tetrahedral/triangular meshes as the computational time required for generating hexahedral meshes would be less compared to

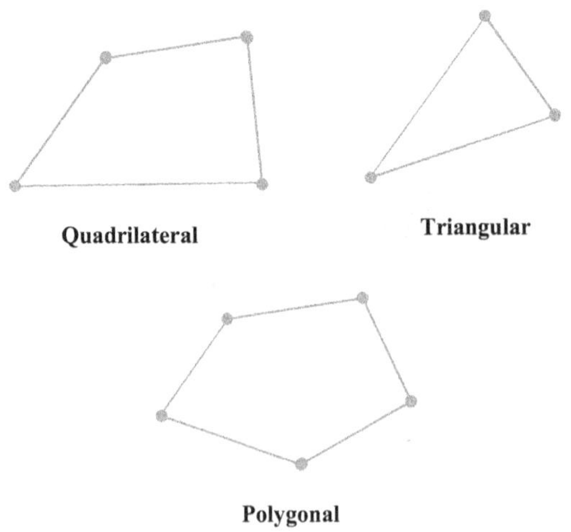

FIGURE 9.10 Two-dimensional meshes.

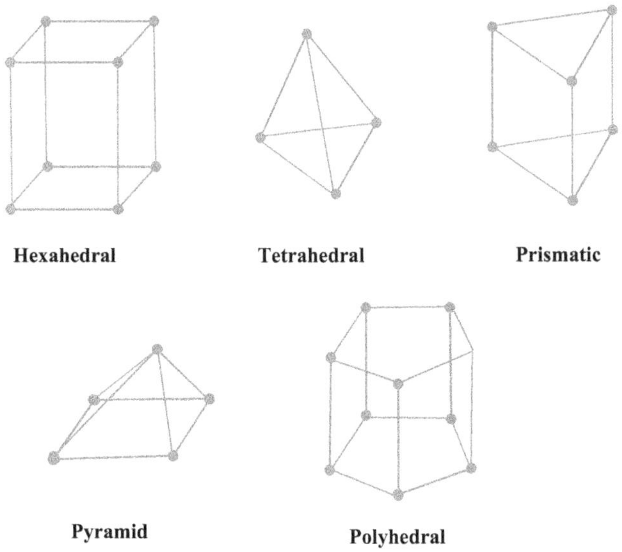

FIGURE 9.11 Three-dimensional meshes.

tetrahedral. One way of generating hexagonal meshes is based on multiblock topology discussed in later section. In such case, the geometry will be partitioned into multiple hexahedral blocks within which an array of elements will be created. The major advantage of using this technique is that it produces highly structured grids.

9.2.2 Mesh Size Distributions

So far in our presentation of control volume and finite difference discretization methods, we have employed straight forward structured and orthogonal mesh or grid shapes as demonstrated in Figures 5.1, 5.3, 6.9, and 6.10. Use of such orthogonal mesh works very well in discretization or creating a structured mesh or grid distribution for a simpler geometries as shown in Figure 9.12 below.

For simple and regular flow geometries, the most straight forward approach is to employ an orthogonal (90°) grid in a Cartesian co-ordinate system. This type of grid is generally termed as a *structured mesh* where the grid lines follow the coordinate directions. A *structured grid* consists of planar cells with four edges for 2-D and volumetric cells with six faces for 3-D. Cells may be distorted from rectangular, but each cell is numbered according to i, j, k indices [see Figure 9.13a]. Grids are generated by connecting grid points from left side to corresponding nodes on the right and from bottom to top.

Some of the advantages and disadvantages of structured mesh types are summarized below:

Advantages include (a) points of structured mesh can be easily addressed by the indices i, j, k; (b) connectivity is straight forward because the cells adjacent to elemental face are also identified

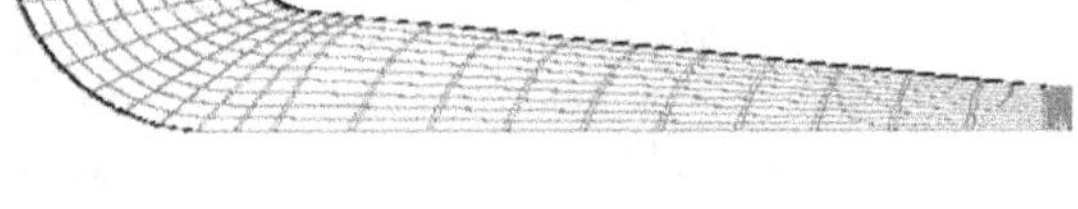

FIGURE 9.12 Structured mesh distribution in a fan blade passage.

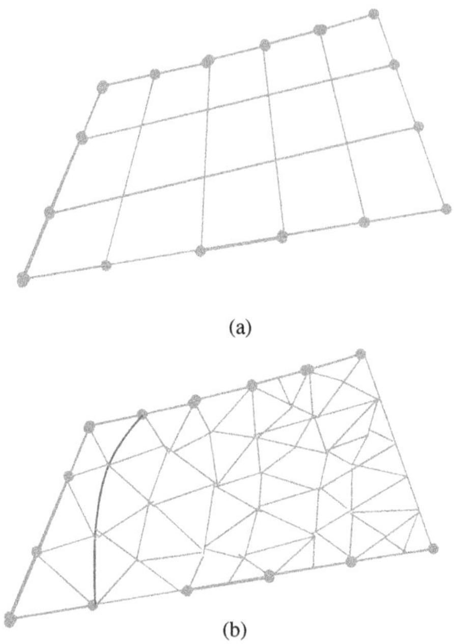

FIGURE 9.13 Structured and unstructured meshes. (a) Structured 5×3 grid with 15. (b) Unstructured triangular grid.

by the indices; (c) cell edges form continuous mesh lines that begin and end on opposite elemental faces; (d) in **two dimensions**, the central cell is connected by four neighboring cells and in **three dimensions** the central cell is connected by six neighboring cells; (e) provides easy data management, connectivity, and programming; and fewer cells are usually generated with structured grid than with an unstructured grid.

Disadvantages (a) For more complex geometry, the nonorthogonality or skewness can cause unrealistic solution; (b) the accuracy and efficiency of the numerical algorithm depend strongly on skewness; (c) additional cost of computation and slower convergence; (d) difficulties in programming.

A **body-fitted grid** that allows the creation of a nonorthogonal mesh with deformed grid cells can be adopted. The cell surfaces within this grid layout are allowed to follow the surface of the domain boundaries. These **cells still retain a regular elemental shape property** in the form of either a **skewed rectangular shape** type of element with four nodal corner points in two-dimensions or a **distorted hexahedral-shape type of element** with eight-nodal corner points in three-dimensions.

Figure 9.14 shows a body-fitted mesh-size distribution at the bottom curve surface of a flow domain and the resulting solution in a Computational Fluid Dynamics (CFD) flow simulation study.

Nonuniform structured mesh size distribution without matching faces is often used in many problems to capture steep variation of dependent variables. Figure 9.15 shows such use of mesh distribution that includes nonmatching mesh faces.

9.2.2.1 Unstructured Mesh

For flow geometry with inclined surfaces or more complex shapes, unstructured meshes are more effective than structured mesh in representing the region more accurately and effectively. Unstructured mesh (See Figure 9.13b) consists of cells of various shapes: typically, **triangles** or **quadrilaterals** for 2-D and **tetrahedrons** or **hexahedrons** for 3-D. **polyhedral meshes** are developed to fill the interior volume domain and found to be advantages over tetrahedral meshing with regard to accuracy and efficiency of the numerical computations. The use of unstructured mesh has become more popular and widespread use in CFD.

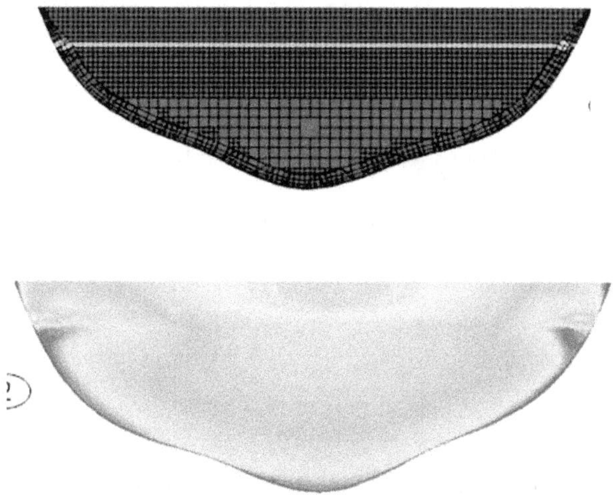

FIGURE 9.14 Body-fitted mesh used in a corrugated flow channel.

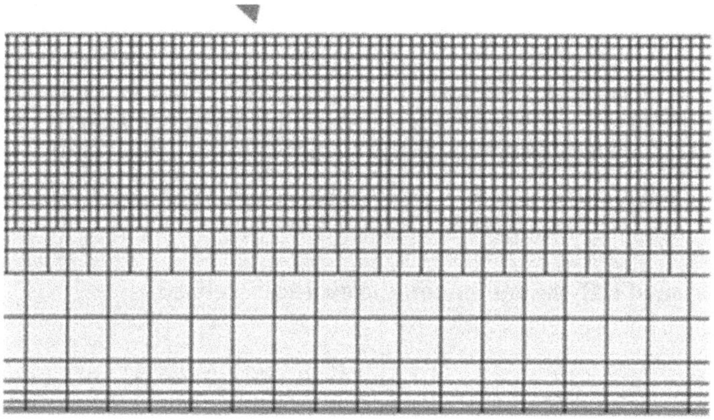

FIGURE 9.15 Nonuniform structured mesh size distribution with nonmatching faces.

Some of the advantages and disadvantages of unstructured mesh types are summarized below:

(a) **Unstructured meshes** are well suited for handling arbitrary shape geometries including high-curvature geometries. (b) Unstructured meshes can be assembled freely within the computational domain. The connectivity information for each face requires appropriate storage in some form of a table. (c) In general, the unstructured meshes do not have the constraint of matching cell faces. This can also be achieved using other types of elements such as triangular or combination of both triangular and quadrilateral elements.

9.2.2.2 Hybrid Mesh

In many problems, hybrid mesh distribution is also used where a variety of different mesh types are used. For example, a combination of structured meshes in far-away uniform region and unstructured meshes in complex or curved region are used.

9.2.2.3 Skewness

It is defined as the amount of departure from the symmetry. Quality of grid is important for both structured and unstructured grid systems. Quality of grid is extremely critical for CFD solutions

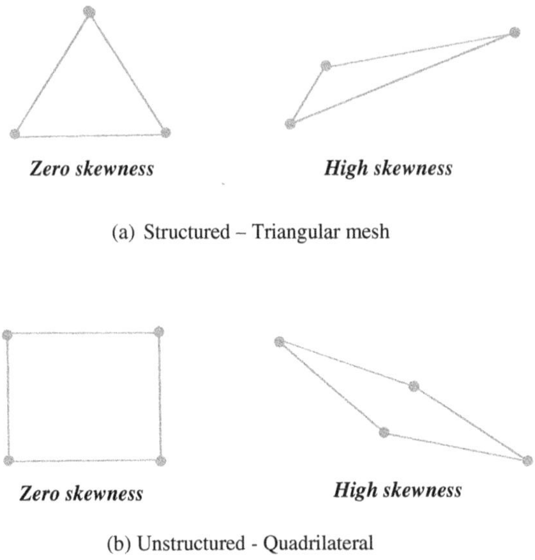

FIGURE 9.16 Level of skewness of meshes. (a) Structured – triangular mesh. (b) Unstructured – quadrilateral.

because of the presence of strong nonlinearity and advection components in a coupled set of differential equations for velocity and scalar components. It is important to ensure that the individual cell is not highly skewed. Otherwise, it can lead to difficulties in convergence and inaccuracies in the numerical results. Figure 9.16 demonstrates the level of skewness in different meshes.

There are different kinds of skewness for both two-dimensional and three-dimensional grid systems.

For **two-dimensional cell**, the **equiangular skewness** is defined as

$$Q_{\text{EAS}} = \text{MAX}\left(\frac{\theta_{\max} - \theta_{\text{equal}}}{180^\circ - \theta_{\text{equal}}}, \frac{\theta_{\text{equal}} - \theta_{\min}}{\theta_{\text{equal}}} \right)$$

where

$\theta_{\max}$, $\theta_{\min}$ = maximum and minimum angles (in degrees) between any two edges of the cell.

θ_{equal} = angle between any two edges of the ideal equilateral cell with the same number of edges.

For any two-dimensional cell, the equiangular skewness can fall in the range of $0 < \text{QEAS} < 1$.

Equilateral triangular cell or square and rectangular cells have zero skewness. Figure 9.17 shows some examples of such level of mesh skewness and quality. A grossly distorted triangular and

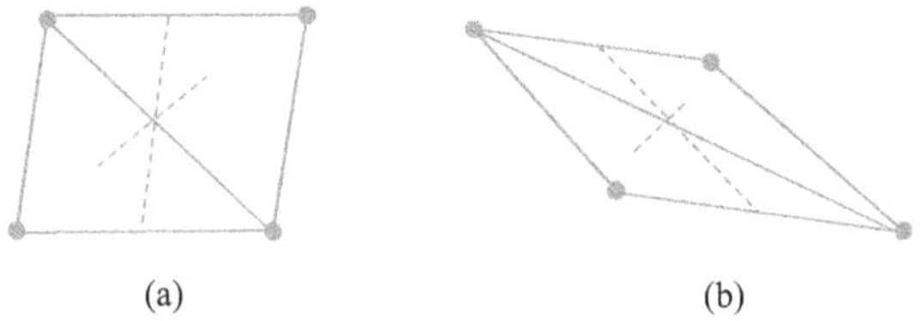

FIGURE 9.17 Skewness and quality for two-dimensional meshes. (a) Lower skewness and higher quality. (b) High skewness and lower quality.

quadrilateral cell may have unacceptably high skewness. Some mesh generation codes use numerical schemes to smooth the meshes to minimize skewness.

Other factors that affect the mesh quality are (a) abrupt changes in grid sizes and so it is necessary to vary the mesh spacing gradually; and (b) meshes with exceptionally large aspect ratios and so try to keep the aspect ratio less than a factor 8–10. High-quality unstructured mesh may be better than a poor-quality structured grid.

9.2.3 Mesh Generation Procedure

A typical mesh generating procedure is demonstrated with an example shown in Figure 9.18 for an application in the simulation study of wind turbine performance. In this analysis, different meshing schemes are used for discretizing the complex flow domains around the wind turbine blades and structures. Both surface meshing and volume meshing are used to capture the complex flow interactions of air flow over the wind turbine blades and structure.

Use of surface meshing is used to divide the whole air flow domain into several surfaces. Different types of mesh shapes used for representing those complex-shaped surfaces and progressively grow into volume mesh cells. Surface meshing helps in representing the curves and blade profile geometry. In the case of volume meshing, polyhedral meshing worked well for some application since it has more correlated nodes and increases the accuracy of the computation. Prism layer meshing has been chosen to handle the boundaries in the flow field, especially near the blade boundaries.

In addition, fine meshing zones are used upstream near the rotor and then near field of downstream of the wind turbine. This feature increases the number of cells in those regions to capture the boundary layer formation over the blades and wake region of the flow.

While the minimum size has been assigned to 10% of the base size on the core volume region; however, the trials and mesh refinement study needs to be conducted to reveal the final choice for the small enough mesh element size for blades, cylinder, and downstream to reach an acceptable mesh quality. Mesh customization in this kind of problem plays an important role since the number of mesh controls the computational time. Also, it can be noticed that considerably large number of meshes are used in the rotational region compared to the outer wind flow region. Prism layers are used to allow the solver to resolve near wall-boundary flow accurately, which is critical in determining not only the forces and heat transfer on walls but also flow features such as separation and recirculating eddies and turbulence quantities for additional studies such as noise generation over a range of frequencies.

Figure 9.19 displays some results of the wind turbine performance in terms of pressure distribution over the wind turbine surfaces and in regions around a blade section.

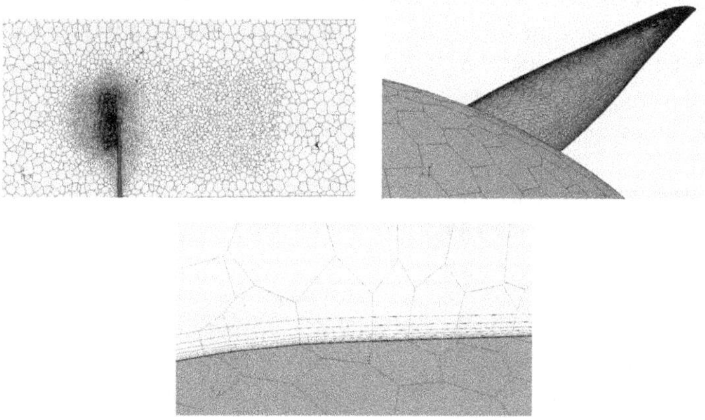

FIGURE 9.18 Meshes and results in simulation of wind turbine blade performance.

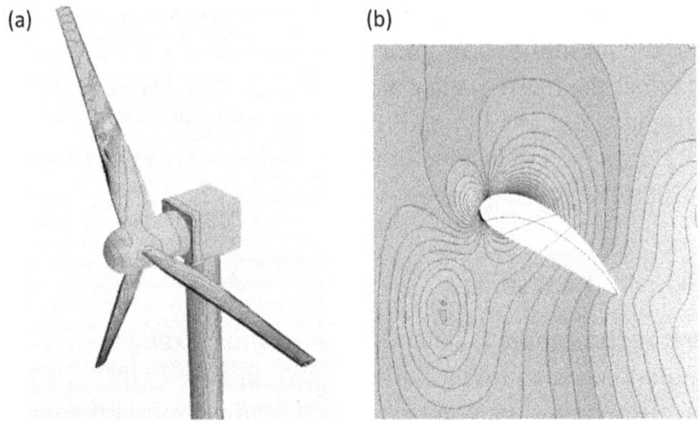

FIGURE 9.19 CFD results based on the meshes used in the wind turbine performance simulation. (a) Static pressure distribution on wind turbine blades. (b) Pressure distributions are wind turbine blades.

9.2.4 Multiblock Mesh System

Multiblock meshing is an efficient way of using different scales of mesh size distribution for select regions of complete geometries. The procedure involves constructing different block regions and configurations and then constructs the mesh with increasing details providing higher resolution to infer greater details to the physics of the problem.

Figure 9.20 shows some examples of multiblock meshing approach. Figure 9.20a shows the use of multiblock meshing near the interfaces of electrode–membrane interfaces and the gas flow channels.

Figure 9.20b shows that in simulating the multiscale transport phenomena in a lithium-ion battery cell, multiple block regions are created with different mesh resolutions in different parts of Li-ion cell including the electrode terminals and adjacent cold plates with integrated cooling channels.

Figure 9.20c shows the use of multiblock meshing with different mesh resolutions at the curve tube surface regions and in the core flow regions to capture the complex flow pattern and nature of heat transfer surrounding the tube surfaces.

9.2.5 Prism Layer

Prism layers are used to provide higher resolutions in thin boundary layers in which the velocity and other scalar quantities experience steep variation experience in the transverse direction near the sold-fluid interfaces as well interfaces of the different media such as at the interface of fluid region over a porous media region. While an exceptionally fine mesh size distribution could be used over the entire solution domain, such an approach is computationally expensive and not highly effective. Alternatively, a finer mesh distribution, normally known as the **prism layer mesh**, could be used near the wall interface regions in the presence of viscous sublayer and progressively increase the mesh sizes to coarser meshes in the core region. This helps in reducing computational time at the same time achieving the same level of accuracy. Prism layers are tightly packed meshes with extremely high aspect ratios like prismatic meshes without increasing the stream-wise resolution.

Accurate prediction of these flow features depends on resolving the velocity and temperature gradients normal to the wall. These gradients are much sharper in the viscous sublayer of turbulent boundary layers than would be implied by taking gradients from a coarse mesh. Using a prism layer mesh allows you to resolve the viscous sublayer directly if the turbulence model supports it (low $y+\sim 1$). Alternatively, for coarser meshes, it allows the code to fit a wall function more accurately for higher $y+$ values.

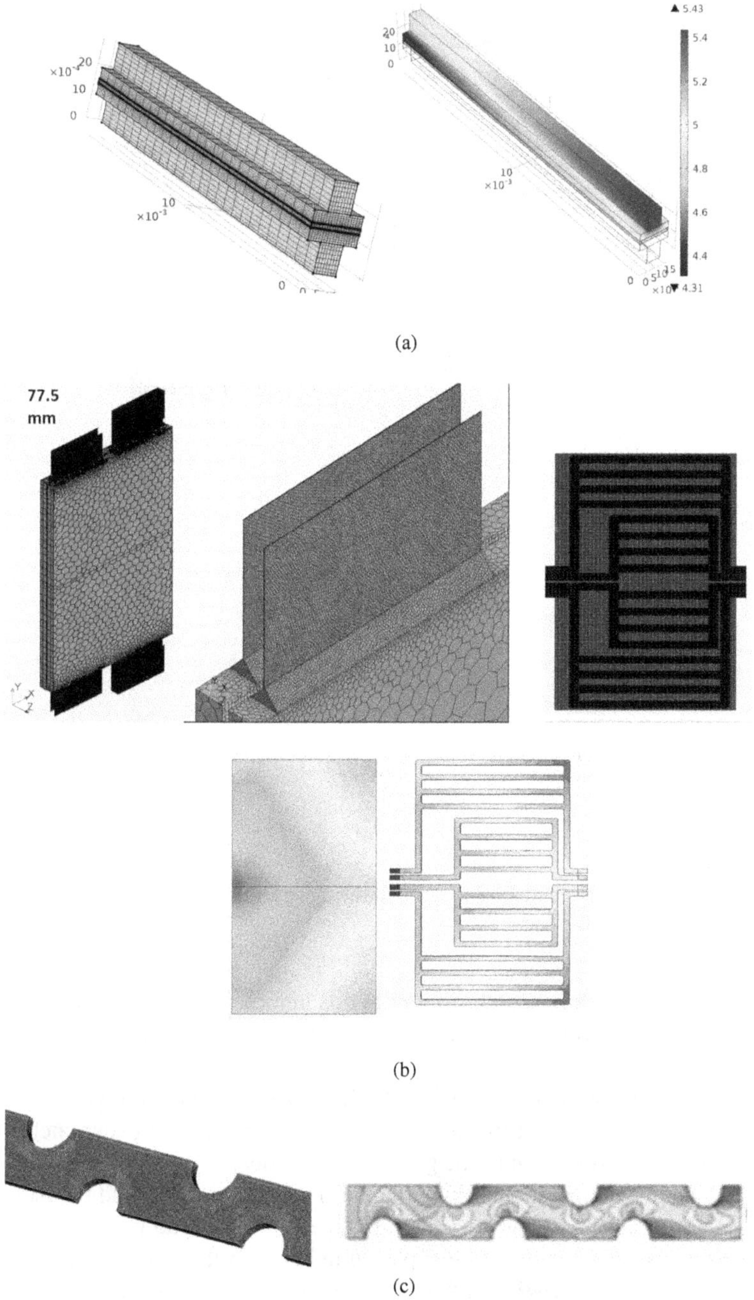

FIGURE 9.20 Multiblock meshing examples. (a) Multiblock meshing in fuel cell and gas flow channels. (b) Block meshing and solution for Li-ion battery pack with cooling plates and cooling plates design. (c) Block meshing and periodic solution for flow over a bundle of tubes.

The number of prism layers needed varies widely depending application and the type of wall function used. The adequate number is always established after conducting some numerical experimentations.

Figure 9.21 shows an example of using prism layers in computational simulation study for open channel flow over a porous sediment bed and bridge deck. The use of prism layers at the

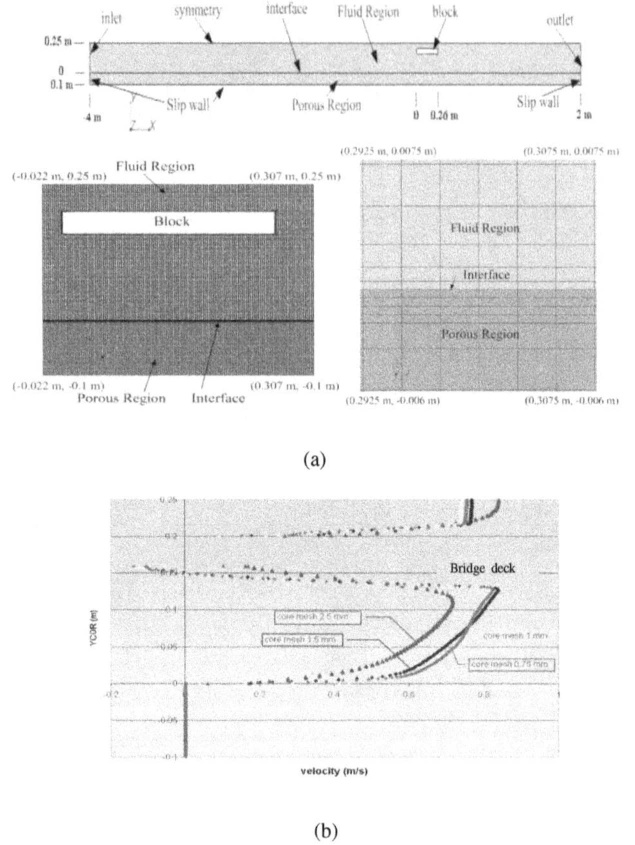

FIGURE 9.21 Open-channel river water flow around the bridge. (a) Velocity profiles in the channel around the pier. (b) Velocity profiles in the channel.

interface of water flow over a porous sediment bed is demonstrated. To capture the steep variation of the axial component velocity in the transverse direction at that interface, 5–6 prism layers are used.

An example of block meshing and use prism layers is demonstrated in Figure 9.22 for flow in an open channel with a cylindrical pier located at the bottom surface. Due to symmetric nature of the geometry and flow, only half section of the geometry is modeled.

In order to reduce the computational time, the mesh is made finer around the pier and the prism layer meshing is used to refine the grid at the interface. The fluid and porous regions are meshed separately, and after meshing, an interface is placed in between them. In the prism layer mesh, the pink region represents the **fluid region** and the blue region represents the **porous region**. The prism grid layer thickness was gradually increased to match the core mesh size. A volumetric grid was used to refine the grid around the cylindrical pier.

Domain Decomposition and Meshing a Region

Depending on the problems, a free or a mapped mesh can be generated using different schemes. While a prescribed control volume shape and regulated order of spacing can be achieved through a mapped meshing, edge meshing can be performed considering the distance from the wall using a gradual nonuniform grading scheme. The mesh at the wall region is resolved by placing the first point from the wall at a distance given by y-plus criterion for turbulent flow.

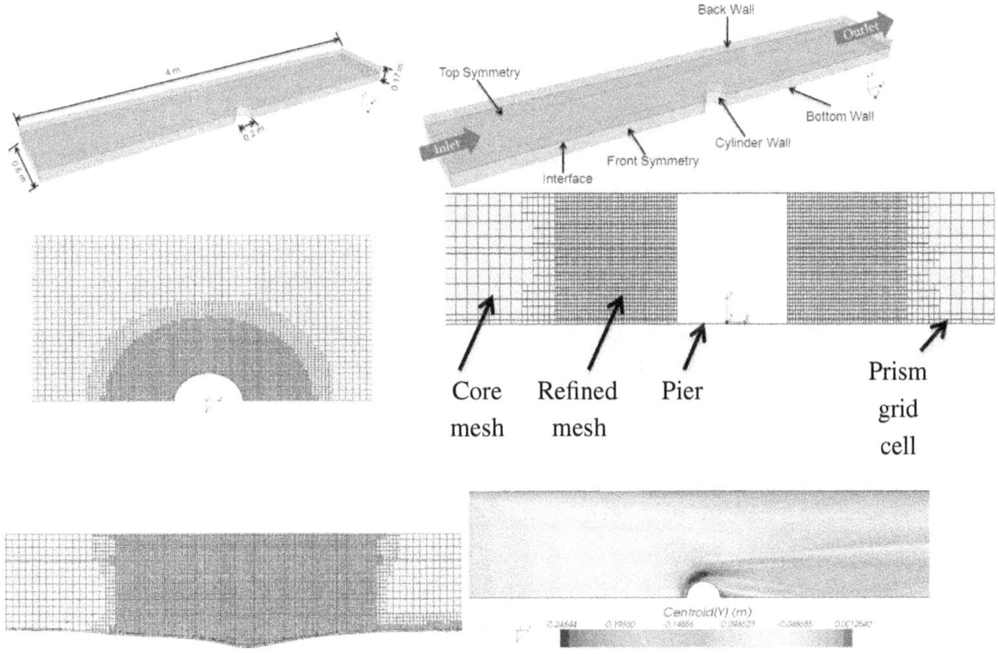

FIGURE 9.22 Example of block meshing around a cylindrical pier and the use of prism layer mesh at interface surfaces.

Example 9.1: Multiblock Mesh Generation for the Centrifugal Fan Impeller

Figure 9.23 depicts ways to use multiblock domain decomposition in creating mesh in the symmetric section of the impeller blade with periodic boundary conditions. The decomposition divides the impeller region into four volumes, each of which could be mapped with hexahedral structured mesh.

It can be seen that the used mesh and the computational physics predict large recirculation regions on the blade suction side. On the pressure side of the blade, low-pressure region is developed at the hub-pressure side corner and on the suction side at the suction-casing corner. The vector plot of relative velocity vectors in the meridional view, as presented, reveals large areas of low-velocity regions on the blade suction side. The flow separation region is developed close to suction-casing corner due to the flow turning from the axial direction to radial direction

9.3 MULTIGRID (MG) METHOD

MG methods are iterative methods which has the rate of convergence independent of the mesh resolution (Nicolaides, 1975,1977). MG methods are used in computational fluid dynamics and heat-transfer problems where regular iterative solvers such as Gauss-Seidel, Successive Over-relaxation (SOR), Conjugate Gradient (CG), and Generalized Minimal Residual (GMRES) methods experience time-consuming and very slow convergence in solving a large system of algebraic equations that are derived based on the grid discretization of equations using control volume/finite difference or finite element methods. The large system of linear equations derived based on the control volume/finite difference method discussed so far requires repetitive use of iterative solvers until convergence is reached, i.e., the residuals for mass continuity, momentum, and other scalar equations and scalar quantities such as temperature, mass species transport, and turbulence quantities fall below the set error tolerance level. The MG method is particularly useful to address the multiple scales of error frequencies that are present in the errors as observed in the residual balance progression toward

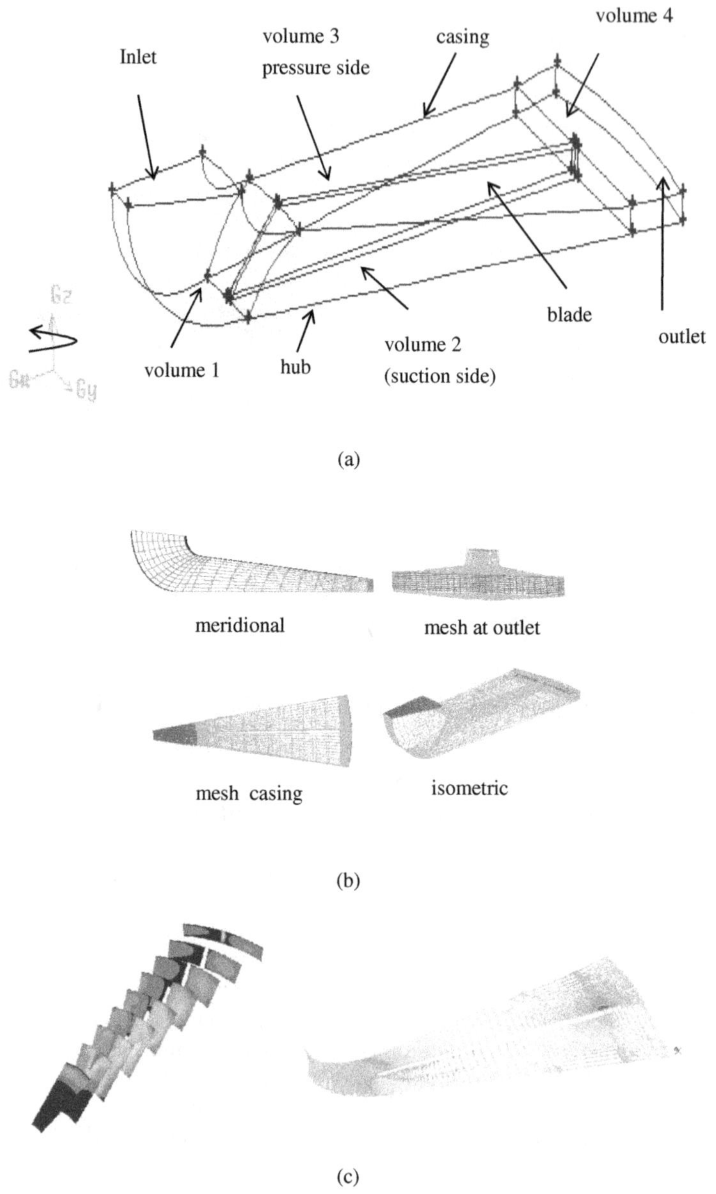

FIGURE 9.23 Multiblock domain decomposition in creating mesh in a symmetric section of impeller blade with periodic boundary conditions. (a) Multiblock regions for mesh generation. (b) Block meshing for different regions of the impeller blade. (c) Velocity magnitude contours in a symmetric section of the impeller blade.

the convergence (see Figure 9.24). Basically, the slow converging residuals on a given mesh or grid level are interpolated to a coarser grid and solved on that grid with better convergence rate. The MG method is an accelerated technique for solving a system of linear equations more efficiently in coordination with the relaxation methods such as iterative solvers such as the Gauss-Seidel, Jacobi, SOR, CG, and GMRES methods.

Figure 9.24 depicts typical progression of residuals for mass, momentum, energy, and turbulence quantities with iteration using one of the relaxation-iterative solver. As we can see that even with the set tolerance limit of 10^{-6}, some of the residuals tend to level off to a much high error limit exhibiting multiple scales of frequencies. Major reason for using MG methods is to enhance the

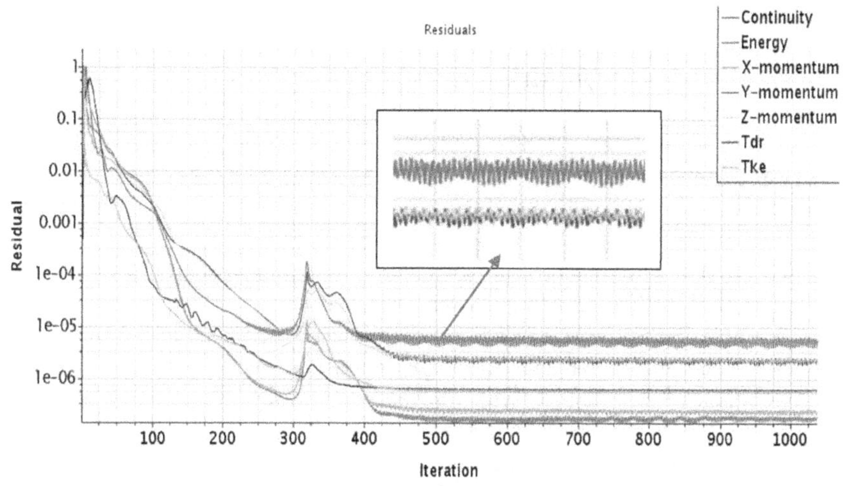

FIGURE 9.24 Typical progression of residuals.

convergence rate. Such a method specifically helps improving the slow converging residuals or error levels in some equations such as in the solution of mass conservation equation or the pressure equation derived based on combining mass continuity and momentum equations. If one can solve system of equations quite efficiently with Gauss-Seidel or SOR or CG methods, then it would be unnecessary and expensive to activate the use of MG methods. MG method is always used along a regular iterative solver and activated during computation process as needed and for specific variables.

MG Method is developed for solving differential equations based on an algorithm that uses a layer of discretization of equations derived based on different mesh or grid levels from finer grid to coarser meshes. The basic idea is that certain error levels or frequencies can be resolved at coarser grid level using iterative relaxation methods or even a direct solver. The original MG Method (Nicolaides, 1975, 1977; Braess and Hackbush, 1983; Bank and Dupont, 1980) is referred to as the **Geometric Multigrid (MGD)** methods. Basic computational methodology in MG method includes several basic steps:

a. Reduce high-frequency errors using number iterations-based relaxation methods. This step is referred to as the **Smoothing**.
b. **Residual error computation** at the end of the smoothing operation.
c. Reduce the grid size distribution to a coarser one. This is referred to as the **Restriction**
d. Interpolate the corrected computation at a coarser grid level to finer grid level. This procedure is referred to as the **Prolongation**.
e. Include prolonged solution obtained at a coarser grid level onto a finer grid level. This procedure is referred to as the **Correction**.

The **MGD** method can be performed following a number of different sweeping patterns such as V-cycle, F-cycle, and W-cycle types, which are categorized based on the nature of sweeping pattern in going down and up from/to different mesh levels.

The algorithm for GMD method following V-cycle sweeping is summarized below:

i. Start computation with a finest mesh size distribution and obtain solution after applying few iterations using one of the relaxation methods.
ii. Using the **presmoothing** step, this approximate solution obtained at finest mesh is than mapped to nest coarser mesh level. Using the results from finest level is then transformed to next coarse level of mesh distribution.

iii. The solution is then corrected at this coarser mesh level using few iterations using the relaxation method. Computation is continued at this coarse level for some relaxation cycles and then **restricted** to next higher coarse level.
iv. This process of successive down restrictions and corrections of the solution to a number of coarse mesh levels is continued until the convergence or the residual error reaches a set tolerance and hence the higher frequency level of error is resolved at the coarsest level.
v. The process is now reversed by moving up from the coarsest level to the finest level after passing through the coarser mesh levels.
vi. Step 3 is continued with transforming the results from current coarse level to even higher coarser level until the highest coarser level is reached.
vii. Results from the highest coarser level are then interpolated back to the finest mesh size distribution. This step is referred to as the **prolongation**.
viii. After some iterative computation with some relaxation steps, the solution from this finer level is than interpolated to next finer level. This is referred to as the **post-MG sweep**.
ix. Steps 1–6 is then repeated until a required level of convergence is reached set by tolerance limit.

The slow convergence rate on a given grid level is interpolated to a coarser grid and solved on that grid with better convergence rate. The corrections obtained on the coarser grid are then extrapolated (**prolongated**) to the finer grid level. This cycling is continued until convergence.

While it has been demonstrated that MG method improves the grid, mathematical theory shows that this strategy procures grid independent rates of convergence; however, additional effort is required in transferring the residuals and corrections between the finer-coarser grid levels. In the present effort, we will incorporate coarsening strategies by link listing points to various levels in a Cartesian box grid.

9.3.1 Algebraic Multigrid Method

Algebraic Multigrid Method (AMG) was proposed to alleviate some difficulties encountered in a using **MGD** especially for some complex geometries [1987]. One may experience difficulties in implementing the coarsening steps, which involve generating a new mesh over the geometry and over which a new set of discretization equations is solved. Such a process not only involves increased computational efforts in processing and storing the coefficient information but may also suffer from loosening mesh quality for some complex shapes.

To overcome such difficulties, the AMG developed with some simplification. In AMG method, the coarsening procedure does not involve creation of a new set and hence a new set of discretization equation, but rather reduces the order of the matrix system by agglomerating the element entries of the matrix and then iteratively solves the new agglomerated reduced matrix system using the relaxation solvers. This process of reducing order of the system by just agglomerating system at coarse level without really creating any new coarse mesh and solving new set of discretization equations not only reduces the computation difficulties but able resolves the higher frequency errors by restricting to coarse level.

In AMG method, the coarsening of a fine grid is done by the strengths of the coefficients rather than by the geometric layout. The coarse level system of equations is generated without the use of the geometry or re-designating grid points to preserve the best convergence. This provides opportunities for coarsening in selected directions and regions such as the wake regions to reach targeted convergence. The corrections obtained on the coarser grid are then extrapolated (prolongated) to the finer grid level while sweeping upward following a sweeping cycle type like in a V-cycle. This cycling is continued until convergence. The strategy secures grid-independent rates of convergence. However, some work is needed in the transfer of residuals and corrections between the grids.

10 Turbulent Flow Modeling

In this chapter, computational modeling of turbulence flow is discussed with a primary focus on the Reynolds Averages Navier Stokes (RANS) turbulence model and associated turbulence closure models. Brief descriptions of various turbulence closure models and some guidelines for choosing a turbulence model are also given. Furthermore, some important aspects related to the turbulent fluid flow computations such as wall function treatments; selection of $y+$ values; and wall and boundary conditions for turbulence scalar variable are outlined.

10.1 PHYSICAL DESCRIPTION OF TURBULENCE

A brief review of the physical description and some major characteristics of turbulence is given in this section. For more comprehensive understanding of turbulence, it is recommended that readers refer to the books by Schlichting and Gersten (2000), Hinze (1975), and White (1991) as well as the seminal article by Klebanoff (1955). One important characteristic of a turbulent flow is that the velocity and pressure may be steady or remain constant at a point but may still exhibit instantaneous fluctuations over the mean or average value. Figure 10.1 shows such instantaneous random fluctuations of a single component of the velocity, $u\ (x, y, z, t)$, in shear layer. In this plot, the mean flow is subtracted from the mean to demonstrate the instantaneous fluctuating component, which may be as high as 5%–10% around the mean for a turbulent flow. Such continuous random fluctuation forms are applicable to all three velocity components u, v, w with different orders of magnitudes, and also to the scalar quantities like p and T.

For the velocity $u\ (t)$ at a point in the flow, we can define the time average value of velocity component as

$$\bar{u} = \frac{1}{\Delta T}\int_{t}^{t+\Delta T} u\,\mathrm{d}t \tag{10.1}$$

Where ΔT is a selected period that is larger than the period of fluctuations.

While the mean fluctuation $\bar{u}'$ of each component is zero, the significant turbulent quantity that characterizes the magnitude of fluctuations or turbulence is the mean-square value which is defined as

$$\bar{u}'^2 = \frac{1}{\Delta T}\int_{t}^{t+\Delta T} \bar{u}'^2\,\mathrm{d}t \tag{10.2a}$$

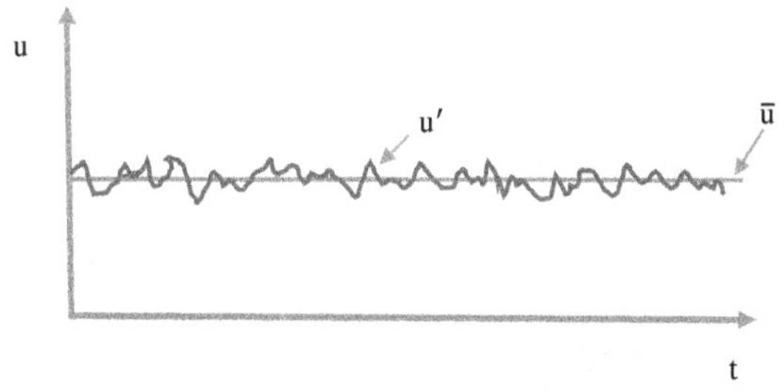

FIGURE 10.1 Instantaneous velocity fluctuations of a velocity component at a point in turbulent flows.

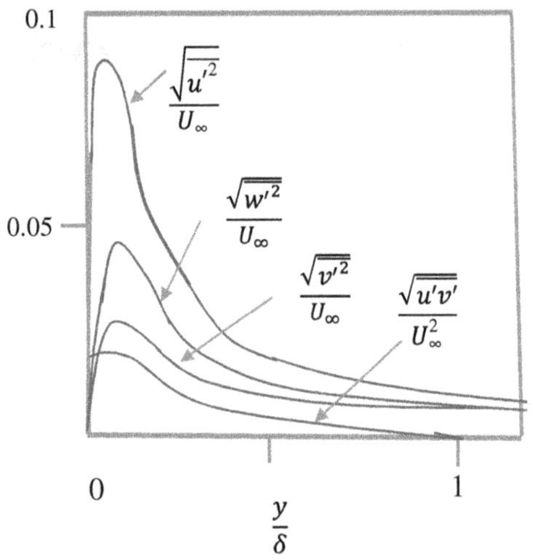

FIGURE 10.2 Representation of velocity fluctuations and turbulent shear stress variations in turbulent boundary layer flow over flat wall surface.

and the **root-mean-square (rms)** value is given as

$$u'_{\text{rms}} = \left(\overline{u'^2}\right)^{1/2} \tag{10.2b}$$

Experimental data for the rms values for all three velocity components and the covariance quantity $\overline{u'v'}$ referred to as the apparent turbulent shear stress show significantly higher values in region near the wall (Klebanoff, 1955; Hinze, 1975; White, 1991). While these fluctuations exist for all three components of the velocity with different orders, all three components vanish at the wall surface ($u' = v' = w' = 0$) at the wall due to no-slip condition.

However, as demonstrated in Figure 10.2 for turbulent boundary layer flow over flat wall surface, these fluctuations are quite dominant against the wall dampening in the boundary layer and demonstrate isotropic nature until it approaches very close to the wall surface where all three fluctuations approach anisotropic nature until reaching a zero value satisfying no slip condition. Another important aspect is that the highest order fluctuation is associated with the u-component velocity, i.e., the longitudinal or streamwise component, and to a lesser order for the lateral component, w, and least for the normal component, v. The near-wall region that shows peak and anisotropic nature of the fluctuating velocity components is associated with most of the production and dissipation of turbulence energy.

The fluid elements which carry out fluctuations both in the streamwise direction of main flow and in transverse directions to flow are not individual molecules, but rather are lumps of fluid of different sizes, known as eddies. The fluctuating component may be a few percent of the mean value, but it strongly influences the turbulent flow and contributes to additional momentum exchange across adjacent layers of fluid and, hence, results in additional stresses on top of the viscous shear stress due to the main flow stream.

Turbulent fluid flows are also composed of three-dimensional and unstable **localized rotational eddies** which forms and dissipates randomly. These eddies have a wide range of sizes or length scales and continuously form and disintegrate within few oscillation periods and, hence, have very small time scales. In general, the frequencies of the unsteadiness and the size of the scales of motion span several orders of magnitude. Another important aspect is that these eddies overlap in space, larger one carrying the smaller ones following a cascading process as demonstrated in Figure 10.3.

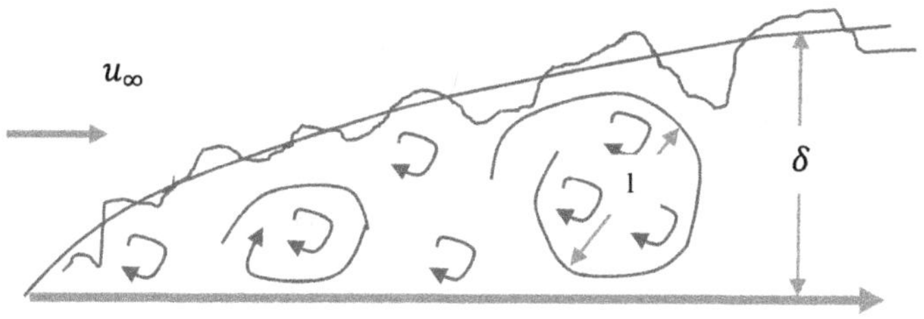

FIGURE 10.3 Eddies in turbulent flows.

The process can be characterized as a cascading process by which the turbulence dissipates its kinetic energy from the larger eddies to the smaller eddies through vortex stretching. The energy is finally dissipated into heat through the action of molecular viscosity in the smallest eddies. These larger eddies randomly stretch the vortex elements that compress the smaller eddies cascading energy to them. The cascading process gives rise to the important features such as apparent stresses and enhanced diffusivity, which are several orders of magnitude larger than those in corresponding laminar flows. The scales of motion or wavelengths usually extend all the way from a maximum size comparable to the characteristic length of the flow channel to a minimum scale corresponding to the smallest eddy fixed by the viscous dissipation. The range of these scales or the ratio of minimum to maximum wavelengths varies with characteristic flow parameter such as Reynolds number of the flow. Turbulence energy spectrum study (Klebanoff, 1955; Hinze, 1991; White, 1991) of the fluctuations for the streamwise velocity component shows that more energy contents are associated with smaller eddies and in closer proximity to wall. On the other hand, the larger eddies dominate in regions away from the wall. Another important aspect is that the fluctuations of streamwise velocity component in the outer layer of the boundary even extend beyond the interface region or superlayer between the outer flow region and the boundary layer region.

10.2 GOVERNING EQUATIONS FOR TURBULENT FLUID FLOW ANALYSIS

To describe and analyze turbulent flow, one needs to resolve the random nature of the fluctuating flow quantities and rely on statistical theory of turbulence. A simpler and more widely used turbulent flow modeling approach relies on the concept of time averaging of the equation of motion such as the Navier-Stokes equations.

Consider the Navier-Stokes equations for incompressible flow with constant viscosity from Equation 1.55

$$\frac{\partial u}{\partial x}+\frac{\partial v}{\partial y}+\frac{\partial w}{\partial z}=0 \tag{1.55a}$$

$$\rho\left(\frac{\partial u}{\partial t}+u\frac{\partial u}{\partial x}+v\frac{\partial u}{\partial y}+w\frac{\partial u}{\partial z}\right)=\rho g_x-\frac{\partial p}{\partial x}+\mu\left(\frac{\partial^2 u}{\partial x^2}+\frac{\partial^2 u}{\partial y^2}+\frac{\partial^2 u}{\partial z^2}\right) \tag{1.55b}$$

$$\rho\left(\frac{\partial v}{\partial t}+u\frac{\partial v}{\partial x}+v\frac{\partial v}{\partial y}+w\frac{\partial v}{\partial z}\right)=\rho g_y-\frac{\partial p}{\partial y}+\mu\left(\frac{\partial^2 v}{\partial x^2}+\frac{\partial^2 v}{\partial y^2}+\frac{\partial^2 v}{\partial z^2}\right) \tag{1.55c}$$

$$\rho\left(\frac{\partial w}{\partial t}+u\frac{\partial w}{\partial x}+v\frac{\partial w}{\partial y}+w\frac{\partial w}{\partial z}\right)=\rho g_z-\frac{\partial p}{\partial z}+\mu\left(\frac{\partial^2 w}{\partial x^2}+\frac{\partial^2 w}{\partial y^2}+\frac{\partial^2 w}{\partial z^2}\right) \tag{1.55d}$$

and in vector notation

$$\nabla \cdot \vec{V} = 0$$

$$\rho \frac{D\vec{V}}{dt} = \rho\vec{g} - \nabla p + \mu\nabla^2\vec{V} \tag{1.55}$$

In principle, the time-dependent three-dimensional Navier-Stokes equations can fully describe all the physics of a turbulent flow. This is due to the fact that turbulence is continuous process which consists of continuous spectrum of scales, ranging from the largest one associated with the largest eddy to the smallest scales associated with the smallest **eddy**, referred as **Kolmogorov microscale**, a concept brought by the theory of turbulence statistics.

10.3 COMPUTATIONAL MODEL FOR TURBULENCE FLOW

In the computational simulation of turbulent flow, it is important to decide **how finely we should resolve** these eddies in the computational model as this directly affects the accuracy of the prediction as well as the computational time. Available methods are (a) **Direct Numerical Simulation (DNS)** based on direct solution of Navier-Stokes equations and (b) **Averaged or Filtered Simulation** based on averaged solution of Navier-Stokes equations. A brief description of these methods is described as follows:

10.3.1 Direct Numerical Simulation

A computational model based on the microscale discretization is called **DNS**. It involves complete resolution of the flow field by a **direct solution of unsteady Navier-Stokes equations** resolving all active scales of motion in the flow field without using any approximation and models for eddies. There are two basic requirements that a DNS model must meet to represent turbulence: (a) It must represent a solution of Navier-Stokes equations **resolving all scales of motion (viscous dissipation scales)** adequately by the computational mesh, and (b) it should provide adequate statistical resolution, i.e., large samples or smaller time steps, of the set of all possible fluid motions allowed by the Navier-Stokes equations. However, these two requirements for a turbulence simulation conflict. The sample improves as the energy moves to smaller scales, but the viscous resolution is degraded. The grid spacing and time steps should be fine enough to capture the dynamics of all scales down to the smallest scale associated with the smallest eddy, which is established by the **Kolmogorov microscale**. The smallest eddy based on **Kolmogorov microscale** decreases with the increase in flow Reynolds number in proportion to the value of $Re^{\frac{9}{4}}$ and could be as small as 0.1–1 mm. Also, the computational domain should be large enough to include the **largest scale** of the flow dynamic, which is established by the characteristic dimension such as the height and width of the flow domain such as the thickness of the boundary or the pipe diameter.

Resolving all scales and frequencies of turbulent eddies based on the **Kolmogorov microscale** requires excessively refined mesh size distribution or large number of nodal points using fine time steps. The discretization of convective terms requires 5–7 points approximation schemes with low truncation error. However, it faces serious obstacles even with the most powerful computer clusters or supercomputers available today due to excessive computational time.

As a result, a DNS model of three-dimensional time-dependent Navier-Stokes equations for all important scales of turbulence has posed a great challenge for computer and numerical techniques in the past due to the requirement of extremely fine mesh size distribution and very small time steps to capture the essential details of the turbulent structures. Because of such requirements and restrictions, the DNS study of fluid flow is only restricted to problems with low Reynolds numbers. A more detail description of DNS method is given by Eswaran and Pope (1988), Rai and Moin (1991), Kim et al. (1987), Rogallo (1981), and Deb and Majumdar (1999).

10.3.2 Averaged or Filtered Simulation

This is a simplified approach to overcome the computational difficulties in terms grid size limitation imposed by **Kolmogorov microscale**. An **averaging** or **filtering** operation is employed over the Navier-Stokes equation to smooth out certain range of high-frequency variations of flow variables or smaller scales of turbulent eddies. This **averaging** or **filtering operation**, also known as **coarse graining**, leads to a new set of governing equations that represent only the larger scale eddies or lower frequencies of flow variables. Because of the smoother variations of the flow variables, the smallest scale is no longer of the order of Kolmogorov microscale, but rather limited by the cut-off scale used in the averaging or filtering method. This results in a considerable reduction in the number of grid points and savings in computational time.

The major features of Averaged or Filtered Simulation methods can be summarized as follows: (a) turbulence modeling is designed to simulate the averaged flow field named as *coarse graining* instead of the original flow field, (b) only large scales of turbulence eddies are resolved, (c) small-scale eddies that are difficult to resolve are neglected, (d) average effects of small-scale eddies on the resolved scales are considered using *statistical average model*, known as *turbulence closure models*.

Types of Averaged or Filtered Simulation Models

Options available for analyzing turbulent flows based on averaged or filtered simulation models are (a) a **time-averaged approach** using **RANS equations** along with turbulence closure models, and (b) a **space-averaged approach using Large Eddy Simulations (LES)** that takes **into account of only large-scale eddies** and uses **turbulence closure model** for the smaller eddies but require large amounts of computational time as well.

10.3.2.1 Large Eddy Simulation

In LES, the unsteady nature of turbulent eddies and only large-scale eddies are resolved. The **large-scale eddies are anisotropic** in nature and responsible for driving physical mechanisms such as **production and major carrier of the turbulent kinetic energy**. The **small-scale eddies are only responsible for viscous dissipation** of small fraction of kinetic energy that they carry. The small-scale eddies are modeled based on **assuming an isotropic or a direction independent nature of eddies** that follow a statistically predictive behavior for all turbulent flows. As small-scale eddies are not resolved, LES methods are computationally less expensive than DNS method. Nevertheless, LES method still requires finer mesh size distributions and computationally more expansive than RANS model. For more comprehensive descriptions of LES, refer books by Fernando et al. (2007).

10.4 REYNOLDS AVERAGED NAVIER-STOKES MODEL

RANS model is the next level of approximation in which no attempts are made to resolve the unsteady nature of any sizes of turbulence eddies. The increased level of mixing and dissipation caused by the turbulent eddies is considered through the turbulence closure models. In this approximation, the turbulence itself is not directly computed, but rather its average effect on mean flow is modeled by describing the turbulent motion in terms of time-averaged form of Navier-Stokes equation, referred to as the RANS.

Instantaneous turbulent flow quantities are composed of two different types of motions: (i) mean motion and (ii) fluctuating motion.

The instantaneous velocity components and pressure are presented as

$$u(t) = \bar{u} + u'(t), v = \bar{v} + v'(t), w = \bar{w} + w'(t), p = \bar{p} + p'(t), T = \bar{T} + T'(t) \tag{10.3}$$

where

$\bar{u}$, $\bar{v}$, and $\bar{w}$ = mean velocity components
u', v', and w' = fluctuating components
$\bar{p}$ and $\bar{T}$ = mean pressure and temperature velocity
p' and T' = fluctuating components

The time mean of a quantity φ (u, v, w, p, and T) is described as

$$\bar{\phi} \equiv \lim_{T_o \to \infty} \frac{1}{T} \int_{t_o}^{t_o+T_o} \phi \, dt \tag{10.4}$$

The velocity fluctuations produce mean rates of momentum transfer in addition to those produced by the mean velocity components.

Substituting all fluctuating flow quantities given by Equation 10.3 into in the Navier-Stokes equations (10.1 and 10.2) and performing the time-averaged integration, the **RANS** is obtained as

$$\frac{\partial \bar{u}}{\partial x} + \frac{\partial \bar{v}}{\partial x} + \frac{\partial \bar{w}}{\partial x} = 0 \tag{10.5a}$$

$$\rho\left(\bar{u}\frac{\partial \bar{u}}{\partial x} + \bar{v}\frac{\partial \bar{u}}{\partial y} + \bar{w}\frac{\partial \bar{u}}{\partial z}\right) = -\frac{\partial \bar{p}}{\partial x} + \mu\left(\frac{\partial^2 \bar{u}}{\partial x^2} + \frac{\partial^2 \bar{u}}{\partial y^2} + \frac{\partial^2 \bar{u}}{\partial z^2}\right) + \left(\frac{\partial \overline{u'^2}}{\partial x} + \frac{\partial \overline{u'v'}}{\partial y} + \frac{\partial \overline{u'w'}}{\partial z}\right) \tag{10.5b}$$

$$\rho\left(\bar{u}\frac{\partial v}{\partial x} + \bar{v}\frac{\partial \bar{v}}{\partial y} + \bar{w}\frac{\partial \bar{v}}{\partial z}\right) = -\frac{\partial \bar{p}}{\partial y} + \mu\left(\frac{\partial^2 \bar{v}}{\partial x^2} + \frac{\partial^2 \bar{v}}{\partial y^2} + \frac{\partial^2 \bar{v}}{\partial z^2}\right) + \left(\frac{\partial \overline{u'v'}}{\partial x} + \frac{\partial \overline{v'^2}}{\partial y} + \frac{\partial \overline{v'w'}}{\partial z}\right) \tag{10.5c}$$

$$\rho\left(\bar{u}\frac{\partial \bar{w}}{\partial x} + \bar{v}\frac{\partial \bar{w}}{\partial y} + \bar{w}\frac{\partial \bar{w}}{\partial z}\right) = -\frac{\partial \bar{p}}{\partial z} + \mu\left(\frac{\partial^2 \bar{w}}{\partial x^2} + \frac{\partial^2 \bar{w}}{\partial y^2} + \frac{\partial^2 \bar{w}}{\partial z^2}\right) + \left(\frac{\partial \overline{u'w'}}{\partial x} + \frac{\partial \overline{v'w'}}{\partial y} + \frac{\partial \overline{w'^2}}{\partial z}\right) \tag{10.5d}$$

Equation 10.5 is very similar to Navier-Stokes equations (1.55) for laminar incompressible flows, except the last term on the right-hand side which represents the additional stress due to turbulent fluctuations; Equation 10.5 can be expressed in an analogy with the derivatives of the stress tensor and we can recast in the following form:

$$\rho\left(\bar{u}\frac{\partial \bar{u}}{\partial x} + \bar{v}\frac{\partial \bar{u}}{\partial y} + \bar{w}\frac{\partial \bar{u}}{\partial z}\right) = -\frac{\partial \bar{p}}{\partial x} + \mu\left(\frac{\partial^2 \bar{u}}{\partial x^2} + \frac{\partial^2 \bar{u}}{\partial y^2} + \frac{\partial^2 \bar{u}}{\partial z^2}\right) + \left(\frac{\partial \sigma'_x}{\partial x} + \frac{\partial \tau'_{xy}}{\partial y} + \frac{\partial \tau'_{xz}}{\partial z}\right) \tag{10.6a}$$

$$\rho\left(\bar{u}\frac{\partial v}{\partial x} + \bar{v}\frac{\partial \bar{v}}{\partial y} + \bar{w}\frac{\partial \bar{v}}{\partial z}\right) = -\frac{\partial \bar{p}}{\partial y} + \mu\left(\frac{\partial^2 \bar{v}}{\partial x^2} + \frac{\partial^2 \bar{v}}{\partial y^2} + \frac{\partial^2 \bar{v}}{\partial z^2}\right) + \left(\frac{\partial \tau'_{xy}}{\partial x} + \frac{\partial \sigma'_y}{\partial y} + \frac{\partial \tau'_{yz}}{\partial z}\right) \tag{10.6b}$$

$$\rho\left(\bar{u}\frac{\partial \bar{w}}{\partial x} + \bar{v}\frac{\partial \bar{w}}{\partial y} + \bar{w}\frac{\partial \bar{w}}{\partial z}\right) = -\frac{\partial \bar{p}}{\partial z} + \mu\left(\frac{\partial^2 \bar{w}}{\partial x^2} + \frac{\partial^2 \bar{w}}{\partial y^2} + \frac{\partial^2 \bar{w}}{\partial z^2}\right) + \left(\frac{\partial \tau'_{xz}}{\partial x} + \frac{\partial \tau'_{yz}}{\partial y} + \frac{\partial \sigma'_z}{\partial z}\right) \tag{10.6c}$$

The time-averaged Navier-Stokes equation is complicated by the inclusion of the **new turbulent shear stress term**

$$T_{ijt} = -\rho \overline{u_i' u_j'} \tag{10.7}$$

where the stress turbulent tensor based on turbulent velocity fluctuation components as

$$T_{ij_t} = \begin{bmatrix} \sigma_x' & \tau_{xy}' & \tau_{xz}' \\ \tau_{xy}' & \sigma_y' & \tau_{yz}' \\ \tau_{xz}' & \tau_{yz}' & \sigma_x' \end{bmatrix} = \rho \begin{bmatrix} \overline{u'^2} & \overline{u'v'} & u'w' \\ \overline{u'v'} & \overline{v'^2} & v'w' \\ u'w' & \overline{v'w'} & w'^2 \end{bmatrix} = -\rho \overline{u_i' u_j'} \tag{10.8a}$$

or

$$T_{ij_t} = -\rho \overline{u_i' u_j'} \tag{10.8b}$$

This addition stress tensor is referred to as the **turbulent apparent stress** and represents **nine additional turbulent shear stress components** caused by the cross-products of the fluctuating velocity components and referred also as the **Reynolds stress** components.

The nine components of Reynolds stress tensor can be summarized by the following way:

$$T_{ij_t} = -\rho \overline{u_i' u_j'} = \rho \begin{vmatrix} \overline{u'^2} & \overline{u'v'} & \overline{u'w'} \\ \overline{u'v'} & \overline{v'^2} & \overline{v'w'} \\ \overline{u'w'} & \overline{v'w'} & \overline{w'^2} \end{vmatrix} \tag{10.9}$$

The **total stress** is written as the sum of laminar viscous stress and apparent turbulent Reynolds stress as

$$\tau_{ij} = \mu \left(\frac{\partial \overline{u_i}}{\partial x_j} + \frac{\partial \overline{v_j}}{\partial x_i} \right) - \rho \overline{u_i' u_j'} \tag{10.10}$$

where the normal and shear stresses on x-plane can be written, as an example, as follows:

$$\text{Normal component: } \sigma_x = -p + 2\mu \frac{\partial \overline{u}}{\partial x} - \rho \overline{u'^2} \tag{10.11a}$$

and

$$\text{Shear component: } \tau_{xy} = \mu \left(\frac{\partial \overline{u}}{\partial y} + \frac{\partial \overline{v}}{\partial x} \right) - \rho \overline{u'v'} \tag{10.11b}$$

In a symbolic compact notation, the **RANS** is written as

$$\frac{\partial}{\partial x_i}(u_i) = 0 \quad \text{or} \quad \nabla.\vec{V} = 0 \tag{10.12a}$$

$$\rho \frac{D\vec{V}}{dt} = \rho \vec{g} - \nabla p + \nabla.\, \tau_{ij} \tag{10.12b}$$

$$\tau_{ij} = \mu\left(\frac{\partial \overline{u_i}}{\partial x_j} + \frac{\partial \overline{v_j}}{\partial x_i}\right) - \rho\overline{u_i' u_j'} \tag{10.12c}$$

Turbulent stress tensor depends not only on the fluid properties but also on the flow conditions such as geometry, velocity, surface roughness, and the upstream conditions defined based on the turbulence structure, which needs to be defined as well.

Energy Equation for Turbulent Flow

Let us now consider the ***energy equation*** for incompressible and constant properties as

$$\rho c_p \frac{DT}{Dt} = \nabla \cdot (k\, \nabla T) + \Phi \tag{10.13}$$

Where φ is the viscous heat dissipation term.

Substituting fluctuating quantities and performing the time averaged integration, we get the **time-averaged energy equation**

$$\rho c_p \frac{D\overline{T}}{Dt} = -\frac{\partial}{\partial x_i}(q_i) + \overline{\Phi} \tag{10.14}$$

where q_i is the local heat flux term considering both molecular laminar heat flux and the turbulent heat flux.

$$q_i = -k\frac{\partial \overline{T}}{\partial x_i} + \rho c_p \overline{u_i' T'} \tag{10.15}$$

Laminar Flux Turbulent Flux

The turbulent dissipation term is expressed for a simpler case like in a two dimensional-boundary layer flow as

$$\overline{\Phi} = \frac{\partial \overline{u}}{\partial y}\left(\mu\frac{\partial \overline{u}}{\partial y} - \rho\, \overline{u'v'}\right) \tag{10.16}$$

10.4.1 Turbulence Kinetic Energy Transport Equation

Before proceeding with the computational modeling of turbulence fluid flow and heat transfer, it is important to have some understanding of the overall balance of the kinetic energy associated with turbulence fluctuations due to the velocity fluctuations. In an analogy with the momentum transport and energy transport equation, a transport equation that addresses the conservation of turbulence considering production, transport, and dissipation terms is derived. In this effort, the **turbulent kinetic energy of fluctuations** is defined as

$$k = \frac{1}{2}\,\overline{u_i' u_i'} = \frac{1}{2}\left(\overline{u'u'} + \overline{v'v'} + \overline{w'w'}\right) = \frac{1}{2}\left(\overline{u'^2} + \overline{v'^2} + \overline{v'^2}\right) \tag{10.17}$$

Following this definition, the turbulence kinetic energy can be related to the **Reynold turbulent stress tensor**. For example, the trace of the stress tensor is used to express the turbulent kinetic energy of fluctuations.

$$\tau_{ii} = = -\,\rho\,\overline{u_i' u_i'} = -\,2\rho k \tag{10.18a}$$

The turbulence kinetic energy, k, is the specific kinetic energy given based on per unit mass.

Based on this definition of turbulence kinetic energy and turbulence length scale, l, the turbulent eddy viscosity is considered as

$$\mu_t = C_\mu \ \rho \ k^{1/2} l \tag{10.18b}$$

where C_μ is a constant that needs to be determined for completing the turbulence closure model.

More comprehensive description of the formulation of the Turbulence Kinetic Energy Transport equation is given in reference books like Wilcox (1993); Schlichting and Gersten (2000); and White (1991). The conservation of turbulence kinetic is derived considering the rate of change of turbulence kinetic energy as a balance of turbulence production, convection and diffusion of turbulence kinetic energy, work done by turbulence kinetic energy, and turbulent viscous dissipation. The conservation of turbulence kinetic equation for incompressible flow is given in a symbolic notation as follows:

$$\rho\left(\frac{\partial k}{\partial t} + u_j \frac{\partial k}{\partial x_j}\right) = \tau_{ij} \frac{\partial \overline{u_i}}{\partial x_j} - \rho\varepsilon + \frac{\partial}{\partial x_i}\left[\left(\mu \frac{\partial k}{\partial x_j} - \frac{1}{2}\rho\overline{u_i' u_i' u_j'} - \overline{p' u_j'}\right)\right] \tag{10.19a}$$

where ε is the rate of dissipation rate with which the turbulence kinetic energy is converted into thermal heat energy due to the work done by the fluctuating component of the strain rate against the corresponding fluctuating viscous stress, and this is expressed as

$$\varepsilon = \overline{\frac{\partial u_i'}{\partial x_k} \frac{\partial u_i'}{\partial x_k}} \tag{10.19b}$$

As we can see that the time-averaged equations (Equations 10.6, 10.15, and 10.19) of momentum, energy, and turbulence kinetic energy include additional time-averaged fluctuating components such as the turbulent shéar stress and turbulent flux and hence require additional empirical-based closure modeling relations before proceeding with the solution of the system equations for turbulence flow and heat transfer equation.

A major challenge is to express Reynolds turbulent stress tensor in terms of mean flow. There are two approaches to evaluate the Reynolds stresses in terms of mean flow variables

1. Boussinesq Eddy viscosity concept and Prandtl Mixing Length model
2. Reynolds Stress Transport Model

10.4.2 Boussinesq Eddy Viscosity Concept and Prandtl Mixing Length Model

We notice that we have nine additional unknowns ($\overline{u'u'}, \overline{u'v'}, \overline{u'w'}, \overline{v'u'}, \overline{v'v'}, \overline{w'v'}, \overline{w'u'}, \overline{w'v'}, \overline{w'w'}$) for the velocity field components and three additional terms ($\overline{u'T'}$, $\overline{v'T'}$ and $\overline{w'T}$) the energy equations. These are the Reynolds stresses which are related to the mean rates of deformation or stain rate as given by Boussinesq as discussed in the section below.

10.4.2.1 Boussinesq Eddy Viscosity

To mathematically describe turbulent stress in terms of mean flow quantities, **Boussinesq**, in 1868, introduced the concept of **eddy viscosity**. **Boussinesq assumption**, which relates Reynolds turbulent stresses to the mean flow and strain rate, is in similarity with the laminar linear shear–stress–strain relation. Based on this assumption, the general expression for the Reynolds turbulent stress-strain is expressed as follows:

$$T_{ijt} = \left[\mu_t\left(\frac{\partial u_i}{\partial x_j} + \frac{\partial u_j}{\partial x_i}\right)\right] - \frac{2}{3}\mu_t \frac{\partial u_k}{\partial x_k}\delta_{ij} \tag{10.20}$$

where μ_t is termed as **the turbulent viscosity or eddy viscosity**.

The total stress is the sum of laminar stress and turbulent stress and is written as

$$\tau_{ij} = \left[\mu_{tot}\left(\frac{\partial u_i}{\partial x_j} + \frac{\partial u_j}{\partial x_i}\right)\right] - \frac{2}{3}\left(\rho k + \mu_{tot}\frac{\partial u_k}{\partial x_k}\right)\delta_{ij} \tag{10.21}$$

where $\mu_{tot} = \mu + \mu_t$= total viscosity, which is the sum of the **molecular dynamic viscosity**, μ and **turbulent or eddy viscosity**, μ_t.

Turbulent or eddy viscosity value is generally several orders of magnitude higher than the **molecular dynamic viscosity** depending on the order of magnitude of the turbulence in the flow. Another important characteristic of **turbulent or eddy viscosity** is that it depends not only on the fluid but it varies throughout the flow domain and depends strongly on fluid flow characteristics such as flow field, geometry, roughness, and upstream conditions.

In an analogous manner to these turbulent stress–strain relations, the three additional turbulent transport terms involving fluctuations of velocity and temperature in energy equation are assumed to be proportional to the gradient of the mean values of the transported quantities as

Turbulent Heat Transport Relation

$$-\rho\overline{u'T'} = \alpha_t \frac{\partial \bar{T}}{\partial x} \tag{10.22a}$$

$$-\rho\overline{v'T'} = \alpha_t \frac{\partial \bar{T}}{\partial y} \tag{10.22b}$$

$$-\rho\overline{w'T'} = \alpha_t \frac{\partial \bar{T}}{\partial z} \tag{10.22c}$$

where α_t = **turbulent or eddy diffusivity**.

Substituting Reynolds turbulent stress–strain relations given by Equation 10.11 and turbulent heat transport relation (10.15) into time-averaged equations (10.12 and 10.14), we get

$$\textbf{Continuity}: \quad \frac{\partial \bar{u}}{\partial x} + \frac{\partial \bar{v}}{\partial y} + \frac{\partial \bar{w}}{\partial z} = 0 \tag{10.23a}$$

$$x\text{-}momentum: \quad \frac{\partial \bar{u}}{\partial t} + \frac{\partial(\overline{uu})}{\partial x} + \frac{\partial(\overline{vu})}{\partial y} + \frac{\partial(\overline{wu})}{\partial z} = -\frac{1}{\rho}\frac{\partial \bar{p}}{\partial x} + \frac{\partial}{\partial x}\left[(\nu+\nu_t)\frac{\partial \bar{u}}{\partial x}\right]$$

$$+\frac{\partial}{\partial y}\left[(\nu+\nu_t)\frac{\partial \bar{v}}{\partial y}\right] + \frac{\partial}{\partial z}\left[(\nu+\nu_t)\frac{\partial \bar{w}}{\partial z}\right] \tag{10.23b}$$

$$y\text{-}momentum: \quad \frac{\partial \bar{v}}{\partial t} + \frac{\partial(\overline{uv})}{\partial x} + \frac{\partial(\overline{vv})}{\partial y} + \frac{\partial(\overline{wv})}{\partial z} = -\frac{1}{\rho}\frac{\partial \bar{p}}{\partial y} + \frac{\partial}{\partial x}\left[(\nu+\nu_t)\frac{\partial v}{\partial x}\right]$$

$$+\frac{\partial}{\partial y}\left[(\nu+\nu_t)\frac{\partial v}{\partial y}\right] + \frac{\partial}{\partial z}\left[(\nu+\nu_t)\frac{\partial v}{\partial z}\right] \tag{10.23c}$$

$$z\text{-}momentum: \quad \frac{\partial \bar{w}}{\partial t} + \frac{\partial \overline{(uw)}}{\partial x} + \frac{\partial \overline{(vw)}}{\partial y} + \frac{\partial \overline{(ww)}}{\partial z} = -\frac{1}{\rho}\frac{\partial \bar{p}}{\partial z} + \frac{\partial}{\partial z}\left[(\nu + \nu_t)\frac{\partial \bar{w}}{\partial x}\right]$$

$$+\frac{\partial}{\partial z}\left[(\nu + \nu_t)\frac{\partial \bar{w}}{\partial y}\right] + \frac{\partial}{\partial z}\left[(\nu + \nu_t)\frac{\partial \bar{w}}{\partial z}\right] \tag{10.23d}$$

Energy Equation

$$\frac{\partial \bar{T}}{\partial t} + \frac{\partial \overline{(uT)}}{\partial x} + \frac{\partial \overline{(vT)}}{\partial y} + \frac{\partial \overline{(wT)}}{\partial z} = \frac{\partial}{\partial x}\left[(\alpha + \alpha_t)\frac{\partial T}{\partial x}\right] + \left[(\alpha + \alpha_t)\frac{\partial T}{\partial y}\right] + \left[(\alpha + \alpha_t)\frac{\partial T}{\partial z}\right] \tag{10.23e}$$

Turbulence Kinetic Energy Transport Equation

To correlate turbulence fluctuation involving velocity and pressure fluctuations to mean flow terms, several assumptions and closure coefficients are introduced. For example, for the turbulence kinetic energy transport, the diffusion term is correlated with the mean flow scalar quantity by assuming a gradient-transport process and is written as follows:

$$\frac{1}{2}\rho\overline{u_i'u_i'u_j'} + \overline{p'u_j'} = -\frac{\mu_t}{\sigma_k}\frac{\partial k}{\partial x_j} \tag{10.24}$$

where σ_k is a turbulence closure coefficient and is used in many turbulence closure model as a constant value.

The additional new quantities like the kinetic energy dissipation quantity, ε in turbulence kinetic energy transport equation (10.19), and the turbulence length scale (l) in Equation 10.18 also need to be defined in order to complete or close the system of turbulence modeling equations. These two quantities and the turbulence kinetic energy are correlated using a proportionality function

$$\varepsilon \approx k^{3/2}/l \tag{10.25}$$

The turbulence length scale (l) and the proportionality constant are assigned through the turbulence closure models as discussed in Section 10.5.

Introducing Equation 10.24 into the turbulence kinetic energy equation (10.19), we get the following form of the turbulence kinetics energy transport equation:

$$\rho\left(\frac{\partial k}{\partial t} + u_j\frac{\partial k}{\partial x_j}\right) = \tau_{ij}\frac{\partial \overline{u_i}}{\partial x_j} - \rho\varepsilon + \frac{\partial}{\partial x_i}\left[(\mu + \mu_t/\sigma_k)\frac{\partial k}{\partial x_j}\right] \tag{10.26}$$

10.4.2.2 Prandtl Mixing Length

Prandtl in 1925 introduced the concept of mixing length l_m theory that closely relates to eddy viscosity concept and formed the basis for all turbulent modeling efforts. The Prandtl mixing length is defined as the average distance travelled by a lump of fluid or the fluid eddy in the normal direction across the flow in similarity with the mean free path length of molecules as demonstrated in Figure 10.4.

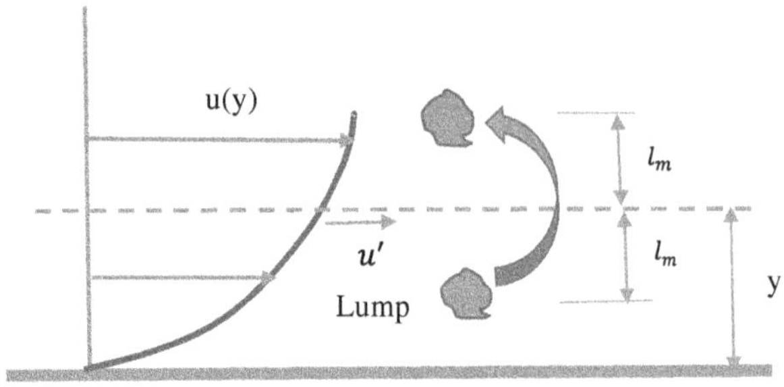

FIGURE 10.4 Prandtl mixing length and turbulent shear stress.

As shown, a fluid lump travels a distance of $2l_m$ across a plane, exchanging momentum and resulting in the turbulent eddy shear stress. The velocities at the two locations $(y+l_m)$ and $(y+l_m)$ are given as

$$u\left(y+l_m\right) \approx u\left(y\right)+l_m\,\frac{\partial u}{\partial y}$$

and

$$u\left(y-l_m\right) \approx u\left(y\right)-l_m\,\frac{\partial u}{\partial y}$$

According to Prandtl postulation, the turbulent velocity fluctuation, u', is proportional to the mean of the velocities $u\ (y+l_m)$ and $u\ (y-l_m)$ at locations $(y+l_m)$ and $(y+l_m)$ across the plane and is written as

$$u' \approx l_m\,\frac{\partial u}{\partial y} \tag{10.27}$$

Based on this concept and assuming ν' of the same order, the turbulent stress and turbulent eddy viscosity are expressed in terms of ***Prandtl mixing length hypothesis*** in the following manner:

$$\tau_t = -\rho\,\overline{u'v'} = \rho\ \mu_t\,\frac{\partial u}{\partial y} = \varrho l_m^2\left(\frac{\partial u}{\partial y}\right)^2 \tag{10.28a}$$

where μ_t is the **turbulent eddy viscosity** given as

$$\mu_t = l_m^2\left(\frac{\partial u}{\partial y}\right)^2 \tag{10.28b}$$

and l_m is the turbulent mixing length.

Prandtl proposed that along with the strong variation of turbulent eddy viscosity, μ_t, within the boundary layer, mixing length, l_m, also varies throughout the boundary layer following the relation:

$$l_m \approx y \quad \text{or} \quad l_m = Ky \tag{10.28c}$$

where K is the constant of proportionality.

Substituting Equation 10.28c for the mixing length into Equation 10.28a, the turbulent eddy viscosity is expressed as

$$\tau_t = -\rho\,\overline{u'v'} = = \tilde{n}\; K^2 y^2 \left(\frac{\partial u}{\partial y}\right)^2 \tag{10.28d}$$

Further assuming that in the near-wall region $\tau_t = \tau_w$, integration of Equation 10.28d resulted in the following velocity distribution profile:

$$u = \frac{1}{K}\sqrt{\frac{\tau_w}{\rho}}\;\ln y + E \tag{10.29}$$

where E is the constant of integration.

Equation 10.29 matched well with experimental data in the near-wall region of the turbulent boundary layer, except in the so-called laminar sublayer region close to the wall where viscous shear is dominant.

10.5 DIFFERENT CLASSES OF TURBULENCE CLOSURE MODELS

Turbulence closure models consider the **statistical average effect of small-scale eddies on the time-averaged mean flow** that only resolves large-scale eddies. To resolve turbulent viscosity and Reynolds turbulent stresses, the eddy viscosity model based on Boussinesq assumptions leads to different order (zero-equation, one-equation, and two-equation) turbulence closure models and Reynolds Stress Model (RSM). To define the turbulent eddy viscosity, it is necessary determine a suitable velocity scale and a length scale. The **mixing length model** is considered as **an algebraic model** or a **zero-equation turbulence model**. Subsequently, **additional class of turbulence** models was developed based on number of additional equations to describe the turbulent viscosity. An ***n*-equation turbulence model** requires solution of n-additional transport equations and additional closure variables and functions used to describe the length and velocity scales used in the estimation of turbulent viscosity and close the system of equations for solving turbulent flow and heat transfer.

Turbulence Models

In **one-equation turbulence model, turbulence kinetic energy (*k*)** was introduced to describe the velocity scale. Subsequently, in **two-equation model**, additional variable like the **rate of dissipation** (ε) of turbulence kinetic energy was introduced to represent the length scale of turbulence.

The turbulence kinetic energy (k) and the dissipation rate of turbulence kinetic energy (ε) are defined in the following manner:

$$k = \frac{1}{2} u_i' u_j' \tag{10.30a}$$

and

$$\varepsilon = \nu_t\, \overline{\left(\frac{\partial u_i'}{\partial x_j}\right)\left(\frac{\partial u_i'}{\partial x_j}\right)} \tag{10.30b}$$

For $i, j = 1, 2$, and 3

The estimation of turbulence viscosity in terms of turbulence kinetic energy (k) and turbulence dissipation rate (ε) is given by the **Prandtl-Kolmogorov relation**:

$$\mu_t = \rho C_\mu k^2/\varepsilon \tag{10.31}$$

where
μ_t=turbulent or eddy viscosity
ν_t=kinematic turbulent or eddy viscosity $\nu_t = \dfrac{\mu_t}{\rho}$
C_μ = empirical turbulence constant

10.6 CLASSIFICATION OF TURBULENCE MODELS

Turbulence closure models vary in complexities in terms of inclusions of number additional closure coefficients, damping functions, and closure functions for turbulence length scales.

10.6.1 Algebraic Turbulence Model or Zero-Equation Models

Algebraic turbulence models are zero-equation turbulence models that do not require the solution of any additional equation, and are calculated directly from the flow variables. Zero-equation models do not consider history effects of the turbulence, such as **convection and diffusion of turbulent energy**, and are often adequate for simpler flow geometries. Some of the most popular algebraic zero-equation turbulence models are (a) **Cebeci-Smith model**, (b) **Baldwin-Lomax model**, and (c) **Johnson-King model**. Algebraic models are simple, quite robust, and computationally less expensive. A major limitation of the algebraic turbulence is the weak physical base because of the semi-empirical form and so not applicable to flow problems that are significantly different from flow problems for which the empirical constant is derived.

10.6.2 One-Equation Model

One-equation turbulence models include (a) **Prandtl–Emmons–Glushko model**; (b) **Baldwin–Barth model**, and (c) **Spalart–Allmaras model**.

10.6.2.1 Prandtl–Emmons–Glushko Model

This is one of the simplest closure model and was originated with the postulation that relates the dissipation with kinetic energy and length scale given by Equation 10.25 with a closure coefficient and assuming that the turbulence length scale is proportional to turbulence mixing length $(l \approx l_m)$ for proportionality as

$$\varepsilon = C_D k^{3/2}/l \tag{10.32}$$

where C_D = closure coefficient.

This one-equation model is written based on Equation 10.26 and using Equation 10.31 as

$$\rho\left(\frac{\partial k}{\partial t} + u_j\frac{\partial k}{\partial x_j}\right) = \tau_{ij}\frac{\partial \overline{u_i}}{\partial x_j} - C_D\rho k^{3/2}/l + \frac{\partial}{\partial x_i}\left[\left(\mu + \mu_t/\sigma_k\right)\frac{\partial k}{\partial x_j}\right] \tag{10.33}$$

where the stress tensor is given by Equation 10.26 and turbulent eddy viscosity by Equation 10.18. The closure coefficients are given as $\sigma_k = 1$ and $C_D = 0.07 \sim 0.09$.

10.6.2.2 Spalart–Allmaras Model

The model proposed by Spalart and Allmaras is a one-equation model that solves for eddy viscosity from the transport equation. It is being widely used in turbomachinery applications.

The transport equation is given by for a viscosity-like variable:

$$\frac{\partial}{\partial t}(\rho\tilde{v})+\frac{\partial}{\partial x_i}(\rho\tilde{v}u_i)=G_v+\frac{1}{\sigma_{\tilde{v}}}\left[\frac{\partial}{\partial x_j}\left\{(\mu+\rho\tilde{v})\frac{\partial\tilde{v}}{\partial x_j}\right\}+C_{b2}\rho\left(\frac{\partial\tilde{v}}{\partial x_j}\right)^2\right]-Y_{\tilde{v}} \tag{10.34a}$$

where

$\tilde{v}$ = viscosity like variable – referred to as ***Spalart–Allmaras variable***

G_v = production of turbulent viscosity

$$G_v = C_{b1}\rho\tilde{S}\tilde{v} \tag{10.34b}$$

$Y_{\tilde{v}}$ = destruction of turbulent viscosity that occurs in the near-wall region

$$Y_{\tilde{v}} = C_{w1}\rho f_w\left(\frac{\tilde{v}}{d}\right)^2 \tag{10.34c}$$

The turbulent viscosity (μ_t) is computed as

$$\mu_t = \rho\tilde{v}f_{v1} \tag{10.34d}$$

f_{v1} = **viscous damping function** given by

$$f_{v1} = \frac{\chi^3}{\chi^3 + C_{v1}^3}, \chi = \frac{\tilde{v}}{v} \tag{10.34e}$$

$\tilde{S}$ = mean rate of rotation tensor and **viscous damping function**.

f_{w1} = function of vorticity

d = distance from the wall.

Boundary Conditions: $\tilde{v} = 0$ on the Wall

This model works well for applications in which the flows are mostly attached with no or minor flow separation.

10.6.2.3 Baldwin–Barth Model

This model brings in more complexity with the addition of many more closure coefficients and two empirical damping functions and functions for defining turbulence length scale. For more detailed description, see book by Wilcox (1993).

10.6.3 Two-Equation Model

Two-equation turbulence models include **two additional transport equations** to represent the turbulent properties of the flow and account for production, dissipation, convection, and diffusion of turbulent energy. Most often one of the transport variables is the turbulent kinetic energy, k. The **second transport variable** varies with different two-equation models. Most common choices are (a) the **turbulent dissipation** and (b) the **specific dissipation**, ω.

While the **turbulent kinetic energy**, k, represents the specific energy in the turbulence or velocity scale, the **turbulent dissipation**, ε, or the **specific dissipation**, ω, represents the turbulence length scale.

Two-equation models that work well for a list of some of the widely used two-equation turbulence models are given here:

Two-equation $k-\varepsilon$ Turbulence Models

i. $k-\varepsilon$ Standard turbulence model
ii. $k-\varepsilon$ Realizable model
iii. $k-\varepsilon$ Renormalization Group (RNG) turbulence model
iv. $k-\varepsilon$ Chen turbulence model
v. $k-\varepsilon$ Standard Quadratic High Reynolds Turbulence model
vi. $k-\varepsilon$ Suga Quadratic High Reynolds Turbulence model

Two-Equation $k-\omega$ Turbulence Models

1. $k-\omega$ Standard High Re
2. $k-\omega$ Shear Stress Transport (SST) High Re
3. $k-\omega$ Standard Low Re
4. $k-\omega$ SST Low Re

Let us briefly discuss several of these two-equation turbulence models.

10.6.3.1 $k-\varepsilon$ Turbulence Model

The Standard Two-Equation $k-\varepsilon$ Turbulence Model

This $k-\varepsilon$ turbulence model is the most widely used one and is also known as $k-\varepsilon$ High Reynolds Turbulence Model. This is one of most popular two-equation $k-\varepsilon$ turbulence models and also referred as the standard $k-\varepsilon$ turbulence model as proposed and developed by Jones and Launder (1972), Launder and Spalding (1974), and Launder and Sharma (1974). It is the first two-equation model used in many early studies of computational fluid dynamics (CFD). It includes two transport equations to define the turbulence scale. The k denotes the turbulent kinetic energy (m^2/s^2), whereas ε denotes the dissipation rate (m^2/s^3). Following are the transport equations for momentum and energy, the transport equation for the **turbulence kinetic energy (k)**, and **dissipation rate of turbulence kinetic energy (ε)**:

Turbulence Kinetic Energy (k)

$$\rho\left(\frac{\partial k}{\partial t}+u_j\frac{\partial k}{\partial x_j}\right)=\tau_{ij}\frac{\partial \overline{u_i}}{\partial x_j}-\rho\,\varepsilon+\frac{\partial}{\partial x_i}\left[\left(\mu+\mu_t/\sigma_k\right)\frac{\partial k}{\partial x_j}\right] \tag{10.35a}$$

or

$$\frac{\partial k}{\partial t}+u\frac{\partial k}{\partial x}+v\frac{\partial k}{\partial y}+w\frac{\partial k}{\partial z}=\frac{\partial}{\partial x}\left(\frac{v_t}{\sigma_k}\frac{\partial k}{\partial x}\right)+\frac{\partial}{\partial y}\left(\frac{v_t}{\sigma_k}\frac{\partial k}{\partial y}\right)+\frac{\partial}{\partial z}\left(\frac{v_t}{\sigma_k}\frac{\partial \bar{k}}{\partial z}\right)+P-D \tag{10.35b}$$

Dissipation Rate of Turbulence Kinetic Energy (ε)

$$\rho\left(\frac{\partial \varepsilon}{\partial t}+u_j\frac{\partial \varepsilon}{\partial x_j}\right)=C_{\varepsilon 1}\frac{\varepsilon}{k}\tau_{ij}\frac{\partial \overline{u_i}}{\partial x_j}-C_{\varepsilon 2}\rho\frac{\varepsilon^2}{k}+\frac{\partial}{\partial x_i}\left[\left(\mu+\mu_t/\sigma_k\right)\frac{\partial k}{\partial x_j}\right] \tag{10.36a}$$

or

$$\frac{\partial \varepsilon}{\partial t}+u\frac{\partial \varepsilon}{\partial x}+v\frac{\partial \varepsilon}{\partial y}+w\frac{\partial \varepsilon}{\partial z}=\frac{\partial}{\partial x}\left(\mu+\frac{v_t}{\sigma_\varepsilon}\frac{\partial \varepsilon}{\partial x}\right)+\frac{\partial}{\partial y}\left(\mu+\frac{v_t}{\sigma_\varepsilon}\frac{\partial \varepsilon}{\partial y}\right)+\frac{\partial}{\partial z}\left(\mu+\frac{v_t}{\sigma_\varepsilon}\frac{\partial \bar{\varepsilon}}{\partial z}\right)$$

$$+\frac{\varepsilon}{k}\left(C_{\varepsilon 1}P-C_{\varepsilon 2}D\right) \tag{10.36b}$$

where P is the **rate of production of turbulence kinetic energy** and is expressed as

$$P=2\,v_t\left[\left(\frac{\partial u}{\partial x}\right)^2+\left(\frac{\partial v}{\partial y}\right)^2+\left(\frac{\partial w}{\partial z}\right)^2\right]+v_t\left[\left(\frac{\partial u}{\partial y}+\frac{\partial v}{\partial x}\right)^2+\left(\frac{\partial v}{\partial z}+\frac{\partial w}{\partial y}\right)^2+\left(\frac{\partial w}{\partial x}+\frac{\partial u}{\partial z}\right)^2\right] \tag{10.36c}$$

and $D=\varepsilon$ is the **rate destruction of turbulence kinetic energy**.
Eddy Viscosity

$$\mu_t=\rho C_\mu k^2/\varepsilon \tag{10.36d}$$

Closure Coefficients

$$C_{\varepsilon 1}=1.44,\ C_{\varepsilon 2}=1.92,\ C_\mu=0.09,\ \sigma_k=1.0 \text{ and } \sigma_\varepsilon=1.3$$

Other Relations

$$l=C_\mu\ k^{3/2}/\varepsilon \quad \text{and} \quad \omega=\varepsilon/(C_\mu k)$$

Often additional terms related to the generation of turbulent kinetic energy due to buoyancy and dissipation due to fluctuating dilatation in compressible turbulence are included. User-defined source terms $S_\kappa$and S_ε can also be included depending on the application problems.

A two-equation standard high-Reynolds number $k-\varepsilon$ model is found to be quite suitable for flow in straight flow channels without the presence of any large-scale flow separations and adverse pressure gradient. Also, in problems where only average parameters are of major interest without the requirement of resolving detailed turbulence quantities such as in many industrial applications. The standard high-Reynolds number $k-\varepsilon$ model also performs poorly in regions involving flow reattachment and recovery following the flow separation.

The RNG $\kappa-\varepsilon$ Model

The RNG κ-ε model was originated from the instantaneous Navier-Stokes equations, utilizing a **mathematical technique called** RNG methods (Yakhot et al., 1992). The RNG $\kappa-\varepsilon$ model has similar form for transport equations for k as in the standard $\kappa-\varepsilon$ model:

$$\frac{\partial k}{\partial t}+u\frac{\partial k}{\partial x}+v\frac{\partial k}{\partial y}+w\frac{\partial k}{\partial z}=\frac{\partial}{\partial x}\left(\frac{v_t}{\sigma_k}\frac{\partial k}{\partial x}\right)+\frac{\partial}{\partial y}\left(\frac{v_t}{\sigma_k}\frac{\partial k}{\partial y}\right)+\frac{\partial}{\partial z}\left(\frac{v_t}{\sigma_k}\frac{\partial \bar{k}}{\partial z}\right)+P-D \tag{10.37}$$

For the transport equation for the dissipation ε, the derivation results in additional source term and functions.

$$\frac{\partial \varepsilon}{\partial t}+u\frac{\partial \varepsilon}{\partial x}+v\frac{\partial \varepsilon}{\partial y}+w\frac{\partial \varepsilon}{\partial z}=\frac{\partial}{\partial x}\left(\mu+\frac{v_t}{\sigma_\varepsilon}\frac{\partial \varepsilon}{\partial x}\right)+\frac{\partial}{\partial y}\left(\mu+\frac{v_t}{\sigma_\varepsilon}\frac{\partial \varepsilon}{\partial y}\right)+\frac{\partial}{\partial z}\left(\mu+\frac{v_t}{\sigma_\varepsilon}\frac{\partial \bar{\varepsilon}}{\partial z}\right)$$

$$+\frac{\varepsilon}{k}(C_{\varepsilon 1}P-C_{\varepsilon 2}D)+R \tag{10.38}$$

The additional source term R is given as

$$R=\frac{C_\mu \eta^3\left(1-\eta/\eta_0\right)}{1+\beta\eta^3}\frac{\varepsilon^2}{k} \tag{10.39}$$

where constants are given as $\beta = 0.015$ and $\eta_0 = 4.38$.

The RNG theory results in a differential equation for turbulent viscosity:

$$d\left(\frac{\rho^2 k}{\sqrt{\varepsilon\mu}}\right)=1.72\frac{\hat{v}}{\sqrt{\hat{v}^3-1+C_v}}d \tag{10.40}$$

where $\hat{v}=\frac{\mu_{\text{eff}}}{\mu}$ and $C_v \approx 100$.

Turbulent quantities change considerably with the effect of swirl in the mean flow direction. To incorporate the **swirl effect**, the turbulent viscosity is calculated from

$$\mu_t=\mu_{t0} f\left(\alpha_s,\Omega,\frac{k}{s}\right) \tag{10.41}$$

where μ_{t0} is the turbulent viscosity calculated without swirl modification from Equation 10.36d, $\alpha_s = 0.05$ for moderately swirl flow, Ω swirl number. For higher Reynolds numbers, equation turbulent viscosity is calculated by (10.36d) and c_μ=0.0845. The default model constants are $C_{1\varepsilon}=1.42, C_{2\varepsilon}=1.68, \sigma_k=0.718$, and $\sigma_\varepsilon=0.718$.

RNG k-ε models have shown substantial improvements over the standard k-ε model where the flow features include strong streamline curvature, vortices, and rotation.

The Realizable $k-\varepsilon$ Model

The realizable $\kappa-\varepsilon$ model proposed by Shih et al. (1995) satisfies certain mathematical constraints consistent with certain physics of the turbulent flow such as the mathematical constraints of normal stresses. It addresses the deficiencies of traditional $\kappa-\varepsilon$ models by adopting a new eddy-viscosity formula with a variable C_μ and a new model equation for dissipation rate ε based on a new source term.

The transport equations in realizable $\kappa-\varepsilon$ model:

Turbulent Kinetic Energy:

$$\frac{\partial k}{\partial t}+u\frac{\partial k}{\partial x}+v\frac{\partial k}{\partial y}+w\frac{\partial k}{\partial z}=\frac{\partial}{\partial x}\left(\frac{v_t}{\sigma_k}\frac{\partial k}{\partial x}\right)+\frac{\partial}{\partial y}\left(\frac{v_t}{\sigma_k}\frac{\partial k}{\partial y}\right)+\frac{\partial}{\partial z}\left(\frac{v_t}{\sigma_k}\frac{\partial \bar{k}}{\partial z}\right)+P-D \tag{10.42}$$

Dissipation Equation:

$$\frac{\partial \varepsilon}{\partial t}+u\frac{\partial \varepsilon}{\partial x}+v\frac{\partial \varepsilon}{\partial y}+w\frac{\partial \epsilon}{\partial z}=\frac{\partial}{\partial x}\left(\mu+\frac{v_t}{\sigma_\varepsilon}\frac{\partial \varepsilon}{\partial x}\right)+\frac{\partial}{\partial y}\left(\mu+\frac{v_t}{\sigma_\varepsilon}\frac{\partial \epsilon}{\partial y}\right)+\frac{\partial}{\partial z}\left(\mu+\frac{v_t}{\sigma_\varepsilon}\frac{\partial \bar{\varepsilon}}{\partial z}\right)$$

$$+\rho C_1 S\varepsilon-\rho C_2\frac{\varepsilon^2}{\kappa+\sqrt{v_t\varepsilon}} \tag{10.43}$$

where $C_1=\max\left[0.43,\frac{\eta}{\eta+5}\right]$, $\eta=S\frac{k}{\varepsilon}$, $C_2=1.9$, $\sigma_k=1.0$, and $\sigma_\varepsilon=1.2$ are the model constants.

The eddy viscosity is calculated as before, but the correlation coefficient is computed based on the following functional relation:

$$C_\mu=\frac{1}{A_0+A_s\frac{kU^*}{\varepsilon}} \tag{10.44}$$

where

$A_0=4.04$ and $A_s=4.04$ and the new parameter U^* is computed based on the following relations:

$$U^*\equiv\sqrt{S_{ij}S_{ij}+\tilde{\Omega}_{ij}\tilde{\Omega}_{ij}} \tag{10.45}$$

where $\tilde{\Omega}_{ij}$ is the mean rate-of-rotation tensor viewed in the rotating reference frame with an angular velocity and S_{ij} is given as

$$S_{ij}=\frac{1}{2}\left(\frac{\partial u_i}{\partial x_j}+\frac{\partial u_j}{\partial x_i}\right) \tag{10.46}$$

Like RNG k-ε model, the realizable k-ε model also has shown substantial improvements over the standard k-ε model where the flow features include strong streamline curvature, vortices, and rotation.

The realizable κ-ε model provides superior performance for flows involving rotation, boundary layers under strong adverse pressure gradients, separation, and recirculation.

The limitations of the realizable κ-ε model are that it produces nonphysical turbulent viscosities in situations when the computational domain contains both rotating and stationary fluid zones like in the use of multiple reference frames or rotating sliding meshes.

$k-\varepsilon$ Chen Turbulence Model

The Chen model has been introduced to have a better response of the energy transfer mechanism of turbulence toward the mean strain rate. It does not consider the compressibility and buoyancy effects explicitly. These effects are modeled in the same way as in the standard $k-\varepsilon$ model.

Equation for k

$$\frac{\partial}{\partial t}(\rho k)+\frac{\partial}{\partial x_j}\left[\rho u_j k-\left(\mu+\frac{\mu_t}{\sigma_k}\right)\frac{\partial k}{\partial x_j}\right]=\mu_t(P+P_B)-\rho\varepsilon-\frac{2}{3}\left(\mu_t\frac{\partial u_i}{\partial x_j}+\rho k\right)\frac{\partial u_i}{\partial x_j} \tag{10.47a}$$

Equation for ε

$$\frac{\partial}{\partial t}(\rho\varepsilon)+\frac{\partial}{\partial x_j}\left[\rho u_j\varepsilon-\left(\mu+\frac{\mu_t}{\sigma_\varepsilon}\right)\frac{\partial\varepsilon}{\partial x_j}\right]=C_{\varepsilon 1}\frac{\varepsilon}{k}\left[\mu_t P-\frac{2}{3}\left(\mu_t\frac{\partial u_i}{\partial x_j}+\rho k\right)\frac{\partial u_i}{\partial x_j}\right]$$

$$+C_{\varepsilon 3}\frac{\varepsilon}{k}\mu_t P_B-C_{\varepsilon 2}\rho\frac{\varepsilon^2}{k}+C_{\varepsilon 4}\rho\varepsilon\frac{\partial u_i}{\partial x_j}+C_{\varepsilon 5}\frac{\mu_t^2}{\rho}\frac{P^2}{k} \quad (10.47b)$$

Suga's High Reynolds Number $k-\varepsilon$ Turbulence Model

This a nonlinear version of the $\boldsymbol{k-\varepsilon}$ Turbulence Model. In Suga's $k-\varepsilon$ model $\tilde{\varepsilon}$ is solved instead of ε, which is the isotropic part of ε and is zero at the wall. The k equation for this Turbulence Model is same as in standard $k-\varepsilon$ and the dissipation equation is given as

Equation for ε

$$\frac{\partial}{\partial t}(\rho\tilde{\varepsilon})+\frac{\partial}{\partial x_j}\left[\rho u_j\tilde{\varepsilon}-\left(\mu+\frac{\mu_t}{\sigma_\varepsilon}\right)\frac{\partial\tilde{\varepsilon}}{\partial x_j}\right]=\rho C_{\varepsilon 1}\frac{\tilde{\varepsilon}}{k}P_k-\rho C_{\varepsilon 2}+C_{\varepsilon 3}\frac{\tilde{\varepsilon}}{k}\mu_t P_B+C_{\varepsilon 4}\rho\tilde{\varepsilon}\frac{\partial u_i}{\partial x_i} \quad (10.48a)$$

where

$$P_B\equiv\frac{g_i}{\sigma_{h,t}}\frac{1}{\rho}\frac{\partial\rho}{\partial x_j},\quad P_k=-\overline{u_i'u_j'}\frac{\partial u_i}{\partial x_j},\quad \tilde{R}_t=\frac{k^2}{\nu\tilde{\varepsilon}},\quad \tilde{\varepsilon}=\varepsilon-2\nu\left[\frac{\partial\sqrt{k}}{\partial x_i}\right]^2 \quad (10.48b)$$

Turbulent viscosity μ_t is defined as

$$\mu_t=f_\mu\frac{C_\mu\rho\cdot k^2}{\varepsilon} \quad (10.48c)$$

where $f_\mu=1-\exp\left[-\left(\frac{\tilde{R}_t}{90}\right)^{\frac{1}{2}}-\left(\frac{\tilde{R}_t}{400}\right)^2\right]$

Quadratic High Reynolds $k-\varepsilon$ Turbulence Model

Many real flows exhibit **anisotropic turbulence** characteristics. Nonlinear turbulence models are introduced to consider the anisotropic turbulence characteristics present in many real flows by adopting **nonlinear relationships between Reynolds stresses and the rate of strain**. For quadratic models, the constitutive relations for the Reynolds stresses are

$$\rho\frac{\overline{u_i'u_j'}}{k}=\frac{2}{3}\left(\frac{\mu_t}{k}\frac{\partial u_k}{\partial x_k}+\rho\right)\delta_{ij}-\frac{\mu_t}{k}S_{ij}+C_1\frac{\mu_t}{\varepsilon}\left[S_{ik}S_{kj}-\frac{1}{3}\delta_{ij}S_{kl}S_{kl}\right]$$

$$+C_2\frac{\mu_t}{\varepsilon}\left[\Omega_{ik}S_{kj}+\Omega_{jk}S_{ki}\right]+C_3\frac{\mu_t}{\varepsilon}\left[\Omega_{ik}\Omega_{ik}-\frac{1}{3}\delta_{ij}\Omega_{kl}\Omega_{kl}\right] \quad (10.49a)$$

where Ω_{ij} is the mean vorticity tensor given by

$$\Omega_{ij}=\frac{\partial u_i}{\partial x_j}-\frac{\partial u_j}{\partial x_i} \quad (10.49b)$$

Coefficients C_1, C_2, and C_3 are defined as

$$C_1 = \frac{c_{\text{NL1}}}{\left(c_{\text{NL6}} + c_{\text{NL7}} S^3\right) C_\mu}, \quad C_2 = \frac{c_{\text{NL2}}}{\left(c_{\text{NL6}} + c_{\text{NL7}} S^3\right) C_\mu}, \quad C_3 = \frac{c_{\text{NL3}}}{\left(c_{\text{NL6}} + c_{\text{NL7}} S^3\right) C_\mu} \tag{10.49c}$$

where

$$C_\mu = \frac{c_{A0}}{c_{A1} + c_{A2} S + c_{A3} \Omega}$$

$$S = \frac{k}{\varepsilon} S^* \text{ and } \Omega = \frac{k}{\varepsilon} \Omega^*,\; S^* = \sqrt{\frac{1}{2} S_{ij} S_{ij}} \text{ AND } \Omega^* = \sqrt{\frac{1}{2} \Omega_{ij} \Omega_{ij}}$$

Empirical coefficients for $k - \varepsilon$ quadratic high Re turbulence model

c_{A0}	c_{A1}	c_{A2}	c_{A3}	c_{NL1}	c_{NL2}	c_{NL3}	c_{NL6}	c_{NL7}
0.667	1.25	1.0	0.9	0.75	3.75	4.75	1000.0	1.0

10.6.3.2 Two-Equation $k - \omega$ Turbulence Models

An alternate approach to the $k - \varepsilon$ model is the $k - \omega$ model, where ω is the specific dissipation rate, which is defined as

$$\omega = \varepsilon / C_\mu k \tag{10.50}$$

This turbulence dissipation parameter ω is referred to as dissipation per unit turbulence kinetic energy and was introduced by Kolmogorov (1942) along with a transport equation for the specific dissipation rate as

$$\rho \frac{\partial \omega}{\partial t} + \rho u_j \frac{\partial \omega}{\partial x_j} = \frac{\partial}{\partial x_j}\left[\sigma \mu_t \frac{\partial \omega}{\partial x_j}\right] - \rho \beta \omega^2 \tag{10.51}$$

where σ and β are the closure coefficients.

Since the first $k - \omega$ model introduced by Kolmogorov, several newer variants of $k - \omega$ models were proposed such as the Wilcox standard $k - \omega$ model, $k - \omega$ SST Turbulence Model, and some variants of low Reynolds number $k - \omega$ models.

Standard $k - \omega$ Turbulence Model

The Wilcox standard $k - \omega$ model along with all closure coefficients and relations is summarized below.

Turbulent Kinetic Energy:

$$\rho\left(\frac{\partial k}{\partial t} + u_j \frac{\partial k}{\partial x_j}\right) = \tau_{ij} \frac{\partial \overline{u_i}}{\partial x_j} - \beta^* \rho k \omega + \frac{\partial}{\partial x_i}\left[\left(\mu + \sigma^* \mu_t\right) \frac{\partial k}{\partial x_j}\right] \tag{10.52a}$$

Specific Dissipation Rate:

$$\rho \frac{\partial \omega}{\partial t} + \rho u_j \frac{\partial \omega}{\partial x_j} = \alpha \frac{\omega}{k} \tau_{ij} \frac{\partial u_i}{\partial x_j} + \frac{\partial}{\partial x_j}\left[\mu + \sigma \mu_t \frac{\partial \omega}{\partial x_j}\right] - \rho \beta \omega^2 \tag{10.52b}$$

Eddy Viscosity:

$$\mu_t = \rho k/\omega \tag{10.52c}$$

Closure Coefficients:

$$\alpha = 5/9,\ \beta = 3/40,\ \beta^* = 9/100,\ \sigma = 1/2,\ \sigma^* = 1/2 \tag{10.52d}$$

Other Auxiliary Equation:

$$\varepsilon = \beta^* k\omega \quad \text{and} \quad l = k^{1/2}/\omega \tag{10.52c}$$

$k-\omega$ SST Turbulence Model

This is a

Equation for k

$$\frac{\partial}{\partial t}(\rho k) + \frac{\partial}{\partial x_j}\left[\rho u_j k - \left(\mu + \frac{\mu_t}{\sigma_k^{\omega}}\right)\frac{\partial k}{\partial x_j}\right] = \mu_t(P + P_B) - \rho\beta^* k\omega \tag{10.53a}$$

Equation for ω

$$\frac{\partial}{\partial t}(\rho\omega) + \frac{\partial}{\partial x_j}\left[\rho u_j \omega - \left(\mu + \frac{\mu_t}{\sigma_\omega^{\omega}}\right)\frac{\partial \omega}{\partial x_j}\right] = \alpha\frac{\omega}{k}\mu_t P - \rho\beta\omega^2 + \rho S_\omega + C_{\varepsilon 3}\mu_t P_B C_\mu \omega \tag{10.53b}$$

where C_μ and $C_{\varepsilon 3}$ are empirical coefficients.

For the $k-\omega$ SST turbulence model, the coefficients are expressed in the following general form:

$$C_\varphi = F_1 C_{\varphi 1} + (1 - F_1)C_{\varphi 2} \tag{10.53c}$$

where $C_{\varphi 1}$ and $C_{\varphi 2}$ are given by two separate coefficient sets and

$$F_1 = \tanh\left(\arg_1^4\right)$$

$$\arg_1 = \min\left[\max\left(\frac{\sqrt{k}}{0.09\omega y}, \frac{500\nu}{y^2\omega}\right), \frac{4\rho k}{\sigma_{\omega 2}^{\omega} CD_{k\omega} y^2}\right]$$

$$CD_{k\omega} = \max\left(\frac{2\rho}{\omega\sigma_{\omega 2}^{\omega}}\frac{\partial k}{\partial x_j}\frac{\partial \omega}{\partial x_j}, 10^{-20}\right)$$

Coefficients for $k-\omega$ SST turbulence model

σ_{k1}^{ω}	$\sigma_{\omega 1}^{\omega}$	β_1	β_1^*	κ
1.176	2.0	0.075	0.09	0.41

With $\alpha_1 = \dfrac{\beta_1}{\beta_1^*} - \dfrac{\kappa^2}{\sigma_{\omega 1}^{\omega}\sqrt{\beta_1^*}}$

Coefficients for $k-\omega$ SST turbulence model

σ_{k2}^{ω}	$\sigma_{\omega 2}^{\omega}$	β_2	β_2^*	κ
1.0	1.168	0.0828	0.09	0.41

where

$$\alpha_2 = \frac{\beta_2}{\beta_2^*} - \frac{\kappa^2}{\sigma_{\omega 2}^{\omega}\sqrt{\beta_2^*}}, S_\omega = 2(1-F_1)\frac{1}{\sigma_{\omega 2}^{\omega}\omega}\frac{\partial k}{\partial x_j}\frac{\partial \omega}{\partial x_j}$$

The **turbulent viscosity** is defined as

$$\mu_t = \rho\frac{a_1 k}{\max\left(a_1\omega, \Omega^* F_2\right)} \tag{10.53d}$$

where $a_1 = 0.31$, $F_2 = \tan h\left(\arg_2^2\right)$, and $\arg_2 = \max\left(2\frac{\sqrt{k}}{0.09\omega y}, \frac{500\nu}{y^2\omega}\right)$.

10.6.3.3 Low Reynolds Turbulence Model

Low Reynolds $k-\varepsilon$ Turbulence Model

The Low-Reynolds number $k-\varepsilon$ turbulence model is developed in order to address the deficiencies of standard high-Reynold number $k-\varepsilon$ model to predict the behavior at lower Reynolds number applications and also in problems involving separated flows (Lam and Bremhorst, 1981; Jones and Launder, 1972; Majumdar and Deb, 2003). The Low-Reynolds number model uses a special near-wall treatment for proper prediction of the flow at near wall, referred to as the laminar sublayer.

As we have discussed that the turbulent kinetic energy distribution reaches to its peak value in the near-wall region. Even though dissipation rate ε also increases in this region, k^2 term in the turbulent viscosity relation given by Equation 10.30 plays a stronger role by changing the μ_t by a huge amount. On the other hand, at a location remarkably close to the wall, k^2 suppresses μ_t. Basically, a high Reynolds number model is unable to predict such sharp variations in turbulence kinetic energy in the region close to the wall surface. To counteract these effects and have appropriate corrections for the viscous effects, two approaches are normally used: (a) use of the **wall-function method** near the wall region where an empirical wall function equation is introduced along with use of high Reynolds number turbulence in rest of the flow region; however, such an approach is not always suitable for many flows; and (b) use of the **low-Reynolds-number turbulence model** along with the wall boundary conditions applied directly to the equations and use some corrections for some low-Reynolds number without introducing any wall function treatments. Among many available low-Reynolds number models, the most popularly ones are (a) Jones and Launder (1972); (b) Launder and Sharma (1974); (c) Lam and Bremhorst (1981).

The turbulence kinetic energy k –equation for this turbulence model is the same as in equation in the standard $k-\varepsilon$ model, except the expression for the turbulent viscosity μ_t, which is defined as

$$\mu_t = f_\mu\frac{C_\mu\rho\cdot k^2}{\varepsilon} \tag{10.54a}$$

where

$$f_\mu = \left[1 - e - 0.0198\ \mathrm{Re}_y\ \right]\left[1 + \frac{5.29}{\mathrm{Re}_y}\right] \tag{10.54b}$$

ε Equation

$$\frac{\partial}{\partial t}(\rho\varepsilon)+\frac{\partial}{\partial x_j}\left[\rho u_j\varepsilon-\left(\mu+\frac{\mu_t}{\sigma_\varepsilon}\right)\frac{\partial\varepsilon}{\partial x_j}\right]=C_{\varepsilon 1}\frac{\varepsilon}{k}\left\{\mu_t\left(P+P_{\mathrm{NL}}+P'\right)-\frac{2}{3}\left(\mu_t\frac{\partial u_i}{\partial x_j}+\rho k\right)\frac{\partial u_i}{\partial x_j}\right\}$$

$$+C_{\varepsilon 3}\frac{\varepsilon}{k}\mu_t P_B-C_{\varepsilon_2}\left(1-0.3e^{-R_t^2}\right)\rho\frac{\varepsilon^2}{k}+C_{\varepsilon_4}\rho\varepsilon\frac{\partial u_i}{\partial x_j} \quad (10.55)$$

The additional term P' is given by

$$P'=1.33\left(1-0.3e^{-R_t^2}\right)\left[P+P_{\mathrm{NL}}+2\frac{\mu}{\mu_t}\frac{k}{y^2}\right]e^{-0.00375\mathrm{Re}_y^2} \quad (10.56)$$

where $\mathrm{Re}_y=\frac{y\sqrt{k}}{\nu}$ and y as a normal distance to the nearest wall.

Turbulent Reynolds Number $=R_t=\frac{k^2}{\nu\varepsilon}$.

As the integration or discretizations of Low Reynold's number model equations are applied all the way to the wall and use no slip condition, much finner mesh resolutions are required and hence are computationally more expensive.

10.7 REYNOLDS STRESS MODEL

RSM uses an alternative approach to solving RANS equation for turbulent flow. This model discards the use of turbulence closure models and the use of eddy viscosity approach. It is a higher level more elaborate turbulent model, which introduces exact Reynolds stress transport equation to compute the Reynolds stresses $T_{ij_t}=-\rho\overline{u_i'u_j'}$ directly by solving conservation equations for each component of the turbulent stress tensor. It accounts for anisotropic nature or the directional effects of Reynolds stress field.

Derivation of the Reynolds stress transport equation is given in books by Wilcox (1993) and White (1991). The time-averaged form of the Reynolds stress transport equation is given as

$$\underbrace{\frac{\partial\tau_{ij}}{\partial t}}_{\mathrm{I}}+\underbrace{u_k\frac{\partial\tau_{ij}}{\partial x_k}}_{\mathrm{II}}=-\underbrace{\left(\tau_{ik}\frac{\partial u_j}{\partial x_k}+\tau_{jk}\frac{\partial u_i}{\partial x_k}\right)}_{\mathrm{III}}+\underbrace{2\nu\overline{\frac{\partial u_i'}{\partial x_k}\frac{\partial u_j'}{\partial x_k}}}_{\mathrm{IV}}$$

$$-\underbrace{\overline{\frac{p'}{\rho}\left(\frac{\partial u_i'}{\partial x_j}+\frac{\partial u_j'}{\partial x_i}\right)}}_{\mathrm{V}}+\underbrace{\frac{\partial}{\partial x_k}\left[\nu\frac{\partial\tau_{ij}}{\partial x_k}+\overline{u_i'u_j'u_k'}+\overline{\frac{p'}{\rho}\left(u_i'\delta_{jk}+u_j'\delta_{ik}\right)}\right]}_{\mathrm{VI}} \quad (10.57)$$

The physical descriptions of the six terms (term I–VI) in the Reynold stress equation (10.57) are as follows: I – rate of change of Reynolds stress; II – convective diffusion; III – production of stress; IV – dissipation (ε_{ij}); V – pressure–strain effect; VI – diffusion of Reynolds stress.

As we can see that this higher moment representation of the RANS equation includes additional differential equations representing all components of Reynolds stress, but resulted in additional unknowns, which need to be approximated in terms of flow properties using correlation coefficients to close the system of equations. Most popular RSM models are (a) RSM/Gibson-Launder (wall Reflection (1: Standard)); (b) RSM/Gibson-Launder (wall Reflection: Craft); (c) RSM/

Speziale, Sarkar, and Gatski. Due to the consideration of all directional nature of transportation equations, RSM model shows superior results with flows **involving anisotropic turbulence**. The model includes **additional equations** and closure coefficients, and accounts for higher accuracy, but requires increased computational resources.

To narrow down the choice of turbulence closure model in terms of stable converging solutions, each class of turbulence models has to be compared among themselves first before comparing all models with the experimental data or DNS/LES.

10.8 NEAR-WALL REGION MODELING

The high Reynolds number $k - \varepsilon$ or k–ω turbulence models are valid in the region away from the wall where the flow is fully turbulent. However, the flow is laminarized near the wall due to reduced velocity dictated by the laminar sublayer profile and no-slip wall condition. In this region, the high Reynolds number turbulence models are not valid. To model the dominance of **laminarization of flow** or **the presence of laminar viscous sublayer near the wall**, two approaches are employed: (a) the use of low Reynolds number model; and (b) the use of **wall function treatments**.

The **first approach** involves modifying the large Reynolds number turbulence models and use of low Reynolds number models as discussed earlier in Section 10.6.3.3 along with using an appropriate fine mesh resolution to resolve the near-wall viscous sublayer region. Before proceeding with the discussion about the computational methodology for modeling turbulent flow near-wall region using the **second approach: wall function treatment**, let us briefly discuss here the well-known concept of **Universal Laws of the Wall** (Schlichting and Gersten, 1996; Fox and McDonald, 1998; Holman, 1990).

The universal laws of the wall are developed to represent the velocity profile in turbulent boundary flows. Figure 10.5 depicts three major regions of turbulent boundary layer for flow over a flat wall surface. The **viscous sublayer** is the very thin fluid layer close to the wall where molecular viscous effects as in laminar flow characteristics are present. The layer farther away is the **buffer layer** where some turbulent effects are present along with the molecular viscous effect. Farther away is the **fully turbulent flow region**.

The universal wall law was originally proposed as an empirical representation of experimental data for the turbulent velocity profile in the near-wall region of Couette flow in a smooth wall pipe. This profile was assumed following the velocity distribution and eddy viscosity concept postulated by Prandtl as discussed in Section 10.4.2.2. This is, however, universally accepted for all fully developed turbulent flows through smooth pipes at higher Reynolds range involving thin boundary layers.

The velocity profile is shown in Figure 10.6 in terms of the dimensionless velocity, u^+, and dimensionless local distance y^+. Both the streamline velocity and the position are normalized using friction velocity and are defined in the following manner:

$$u^+ = \frac{u}{u^*}, \tag{10.58a}$$

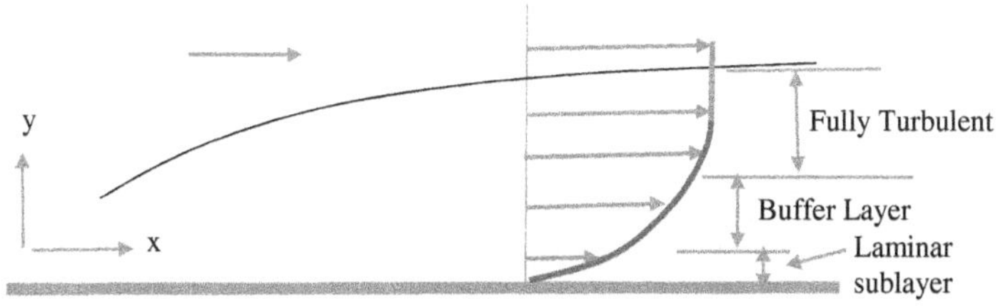

FIGURE 10.5 Flow regions in turbulent flow boundary layer.

and

$$\frac{yu^*}{\nu} = y^+ \tag{10.58b}$$

where

R = pipe radius

y = the local position

u = mean velocity

u^* is the **friction velocity** and is defined as

$$u^* = \left(\tau_w/\rho\right)^{1/2} \tag{10.58c}$$

where τ is the total shear stress across the pipe and $\tau_w = \mu\left(\frac{\partial u}{\partial y}\right)_{y=0}$ is the wall shear stress.

Velocity profile for fully developed turbulent flow in smooth pipes, derived based on the correlation of experimental data, shows three distinct regions: (a) viscous sublayer or near-wall region where viscous shear is dominant; (b) the buffer or transition region between the viscous sublayer and the center core region, where both viscous and turbulent shear are important; and (c) the center core region where turbulent shear stress is dominant. The functional relations for these regions are given as

Viscous Sublayer Velocity profile follows linear viscous relation as

$$u^+ = y^+ \quad \text{for} \quad 0 \le y^+ \le 5 \tag{10.59}$$

Buffer Layer Both viscous and turbulent share are important and the velocity profile follows the logarithmic relation as

$$u^+ = 2.5 \ln y^+ + 5.0 \tag{10.60a}$$

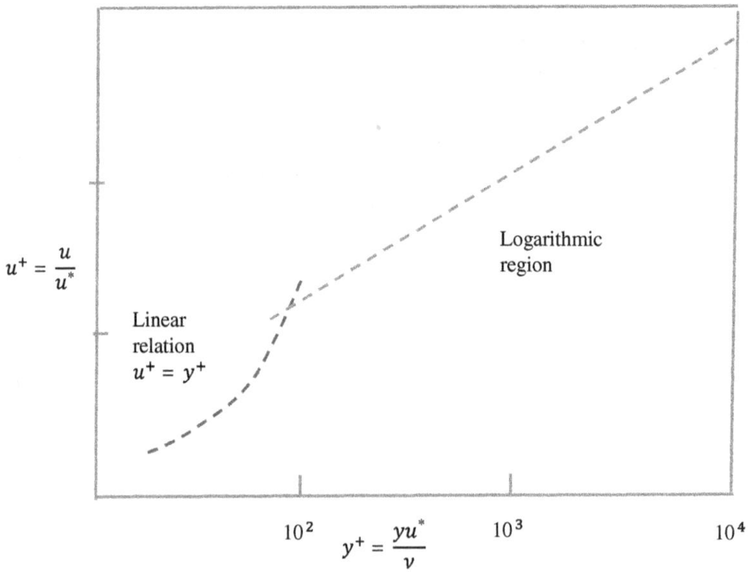

FIGURE 10.6 Velocity profile and wall law for turbulent flow.

or

$$\frac{u}{u^*} = 5.0 \ln \frac{yu^*}{\nu} - 5.0 \quad \text{for} \quad 5 \le y^+ \le 30 \tag{10.60b}$$

where $K=0.25$ and $E=-3.05$.

Turbulent Layer This is the further outer layer or center core region where turbulent shear stress is dominant

$$\frac{u}{u^*} = 2.5 \ln \frac{yu^*}{\nu} + 5.0 \quad \text{for} \quad 5 \le y^+ \le 30 \tag{10.60c}$$

where $K=0.4$ and $E=5.5$.

As we have discussed in Section 10.1 and as demonstrated in Figure 10.2 that the turbulence kinetic energy associated with the velocity fluctuations is dominant at the near-wall region. This also results in a strong apparent turbulent shear stress variation in that region. Figure 10.7 shows the turbulent shear stress variation with dimensionless distance y/R for fully developed turbulent flow in a pipe of radius R.

Turbulence modeling in this near-wall region requires to address these near-wall turbulence quantities such as the turbulence kinetic energy production, dissipation rate, and the resulting turbulence viscosity.

Wall Functions

Important aspect of computational turbulence flow modeling is to take into account the near-wall viscous sublayer region where viscous shear force is dominant and the shear stress remains constant with a value set by the wall shear stress, τ_w. The mean velocity profile follows a linear viscous relation given by the Equation 10.59.

In the second approach, the viscosity-affected region is not resolved and rather **wall functions** are used to bridge the viscosity-affected region between the wall and the fully turbulent region.

There are number of different wall functions considered. standard wall functions, two-layer nonequilibrium wall functions, and enhanced wall functions. These wall function treatment formulations are implemented in many commercial codes such as ANSYS-Fluent and Star CCM+. A brief discussion of these wall function treatments is discussed here. For more comprehensive descriptions, readers are recommended to refer to Launder and Spalding (1974), Kim and Choudhury (1995), ANSYS-Fluent user guide, Star=CCM+ user guide.

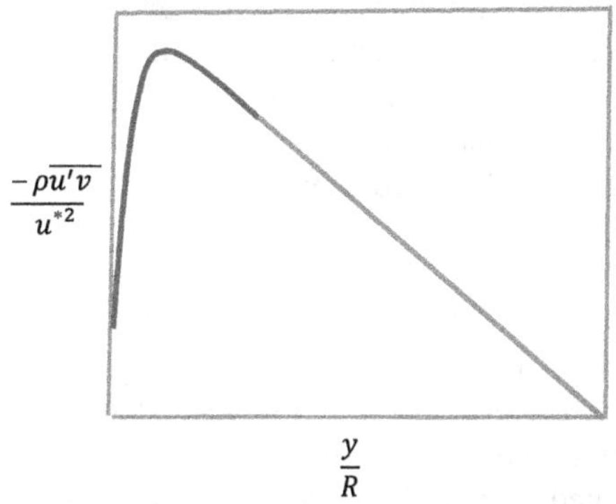

FIGURE 10.7 Variation of turbulent shear stress in fully developed turbulent flow in pipe.

Standard Wall Function:

In the **standard wall function method**, the **Universal Laws of the Wall** are considered in many turbulence modeling studies for wall function treatment. The wall function treatment proposed by Launder and Spalding (1972) is applied successfully in many industrial application studies. This standard wall function treatment is summarized as follows:

Viscous Sublayer:

$$u^+ = y^+ \quad \text{for} \quad 0 \le y^+ \le 5 \tag{10.61}$$

Log-layer Layer

Outside the viscous sublayer, the turbulent diffusion effect is present and the following log-law is applied

$$u^+ = \frac{1}{K} \, ln\left(E\, y^+\right) \quad \text{for} \quad 30 \le y^+ \le 300 \tag{10.62}$$

where $K=0.41$ and $E=9.793$.

In many computational analyses and computational codes, a lower limit of $10 \le y^+$ is often used.

Along with the near-wall function treatment, the standard high Reynolds number turbulence models are used outside this region without any modification.

The **standard wall functions** are generally used for high Reynolds number wall-bounded flows and for satisfying equilibrium conditions for production and dissipation of turbulent kinetic energy at the wall.

One of the limitations of the standard wall function is the equilibrium turbulence condition based on which the universal wall law was derived involving with small pressure gradient as in the original Couette flow experiments. Because of this restriction, standard wall functions are used cautiously for problem strong pressure gradient or adverse pressure gradient such as in the applications involving flow separation, reattachment, and flow developments.

Nonequilibrium Wall Functions

For flows that involve adverse pressure gradient, rotation, and strong streamline curvature, the flow conditions depart from equilibrium. The nonequilibrium wall functions consider the effects of pressure gradient and departure from equilibrium conditions and are generally involve severe pressure gradient. In nonequilibrium wall function method, two major considerations are given. One is to include the pressure gradient effects through sensitizing of the log-law profile of the standard wall function. Second, the two-layers concept is used to take into account of the difference in the production of the turbulence kinetic energy (k) with the dissipation rate (ε) of the turbulence kinetic energy by applying a cell averaged k and ε is to wall adjacent mesh cells. This average value is estimated by taking a volume average over a depth covering number of adjacent cells consisting of viscous sublayer and the fully turbulent layer. The nonequilibrium wall function is recommended for problems that involve flow separation, reattachment, and jet impingement.

Enhanced Wall Functions

This is another optional wall function treatment method that uses two-layer model along an enhanced wall function. The concept is to use fine enough mesh for the sublayer with a near wall node at y^+ close to 1 and then use a standard two-layer formulation without sacrificing the accuracy for coarser wall function mesh usage. The two-layer formulation in the viscosity dominated near-wall region is blended smoothly with the high-Reynolds number model for the outer fully turbulent region.

In the two-layer formulation, the solution domain is divided into two regions: fully turbulent outer layer and the viscosity dominated near-wall region. The division is based on a wall-distance-based Reynolds number defined as

$$\mathrm{Re}_y = \frac{\rho y \sqrt{k}}{\mu} \tag{10.63}$$

and using a limiting value of this Reynolds as $\mathrm{Re}_y^* = 200$.

In outer fully developed region with $\mathrm{Re}_y > \mathrm{Re}_y^*$, the high Reynolds number model is used, and for the near wall region with $\mathrm{Re}_y < \mathrm{Re}_y^*$, a one-equation model is used.

The enhanced wall function formulation then uses a single-wall law for the entire wall region.

Covering the near-wall and the outer flow region by using the following weighted-average values for the dimensionless velocity, velocity gradient, and the turbulent viscosity:

$$u_{\text{en}}^+ = \lambda_e u_{\text{lam}}^+ + (1-\lambda_e) u_{\text{turb}}^+ \tag{10.64a}$$

and

$$\frac{\partial u^+}{\partial y^+} = \lambda_e \frac{\partial u_{\text{lam}}^+}{\partial y^+} + (1-\lambda_e) \frac{\partial u_{\text{turb}}^+}{\partial y^+} \tag{10.64b}$$

The turbulent viscosity in the enhanced wall function treatment is computed as a weighted average given as

$$\mu_{t,\text{enh}} = \lambda_e \mu_t + (1-\lambda_e) \mu_{t,\text{2-layer}} \tag{10.65}$$

where μ_t is the based on high Reynolds number model for the outer fully turbulent region; $\mu_{t,\text{2-layer}}$ is based on the one-equation turbulence model; λ_e is a blending function defined as

$$\lambda_e = \frac{1}{2}\left[1 + \tan h\left(\frac{\mathrm{Re}_y - R_y^*}{A}\right)\right] \tag{10.66}$$

Constant A in Equation 10.66 represents a constant factor to keep the width of the blending function within 1% of the far-field value.

10.9 ESTIMATION OF *Y*-PLUS

The refined mesh size distribution near the wall is limited by satisfying y^+ requirement, which defines the minimum distance of the computational cell from the wall boundary in both approaches. Having the correct y^+ value for the cells next to the wall is extremely important to obtain the correct velocity, pressure, and shear stress values. For example, in the study using the low Reynolds number turbulence closure models, the y^+ value is required to be kept within $y^+ < 5$.

For using turbulence models with wall functions, the y^+ value of the near-wall cells is a basic requirement that has to be satisfied. For example, in the study of $k-\varepsilon$ high Reynolds number turbulence closure model with standard wall function, the y^+ value is required to be kept within 30–120.

All computational turbulence studies start with an initial search for the correct cell size to satisfy the y^+ requirement for the turbulence model used.

10.9.1 Procedure to Estimate y^+ in Wall Function Treatment

To overcome this situation, several different wall function treatments using the wall law profile are used. The basic requirement is that the grid refinement near the wall must satisfy a y^+value in the range of 30–120 for the $k-\varepsilon$ high Reynolds number turbulence closure model with standard wall function. The cell size near the wall or the distance of the centroid of cell size nearest to the wall is calculated based on the following formula as follows:

$$y^+ = \frac{u^* \cdot y}{\upsilon} = \frac{u^* \cdot y \cdot \rho}{\mu} \tag{10.67a}$$

where

$$u^* = \sqrt{\frac{\tau_w}{\rho}} \tag{10.67b}$$

This is initiated by searching for the correct cell size to satisfy the y^+ requirement for the turbulence model used. Notice that y^+ depends on the distance of the centroid of the nearest cell from the wall (y_p) as well as on the wall shear stress (τ_w) for a specific problem. So, the grid refinement near the wall must be done by a trial and error basis. A numerical experiment has to be conducted to check with different near wall cell size to see the effect of cell size on the y^+ value

Near-wall mesh size distribution and the value of y^+ can be computed based on following steps:

1. Select a grid size near the wall.
2. Solve the CFD problem
3. Estimate wall shear stress
4. Calculate y^+ from Equation (10.67a) along the wall surface
5. Check if it is within 30–120
6. Adjust near-wall cell size accordingly

10.10 BOUNDARY CONDITION FOR TURBULENCE QUANTITIES

10.10.1 Inlet Turbulence

Solution of turbulent flow requires specification of turbulence intensity or turbulence quantities such as turbulence kinetic energy and turbulence dissipation rate at the inlet.

In general, the inlet turbulence is a function of the upstream flow conditions unless some form of turbulence control is applied.

The specified values may have a significant effect on the resulting flow solutions. However, the inlet turbulence conditions are mostly unknown. The assignment of inlet turbulence is particularly important if the computational region of interest is close to the inlet. Solutions are less sensitive of inlet turbulence condition when inlet is sufficiently far away from the region of interest. Because of this reason, extra entrance length is often included in the computational analysis of fluid flow and heat transfer.

Specification of turbulence quantities such as turbulence kinetic energy and turbulence dissipation rate at inlet can be quite difficult and often rely on engineering judgments.

One approach is to directly assign the values of turbulent kinetic energy (k) and turbulent dissipation rate (ε). It is always preferred to assign experimentally measured values of such turbulence quantities. If such data are not available, then values can be prescribed based on engineering judgements and assumptions, and a sensitivity study must be performed while varying the assigned turbulence conditions. For the specification of the turbulent kinetic energy, appropriate values can be specified through **turbulence intensity** (I), which is defined by the **ratio of the fluctuating**

components of the velocity to the mean velocity. Inlet turbulence condition is often also assigned in form of **length of turbulence scale (*L*)**.

Approximate values of the inlet turbulence quantities can be determined according to the following relationships:

Turbulent Kinetic Energy:

$$k_{\text{inlet}} = \frac{3}{2}(U_{\text{inlet}}\, I)^2 \tag{10.64}$$

Dissipation Rate:

$$\varepsilon_{\text{inlet}} = \frac{C_\mu^{3/4}\, k_{\text{inlet}}^{3/2}}{L} \quad \text{for } k-\varepsilon \text{ models} \tag{10.65}$$

or

$$\omega_{\text{inlet}} = \frac{k_{\text{inlet}}^{3/2}}{C_\mu^{1/4} L} \quad \text{for } k-\omega \text{ models} \tag{10.66}$$

where $C_\mu = 0.09$

Typical recommended values of turbulence intensity (*I*) are in the range of 3%–10%. For internal flows, the recommended values are in the range of 1%–10%. For atmospheric boundary layer flows, the level can be as high as two orders of magnitude – 30%.

Estimated values of inlet turbulence length scale (*L*) are often estimated based on the following the geometric characteristics dimension with a scale down factor of 1–10. In an internal channel flow, the inlet turbulent length scale can be assigned as

$$L = D_h/10 \tag{10.67}$$

Another common practice is to assign inlet turbulence intensity at the core of a **fully developed duct flow** estimated from the following formula:

$$\text{I} = 0.16\left(Re_{D_h}\right)^{1/8} \tag{10.68}$$

where Re_{D_h} is Reynolds number defined based on the hydraulic diameter.

10.10.2 Wall Boundary Condition

For use with wall-function treatment:

Normal flux component for turbulent kinetic energy is zero; $\frac{\partial k}{\partial n} = 0$.

For dissipation rate of near-wall node value is specified as follows:

$$\varepsilon_P = \frac{C_\mu^{3/4}\, k_P^{3/2}}{K y_p}$$

where y_P is *y*-distance of the near-wall boundary node-*P* from the wall.

For use with low Reynolds number models:

$$k = 0 \quad \text{and} \quad \frac{\partial \varepsilon}{\partial n} = 0$$

Outlet and Symmetric Boundary:

$$\frac{\partial k}{\partial n} = 0 \quad \text{and} \quad \frac{\partial \varepsilon}{\partial n} = 0$$

Far-Away Boundary in Free Stream Flow:

$$k = 0 \quad \text{and} \quad \varepsilon = 0$$

Part III

Finite Element Method

11 Introduction and Basic Steps in Finite Element Method

As we have discussed in previous chapters, in the **finite difference methods**, the governing equations are approximated by a point-wise discretization scheme where derivatives are replaced by difference formulas that involve the unknown values at the nodal points. Other commonly used approximation methods are the variational methods and the finite element methods (FEMs).

In the solution of a mathematical model by the **variational method**, the governing equation is transformed into an equivalent integral form. The approximate solution to this integral form of the governing equation is obtained over the entire domain. The approximate solution is assumed to be a linear combination of an appropriately chosen function $\psi_i(x)$ and undetermined coefficient a_i as

$$\phi = \sum_{i=1} a_i \psi_i \tag{11.1}$$

Coefficients a_i are determined such that the integral statement equivalent to the original differential equation is satisfied. There are different kinds of variational methods, for example, **Rayleigh–Ritz**, **collocation**, **least-squared**, and **Galerkin methods**. These different types of variational methods differ from each other based on the type of integral form used and selection of weighted functions. The major disadvantages of the variational method are the difficulty in selecting an approximate function that is valid over the entire solution domain satisfying the continuity and boundary conditions, and the lack of any systematic procedure for constructing them. The situation becomes even more difficult when the solution domain is geometrically complex. Due to these disadvantages, the variational methods of approximations are not as widely used as the finite difference–control volume methods or the FEM.

The FEM, like the finite difference method, is a numerical procedure based on discretization of the solution domain. This method overcomes the disadvantages associated with the variational method by providing a systematic procedure for discretizing the solution domain into simple shaped subregions called **finite elements**. This is followed by deriving the approximate function or solution over each of these elements, using one of the previously mentioned variational methods of approximation such as **Rayleigh–Ritz** or **Galerkin** method. The total solution over the whole domain is then generated by linking together or assembling the individual element solutions satisfying the continuity at the interelement boundaries. Hence, the FEM can be viewed as a **piecewise** or **element-wise** application of the variational methods. Also, a FEM is named based on the type of variational method being used such as the Rayleigh–Ritz FEM or the Galerkin FEM.

The **FEM** is one of the most widely used solution methods in engineering. It was originally developed for solving problems in structural mechanics. However, it became increasing popular in all branches of engineering including heat transfer and fluid mechanics. It has also become a primary analysis tool for many computer-aided design programs. The advantages of the FEM are quite clear. The method is easily applied to irregular-shaped objects; the medium composed of several different materials and having mixed boundary conditions.

11.1 COMPARISON BETWEEN FINITE DIFFERENCE–CONTROL VOLUME METHOD AND FEM

We have discussed in Chapters 5–8 that in the finite difference–control volume method the solution domain is divided into a grid of discrete points, called nodes. The governing mathematical equations are then written at each node and its derivatives are expressed by the finite difference formulas, which involve unknown values at discrete grid or nodal points of the domain. This discretization procedure is referred to as **pointwise approximation**. The system of equations resulting from all nodes including the boundary nodes is solved for the unknown values at the nodal points.

The major disadvantages of the finite difference method are (a) difficulty in accurately representing a geometrically complex domain, (b) difficulty in imposing the boundary conditions along a nonstraight boundary, and (c) inability to employ nonrectangular mesh size distribution.

In the FEM, the solution domain is subdivided into a mesh of interconnected subregions referred to as finite elements. A finite element model of a problem gives a **piecewise approximation** to the governing equations. This method has some features that account for its superiority over other numerical methods. These are as follows:

1. A geometrically complex domain of the problem is represented as a collection of geometrically simple subdomains, called **finite elements**.
2. The approximate function is derived over each finite element by using the basic idea that any continuous function can be represented by a linear combination of interpolation functions.
3. The algebraic relations among the undetermined coefficients are obtained by satisfying the governing equations either in a **weighted-residual integral sense** or an equivalent **variational form** of the problem over each element.

One important point we should remember is that both finite difference and finite element methods are used to obtain an approximate solution of a mathematical model, which is an idealization of a physical problem subject to some assumptions. Hence the selection of the mathematical model is especially important, and we should not expect any additional information regarding the physical phenomena than what is represented by the mathematical model. Also, it is necessary for us to assess the solution accuracy. If the accuracy criterion is not met, then the numerical solution must be repeated with refined grids or mesh size distributions and/or with higher order elements until sufficient accuracy is reached. Once the mathematical model has been solved accurately and the results have been interpreted, we may decide to consider a more refined mathematical model to get additional insight of the physical problem. This will lead to higher-order approximations and additional grid or mesh size distributions.

11.2 BASIC STEPS IN FEMs

The general procedure in the finite element analysis usually involves some basic steps. Details of each step may vary from problem to problem. In this section, we will briefly discuss the basic steps employed in the formulation and application of the FEM. The details of each one of these steps will be developed and described fully in the subsequent sections and chapters. Although the particulars may vary, the implementation of the FEM can be subdivided into the following basic steps:

1. **Discretization of the Solution Domain** This step involves dividing the solution domain into a finite number of subdomains. Each subdomain is called a **finite element**. The collection of all elements is called the **finite element mesh**. The shape, size, number, and orientation of the elements are selected in such a way that the solution domain is represented closely and accurately, and solution is obtained with least computational difficulties. The final selection is

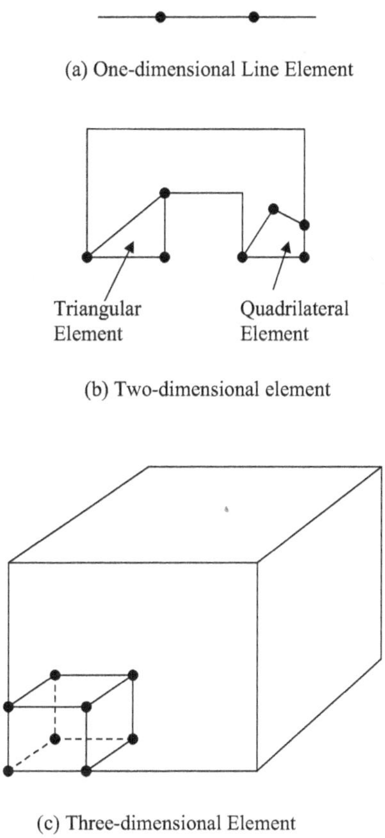

FIGURE 11.1 Examples of finite elements in one, two, and three dimensions.

established through a systematic mesh refinement study until desirable accuracy is achieved. Figure 11.1 shows some examples of elements employed in one, two, and three dimensions.

The points of intersection of the lines that make up the sides of the elements are referred to as **nodes** or **nodal points** and the sides themselves are called **nodal lines** or **planes**. This step also includes locating and numbering the nodal points, as well as specifying their coordinate values. A detail description of the type, selection, orientation of finite elements, and the numbering scheme is given in Chapter 12.

2. **Integral Formulation of the Governing Equation** In this step, the governing differential equation is transformed into an integral form, which represents an optimum fit of the approximating function to the solution of the equation. There are several methods that are available for this purpose. The most common ones are the **method of weighted-residual integral** and the **variational formulation**. These methods specify relationships between the unknowns in the approximation function that satisfy the underlying governing differential equation in an optimal manner. The **method of weighted-residual integral** is based on the minimization of a residual term R, which is obtained as an approximate or a trial solution is substituted into the governing differential equation. In the **variational formulation**, the minimization is sought for a **variational form** or the equivalent **functional form** I of the governing equation. The variational form or the functional I could be obtained either by using the principles of variational calculus or by deriving a weak form by evaluating the weighted-residual integral. Both methods, however, use **classical variational method of approximations** to solve the integral equations. Detailed descriptions of these methods are given in subsequent sections.

3. **Selection of Approximation Function** In this step, we develop equations to approximate the solutions for each element. This involves two steps. First, we must choose an appropriate function with unknown coefficients that will be used to approximate the solution. Second, we evaluate the coefficients so that the chosen function approximates the solution in an optimal fashion.

   Several mathematical functions such as polynomials or trigonometric series can be used for this purpose. Polynomials are often selected for this purpose, as they are easy to manipulate mathematically. The approximate solution is assumed to be a linear combination of appropriately selected functions $\psi_i(x)$ and undetermined coefficients a_i as expressed by Equation 11.1. For example, in one-dimensional problems, the simplest choice is a first-order polynomial or a straight line such as

$$\phi(x) = a_0 + a_1 x \tag{11.2}$$

   where $\phi(x)$ is the dependent variable, a's are constants, and x is the independent variable and represents the function $\psi_1(x)$. Figure 11.1 shows a one-dimensional line element with length L and two nodal points i and j. The coefficients in Equation 11.2 can be found by satisfying two nodal points, i.e., $\phi_i(x_i)$ and $\phi_j(x_j)$. Substituting these two nodal points in Equation 11.1, we have

$$\phi_i(x) = a_0 + a_1 x_i \tag{11.3a}$$

   and

$$\phi_j(x) = a_0 + a_1 x_j \tag{11.3b}$$

   These equations can be solved using Cramer's rule for

$$a_0 = \frac{\phi_i x_j - \phi_j x_i}{x_j - x_i} \tag{11.4a}$$

   and

$$a_1 = \frac{\phi_j - \phi_i}{x_j - x_i} \tag{11.4b}$$

   Substituting Equation 11.4 into Equation 11.2, we get

$$\phi = \left(\frac{x_j - x}{x_j - x}\right)\phi_i + \left(\frac{x - x_i}{x_j - x_i}\right)\phi_j \tag{11.5}$$

   or

$$\phi = N_i \phi_i + N_j \phi_j \tag{11.6}$$

   where

$$N_i = \frac{x_j - x}{x_j - x_i} \tag{11.7a}$$

and

$$N_j = \frac{x - x_i}{x_j - x_i} \tag{11.7b}$$

Equation 11.6 is called an **approximation function**, which is a sum of nodal values multiplied by linear interpolation functions of x. It also provides a means to predict intermediate values between given values of ϕ at the nodal points. The interpolation functions expressed as N, given by Equation 11.7, are called **interpolation or shape functions**. The approximation Equation 11.6 can be written in matrix form as

$$\phi = [N]\{\phi\} \tag{11.8}$$

where

$$[N] = \left[N_i, N_j\right] = \text{row vector of shape functions}$$

$$\{\phi\} = \begin{Bmatrix} \phi_i \\ \phi_j \end{Bmatrix} = \text{column vector of unknown values at the nodal points}$$

In a general form, the approximation function can be written as

$$\phi = N_i\phi_i + N_2\phi_2 \cdots N_n\phi_n \tag{11.9a}$$

or

$$\phi = \left[N\right]\{\phi\} \tag{11.9b}$$

where

$$\{\phi\} = \begin{Bmatrix} \phi_1 \\ \phi_2 \\ \vdots \\ \phi_n \end{Bmatrix} = \text{a column vector of unknown nodal values}$$

$$[N] = \left[N_1, N_2, \ldots, N_n\right] = \text{a row vector of shape functions}$$

4. **Formation of Element Characteristics Equation** After the selection of the approximation function, we need to develop an equation which governs the behavior of the element. A direct substitution of the approximation function into the integral form of the element governing equation results in the **element characteristics equation**, which is also referred to as element stiffness equation. Mathematically, the resulting element equation forms a system of linear algebraic equations that can be expressed in matrix form as

$$\left[k^e\right]\{\phi\} = \left\{f^e\right\} \tag{11.10}$$

where

$\left[k^e\right]$ = an element characteristics matrix or element stiffness matrix

$\{\phi\}$ = a column vector of unknown values at the nodal points of the element

$\left\{f^e\right\}$ = a column vector of element nodal forcing parameters that represents the effect of any external influences applied at the nodal points

This equation represents the optimum fit of the function to the solution of the governing differential equation. It can be mentioned here again that the procedure for obtaining the element characteristics equation is based on element-wise application of the **classical variational method of approximation**.

5. **Assembly of Element Equations** The individual element equations are linked together or assembled to characterize the unified behavior of the entire system. The assembly process is governed by the concept of continuity. That is, the solutions for contiguous elements are matched so that the unknown values and/or its derivatives at their common nodes are equivalent. This will cause a continuous solution in the domain. When all the individual element equations are finally assembled, the entire system of equations, known as the global characteristic equation or global stiffness equation, can be expressed in matrix form as

$$[K]\{\Phi\}=\{F\} \tag{11.11}$$

where

K = global characteristic matrix or stiffness matrix

Φ = column vector for unknowns

F = column vector for external forces

Φ and F are the collection of all vectors ϕ and f associated with all individual elements, respectively.

6. **Implementation of Boundary Conditions** In this step, the boundary conditions are introduced directly in the global system of equations. This may also require rearrangement of the system of equations, particularly for cases with assigned boundary conditions of the second or third kind.
7. **Solution of System of Equations** In this step, the appropriate solver is selected and solutions to the system of equations are obtained. In many cases, the elements can be reconfigured so that the resulting equations are banded, and highly efficient solution algorithms could be employed.
8. **Error Estimate and Convergence Check** An error analysis study is conducted in this step. Errors associated with the approximate solutions are estimated and checked for convergence. The errors in the FEM include (a) the truncation error due to the discretization of the solution domain and error associated with selection of the approximate solution as a linear combination of interpolation functions, and (b) truncation and round-off errors associated with the numerical evaluation of integrals or selection of quadrature formulas, and solutions of systems of equations on the computer.

A mesh refinement study is conducted to improve the solution and obtain the desired accuracy or convergence. This may also, in some cases, require the selection of higher-order approximation solution.

9. **Postprocessing** Once the converged solution is obtained, other quantities or secondary variables of interest are calculated. These quantities are usually related to the derivative of the variables and include quantities such as the volume flow, shear stress, and heat transfer rate.

11.3 INTEGRAL FORMULATION

We have discussed in the previous section that in a FEM we start with an integral form of the governing equation. One of the important steps in the FEM is to transform the governing differential equation into an equivalent integral form using one of the integration formulation techniques, namely, **variational formulation**, or **weighted residual integration formulation**. Let us now briefly describe these two integral formulation procedures.

11.3.1 Variational Formulation

In this approach, an integral statement of the problem is obtained by deriving an equivalent **variational form** of the problem using the principles of **calculus of variations**. A brief discussion of the principles of **calculus of variations** is given in the next subsection. This **variational form** involves a **functional**, I, such that finding a minimized or stationary or extremized value of the functional, I, is equivalent to finding the solution of the original differential statement of the problem. A general expression of this functional form is given as

$$I = \int_V F\left(\tilde{x}, \phi, \frac{\partial \phi}{\partial \tilde{x}} \ldots\right) \mathrm{d}V \tag{11.12}$$

where V is the solution domain, $\tilde{x}$ is the vector for independent variables, and $\{\phi\}$ is the unknown solution vector. The condition of minimum value of the functional with respect to the unknown solution vector is given as

$$\frac{\partial I}{\partial \{\phi\}} = \begin{Bmatrix} \dfrac{\partial I}{\partial \phi_1} \\ \dfrac{\partial I}{\partial \phi_2} \\ \vdots \\ \dfrac{\partial I}{\partial \phi_M} \end{Bmatrix} = 0 \tag{11.13}$$

where M indicates the number of unknown variables. In the case of a finite element formulation, the functional over the whole solution domain can be written as a summation of all elemental contribution, $I^{(e)}$, as

$$I = \sum_{e=1}^{n} I^{(e)} \tag{11.14}$$

Combining Equations 11.13 and 11.14, we get the condition of minimization of the functional over the entire solution domain as

$$\sum_{e=1}^{n} \frac{\partial I^{(e)}}{\partial \phi^{(e)}} = 0 \tag{11.15}$$

As we evaluate the elemental contribution $\partial I^{(e)}/\partial \phi^{(e)}$, we obtain the element characteristic equation as

$$\frac{\partial I^{(e)}}{\partial \phi^{(e)}} = \left[K^{(e)}\right]\left\{\phi^{(e)}\right\} - \left\{f^{(e)}\right\} = 0 \tag{11.16}$$

Equation 11.16 is applied successively to all elements, and the resulting elemental equations are then assembled to form the global system of equations.

Principles of Calculus of Variations We have mentioned that one of the important steps in the formation of an integral statement using the variational method of approximation, as well as the FEM, is to determine an equivalent **variational** or **functional** form using the **calculus of variations**. Lagrange introduced an operator δ, which is like the differentiation operation used in 1760 (Bliss, 1946). This operator is referred to as variation, and all techniques and theory developed using this operator are referred to as the **calculus of variation**. Application of the calculus of variation is mainly concerned with the determination of the maximum and minimum values or stationery of a certain expression known as **functional**. A brief review of the variational form and notation and a procedure for obtaining the equivalent variational or functional forms are given here. For a more detailed description of this method, readers are suggested to read other reference books (Bliss, 1946; Hilderbrand, 1965; Ewing 1985).

The Variational Forms and Notation In order to establish the techniques, we will introduce the definition of a functional and various notation to the calculus of variation.

Functional A quantity such as

$$I = \int_{x_1}^{x_2} F\left(x, \phi(x), \frac{d\phi}{dx}, \ldots\right) dx \tag{11.17}$$

is called a functional if for any function $\phi(x)$ the quantity becomes a definite numerical value.

δ Operator For the integrand, $F\,(x, \phi(x), d\phi/dx, \ldots)$, we change the function $\phi(x)$ to be determined as a new function $\phi(x) + \varepsilon\eta(x)$. The change $\varepsilon\eta(x)$ in $\phi(x)$ is called the variation of ϕ and it is denoted as $\delta\phi$.

Laws of Variation Laws of variation of sums, products, ratios, powers, and other similar operations are completely analogous to the corresponding laws of differentiations. For example,

$$\text{Sum:} \quad \delta(\phi_1 + \phi_2) = \delta\phi_1 + \delta\phi_2$$

$$\text{Products:} \quad \delta(\phi_1\phi_2) = \phi_1\delta\phi_2 + \phi_2\delta\phi_1$$

$$\text{Ratio:} \quad \delta\left(\frac{\phi_1}{\phi_2}\right) = \frac{\phi_2\delta\phi_1 - \phi_1\delta\phi_2}{\phi_2^2}$$

$$\text{Power:} \quad \delta\left[(\phi_1)^n\right] = n\phi_1^{n-1}\delta\phi_1$$

Operator δ is commutative with differential and integral operators, i.e.,

$$\frac{\partial}{\partial x}(\delta\phi) = \delta\left(\frac{\partial \phi}{\partial x}\right)$$

$$\delta\int_{x_1}^{x_2} \phi(x)dx = \int_{x_1}^{x_2} \delta\phi(x)dx$$

and

$$\int_{x_1}^{x_2} \phi\,\delta\phi\,dx = \delta\int_{x_1}^{x_2} \tfrac{1}{2}\phi^2\,dx$$

Determination of a Functional or the Variational Form Let us demonstrate a general procedure for determining the functional, I, or the variational form of a problem that involves governing differential equations and boundary conditions. Consider the governing differential equation of the form

$$\frac{\mathrm{d}}{\mathrm{d}x}\left(\Gamma_x\frac{\mathrm{d}\phi}{\mathrm{d}x}\right) + S_1\phi + S_0 = 0 \tag{11.18}$$

The associated boundary conditions will be prescribed as we proceed. To derive a corresponding variational form, we first multiply both sides of Equation 11.18 by a variation $\delta\phi$ and integrate it over the solution domain (0–L)

$$\int_0^L \frac{\mathrm{d}}{\mathrm{d}x}\left(\Gamma_x\frac{\mathrm{d}\phi}{\mathrm{d}x}\right)\delta\phi\mathrm{d}x + \int_0^L S_1\phi\delta\phi\mathrm{d}x + \int_0^L S_0\delta\phi\mathrm{d}x = 0 \tag{11.19}$$

Using rules of variation, the second and third integrands are evaluated as

$$\int_0^L S_1\phi\delta\phi\mathrm{d}x = \delta\int_0^L \tfrac{1}{2}S_1\phi^2\,\mathrm{d}x \tag{11.20}$$

and

$$\int_0^L S_0\delta\phi\mathrm{d}x = \delta\int_0^L S_0\phi\mathrm{d}x \tag{11.21}$$

The first integral is evaluated using integration by parts and rules of variation as

$$\int_0^L \frac{\mathrm{d}}{\mathrm{d}x}\left(\Gamma_x\frac{\mathrm{d}\phi}{\mathrm{d}x}\right)\delta\phi\mathrm{d}x = \left[\Gamma_x\frac{\mathrm{d}\phi}{\mathrm{d}x}\delta\phi\right]_0^L - \int_0^L \Gamma_x\frac{\mathrm{d}\phi}{\mathrm{d}x}\delta\left(\frac{\mathrm{d}\phi}{\mathrm{d}x}\right)\mathrm{d}x$$

$$= \Gamma_x\frac{\mathrm{d}\phi}{\mathrm{d}x}\delta\phi\Big|_0^L - \delta\int_0^L \frac{1}{2}\Gamma_x\left(\frac{\mathrm{d}\phi}{\mathrm{d}x}\right)^2\mathrm{d}x \tag{11.22}$$

Substituting Equations 11.20–11.22 into Equation 11.19, we have

$$\delta\int_0^L\left[-\frac{1}{2}\Gamma_x\left(\frac{\mathrm{d}\phi}{\mathrm{d}x}\right)^2 + \frac{1}{2}S_1\phi^2 + S_0\phi\right]\mathrm{d}x + \left[\Gamma_x\frac{\mathrm{d}\phi}{\mathrm{d}x}\delta\phi\right]_0^L = 0 \tag{11.23}$$

Let us now consider different cases involving different types of boundary conditions.

Case 1: Constant Surface Value

For a constant surface value or boundary condition of the first kind, let us assume $\phi=\phi_0$ at $x=0$ and $\phi=\phi_L$ at $x=L$. Since the values are fixed at the boundaries, $\delta\phi$ becomes zero at $x=0$ and $x=L$, and the last integrated term $\left[\Gamma(\mathrm{d}\phi/\mathrm{d}x)\delta\phi\right]_0^L$ vanishes. This reduces Equation 11.23 to the final variational form as

$$-\delta\int_0^L\left[\frac{1}{2}\Gamma_x\left(\frac{\mathrm{d}\phi}{\mathrm{d}x}\right)^2-\frac{1}{2}S_1\phi^2-S_0\phi\right]\mathrm{d}x=0 \tag{11.24}$$

and the **functional** is given as

$$I=\int_0^L\left[-\frac{1}{2}\Gamma_x\left(\frac{\mathrm{d}\phi}{\mathrm{d}x}\right)^2+\frac{1}{2}S_1\phi^2+S_0\phi\right]\mathrm{d}x \tag{11.25}$$

with essential boundary condition $\phi(0)=\phi_0$ and $\phi(L)=\phi_L$.

Case 2: Constant Surface Flux

The mathematical statement of the constant surface flux or the boundary conditions of the second kind, as depicted in Figure 11.2, is given as

$$\text{at}\,x=0,\quad -\Gamma_x\frac{\mathrm{d}\phi}{\mathrm{d}x}=f_{\mathrm{ls}}'' \tag{11.26}$$

and

$$\text{at}\,x=L,\quad \Gamma_x\frac{\mathrm{d}\phi}{\mathrm{d}x}=f_{\mathrm{rs}}'' \tag{11.27}$$

With the substitution of these conditions, Equation 11.23 takes the form

$$\delta\int_0^L\left[-\frac{1}{2}\Gamma_x\left(\frac{\mathrm{d}\phi}{\mathrm{d}x}\right)^2+\frac{1}{2}S_1\phi^2+S_0\phi\right]\mathrm{d}x+f_{\mathrm{rs}}''\delta\phi\Big|_{x=L}+f_{\mathrm{ls}}''\delta\phi\Big|_{x=0}=0 \tag{11.28}$$

Using rules of variation, we get the equivalent variational form of the problem as

$$\delta\left[\int_0^L\left\{-\frac{1}{2}\Gamma_x\left(\frac{\mathrm{d}\phi}{\mathrm{d}x}\right)^2+\frac{1}{2}S_1\phi^2+S_0\phi\right\}\mathrm{d}x+f_{\mathrm{rs}}''\phi\Big|_{x=L}+f_{\mathrm{ls}}''\phi\Big|_{x=0}\right]=0 \tag{11.29a}$$

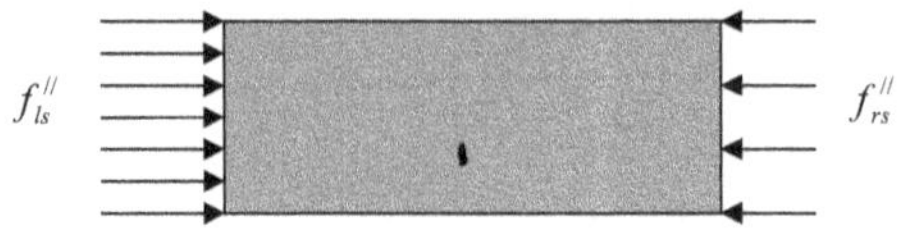

FIGURE 11.2 Constant surface boundary conditions on a plane slab.

and the corresponding **functional** is given as

$$I = \int_0^L \left\{ -\frac{1}{2}\Gamma_x \left(\frac{d\phi}{dx} \right)^2 + \frac{1}{2} S_1 \phi^2 + S_0 \phi \right\} dx + f_{rs}'' \phi \Big|_{x=L} + f_{ls}'' \phi \Big|_{x=0} \tag{11.29b}$$

It can also be noted that for the case of zero surface flux condition, i.e.,

$$\Gamma_x \frac{d\phi}{dx}\Big|_{x=0} = 0 \quad \text{and} \quad \Gamma_x \frac{d\phi}{dx}\Big|_{x=L} = 0$$

and the equivalent variational form and the functional are given as

$$\delta \left[\int_0^L \left\{ -\frac{1}{2}\Gamma_x \left(\frac{d\phi}{dx} \right)^2 + \frac{1}{2} S_1 \phi^2 + S_0 \phi \right\} dx \right] = 0 \tag{11.30}$$

and

$$I = \int_0^L \left\{ -\frac{1}{2}\Gamma_x \left(\frac{d\phi}{dx} \right)^2 + \frac{1}{2} S_1 \phi^2 + S_0 \phi \right\} dx \tag{11.31}$$

Case 3: Convective Boundary Condition

Let us now consider the convective boundary condition or the boundary condition of the third kind at the two boundaries shown in Figure 11.3.

The corresponding mathematical statements for the boundary conditions are

$$\text{at}\, x = 0, \quad \Gamma_x \frac{d\phi}{dx}\Big|_{x=0} = h_{ls} \left(\phi\big|_{x=0} - \phi_{l\infty} \right) A \tag{11.32}$$

and

$$\text{at } x = L, \quad -\Gamma_x \frac{d\phi}{dx}\Big|_{x=L} = h_{rs} \left(\phi\big|_{x=L} - \phi_{r\infty} \right) \tag{11.33}$$

Let us now evaluate the last integrated term in Equation 11.23 as

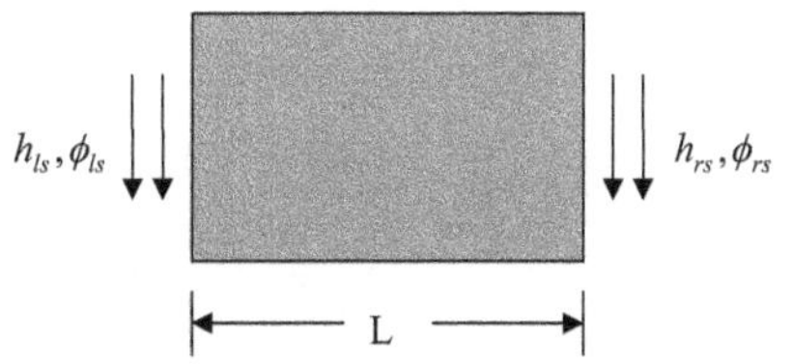

FIGURE 11.3 Convective boundary conditions on plane slab.

$$\left[\Gamma_x \frac{\mathrm{d}\phi}{\mathrm{d}x}\delta\phi\right]_0^L = \left[\Gamma_x \frac{\mathrm{d}\phi}{\mathrm{d}x}\delta\phi\right]_{x=L} - \left[\Gamma_x \frac{\mathrm{d}\phi}{\mathrm{d}x}\delta\phi\right]_{x=0}$$

$$= h_{\mathrm{rs}}\left(\phi_{r\infty} - \phi\big|_{x=L}\right)\delta\phi\big|_{x=L} - h_{\mathrm{ls}}\left(\phi\big|_{x=0} - \phi_{l\infty}\right)\delta\phi\big|_{x=0}$$

$$= \delta\left[h_{\mathrm{rs}}\phi_{r\infty}\,\phi\big|_{x=L} - \frac{1}{2}h_{\mathrm{rs}}\left(\phi\big|_{x=L}\right)^2\right] - \delta\left[\frac{1}{2}h_{\mathrm{ls}}\left(\phi\big|_{x=0}\right)^2 - h_{\mathrm{ls}}\phi_{l\infty}\,\phi\big|_{x=0}\right]$$

$$= \delta\left[h_{\mathrm{rs}}\phi_{r\infty}\,\phi\big|_{x=L} - \frac{1}{2}h_{\mathrm{rs}}\left(\phi\big|_{x=L}\right)^2 - \frac{1}{2}h_{\mathrm{ls}}\left(\phi\big|_{x=0}\right)^2 + h_{\mathrm{ls}}\phi_{l\infty}\,\phi\big|_{x=0}\right] \tag{11.34}$$

Now substituting Equation 11.34 into Equation 11.23, we get the variational form as

$$\delta\left[\int_0^L \left\{-\frac{1}{2}\Gamma_x\left(\frac{\mathrm{d}\phi}{\mathrm{d}x}\right)^2 + \frac{1}{2}S_1\phi^2 + S_0\phi\right\} + \left\{h_{\mathrm{rs}}\phi_{r\infty}\,\phi\big|_{x=L} - \frac{1}{2}h_{\mathrm{rs}}\left(\phi\big|_{x=L}\right)^2 - \frac{1}{2}h_{\mathrm{ls}}\left(\phi\big|_{x=0}\right)^2 + h_{\mathrm{ls}}\phi_{l\infty}\,\phi\big|_{x=0}\right\}\right] = 0 \tag{11.35}$$

and the functional is given as

$$I = \int_0^L \left\{-\frac{1}{2}\Gamma_x\left(\frac{\mathrm{d}\phi}{\mathrm{d}x}\right)^2 + \frac{1}{2}S_1\phi^2 + S_0\phi\right\}\mathrm{d}x + \left\{h_{\mathrm{rs}}\phi_{r\infty}\,\phi\big|_{x=L} - \frac{1}{2}h_{\mathrm{rs}}\left(\phi\big|_{x=L}\right)^2 - \frac{1}{2}h_{\mathrm{ls}}\left(\phi\big|_{x=0}\right)^2 + h_{\mathrm{ls}}\phi_{l\infty}\,\phi\big|_{x=0}\right\} \tag{11.36}$$

Example 11.1: One-Dimensional Variational Formulation

Determine the variational form and functional form for the following one-dimensional fin problem using the **calculus of variation**.

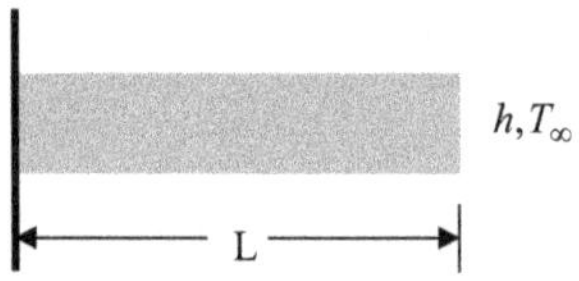

Governing Equation:

$$\frac{\mathrm{d}}{\mathrm{d}x}\left(kA\frac{\mathrm{d}T}{\mathrm{d}x}\right) - h_c P(T - T_\infty) = 0 \tag{11.37}$$

Boundary Conditions:

$$1.\ x = 0, \quad T = T_0 \tag{11.38a}$$

$$2.\ x = L, \quad -k\left.\frac{\mathrm{d}T}{\mathrm{d}x}\right|_{x=L} = h_c\left(T|_{x=L} - T_\infty\right) \tag{11.38b}$$

Solution:

To derive the variational form, we multiply both sides of the governing equation by a variation δT and integrate over the length of the fin

$$\int_0^L \frac{\mathrm{d}}{\mathrm{d}x}\left(kA\frac{\mathrm{d}T}{\mathrm{d}x}\right)\delta T\,\mathrm{d}x - \int_0^L h_c PT\,\delta T\,\mathrm{d}x + \int_0^L h_c PT_\infty\,\delta T\,\mathrm{d}x = 0 \tag{11.39}$$

Evaluating the first term by integration-by-parts

$$\left[kA\frac{\mathrm{d}T}{\mathrm{d}x}\delta T\right]_0^L - \int_0^L kA\frac{\mathrm{d}T}{\mathrm{d}x}\frac{\mathrm{d}}{\mathrm{d}x}(\delta T)\mathrm{d}x - \int_0^L h_c PT\,\delta T\,\mathrm{d}x + \int_0^L h_c PT_\infty\,\delta T\,\mathrm{d}x = 0 \tag{11.40}$$

Using rules of variational calculus

$$\begin{aligned} &\left.kA\frac{\mathrm{d}T}{\mathrm{d}x}\delta T\right|_{x=L} - \left.kA\frac{\mathrm{d}T}{\mathrm{d}x}\delta T\right|_{x=0} \\ &-\frac{1}{2}\delta\int_0^L kA\left(\frac{\mathrm{d}T}{\mathrm{d}x}\right)^2\mathrm{d}x - \frac{1}{2}\delta\int_0^L h_c PT^2\,\mathrm{d}x + \delta\int_0^L h_c PT_\infty T\,\mathrm{d}x = 0 \end{aligned} \tag{11.41}$$

We have discussed before that for the **boundary condition of the first kind**, i.e., of constant surface temperature value, δT must vanish at that boundary surface. So, in this problem, we set $\delta T|_{x=0} = 0$. At the tip of the fin we have the **boundary condition of the third kind**, i.e., heat transfer rate is specified as convection rate and we set $-kA(\mathrm{d}T/\mathrm{d}x)|_{x=L} = hA\left(T|_{x=L} - T_\infty\right)$. Substituting these conditions directly in Equation 11.41, we get

$$\begin{aligned} &h_c A\left(T|_{x=L} - T_\infty\right)\delta T|_{x=L} - \frac{1}{2}\delta\int_0^L kA\left(\frac{\mathrm{d}T}{\mathrm{d}x}\right)^2\mathrm{d}x - \frac{1}{2}\delta \\ &\int_0^L h_c PT^2\,\mathrm{d}x + \delta\int_0^L h_c PT_\infty T\,\mathrm{d}x = 0 \end{aligned} \tag{11.42}$$

Further using the rules of calculus of variation in the first term of the equation, we have the **variational form** of the problem as

$$\begin{aligned} \delta\Bigg[&-\frac{1}{2}\int_0^L kA\left(\frac{\mathrm{d}T}{\mathrm{d}x}\right)^2\mathrm{d}x - \frac{1}{2}\int_0^L h_c PT^2\,\mathrm{d}x \\ &+\int_0^L h_c PT_\infty T\,\mathrm{d}x + \frac{1}{2}hA\left(T|_{x=L}\right)^2 - h_c AT_\infty\,\delta T|_{x=L}\Bigg] = 0 \end{aligned} \tag{11.43}$$

and the corresponding **functional** is given as

$$I = -\frac{1}{2}\int_0^L kA\left(\frac{dT}{dx}\right)^2 dx - \frac{1}{2}\int_0^L h_c P T^2 dx$$

$$+\int_0^L h_c P T_\infty T\,dx + \frac{1}{2}h_c A\left(T|_{x=L}\right)^2 - h_c A T_\infty T|_{x=L} \qquad (11.44)$$

with essential condition $T(0) = T_0$.

Variational Formulation for Multidimensional Problems Let us consider the derivation of the variational form for a three-dimensional problem such as the one given by

$$\frac{d}{dx}\left(\Gamma_x \frac{d\phi}{dx}\right) + \frac{d}{dy}\left(\Gamma_y \frac{d\phi}{dy}\right) + \frac{d}{dz}\left(\Gamma_z \frac{d\phi}{dz}\right) + S_1\phi + S_0 = 0 \quad \text{over volume } \mathbf{V} \qquad (11.45)$$

subject to the boundary conditions

$$\phi = \phi_s \qquad \text{on } A_1 \qquad (11.46a)$$

$$\Gamma_x \frac{\partial \phi}{\partial x} n_x + \Gamma_y \frac{\partial \phi}{\partial y} n_y + \Gamma_z \frac{\partial \phi}{\partial z} n_z + q_s'' = 0 \qquad \text{on } A_2 \qquad (11.46b)$$

$$\Gamma_x \frac{\partial \phi}{\partial x} n_x + \Gamma_y \frac{\partial \phi}{\partial y} n_y + \Gamma_z \frac{\partial \phi}{\partial z} n_z + h_s\left(\phi - \phi_\infty\right) = 0 \qquad \text{on } A_3 \qquad (11.46c)$$

where n_x, n_y, and n_z are outward direction cosines drawn normal to the surface; A_1 is the surface area on which the constant surface value ϕ_s is specified; A_2 is the surface area on which the constant surface flux ϕ_s'' is specified; and A_3 is the surface area on which the mixed or the convective boundary condition is specified.

To obtain the variational form of this problem, multiply the equation by $\delta\phi$

$$\iiint_V \frac{d}{dx}\left(\Gamma_x \frac{d\phi}{dx}\right)\delta\phi dV + \iiint_V \frac{d}{dy}\left(\Gamma_x \frac{d\phi}{dy}\right)\delta\phi dV + \iiint_V \frac{d}{dz}\left(\Gamma_z \frac{d\phi}{dz}\right)\delta\phi dV$$

$$+\iiint_V S_1\phi\delta\phi dV + \iiint_V S_0\delta\phi dV = 0 \qquad (11.47)$$

The first integrand term can be written using differentiation rule as

$$\iiint_V \frac{d}{dx}\left(\Gamma_x \frac{d\phi}{dx}\right)\delta\phi dV = \iiint_V \frac{d}{dx}\left(\Gamma_x \frac{d\phi}{dx}\delta\phi\right)dV - \iiint_V \Gamma_x \frac{d\phi}{dx}\frac{d}{dx}(\delta\phi)dV \qquad (11.48)$$

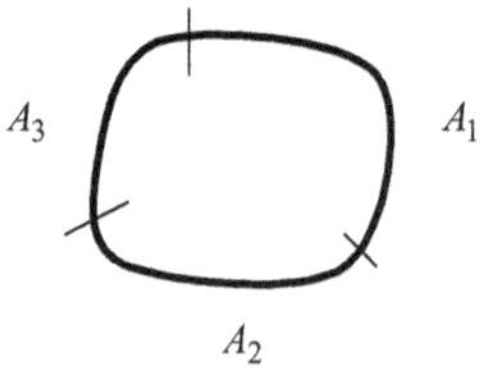

FIGURE 11.4 Control volume with boundary surfaces.

Using the extended divergence theorem (see Appendix B) for the first term and rules of calculus of variation for the second term on the right-hand side of the Equation 11.48, we get

$$\iiint_V \frac{d}{dx}\left(\Gamma_x \frac{d\phi}{dx}\right)\delta\phi dV = \oiint_A \Gamma_x \frac{d\phi}{dx}\delta\phi n_x dA - \delta \iiint_V \tfrac{1}{2}\Gamma_x\left(\frac{d\phi}{dx}\right)^2 dV \tag{11.49a}$$

Similarly, we can express the second and third integrand terms of Equation 11.47 as

$$\iiint_V \frac{d}{dy}\left(\Gamma_y \frac{d\phi}{dy}\right)\delta\phi dV = \oiint_A \Gamma_y \frac{d\phi}{dy}\delta\phi n_y dA - \delta \iiint_V \tfrac{1}{2}\Gamma_y\left(\frac{d\phi}{dy}\right)^2 dV \tag{11.49b}$$

and

$$\iiint_V \frac{d}{dz}\left(\Gamma_z \frac{d\phi}{dz}\right)\delta\phi dV = \iint_A \Gamma_z \frac{d\phi}{dz}\delta\phi n_z dA - \delta \iiint_V \tfrac{1}{2}\Gamma_z\left(\frac{d\phi}{dz}\right)^2 dV \tag{11.49c}$$

Substituting Equation 11.49 into Equation 11.48

$$-\delta \iiint_V \left[\frac{1}{2}\left(\Gamma_x\left(\frac{d\phi}{dx}\right)^2 + \Gamma_x\left(\frac{d\phi}{dx}\right)^2 + \Gamma_x\left(\frac{d\phi}{dx}\right)^2\right) - \frac{1}{2}S_1\phi^2 - S_0\phi\right] dV$$
$$+ \oiint_A \left(\Gamma_x \frac{d\phi}{dx} n_x + \Gamma_y \frac{d\phi}{dy} n_y + \Gamma_x \frac{d\phi}{dz} n_z\right)\delta\phi dA \tag{11.50}$$

The boundary integral term can now be evaluated by considering the specified boundary conditions given by Equation 11.46 on different segments as

$$\oiint_A \left(\Gamma_x \frac{d\phi}{dx} n_x + \Gamma_y \frac{d\phi}{dy} n_y + \Gamma_x \frac{d\phi}{dz} n_z\right)\delta\phi dA$$
$$= \oiint_{A_1} \left(\Gamma_x \frac{d\phi}{dx} n_x + \Gamma_y \frac{d\phi}{dy} n_y + \Gamma_x \frac{d\phi}{dz} n_z\right)\delta\phi dA$$
$$+ \oiint_{A_2} \left(\Gamma_x \frac{d\phi}{dx} n_x + \Gamma_y \frac{d\phi}{dy} n_y + \Gamma_x \frac{d\phi}{dz} n_z\right)\delta\phi dA$$
$$+ \oiint_{A_3} \left(\Gamma_x \frac{d\phi}{dx} n_x + \Gamma_y \frac{d\phi}{dy} n_y + \Gamma_x \frac{d\phi}{dz} n_z\right)\delta\phi dA \tag{11.51}$$

The first boundary integral on the right-hand side vanishes as $\delta\phi=0$ since ϕ is specified on this segment of the boundary. The remaining two boundary integrals are evaluated by directly substituting the boundary conditions given by Equations 11.46b and 11.46c. This simplifies the boundary integral as

$$\oiint_A \left(\Gamma_x \frac{d\phi}{dx} n_x + \Gamma_y \frac{d\phi}{dx} n_y + \Gamma_x \frac{d\phi}{dx} n_z \right) \delta\phi dA = -\oiint_{A_2} q_s'' \delta\phi dA - \oiint_{A_3} h_s (\phi - \phi_\infty) \delta\phi dA \qquad (11.52)$$

Substituting Equation 11.52 into Equation 11.50, we obtain the *variational form* of the problem as

$$-\delta \left[\iiint_V \left\{ \frac{1}{2} \left(\Gamma_x \left(\frac{d\phi}{dx} \right)^2 + \Gamma_x \left(\frac{d\phi}{dy} \right)^2 + \Gamma_x \left(\frac{d\phi}{dz} \right)^2 \right) - \frac{1}{2} S_1 \phi^2 - S_0 \phi \right\} dV \right.$$

$$\left. + \oiint_{A_2} q_s'' \phi dA + \oiint_{A_3} h_s \left(\frac{1}{2} \phi^2 - \phi \phi_s \right) dA \right] \qquad (11.53)$$

and the corresponding *functional* as

$$I = \iiint_V \left\{ \frac{1}{2} \left(\Gamma_x \left(\frac{d\phi}{dx} \right)^2 + \Gamma_x \left(\frac{d\phi}{dx} \right)^2 + \Gamma_x \left(\frac{d\phi}{dx} \right)^2 \right) - \frac{1}{2} S_1 \phi^2 - S_0 \phi \right\} dV$$

$$+ \oiint_{A_2} q_s'' \phi dA + \oiint_{A_3} h \left(\frac{1}{2} \phi^2 - \phi \phi_\infty \right) dA \qquad (11.54)$$

with essential condition $\phi = \phi_s$ on A_1.

Example 11.2: Two-dimensional Variational Formulation

Consider a two-dimensional steady state conduction in a rectangular plate with constant surface heat flux q_0'' in the mid-section of the top surface due to an incident high-energy laser beam. The figure shows the schematic of the problem with boundary conditions specified at the surfaces.

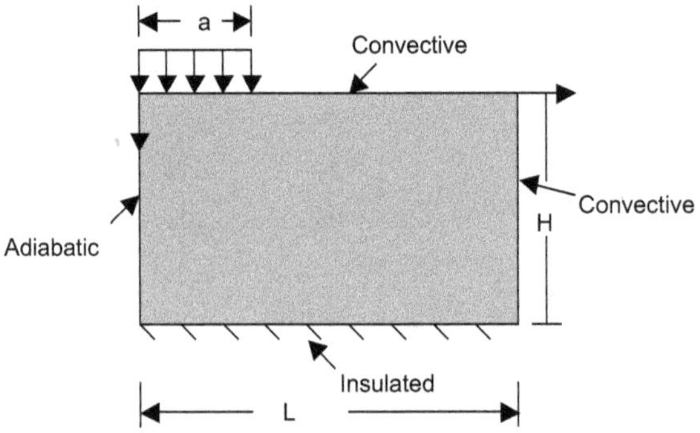

The mathematical statement of the problem is given as follows.

Governing Equation:

$$\frac{\partial}{\partial x}\left(k_x \frac{\partial T}{\partial x} \right) + \frac{\partial}{\partial y}\left(k_y \frac{\partial T}{\partial y} \right) = 0 \qquad (11.55)$$

Boundary Conditions:

Top Surface

$$y = 0, \quad 0 < x \le a, \quad -k_y \frac{\partial T}{\partial y} + q_0'' = h_c(T - T_\infty) \tag{11.56a}$$

$$y = 0, \quad a < x \le L, \quad -k_y \frac{\partial T}{\partial y} = h_c(T - T_\infty) \tag{11.56b}$$

Bottom Surface

$$y = H, \quad \frac{\partial T}{\partial y} = 0 \tag{11.57}$$

Left Surface

$$x = 0, \quad \frac{\partial T}{\partial x} = 0 \tag{11.58}$$

Right Surface

$$x = L, \quad -k_x \frac{\partial T}{\partial x} = h_c(T - T_\infty) \tag{11.59}$$

Solution

To obtain the variational form of this problem, multiply the equation by δT and integrate over the two-dimensional solution domain A

$$\iint_A \frac{\partial}{\partial x}\left(k_x \frac{\partial T}{\partial x}\right)\delta T\, dA + \iint_A \frac{\partial}{\partial y}\left(k_y \frac{\partial T}{\partial y}\right)\delta T\, dA \tag{11.60}$$

The first integrand term can be written using the differentiation rule as

$$\iint_A \frac{\partial}{\partial x}\left(k_x \frac{dT}{dx}\right)\delta T\, dA = \iint_A \frac{\partial}{\partial x}\left(k_x \frac{\partial T}{\partial x}\delta T\right) dA - \iint_A k_x \frac{\partial T}{\partial x}\frac{\partial}{\partial x}(\delta T)\, dA \tag{11.61}$$

Using the divergence theorem (see Appendix B) for the first term and rules of calculus of variation for the second term on the right-hand side of Equation 11.61, we get

$$\iint_A \frac{\partial}{\partial x}\left(k_x \frac{dT}{dx}\right)\delta T\, dA = \oint_S k_x \frac{\partial T}{\partial x}\delta T n_x\, ds - \delta \iint_A \tfrac{1}{2} k_x \left(\frac{\partial T}{\partial x}\right)^2 dA \tag{11.62a}$$

Similarly, we can express the second term of Equation 11.60 as

$$\iint_A \frac{\partial}{\partial y}\left(k_y \frac{dT}{dy}\right)\delta T\, dA = \oint_S k_y \frac{\partial T}{\partial y}\delta T n_y\, ds - \delta \iint_A \frac{1}{2} k_y \left(\frac{\partial T}{\partial y}\right)^2 dA \tag{11.62b}$$

Substituting Equation 11.62 into Equation 11.60

$$\delta \iint_A \frac{1}{2}\left(k_x \left(\frac{\partial T}{\partial x}\right)^2 + k_y \left(\frac{\partial T}{\partial y}\right)^2\right) dA - \oint_S \left(k_x \frac{\partial T}{\partial x} n_x + k_y \frac{\partial T}{\partial y} n_y\right)\delta T\, ds = 0 \tag{11.63}$$

The boundary integral term in Equation 11.63 can now be evaluated by considering the specified boundary conditions given by Equations 11.56–11.59 on different surfaces as

$$\oint_S \left(k_x \frac{\partial T}{\partial x} n_x + k_y \frac{\partial T}{\partial y} n_y\right)\delta T\, ds = \oint_{\substack{\text{top}\\\text{surface}}} \left(k_x \frac{\partial T}{\partial x} n_x + k_y \frac{\partial T}{\partial y} n_y\right)\delta T\, ds + \oint_{\substack{\text{right}\\\text{surface}}} \left(k_x \frac{\partial T}{\partial x} n_x + k_y \frac{\partial T}{\partial y} n_y\right)\delta T\, ds + \oint_{\substack{\text{bottom}\\\text{surface}}} \left(k_x \frac{\partial T}{\partial x} n_x + k_y \frac{\partial T}{\partial y} n_y\right)\delta T\, ds + \oint_{\substack{\text{left}\\\text{surface}}} \left(k_x \frac{\partial T}{\partial x} n_x + k_y \frac{\partial T}{\partial y} n_y\right)\delta T\, ds \tag{11.64}$$

For the first boundary integral on the top surface $n_x = 0$ and $n_y = -1$, and as we substitute the boundary condition given by Equation 11.56, we get

$$\oint_{\substack{\text{top}\\\text{surface}}} \left(k_x \frac{\partial T}{\partial x} n_x + k_y \frac{\partial T}{\partial y} n_y\right)\delta T\, ds = \int_0^a \left\{h(T(x,0) - T_\infty) - q_0''\right\}\delta T\, dx + \int_a^L h(T(x,0) - T_\infty)\delta T\, dx \tag{11.65a}$$

$$= -\int_0^a q_0''\ \delta T\, dx + \int_0^L \left\{h(T(x,0) - T_\infty)\right\}\delta T\, dx$$

$$= -\delta \int_0^a q_0'' T(x,0)\, dx + \frac{1}{2}\delta \int_0^L h\left[T(x,0)\right]^2 dx - \delta \int_0^L hT_\infty T(x,0)\, dx$$

For the second boundary integral on the right surface $n_x = 1$ and $n_y = 0$, and as we substitute the boundary condition given by Equation 11.59, we get,

$$\oint_{\substack{\text{right}\\\text{surface}}} \left(k_x \frac{\partial T}{\partial x} n_x + k_y \frac{\partial T}{\partial y} n_y\right)\delta T\, ds = -\int_0^H \left\{h\left(T(L,y) - T_\infty\right)\right\}\delta T\, dy \tag{11.65b}$$

$$= -\frac{1}{2}\delta \int_0^H h\left[T(L,y)\right]^2 dy + \delta \int_0^H hT_\infty T(L,y)\, dy$$

For the third boundary integral on the bottom surface $n_x = 0$ and $n_y = 1$, and as we substitute the boundary condition given by Equation 11.61, we get,

$$\oint_{\substack{\text{bottom}\\\text{surface}}} \left(k_x \frac{\partial T}{\partial x} n_x + k_y \frac{\partial T}{\partial y} n_y\right)\delta T\, ds = 0 \tag{11.65c}$$

For the fourth boundary integral on the left surface $n_x = -1$ and $n_y = 0$, and as we substitute the boundary condition given by Equation 11.57, we get

$$\oint_{\substack{\text{left}\\\text{surface}}} \left(k_x \frac{\partial T}{\partial x} n_x + k_y \frac{\partial T}{\partial y} n_y\right) \delta T\, ds = 0 \tag{11.65d}$$

Substituting Equation 11.65 into Equation 11.64, the boundary integral transforms into

$$\oint_S \left(k_x \frac{\partial T}{\partial x} n_x + k_y \frac{\partial T}{\partial y} n_y\right) \delta T\, ds$$
$$= -\delta \int_0^a q_0'' T(x,0)\, dx + \frac{1}{2}\delta \int_0^L h\left[T(x,0)\right]^2 dx - \delta \int_0^L hT_\infty T(x,0)\, dx$$
$$-\frac{1}{2}\delta \int_0^H h\left[T(L,y)\right]^2 dy + \delta \int_0^H hT_\infty T(L,y)\, dy \tag{11.66}$$

Substituting Equation 11.66 into Equation 11.63, we obtain the variational form of the problem as

$$\delta\left[\iint_A \frac{1}{2}\left(k_x\left(\frac{\partial T}{\partial x}\right)^2 + k_y\left(\frac{\partial T}{\partial y}\right)^2\right) dxdy\right.$$
$$+ \int_0^a q_0'' T(x,0) dx - \frac{1}{2}\int_0^L h\left[T(x,0)\right]^2 dx + \int_0^L hT_\infty T(x,0) dx$$
$$\left. + \frac{1}{2}\int_0^H h\left[T(L,y)\right]^2 dy - \int_0^H hT_\infty T(L,y) dy\right] = 0 \tag{11.67}$$

and the corresponding *functional* as

$$I = \iint_A \frac{1}{2}\left(k_x\left(\frac{\partial T}{\partial x}\right)^2 + k_y\left(\frac{\partial T}{\partial y}\right)^2\right) dxdy$$
$$+ \int_0^a q_0'' T(x,0) dx - \frac{1}{2}\int_0^L h\left[T(x,0)\right]^2 dx + \int_0^L hT_\infty T(x,0) dx$$
$$+ \frac{1}{2}\int_0^H h\left[T(L,y)\right]^2 dy - \int_0^H hT_\infty T(L,y) dy = 0 \tag{11.68}$$

11.3.2 Method of Weighted Residuals

The method of weighted residuals is based on the minimization of a residual term that results as an approximate or trial solution is substituted into the governing differential equation. Let us consider the differential (in the form of an operator) equation

$$L\phi = f(\tilde{x}) \tag{11.69}$$

where L is the differential operator, ϕ is the unknown dependent variable, $\tilde{x}$ is the dependent variable, and $f(\tilde{x})$ is the source function.

As we seek an approximate solution to the above equation, let us consider an approximate or trial solution ϕ^* for ϕ as

$$\phi^* = \sum_{i=1}^{N} a_i \psi_i \tag{11.70}$$

Since the approximate solution does not satisfy the governing differential Equation 11.69 exactly, a residual term results when we substitute the approximate solution into the governing equation as

$$R(\phi,\tilde{x}) = L\phi - f(\tilde{x}) = L\left(\sum_{i=1}^{N} a_i \psi_i\right) - f(\tilde{x}) \tag{11.71}$$

We note here that the residual term becomes zero when the approximate or trial solution is exactly equal to the true solution, i.e., $\phi = \phi^*$.

In the method of weighted residuals, the objective is to select the unknown parameters a_i such that the residual $R(\phi,\tilde{x})$ is minimized along with a weighting function, W_i, as

$$\int_V W_i R(\phi,\tilde{x})\,\mathrm{d}V = 0, \quad i = 1,2,\ldots,n \tag{11.72}$$

where V represents the solution domain and W_i is the weighting function. The number of weighting functions equals the number of unknown coefficients in the approximate solution. There are several choices for the weighting function, and some of the most popular choices lead to different names for the **weighted residual method** such as the **Galerkin method**, the **subdomain method**, the **collocation method**, the **least-square method**, and the **Petrov–Galerkin method**. A brief description of these methods is given in Section 11.4 while describing variational methods for solving the integral form of the problem.

It can be noted here that the **weighted-residual integral form** given by Equation 11.72 does not involve any boundary conditions of the problem, and it can be derived for any form of differential equation.

Weak Integral Form A **weak integral form** can be derived from the weighted residual integral form by evaluating the integral 11.72 using integration by parts and imposing the boundary conditions of the problem. To satisfy a specified boundary condition of the first kind or an essential boundary condition, it is assumed that the weighting function W_i represents a variation of the variable ϕ and must vanish at the boundary. The resulting integral form is called the **weak form** because the order of the derivative has been reduced, and hence requires less continuity of the dependent variable. One important aspect of the weak form is that it can be obtained for both linear and nonlinear problems. But it can only be obtained for second- or higher-order equations that contain derivatives of the order of two or higher. Further, it has been shown by Reddy (1984) that for linear problems involving even-order derivatives, the weak form will have a symmetric bilinear form. Such weak forms can be transformed into a **variational form** or a **functional** by simply substituting the weighting function as a variation of the variable $\delta\phi$. It must be remembered that for problems dealing with the first-order equation the weighted-residual integral equation 11.72 is the only choice for the variational method of approximation as well as for the FEM.

Example 11.3: Weighted-Residual Integral Formulation

The exposed surface ($x=0$) of a plane slab is subjected to radiation that causes volumetric heat generation to vary as $\dot{q} = q_0'' a\, \mathrm{e}^{-ax}$, where a is the absorption coefficient, and q_0'' is the incident beam power intensity at the surface. The exposed surface at $x=0$ is also exposed to a fluid at temperature T_∞ and convection coefficient film h. The inner surface ($x=L$) is maintained at a constant temperature T_L.

The mathematical statement of the problem is given as

Governing Equation:

$$\frac{\mathrm{d}}{\mathrm{d}x}\left(kA\frac{\mathrm{d}T}{\mathrm{d}x}\right)+q_0''a\,\mathrm{e}^{-ax}=0 \tag{11.73}$$

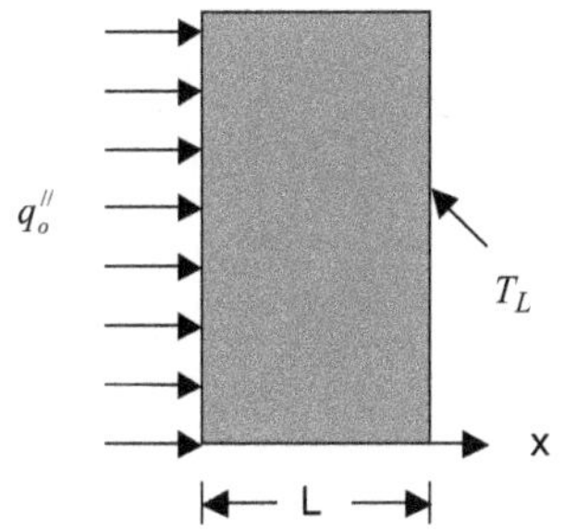

Boundary Conditions:

$$1.\quad x=0,\quad k\left.\frac{\mathrm{d}T}{\mathrm{d}x}\right|_{x=0}=h\left(T|_{x=0}-T_\infty\right) \tag{11.74a}$$

$$2.\quad x=L,\quad T|_{x=L}=T_L \tag{11.74b}$$

Obtain the weighted-residual integral form, the weak form, and, if possible, the variational form or the functional of the problem.

Solution

To derive the **weighted-residual integral form**, we multiply the residual term of the differential equation by a weighting function W_i and integrate over the solution domain (0, L)

$$\int_0^L W_i\left[\frac{\mathrm{d}}{\mathrm{d}x}\left(kA\frac{\mathrm{d}T}{\mathrm{d}x}\right)+q_0''a\,e^{-ax}\right]\mathrm{d}x=0 \tag{11.75}$$

To develop the **weak form**, Equation 11.75 is evaluated by using the differentiation rule as

$$\int_0^L \frac{\mathrm{d}}{\mathrm{d}x}\left[W_i kA\frac{\mathrm{d}T}{\mathrm{d}x}\right]\mathrm{d}x-\int_0^L kA\frac{\mathrm{d}T}{\mathrm{d}x}\frac{\mathrm{d}W_i}{\mathrm{d}x}\mathrm{d}x+\int_0^L W_i q_0''a\,\mathrm{e}^{-ax}\,\mathrm{d}x=0$$

$$\left[W_i kA\frac{\mathrm{d}T}{\mathrm{d}x}\right]_0^L-\int_0^L kA\frac{\mathrm{d}T}{\mathrm{d}x}\frac{\mathrm{d}W_i}{\mathrm{d}x}\mathrm{d}x+\int_0^L W_i q_0''a\,\mathrm{e}^{-ax}\,\mathrm{d}x=0 \tag{11.76}$$

or

$$W_i kA\left.\frac{\mathrm{d}T}{\mathrm{d}x}\right|_{x=L}-W_i kA\left.\frac{\mathrm{d}T}{\mathrm{d}x}\right|_{x=0}-\int_0^L kA\frac{\mathrm{d}T}{\mathrm{d}x}\frac{\mathrm{d}W_i}{\mathrm{d}x}\mathrm{d}x+\int_0^L W_i q_0''a\,\mathrm{e}^{-ax}\mathrm{d}x=0 \tag{11.77}$$

We have discussed before that for the **boundary condition of the first kind**, i.e., of constant surface temperature value, W_i must vanish at that boundary surface. So, in this problem, we set $W_i|_{x=L}=0$ at the right surface of the plane slab. At the left surface, we have the **boundary condition of the third kind**, i.e., heat transfer rate is specified by the convective cooling, and we set $kA(\mathrm{d}T/\mathrm{d}x)|_{x=0}=hA\left(T|_{x=0}-T_\infty\right)$. Substituting these conditions directly into Equation 11.77, we have the **weak form** as

$$-W_i\big|_{x=0} hA\left(T\big|_{x=0} - T_\infty\right) - \int_0^L kA\frac{dT}{dx}\frac{dW_i}{dx}dx + \int_0^L W_i q_0'' a\, e^{-ax}\, dx = 0 \tag{11.78}$$

Further, setting $W_i = \delta T$ in the equation, we have the **variational form** of the problem as

$$\delta\left[-\frac{1}{2}\int_0^L kA\left(\frac{dT}{dx}\right)^2 dx + \int_0^L Tq_0''\; a\, e^{-ax}\, dx - \frac{1}{2}hA\left(T\big|_{x=0}\right)^2 + hAT_\infty\, \delta T\big|_{x=0}\right] = 0 \tag{11.79}$$

and the corresponding **functional** is given as

$$I = -\frac{1}{2}\int_0^L kA\left(\frac{dT}{dx}\right)^2 dx + \int_0^L Tq_0''\; a\, e^{-ax}\, dx - \frac{1}{2}hA\left(T\big|_{x=0}\right)^2 + hAT_\infty\, \delta T\big|_{x=0} \tag{11.80}$$

11.4 VARIATIONAL METHODS

We have mentioned before that the FEM is a piece-wise application of classical variational methods. In the variational methods, the governing differential problem is transformed into an equivalent integral form, and the approximate solution is assumed to be a linear combination of appropriately selected functions ψ_i and undetermined coefficients a_i given as

$$\phi = a_0\psi_0 + a_1\psi_1 + \cdots + a_n\psi_n = \sum_{i=1}^{n} a_i\psi_i \tag{11.81}$$

The coefficient a's are determined by satisfying the equivalent integral statement of the problem. Also, in the variational method, it is assumed that the approximate solution is valid over the entire solution domain.

Variational methods are classified into two primary categories: (a) **Rayleigh–Ritz Method** and (b) the **weighted residual methods**, which include the **Galerkin method, the Collocation method**, the **least square method**, etc. These methods differ from each other based on the selection of the equivalent integral form such as the variational form or the weighted residual integral form, the weighting function in weighted residual integral form, and the approximation functions. Let us present a brief description of some these classical variational methods.

11.4.1 THE RAYLEIGH–RITZ VARIATIONAL METHOD

This method is associated with the use of a functional (I) or the variational form of integral statement obtained by using the calculus of variation. The functions $\psi_0(x), \psi_1(x), \ldots, \psi_n(x)$ in the approximation solution, Equation 11.81, are selected in such a way that the specified boundary conditions are satisfied for any values of a. For example, in problems with constant surface value boundary conditions, the function $\psi_0(x)$ is selected such that it becomes equal to the constant surface values at the boundaries, and the rest of the functions, $\psi_i(x)$, become zero at the boundaries. As we substitute the approximation solution in the functional I, it becomes a function of the unknowns a_i. The necessary conditions for the functional to be stationary are given by

$$\frac{\partial I}{\partial a_i} = 0 \quad \text{for } i = 0,1,2,\ldots,n \tag{11.82}$$

and this leads to a system of n equations, which is solved for the n unknown a's.

It is quite evident that the accuracy of the solution depends on the selection of the approximation function, $\psi_i(x)$. The most used functions are the **polynomials**, which can be used with successively increasing degree and which satisfy the boundary values. However, other special functions such as sine, cosine, Bessel, etc., may also be used and in some cases may exhibit computational advantages. Also, the accuracy of the solution depends on the number of approximation functions retained in the approximation solution. A general procedure involves a sequence of approximations, starting with the first approximation as

$$\phi = a_0^{(1)}\psi_0 + a_1^{(1)}\psi_1 \tag{11.83a}$$

where the superscript indicates that the a's are obtained in the first stage of approximation. Additional functions are then added successively in the subsequent stages as

$$\phi = a_0^{(2)}\psi_0 + a_1^{(2)}\psi_1 + a_2^{(2)}\psi_2 \tag{11.83b}$$

$$\vdots$$

$$\phi = a_0^{(k)}\psi_0 + a_1^{(k)}\psi_1 + a_2^{(k)}\psi_2 + \cdots + a_n^{(k)}\psi_n \tag{11.83c}$$

This successive approximation process is continued until convergence, i.e., the deviations in unknown coefficients a's between two successive approximation stages are less than a specified tolerance value as

$$\left| \frac{a_i^k - a_i^{k-1}}{a_i^k} \right| \le \varepsilon_s \tag{11.84}$$

It can be realized that with these successive stages of approximation, the functional I is monotonically converged to the stationary value.

Example 11.4: Rayleigh–Ritz Variational Method

The exposed surface ($x=L$) of a plane slab is subjected to a constant surface heat flux q_L''. The inner surface ($x=0$) is maintained at a constant temperature T_0.

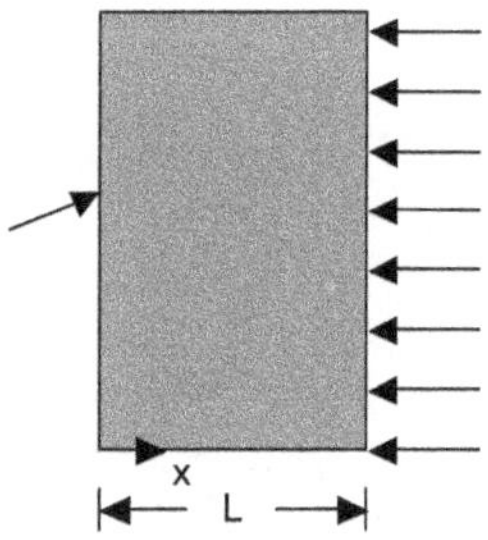

The mathematical statement of the problem is given as follows:

Governing Equation:

$$\frac{\mathrm{d}}{\mathrm{d}x}\left(k\frac{\mathrm{d}T}{\mathrm{d}x}\right) = 0 \tag{11.85}$$

Boundary Conditions:

3. $x = 0, \quad T|_{x=0} = T_0$ (11.86a)

$$4. \quad x = L, \quad k\left.\frac{dT}{dx}\right|_{x=L} = q''_L \tag{11.86b}$$

Obtain the functional form of the differential problem. Use the Rayleigh–Ritz method to obtain a solution to the equivalent variational problem.

Solution:

Step 1: Obtain the Variation Form and the Functional In order to derive the variational form, we multiply both sides of the governing equation by a variation δT and integrate over the length of the slab

$$\int_0^L \frac{d}{dx}\left(k\frac{dT}{dx}\right)\delta T\,dx = 0 \tag{11.87}$$

Evaluating the first term by integration-by-parts

$$\left[k\frac{dT}{dx}\delta T\right]_0^L - \int_0^L k\frac{dT}{dx}\frac{d}{dx}(\delta T)dx = 0 \tag{11.88}$$

Using rules of variational calculus

$$k\frac{dT}{dx}\delta T\bigg|_{x=L} - k\frac{dT}{dx}\delta T\bigg|_{x=0} - \frac{1}{2}\delta\int_0^L k\left(\frac{dT}{dx}\right)^2 dx = 0 \tag{11.89}$$

Since at $x=0$ we have the **boundary condition of the first kind**, i.e., of constant surface temperature value, δT must vanish at that boundary surface. So, in this problem, we set $\delta T|_{x=0} = 0$. At $x=L$, we have the **boundary condition of the second kind**, i.e., constant surface heat flux and we set $k(dT/dx)|_{x=0} = q''_0$. Substituting these conditions directly in Equation 11.89, we get

$$q''_L\,\delta T|_{x=L} - \frac{1}{2}\delta\int_0^L k\left(\frac{dT}{dx}\right)^2 dx = 0 \tag{11.90}$$

Further, using the rules of the calculus of variation in the first term of the equation, we have the **variational form** of the problem as

$$\delta\left[-\frac{1}{2}\int_0^L k\left(\frac{dT}{dx}\right)^2 dx + q''_L\,T|_{x=L}\right] = 0 \tag{11.91}$$

and the corresponding **functional** is given as

$$I = -\frac{1}{2}\int_0^L k\left(\frac{dT}{dx}\right)^2 dx + q''_L\,T|_{x=L} \tag{11.92}$$

with the essential boundary condition $T|_{x=0} = T_0$.

Step 2: Obtain Approximate Solution using the Rayleigh–Ritz Method Let us assume the approximate solution of the form

$$T = \sum_{i=0}^{2} a_i\psi_i = a_0 + a_1 x + a_2 x^2 \tag{11.93}$$

where $\psi_0(x)=1$, $\psi_1(x)=x$, and $\psi_2(x)=x^2$.

Substituting the approximate solution Equation 11.93 into Equation 11.92, we obtain

$$I = -\frac{1}{2}\int_0^L k\left[a_1^2 + 4a_1a_2x + 4a_2^2x^2\right]dx + q_L''\ (a_0 + a_1L + a_2L) \tag{11.94a}$$

or

$$I = -\frac{1}{2}k\left(a_1^2L + 2a_1a_2L^2 + \frac{4}{3}a_2^2L^3\right) + q_L''\left(a_0 + a_1L + a_2L^2\right) \tag{11.94b}$$

Using the condition of stationary functional I, i.e., $\partial I / \partial a_i = 0,$ we get

$$\text{for } i = 0 \quad \frac{\partial I}{\partial a_0} = 0, \quad q_L'' = 0 \tag{11.95a}$$

$$\text{for } i = 1 \quad \frac{\partial I}{\partial a_1} = 0, \quad a_1 + a_2L = \frac{q_L''}{k} \tag{11.95b}$$

$$\text{for } i = 2 \quad \frac{\partial I}{\partial a_2} = 0, \quad a_1 + \frac{4}{3}a_2L = \frac{q_L''}{k} \tag{11.95c}$$

Equations 11.95a–11.95c form a system of equations for a_i, and we can write this in matrix form as

$$\begin{bmatrix} 0 & 0 & 0 \\ 0 & 1 & L \\ 0 & 1 & \frac{4}{3}L \end{bmatrix} \begin{Bmatrix} a_0 \\ a_1 \\ a_2 \end{Bmatrix} = \begin{Bmatrix} q_L'' \\ q_L''/k \\ q_L''/k \end{Bmatrix} \tag{11.96}$$

The solution of this system gives

$$a_1 = \frac{q_L''}{k} \quad \text{and} \quad a_2 = 0 \tag{11.97}$$

We now impose the essential boundary condition, i.e., $T|_{x=0} = T_0$, on the approximate solution

$$a_0 = T_0 \tag{11.98}$$

The approximate solution is then

$$T = T_0 + \frac{q_L''}{k}x \tag{11.99}$$

11.4.2 Weighted Residual Variational Methods

The weighted residual method is used along with the weighted residual integral form of a problem given by Equation 11.72. The approximate solution is of the same form as in the case of the Rayleigh–Ritz, i.e., Equation 11.81. The weighting function $W_i(\tilde{x})$ could be selected from a set of independent functions that may be different from the approximate functions Ψ_i. Another important criterion of this method is that the selected approximate functions Ψ_i need to satisfy all specified

essential as well as **natural** boundary conditions. This is because the weighted residual form of the integral Equation 11.72 does not include any boundary conditions. To satisfy this requirement, the selected approximate functions Ψ_i should have order higher than that used in the Rayleigh–Ritz method. In other words, the functions must have nonzero derivatives up to the order of the governing differential equation.

The **weighted-residual variational methods** can be classified into different kinds depending on the selection of the weighting functions. Some of the popular weighted residual variational methods are the **Galerkin method**, the c**ollocation method**, the **subdomain method**, the **least-square method**, and the **Petrov–Galerkin method**. A brief description of these methods is given in the following section.

Galerkin's Method In this method, the weighting function, $W_i(\tilde{x})$, is assumed to be the same as the approximation function, $\psi_i(\tilde{x})$, used in the approximate solution given by Equation 11.81, and the integral statement is given as

$$\int_V \psi_i(\tilde{x}) R(\phi,\tilde{x}) \mathrm{d}v, \quad i = 1,2,\ldots,n \tag{11.100}$$

Since the number of weighting functions is the same as the number of unknown coefficients, a_i, in the approximate solution, evaluation of the integral Equation 11.86 leads to a system of equations involving unknown coefficients $a_0, a_1, \ldots, a_n$.

Collocation Method In this method, the weighting function is assumed to be a **direct delta function** of the form

$$W_i(\tilde{x}) = \delta(\tilde{x} - \tilde{x}_i) \tag{11.101}$$

where $\tilde{x}_i$ represents a set of selected points, the number of which is equal to the number of unknown coefficients. So, the weighting function is $W_i = 1$ at the point $x = x_i$ and zero everywhere else in the solution domain. The equivalent integral statement is given by

$$\int_V \delta(\tilde{x} - \tilde{x}_i) R(\phi,\tilde{x}_i) \mathrm{d}v \tag{11.102}$$

Such a selection of weighting functions causes the residual integral to vanish everywhere else in the solution domain, except at the selected points, and this leads to a system of equations given as

$$R(\phi,\tilde{x}_i) = 0, \quad i = 1,2,\ldots,n \tag{11.103}$$

Subdomain Method In this method, the weighting function is assumed to be unity over a specific subregion, ΔV_i, and subsequently, the integral of the residual vanishes over that subregion or the integration interval. The number of subregions or integration intervals equals the number of unknown coefficients. The equivalent integral statement becomes

$$\int_{\Delta V_i} R(\phi,\tilde{x}) \mathrm{d}V, \quad i = 1,2,\ldots,n \tag{11.104}$$

Least Square Method In this method, the weighting function is assumed to be the same as the **residual term**, i.e., $W_i(\tilde{x}) = R(\phi,\tilde{x})$, and an error term is evaluated as

$$E = \int_V \left[R(\phi,\tilde{x}) \right]^2 \tag{11.105}$$

The unknown coefficients a_i are determined by minimizing the error with respect to the unknown coefficients as

$$\frac{\partial E}{\partial a_i} = 0, \quad i = 1, 2, \ldots, n \tag{11.106}$$

Equation 11.92 leads to a system of equations, and solutions which give the unknown coefficients a_i.

The Petrov–Galerkin Method In this method, the weighting functions are selected as independent functions, which are different from the approximate functions used in the approximate solution given by Equation 11.81, i.e., $W_i(\tilde{x}) = f_i(\tilde{x}) \neq \psi_i(\tilde{x})$. The equivalent integral form is written as

$$\int_V f_i(\tilde{x}) R(\phi, \tilde{x}) \mathrm{d}V = 0, \quad i = 1, 2, \ldots, n \tag{11.107}$$

This equation represents a system of equations, the solution of which give the unknown coefficients a_i.

Example 11.5: Weighted-Residual Variational Method

Consider the uniform heat generation $\dot{q}$ in a plane slab with exposed surfaces ($x=0$) and ($x=L$) maintained at a constant temperature $T_0=0$.

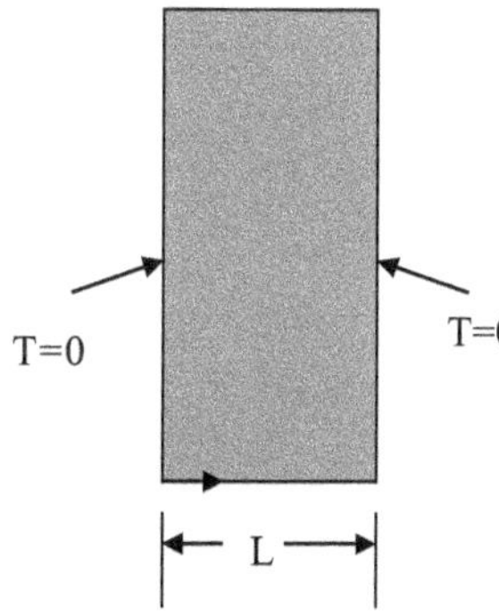

The mathematical statement of the problem is given as

Governing Equation:

$$\frac{\mathrm{d}^2T}{\mathrm{d}x^2} + \frac{\dot{q}}{k} = 0 \tag{11.108}$$

Boundary Conditions:

5. $x = 0, \quad T|_{x=0} = 0$ (11.109a)

6. $x = L, \quad T|_{x=L}, = 0$ (11.109b)

Determine the solution using Galerkin's weighted residual method.

Solution

Let us assume an approximate solution for temperature of the form

$$T^* = a_1 \sin\frac{\pi x}{L} \tag{11.110}$$

with $\psi_1(x) = \sin(\pi x/L)$. Note that the function is selected in such a way that it satisfies the two boundary conditions. In Galerkin's method, we selected the weighting function $W_1(x) = \sin(\pi x/L)$. With the substitution of the weighting function, the residual Equation 11.100 is evaluated as

$$\int_0^L W_1\left(\frac{d^2T^*}{dx^2} + \frac{\dot{q}}{k}\right)dx = 0 \tag{11.111}$$

or

$$\int_0^L \sin\frac{\pi x}{L}\left(-a_1\frac{\pi^2}{L^2}\sin\frac{\pi x}{L} + \frac{\dot{q}}{k}\right)dx = 0$$

or

$$-ka_1\frac{\pi^2}{L^2}\int_0^L \sin^2\frac{\pi x}{L}dx + \frac{\dot{q}}{k}\int_0^L \sin\frac{\pi x}{L}dx = 0$$

$$ka_1\frac{\pi^2}{L^2}\frac{1}{2}\int_0^L\left(1-\cos\frac{2\pi x}{L}\right)dx + \frac{\dot{q}}{k}\int_0^L \sin\frac{\pi x}{L}dx = 0$$

$$a_1\frac{\pi^2}{2L} - \frac{2\dot{q}L}{\pi k} = 0$$

Solving

$$a_1 = \frac{4L^2}{\pi^3 k}\dot{q} \tag{11.112}$$

the approximate temperature solution is

$$T = \frac{4\dot{q}L^2}{\pi^3 k}\sin\frac{\pi x}{L}$$

PROBLEMS

11.1 Determine the variational form and functional for the following one-dimensional conduction heat transfer problem using calculus of variation.

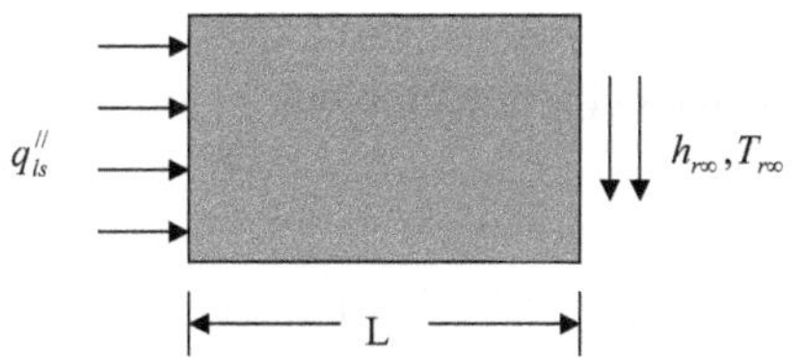

Governing Equation:

$$\frac{\partial}{\partial x}\left(k_x \frac{\partial T}{\partial x}\right) + \frac{\partial}{\partial y}\left(k_y \frac{\partial T}{\partial y}\right) = 0$$

Boundary Conditions:

1. $x = 0, \quad -k\left.\frac{dT}{dx}\right|_{x-0} = q_s''$
2. $x = L, \quad -k\left.\frac{dT}{dx}\right|_{x=L} = h\left(T\big|_{x=L} - T_\infty\right)$

11.2 Determine the variational form and functional for the following one-dimensional fin problem using calculus of variation.

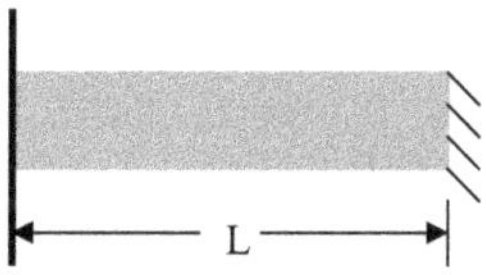

Governing Equation:

One-dimensional Fin $\dfrac{d}{dx}\left(kA\dfrac{dT}{dx}\right) - hP(T - T_\infty) = 0$

Boundary Conditions:

3. $x = 0, \quad T = T_0$
4. $x = L, \quad \left.\frac{dT}{dx}\right|_{x=L} = 0$

11.3 Consider two-dimensional steady-state conduction in a rectangular plate with a cutting groove created by a constant surface heat flux q_0'' in the mid-section of the top surface due to an incident high-energy laser beam. The figure shows the schematic of the problem with boundary conditions specified at the surfaces.

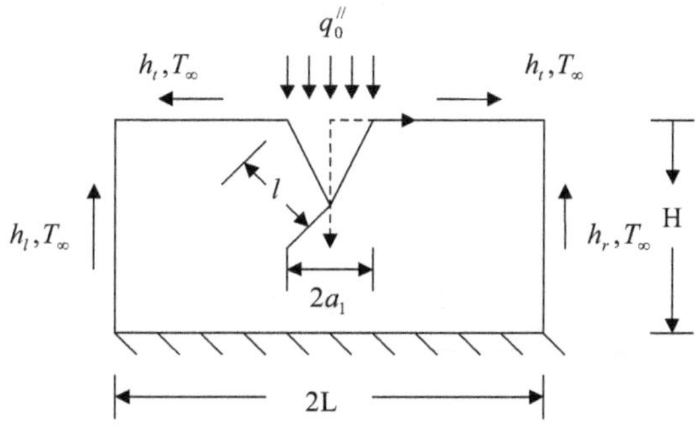

Use calculus of variation to determine the variational form and functional considering the following mathematical statement of the problem.

Governing Equation:

$$\frac{\partial}{\partial x}\left(k_x \frac{\partial T}{\partial x}\right) + \frac{\partial}{\partial y}\left(k_y \frac{\partial T}{\partial y}\right) = 0$$

Boundary Conditions:

Top Surface:

$$y = 0, \quad -L < x \leq -a \quad k_y \frac{\partial T}{\partial y} = h(T - T_\infty),$$

$$y = 0, \quad a < x \leq L, \quad k_y \frac{\partial T}{\partial y} = h(T - T_\infty),$$

On the Groove Surface:

$$-k_y \frac{\partial T}{\partial y} = q_0''$$

Bottom Surface: $y = H, \quad \frac{\partial T}{\partial y} = 0$

Left surface: $x = 0, \quad k_x \frac{\partial T}{\partial x} = h(T - T_\infty)$

Right Surface: $x = L, \quad -k_x \frac{\partial T}{\partial x} = h(T - T_\infty)$

11.4 Determine the weak form of the Problem 11.3 using Galerkin's method

11.5 Consider the problem of cooling an electronic chip that is mounted on a substrate.

Assume uniform volumetric heat generation in the chip. The top and bottom surfaces are convectively cooled. The two side surfaces are assumed to be at constant temperature T_0. The conductivities of the substrate and chip materials are k_s and k_c, respectively.

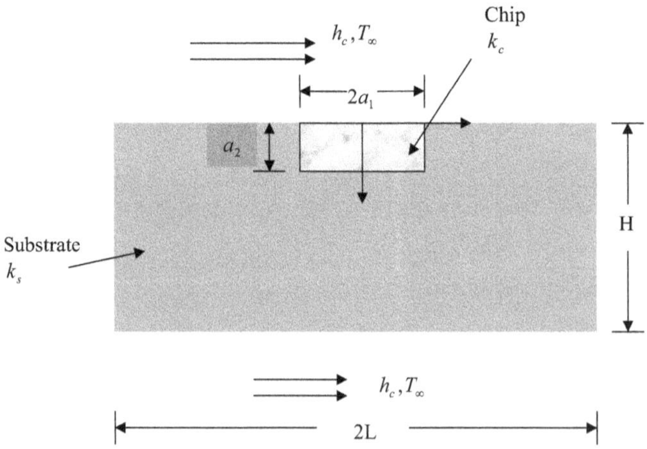

Use **calculus of variation** to determine the variational form and functional considering the following mathematical statement of the problem:

Governing Equation:

$$\frac{\partial}{\partial x}\left(k_x\frac{\partial T}{\partial x}\right)+\frac{\partial}{\partial y}\left(k_y\frac{\partial T}{\partial y}\right)+\dot{q}=0$$

where $\dot{q}=0$ in the substrate.

Boundary Conditions:

Top Surface:

$$y = 0, \quad -L<x<-a_1, \quad a_1<x<L,$$

$$k_s\frac{\partial T}{\partial y}=h(T-T_\infty)$$

$$y=0, \quad -a_1<x<a_1,$$

$$k_c\frac{\partial T}{\partial y}=h(T-T_\infty)$$

Bottom Surface: $y=H, \quad -k_s\dfrac{\partial T}{\partial x}=h(T-T_\infty)$

Left surface: $x=0, \quad T=T_0$

Right Surface: $x = L, \quad T=T_0$

11.6 Determine the integral or weak form of the three-dimensional Navier-Stokes equations for viscous incompressible flows given by the following governing equations and boundary conditions using Galerkin's method.

$$\frac{\partial u}{\partial x}+\frac{\partial v}{\partial y}+\frac{\partial w}{\partial y}=0$$

$$\rho\left(\frac{\partial u}{\partial t}+u\frac{\partial u}{\partial x}+v\frac{\partial u}{\partial y}+w\frac{\partial u}{\partial z}\right)=b_x-\frac{\partial p}{\partial x}+\mu\left(\frac{\partial^2 u}{\partial x^2}+\frac{\partial^2 u}{\partial y^2}+\frac{\partial^2 u}{\partial z^2}\right)$$

$$\rho\left(\frac{\partial v}{\partial t}+u\frac{\partial v}{\partial x}+v\frac{\partial v}{\partial y}+w\frac{\partial v}{\partial z}\right)=b_y-\frac{\partial p}{\partial y}+\mu\left(\frac{\partial^2 v}{\partial x^2}+\frac{\partial^2 v}{\partial y^2}+\frac{\partial^2 v}{\partial z^2}\right)$$

$$\rho\left(\frac{\partial w}{\partial t}+u\frac{\partial w}{\partial x}+v\frac{\partial w}{\partial y}+ww\right)=b_z-\frac{\partial p}{\partial z}+\mu\left(\frac{\partial^2 w}{\partial x^2}+\frac{\partial^2 w}{\partial y^2}+\frac{\partial^2 w}{\partial z^2}\right)$$

over volume V

With boundary conditions given in a general form as

1. $u=u_s$, $v=v_s$ and $w=w_s$ on the boundary A_1

$$\mu\left(\frac{\partial u}{\partial x}\hat{n}_x+\frac{\partial u}{\partial y}\hat{n}_y+\frac{\partial u}{\partial z}\hat{n}_z\right)-p\hat{n}_z=f_{sx}$$

2. $$\mu\left(\frac{\partial v}{\partial x}\hat{n}_x+\frac{\partial v}{\partial y}\hat{n}_y+\frac{\partial v}{\partial z}\hat{n}_z\right)-p\hat{n}_z=f_{sy}$$

$$\mu\left(\frac{\partial w}{\partial x}\hat{n}_x+\frac{\partial w}{\partial y}\hat{n}_y+\frac{\partial w}{\partial z}\hat{n}_z\right)-p\hat{n}_z=f_{sz}$$

on the boundary A_2

11.7 Consider the mathematical statement of a problem given as

Governing Equation:

$$\frac{\mathrm{d}}{\mathrm{d}x}\left(\Gamma\frac{\mathrm{d}\phi}{\mathrm{d}x}\right)+S_0\frac{x}{L}=0$$

Boundary Conditions:

1. $x=0,\quad \phi|_{x=0}=0$
2. $x=L,\quad \phi|_{x=L}=\phi_L$

a. Obtain the functional form using variational method.
b. Obtain approximate solution using Rayleigh-Ritz method, assuming the approximate solution of the form

$$\phi=\psi_0+a_1\psi_1$$

where $\psi_0(x)=\dfrac{\phi_0}{L}x$ and $\psi_1(x)=x(x-L)$ that satisfies the boundary conditions.

11.8 Obtain approximate solution to the problem 11.7 using weighted residual method, assuming the approximate solution of the form

$$\phi=a_1\sin\frac{\pi x}{L}$$

Evaluate a_1 using (a) Galerkin method, (b) collocation method, (c) subdomain method, and (d) least-square method.

11.9 Solve the problem outlined in Example 11.5 using (a) collocation method, (b) subdomain method, and (c) least-square method.

12 Element Shape Functions

We have mentioned in the previous chapter that one of the important steps in the finite element method is the selection of a simple function that is used to approximate the solution over each discretized subregion or finite element. These simple approximate functions are also referred to as **shape functions** or **interpolation functions**. They are mostly polynomial as they are easier to manipulate in mathematical operations and computer implementation compared to other functions such as trigonometric functions. Also, the accuracy of the results can be improved by increasing the order of the polynomial function.

In the following sections, we present the development of some of the common one-, two-, and three-dimensional shape functions for the finite elements.

12.1 ONE-DIMENSIONAL ELEMENT

The general form of the polynomial function in one dimension is given as

$$\phi(x) = a_0 + a_1 x + a_2 x^2 + \cdots + a_n x^n \tag{12.1}$$

where n is the order of the polynomial. Some examples of these functions are depicted in Figure 12.1. The simplest and most widely used function is the polynomial function of order one (Figure 12.1a), which is known as a linear element involving two nodes and is given as

$$\phi(x) = a_0 + a_1 x \tag{12.2}$$

A quadratic element (Figure 12.1b) consists of three nodes and it is given by a second-order polynomial as

$$\phi(x) = a_0 + a_1 x + a_2 x^2 \tag{12.3}$$

Similarly, for the cubic element, the polynomial is of order three and is expressed as

$$\phi(x) = a_0 + a_1 x + a_2 x^2 + a_3 x^3 \tag{12.4}$$

The higher the order of the polynomial, the higher the accuracy in the approximation solution as demonstrated in Figure 12.2.

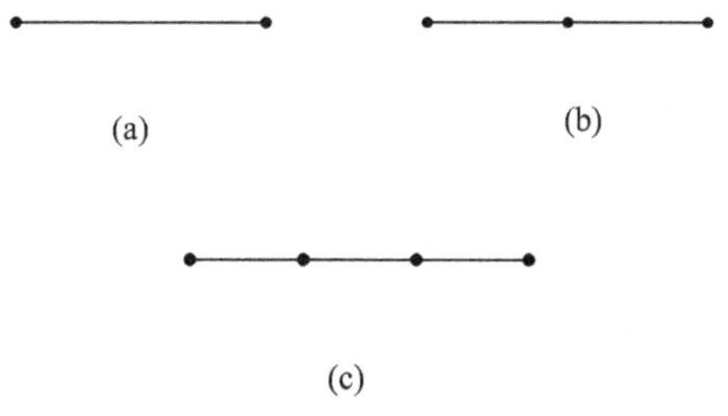

FIGURE 12.1 Examples of polynomial interpolation functions: (a) Linear element, (b) quadratic element, and (c) cubic element.

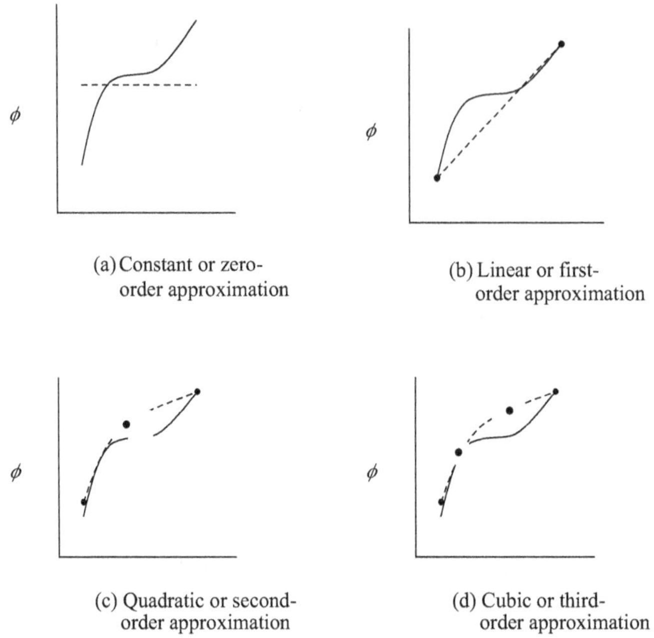

FIGURE 12.2 Polynomials of different order of approximation.

We have mentioned that linear elements are generally preferred due to their simplicity. Accuracy of the results is improved further by using an increased number of elements, i.e., using finer mesh size distributions. On the other hand, one can also choose higher-order elements to achieve similar accuracies with enhanced convergence rate and using coarser mesh size. However, this will bring in additional mathematical complexity and operations involving bigger bandwidth matrices. The choice of linear or higher-order elements is eventually justified through computational savings.

12.1.1 One-Dimensional Linear Element

A one-dimensional linear element is a line of length L_e and consists of two nodes i and j, one at each end as shown in Figure 12.3. The unknown values at these nodes are ϕ_i and ϕ_j.

The approximating function over the element is given as

$$\phi(x) = a_0 + a_1 x \tag{12.5a}$$

The two coefficient values a_0 and a_1 are determined by using two nodal values, i.e.,

$$1.\quad x = x_i, \qquad \phi = \phi_i \tag{12.5b}$$

and

$$2.\quad x = x_j, \qquad \phi = \phi_j \tag{12.5c}$$

Substituting these two conditions, we have

$$\phi_i = a_0 + a_1 x_i$$
$$\phi_j = a_0 + a_1 x_j$$

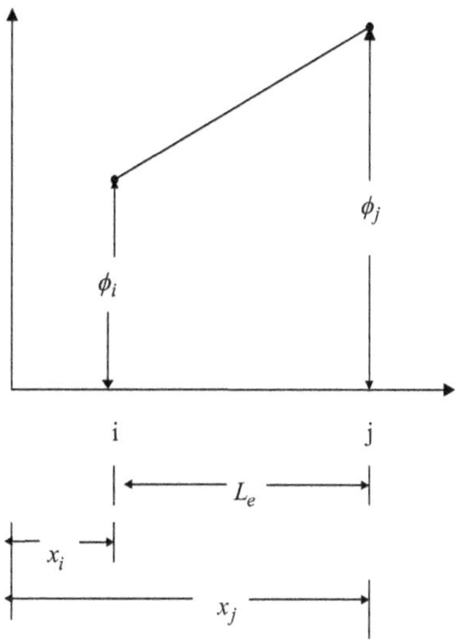

FIGURE 12.3 Linear line element.

and in matrix form as

$$\begin{bmatrix} 1 & x_i \\ 1 & x_j \end{bmatrix} \begin{Bmatrix} a_0 \\ a_1 \end{Bmatrix} = \begin{Bmatrix} \phi_i \\ \phi_j \end{Bmatrix} \tag{12.6}$$

Solving Equation 12.6 using Cramer's rule

$$a_0 = \frac{\begin{vmatrix} \phi_i & x_i \\ \phi_j & x_j \end{vmatrix}}{\begin{vmatrix} 1 & x_i \\ 1 & x_j \end{vmatrix}} = \frac{\phi_i x_j - \phi_j x_i}{x_j - x_i} \tag{12.7a}$$

and

$$a_0 = \frac{\begin{vmatrix} 1 & \phi_i \\ 1 & \phi_j \end{vmatrix}}{\begin{vmatrix} 1 & x_i \\ 1 & x_j \end{vmatrix}} = \frac{\phi_j - \phi_i}{x_j - x_i} \tag{12.7b}$$

Substituting Equation 12.7 into Equation 12.5a, we get the approximating function as

$$\phi = \frac{\phi_i x_j - \phi_j x_i}{x_j - x_i} + \frac{\phi_j - \phi_i}{x_j - x_i} x \tag{12.8}$$

Rearranging

$$\phi = \frac{x_j - x}{x_j - x_i} \phi_i + \frac{x - x_i}{x_j - x_i} \phi_j \tag{12.9a}$$

or

$$\phi = N_i\phi_i + N_j\phi_j = \left[N\right]\{\phi\} \tag{12.9b}$$

where N_i and N_j are shape functions, which are defined as

$$N_i = \frac{x_j - x}{x_j - x_i}, \qquad N_j = \frac{x - x_i}{x_j - x_i} \tag{12.10}$$

and

$$\left[N\right] = \left[\begin{array}{cc} N_i & N_j \end{array}\right] = \text{row vector of shape functions}$$

$$\{\phi\} = \left\{\begin{array}{c} \phi_i \\ \phi_j \end{array}\right\} = \text{column vector of unknown nodal values}$$

It can be noted that the shape functions given by Equation 12.10 are of the same polynomial type as the initial approximation Equation 12.5a. The shape functions also possess two additional properties, given as follows.

1. Each shape function has a value of one at the same node and zero at the other node, i.e.,

$$N_i(x_i) = 1, \qquad N_i(x_j) = 0 \tag{12.11a}$$

   and

$$N_j(x_i) = 0, \qquad N_j(x_j) = 1 \tag{12.11b}$$

   Variations of these shape functions are depicted in Figure 12.4.
2. The sum of the derivatives of the shape functions with respect to x is zero, i.e.,

$$\frac{dN_i}{dx} + \frac{dN_j}{dx} = 0 \tag{12.12a}$$

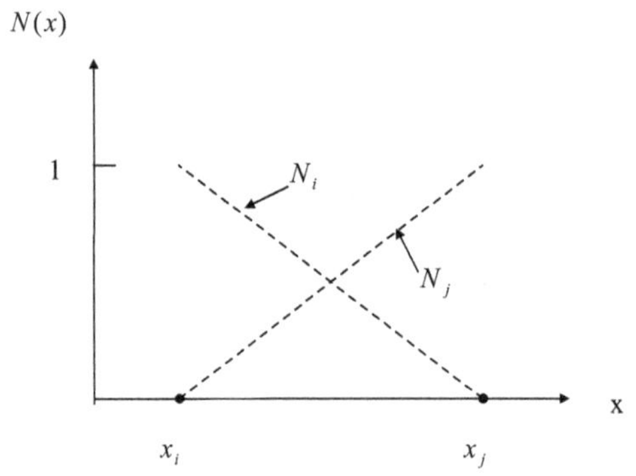

FIGURE 12.4 Variation of shape function for linear line element.

or

$$\sum_{I=i,j} \frac{\mathrm{d}N_I}{\mathrm{d}x} = 0 \qquad (12.12b)$$

The approximation function can be written in a general form for any type of element as

$$\phi = \sum_{I=i,j...} N_I \phi_I, \qquad I = i, j, .., N_n \qquad (12.13)$$

where N_n represents the number of nodes in the element.

Shape Function in Local Coordinate System So far, we have considered shape functions in a global coordinate system. In many situations, it is more convenient to deal with shape functions using the local coordinate systems, particularly in evaluating integrals involving shape functions. Let us consider two such local coordinate systems for a one-dimensional line element. In the first one, the origin is located at node i or the left-hand corner node of the element as shown in Figure 12.5a.

The shape functions for the linear line element in this local coordinate system are obtained by simply substituting x by $x = x_i + \tilde{x}$ in Equation 12.10 as

$$N_i(\tilde{x}) = 1 - \frac{\tilde{x}}{L_e}, \qquad N_j(\tilde{x}) = \frac{\tilde{x}}{L_e} \qquad (12.14)$$

where the coordinate $\tilde{x}$ varies from 0 to L_e.

The second local coordinate system is defined based on its origin located at the center of the element, as shown in Figure 12.5b. The shape functions based on this local coordinate system are obtained by substituting $x = x_i + (L_e / 2) + x'$ and expressed as

$$N_i(x') = \left(\frac{1}{2} - \frac{x'}{L_e}\right), \quad N_j(x') = \left(\frac{1}{2} + \frac{x'}{L_e}\right) \qquad (12.15)$$

where the coordinate x' varies from $-L_e / 2$ to $L_e / 2$.

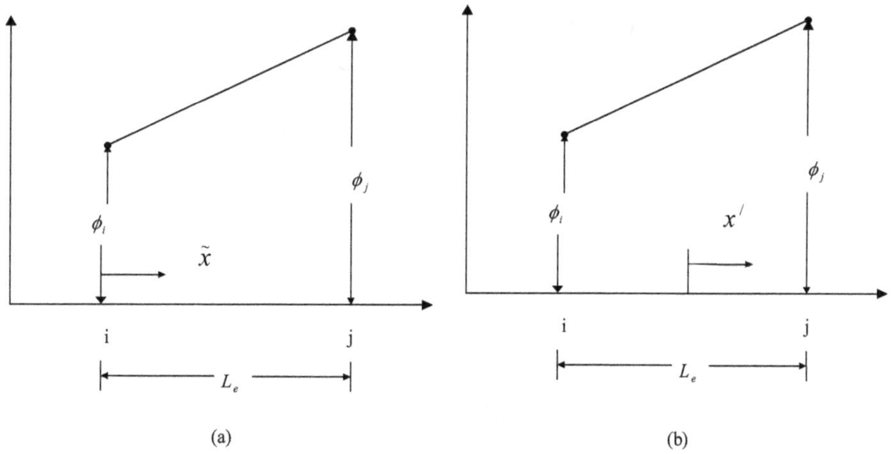

FIGURE 12.5 Linear line element in local coordinate system.

Shape Function in Natural Coordinate System Use of the natural coordinate system is primarily preferred for cases where the integrals involving shape function can easily be evaluated analytically using the integral formulas or by numerically using Gauss–Legendre quadrature formulas. A natural coordinate system is a local coordinate system in which coordinate variables are expressed in a dimensionless form such as $\xi = x / L_e$ or $\xi = x / (L_e / 2)$. In such natural coordinate systems, the coordinate variable ξ varies from 0 to 1, as in Figure 12.6a, or −1 to 1, as shown in Figure 12.6b.

Let us consider the derivation of the shape function for the natural co-ordinate system with the origin at the center of the element as in Figure 12.6b. The approximating function is assumed as

$$\phi(\xi) = a_0 + a_1\xi \tag{12.16}$$

The two coefficient values a_0 and a_1 are determined by using two nodal values, i.e.,

$$1. \quad \xi = -1, \quad \phi = \phi_i \tag{12.17a}$$

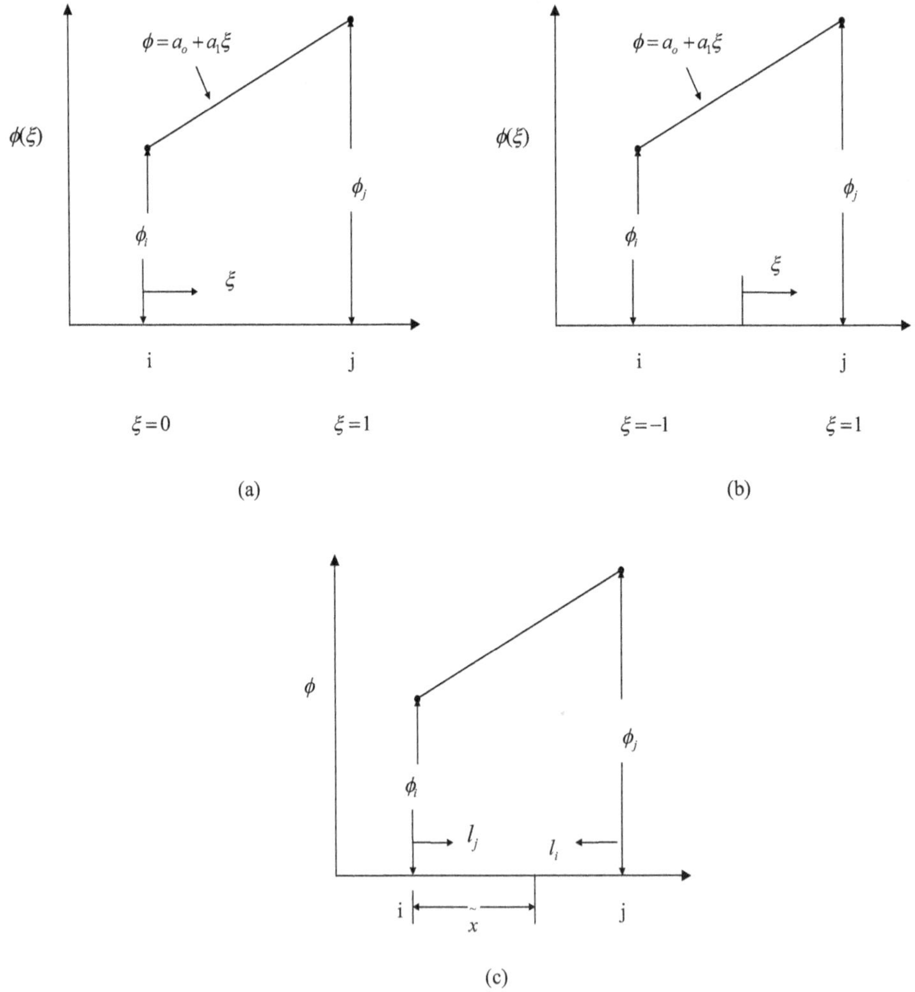

FIGURE 12.6 Shape function for linear line element in natural co-ordinate system.

and

$$2.\quad \xi = 1, \quad \phi = \phi_j \tag{12.17b}$$

Substituting these two conditions, we have

$$\phi_i = a_0 + a_1(-1) \tag{12.18a}$$

$$\phi_j = a_0 + a_1(1) \tag{12.18b}$$

Solving, we get

$$a_0 = \frac{\phi_i + \phi_j}{2} \quad \text{and} \quad a_1 = \frac{\phi_j - \phi_i}{2} \tag{12.19}$$

Substituting Equation 12.19 into Equation 12.16, we get

$$\phi(\xi) = \frac{\phi_i + \phi_j}{2} + \frac{\phi_j - \phi_i}{2}\xi \tag{12.20}$$

Rearranging

$$\phi(\xi) = \tfrac{1}{2}(1-\xi)\phi_i + \tfrac{1}{2}(1+\xi)\phi_j \tag{12.21}$$

or

$$\phi(\xi) = N_i\phi_i + N_j\phi_j \tag{12.22}$$

where the shape functions in the natural coordinate system are given as

$$N_i(\xi) = \tfrac{1}{2}(1-\xi) \quad \text{and} \quad N_j(\xi) = \tfrac{1}{2}(1+\xi) \tag{12.23}$$

The natural coordinate system and shape functions are used along with the coordinate transformation relation

$$x = N_i(\xi)x_i + N_j(\xi)x_j = \sum_{I=i,j} N_I x_I \tag{12.24a}$$

Additionally, we need to transform some integral expressions usually encountered in finite element formulation into natural coordinates using the following relations:

$$\frac{\mathrm{d}}{\mathrm{d}x} = \frac{\mathrm{d}}{\mathrm{d}\xi}\frac{\mathrm{d}\xi}{\mathrm{d}x} = \frac{\mathrm{d}}{\mathrm{d}\xi}\frac{2}{L_e} \tag{12.24b}$$

$$\mathrm{d}x = \frac{L_e}{2}\mathrm{d}\xi \tag{12.24c}$$

Line Coordinates as a Natural Coordinate Another useful natural coordinate system is given in terms of two length ratios as

$$l_i = \frac{L_e - \tilde{x}}{L_e} \quad \text{and} \quad l_j = \frac{\tilde{x}}{L_e} \tag{12.25}$$

where $\tilde{x}$ is the distance of point on the line element as shown in Figure 12.6c. Note that the two natural coordinates l_i and l_j are not independent, but related by the equation

$$l_i + l_j = 1 \tag{12.26}$$

It can be noted that these natural coordinates are identical to the shape functions given by Equation 12.14 for linear line elements in a local coordinate system, i.e.,

$$l_i = N_i(\tilde{x}) \quad \text{and} \quad l_j = N_j(\tilde{x}) \tag{12.27}$$

The main reason for expressing the shape function in terms of natural co-ordinates is the simplicity in evaluating some integrals involving shape functions through use of integral formulas or use Gauss–Legendre quadrature as discussed in Chapter 4. One such common integral formula is given as

$$I = \int_0^1 l_i^a l_j^b \, dl_j = \frac{a!b!}{(a+b+1)} \tag{12.28}$$

For example, the integral expression of the form

$$I = \int_0^{L_e} N_i(\tilde{x}) N_j(\tilde{x}) d\tilde{x} \tag{12.29}$$

can be evaluated using Equations 12.25, 12.27, and 12.28 as

$$I = \int_0^{L_e} N_i(\tilde{x}) N_j(\tilde{x}) d\tilde{x} = \int_0^1 l_i^1 l_j^1 L_e \, dl_j$$

$$= \frac{1!1!}{(1+1+1)} L_e = \frac{L_e}{3} \tag{12.30}$$

12.1.2 One-Dimensional Quadratic Line Element

A quadratic line element has three nodes: one at each end and one at the midpoint as shown in Figure 12.7.

The quadratic approximation equation of this element is given as

$$\phi = a_0 + a_1 x + a_2 x^2 \tag{12.31}$$

The three coefficients a_0, a_1, and a_2 are determined by substituting the three nodal values as

$$\phi_i = a_0 + a_1 x_i + a_2 x_i^2 \tag{12.32a}$$

$$\phi_j = a_0 + a_1 x_j + a_2 x_j^2 \tag{12.32b}$$

$$\phi_k = a_0 + a_1 x_k + a_2 x_k^2 \tag{12.32c}$$

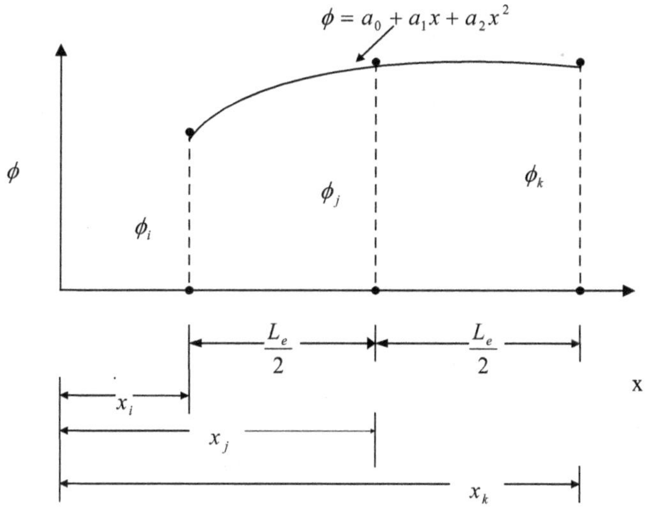

FIGURE 12.7 Quadratic line element.

The system of equation is written in matrix form as

$$\begin{bmatrix} 1 & x_i & x_i^2 \\ 1 & x_j & x_j^2 \\ 1 & x_k & x_k^2 \end{bmatrix} \begin{Bmatrix} a_0 \\ a_1 \\ a_2 \end{Bmatrix} = \begin{Bmatrix} \phi_i \\ \phi_j \\ \phi_k \end{Bmatrix} \tag{12.33}$$

Equation 12.33 is solved by the Cramer rule as

$$a_0 = \frac{\begin{bmatrix} \phi_i & x_i & x_i^2 \\ \phi_j & x_j & x_j^2 \\ \phi_k & x_k & x_k^2 \end{bmatrix}}{\begin{bmatrix} 1 & x_i & x_i^2 \\ 1 & x_j & x_j^2 \\ 1 & x_k & x_k^2 \end{bmatrix}} = \frac{1}{D}\begin{bmatrix} \phi_i(x_jx_k^2 - x_kx_j^2) + \phi_j(x_kx_i^2 - x_ix_k^2) \\ +\phi_k(x_ix_j^2 - x_jx_i^2) \end{bmatrix} \tag{12.34a}$$

$$a_1 = \frac{\begin{bmatrix} 1 & \phi_i & x_i^2 \\ 1 & \phi_j & x_j^2 \\ 1 & \phi_k & x_k^2 \end{bmatrix}}{\begin{bmatrix} 1 & x_i & x_i^2 \\ 1 & x_j & x_j^2 \\ 1 & x_k & x_k^2 \end{bmatrix}} = \frac{1}{D}\left[\phi_i(x_j^2 - x_k^2) + \phi_j(x_k^2 - x_i^2) + \phi_k(x_i^2 - x_j^2)\right] \tag{12.34b}$$

$$a_2 = \frac{\begin{bmatrix} 1 & x_i & \phi_i \\ 1 & x_j & \phi_j \\ 1 & x_k & \phi_k \end{bmatrix}}{\begin{bmatrix} 1 & x_i & x_i^2 \\ 1 & x_j & x_j^2 \\ 1 & x_k & x_k^2 \end{bmatrix}} = \frac{1}{D}\left[\phi_i(x_k - x_j) + \phi_j(x_i - x_k) + \phi_k(x_j - x_i)\right] \tag{12.34c}$$

where

$$D = (x_j x_k^2 - x_k x_j^2) + (x_k x_i^2 - x_i x_k^2) + (x_i x_j^2 - x_j x_i^2)$$

Substituting Equation 12.34 into Equation 12.31 and rearranging, we get

$$\phi = N_i\phi_i + N_j\phi_j + N_k\phi_k \tag{12.35a}$$

or

$$\phi = \sum_{I=i,j,k} N_I\phi_I, \tag{12.35b}$$

The shape functions for the quadratic line element are written in a compact form as

$$N_I(x) = \frac{1}{D}\left(\alpha_I + \beta_I x + \gamma_I x^2\right), \qquad I = i, j \text{ and } k \tag{12.36}$$

where

$$\alpha_I = (x_j x_k^2 - x_k x_j^2) + (x_k x_i^2 - x_i x_k^2) + (x_i x_j^2 - x_j x_i^2) \tag{12.37a}$$

$$\beta_I = (x_j^2 - x_k^2) + (x_k^2 - x_i^2) + (x_i^2 - x_j^2) \tag{12.37b}$$

$$\gamma_I = (x_k - x_j) + (x_i - x_k) + (x_j - x_i) \tag{12.37c}$$

Quadratic Line Element in Local Coordinate The quadratic shape function can be expressed in the local coordinate system with the origin located at the left-hand node as shown in Figure 12.8.

Substituting the coordinate points into the approximating function given by the quadratic Equation 12.31, in the local coordinate ($\tilde{x}$), we get

$$\phi_i = a_0 + a_1 \cdot 0 + a_2 \cdot 0 \tag{12.38a}$$

$$\phi_j = a_0 + a_1 \frac{L_e}{2} + a_2\left(\frac{L_e}{2}\right)^2 \tag{12.38b}$$

$$\phi_k = a_0 + a_1 L_e + a_2 (L_e)^2 \tag{12.38c}$$

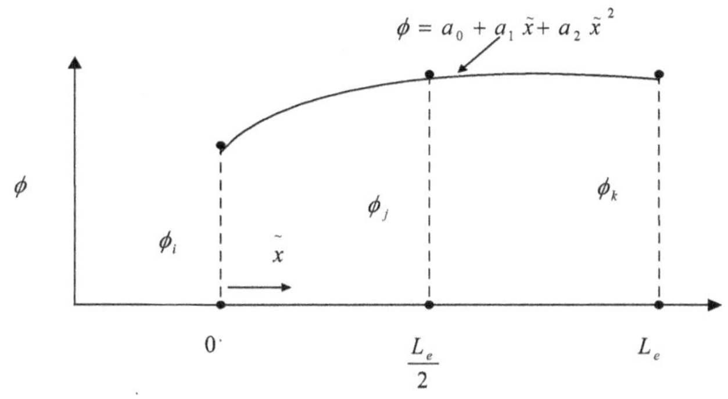

FIGURE 12.8 Quadratic line element in a local co-ordinate system.

The system of equation is written in matrix form as

$$\begin{bmatrix} 1 & 0 & 0 \\ 1 & L_e/2 & L_e^2/4 \\ 1 & L_2 & L_e^2 \end{bmatrix} \begin{Bmatrix} a_0 \\ a_1 \\ a_2 \end{Bmatrix} = \begin{Bmatrix} \phi_i \\ \phi_j \\ \phi_k \end{Bmatrix} \tag{12.39}$$

Solution of this system gives

$$a_0 = \phi_i \tag{12.40a}$$

$$a_1 = \frac{1}{L_e}\left(-3\phi_i + 4\phi_j - \phi_k\right) \tag{12.40b}$$

$$a_2 = \frac{2}{L_e^2}\left(\phi_i - 2\phi_j + \phi_k\right) \tag{12.40c}$$

Substituting Equation 12.40 into Equation 12.31 and rearranging, we get

$$\phi(\tilde{x}) = \left(1 - \frac{\tilde{x}}{L_e}\right)\left(1 - 2\frac{\tilde{x}}{L_e}\right)\phi_i + 4\frac{\tilde{x}}{L_e}\left(1 - \frac{\tilde{x}}{L_e}\right)\phi_j + \frac{\tilde{x}}{L_e}\left(2\frac{\tilde{x}}{L_e} - 1\right)\phi_j \tag{12.41a}$$

or

$$\phi = N_i\phi_i + N_j\phi_j + N_k\phi_k \tag{12.41b}$$

where the shape functions are given as

$$N_i = \left(1 - \frac{\tilde{x}}{L_e}\right)\left(1 - 2\frac{\tilde{x}}{L_e}\right) \tag{12.42a}$$

$$N_j = 4\frac{\tilde{x}}{L_e}\left(1 - \frac{\tilde{x}}{L_e}\right) \tag{12.42b}$$

$$N_k = \frac{\tilde{x}}{L_e}\left(2\frac{\tilde{x}}{L_e} - 1\right) \tag{12.42c}$$

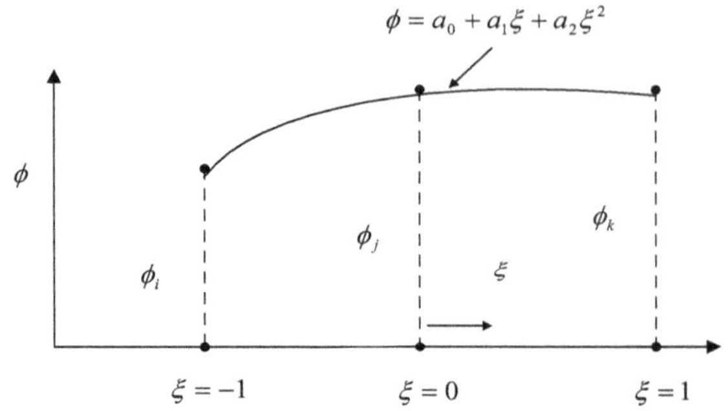

FIGURE 12.9 One-dimensional quadratic element in natural co-ordinate system.

Quadratic Element in Natural Coordinate System The approximation function for the quadratic line element in a natural coordinate system with the origin located at the midpoint, as shown in Figure 12.9, is assumed as

$$\phi(\xi) = a_0 + a_1\xi + a_2\xi^2 \tag{12.43}$$

The coefficients are determined by substituting the coordinate points, and the approximating function is again obtained as

$$\phi = N_i\phi_i + N_j\phi_j + N_k\phi_k \tag{12.44}$$

with shape functions in natural coordinates given as

$$N_i(\xi) = \frac{\xi}{2}(\xi - 1) \tag{12.45a}$$

$$N_j(\xi) = -(\xi + 1)(\xi - 1) \tag{12.45b}$$

$$N_k(\xi) = \frac{\xi}{2}(\xi + 1) \tag{12.45c}$$

The transformation relationship between the local coordinate and natural coordinate systems is given by

$$x = \sum_{I=i,j,k} N_I x_I = N_i(\xi)x_i + N_j(\xi)x_j + N_k(\xi)x_k \tag{12.46a}$$

Additionally, we need the following relations for transforming integral relations into natural coordinate systems

$$\frac{\mathrm{d}}{\mathrm{d}x} = \frac{1}{\mathrm{d}x/\mathrm{d}\xi}\frac{\mathrm{d}}{\mathrm{d}\xi} \tag{12.46b}$$

where

$$\frac{\mathrm{d}x}{\mathrm{d}\xi} = \frac{\mathrm{d}N_i}{\mathrm{d}\xi}x_i + \frac{\mathrm{d}N_j}{\mathrm{d}\xi}x_j + \frac{\mathrm{d}N_k}{\mathrm{d}\xi}x_k \tag{12.46c}$$

12.1.3 One-Dimensional Cubic Element

The one-dimensional cubic element in a local coordinate with the origin located at the left-hand corner node is considered as shown in Figure 12.10, and the approximating function given as

$$\phi = a_0 + a_1\tilde{x} + a_2\tilde{x}^2 + a_3\tilde{x}^3 \tag{12.47}$$

The coefficients are derived using the nodal values at the four nodal points shown in the figure, and the approximation function derived as

$$\phi(\tilde{x}) = N_i\phi_i + N_j\phi_j + N_k\phi_k + N_l\phi_l = \sum_{i=i,j,k,l} N_I\phi_I \tag{12.48}$$

where the shape functions are given as

$$N_i(\tilde{x}) = \left(1 - \frac{3\tilde{x}}{L_e}\right)\left(1 - \frac{3\tilde{x}}{2L_e}\right)\left(1 - \frac{\tilde{x}}{L_e}\right) \tag{12.49a}$$

$$N_j(\tilde{x}) = \frac{9\tilde{x}}{L_e}\left(1 - \frac{3\tilde{x}}{2L_e}\right)\left(1 - \frac{\tilde{x}}{L_e}\right) \tag{12.49b}$$

$$N_k(\tilde{x}) = -\frac{9\tilde{x}}{2L_e}\left(1 - \frac{3\tilde{x}}{2L_e}\right)\left(1 - \frac{\tilde{x}}{L_e}\right) \tag{12.49c}$$

$$N_l(\tilde{x}) = \frac{\tilde{x}}{L}\left(1 - \frac{3\tilde{x}}{L_e}\right)\left(1 - \frac{3\tilde{x}}{2L_e}\right) \tag{12.49d}$$

One-Dimensional Cubic Element in Natural Coordinate The one-dimensional approximation function for a one-dimensional cubic element, as shown in Figure 12.11, is assumed as

$$\phi = a_0 + a_1\xi + a_2\xi^2 + a_3\xi^3 \tag{12.50}$$

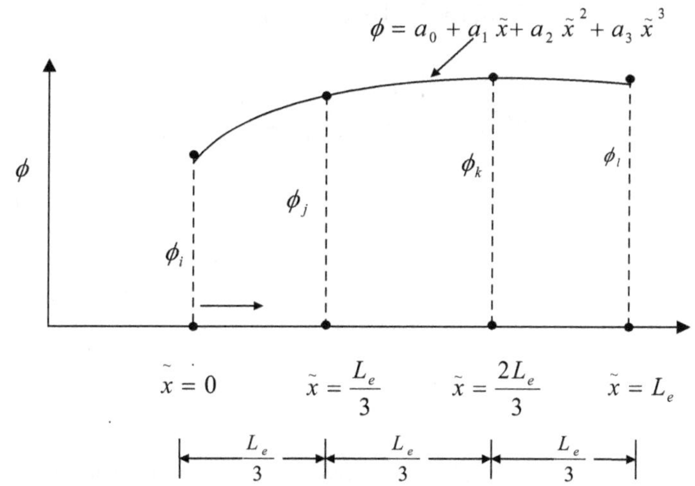

FIGURE 12.10 One-dimensional quadratic element in local co-ordinate system.

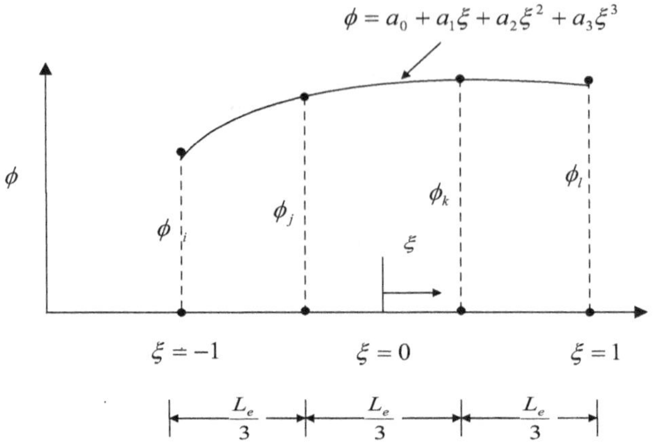

FIGURE 12.11 One-dimensional cubic elements in natural co-ordinate.

Substituting the coordinates of the four nodal points *i*, *j*, *k*, and *l*, we obtain the approximating function as

$$\xi = N_i\phi_i + N_j\phi_j + N_k\phi_k + N_l\phi_l = \sum_{i=i,j,k,l} N_I\phi_I \tag{12.51}$$

where the shape functions for the one-dimensional cubic element in a natural coordinate system are given as

$$N_i(\xi) = -\frac{9}{16}(1-\xi)\left(\frac{1}{3}+\xi\right)\left(\frac{1}{3}-\xi\right) \tag{12.52a}$$

$$N_j(\xi) = \frac{27}{16}(1+\xi)(1-\xi)\left(\frac{1}{3}-\xi\right) \tag{12.52b}$$

$$N_k(\xi) = \frac{27}{16}(1+\xi)(1-\xi)\left(\frac{1}{3}+\xi\right) \tag{12.52c}$$

$$N_l(\xi) = -\frac{9}{16}(1+\xi)\left(\frac{1}{3}+\xi\right)\left(\frac{1}{3}-\xi\right) \tag{12.52d}$$

Example 12.1: One-Dimensional Element

The coordinates of the two nodes of a one-dimensional linear element are $x_i = 0.03$ m and $x_j = 0.04$ m in a media with thermal conductivity $k = 12$ W/m°C. Based on the finite element analysis, the nodal temperature values are found to be $T_i = 85$°C and $T_j = 80$°C. Determine the temperature at a distance x=0.032 m and the heat flux within the element.

Solution

The temperature distribution within a linear one-dimensional element is given by Equation 12.9 as

$$T = N_iT_i + N_jT_j$$

where the shape functions N_i and N_j are given as

$$N_i = \frac{x_j - x}{x_j - x_i}, \qquad N_j = \frac{x - x_i}{x_j - x_i}$$

Using the coordinate values, we get

$$N_i = \frac{x_j - x}{x_j - x_i} = \frac{0.04 - x}{0.04 - 0.03} = 100(0.04 - x)$$

and

$$N_j = \frac{x - x_i}{x_j - x_i} = \frac{x - 0.03}{0.04 - 0.03} = 100(x - 0.03)$$

with the substitution of the shape function expressions, the temperature distribution within the element is given as

$$T = 100(0.04 - x)T_i + 100(x - 0.03)T_j$$

Now, the temperature at a distance $x=0.032$ within the element is given as

$$T = 100 \times (0.04 - 0.032) \times 85.0 + 100 \times (0.032 - 0.03) \times 80.0$$
$$= 84°\text{C}$$

The heat flux rate within the element is

$$q'' = -k\frac{dT}{dx} = -k\left[\left(-\frac{1}{x_j - x_i}\right)T_i + \left(\frac{1}{x_j - x_i}\right)T_j\right]$$
$$= -k\left(\frac{T_j - T_i}{x_j - x_i}\right) = -12 \times \left(\frac{80.0 - 85.0}{0.01}\right)$$
$$q'' = 6000 \text{ W/m}^2$$

12.2 TWO-DIMENSIONAL ELEMENT

Two-dimensional elements are used to discretize a two-dimensional region. The most common polygonal shapes that are being used for two-dimensional elements are **triangular** and **quadrilateral** shapes or a variation of these elements as shown in Figure 12.12.

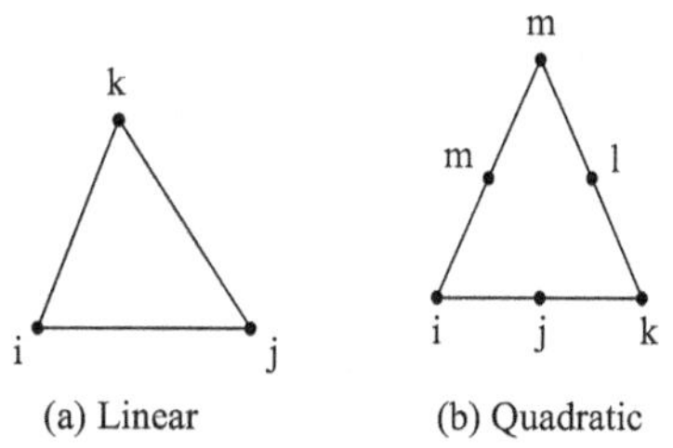

FIGURE 12.12 Triangular elements: (a) linear and (b) quadratic.

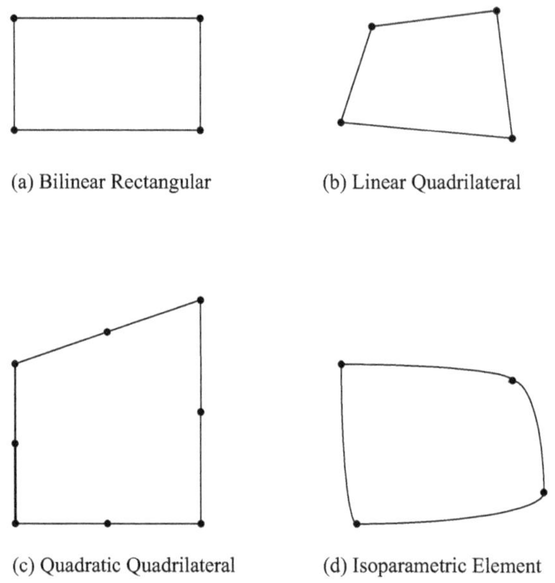

FIGURE 12.13 Rectangular and quadrilateral elements: (a) bilinear rectangular, (b) linear quadrilateral elements, (c) isoparametric.

The triangular elements (Figure 12.12) can take up any orientation and satisfy the continuity requirements involving adjacent elements and gives closer approximation of irregular surfaces.

On the other hand, a bilinear rectangular element (Figure 12.13a) can only take up an orientation that is parallel to an x–y coordinate system, and typically used when the solution domain is rectangular in shape. However, the four-sided quadrilateral element (Figure 12.13b), other than the rectangular element, can have nonorthogonal sides, i.e., each side may have different slopes, which gives it the capability of having a closer fit of the irregular shaped solution domain. In fact, the quadrilateral element is becoming increasingly popular in discretizing the irregular solution domain.

The elements could be linear or cubic depending on the order of the polynomial used to represent the approximate function. For example, a polynomial of the form $\phi = a_0 + a_1x + a_2y$ is linear in x and y, and its coefficients a_0, a_1, and a_2 are expressed in terms of nodal values at the three vertices of the triangle (Figure 12.12a). A quadratic polynomial $\phi = a_0 + a_1x + a_2y + a_3xy + a_4x^2 + a_5y^2$ with six terms can be used to form a geometrical shape or element with six nodes, such as in a triangular element with nodes at three vertices and nodes at mid-points of the three sides of the triangle (Figure 12.12b). A polynomial of the form $\phi = a_0 + a_1x + a_2y + a_3xy$ has linear terms in x and y, and a bilinear term in x and y, and it requires a geometrical shape with four nodes such as the linear rectangular, which is a special case of the quadrilateral element with nodes at the vertices.

The procedure for deriving the shape functions or the interpolation functions for two-dimensional elements is the same as those given for one-dimensional elements presented in the previous sections. In this section, a brief discussion on the derivation of some of the two-dimensional elements in local and natural coordinate systems is given.

12.2.1 Linear Triangular Element

Let us consider a linear triangular element with three nodal values ϕ_i, ϕ_j, and ϕ_k with nodal coordinate (x_i, y_i), (x_j, y_j), and (x_k, y_k) as shown in Figure 12.14.

We have mentioned that the appropriate approximation is the first-order polynomial, i.e., by

$$\phi = a_0 + a_1x + a_2y \tag{12.53}$$

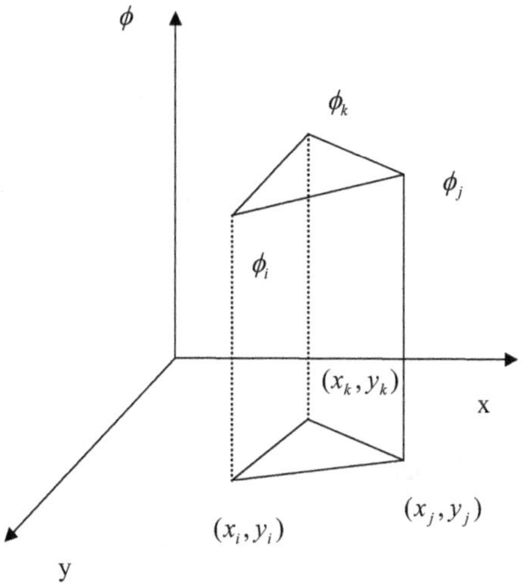

FIGURE 12.14 Linear triangular elements.

Substituting the nodal values, we get

$$\phi_i = a_0 + a_1 x_i + a_2 y_i \tag{12.54a}$$

$$\phi_j = a_0 + a_1 x_j + a_2 y_j \tag{12.54b}$$

$$\phi_k = a_0 + a_1 x_k + a_2 y_k \tag{12.54c}$$

The system of Equation 12.54 is written in a matrix notation as

$$\begin{bmatrix} 1 & x_i & y_i \\ 1 & x_j & y_j \\ 1 & x_k & y_k \end{bmatrix} \begin{Bmatrix} a_0 \\ a_1 \\ a_2 \end{Bmatrix} = \begin{Bmatrix} \phi_i \\ \phi_j \\ \phi_k \end{Bmatrix} \tag{12.55}$$

Solving by the Cramer rule

$$a_0 = \frac{\begin{vmatrix} \phi_i & x_i & y_i \\ \phi_j & x_j & y_j \\ \phi_k & x_k & y_k \end{vmatrix}}{2A} \tag{12.56}$$

or

$$a_0 = \frac{1}{2A}\left[\phi_i\left(x_j y_k - x_k y_j\right) + \phi_j\left(x_k y_i - x_i y_k\right) + \phi_k\left(x_i y_j - x_j y_i\right)\right] \tag{12.57a}$$

Similarly

$$a_1 = \frac{1}{2A}\left[\phi_i\left(y_j - y_k\right) + \phi_j\left(y_k - y_i\right) + \phi_k\left(y_i - y_j\right)\right] \tag{12.57b}$$

and

$$a_2 = \frac{1}{2A}\left[\phi_i\left(x_k - x_j\right) + \phi_j\left(x_i - x_k\right) + \phi_k\left(x_j - x_i\right)\right] \tag{12.57c}$$

where A is the area of the triangle and is given by the determinant

$$A = \frac{1}{2}\begin{vmatrix} 1 & x_i & y_i \\ 1 & x_j & y_j \\ 1 & x_k & y_k \end{vmatrix} \tag{12.58}$$

Substituting the coefficients into Equation 12.53 and rearranging, we get

$$\phi = N_i\phi_i + N_j\phi_j + N_k\phi_k \tag{12.59}$$

or

$$\phi = \left[N\right]\left\{\phi^{(e)}\right\} \tag{12.60}$$

where

$$N_i = \frac{1}{2A}\left(\alpha_i + \beta_i x + \gamma_i y\right) \tag{12.61a}$$

$$N_j = \frac{1}{2A}\left(\alpha_j + \beta_j x + \gamma_j y\right) \tag{12.61b}$$

$$N_k = \frac{1}{2A}\left(\alpha_k + \beta_k x + \gamma_k y\right) \tag{12.61c}$$

and

$$\alpha_i = x_j y_k - x_k y_j, \quad \beta_i = y_j - y_k, \quad \gamma_i = x_k - x_j \tag{12.62a}$$

$$\alpha_j = x_k y_i - x_i y_k, \quad \beta_j = y_k - y_i, \quad \gamma_j = x_i - x_k \tag{12.62b}$$

$$\alpha_k = x_i y_j - x_j y_i, \quad \beta_k = y_i - y_j, \quad \gamma_k = x_j - x_i \tag{12.62c}$$

The properties of a two-dimensional triangular element are like those for one-dimensional elements. Each shape function $N_I, I = i, j, k$ has a value of **one** at its own node and is **zero** at the other nodes as depicted in Figure 12.15. Also, the shape function varies linearly (for the linear element) along two adjacent sides of the node and is zero along the side opposite to the node. For example, the shape function N_i has a value of one at the node i and has a value of zero at nodes j and k. It varies linearly along ij and ik, and zero along jk.

Other relevant properties are

$$\sum_{I=i,j,k} N_I = 1, \quad \sum_{I=i,j,k} \frac{\partial N_I}{\partial x} = 0 \quad \text{and} \quad \sum_{I=i,j,k} \frac{\partial N_J}{\partial y} = 0 \tag{12.63}$$

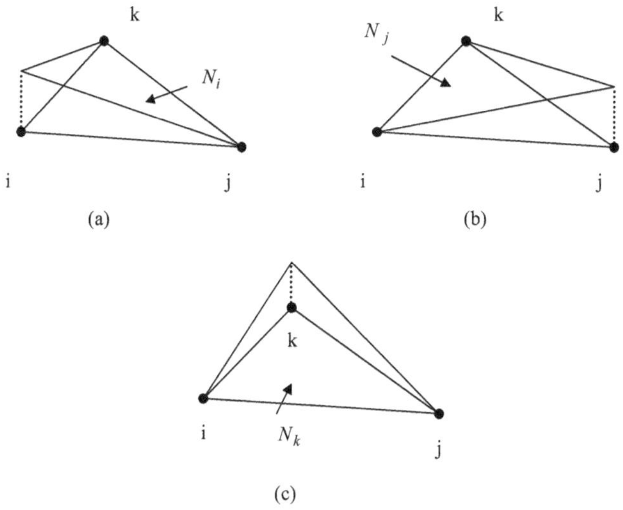

FIGURE 12.15 Linear triangular shape functions.

Linear Triangular Element in Natural Coordinate It was mentioned before that a natural coordinate system simplifies evaluation of integral expressions involving shape functions through use of integral formulas or numerical quadratures such as the Gauss–Legendre formula. A natural coordinate system for a two-dimensional triangular element is known as area coordinate system. In this coordinate system three coordinate variables L_i, L_j, and L_k are defined in dimensionless form as

$$L_i = \frac{s_i}{h}, \quad L_j = \frac{s_j}{h}, \quad L_k = \frac{s_k}{h} \tag{12.64}$$

where s is the perpendicular distance of a point P from the side of the triangle and h is the altitude of the same side as shown in Figure 12.16a.

The lines of constant L_i run parallel to the opposite side of the triangle, i.e., side jk. Similarly, the lines of constant L_j and L_k run parallel to ik and ij, respectively. Also, the line coordinates, L, are called area coordinates as these are equal to the ratio of the partial area to the total area of the triangle as given by

$$L_i = \frac{A_i}{A}, \quad L_j = \frac{A_j}{A}, \quad L_k = \frac{A_k}{A} \tag{12.65}$$

where A_i, A_j, and A_k are the partial areas of the triangular element area. These partial areas are defined by joining a point P in the triangle with three vertices or nodal points i, j, and k, as shown in Figure 12.16b.

Since the sum of Ai, Aj, and Ak is equal to A, the sum of three-line coordinates is equal to one, i.e.,

$$L_i + L_j + L_k = 1 \tag{12.66}$$

Considering the coordinate of the point P as (x, y) and noting that the area of the triangle A is given by Equation 12.58, we can write

$$A_i = \frac{1}{2}\begin{vmatrix} 1 & x & y \\ 1 & x_j & y_j \\ 1 & x_k & y_k \end{vmatrix} \tag{12.67a}$$

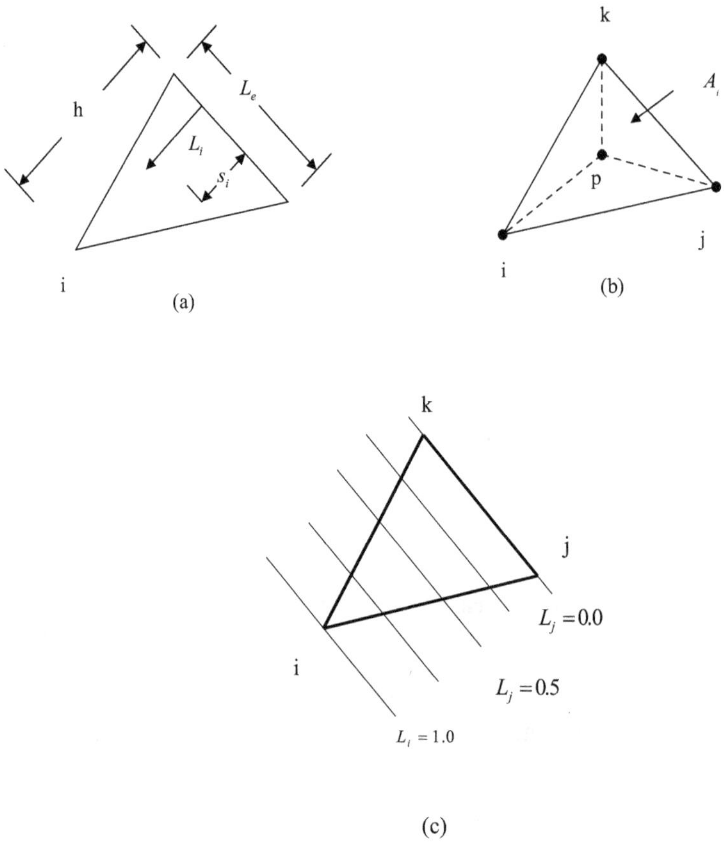

FIGURE 12.16 Natural or area co-ordinates for linear triangular element.

or

$$A_i = \frac{1}{2}\left[\left(x_j y_k - x_k y_j\right) + \left(y_j - y_k\right)x + \left(x_k - x_j\right)y\right] \tag{12.67b}$$

Now, substituting Equation 12.67b into the defining equation of L_i, i.e., Equation 12.65 and comparing with Equation 12.60a we get

$$L_i = N_i \tag{12.68a}$$

Similarly, we can show that

$$L_j = N_j \tag{12.68b}$$

and

$$L_k = N_k \tag{12.68c}$$

So, we can express the approximating function in terms of line coordinate s as

$$\phi = L_i\phi_i + L_j\phi_j + L_k\phi_k \tag{12.69}$$

Also, the relationship between the Cartesian and line coordinates is given by

$$x = L_i x_i + L_j x_j + L_k x_k \tag{12.70a}$$

and

$$y = L_i y_i + L_j y_j + L_k y_k \tag{12.70b}$$

We have mentioned before that the main reason for expressing the two-dimensional shape function in terms of the area coordinate is the simplicity in evaluating some integrals involving shape functions through use of integral formulas. One such common integral formula is given as

$$I = \int_A L_i^a L_j^b L_k^c \, \mathrm{d}A = \frac{a!b!c!}{(a+b+c+2)!} 2A \tag{12.71}$$

For example, the integral expression of the form

$$I = \int_A N_i N_j \mathrm{d}A \tag{12.72}$$

can be evaluated using Equation 12.68 and 12.71 as

$$I = \int_A N_i N_j \mathrm{d}A = \int_A L_i^1 L_j^1 L_k^0 \mathrm{d}A \tag{12.73}$$

$$= \frac{1!1!0!}{(1+1+0+2)!} 2A = \frac{A}{12} \tag{12.74}$$

We will also see in later chapters that the finite element formulations require evaluation of integrals over each element, and these integrals may involve derivatives of the shape functions. To evaluate these integrals in natural coordinate systems, it is also required to express the derivatives in natural coordinates. We can express the derivatives in the natural coordinate system assuming that L_i and L_j are independent, L_k is given by Equation 12.66, and using the chain rule as

$$\frac{\partial N_i}{\partial L_i} = \frac{\partial N_i}{\partial x}\frac{\partial x}{\partial L_i} + \frac{\partial N_i}{\partial y}\frac{\partial y}{\partial L_i} \tag{12.75a}$$

$$\frac{\partial N_i}{\partial L_j} = \frac{\partial N_i}{\partial x}\frac{\partial x}{\partial L_j} + \frac{\partial N_i}{\partial y}\frac{\partial y}{\partial L_j} \tag{12.75b}$$

Writing this set in a matrix form, we get

$$\begin{Bmatrix} \dfrac{\partial N_i}{\partial L_i} \\ \dfrac{\partial N_i}{\partial L_j} \end{Bmatrix} = \begin{bmatrix} \dfrac{\partial x}{\partial L_i} & \dfrac{\partial y}{\partial L_i} \\ \dfrac{\partial x}{\partial L_j} & \dfrac{\partial y}{\partial L_j} \end{bmatrix} \begin{Bmatrix} \dfrac{\partial N_i}{\partial x} \\ \dfrac{\partial N_i}{\partial y} \end{Bmatrix} = J \begin{Bmatrix} \dfrac{\partial N_i}{\partial x} \\ \dfrac{\partial N_i}{\partial y} \end{Bmatrix} \tag{12.75c}$$

where J is the **Jacobian matrix**, which is defined here as

$$J = \begin{bmatrix} \dfrac{\partial x}{\partial L_i} & \dfrac{\partial y}{\partial L_i} \\ \dfrac{\partial x}{\partial L_j} & \dfrac{\partial y}{\partial L_j} \end{bmatrix} \tag{12.76}$$

Equation 12.75c can be inverted to express the derivatives in x-and y-coordinates as

$$\begin{Bmatrix} \dfrac{\partial N_i}{\partial x} \\ \dfrac{\partial N_i}{\partial y} \end{Bmatrix} = J^{-1} \begin{Bmatrix} \dfrac{\partial N_i}{\partial L_i} \\ \dfrac{\partial N_i}{\partial L_j} \end{Bmatrix} \tag{12.77}$$

Equation 12.77 can further be simplified by using the defining equation for the inverse of a matrix (see Appendix A) as

$$\begin{Bmatrix} \dfrac{\partial N_i}{\partial x} \\ \dfrac{\partial N_i}{\partial y} \end{Bmatrix} = \frac{1}{|J|} \begin{bmatrix} \dfrac{\partial y}{\partial L_j} & -\dfrac{\partial y}{\partial L_i} \\ -\dfrac{\partial x}{\partial L_j} & \dfrac{\partial x}{\partial L_i} \end{bmatrix} \begin{Bmatrix} \dfrac{\partial N_i}{\partial L_i} \\ \dfrac{\partial N_i}{\partial L_j} \end{Bmatrix} \tag{12.78}$$

or

$$\begin{Bmatrix} \dfrac{\partial N_i}{\partial x} \\ \dfrac{\partial N_i}{\partial y} \end{Bmatrix} = \frac{1}{|J|} \begin{Bmatrix} \dfrac{\partial y}{\partial L_j}\dfrac{\partial N_i}{\partial L_i} - \dfrac{\partial y}{\partial L_i}\dfrac{\partial N_i}{\partial L_j} \\ -\dfrac{\partial x}{\partial L_j}\dfrac{\partial N_i}{\partial L_i} + \dfrac{\partial x}{\partial L_i}\dfrac{\partial N_i}{\partial L_j} \end{Bmatrix} \tag{12.79}$$

where $|J|$ is the determinant of the Jacobian matrix and it is given as

$$|J| = \frac{\partial x}{\partial L_i}\frac{\partial y}{\partial L_j} - \frac{\partial x}{\partial L_j}\frac{\partial y}{\partial L_i} \tag{12.80}$$

Another important use of the line or area coordinate system is in evaluating an integral along one side of the element (Figure 12.17).

Consider a point p on the side j–k of the triangular element. For this point, the line coordinates are defined as

$$L_j = \frac{A_j}{A} = \frac{\frac{1}{2}h(L_e - \tilde{x})}{\frac{1}{2}hL_e} = 1 - \frac{\tilde{x}}{L_e}, \qquad L_k = \frac{\tilde{x}}{L_e} \tag{12.81}$$

Comparing these with the shape functions for one-dimensional natural coordinates, we can write

$$L_i = 0, \quad L_j = l_i \quad \text{and} \quad L_k = l_j \quad \text{for the side } j\text{–}k \tag{12.82a}$$

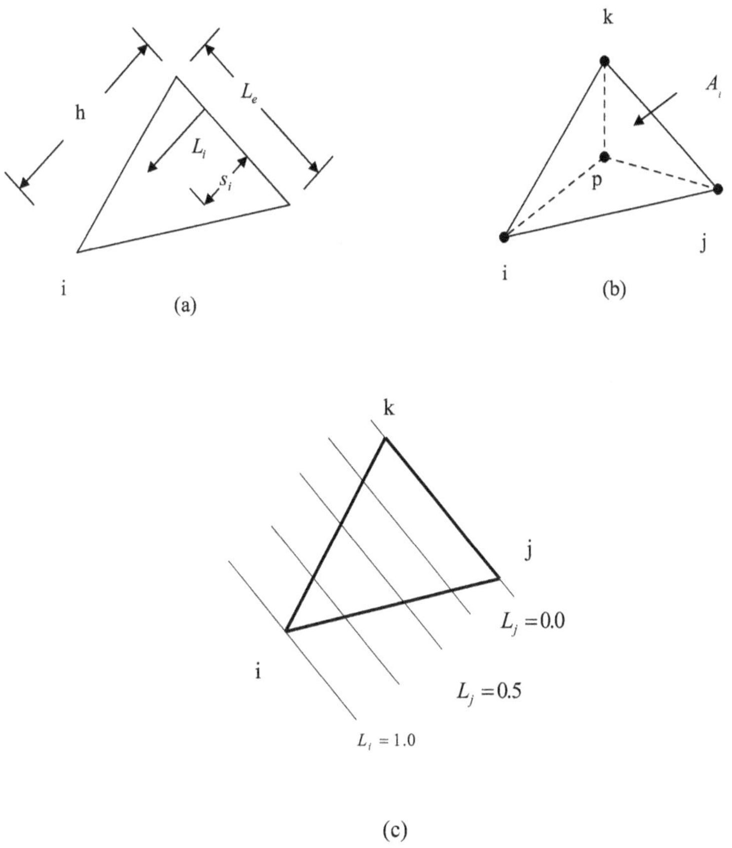

FIGURE 12.17 Natural or area co-ordinates for linear triangular element.

$$L_i = l_j, \quad L_j = 0 \quad \text{and} \quad L_k = l_i \quad \text{for the side } k\text{–}i \tag{12.82b}$$

$$L_i = l_i, \quad L_i = l_j \quad \text{and} \quad L_k = 0 \quad \text{for the side } i\text{–}j \tag{12.82c}$$

12.2.2 Quadratic Triangular Element

A quadratic triangular element has six nodes with three nodes at the three vertices and three at midpoints of the sides of the triangle as shown in Figure 12.18.

The shape function for the quadratic triangular element in the area coordinate is expressed as

$$\begin{aligned} &N_i = L_i(2L_i - 1), \quad N_j = L_j(2L_j - 1), \quad N_k = L_k(2L_k - 1) \\ &N_l = 4L_jL_k, \quad N_m = 4L_iL_k, \quad N_n = 4L_iL_j \end{aligned} \tag{12.83}$$

12.2.3 Two-Dimensional Quadrilateral Elements

We have mentioned before that quadrilateral elements are four-sided polygons. The sides of the element may be straight and orthogonal as in a bilinear rectangular element (Figure 12.13a) or each side may have different slopes (Figure 12.13b) or arbitrarily oriented curved sides as in isoparametric elements (Figure 12.13d).

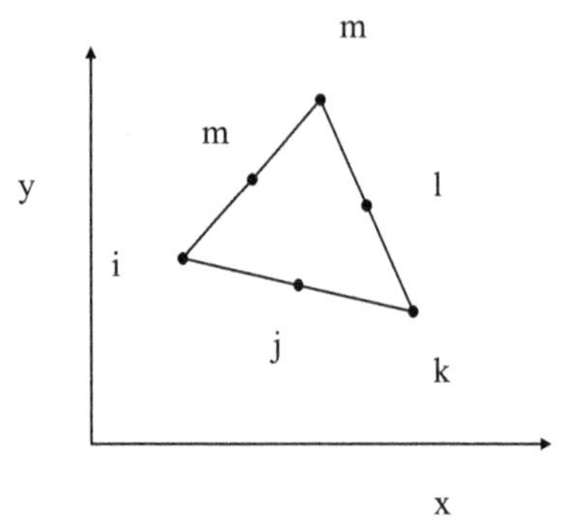

FIGURE 12.18 Quadratic triangular element.

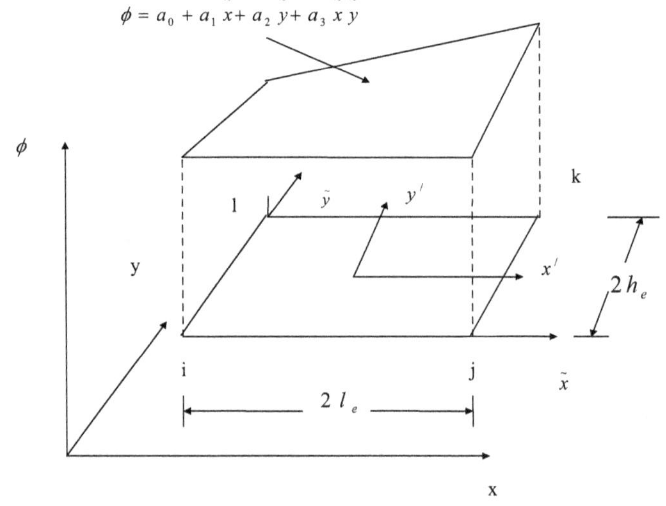

FIGURE 12.19 Bilinear rectangular element.

Bilinear Rectangular The bilinear rectangular element is a special class of quadrilateral element, in which the sides of the element are parallel to the coordinate axes. Let us consider a bilinear rectangular element of length $2l_e$ and height $2h_e$, as shown in Figure 12.19, in a local coordinate system $(\tilde{x}, \tilde{y})$ with the origin located at the left-hand corner node.

The approximation function is assumed as

$$\phi(\tilde{x}, \tilde{y}) = a_0 + a_1\tilde{x} + a_2\tilde{y} + a_3\tilde{x}\tilde{y} \tag{12.84}$$

The coefficients a_0, a_1, a_2, and a_3 are obtained by using the nodal values and coordinates

$$\phi_i = a_0 \tag{12.85a}$$

$$\phi_j = a_0 + 2l_e a_1 \tag{12.85b}$$

$$\phi_k = a_0 + (2l_e)a_1 + (2h_e)a_2 + (4l_e h_e)a_3 \tag{12.85c}$$

$$\phi_l = a_0 + (2h_e)a_2 \tag{12.85d}$$

Solving the set of equations, we obtain the coefficients as

$$a_0 = \phi_i, \quad a_1 = \frac{1}{2l_e}(\phi_j - \phi_i)$$

$$a_2 = \frac{1}{2h_e}(\phi_l - \phi_i) \tag{12.86}$$

$$a_3 = \frac{1}{4l_e h_e}(\phi_i - \phi_j + \phi_k - \phi_l)$$

Substituting Equation 12.86 into Equation 12.84, we get the approximation function for the bilinear rectangular element in the local coordinate system as

$$\phi(\tilde{x},\tilde{y}) = N_i\phi_i + N_j\phi_j + N_k\phi_k + N_l\phi_l = \sum_{I=i,j,k,l} N_I\phi_I \tag{12.87}$$

with shape functions given as

$$N_i = \left(1 - \frac{\tilde{x}}{2l_e}\right)\left(1 - \frac{\tilde{y}}{2h_e}\right), \quad N_j = \frac{\tilde{x}}{2l_e}\left(1 - \frac{\tilde{y}}{2h_e}\right)$$
$$N_k = \frac{\tilde{x}\tilde{y}}{4l_e h_e}, \quad N_l = \frac{\tilde{y}}{2h_e}\left(1 - \frac{\tilde{x}}{2l_e}\right) \tag{12.88}$$

The shape functions for bilinear rectangular element have properties like the triangular elements. Each shape function varies linearly along the adjacent sides of its node and it is zero along the other two sides opposite to its node. For example, the shape function N_i varies linearly along the adjacent sides *ij* and *il*, and zero along *jk* and *kl*.

Rectangular Element in Alternate Local Coordinate System The shape functions given by Equation 12.88 can be transformed for a local coordinate system (x', y') with its origin located at the center of the element as

$$N_i = \frac{1}{4}\left(1 - \frac{x'}{l_e}\right)\left(1 - \frac{y'}{h_e}\right) \tag{12.89a}$$

$$N_j = \frac{1}{4}\left(1 + \frac{x'}{l_e}\right)\left(1 - \frac{y'}{h_e}\right) \tag{12.89b}$$

$$N_k = \frac{1}{4}\left(1 + \frac{x'}{l_e}\right)\left(1 + \frac{y'}{h_e}\right) \tag{12.89c}$$

$$N_l = \frac{1}{4}\left(1 - \frac{x'}{l_e}\right)\left(1 + \frac{y'}{h_e}\right) \tag{12.89d}$$

Quadrilateral Element in Natural Coordinate System In the natural coordinate system for the two-dimensional quadrilateral element, the coordinates are expressed in a dimensionless form as $\xi = x' / l_e$ and $\eta = y' / h_e$, and the range of coordinate variables is given as $-1 \le \xi \le 1$ and $-1 \le \eta \le 1$, respectively, as shown in Figure 12.20.

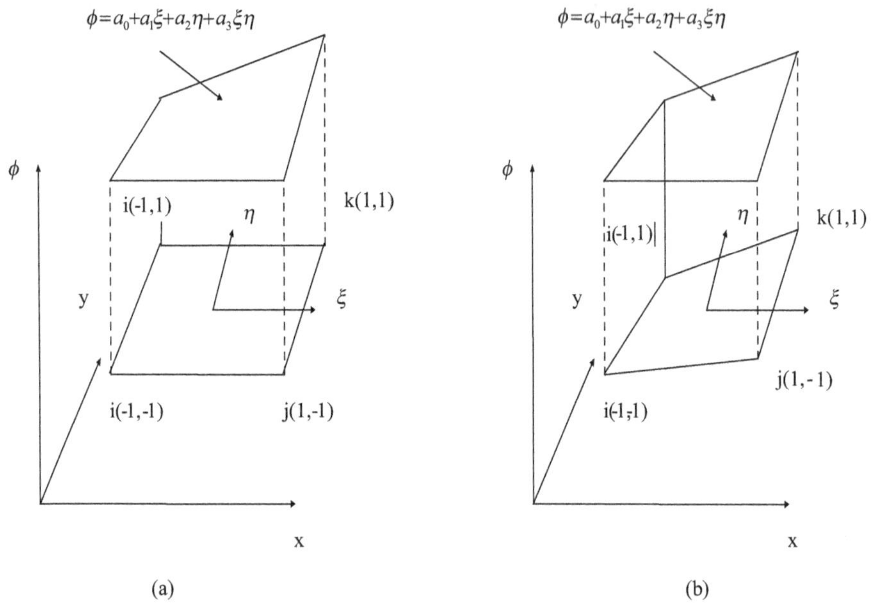

FIGURE 12.20 Quadrilateral element in natural co-ordinate system.

With the definition of these natural coordinates, the sides of the quadrilateral element are no longer required to be parallel to the coordinate axes as in the case of a rectangular element. The shape functions for the quadrilateral element in the natural coordinate system can simply be obtained by substituting the dimensionless coordinate variable in the approximation function of bilinear form (Equation 12.84) in natural coordinates, and expressed as

$$N_i = \frac{1}{4}(1-\xi)(1-\eta) \tag{12.90a}$$

$$N_j = \frac{1}{4}(1+\xi)(1-\eta) \tag{12.90b}$$

$$N_k = \frac{1}{4}(1+\xi)(1+\eta) \tag{12.90c}$$

$$N_l = \frac{1}{4}(1-\xi)(1+\eta) \tag{12.90d}$$

The geometry of the quadrilateral element with arbitrarily oriented slopes (Figure 12.20b) can be represented as

$$x = N_i x_i + N_j x_j + N_k x_k + N_l x_l = \sum_{I=i}^{N_p} N_I x_I \tag{12.91a}$$

and

$$y = N_i y_i + N_j y_j + N_k y_k + N_l y_l = \sum_{I=i}^{N_p} N_I x_I \tag{12.91b}$$

Note that the approximation function ϕ and the geometry (x, y) are expressed by the shape functions of the same order.

Using the procedure outlined for triangular element in Section 12.2.1, we can express the derivative of the shape functions in natural coordinates as

$$\begin{Bmatrix} \dfrac{\partial N_i}{\partial \xi} \\ \dfrac{\partial N_i}{\partial \eta} \end{Bmatrix} = \begin{bmatrix} \dfrac{\partial x}{\partial \xi} & \dfrac{\partial y}{\partial \xi} \\ \dfrac{\partial x}{\partial \eta} & \dfrac{\partial y}{\partial \eta} \end{bmatrix} \begin{Bmatrix} \dfrac{\partial N_i}{\partial x} \\ \dfrac{\partial N_i}{\partial y} \end{Bmatrix} = J \begin{Bmatrix} \dfrac{\partial N_i}{\partial x} \\ \dfrac{\partial N_i}{\partial y} \end{Bmatrix} \tag{12.92}$$

where the Jacobian matrix is defined as

$$J = \begin{bmatrix} \dfrac{\partial x}{\partial \xi} & \dfrac{\partial y}{\partial \xi} \\ \dfrac{\partial x}{\partial \eta} & \dfrac{\partial y}{\partial \eta} \end{bmatrix} \tag{12.93}$$

The expression of the derivatives in x- and y-coordinates is obtained by taking the inversion of Equation 12.92 as

$$\begin{Bmatrix} \dfrac{\partial N_i}{\partial x} \\ \dfrac{\partial N_i}{\partial y} \end{Bmatrix} = J^{-1} \begin{Bmatrix} \dfrac{\partial N_i}{\partial \xi} \\ \dfrac{\partial N_i}{\partial \eta} \end{Bmatrix} \tag{12.94}$$

Evaluating the inversion of the Jacobian matrix, we get

$$\begin{Bmatrix} \dfrac{\partial N_i}{\partial x} \\ \dfrac{\partial N_i}{\partial y} \end{Bmatrix} = \frac{1}{|J|} \begin{bmatrix} \dfrac{\partial y}{\partial \eta} & -\dfrac{\partial y}{\partial \xi} \\ -\dfrac{\partial x}{\partial \eta} & \dfrac{\partial x}{\partial \xi} \end{bmatrix} \begin{Bmatrix} \dfrac{\partial N_i}{\partial \xi} \\ \dfrac{\partial N_i}{\partial \eta} \end{Bmatrix} \tag{12.95}$$

or

$$\begin{Bmatrix} \dfrac{\partial N_i}{\partial x} \\ \dfrac{\partial N_i}{\partial y} \end{Bmatrix} = \frac{1}{|J|} \begin{Bmatrix} \dfrac{\partial y}{\partial \eta}\dfrac{\partial N_i}{\partial \xi} - \dfrac{\partial y}{\partial \xi}\dfrac{\partial N_i}{\partial \eta} \\ -\dfrac{\partial x}{\partial \eta}\dfrac{\partial N_i}{\partial \xi} + \dfrac{\partial x}{\partial \xi}\dfrac{\partial N_i}{\partial \eta} \end{Bmatrix} \tag{12.96}$$

where $|J|$, the determinant of the Jacobian matrix, is given as

$$|J| = \frac{\partial x}{\partial \xi}\frac{\partial y}{\partial \eta} - \frac{\partial x}{\partial \eta}\frac{\partial y}{\partial \xi} \tag{12.97}$$

Also, $|J|$ relates the differential area $dA(\xi,\eta) = d\xi d\eta$ and differential area $dA(x, y)=dx\, dy$ as

$$dA(x,y) = |J| dA(\xi,\eta)$$

The natural coordinate system helps in simplifying the integral expressions and enables us to use Gauss–Legendre quadrature formulas. For example

$$A = \iint_A dx dy = \int_{-1}^{1}\int_{-1}^{1} |J| d\xi d\eta$$

$$I = \iint_A N_i N_j dx dy = \int_{-1}^{1}\int_{-1}^{1} N_i N_j |J| d\xi d\eta$$

and

$$I = \iint_A \frac{\partial N_i}{\partial x}\frac{\partial N_j}{\partial x} dx dy = \int_{-1}^{1}\int_{-1}^{1} \frac{\partial N_i}{\partial x}\frac{\partial N_j}{\partial y} |J| d\xi d\eta$$

Note that these integrals are now in a format suitable for numerical integration by n-term Gauss–Legendre quadrature given by Equation 4.40

$$I \cong \sum_{i=1}^{n}\sum_{j=1}^{n} w_i w_j g(x_{i'}, y_{j'})$$

The number of points n of the quadrature formula will be one less than the order of the polynomial function g.

Quadratic Rectangular Element A higher-order quadratic element contains eight nodes, one at each corner and one at the mid-point of each side, as shown with a local coordinate system in Figure 12.21.

The approximation function for the eight-noded quadratic rectangular element is given by

$$\phi = a_0 + a_1 x' + a_2 y' + a_3 x'y' + a_4 x'^2 + a_5 y'^2 + a_6 x'^2 y' + a_7 x' y'^2 \tag{12.98}$$

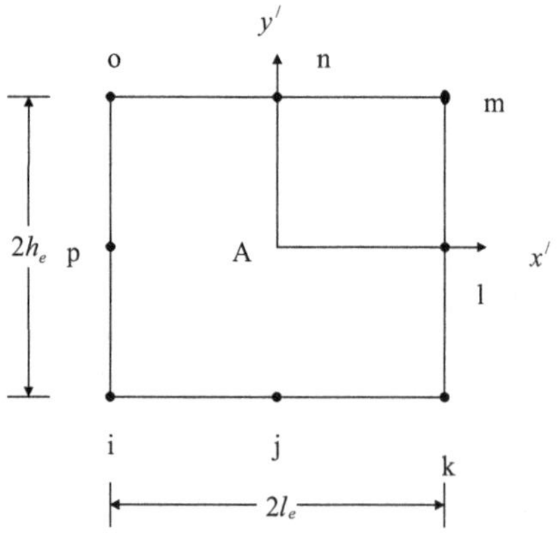

FIGURE 12.21 Quadratic rectangular element.

Substituting the nodal values and coordinates, and solving the resulting system of equations, we get

$$\phi = N_i\phi_i + N_j\phi_j + N_k\phi_k + N_l\phi_l + N_m\phi_m + N_n\phi_n + N_o\phi_o + N_p\phi_p = \sum_{I=i}^{Np} N_I\phi_I \tag{12.99}$$

where the shape functions are given as

$$N_i = -\frac{1}{4l_e^2h_e^2}(h_e - x)(l_e - y)(l_eh_e + xl_e + yh_e) \tag{12.100a}$$

$$N_j = \frac{1}{2l_eh_e^2}(h_e^2 - x^2)(l_e - y) \tag{12.100b}$$

$$N_k = \frac{1}{4l_e^2h_e^2}(h_e + x)(l_e - y)(xl_e - yh_e - l_eh_e) \tag{12.100c}$$

$$N_l = \frac{1}{2l_eh_e^2}(h_e^2 - x^2)(l_e + y) \tag{12.100d}$$

$$N_m = -\frac{1}{4l_e^2h_e^2}(h_e + x)(l_e + y)(xl_e + yh_e - l_eh_e) \tag{12.100e}$$

$$N_n = \frac{1}{2l_eh_e^2}(h_e^2 - x^2)(l_e + y) \tag{12.100f}$$

$$N_o = -\frac{1}{4l_e^2h_e^2}(h_e - x)(l_e + y)(l_eh_e + xl_e - yh_e) \tag{12.100g}$$

$$N_p = \frac{1}{2l_e^2h_e}(h_e - x)(l_e^2 - y^2) \tag{12.100h}$$

In the natural coordinate system, the approximation function is given by Equation 12.99, with shape functions given as

$$N_i = -\frac{1}{4}(1-\xi)(1-\eta)(1+\xi+\eta) \tag{12.101a}$$

$$N_j = \frac{1}{2}(1-\xi^2)(1-\eta) \tag{12.101b}$$

$$N_k = \frac{1}{4}(1+\xi)(1-\eta)(\xi-\eta-1) \tag{12.101c}$$

$$N_l = \frac{1}{2}(1+\xi)(1-\eta^2) \tag{12.101d}$$

$$N_m = \frac{1}{2}(1+\xi)(1+\eta)(\xi+\eta-1) \tag{12.101e}$$

$$N_n = \frac{1}{2}(1-\xi^2)(1+\eta) \tag{12.101f}$$

$$N_o = -\frac{1}{4}(1-\xi)(1+\eta)(1+\xi-\eta) \tag{12.101g}$$

$$N_p = \frac{1}{2}(1-\xi)(1-\eta^2) \tag{12.101h}$$

Example 12.2: Two-Dimensional Quadratic Element

Derive the shape functions for the eight-noded quadratic rectangular element in the natural coordinate system with origin at the point $A(0, 0)$ as shown in the figure.

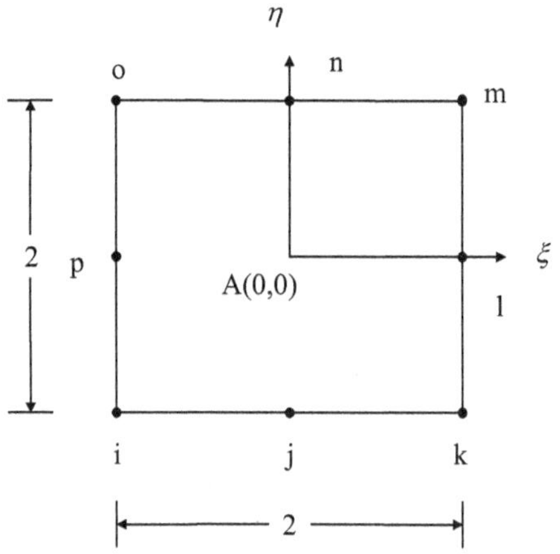

The approximate function for the eight-noded quadratic element is assumed as

$$\phi = a_0 + a_1\xi + a_2\eta + a_3\xi\eta + a_4\xi^2 + a_5\eta^2 + a_6\xi^2\eta + a_7\xi\eta^2 \tag{E.12.2.1}$$

Based on the origin located at A, the coordinates of the nodes are $i(-1, -1)$, $j(0, -1)$, $k(1,-1), l(1,0), m(1,1), n(0,1), o(-1,1), p(-1,0)$. Now, substituting the nodal values and coordinates in the approximation function given by Equation E.12.2.1, we get

$$\begin{aligned}
\phi_i &= a_0 - a_1 - a_2 + a_3 + a_4 + a_5 - a_6 - a_7 \\
\phi_j &= a_0 - a_2 + a_5 \\
\phi_k &= a_0 + a_1 - a_2 - a_3 + a_4 + a_5 - a_6 + a_7 \\
\phi_l &= a_0 + a_1 + a_4 \\
\phi_m &= a_0 + a_1 + a_2 + a_3 + a_4 + a_5 + a_6 + a_7 \\
\phi_n &= a_0 + a_2 + a_5 \\
\phi_o &= a_0 - a_1 + a_2 - a_3 + a_4 + a_5 + a_6 - a_7 \\
\phi_p &= a_0 - a_1 + a_4
\end{aligned} \tag{E.12.2.2}$$

Equation (E.12.2.2) can be expressed in matrix form as

$$\begin{bmatrix} 1 & -1 & -1 & 1 & 1 & 1 & -1 & -1 \\ 1 & -1 & 0 & 0 & 0 & 1 & 0 & 0 \\ 1 & 1 & -1 & -1 & 1 & 1 & -1 & 1 \\ 1 & 1 & 0 & 0 & 1 & 0 & 0 & 0 \\ 1 & 1 & 1 & 1 & 1 & 1 & 1 & 1 \\ 1 & 0 & 1 & 0 & 0 & 1 & 0 & 0 \\ 1 & -1 & 1 & -1 & 1 & 1 & 1 & 1 \\ 1 & -1 & 0 & 0 & 1 & 0 & 0 & 0 \end{bmatrix} \begin{Bmatrix} a_0 \\ a_1 \\ a_2 \\ a_3 \\ a_4 \\ a_5 \\ a_6 \\ a_7 \end{Bmatrix} = \begin{bmatrix} \phi_i \\ \phi_j \\ \phi_k \\ \phi_l \\ \phi_m \\ \phi_n \\ \phi_o \\ \phi_p \end{bmatrix} \tag{E.12.2.3}$$

Solving the system for unknown coefficients *a*, we get

$$\begin{aligned}
a_0 &= \tfrac{1}{2}\phi_j + \tfrac{1}{2}\phi_n + \tfrac{1}{2}\phi_p - \tfrac{1}{4}\phi_o + \tfrac{1}{2}\phi_l - \tfrac{1}{4}\phi_i - \tfrac{1}{4}\phi_m - \tfrac{1}{4}\phi_k \\
a_1 &= -\tfrac{1}{2}\phi_p + \tfrac{1}{2}\phi_l \\
a_2 &= -\tfrac{1}{2}\phi_j + \tfrac{1}{2}\phi_n \\
a_3 &= -\tfrac{1}{4}\phi_o + \tfrac{1}{4}\phi_i + \tfrac{1}{4}\phi_m - \tfrac{1}{4}\phi_k \\
a_4 &= -\tfrac{1}{2}\phi_j - \tfrac{1}{2}\phi_n + \tfrac{1}{4}\phi_o + \tfrac{1}{4}\phi_i + \tfrac{1}{4}\phi_m + \tfrac{1}{4}\phi_k \\
a_5 &= -\tfrac{1}{2}\phi_p + \tfrac{1}{4}\phi_o - \tfrac{1}{2}\phi_l + \tfrac{1}{4}\phi_i + \tfrac{1}{4}\phi_m + \tfrac{1}{4}\phi_k \\
a_6 &= \tfrac{1}{2}\phi_j - \tfrac{1}{2}\phi_n + \tfrac{1}{4}\phi_o - \tfrac{1}{4}\phi_i + \tfrac{1}{4}\phi_m - \tfrac{1}{4}\phi_k \\
a_7 &= \tfrac{1}{2}\phi_p - \tfrac{1}{4}\phi_o - \tfrac{1}{2}\phi_l - \tfrac{1}{4}\phi_i + \tfrac{1}{4}\phi_m + \tfrac{1}{4}\phi_k
\end{aligned} \tag{E.12.2.4}$$

Substituting *a*'s into Equation (E.12.2.1) and rearranging, we get

$$\phi = \phi_i N_i + \phi_j N_j + \phi_k N_k + \phi_l N_l + \phi_m N_m + \phi_n N_n + \phi_o N_o + \phi_p N_p \tag{E.12.2.5}$$

where the shape functions are expressed as

$$\begin{aligned}
N_i &= \frac{1}{4}\eta\xi + \frac{1}{4}\xi^2 + \frac{1}{4}\eta^2 - \frac{1}{4}\xi^2\eta - \frac{1}{4} - \frac{1}{4}\xi\eta^2 = -\frac{(1-\xi)(1-\eta)(1+\xi+\eta)}{4} \\
N_j &= \frac{1}{2} - \frac{1}{2}\xi^2 - \frac{1}{2}\eta + \frac{1}{2}\xi^2\eta = \frac{(1-\xi^2)(1-\eta)}{2} \\
N_k &= \frac{1}{4}\xi\eta^2 + \frac{1}{4}\eta^2 + \frac{1}{4}\xi^2 - \frac{1}{4}\xi\eta - \frac{1}{4} - \frac{1}{4}\xi^2\eta = \frac{(1+\xi)(1-\eta)(\xi-\eta-1)}{4} \\
N_l &= -\frac{1}{2}\xi\eta^2 + \frac{1}{2}\xi - \frac{1}{2}\eta^2 + \frac{1}{2} = \frac{(1+\xi)(1-\eta^2)}{2} \\
N_m &= -\frac{1}{4} + \frac{1}{4}\xi\eta^2 + \frac{1}{4}\xi\eta + \frac{1}{4}\eta^2 + \frac{1}{4}\xi^2 + \frac{1}{4}\xi^2\eta = \frac{(1+\xi)(1+\eta)(\xi+\eta-1)}{4} \\
N_n &= \frac{1}{2} - \frac{1}{2}\xi^2 + \frac{1}{2}\eta - \frac{1}{2}\xi^2\eta = \frac{(1-\xi^2)(1+\eta)}{2} \\
N_o &= -\frac{1}{4}\xi\eta^2 - \frac{1}{4} + \frac{1}{4}\xi^2 - \frac{1}{4}\xi\eta + \frac{1}{4}\eta^2 + \frac{1}{4}\xi^2\eta = \frac{(1-\xi)(1+\eta)(1+\xi-\eta)}{4} \\
N_p &= -\frac{1}{2}\xi + \frac{1}{2}\xi\eta^2 + \frac{1}{2} - \frac{1}{2}\eta^2 = \frac{(1-\xi)(1-\eta^2)}{2}
\end{aligned} \tag{E.12.2.6}$$

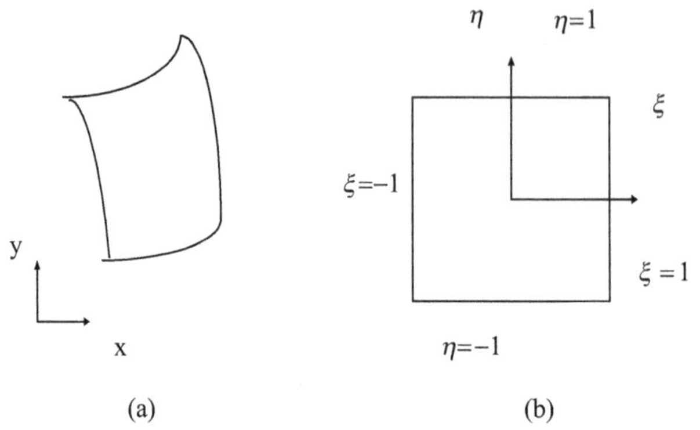

FIGURE 12.22 Two-dimension isoparametric element.

Isoparametric Element The straight-sided triangular and quadrilateral elements are not quite suitable or accurate for approximating shapes with curved surfaces. This inaccuracy is reduced by using elements with arbitrarily oriented curved sides. One such category of these elements with curved surfaces that is increasingly being used in many commercial codes is called the isoparametric element. Basically, in such elements, a procedure involving transformation of coordinates or mapping of straight regions into a curved region is used. The procedure is like that used in the case of quadrilateral elements in natural coordinate systems except for getting an arbitrarily oriented curved surface rather than an arbitrarily oriented slope. Another important characteristic of the isoparametric element is that the same shape functions of the same order are used for both the approximation function ϕ and geometric variables (x, y). The main idea is to transform or map a simple geometric shape with straight edges in some natural coordinate system into distorted shapes with curved edges in the global coordinate system, as depicted in Figure 12.22 for the case of a quadric (eight-nodes) quadrilateral element.

The quadratic variation of the approximation function is given in coordinate system as

$$\phi(\xi,\eta) = \sum_{I=i}^{N_p} N_I(\xi,\eta)\phi_I \tag{12.102}$$

The eight nodes in the natural coordinate system can be mapped into corresponding nodes in the global coordinate plane by the transformation relation

$$x = \sum_{I=i}^{N_p} N_I(\xi,\eta)x_I \tag{12.103a}$$

and

$$y = \sum_{I=i}^{N_p} N_I(\xi,\eta)y_I \tag{12.103b}$$

Equation 12.103 results in a curve-sided quadrilateral element.

A similar procedure can be used for one-dimensional and three-dimensional elements and shapes.

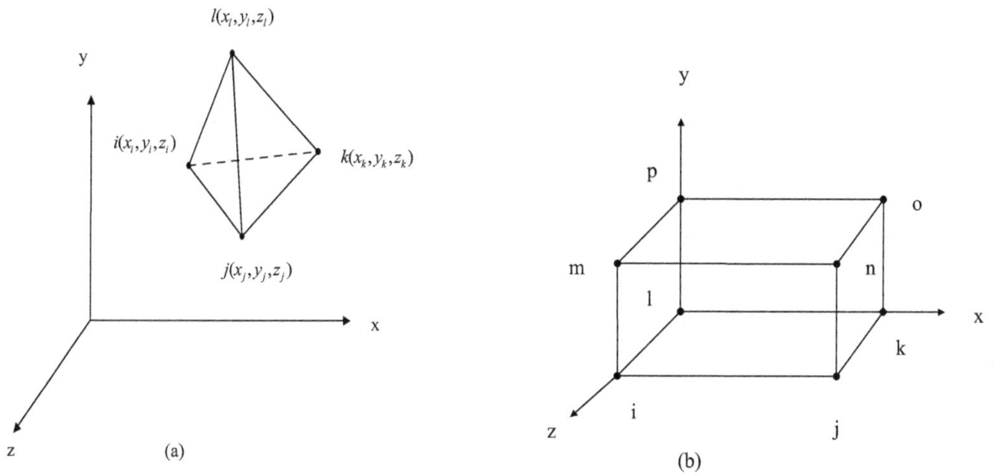

FIGURE 12.23 Three-dimensional elements: (a) tetrahedron element and (b) hexahedron element.

12.3 THREE-DIMENSIONAL ELEMENT

The most widely used three-dimensional elements are the **tetrahedron** and **hexahedron** or **brick** element as shown in Figure 12.23.

12.3.1 Three-Dimensional Tetrahedron Element

The simplest tetrahedron element is the linear flat-faced tetrahedron with four nodes, one at each vertex as shown in Figure 12.23a. The approximation function is assumed as

$$\phi(x,y,z) = a_0 + a_1 x + a_2 y + a_3 z \tag{12.104}$$

Substituting the nodal values ϕ_i, ϕ_j, ϕ_k, ϕ_l, and their nodal values in Equation 12.104, we get

$$\phi_i = a_0 + a_1 x_i + a_2 y_i + a_3 z_i \tag{12.105a}$$

$$\phi_j = a_0 + a_1 x_j + a_2 y_j + a_3 z_j \tag{12.105b}$$

$$\phi_k = a_0 + a_1 x_k + a_2 y_k + a_3 z_k \tag{12.105c}$$

$$\phi_l = a_0 + a_1 x_l + a_2 y_l + a_3 z_l \tag{12.105d}$$

Writing the equations in matrix notation, we have

$$\begin{bmatrix} 1 & x_i & y_i & z_i \\ 1 & x_j & y_j & z_j \\ 1 & x_k & y_k & z_k \\ 1 & x_l & y_l & z_l \end{bmatrix} \begin{Bmatrix} a_0 \\ a_1 \\ a_2 \\ a_3 \end{Bmatrix} = \begin{Bmatrix} \phi_i \\ \phi_j \\ \phi_k \\ \phi_l \end{Bmatrix} \tag{12.106}$$

Solving the system of equations for the coefficients a as

$$a_0 = \frac{\begin{vmatrix} \phi_i & x_i & y_i & z_i \\ \phi_j & x_j & y_j & z_j \\ \phi_k & x_k & y_k & z_k \\ \phi_l & x_l & y_l & z_l \end{vmatrix}}{\begin{vmatrix} 1 & x_i & y_i & z_i \\ 1 & x_j & y_j & z_j \\ 1 & x_k & y_k & z_k \\ 1 & x_l & y_l & z_l \end{vmatrix}} = \frac{1}{6V}\left(\alpha_i\phi_i + \alpha_j\phi_j + \alpha_k\phi_k + \alpha_l\phi_l\right) \tag{12.107a}$$

where

$$\alpha_i = \begin{vmatrix} x_j & y_j & z_j \\ x_k & y_k & z_k \\ x_l & y_l & z_l \end{vmatrix}, \quad \alpha_j = -\begin{vmatrix} x_i & y_i & z_i \\ x_k & y_k & z_k \\ x_l & y_l & z_l \end{vmatrix},$$

$$\alpha_k = \begin{vmatrix} x_i & y_i & z_i \\ x_j & y_j & z_j \\ x_l & y_l & z_l \end{vmatrix}, \quad \alpha_l = -\begin{vmatrix} x_i & y_i & z_i \\ x_j & y_j & z_j \\ x_k & y_k & z_k \end{vmatrix} \tag{12.107b}$$

$$a_1 \frac{\begin{vmatrix} 1 & \phi_i & y_i & z_i \\ 1 & \phi_j & y_j & z_j \\ 1 & \phi_k & y_k & z_k \\ 1 & \phi_l & y_l & z_l \end{vmatrix}}{\begin{vmatrix} 1 & x_i & y_i & z_i \\ 1 & x_j & y_j & z_j \\ 1 & x_k & y_k & z_k \\ 1 & x_l & y_l & z_l \end{vmatrix}} = \frac{1}{6V}\left(\beta_i\phi_i + \beta_j\phi_j + \beta_k\phi_k + \beta_l\phi_l\right) \tag{12.108a}$$

where

$$\beta_i = \begin{vmatrix} 1 & y_j & z_j \\ 1 & y_k & z_k \\ 1 & y_l & z_l \end{vmatrix}, \quad \beta_j = -\begin{vmatrix} 1 & y_i & z_i \\ 1 & y_k & z_k \\ 1 & y_l & z_l \end{vmatrix},$$

$$\beta_k = \begin{vmatrix} 1 & y_i & z_i \\ 1 & y_j & z_j \\ 1 & y_l & z_l \end{vmatrix}, \quad \beta_l = -\begin{vmatrix} 1 & y_i & z_i \\ 1 & y_j & z_j \\ 1 & y_k & z_k \end{vmatrix} \tag{12.108b}$$

$$a_2 = \frac{\begin{vmatrix} 1 & x_i & \phi_i & z_i \\ 1 & x_j & \phi_j & z_j \\ 1 & x_k & \phi_k & z_k \\ 1 & x_l & \phi_l & z_l \end{vmatrix}}{\begin{vmatrix} 1 & x_i & y_i & z_i \\ 1 & x_j & y_j & z_j \\ 1 & x_k & y_k & z_k \\ 1 & x_l & y_l & z_l \end{vmatrix}} = \frac{1}{6V}\left(\gamma_i\phi_i + \gamma_j\phi_j + \gamma_k\phi_k + \gamma_l\phi_l\right) \tag{12.109a}$$

where

$$\gamma_i = \begin{vmatrix} 1 & x_j & z_j \\ 1 & x_k & z_k \\ 1 & x_l & z_l \end{vmatrix}, \quad \gamma_j = -\begin{vmatrix} 1 & x_i & z_i \\ 1 & x_k & z_k \\ 1 & x_l & z_l \end{vmatrix},$$

$$\gamma_k = \begin{vmatrix} 1 & x_i & z_i \\ 1 & x_j & z_j \\ 1 & x_l & z_l \end{vmatrix}, \quad \gamma_l = -\begin{vmatrix} 1 & x_i & z_i \\ 1 & x_j & z_j \\ 1 & x_k & z_k \end{vmatrix} \tag{12.109b}$$

$$a_3 = \frac{\begin{vmatrix} 1 & x_i & y_i & \phi_i \\ 1 & x_j & y_j & \phi_j \\ 1 & x_k & y_k & \phi_k \\ 1 & x_l & y_l & \phi_l \end{vmatrix}}{\begin{vmatrix} 1 & x_i & y_i & z_i \\ 1 & x_j & y_j & z_j \\ 1 & x_k & y_k & z_k \\ 1 & x_l & y_l & z_l \end{vmatrix}} = \frac{1}{6V}\left(\delta_i\phi_i + \delta_j\phi_j + \delta_k\phi_k + \delta_l\phi_l\right) \tag{12.110a}$$

where

$$\delta_i = \begin{vmatrix} 1 & x_j & y_j \\ 1 & x_k & y_k \\ 1 & x_l & y_l \end{vmatrix}, \quad \delta_j = -\begin{vmatrix} 1 & x_i & y_i \\ 1 & x_k & y_k \\ 1 & x_l & y_l \end{vmatrix},$$

$$\delta_k = \begin{vmatrix} 1 & x_i & y_i \\ 1 & x_j & y_j \\ 1 & x_l & y_l \end{vmatrix}, \quad \delta_l = -\begin{vmatrix} 1 & x_i & y_i \\ 1 & x_j & y_j \\ 1 & x_k & y_k \end{vmatrix} \tag{12.110b}$$

and V is the volume of the tetrahedron given by

$$V = \frac{1}{6}\begin{vmatrix} 1 & x_i & y_i & z_i \\ 1 & x_j & y_j & z_j \\ 1 & x_k & y_k & z_k \\ 1 & x_l & y_l & z_l \end{vmatrix} \tag{12.111}$$

Substituting Equations 12.107–12.110 into Equation 12.104, we obtain the approximation function as

$$\phi(x,y,z) = N_i\phi_i + N_j\phi_j + N_k\phi_k + N_l\phi_l = [N]^{\backslash T}\{\phi\} \tag{12.112}$$

where

$$[N] = \begin{bmatrix} N_i(x,y,z) \\ N_j(x,y,z) \\ N_k(x,y,z) \\ N_l(x,y,z) \end{bmatrix} \tag{12.113}$$

$$N_i(x,y,z) = \frac{1}{6V}(\alpha_i + \beta_i x + \gamma_i y + \delta_i z) \tag{12.114a}$$

$$N_j(x,y,z) = \frac{1}{6V}(\alpha_j + \beta_j x + \gamma_j y + \delta_j z) \tag{12.114b}$$

$$N_k(x,y,z) = \frac{1}{6V}(\alpha_k + \beta_k x + \gamma_k y + \delta_k z) \tag{12.114c}$$

$$N_l(x,y,z) = \frac{1}{6V}(\alpha_l + \beta_l x + \gamma_l y + \delta_l z) \tag{12.114d}$$

Volume Coordinate System One of the natural coordinate systems for the three-dimensional tetrahedron element is the volume coordinate system, which is developed in the same way as in one-dimensional line coordinate systems and two-dimensional area coordinate systems. In this case, the volume coordinates are the distance ratios L_i, L_j, L_k, and L_l, which are defined by considering a point p normal to each side, as shown in Figure 12.24, and taking the ratio of volumes as

$$L_i = \frac{V_i}{V}, \quad L_j = \frac{V_j}{V}, \quad L_k = \frac{V_k}{V}, \quad \text{and} \quad L_l = \frac{V_l}{V} \tag{12.115}$$

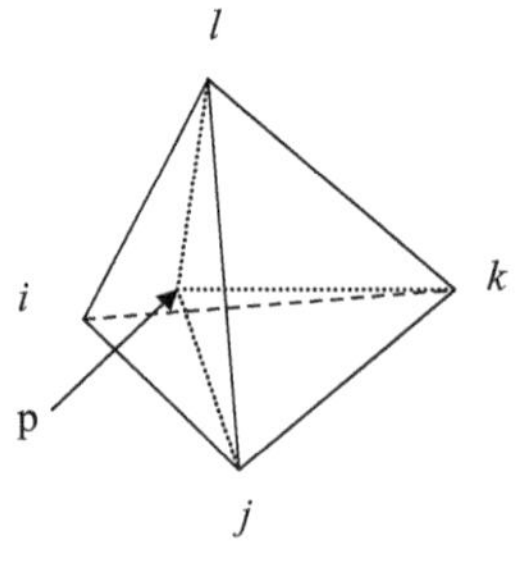

FIGURE 12.24 Volume co-ordinate system for tetrahedron element.

where the volume $V_I\left(I=i,j,k,l\right)$ is the volume of the tetrahedron formed by the point p and the vertices other than the vertex I.

Based on this coordinate system, the shape functions are defined as

$$N_i = L_i,\quad N_j = L_j,\quad N_k = L_k \tag{12.116}$$

Additionally, the volume coordinates are related by

$$L_i + L_j + L_k + L_l = 1 \tag{12.117}$$

and the relationship between the volume or natural coordinates and Cartesian coordinates is given by

$$x = L_i x_i + L_j x_j + L_k x_k + L_l x_l \tag{12.118a}$$

$$y = L_i y_i + L_j y_j + L_k y_k + L_l y_l \tag{12.118b}$$

$$z = L_i z_i + L_j z_j + L_k z_k + L_l z_l \tag{12.118c}$$

The integral formula that can be used along with the three-dimensional integrals in a natural coordinate system is

$$I = \int_V L_i^a L_j^b L_k^c L_l^d \,\mathrm{d}V = \frac{a!b!c!d!}{(a+b+c+d+3)!}6V \tag{12.119}$$

For example, the integral expression of the form

$$I = \int_V N_i N_j N_k \,\mathrm{d}V \tag{12.120}$$

can be evaluated using Equations 12.116 and 12.119 as

$$\begin{aligned} I &= \int_V N_i N_j N_k \,\mathrm{d}V = \int_V L_i^1 L_j^1 L_k^1 L_l^0 \,\mathrm{d}V \\ &= \frac{1!1!1!0!}{(1+1+1+0+3)!}6V = \frac{V}{120} \end{aligned} \tag{12.121}$$

For a more detailed discussion of higher-order tetrahedral elements, refer to the books by Zienkiewicz and Taylor (1986) and Reddy and Gartling (1994).

12.3.2 Three-Dimensional Hexahedron Element

A simple and more accurate three-dimensional element, which is easier to manipulate, is the **hexahedron** or **brick element**. The simplest of the hexahedron elements is the trilinear eight-noded hexahedron element as shown in Figure 12.23b. The approximation function is assumed to be

$$\phi(x,y,z) = a_0 + a_1 x + a_2 y + a_3 z + a_4 xy + a_5 xz + a_6 yz + a_7 xyz \tag{12.122}$$

The a-coefficients are determined by substituting the nodal values and nodal coordinates in Equation 12.122 and by solving the system of equation

$$\begin{bmatrix} 1 & x_i & y_i & z_i & x_iy_i & x_iz_i & y_iz_i & x_iy_iz_i \\ 1 & x_j & y_j & z_j & x_iy_i & x_iz_i & y_iz_i & x_iy_iz_i \\ 1 & x_k & y_k & z_k & x_ky_k & x_kz_k & y_kz_k & x_ky_kz_k \\ 1 & x_l & y_l & z_l & x_ly_l & x_lz_l & y_lz_l & x_ly_lz_l \\ 1 & x_m & y_m & z_m & x_my_m & x_mz_m & y_mz_m & x_my_mz_m \\ 1 & x_n & y_n & z_n & x_ny_n & x_nz_n & y_nz_n & x_ny_nz_n \\ 1 & x_o & y_o & z_o & x_oy_o & x_oz_o & y_oz_o & x_oy_oz_o \\ 1 & x_p & y_p & z_p & x_py_p & x_pz_p & y_pz_p & x_py_pz_p \end{bmatrix} \begin{Bmatrix} a_0 \\ a_1 \\ a_2 \\ a_3 \\ a_4 \\ a_5 \\ a_6 \\ a_7 \end{Bmatrix} = \begin{Bmatrix} \phi_i \\ \phi_j \\ \phi_k \\ \phi_l \\ \phi_m \\ \phi_n \\ \phi_o \\ \phi_p \end{Bmatrix} \quad (12.123a)$$

or

$$X\{a\} = \{\phi\} \quad (12.123b)$$

Since the matrix dimension is large, the use of Cramer's rule leads to very tedious algebraic manipulation. So, the a-coefficients are determined more conveniently by using the inverse of the coefficient matrix as

$$\{a\} = X^{-1}\{\phi\} \quad (12.124)$$

Substituting Equation 12.124 into Equation 12.122, we obtain the approximation function as

$$\phi = \begin{bmatrix} 1 & x & y & z & xy & xz & yz & xyz \end{bmatrix} X^{-1}\{\phi\} \quad (12.125)$$

The shape functions for the hexahedron element can be deduced from Equation 12.125 as

$$[N] = \begin{bmatrix} 1 & x & y & z & xy & xz & yz & xyz \end{bmatrix} X^{-1} \quad (12.126)$$

Natural Coordinate System for Hexahedron Element The natural coordinate system for the three-dimensional hexahedron element is defined in terms of the dimensionless form of the local coordinate system with the origin located at the center of the hexahedron (Figure 12.25) as

$$\xi = \frac{x}{l_e}, \quad \eta = \frac{y}{h_e}, \quad \text{and} \quad \zeta = \frac{z}{w_e} \quad (12.127)$$

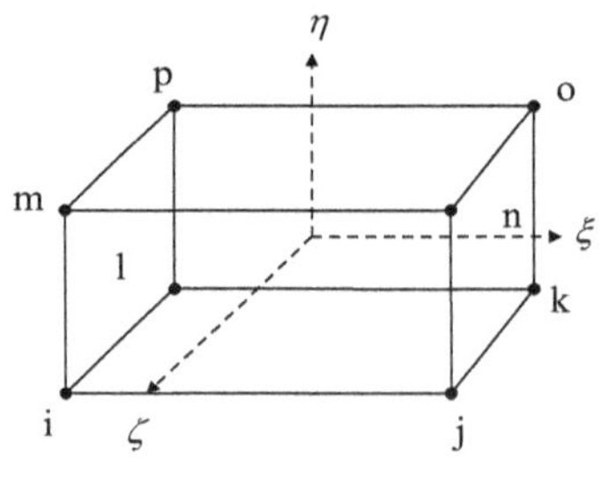

FIGURE 12.25 Hexahedron element in natural co-ordinate.

With this definition of the natural coordinate system, the shape functions are defined as

$$N_i = \frac{1}{8}(1-\xi)(1-\eta)(1-\zeta), \quad N_j = \frac{1}{8}(1+\xi)(1-\eta)(1-\zeta) \tag{12.128a}$$

$$N_k = \frac{1}{8}(1+\xi)(1+\eta)(1-\zeta), \quad N_l = \frac{1}{8}(1-\xi)(1+\eta)(1-\zeta) \tag{12.128b}$$

$$N_m = \frac{1}{8}(1-\xi)(1-\eta)(1+\zeta), \quad N_n = \frac{1}{8}(1+\xi)(1-\eta)(1+\zeta) \tag{12.128c}$$

$$N_o = \frac{1}{8}(1+\xi)(1+\eta)(1+\zeta), \quad N_p = \frac{1}{8}(1-\xi)(1+\eta)(1+\zeta) \tag{12.128d}$$

The coordinate transformation relationship between the Cartesian and the natural coordinate systems

$$x = \sum x_I N_I(\xi,\eta,\zeta), \quad y = \sum y_I N_I(\xi,\eta,\zeta), \quad \text{and} \quad z = \sum z_I N_I(\xi,\eta,\zeta) \tag{12.129}$$

Additionally, by using the procedure outlined in this section, we can express the derivative of a shape function in the natural coordinate system by following the procedure outlined for two-dimensional elements as

$$\begin{Bmatrix} \dfrac{\partial N_I}{\partial x} \\ \dfrac{\partial N_I}{\partial y} \\ \dfrac{\partial N_I}{\partial z} \end{Bmatrix} = J^{-1} \begin{Bmatrix} \dfrac{\partial N_I}{\partial \xi} \\ \dfrac{\partial N_I}{\partial \eta} \\ \dfrac{\partial N_I}{\partial \zeta} \end{Bmatrix} \tag{12.130}$$

where the **Jacobian** matrix J is given as

$$J = \begin{bmatrix} \dfrac{\partial x}{\partial \xi} & \dfrac{\partial y}{\partial \xi} & \dfrac{\partial z}{\partial \xi} \\ \dfrac{\partial x}{\partial \eta} & \dfrac{\partial y}{\partial \eta} & \dfrac{\partial z}{\partial \eta} \\ \dfrac{\partial x}{\partial \zeta} & \dfrac{\partial y}{\partial \psi} & \dfrac{\partial z}{\partial \zeta} \end{bmatrix} \tag{12.131}$$

The differential volume can be transformed into a natural coordinate system using

$$dx\,dy\,dz = |J|\,d\xi\,d\eta\,d\zeta \tag{12.132}$$

For more discussions on higher-order **hexahedron** or **brick elements**, refer to books by Zienkiewicz and Taylor (1986) and Reddy and Gartling (1994).

PROBLEMS

12.1 Evaluate the following integrals for a linear line element:

a. $\int_{x_i}^{x_j} N_j \,\mathrm{d}x$, (b) $\int_{x_i}^{x_j} \frac{\mathrm{d}N_i}{\mathrm{d}x}\frac{\mathrm{d}N_j}{\mathrm{d}x}\mathrm{d}x$, and (c) $\int_{x_i}^{x_j} N_i^2 \,\mathrm{d}x$

12.2 Evaluate the following integrals using the quadratic line element with shape functions given by Equation 12.42

a. $\int_{x_i}^{x_j} N_i N_j \,\mathrm{d}x$ and (b) $\int_{x_i}^{x_j} \frac{\mathrm{d}N_i}{\mathrm{d}x}\frac{\mathrm{d}N_j}{\mathrm{d}x}\mathrm{d}x$

12.3 Consider a local coordinate system for a triangular element as shown. The origin is at the node *i* and the *x*-axis passes though the node *j*. Use the coordinates of the nodes as (x_i, y_i), (x_j, y_j), and (x_k, y_k).

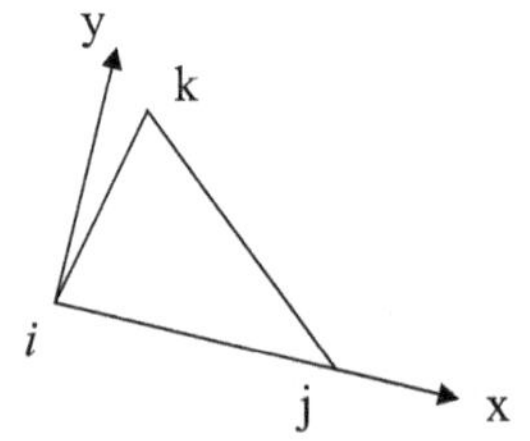

Consider the first-order polynomial of the form $\phi = a_0 + a_1 x + a_2 y$, and determine the shape functions for the linear triangular element.

12.4 Consider a linear triangular element with global coordinates of the three nodes given as (0.125, 0), (0.25, 0), and (0.25, 0.15). Determine the area of the triangle and the expressions for the shape functions. If the nodal values for this element are estimated as $\phi_i = 0.002$, $\phi_j = 0.0025$, and $\phi_j = 0.0031$, then determine the value of ϕ at the point (0.22, 0.125).

12.5 Evaluate the following integrals using the natural coordinate system for the linear triangular element: (a) $\int_A N_k^2 \,\mathrm{d}A$ and (b) $\int_A N_j N_k \,\mathrm{d}A$.

12.6 Use the local coordinate system in a bilinear rectangular element with shape functions given by Equation 12.88 to evaluate the integral

$$I = \int_A N_i N_j \,\mathrm{d}A$$

12.7 Consider a quadrilateral element shown in the figure with nodal coordinates given as (1, 1), (4, −1), (3.7, 2.8), and (0.8, 2.5).

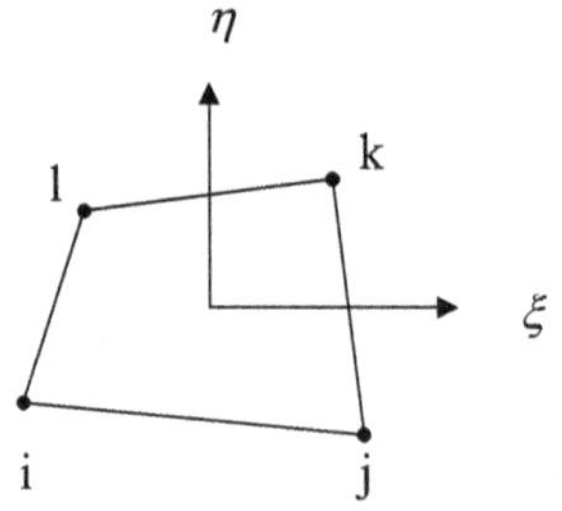

Use the natural coordinate system with shape function given by Equation 12.90 to determine the Jacobian matrix J, the inverse of the Jacobian matrix, the transformation coordinates given by Equation 12.91, and $\partial N_i/\partial x$ and $\partial N_i/\partial y$.

12.8 Use the natural coordinate system relations derived in Problem 12.7 to evaluate the following integral using Gauss–Legendre quadrature formula:

$$I = \int_A \frac{\partial N_i}{\partial x}\frac{\partial N_j}{\partial x}\mathrm{d}A = \int_{-1}^{1}\int_{-1}^{1}\frac{\partial N_i}{\partial x}\frac{\partial N_j}{\partial x}|J|\mathrm{d}\xi\mathrm{d}\eta$$

12.9 Evaluate the following integral numerically using the quadratic line element given by Equation 12.45 and three-point Gauss–Legendre quadrature:

$$I = \int_{-1}^{1}[N]^{\mathrm{T}}[N]\mathrm{d}\xi$$

13 Finite Element Method
One-Dimensional Steady State Problems

In this chapter, we will consider formulation and application of finite element method using a general one-dimensional steady-state equation containing diffusion, source, and surface convection terms, and considering different types of boundary conditions. This will be followed by the presentation of number of classical example problems. As we have mentioned in Chapter 10 that there are usually two most popular approaches in the development and formulation of an integral form of the governing equation and, subsequently, the formation of the element equation in the finite element method. These are (a) **Galerkin-based weighted residual approach** and (b) **variational approach**. So, we will describe and discuss both approaches for forming the integral or weak form while describing the finite element formulation.

13.1 FINITE ELEMENT FORMULATION USING GALERKIN METHOD

Let us consider a general one-dimensional steady-state transport process in a one-dimensional media as shown in Figure 13.1 below:

The general mathematical statement of this transport process is given as:

$$\frac{\mathrm{d}}{\mathrm{d}x}\left(\Gamma_x \frac{\mathrm{d}\phi}{\mathrm{d}x}\right) + S = 0 \tag{13.1}$$

Boundary Conditions

1. $x = 0, \phi(0) = \phi_l$ (13.2a)

2. $x = L, \phi(L) = \phi_r$ (13.2a)

This equation contains a diffusion term and a source term, and the boundary conditions are of the first kind, i.e., of constant surface value. Let us demonstrate the application of finite element

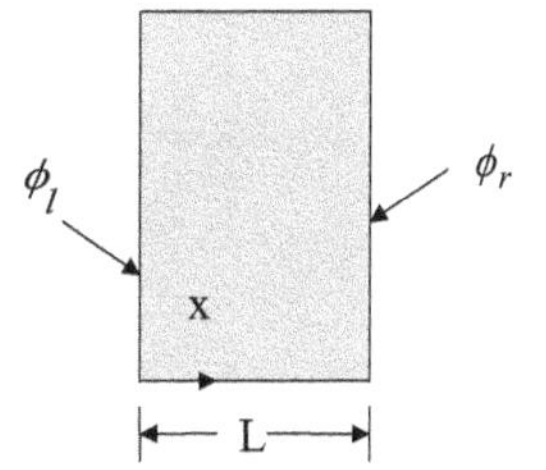

FIGURE 13.1 Steady-state transport process in a plane slab.

method to this problem through detail descriptions of several steps that constitute the finite element method.

Step 1 Discretization of the Solution Domain

The solution domain is discretized into a mesh size distribution using one-dimensional line elements as shown in Figure 13.2 below.

Let us consider the elements as linear two-point line elements with a typical element shown in Figure 13.3. The approximate solution function, $\phi \approx \phi^*$, over this element is assumed as

$$\phi^* = N_i\phi_i + N_j\phi_j = [N]\{\phi^{(e)}\} \tag{13.3a}$$

where N_i and N_j are the interpolation or shape functions given as

$$N_i = \frac{x_j - x}{x_j - x_i}, N_j = \frac{x - x_i}{x_j - x_i} \tag{13.3b}$$

Step 2 Formation Integral Statement of the Problem

Let us demonstrate the formation of the integral statement of the problem using **Galerkin's-based Weighted Residual Method**. As we substitute the approximate solution given by Equation 13.3 into the governing differential equation, Equation 13.1, a residual term will be obtained as

$$R^{(e)} = \frac{d}{dx}\left(\Gamma_x \frac{d\phi^*}{dx}\right) + S \tag{13.4}$$

The method of weighted residual involves finding a minimum for the residual over the solution domain as follows:

$$\int_{x_i}^{x_j} W_I R^{(e)} dx \tag{13.5}$$

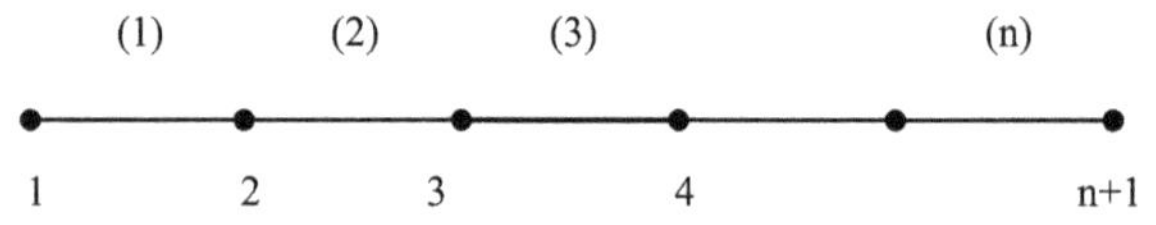

FIGURE 13.2 Mesh size distributions.

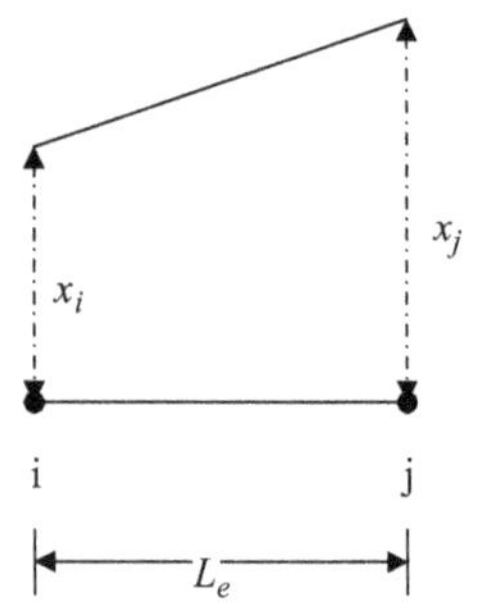

FIGURE 13.3 A typical two-point line element.

where

W_I = weighting factors

I = index for number of weighting factor or number of nodal points in an element such as $i, j \ldots N_n$

N_n = number of weighting factors or number of nodal points in element

As we have mentioned before that in the **Galerkin's-based weighted residual method**, the **interpolation** or **shape functions** are used as the weighting factors. So, in the case with two-point line element, the weighting factors are N_i and N_j. Using these weighting factors, Equation 13.5 leads to

For $I=i$

$$R_i^{(e)} = \int_{x_i}^{x_j} N_i(x)\left(\frac{\mathrm{d}}{\mathrm{d}x}\left(\Gamma_x \frac{\mathrm{d}\phi^*}{\mathrm{d}x}\right) + S\right)\mathrm{d}x = 0 \tag{13.6a}$$

and

For $I=j$

$$R_j^{(e)} = \int_{x_i}^{x_j} N_j(x)\left(\frac{\mathrm{d}}{\mathrm{d}x}\left(\Gamma_x \frac{\mathrm{d}\phi^*}{\mathrm{d}x}\right) + S\right)\mathrm{d}x = 0 \tag{13.6b}$$

Equations 13.6a and 13.6b can also be written in a compact form as

$$\left\{R^{(e)}\right\} = \int_{x_i}^{x_j} [N]^T\left(\frac{\mathrm{d}}{\mathrm{d}x}\left(\Gamma_x \frac{\mathrm{d}\phi^*}{\mathrm{d}x}\right) + S\right)\mathrm{d}x = 0 \tag{13.6c}$$

Let us now evaluate the first terms of Equations 13.6a and 13.6b using integration-by-parts as follows:

$$\int_{x_i}^{x_j} N_i(x)\left(\frac{\mathrm{d}}{\mathrm{d}x}\left(\Gamma_x \frac{\mathrm{d}\phi^*}{\mathrm{d}x}\right)\right)\mathrm{d}x = \left(N_i(x)\Gamma_x \frac{\mathrm{d}\phi^*}{\mathrm{d}x}\right)\Bigg|_{x_i}^{x_j} - \int_{x_i}^{x_j} \Gamma_x \frac{\mathrm{d}N_i}{\mathrm{d}x}\frac{\mathrm{d}\phi^*}{\mathrm{d}x}\mathrm{d}x$$

$$= N_i(x_j)\left(\Gamma_x \frac{\mathrm{d}\phi^*}{\mathrm{d}x}\right)\Bigg|_{x_j} - N_i(x_i)\left(\Gamma_x \frac{\mathrm{d}\phi^*}{\mathrm{d}x}\right)\Bigg|_{x_i} - \int_{x_i}^{x_j} \Gamma_x \frac{\mathrm{d}N_i}{\mathrm{d}x}\frac{\mathrm{d}\phi^*}{\mathrm{d}x} \tag{13.7a}$$

and

$$\int_{x_i}^{x_j} N_j(x)\left(\frac{\mathrm{d}}{\mathrm{d}x}\left(\Gamma_x \frac{\mathrm{d}\phi^*}{\mathrm{d}x}\right)\right)\mathrm{d}x = \left(N_j(x)\Gamma_x \frac{\mathrm{d}\phi^*}{\mathrm{d}x}\right)\Bigg|_{x_i}^{x_j} - \int_{x_i}^{x_j} \Gamma_x \frac{\mathrm{d}N_j}{\mathrm{d}x}\frac{\mathrm{d}\phi^*}{\mathrm{d}x}\mathrm{d}x$$

$$= N_j(x_j)\left(\Gamma_x \frac{\mathrm{d}\phi^*}{\mathrm{d}x}\right)\Bigg|_{x_j} - N_j(x_i)\left(\Gamma_x \frac{\mathrm{d}\phi^*}{\mathrm{d}x}\right)\Bigg|_{x_i} - \int_{x_i}^{x_j} \Gamma_x \frac{\mathrm{d}N_j}{\mathrm{d}x}\frac{\mathrm{d}\phi^*}{\mathrm{d}x} \tag{13.7b}$$

Note that while evaluating integrals by integration-by-parts, we should select the two terms of the integration in such a way that the resulting integral will be easier to evaluate than the original one.

If we now use the properties of the interpolation or shape functions, i.e., for a two-points line element, $N_i(x_i)=1$, $N_i(x_j)=0$ and $N_j(x_i)=0$, $N_j(x_j)=1$, then Equations 13.7a and 13.7b transform into

$$\int_{x_i}^{x_j} N_i(x)\left(\frac{\mathrm{d}}{\mathrm{d}x}\left(\Gamma_x \frac{\mathrm{d}\phi^*}{\mathrm{d}x}\right)\right)\mathrm{d}x = -\left(\Gamma_x \frac{\mathrm{d}\phi^*}{\mathrm{d}x}\right)\Bigg|_{x_i} - \int_{x_i}^{x_j} \Gamma_x \frac{\mathrm{d}N_i}{\mathrm{d}x}\frac{\mathrm{d}\phi^*}{\mathrm{d}x}\mathrm{d}x \tag{13.8a}$$

And

$$\int_{x_i}^{x_j} N_j(x)\left(\frac{\mathrm{d}}{\mathrm{d}x}\left(\Gamma_x \frac{\mathrm{d}\phi^*}{\mathrm{d}x}\right)\right)\mathrm{d}x = \left(\Gamma_x \frac{\mathrm{d}\phi^*}{\mathrm{d}x}\right)\Bigg|_{x_j} - \int_{x_i}^{x_j} \Gamma_x \frac{\mathrm{d}N_j}{\mathrm{d}x}\frac{\mathrm{d}\phi^*}{\mathrm{d}x}\mathrm{d}x \tag{13.8b}$$

Finally, we substitute Equations 13.8a and 13.8b into Equations 13.6a and 13.6b, respectively, to obtain the integral statement of the governing equation as

$$R_i^{(e)} = -\left(\Gamma_x \frac{\mathrm{d}\phi^*}{\mathrm{d}x}\right)\Bigg|_{x_i} - \int_{x_i}^{x_j} \Gamma_x \frac{\mathrm{d}N_i}{\mathrm{d}x}\frac{\mathrm{d}\phi^*}{\mathrm{d}x}\mathrm{d}x + \int_{x_i}^{x_j} SN_i(x)\mathrm{d}x = 0 \tag{13.9a}$$

and

$$R_j^{(e)} = \left(\Gamma_x \frac{\mathrm{d}\phi^*}{\mathrm{d}x}\right)\Bigg|_{x_j} - \int_{x_i}^{x_j} \Gamma_x \frac{\mathrm{d}N_j}{\mathrm{d}x}\frac{\mathrm{d}\phi^*}{\mathrm{d}x}\mathrm{d}x + \int_{x_i}^{x_j} SN_j(x)\mathrm{d}x = 0 \tag{13.9b}$$

Or in vector form as

$$\left\{R^{(e)}\right\} = \left\{\begin{bmatrix} -\Gamma_x \dfrac{\mathrm{d}\phi}{\mathrm{d}x}\Big|_{x_i} \\ \Gamma_x \dfrac{\mathrm{d}\phi}{\mathrm{d}x}\Big|_{x_j} \end{bmatrix}\right\} - \int_{x_i}^{x_j} \Gamma_x \frac{\mathrm{d}[N]^T}{\mathrm{d}x}\frac{\mathrm{d}\phi^*}{\mathrm{d}x}\mathrm{d}x + \int_{x_i}^{x_j} S[N]^T\mathrm{d}x = 0 \tag{13.9c}$$

Equation 13.9 represents the **integral statement** of the governing differential Equation 13.1. This is also known as the **weak formulation** as mathematical manipulation is used to lower the highest-order derivative term from a second-order to a first-order derivative.

Step 3 Formation of Element Characteristics Equation

In this step the element characteristics equation is obtained by evaluating the integrals by substituting the approximate solution for the chosen element type. Substituting $\phi^* = [N]\left\{\phi^{(e)}\right\}$ in Equation 13.9c, we get

$$\left\{R^{(e)}\right\} = \left\{\begin{bmatrix} -\Gamma_x \dfrac{\mathrm{d}\phi}{\mathrm{d}x}\Big|_{x_i} \\ \Gamma_x \dfrac{\mathrm{d}\phi}{\mathrm{d}x}\Big|_{x_j} \end{bmatrix}\right\} - \left(\int_{x_i}^{x_j} \Gamma_x \frac{\mathrm{d}[N]^T}{\mathrm{d}x}\frac{\mathrm{d}[N]}{\mathrm{d}x}\mathrm{d}x\right)\left\{\phi^{(e)}\right\} + \int_{x_i}^{x_j} S[N]^T\mathrm{d}x = 0 \tag{13.10a}$$

or

$$\left[K_\Gamma^{(e)}\right]\left\{\phi^{(e)}\right\}=\left\{I^{(e)}\right\}+\left\{f^{(e)}\right\} \tag{13.10b}$$

where

$$\left\{\phi^{(e)}\right\}=\begin{Bmatrix}\phi_i\\ \phi_j\end{Bmatrix}=\text{vector of element nodal values} \tag{13.11a}$$

$$\left[K_\Gamma^{(e)}\right]=\left(\int_{x_i}^{x_j}\Gamma_x\frac{\mathrm{d}[N]^T}{\mathrm{d}x}\frac{\mathrm{d}[N]}{\mathrm{d}x}\mathrm{d}x\right)\left\{\phi^{(e)}\right\}=\text{element characteristic or stiffness matrix} \tag{13.11b}$$

$$\left\{f^{(e)}\right\}=\int_{x_i}^{x_j}S[N]^T\mathrm{d}x=\text{element source or force vector} \tag{13.11c}$$

$$\left\{I^{(e)}\right\}=\begin{Bmatrix}-\Gamma_x\left.\dfrac{\mathrm{d}\phi}{\mathrm{d}x}\right|_{x_i}\\ \Gamma_x\left.\dfrac{\mathrm{d}\phi}{\mathrm{d}x}\right|_{x_j}\end{Bmatrix}=\text{vector for element contribution to inter-element conditions} \tag{13.11d}$$

Case: Linear Line Element

From the approximate solution, Equation 13.3, selected for the linear line element we can evaluate the following derivatives:

$$\frac{\mathrm{d}\phi^*}{\mathrm{d}x}=\frac{1}{x_j-x_i}(-\phi_i)+\frac{1}{x_j-x_i}(\phi_j)=\frac{1}{L_e}(-\phi_i+\phi_j) \tag{13.12a}$$

and

$$\frac{\mathrm{d}N_i}{\mathrm{d}x}=-\frac{1}{x_j-x_i}=-\frac{1}{L_e},\quad \frac{\mathrm{d}N_j}{\mathrm{d}x}=\frac{1}{x_j-x_i}=\frac{1}{L_e} \tag{13.12b}$$

Using Equation 13.12, we can evaluate the first integrals term in Equation 13.9 as follows:
For $I=i$

$$\int_{x_i}^{x_j}\Gamma_x\frac{\mathrm{d}N_i}{\mathrm{d}x}\frac{\mathrm{d}\phi^*}{\mathrm{d}x}\mathrm{d}x=\int_{x_i}^{x_j}\Gamma_x\left(-\frac{1}{L_e}\right)\frac{\phi_j-\phi_i}{L_e}\cdot\mathrm{d}x=\frac{\Gamma_x}{L_e}(\phi_i-\phi_j) \tag{13.13a}$$

For $I=j$

$$\int_{x_i}^{x_j}\Gamma_x\frac{\mathrm{d}N_j}{\mathrm{d}x}\frac{\mathrm{d}\phi^*}{\mathrm{d}x}\mathrm{d}x=\int_{x_i}^{x_j}\Gamma_x\left(\frac{1}{L_e}\right)\cdot\frac{-\phi_i+\phi_j}{L_e}\mathrm{d}x=\frac{\Gamma_x}{L_e}(-\phi_i+\phi_j) \tag{13.13b}$$

Equations 13.13a and 13.13b can also be written in a matrix form as

$$\int_{x_i}^{x_j} \Gamma_x \frac{\mathrm{d}[N]^T}{\mathrm{d}x}\frac{\mathrm{d}\phi^*}{\mathrm{d}x}\mathrm{d}x = \left(\int_{x_i}^{x_j} \Gamma_x \frac{\mathrm{d}[N]^T}{\mathrm{d}x}\frac{\mathrm{d}[N]}{\mathrm{d}x}\mathrm{d}x\right)\left\{\phi^{(e)}\right\} = \frac{\Gamma_x}{L_e}\begin{bmatrix} 1 & -1 \\ -1 & 1 \end{bmatrix}\begin{Bmatrix} \phi_i \\ \phi_j \end{Bmatrix} \quad (13.13c)$$

While evaluating the integrals, we assume a stepwise constant value for the transport property, Γ. For the second **integral term**, let us assume a step-wise constant value $S = \bar{S}$ for the source in the media for this demonstration and evaluate the integrals as follows:

$$\int_{x_i}^{x_j} SN_i(x)\mathrm{d}x = \int_{x_i}^{x_j} \bar{S}\frac{x_j - x}{x_j - x_i}\mathrm{d}x = \frac{\bar{S}}{x_j - x_i}\left(x_j\, x\Big|_{x_i}^{x_j} - \frac{x^2}{2}\Big|_{x_i}^{x_j}\right)$$

$$= \frac{\bar{S}}{x_j - x_i}\left(x_j\left(x_j - x_i\right) - \frac{x_j^2 - x_i^2}{2}\right)$$

or

$$\int_{x_i}^{x_j} SN_i(x)\mathrm{d}x = \frac{\bar{S}L_e}{2} \quad (13.14a)$$

and similarly,

$$\int_{x_i}^{x_j} SN_j(x)\mathrm{d}x = \int_{x_i}^{x_j} \bar{S}\frac{x - x_i}{x_j - x_i}\mathrm{d}x = \frac{\bar{S}L_e}{2} \quad (13.14b)$$

Equations 13.14a and 13.14b can be written in vector form as

$$\int_{x_i}^{x_j} S[N]^T\mathrm{d}x = \frac{\bar{S}L_e}{2}\begin{Bmatrix} 1 \\ 1 \end{Bmatrix} \quad (13.14c)$$

Substituting Equations 13.13 and 13.14 into Equation 13.9, we obtain the element characteristics equations as follows:

$$R_i^{(e)} = \left(\Gamma_x \frac{\mathrm{d}\phi^*}{\mathrm{d}x}\right)\Bigg|_{x_i} + \frac{\Gamma_x}{L_e}\left(\phi_i - \phi_j\right) - \frac{\bar{S}L_e}{2} = 0 \quad (13.15a)$$

and

$$R_j^{(e)} = -\left(\Gamma_x \frac{\mathrm{d}\phi^*}{\mathrm{d}x}\right)\Bigg|_{x_j} + \frac{\Gamma_x}{L_e}\left(-\phi_i + \phi_j\right) - \frac{\bar{S}L_e}{2} = 0 \quad (13.15b)$$

Equations 13.14a and 13.14b can be written in matrix form as follows:

$$\left\{\begin{array}{c} R_i^{(e)} \\ R_j^{(e)} \end{array}\right\} = \left\{\begin{array}{c} \Gamma_x \left.\frac{d\phi}{dx}\right|_{x_i} \\ -\Gamma_x \left.\frac{d\phi}{dx}\right|_{x_j} \end{array}\right\} + \frac{\Gamma_x}{L_e}\left[\begin{array}{cc} 1 & -1 \\ -1 & 1 \end{array}\right]\left\{\begin{array}{c} \phi_i \\ \phi_j \end{array}\right\} - \frac{\bar{S}L_e}{2}\left\{\begin{array}{c} 1 \\ 1 \end{array}\right\} = 0 \quad (13.16a)$$

or

$$\left[K_\Gamma^{(e)}\right]\left\{\phi^{(e)}\right\} = \left\{I^{(e)}\right\} + \left\{f^{(e)}\right\} \quad (13.16b)$$

where

$$\left\{\phi^{(e)}\right\} = \left\{\begin{array}{c} \phi_i \\ \phi_j \end{array}\right\} = \text{vector of element nodal values} \quad (13.17a)$$

$$\left[K_\Gamma^{(e)}\right] = \frac{\Gamma_x}{L_e}\left[\begin{array}{cc} 1 & -1 \\ -1 & 1 \end{array}\right] = \left[\begin{array}{cc} k_{11}^{(e)} & k_{12}^{(e)} \\ k_{21}^{(e)} & k_{22}^{(e)} \end{array}\right] = \text{element characteristic or stiffness matrix} \quad (13.17b)$$

$$\left\{f^{(e)}\right\} = \left\{\begin{array}{c} \frac{\bar{S}L_e}{2} \\ \frac{\bar{S}L_e}{2} \end{array}\right\} = \left\{\begin{array}{c} f_i^{(e)} \\ f_j^{(e)} \end{array}\right\} = \text{element source or force vector} \quad (13.17c)$$

$$\left\{I^{(e)}\right\} = \left\{\begin{array}{c} -\Gamma_x \left.\frac{d\phi}{dx}\right|_{x_i} \\ \Gamma_x \left.\frac{d\phi}{dx}\right|_{x_j} \end{array}\right\} = \left\{\begin{array}{c} I_i^{(e)} \\ I_j^{(e)} \end{array}\right\} = \text{vector for element contribution to inter-element conditions} \quad (13.17d)$$

Equation 13.16 is a system of equations with number of equations equal to the number nodal points in the element, and it is referred to as the **element characteristic equation**. The vector for **interelement contributions** represents the boundary conditions and interelement continuity conditions. We can now apply Equations 13.16 and 13.17 repeatedly to all elements to obtain corresponding element equations.

In a similar manner, we can derive the element characteristics equation using quadratic and cubic line elements.

Step 4 Assembly of Element Equations to Form the Global System

In this step, the element equations are assembled to form a global system of equations, which consist of all nodal values in the mesh as unknowns. However, a systematic procedure is required for this assembly process, particularly for efficient computer implementations. For this purpose, a numbering or connectivity scheme is used as described in Table 13.1 below:

TABLE 13.1
Node Numbering or Connectivity Scheme

	Node Numbers	
Element Number (e)	Local System	Global System
1	$i=1$	$i=1$
	$j=2$	$j=2$
2	$i=1$	$i=2$
	$j=2$	$j=3$
3	$i=1$	$i=3$
	$j=2$	$j=4$
.		
$n-1$	$i=1$	$i=n-1$
	$j=2$	$j=n$
N	$i=1$	$i=n$
	$j=2$	$j=n+1$

Such a connectivity scheme may look trivial, particularly for a one-dimensional problem. However, it is an essential feature in the finite element formulations while solving multidimensional problems. Let us show this assembly process by considering the element equations one at a time in terms of global co-ordinates or global numbering scheme, and then substituting them in the global system following the connectivity rules outlined in Table 13.1.

Element Equations

Element # 1

$$\begin{bmatrix} k_{11}^{(1)} & k_{12}^{(1)} \\ k_{21}^{(1)} & k_{22}^{(1)} \end{bmatrix} \begin{Bmatrix} \phi_1^{(1)} \\ \phi_2^{(1)} \end{Bmatrix} = \begin{Bmatrix} I_1^{(1)} \\ I_2^{(1)} \end{Bmatrix} + \begin{Bmatrix} f_1^{(1)} \\ f_2^{(1)} \end{Bmatrix} \tag{13.18a}$$

Element # 2

$$\begin{bmatrix} k_{11}^{(2)} & k_{12}^{(2)} \\ k_{21}^{(2)} & k_{22}^{(2)} \end{bmatrix} \begin{Bmatrix} \phi_2^{(2)} \\ \phi_3^{(2)} \end{Bmatrix} = \begin{Bmatrix} I_2^{(2)} \\ I_3^{(2)} \end{Bmatrix} + \begin{Bmatrix} f_2^{(2)} \\ f_3^{(2)} \end{Bmatrix} \tag{13.18b}$$

Element # 3

$$\begin{bmatrix} k_{11}^{(3)} & k_{12}^{(3)} \\ k_{21}^{(3)} & k_{22}^{(3)} \end{bmatrix} \begin{Bmatrix} \phi_3^{(3)} \\ \phi_4^{(3)} \end{Bmatrix} = \begin{Bmatrix} I_3^{(3)} \\ I_4^{(3)} \end{Bmatrix} + \begin{Bmatrix} f_3^{(3)} \\ f_4^{(3)} \end{Bmatrix} \tag{13.18c}$$

Element # (n − 1)

$$\begin{bmatrix} k_{11}^{(n-1)} & k_{12}^{(n-1)} \\ k_{21}^{(n-1)} & k_{22}^{(n-1)} \end{bmatrix} \begin{Bmatrix} \phi_{n-1}^{(n-1)} \\ \phi_{n}^{(n-1)} \end{Bmatrix} = \begin{Bmatrix} I_{n-1}^{(n-1)} \\ I_{n}^{(n-1)} \end{Bmatrix} + \begin{Bmatrix} f_{n-1}^{(n-1)} \\ f_{n}^{(n-1)} \end{Bmatrix} \quad (13.18\text{n-1})$$

Element # n

$$\begin{bmatrix} k_{11}^{(n)} & k_{12}^{(n)} \\ k_{21}^{(n)} & k_{22}^{(n)} \end{bmatrix} \begin{Bmatrix} \phi_{n}^{(n)} \\ \phi_{n+1}^{(n)} \end{Bmatrix} = \begin{Bmatrix} I_{n}^{(n)} \\ I_{n+1}^{(n)} \end{Bmatrix} + \begin{Bmatrix} f_{n}^{(n)} \\ f_{n+1}^{(n)} \end{Bmatrix} \quad (13.18\text{n})$$

Let us now form the global system by including element equations one at a time.

The inclusion of **first** element equation, Equation 13.18a, gives the global system as

$$\begin{bmatrix} k_{11}^{(1)} & k_{12}^{(1)} & 0 & 0 & \cdots & 0 & 0 \\ k_{21}^{(1)} & k_{22}^{(1)} & 0 & 0 & \cdots & 0 & 0 \\ 0 & 0 & 0 & 0 & \cdots & 0 & 0 \\ 0 & 0 & 0 & 0 & \cdots & 0 & 0 \\ \vdots & & & & & & \\ 0 & 0 & 0 & 0 & \cdots & 0 & 0 \\ 0 & 0 & 0 & 0 & \cdots & 0 & 0 \end{bmatrix} \begin{Bmatrix} \phi_1 \\ \phi_2 \\ \phi_3 \\ \\ \vdots \\ \phi_n \\ \phi_{n+1} \end{Bmatrix} = \begin{Bmatrix} I_1^{(1)} \\ I_2^{(1)} \\ \\ \\ \\ \\ \\ \end{Bmatrix} + \begin{Bmatrix} f_1^{(1)} \\ f_2^{(1)} \\ \\ \\ \\ \\ \\ \end{Bmatrix}$$

Inclusion of the **second** element equation, Equation 13.18b, gives

$$\begin{bmatrix} k_{11}^{(1)} & k_{12}^{(1)} & 0 & 0 & \cdots & 0 & 0 \\ k_{21}^{(1)} & k_{22}^{(1)} + k_{11}^{(2)} & k_{12}^{(2)} & 0 & \cdots & 0 & 0 \\ 0 & k_{21}^{(2)} & k_{22}^{(2)} & 0 & \cdots & 0 & 0 \\ 0 & 0 & 0 & 0 & \cdots & 0 & 0 \\ \vdots & & & & & & \\ 0 & 0 & 0 & 0 & \cdots & 0 & 0 \\ 0 & 0 & 0 & 0 & \cdots & 0 & 0 \end{bmatrix} \begin{Bmatrix} \phi_1 \\ \phi_2 \\ \phi_3 \\ \\ \vdots \\ \phi_n \\ \phi_{n+1} \end{Bmatrix}$$

$$= \begin{Bmatrix} I_1^{(1)} \\ I_2^{(1)} + I_1^{(2)} \\ I_2^{(2)} \\ \\ \\ \\ \\ \end{Bmatrix} + \begin{Bmatrix} f_1^{(1)} \\ f_2^{(1)} + f_1^{(2)} \\ f_2^{(2)} \\ \\ \\ \\ \\ \end{Bmatrix}$$

Inclusion of the **third** element equation, Equation 13.18c, gives:

$$
\begin{bmatrix}
k_{11}^{(1)} & k_{12}^{(1)} & 0 & 0 & \cdots & 0 & 0 \\
k_{21}^{(1)} & k_{22}^{(1)} + k_{11}^{(2)} & k_{12}^{(2)} & 0 & \cdots & 0 & 0 \\
0 & k_{21}^{(2)} & k_{22}^{(2)} + k_{11}^{(3)} & k_{12}^{(3)} & \cdots & 0 & 0 \\
0 & & k_{21}^{(3)} & k_{22}^{(3)} & \cdots & 0 & 0 \\
 & & & & & & \\
0 & 0 & 0 & 0 & \cdots & 0 & 0 \\
0 & 0 & 0 & 0 & \cdots & 0 & 0
\end{bmatrix}
\begin{Bmatrix}
\phi_1 \\ \phi_2 \\ \phi_3 \\ \\ \vdots \\ \phi_n \\ \phi_{n+1}
\end{Bmatrix}
$$

$$
= \begin{Bmatrix}
I_1^{(1)} \\ I_2^{(1)} + I_1^{(2)} \\ I_2^{(2)} + I_1^{(3)} \\ I_2^{(3)} \\ \\ 0 \\ 0
\end{Bmatrix}
+ \begin{Bmatrix}
f_1^{(1)} \\ f_2^{(1)} + f_1^{(2)} \\ f_2^{(2)} + f_2^{(3)} \\ I_2^{(3)} \\ \\ 0 \\ 0
\end{Bmatrix}
$$

This process of including the element equations is continued, and with the inclusion of the nth element equation, Equation 13.18n gives the final ***global system*** as

$$
\begin{bmatrix}
k_{11}^{(1)} & k_{21}^{(1)} & 0 & 0 & \cdots & 0 & 0 \\
k_{21}^{(1)} & k_{22}^{(1)} + k_{11}^{(2)} & k_{12}^{(2)} & 0 & \cdots & 0 & 0 \\
0 & k_{21}^{(2)} & k_{22}^{(2)} + k_{11}^{(3)} & k_{12}^{(3)} & \cdots & 0 & 0 \\
0 & 0 & k_{21}^{(3)} & k_{22}^{(3)} & \cdots & 0 & 0 \\
\vdots & & & & & 0 & 0 \\
0 & 0 & 0 & 0 & \cdots & k_{22}^{((n01)} + k_{11}^{(n)} & k_{12}^{(n)} \\
0 & 0 & 0 & 0 & \cdots & k_{21}^{(n)} & k_{22}^{(n)}
\end{bmatrix}
\begin{Bmatrix}
\phi_1 \\ \phi_2 \\ \phi_3 \\ \\ \vdots \\ \phi_n \\ \phi_{n+1}
\end{Bmatrix}
$$

$$
= \begin{Bmatrix}
I_1^{(1)} \\ I_2^{(1)} + I_1^{(2)} \\ I_2^{(2)} + I_1^{(3)} \\ I_2^{(3)} \\ \\ I_2^{(n-1)} + I_1^{(n)} \\ I_2^{(n)}
\end{Bmatrix}
+ \begin{Bmatrix}
f_1^{(1)} \\ f_2^{(1)} + f_1^{(2)} \\ f_2^{(2)} + f_2^{(3)} \\ I_2^{(3)} \\ \\ f_2^{(n-1)} + f_1^{(n)} \\ f_2^{(n)}
\end{Bmatrix}
\tag{13.19}
$$

Let us now introduce the expressions for the element matrix, force vector and the interelement vector as given by Equations 13.17b–13.17d into Equation 13.19 to obtain the global system as follows:

$$
\begin{bmatrix}
\dfrac{\Gamma_x}{L_e} & -\dfrac{\Gamma_x}{L_e} & & & & & \\[2mm]
-\dfrac{\Gamma_x}{L_e} & \dfrac{\Gamma_x}{L_e}+\dfrac{\Gamma_x}{L_e} & -\dfrac{\Gamma_x}{L_e} & & & & \\[2mm]
 & -\dfrac{\Gamma_x}{L_e} & \dfrac{\Gamma_x}{L_e}+\dfrac{\Gamma_x}{L_e} & -\dfrac{\Gamma_x}{L_e} & & & \\[2mm]
 & & -\dfrac{\Gamma_x}{L_e} & \dfrac{\Gamma_x}{L_e}+\dfrac{\Gamma_x}{L_e} & -\dfrac{\Gamma_x}{L_e} & & \\[2mm]
\vdots & & & \ddots & \ddots & \ddots & \\[2mm]
 & & & -\dfrac{\Gamma_x}{L_e} & \dfrac{\Gamma_x}{L_e}+\dfrac{\Gamma_x}{L_e} & -\dfrac{\Gamma_x}{L_e} \\[2mm]
 & & & & -\dfrac{\Gamma_x}{L_e} & \dfrac{\Gamma_x}{L_e}
\end{bmatrix}
\begin{Bmatrix}
\phi_1 \\ \phi_2 \\ \phi_3 \\ \vdots \\ \vdots \\ \phi_{n-1} \\ \phi_n
\end{Bmatrix}
$$

$$
=
\begin{Bmatrix}
-\Gamma_x \dfrac{d\phi}{dx}\Big|_{x_1} \\[3mm]
\Gamma_x \dfrac{d\phi}{dx}\Big|_{x_2} -\Gamma_x \dfrac{d\phi}{dx}\Big|_{x_2} \\[3mm]
\Gamma_x \dfrac{d\phi}{dx}\Big|_{x_3} -\Gamma_x \dfrac{d\phi}{dx}\Big|_{x_3} \\[3mm]
\vdots \\[3mm]
\vdots \\[3mm]
\Gamma_x \dfrac{d\phi}{dx}\Big|_{x_n} -\Gamma_x \dfrac{d\phi}{dx}\Big|_{x_n} \\[3mm]
+\Gamma_x \dfrac{d\varphi}{dx}\Big|_{x_{n+1}}
\end{Bmatrix}
\begin{Bmatrix}
\dfrac{\overline{S}L_e}{2} \\[3mm]
\dfrac{\overline{S}L_e}{2}+\dfrac{\overline{S}L_e}{2} \\[3mm]
\dfrac{\overline{S}L_e}{2}+\dfrac{\overline{S}L_e}{2} \\[3mm]
\vdots \\[3mm]
\vdots \\[3mm]
\dfrac{\overline{S}L_e}{2}+\dfrac{\overline{S}L_e}{2} \\[3mm]
\dfrac{\overline{S}L_e}{2}
\end{Bmatrix}
\qquad (13.20)
$$

Sampling and canceling terms due to interelement continuity conditions, we get final **global system** as follows:

$$
\begin{bmatrix}
\dfrac{\Gamma_x}{L_e} & -\dfrac{\Gamma_x}{L_e} & & & & & \\[2mm]
-\dfrac{\Gamma_x}{L_e} & \dfrac{2\Gamma_x}{L_e} & -\dfrac{\Gamma_x}{L_e} & & & & \\[2mm]
& -\dfrac{\Gamma_x}{L_e} & \dfrac{2\Gamma_x}{L_e} & -\dfrac{\Gamma_x}{L_e} & & & \\[2mm]
& & -\dfrac{\Gamma_x}{L_e} & \dfrac{2\Gamma_x}{L_e} & -\dfrac{\Gamma_x}{L_e} & & \\[2mm]
\vdots & & & \ddots & \ddots & \ddots & \\[2mm]
& & & & -\dfrac{\Gamma_x}{L_e} & \dfrac{2\Gamma_x}{L_e} & -\dfrac{\Gamma_x}{L_e} \\[2mm]
& & & & & -\dfrac{\Gamma_x}{L_e} & \dfrac{\Gamma_x}{L_e}
\end{bmatrix}
\begin{Bmatrix}
\phi_1 \\ \phi_2 \\ \phi_3 \\ \vdots \\ \vdots \\ \phi_{n-1} \\ \phi_n
\end{Bmatrix}
$$

$$
= \begin{Bmatrix}
-\Gamma_x \left.\dfrac{d\phi}{dx}\right|_{x_1} \\[3mm]
0 \\[2mm]
0 \\[2mm]
\vdots \\[2mm]
\vdots \\[2mm]
0 \\[2mm]
\Gamma_x \left.\dfrac{d\phi}{dx}\right|_{x_{n+1}}
\end{Bmatrix}
+ \begin{Bmatrix}
\dfrac{\bar{S}L_e}{2} \\[3mm]
\bar{S}L_e \\[2mm]
\bar{S}L_e \\[2mm]
\vdots \\[2mm]
\vdots \\[2mm]
\bar{S}L_e \\[2mm]
\dfrac{\bar{S}L_e}{2}
\end{Bmatrix}
\qquad (13.21)
$$

It can be noticed that in the assembled global system, the interelement conditions are already implemented by satisfying the continuous flux quantity at all interelement nodes. Again, the global system is a set of algebraic equations with number of unknowns same as the number of unknown nodal values in the selected mesh distribution system. The boundary conditions now can be directly introduced in the global system before solving the system of equations.

The global system can be written in a compact form as follows:

$$
[K]\{\Phi\} = \{F\} \qquad (13.22)
$$

where

$$[K]=\begin{bmatrix} \frac{\Gamma_x}{L_e} & -\frac{\Gamma_x}{L_e} & & & & & \\ -\frac{\Gamma_x}{L_e} & \frac{2\Gamma_x}{L_e} & -\frac{\Gamma_x}{L_e} & & & & \\ & -\frac{\Gamma_x}{L_e} & \frac{2\Gamma_x}{L_e} & -\frac{\Gamma_x}{L_e} & & & \\ & & -\frac{\Gamma_x}{L_e} & \frac{2\Gamma_x}{L_e} & -\frac{\Gamma_x}{L_e} & & \\ \vdots & & & \ddots & \ddots & \ddots & \\ & & & & -\frac{\Gamma_x}{L_e} & \frac{2\Gamma_x}{L_e} & -\frac{\Gamma_x}{L_e} \\ & & & & & -\frac{\Gamma_x}{L_e} & \frac{\Gamma_x}{L_e} \end{bmatrix} = \text{global characteristics matrix} \tag{13.23a}$$

$$\{\Phi\}=\begin{Bmatrix} \phi_1 \\ \phi_2 \\ \phi_3 \\ \vdots \\ \vdots \\ \phi_{n-1} \\ \phi_n \end{Bmatrix} = \text{Vector for unknown nodal values in the solution domain} \tag{13.23b}$$

$$\{F\}=\begin{Bmatrix} -\Gamma_x \left.\frac{d\phi}{dx}\right|_{x_1} \\ 0 \\ 0 \\ \vdots \\ \vdots \\ 0 \\ \Gamma_x \left.\frac{d\phi}{dx}\right|_{x_{n+1}} \end{Bmatrix} + \begin{Bmatrix} \frac{\bar{S}L_e}{2} \\ \bar{S}L_e \\ \bar{S}L_e \\ \vdots \\ \vdots \\ \bar{S}L_e \\ \frac{\bar{S}L_e}{2} \end{Bmatrix} = \text{Source or force vector} \tag{13.23c}$$

Step 6 Implementation of Boundary Conditions

In this step, the boundary conditions are introduced directly in the in the global system of equations. For boundary conditions of the **first kind**, i.e., of constant surface value as defined by Equations 13.2, we simply assign the given values to the boundary nodes as $\phi_1 = \phi_l$ and $\phi_{n+1} = \phi_r$, and rearranged the system so that the natural boundary conditions $\left.\frac{d\phi}{dx}\right|_{x1}$ and $\left.\frac{d\phi}{dx}\right|_{x_{n+1}}$ appear as one of the unknowns as follows:

$$\begin{bmatrix} -\Gamma_x & -\frac{\Gamma_x}{L_e} & & & & & \\ 0 & \frac{2\Gamma_x}{L_e} & -\frac{\Gamma_x}{L_e} & & & & \\ & -\frac{\Gamma_x}{L_e} & \frac{2\Gamma_x}{L_e} & -\frac{\Gamma_x}{L_e} & & & \\ & & -\frac{\Gamma_x}{L_e} & \frac{2\Gamma_x}{L_e} & -\frac{\Gamma_x}{L_e} & & \\ \vdots & & & \ddots & \ddots & \ddots & \\ & & & & -\frac{\Gamma_x}{L_e} & \frac{2\Gamma_x}{L_e} & 0 \\ & & & & & -\frac{\Gamma_x}{L_e} & -\Gamma_x \end{bmatrix} \begin{Bmatrix} \left.\frac{d\phi}{dx}\right|_{x1} \\ \phi_2 \\ \phi_3 \\ \vdots \\ \vdots \\ \phi_n \\ \left.\frac{d\phi}{dx}\right|_{x_{n+1}} \end{Bmatrix} = \begin{Bmatrix} -\frac{\Gamma_x\phi_l}{L_e} + \frac{\bar{S}L_e}{2} \\ \frac{\Gamma_x\phi_l}{L_e} + \bar{S}L_e \\ \bar{S}L_e \\ \vdots \\ \vdots \\ \frac{\Gamma_x\phi_r}{L_e} + \bar{S}L_e \\ -\frac{\Gamma_x\phi_r}{L_e} + \frac{\bar{S}L_e}{2} \end{Bmatrix} \tag{13.24}$$

The global system can be written in a compact form as follows:

$$[\mathrm{K}]\{\Phi\} = \{F\} \tag{13.25}$$

where

$$[K] = \begin{bmatrix} -\Gamma_x & -\frac{\Gamma_x}{L_e} & 0 & 0 & 0 & 0 & 0 \\ 0 & \frac{2\Gamma_x}{L_e} & -\frac{\Gamma_x}{L_e} & 0 & 0 & 0 & 0 \\ 0 & -\frac{\Gamma_x}{L_e} & \frac{2\Gamma_x}{L_e} & -\frac{\Gamma_x}{L_e} & 0 & 0 & 0 \\ 0 & 0 & -\frac{\Gamma_x}{L_e} & \frac{2\Gamma_x}{L_e} & -\frac{\Gamma_x}{L_e} & 0 & 0 \\ \vdots & & & \ddots & \ddots & \ddots & \\ 0 & 0 & 0 & 0 & -\frac{\Gamma_x}{L_e} & \frac{2\Gamma_x}{L_e} & 0 \\ 0 & 0 & 0 & 0 & & -\frac{\Gamma_x}{L_e} & -\Gamma_x \end{bmatrix} = \text{global characteristics matrix} \tag{13.26a}$$

$$\{\Phi\} = \left\{ \begin{array}{c} \left.\dfrac{d\phi}{dx}\right|_{x1} \\ \phi_2 \\ \phi_3 \\ \vdots \\ \vdots \\ \phi_n \\ \left.\dfrac{d\phi}{dx}\right|_{x_{n=1}} \end{array} \right\} = \text{Vector for unknown nodal values in the solution domain} \tag{13.26b}$$

$$\{F\} = \left\{ \begin{array}{c} -\dfrac{\Gamma_x \phi_l}{L_e} + \dfrac{\bar{S}L_e}{2} \\ \dfrac{\Gamma_x \phi_l}{L_e} + \bar{S}L_e \\ \bar{S}L_e \\ \vdots \\ \vdots \\ \dfrac{\Gamma_x \phi_r}{L_e} + \bar{S}L_e \\ -\dfrac{\Gamma_x \phi_r}{L_e} + \dfrac{\bar{S}L_e}{2} \end{array} \right\} \tag{13.26c}$$

Step 5 Solution of the Global System of Equations

The global system of equations is now solved using appropriate solver as discussed in Chapter 3.

Example 13.1: One-Dimensional Steady-State Conduction

Consider one-dimensional steady-state conduction without heat generation in a plane slab as shown. The boundary surfaces at $x=0$ and $x=L$ are maintained at constant temperatures T_l and T_r, respectively. Determine the temperature distribution in the slab and heat transfer rates at the surfaces. Use $k=1.2\ \dfrac{\text{W}}{\text{cm}\cdot\text{C}}$, $L=10$ cm, $T_l=30°\text{C}$, $T_r=100°\text{C}$.

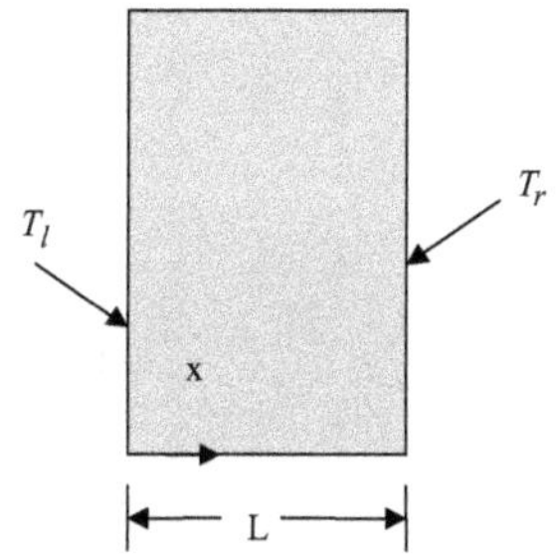

The mathematical statement of the problem is

Governing Equation

$$\frac{d}{dx}\left(k\frac{dT}{dx}\right)=0 \tag{E.13.1.1}$$

Boundary Condition

$$1.\quad x=0, T=T_L \tag{E.13.1.2a}$$

$$2.\quad x=L, T=T_R \tag{E.13.1.2b}$$

Solution

In the **first step**, let us discretize the domain using five uniform line elements as shown in the figure.

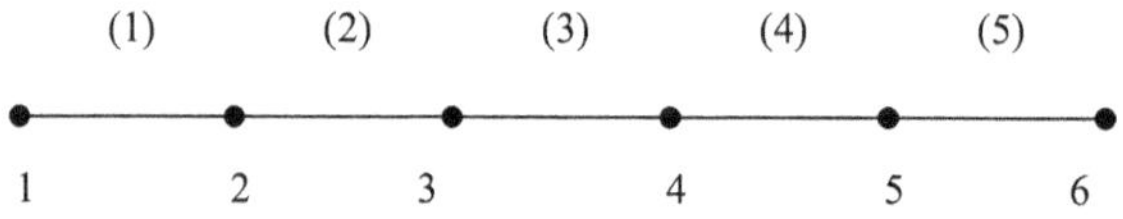

With this mesh size distribution, we have $L_e = L/ = 2.0\,\text{cm}$ and number of elements $N=5$ with node $i=1$ and $i=6$ located at $x=0$ and $x=10$.

In the **second step**, we assume that the temperature distribution in each element be represented by the approximate function

$$T^* = N_iT_i + N_jT_j \tag{E.13.1.3}$$

In the **third step**, we write the weak form of the governing equation following Equation 13.9, which is obtained by Galerkin's weighted residual method as

$$\begin{Bmatrix} k\left.\dfrac{dT}{dx}\right|_{x_i} \\ -k\left.\dfrac{dT}{dx}\right|_{x_j} \end{Bmatrix} + \int_{x}^{x_j} \Gamma_x \frac{d[N]^T}{dx}\frac{dT^*}{dx}dx = 0 \tag{E.13.1.4}$$

In the **fourth step**, the element characteristics equation is obtained following Equations 13.16 and 13.17:

$$\left[\kappa_k^{(e)}\right]\left\{\varphi^{(e)}\right\} = \left\{I^{(e)}\right\} \tag{E.13.1.5a}$$

where

$$\left\{T^{(e)}\right\} = \begin{Bmatrix} T_i \\ T_j \end{Bmatrix},\quad \left[\kappa_k^{(e)}\right] = \frac{k}{L}\begin{bmatrix} 1 & -1 \\ -1 & 1 \end{bmatrix},\quad \left\{I^{(e)}\right\} = \begin{Bmatrix} -k\left.\dfrac{dT}{dx}\right|_{x_i} \\ k\left.\dfrac{dT}{dx}\right|_{x_j} \end{Bmatrix} \tag{E.13.1.5b}$$

Since the material is homogeneous and the mesh is uniform, the element characteristics equation is identical for all elements.

In the **fifth step**, we assemble all element equations and obtain the global system following Equation 13.21

$$\begin{bmatrix} 1 & -1 & & & & \\ -1 & 2 & -1 & & & \\ & -1 & 2 & -1 & & \\ & & -1 & 2 & -1 & \\ & & & -1 & 2 & -1 \\ & & & & -1 & 1 \end{bmatrix} \begin{Bmatrix} T_1 \\ T_2 \\ T_3 \\ T_4 \\ T_5 \\ T_6 \end{Bmatrix} = \begin{Bmatrix} -\left.\dfrac{dT}{dx}\right|_{x1} \cdot L \\ 0 \\ 0 \\ 0 \\ 0 \\ \left.\dfrac{dT}{dx}\right|_{x6} \cdot L \end{Bmatrix} \tag{E.13.1.6}$$

In the **sixth step**, we incorporate the boundary conditions in the global system. *Since* the boundary temperatures are known in this problem, we substitute $T_1 = T_l = 30°C$ and $T_6 = T_r = 100°C$ and rearrange the global system as follows:

$$\begin{bmatrix} -L & -1 & 0 & 0 & 0 & 0 \\ 0 & 2 & -1 & 0 & 0 & 0 \\ 0 & -1 & 2 & -1 & 0 & 0 \\ 0 & 0 & -1 & 2 & -1 & 0 \\ 0 & 0 & 0 & -1 & 2 & 0 \\ 0 & 0 & 0 & 0 & -1 & -L \end{bmatrix} \begin{Bmatrix} \left.\dfrac{dT}{dx}\right|_{x1} \\ T_2 \\ T_3 \\ T_4 \\ T_5 \\ \left.\dfrac{dT}{dx}\right|_{x6} \end{Bmatrix} = \begin{Bmatrix} -T_l \\ T_l \\ 0 \\ 0 \\ T_r \\ -T_r \end{Bmatrix} \tag{E.13.1.7}$$

In the **final step**, we substitute $L = 2$ cm, $T_l = 30°C$, $T_r = 100°C$, and choose an appropriate solver for the global system of equations. Using the Gaussian elimination program, we can get the numerical solution as

$$\left.\frac{dT}{dx}\right|_{x1} = -7.0\,°\text{C/cm}$$

$$T_2 = 44°\text{C}$$

$$T_3 = 58°\text{C}$$

$$T_4 = 72°\text{C}$$

$$T_5 = 86°\text{C}$$

$$\left.\frac{dT}{dx}\right|_{x6} = -7.0\,°\text{C/cm}$$

The heat transfer rate at $x=0$ is estimated as

$$q''_{x=0} = -k\left.\frac{dT}{dx}\right|_{x1} = 1.2 \times (-7.0)$$

$$q''_{x=0} = -8.40\frac{\text{W}}{\text{cm}^2}$$

The heat transfer rate at $x=L$ is estimated as

$$q''_{x=L} = -k\left.\frac{dT}{dx}\right| = 1.2 \times (-7.0)$$

$$q''_{x=L} = -8.40\frac{\text{W}}{\text{cm}^2}$$

13.2 FINITE ELEMENT FORMULATION USING VARIATIONAL APPROACH

Most of the steps outlined for **Galerkin-based finite element formulation** are also applicable to finite element formulation based on variational approach with the exception of steps 3 and 4 in which an integral statement of the problem and the element characteristics equations are obtained. So, while discussing the finite element formulation based on variational approach, we will focus on the formulation of the integral statement and the element characteristic equation in steps 3 and 4 only.

In this approach, the governing differential equation is transformed into an integral form by evaluating an equivalent **variational form** of the problem. This variational form primarily involves a **functional, *I***, such that finding a minimized or extremized value of the functional (I) is equivalent to finding the solution of the original differential equation. In the next step, the functional is extremized with respect to the unknown solution vector, $\{\Phi\}$, which constitute all unknown nodal values of the mesh, i.e.,

$$\frac{\partial I}{\partial \Phi} = 0 \tag{13.27}$$

where the functional I over the whole solution domain can be written as the summation of elemental contribution $I^{(e)}$ as

$$I = \sum_{e=1}^{n} I^{(e)} \tag{13.28}$$

Combining Equations 13.27 and 13.28, we get

$$\sum_{e=1}^{n} \frac{\partial I^{(e)}}{\partial \phi^{(e)}} = 0 \tag{13.29}$$

As we evaluate the elemental contribution, $\dfrac{\partial I^{(e)}}{\partial \phi^{(e)}}$, we obtain the element characteristic equation as

$$\frac{\partial I^{(e)}}{\partial \phi^{(e)}} = \left[\kappa^{(e)}\right]\left\{\phi^{(e)}\right\} - \left\{f^{(e)}\right\} \tag{13.30}$$

Equation 13.30 is applied successively to all elements, and the resulting element characteristic equations are then assembled through Equation 13.29 and using global connectivity system (Table 13.1) to obtain the global system of equations:

$$[K]\{\Phi\} = \{F\} \tag{13.31}$$

A brief description of the variational calculus as well as the procedure for obtaining the functional (I) of a differential problem is given in Chapter 9. Let us now consider the one-dimensional steady-state transport equation given by Equation 13.1 and constant surface boundary conditions given by Equation 13.2, and demonstrate the formulations of the integral form and subsequently the element characteristic equation using the variational formulation. We will consider the mesh size distribution given in Figure 13.2, and approximate solution and linear line element given by Equation 13.3.

Step 3 Formation of the Integral Statement or Variational Form

Multiply both sides of Equation 13.1 by a variation, $\delta\varphi$, and integrate over an element (to get,

$$\int_{x_i}^{x_j}\left(\frac{\mathrm{d}}{\mathrm{d}x}\left(\Gamma_x \frac{\mathrm{d}\phi}{\mathrm{d}x}\right) + S\right)\delta\phi\,\mathrm{d}x = 0$$

or

$$\int_{x_i}^{x_j}\frac{\mathrm{d}}{\mathrm{d}x}\left(\Gamma_x \frac{\mathrm{d}\phi}{\mathrm{d}x}\right)\delta\phi\,\mathrm{d}x + \int_{x_i}^{x_j}S\,\delta\phi\,\mathrm{d}x = 0 \tag{13.32}$$

Evaluating the first integral term using integration-by-parts, we get

$$\left(\Gamma_x \frac{\mathrm{d}\phi}{\mathrm{d}x}\delta\varphi\right)\Bigg|_{x_i}^{x_j} - \int_{x_i}^{x_j}\Gamma_x \frac{\mathrm{d}\phi}{\mathrm{d}x}\delta\left(\frac{\mathrm{d}\phi}{\mathrm{d}x}\right)\mathrm{d}x + \int_{x_i}^{x_j}S\,\delta\phi\,\mathrm{d}x = 0$$

or

$$\left(\Gamma_x \frac{\mathrm{d}\phi}{\mathrm{d}x}\delta\phi\right)_{x=x_j} - \left(\Gamma_x \frac{\mathrm{d}\phi}{\mathrm{d}x}\delta\phi\right)_{x_i} - \int_{x_i}^{x_j}\Gamma_x \frac{1}{2}\delta\left(\frac{\mathrm{d}\phi}{\mathrm{d}x}\right)^2 \mathrm{d}x + \int_{x_i}^{x_j}S\,\delta\varphi\,\mathrm{d}x = 0 \tag{13.33}$$

It can be noted that the first two integrated terms represent flux conditions at the boundaries. These terms vanish if no variation of ϕ is permitted at the boundary, i.e., $\delta\phi = 0$ or flux is zero at the boundary such as in an adiabatic or symmetric boundaries. We further reformulate the rest of the terms in Equation 13.33 by bringing the δ operator outside the integrals as follows:

$$\delta \int_0^L \left(-\frac{1}{2}\Gamma_x \left(\frac{\mathrm{d}\phi}{\mathrm{d}x} \right)^2 + S\phi \right) \mathrm{d}x = 0 \tag{13.34}$$

Now, Equation 13.34 represents the variation of a functional, i.e., $\delta I = 0$. So, the functional, I, corresponding to the governing differential problem is

$$I = \int_{x_i}^{x_j} \left(\frac{1}{2}\Gamma_x \left(\frac{\mathrm{d}\phi}{\mathrm{d}x} \right)^2 - S\phi \right) \mathrm{d}x \tag{13.35}$$

Minimization of the functional leads to the solution, which is also equivalent to the solution of the governing differential problem.

However, for any other types of boundary conditions such as mixed boundary condition or constant flux conditions, the integrated terms do not vanish. Such nonzero-integrated terms may also arise for any interior nodes that are located at the interface of two elements. In such situations, the integrated terms are reformulated before forming the functional as follows:

$$\left(\Gamma_x \left.\frac{\mathrm{d}\phi}{\mathrm{d}x}\right|_{x_j} \delta\phi(x_j) \right) - \left(\Gamma_x \left.\frac{\mathrm{d}\phi}{\mathrm{d}x}\right|_{x_i} \delta\phi(x_i) \right) - \int_{x_i}^{x_j} \Gamma_x \frac{1}{2} \delta \left(\frac{\mathrm{d}\phi}{\mathrm{d}x} \right)^2 \mathrm{d}x + \int_{x_i}^{x_j} S\,\delta\phi\,\mathrm{d}x = 0$$

or

$$\delta \left\{ \left(\Gamma_x \left.\frac{\mathrm{d}\phi}{\mathrm{d}x}\right|_{x_j} \phi(x_j) \right) - \left(\Gamma_x \left.\frac{\mathrm{d}\phi}{\mathrm{d}x}\right|_{x_i} \phi(x_i) \right) - \int_{x_i}^{x_j} \Gamma_x \frac{1}{2} \left(\frac{\mathrm{d}\phi}{\mathrm{d}x} \right)^2 \mathrm{d}x + \int_{x_i}^{x_j} S\phi\,\mathrm{d}x \right\} = 0$$

The corresponding functional is given as

$$I = -\left(\Gamma_x \left.\frac{\mathrm{d}\phi}{\mathrm{d}x}\right|_{x_j} \phi(x_j) \right) + \left(\Gamma_x \left.\frac{\mathrm{d}\phi}{\mathrm{d}x}\right|_{x_i} \phi(x_i) \right) + \int_{x_i}^{x_j} \Gamma_x \frac{1}{2} \left(\frac{\mathrm{d}\phi}{\mathrm{d}x} \right)^2 \mathrm{d}x - \int_{x_i}^{x_j} S\phi\,\mathrm{d}x$$

Substituting $\phi(x_j) = \phi_j$ and $\phi(x_i) = \phi_i$, we get

$$I = -\left(\Gamma_x \left.\frac{\mathrm{d}\phi}{\mathrm{d}x}\right|_{x_j} \phi_j \right) + \left(\Gamma_x \left.\frac{\mathrm{d}\phi}{\mathrm{d}x}\right|_{x_i} \phi_i \right) + \int_{x_i}^{x_j} \Gamma_x \frac{1}{2} \left(\frac{\mathrm{d}\phi}{\mathrm{d}x} \right)^2 \mathrm{d}x - \int_{x_i}^{x_j} S\phi\,\mathrm{d}x$$

Step 4 Formation of Element Characteristic Equation

In this step, element characteristic equation is obtained by evaluating the elemental functional quantity, $I^{(e)}$, given as

$$I^{(e)} = \int_{x_i}^{x_j} \left(\frac{1}{2}\Gamma_x \left(\frac{\mathrm{d}\phi}{\mathrm{d}x} \right)^2 + S\phi \right) \mathrm{d}x \tag{13.36}$$

Let us first evaluate the derivative from Equation 13.3 as follows:
From Equation 13.3a,

$$\frac{d\phi}{dx} = \frac{dN_i}{dx}\phi_i + \frac{dN_j}{dx}\phi_j \tag{13.37}$$

Substituting Equations 13.3a and 13.37 into Equation 13.36, we get

$$I^{(e)} = -\left(\Gamma_x \frac{d\phi}{dx}\bigg|_{x_j}\phi_j\right) + \left(\Gamma_x \frac{d\phi}{dx}\bigg|_{x_i}\phi_i\right)$$

$$+\frac{1}{2}\Gamma_x \int_{x_i}^{x_j}\left[\left(\frac{dN_i}{dx}\right)^2\phi_i^2 + \left(\frac{dN_j}{dx}\right)^2\phi_j^2 + 2\frac{dN_i}{dx}\frac{dN_i}{dx}\phi_i\phi_j\right]dx - \int_{x_i}^{x_j}S\left[N_i\phi_i + N_j\phi_j\right]dx \tag{13.38}$$

To minimize the functional, we evaluate the derivative given by Equation 13.29 as follows:
For $I=i$,

$$\frac{dI^{(e)}}{d\phi_i} = \Gamma_x \frac{d\phi}{dx}\bigg|_{x_i} + \frac{\Gamma_{\bar{x}}}{2}\int_{x_i}^{x_j}\left[\left(\frac{dN_i}{dx}\right)^2 2\phi_i + 2\frac{dN_i}{dx}\frac{dN_j}{dx}\phi_j\right]dx - \bar{S}\int_{x_i}^{x_i}N_i\,dx = 0 \tag{13.39a}$$

For $I=j$,

$$\frac{dI^{(e)}}{d\phi_i} = -\Gamma_x \frac{d\phi}{dx}\bigg|_{x_j} + \frac{\Gamma_{\bar{x}}}{2}\int_{x_i}^{x_j}\left[2\frac{dN_i}{dx}\frac{dN_i}{dx}\phi_j + \left(\frac{dN_j}{dx}\right)^2 2\phi_j\right]dx - \bar{S}\int_{x_i}^{x_i}SN_j\,dx = 0 \tag{13.39b}$$

Notice that we have assumed step-wise constant values for the transport property, $\Gamma_x = \Gamma_{\bar{x}}$, and for the source term as $S = \bar{S}$. Equation 13.39 is the **element characteristic equation**, which can also be written in a matrix form as

$$\left\{\frac{dI^{(e)}}{d\phi^{(e)}}\right\} = \left\{\begin{array}{c}\Gamma_x \frac{d\phi}{dx}\big|_{x_i} \\[2mm] -\Gamma_x \frac{d\phi}{dx}\big|_{x_j}\end{array}\right\} + \Gamma_{\bar{x}}\begin{bmatrix}\int_{x_i}^{x_j}\left(\frac{dN_i}{dx}\right)^2 dx & \int_{x_i}^{x_j}\frac{dN_i}{dx}\frac{dN_j}{dx}dx \\[4mm] \int_{x_i}^{x_j}\frac{dN_i}{dx}\frac{dN_j}{dx}dx & \int_{x_i}^{x_j}\left(\frac{dN_j}{dx}\right)^2 dx\end{bmatrix}\left\{\begin{array}{c}\phi_i \\ \phi_j\end{array}\right\} - \bar{S}\left\{\begin{array}{c}\int_{x_i}^{x_j}N_i\,dx \\[4mm] \int_{x_i}^{x_j}N_j\,dx\end{array}\right\} = 0 \tag{13.40}$$

or

$$\left[\kappa_\Gamma^{(e)}\right]\left\{\phi^{(e)}\right\} = \left\{I^{(e)}\right\} + \left\{f^{(e)}\right\} \tag{13.41}$$

where

$$\left[\kappa_{\Gamma}^{(x)}\right]=\Gamma_{\bar{x}}\begin{bmatrix}\int_{x_i}^{x_j}\left(\frac{dN_i}{dx}\right)^2 dx & \int_{x_i}^{x_j}\frac{dN_i}{dx}\frac{dN_j}{dx}dx \\ \int_{x_i}^{x_j}\frac{dN_i}{dx}\frac{dN_j}{dx}dx & \int_{x_i}^{x_j}\left(\frac{dN_j}{dx}\right)^2 dx\end{bmatrix} = \text{element Characteristic matrix} \tag{13.42}$$

$$\left\{f^{(e)}\right\}=\bar{S}\begin{Bmatrix}\int_{x_i}^{x_j} N_i\,dx \\ \int_{x_i}^{x_j} N_j\,dx\end{Bmatrix} = \text{element source vector} \tag{13.43}$$

$$\left\{I^{(e)}\right\}=\begin{Bmatrix}-\Gamma_x\left.\frac{d\phi}{dx}\right|_{x_i} \\ \Gamma_x\left.\frac{d\phi}{dx}\right|_{x_j}\end{Bmatrix}=\begin{Bmatrix}I_i^{(e)} \\ I_j^{(e)}\end{Bmatrix}$$

$$\left\{\phi^{(e)}\right\}=\begin{Bmatrix}\phi_i \\ \phi_j\end{Bmatrix} = \text{vector for element nodal values} \tag{13.44}$$

Let us know evaluate the elements of the characteristic matrix as follows:

From Equation 13.3b

$$\frac{dN_i}{dx}=-\frac{1}{L_e},\frac{dN_j}{dx}=\frac{1}{L_e} \tag{13.45}$$

Using these we have

$$\int_{x_i}^{x_j}\left(\frac{dN_i}{dx}\right)^2 dx=\int_{x_i}^{x_j}\frac{1}{L_e^2}dx=\frac{1}{L_e} \tag{13.46a}$$

$$\int_{x_i}^{x_j}\left(\frac{dN_j}{dx}\right)^2 dx=\int_{x_i}^{x_j}\frac{1}{L_e^2}dx=\frac{1}{L_e} \tag{13.46b}$$

$$\int_{x_i}^{x_j}\left(\frac{dN_i}{dx}\right)\left(\frac{dN_j}{dx}\right)dx=\int_{x_i}^{x_j}-\frac{1}{L_e^2}dx=-\frac{1}{L_e} \tag{13.46c}$$

The elements of the source vector are evaluated in Section 13.1, and we use

$$\int_{x_i}^{x_j} N_i \mathrm{d}x = \frac{L_e}{2} \tag{13.47a}$$

$$\int_{x_i}^{x_j} N_j \mathrm{d}x = \frac{L_e}{2} \tag{13.47b}$$

Substituting Equations 13.46 and 13.47 into Equations 13.42 and 13.43, we compute the element characteristic matrix and source vectors as follows:

$$\left[K_{\Gamma}^{(e)}\right] = \frac{\Gamma_{\bar{x}}}{L_e}\begin{bmatrix} 1 & -1 \\ -1 & 1 \end{bmatrix} \tag{13.48a}$$

$$\left\{I^{(e)}\right\} = \begin{Bmatrix} -\Gamma_x \left.\dfrac{\mathrm{d}\phi}{\mathrm{d}x}\right|_{x_i} \\ \Gamma_x \left.\dfrac{\mathrm{d}\phi}{\mathrm{d}x}\right|_{x_j} \end{Bmatrix} \tag{13.48b}$$

and

$$\left\{f^{(e)}\right\} = \begin{Bmatrix} \dfrac{\bar{S}L_e}{2} \\ \dfrac{\bar{S}L_e}{2} \end{Bmatrix} \tag{13.48c}$$

Equations 13.41, 13.44, and 13.48 are now applied successively to all elements in the mesh and assembled through Equation 13.29 and using the global connectivity system given in Table 13.1 to form the global system of equations given by Equations 13.22 and 13.23.

Example 13.2: One-Dimensional Steady State Mass Transfer in a Adsorbing Desiccant Material

Consider one-dimensional steady state diffusion of moisture in an adsorbing desiccant felt with top surface maintained at constant moisture concentration of C_0, by subjecting it to a moist air stream. The desiccant felt is supported at the bottom by an aluminum plate. Assume constant rate of moisture adsorption, m'''_{ad} and determine the moisture concentration profile in the felt using variational-based finite element formulation.

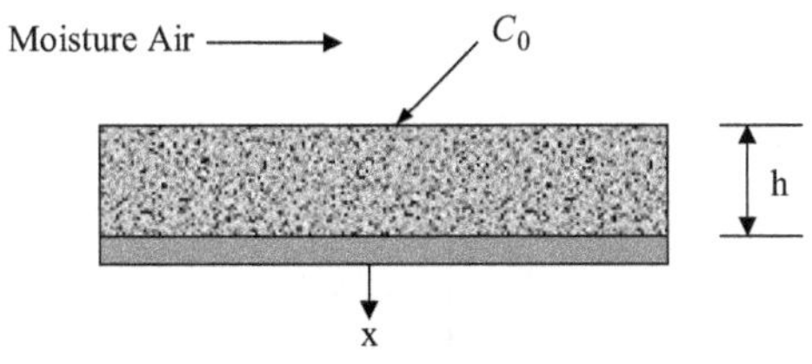

The mathematical statement of the problem is given as:

Governing Equation:

$$\frac{d}{dx}\left(D\frac{dC}{dx}\right)-m_{ad}^{///}=0 \tag{E13.2.1}$$

Boundary Conditions:

$$1.\ x=0,\ C=C_0 \tag{E.13.2.2}$$

$$2.\ x=h,\ \frac{dC}{dx}=0 \quad \text{due to non-permeable metal plate the bottom} \tag{E13.2.3}$$

Solution:

Let us discretize the solution domain into a mesh size distribution using one-dimensional line element as shown in the figure below

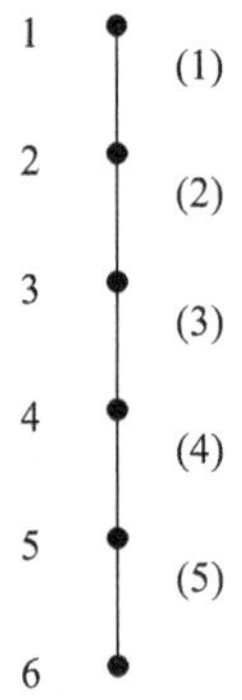

Application of the procedure outlined in Section 13.3 leads to the element characteristic equations as

$$\left[K_D^{(e)}\right]\left\{C^{(e)}\right\}=\left\{I^{(e)}\right\}+\left\{f^{(e)}\right\}$$

Where the element characteristic matrix and source vectors are given as follows:

$$\left[K_D^{(e)}\right]=\frac{D}{L_e}\begin{bmatrix} 1 & -1 \\ -1 & 1 \end{bmatrix} \tag{13.48a}$$

$$\left\{I^{(e)}\right\}=\begin{Bmatrix} -D\left.\dfrac{dC}{dx}\right|_{x_i} \\ D\left.\dfrac{dC}{dx}\right|_{x_j} \end{Bmatrix} \tag{13.48b}$$

$$\left\{f^{(e)}\right\}=\begin{Bmatrix} -\dfrac{m_{ad}^{///}L_e}{2} \\ -\dfrac{m_{ad}^{///}L_e}{2} \end{Bmatrix} \tag{13.48c}$$

Assembling all element equations and applying inter-element continuity conditions, we get final **global system** as follows:

$$\frac{D}{L_e}\begin{bmatrix} 1 & -1 & 0 & 0 & 0 & 0 \\ -1 & 2 & -1 & 0 & 0 & 0 \\ 0 & -1 & 2 & -1 & 0 & 0 \\ 0 & 0 & -1 & 2 & -1 & 0 \\ 0 & 0 & 0 & -1 & 2 & -1 \\ 0 & 0 & 0 & 0 & -1 & 1 \end{bmatrix}\begin{Bmatrix} C_1 \\ C_2 \\ C_3 \\ C_4 \\ C_5 \\ C_6 \end{Bmatrix} = \begin{Bmatrix} -D\left.\dfrac{dC}{dx}\right|_{x=0} - \dfrac{m'''_{ad} L_e}{2} \\ -m'''_{ad} L_e \\ -m'''_{ad} L_e \\ -m'''_{ad} L_e \\ -m'''_{ad} L_e \\ D\left.\dfrac{dC}{dx}\right|_{x=h} - \dfrac{m'''_{ad} L_e}{2} \end{Bmatrix}$$

With the application of the boundary conditions, i.e. $C_1 = C_o$ and $\left.\dfrac{dC}{dx}\right|_{x=h} = 0$, the global system reduces to

$$\frac{D}{L_e}\begin{bmatrix} L_e & -1 & 0 & 0 & 0 & 0 \\ 0 & 2 & -1 & 0 & 0 & 0 \\ 0 & -1 & 2 & -1 & 0 & 0 \\ 0 & 0 & -1 & 2 & -1 & 0 \\ 0 & 0 & 0 & -1 & 2 & -1 \\ 0 & 0 & 0 & 0 & -1 & 1 \end{bmatrix}\begin{Bmatrix} \left.\dfrac{dC}{dx}\right|_{x=0} \\ C_2 \\ C_3 \\ C_4 \\ C_5 \\ C_6 \end{Bmatrix} = \begin{Bmatrix} -\dfrac{DC_0}{L_e} - \dfrac{m'''_{ad} L_e}{2} \\ \dfrac{DC_0}{L_e} - m'''_{ad} \\ -m'''_{ad} \\ -m'''_{ad} \\ -m'''_{ad} \\ -\dfrac{m'''_{ad} L_e}{2} \end{Bmatrix}$$

13.3 BOUNDARY CONDITIONS

In the previous section, we have considered problems with boundary conditions of the first kind, i.e., of constant surface value. For such problems, we simply assign the given boundary values to the unknown variable at the boundary nodes and solved the global system for rest of the unknown nodal values along with the natural boundary conditions such as the flux quantities at the boundary nodes.

13.3.1 Boundary Condition of the Second Kind or Constant Surface Flux

Let us consider the one-dimensional problem with constant surface flux boundary conditions as shown in Figure 13.4. The mathematical statement of the problem is as follows:

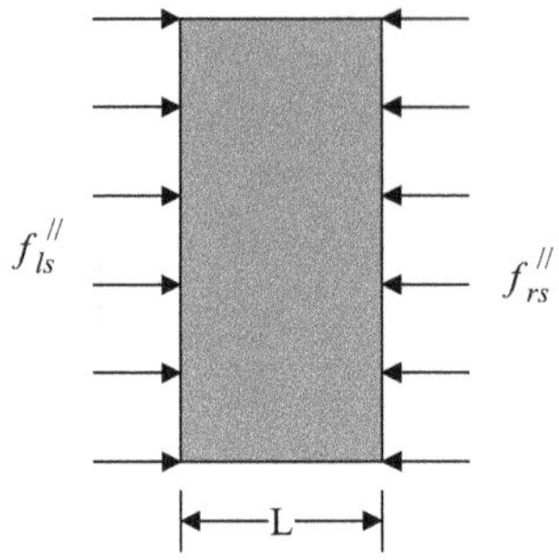

FIGURE 13.4 Constant surface flux boundary condition.

Governing Equation:

$$\frac{d}{dx}\left(\Gamma_x \frac{d\phi}{dx}\right) + \bar{S} = 0 \tag{13.49}$$

Boundary Conditions:

$$1. \quad x = 0, \quad -\Gamma_x \left.\frac{d\phi}{dx}\right|_{x=0} = f''_{ls} \tag{13.50a}$$

$$2. \quad x = 0, \quad -\Gamma_x \left.\frac{d\phi}{dx}\right|_{x=0} = f''_{ls} \tag{13.50b}$$

The Galerkin-based finite element formulation will lead to the same element characteristic equation as in Section 13.1, i.e.,

$$\left[k^{(e)}\right]\left\{\phi^{(e)}\right\} = \left\{I^{(e)}\right\} + \left\{f^{(e)}\right\} \tag{13.16b}$$

where the interelement vectors for element #1 and element #2 are modified based on using the boundary conditions given by Equations 13.50a and 13.50b as follows:

Element #1:

$$\left\{I^{(1)}\right\} = \begin{Bmatrix} \left.\left(-\Gamma_x \dfrac{d\phi}{dx}\right)^{(1)}\right|_{x_i} \\ \left.\left(\Gamma_x \dfrac{d\phi}{dx}\right)^{(1)}\right|_{x_j} \end{Bmatrix} = \begin{Bmatrix} f''_{ls} \\ \left.\left(\Gamma_x \dfrac{d\phi}{dx}\right)^{(1)}\right|_{x_j} \end{Bmatrix} \tag{13.51}$$

and

Element #*n*:

$$\left\{I^{(n)}\right\} = \begin{Bmatrix} \left.\left(-\Gamma_x \dfrac{d\phi}{dx}\right)^{(n)}\right|_{x_i} \\ \left.\left(\Gamma_x \dfrac{d\phi}{dx}\right)^{(n)}\right|_{x_j} \end{Bmatrix} = \begin{Bmatrix} \left.\left(-\Gamma_x \dfrac{d\phi}{dx}\right)^{(n)}\right|_{x_j} \\ f''_{rs} \end{Bmatrix} \tag{13.52}$$

The expressions for the rest of the interelement vectors as well as the force vectors and the characteristics equations will remain the same as that given by in Equations 13.18b to 13.18n − 1. The assembly of all element equations will lead to the global system as equations will lead to the global system as

$$\begin{bmatrix} \frac{\Gamma_x}{L_e} & -\frac{\Gamma_x}{L_e} & & & & \\ -\frac{\Gamma_x}{L_e} & \frac{2\Gamma_x}{L_e} & -\frac{\Gamma_x}{L_e} & & & \\ & -\frac{\Gamma_x}{L_e} & \frac{2\Gamma_x}{L_e} & -\frac{\Gamma_x}{L_e} & & \\ & & -\frac{\Gamma_x}{L_e} & \frac{2\Gamma_x}{L_e} & & \\ \vdots & & & & & \\ & & & & \frac{2\Gamma_x}{L_e} & -\frac{\Gamma_x}{L_e} \\ & & & & -\frac{\Gamma_x}{L_e} & \frac{\Gamma_x}{L_e} \end{bmatrix} \begin{Bmatrix} \phi_1 \\ \phi_2 \\ \phi_3 \\ \vdots \\ \vdots \\ \phi_{n-1} \\ \phi_n \end{Bmatrix} = \begin{Bmatrix} f_{ls}^{//} \\ 0 \\ 0 \\ \vdots \\ \vdots \\ 0 \\ f_{rs}^{//} \end{Bmatrix} + \begin{Bmatrix} \frac{\bar{S}L_e}{2} \\ \bar{S}L_e \\ \bar{S}L_e \\ \vdots \\ \vdots \\ \bar{S}L_e \\ \frac{\bar{S}L_e}{2} \end{Bmatrix} \tag{13.53}$$

The global system can be written in a compact form as follows:

$$[K]\{\Phi\} = \{F\} \tag{13.54}$$

where

$$K = \begin{bmatrix} \frac{\Gamma_x}{L_e} & -\frac{\Gamma_x}{L_e} & & & & \\ -\frac{\Gamma_x}{L_e} & \frac{2\Gamma_x}{L_e} & -\frac{\Gamma_x}{L_e} & & & \\ & -\frac{\Gamma_x}{L_e} & \frac{2\Gamma_x}{L_e} & -\frac{\Gamma_x}{L_e} & & \\ & & -\frac{\Gamma_x}{L_e} & \frac{2\Gamma_x}{L_e} & & \\ \vdots & & & & & \\ & & & & \frac{2\Gamma_x}{L_e} & -\frac{\Gamma_x}{L_e} \\ & & & & -\frac{\Gamma_x}{L_e} & \frac{\Gamma_x}{L_e} \end{bmatrix} = \text{global characteristics matrix} \tag{13.55a}$$

$$\{\Phi\} = \begin{Bmatrix} \phi_1 \\ \phi_2 \\ \phi_3 \\ \vdots \\ \vdots \\ \phi_n \\ \phi_{n+1} \end{Bmatrix} = \text{Vector for unknown nodal values in the solution domain} \quad (13.55b)$$

$$\{F\} = \begin{Bmatrix} f_{ls}^{//} \\ \\ \\ \\ \\ \\ f_{rs}^{//} \end{Bmatrix} + \begin{Bmatrix} \frac{\bar{S}L_e}{2} \\ \bar{S}L_e \\ \bar{S}L_e \\ \vdots \\ \vdots \\ \bar{S}L_e \\ \frac{\bar{S}L_e}{2} \end{Bmatrix} = \begin{Bmatrix} f_{ls}^{//} + \frac{\bar{S}L_e}{2} \\ \bar{S}L_e \\ \bar{S}L_e \\ \vdots \\ \vdots \\ \bar{S}L_e \\ f_{rs}^{//} + \frac{\bar{S}L_e}{2} \end{Bmatrix} = \text{Source or force vector} \quad (13.55c)$$

Special Case

For the case of a zero-surface flux condition or at a symmetric boundary as shown (Figure 13.5) on the right surface of the plane slab, the mathematical statement of the problem is given as

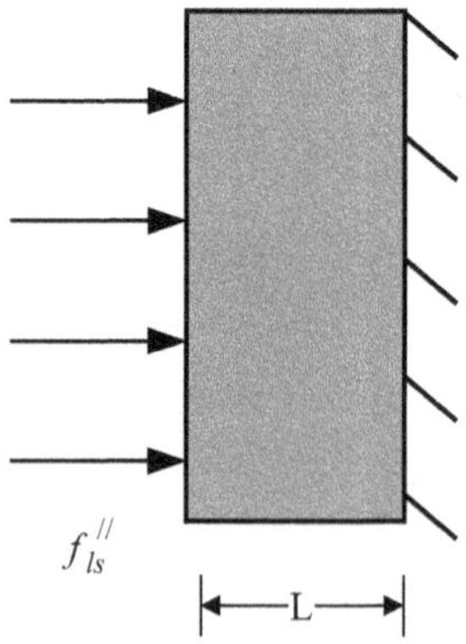

FIGURE 13.5 Symmetric boundary condition.

Governing Equation:

$$\frac{d}{dx}\left(\Gamma_x \frac{d\phi}{dx}\right) + \bar{S} = 0 \tag{13.56}$$

Boundary Conditions:

$$1. \quad x = 0, -\Gamma_x \left.\frac{d\varphi}{dx}\right|_{x=0} = f_{ls}^{//} \tag{13.57a}$$

$$2. \quad x = L, \left.\frac{d\phi}{dx}\right|_{x=L} = 0 \tag{13.57b}$$

The global system for this problem is as same as that given by Equations 13.22 and 13.23a, except for the force vector, which is modified as follows:

$$\{F\} = \begin{Bmatrix} f_{ls}^{//} + \dfrac{\bar{S}L_e}{2} \\ \bar{S}L_e \\ \bar{S}L_e \\ \vdots \\ \vdots \\ \bar{S}L_e \\ f_{rs}^{//} + \dfrac{\bar{S}L_e}{2} \end{Bmatrix} \tag{13.58}$$

Example 13.3: Motion of Falling Liquid Film Down an Inclined Surface

Consider the flow of water film in steady and laminar motion, falling down an inclined surface with slope $\theta = 45°$from the horizontal. The thickness, d of the film is constant along the length. Assume the flow as fully developed and at zero-pressure gradient, i.e. $\frac{dP}{dx} = 0$.

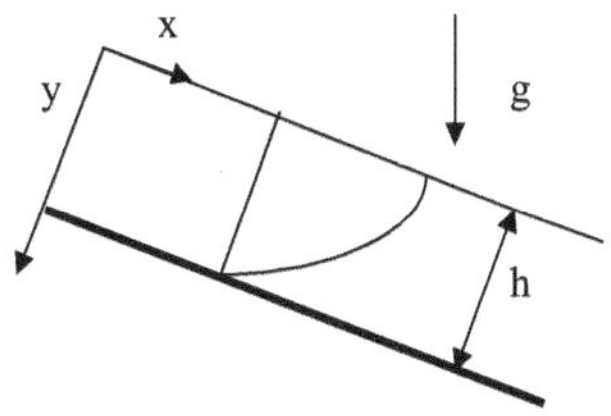

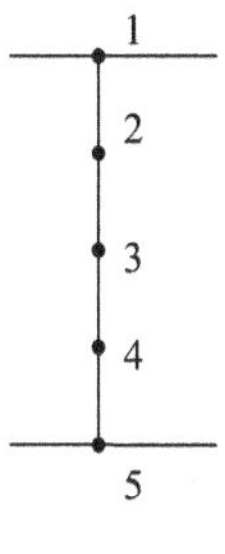

The mathematical statement of the problem can be derived from mass and momentum equations of fluid flow and it is given as follows:

Governing Equation:

$$\frac{d}{dx}\left(\mu\frac{du}{dx}\right)+\rho g\cos\theta=0 \tag{E13.3.1}$$

Boundary Conditions:

$$1.\ y=0,\ \frac{du}{dx}=0 \quad \text{zero shear stress at the free surface of the film} \tag{E13.3.2}$$

$$2.\ y=h,\ u=0, \quad \text{no-slip condition at the plate} \tag{E13.3.3}$$

Use following data for the calculation: h = 10.0 mm; $\rho=1000\,\text{kg/m}^3$; $\mu=0.0005\,\text{kg/m}\cdot\text{s}$; g = 9.8 m/s^2, and determine (a) the velocity distribution in the film, and (b) the shear stress at the plate and (c) validate the results by comparing the results with that given by the exact solution:

$$u=\frac{\rho g}{\mu}\cos\theta\left(D(h-y)-\frac{(h-y)^2}{2}\right) \tag{E13.3.4}$$

Solution

In the first step, let us discretize the solution domain using four uniform line elements as shown in the figure. With this mesh size distribution, we have $L_e=\frac{h}{N_x}=\frac{0.01\,\text{m}}{4}=0.0025\,\text{mm}$ and number of nodal points, N = 5. The nodal points i =1 and i = 6 are located at y = 0 and y = 0.01 m, respectively. Following the Galerkin's finite element formulation and using two-points line element, we obtain the element characteristics equation same as in Equation 13.21, i.e.

$$\frac{\mu}{L_e}\begin{bmatrix}1&-1&0&0&0\\-1&2&-1&0&0\\0&-1&2&-1&0\\0&0&-1&2&-1\\0&0&0&-1&1\end{bmatrix}\begin{Bmatrix}u_1\\u_2\\u_3\\u_4\\u_5\end{Bmatrix}=\begin{Bmatrix}-\mu\left.\frac{du}{dy}\right|_{y=0}+\frac{\rho g\cos\theta L_e}{2}\\ \rho g\cos\theta L_e\\ \rho g\cos\theta L_e\\ \rho g\cos\theta L_e\\ \mu\left.\frac{du}{dy}\right|_{y=h}+\rho g\cos\theta\frac{L_e}{2}\end{Bmatrix} \tag{E13.3.5}$$

Applying boundary conditions given by Equations E13.3.2 and E13.3.3, and rearranging, we get

$$\frac{\mu}{L_e}\begin{bmatrix}1&-1&0&0&0\\-1&2&-1&0&0\\0&-1&2&-1&0\\0&0&-1&2&0\\0&0&0&-1&\frac{L_e}{\mu}\end{bmatrix}\begin{Bmatrix}u_1\\u_2\\u_3\\u_4\\ \mu\left.\frac{du}{dy}\right|_{y=h}\end{Bmatrix}=\begin{Bmatrix}\frac{\rho g\cos\theta L_e}{2}\\ \rho g\cos\theta L_e\\ \rho g\cos\theta L_e\\ \rho g\cos\theta L_e\\ \frac{\rho g\cos\theta L_e}{2}\end{Bmatrix} \tag{E13.3.6}$$

Substituting $\frac{\mu}{L_e} = 0.2$ and $\rho g \cos\theta L_e = 1000 \times 9.8 \times \cos 45 \times 0.0025 = 17.3242$, the system of equation become

$$\begin{bmatrix} 0.2 & -0.2 & 0 & 0 & 0 \\ -0.2 & 0.4 & -0.2 & 0 & 0 \\ 0 & -0.2 & 0.4 & -0.2 & 0 \\ 0 & 0 & -0.2 & 0.4 & 0 \\ 0 & 0 & 0 & -0.2 & 1 \end{bmatrix} \begin{Bmatrix} u_1 \\ u_2 \\ u_3 \\ u_4 \\ \mu \left.\frac{du}{dy}\right|_{y=h} \end{Bmatrix} = \begin{Bmatrix} 8.6621 \\ 17.3242 \\ 17.3242 \\ 17.3242 \\ 8.6621 \end{Bmatrix} \quad \text{(E13.3.7)}$$

Solution to the system of equations is given as

$$\begin{Bmatrix} u_1 \\ u_2 \\ u_3 \\ u_4 \\ \mu \left.\frac{du}{dy}\right|_{y=h} \end{Bmatrix} = \begin{Bmatrix} 692.9680 \\ 649.6575 \\ 519.1725 \\ 303.1735 \\ -69.2968 \end{Bmatrix}$$

The wall shear stress at the wall is

$$\tau_w = \mu \left.\frac{du}{dy}\right|_{y=h} = -69.2968$$

Comparison of the finite element solution with the analytical solution is given in the table below.

Nodal Point	**Analytical solution (m/s)**	**Finite Element Solution (m/s)**
u_1	692.9646	692.9680
u_2	649.6543	549.6575
u_3	519.723	519.7260
u_4	303.1720	303.1725
u_5	0	0

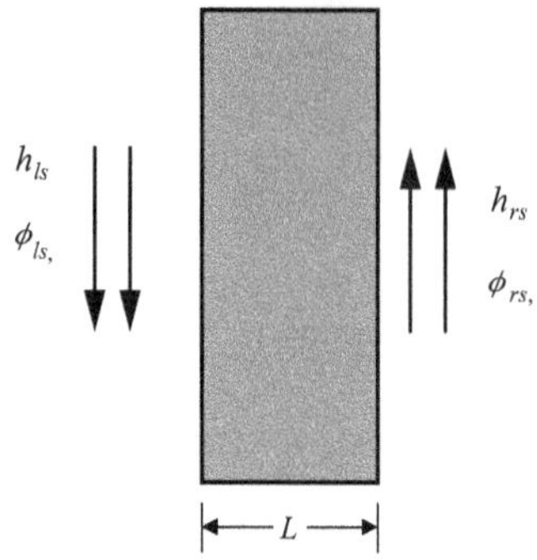

FIGURE 13.6   Convective boundary conditions.

13.3.2 MIXED BOUNDARY CONDITIONS

Let us consider the same one-dimensional problem with the boundary condition of the third kind or the mixed boundary condition at left and right boundary of the plane slab as shown in Figure 13.6. The mathematical statement of the problem is as follows:

Governing Equation

$$\frac{d}{dx}\left(\Gamma_x \frac{d\phi}{dx}\right) + \bar{S} = 0 \tag{13.59}$$

Boundary Conditions

$$1. \quad x = 0, \quad \Gamma_x \frac{d\phi}{dx}\bigg|_{x=0} = h_{ls}\left(\phi - \phi_{ls,\infty}\right) \tag{13.60a}$$

$$2. \quad x = L, -\Gamma_x \frac{d\phi}{dx}\bigg|_{x=L} = h_{rs}\left(\phi - \phi_{rs,\infty}\right) \tag{13.60b}$$

The **Galerkin-based finite element formulation** will lead to the same element characteristic equation as in Section 13.1, i.e.,

$$\left[\kappa^{(e)}\right]\left\{\phi^{(e)}\right\} = -\left\{I^{(e)}\right\} + \left\{f^{(e)}\right\} \tag{13.61}$$

where the interelement vectors for the element #1 and element #n need to be modified to implement the mixed boundary conditions. To do that, these interelement vectors are split into two vectors as follows:

Element #1

$$\left\{I^{(1)}\right\} = \left\{ \begin{array}{c} \left(\Gamma_x \frac{d\phi}{dx}\right)^{(1)}\bigg|_{x_i} \\ \left(-\Gamma_x \frac{d\phi}{dx}\right)^{(1)}\bigg|_{x_j} \end{array} \right\} = \left\{ \begin{array}{c} \left(\Gamma_x \frac{d\phi}{dx}\right)^{(1)}\bigg|_{x_i} \\ 0 \end{array} \right\} + \left\{ \begin{array}{c} 0 \\ \left(-\Gamma_x \frac{d\phi}{dx}\right)^{(1)}\bigg|_{x_j} \end{array} \right\} \tag{13.62a}$$

or

$$\left\{I^{(1)}\right\} = \left\{I_{b,ls}^{(1)}\right\} + \left\{I_j^{(1)}\right\} \tag{13.62b}$$

where $\left\{I_{b,ls}^{(1)}\right\}$ and $\left\{I_j^{(1)}\right\}$ are associated with the boundary conditions at node i located on the left surface and interelement requirement at node j, respectively.

Element #*n*

$$\left\{I^{(n)}\right\}=\left\{\begin{array}{c}\left(\Gamma_x\dfrac{\mathrm{d}\phi}{\mathrm{d}x}\right)^{(n)}\Bigg|_{x_i}\\ \left(-\Gamma_x\dfrac{\mathrm{d}\varphi}{\mathrm{d}x}\right)^{(n)}\Bigg|_{x_j}\end{array}\right\}=\left\{\begin{array}{c}\left(\Gamma_x\dfrac{\mathrm{d}\phi}{\mathrm{d}x}\right)^{(n)}\Bigg|_{x_i}\\ 0\end{array}\right\}+\left\{\begin{array}{c}0\\ \left(-\Gamma_x\dfrac{\mathrm{d}\phi}{\mathrm{d}x}\right)^{(n)}\Bigg|_{x_j}\end{array}\right\} \tag{13.63a}$$

or

$$\left\{I^{(n)}\right\}=\left\{I_i^{(n)}\right\}+\left\{I_{b,\mathrm{rs}}^{(n)}\right\} \tag{13.63b}$$

where $\left\{I_i^{(n)}\right\}$ and $\left\{I_{b,\mathrm{rs}}^{(n)}\right\}$ are the components of interelement vector associated with the interelement requirement at node i and the boundary conditions at node j located on the right surface, respectively.

In the next step, we implement the mixed boundary conditions (Equations 13.60) directly into the interelement vectors for boundary nodes as follows:

Element #1

$$\left\{I_{b,\mathrm{ls}}^{(1)}\right\}=\left\{\begin{array}{c}h_{\mathrm{ls}}(\phi_i-\phi_{\mathrm{ls},\infty})\\ 0\end{array}\right\}=\left\{\begin{array}{c}h_{\mathrm{ls}}\phi_i\\ 0\end{array}\right\}-\left\{\begin{array}{c}h_{\mathrm{ls}}\phi_{\mathrm{ls},\infty})\\ 0\end{array}\right\} \tag{13.64a}$$

or

$$\left\{I_{b,\mathrm{ls}}^{(1)}\right\}=\left[\begin{array}{cc}h_{\mathrm{ls}} & 0\\ 0 & 0\end{array}\right]\left\{\begin{array}{c}\phi_i\\ \phi_j\end{array}\right\}-\left\{\begin{array}{c}h_{\mathrm{ls}}\phi_{\mathrm{ls},\infty})\\ 0\end{array}\right\} \tag{13.64b}$$

or

$$\left\{I_{b,\mathrm{ls}}^{(1)}\right\}=\left[K_{b,\mathrm{ls}}^{(1)}\right]\left\{\phi^{(1)}\right\}-\left\{f_{b,\mathrm{ls}}^{((1))}\right\} \tag{13.64c}$$

where

$\left[K_{b,\mathrm{ls}}^{(1)}\right]=\left[\begin{array}{cc}h_{\mathrm{ls}} & 0\\ 0 & 0\end{array}\right]$ = component of element characteristics matrix due to the mixed boundary condition at the left boundary surface (13.65a)

$\left\{f_{b,\mathrm{ls}}^{(n)}\right\}=\left\{\begin{array}{c}h_{\mathrm{ls}}\phi_{\mathrm{ls},\infty}\\ 0\end{array}\right\}$ = component of element source vector due to the mixed boundary conditions at the left boundary surface (13.65b)

Similarly, for

Element #*n*

$$\left\{I_{b,\mathrm{rs}}^{(n)}\right\}=\left\{\begin{array}{c}0\\ h_{\mathrm{rs}}(\phi_i-r_{\mathrm{ls},\infty})\end{array}\right\}=\left\{\begin{array}{c}0\\ h_{\mathrm{rs}}\phi_i\end{array}\right\}-\left\{\begin{array}{c}0\\ h_{\mathrm{rs}}\varphi_{\mathrm{rs},\infty}\end{array}\right\} \tag{13.66a}$$

or

$$\left\{I_{b,\mathrm{rs}}^{(n)}\right\}=\left[\begin{array}{cc}0 & 0\\ 0 & h_{\mathrm{rs}}\end{array}\right]\left\{\begin{array}{c}\phi_i\\ \phi_j\end{array}\right\}-\left\{\begin{array}{c}0\\ h_{\mathrm{rs}}\varphi_{\mathrm{rs},\infty}\end{array}\right\} \tag{13.66b}$$

or

$$\left\{I_{b,\mathrm{rs}}^{(n)}\right\}=\left[K_{b,\mathrm{rs}}^{(n)}\right]\left\{\phi^{(n)}\right\}-\left\{f_{b,\mathrm{rs}}^{(n)}\right\} \tag{13.66c}$$

where

$$\left[K_{b,\mathrm{rs}}^{(n)}\right]=\left[\begin{array}{cc}0 & 0\\ 0 & h_{\mathrm{rs}}\end{array}\right] \tag{13.67a}$$

= component of element characteristics matrix due to the boundary condition at the right boundary surface

$$\left\{f_{b,\mathrm{rs}}^{(n)}\right\} = \left\{\begin{array}{c} 0 \\ h_{\mathrm{rs}}\phi_{\mathrm{rs},\infty} \end{array}\right\} = \text{component of element source vector due to the mixed boundary condition at the right boundary surface} \quad (13.67b)$$

We now substitute Equations 13.64–13.67 into Equation 13.63 to obtain the interelement vectors as follows:

Element #1:

$$\left\{I^{(1)}\right\} = \left[K_{b,\mathrm{ls}}^{(1)}\right]\left\{\phi^{(1)}\right\} - \left\{f_{b,\mathrm{ls}}^{(1)}\right\} + \left\{I_j^{(1)}\right\} \quad (13.68a)$$

and

Element #*n*:

$$\left\{I^{(n)}\right\} = \left\{I_i^{(n)}\right\} + \left[K_{b,\mathrm{rs}}^{(n)}\right]\left\{\phi^{(n)}\right\} - \left\{f_{b,\mathrm{rs}}^{(n)}\right\} \quad (13.68b)$$

Finally, we substitute Equation 13.68 into Equation 13.61 to obtain the **element characteristic equations** for the **element #1 and #*n*** located at the boundaries as follows:

Element #1:

$$\left(\left[K_\Gamma^{(1)}\right] + \left[K_{b,\mathrm{ls}}^{(1)}\right]\right)\left\{\phi^{(1)}\right\} = -\left\{I_j^{(1)}\right\} + \left\{f_s^{(1)}\right\} + \left\{f_{b,\mathrm{ls}}^{(1)}\right\} \quad (13.69a)$$

and

Element #*n*:

$$\left(\left[K_\Gamma^{(n)}\right] + \left[K_{b,\mathrm{rs}}^{(n)}\right]\right)\left\{\phi^{(n)}\right\} = -\left\{I_i^{(n)}\right\} + \left\{f_s^{(n)}\right\} + \left\{f_{b,\mathrm{ls}}^{(n)}\right\} \quad (13.69b)$$

It can be noticed that all elements, which have a boundary node, will have additional components for the element characteristic matrix and the source vector. The characteristic equations for rest of the elements will remain the same as that given by Equations 13.18b–13.18n.

The assembly of all element equations will lead to the global system as follows:

$$\begin{bmatrix} \frac{\Gamma_x}{L_e} + h_{ls} & -\frac{\Gamma_x}{L_e} & & & & \\ -\frac{\Gamma_x}{L_e} & \frac{\Gamma_x}{L_e} + \frac{\Gamma_x}{L_e} & -\frac{\Gamma_x}{L_e} & & & \\ & -\frac{\Gamma_x}{L_e} & \frac{\Gamma_x}{L_e} + \frac{\Gamma_x}{L_e} & -\frac{\Gamma_x}{L_e} & & \\ & & -\frac{\Gamma_x}{L_e} & \frac{\Gamma_x}{L_e} + \frac{\Gamma_x}{L_e} & & \\ \vdots & & & & & \\ & & & & \frac{\Gamma_x}{L_e} + \frac{\Gamma_x}{L_e} & -\frac{\Gamma_x}{L_e} \\ & & & & -\frac{\Gamma_x}{L_e} & \frac{\Gamma_x}{L_e} + h_{rs} \end{bmatrix} \left\{\begin{array}{c} \phi_1 \\ \phi_2 \\ \phi_3 \\ \vdots \\ \vdots \\ \phi_{n-1} \\ \phi_n \end{array}\right\}$$

$$
= \begin{Bmatrix} 0 \\ \Gamma_x \left.\dfrac{d\phi}{dx}\right|_{x2} - \Gamma_x \left.\dfrac{d\phi}{dx}\right|_{x2} \\ \Gamma_x \left.\dfrac{d\phi}{dx}\right|_{x3} - \Gamma_x \left.\dfrac{d\phi}{dx}\right|_{x3} \\ \vdots \\ \vdots \\ \Gamma_x \left.\dfrac{d\phi}{dx}\right|_{x_n} - \Gamma_x \left.\dfrac{d\phi}{dx}\right|_{x_n} \\ 0 \end{Bmatrix} + \begin{Bmatrix} \dfrac{\bar{S}L_e}{2} + h_{\text{ls}}\phi_{b,\text{ls}} \\ \dfrac{\bar{S}L_e}{2} + \dfrac{\bar{S}L_e}{2} \\ \dfrac{\bar{S}L_e}{2} + \dfrac{\bar{S}L_e}{2} \\ \vdots \\ \vdots \\ \dfrac{\bar{S}L_e}{2} + \dfrac{\bar{S}L_e}{2} \\ \dfrac{\bar{S}L_e}{2} + h_{\text{rs}}\phi_{b,\text{rs}} \end{Bmatrix} + \begin{Bmatrix} h_{\text{ls}}\phi_{b,\text{ls}} \\ 0 \\ 0 \\ \vdots \\ \vdots \\ 0 \\ h_{\text{rs}}\phi_{b,\text{rs}} \end{Bmatrix} \tag{13.70}
$$

Simplifying and canceling terms due to interelement continuity conditions, we get final **global system** as follows:

$$
\begin{bmatrix} \dfrac{\Gamma_x}{L_e} + h_{\text{ls}} & -\dfrac{\Gamma_x}{L_e} & & & & \\ -\dfrac{\Gamma_x}{L_e} & \dfrac{2\Gamma_x}{L_e} & -\dfrac{\Gamma_x}{L_e} & & & \\ & -\dfrac{\Gamma_x}{L_e} & \dfrac{2\Gamma_x}{L_e} & -\dfrac{\Gamma_x}{L_e} & & \\ & & -\dfrac{\Gamma_x}{L_e} & \dfrac{2\Gamma_x}{L_e} & & \\ \vdots & & & & & \\ & & & & \dfrac{2\Gamma_x}{L_e} & -\dfrac{\Gamma_x}{L_e} \\ & & & & -\dfrac{\Gamma_x}{L_e} & \dfrac{\Gamma_x}{L_e} + h_{\text{rs}} \end{bmatrix} \begin{Bmatrix} \phi_1 \\ \phi_2 \\ \phi_3 \\ \vdots \\ \vdots \\ \phi_{n-1} \\ \phi_n \end{Bmatrix} = \begin{Bmatrix} \dfrac{\bar{S}L_e}{2} + h_{\text{ls}}\phi_{b,\text{ls}} \\ \bar{S}L_e \\ \bar{S}L_e \\ \vdots \\ \vdots \\ \bar{S}L_e \\ \dfrac{\bar{S}L_e}{2} + h_{\text{rs}}\phi_{b,\text{rs}} \end{Bmatrix} \tag{13.71}
$$

The global system can be written in a compact form as follows:

$$
[\mathrm{K}]\{\Phi\} = \{F\} \tag{13.72}
$$

where

$$[K]=\begin{bmatrix} \frac{\Gamma_x}{L_e}+h_{\mathrm{ls}} & -\frac{\Gamma_x}{L_e} & & & & \\ -\frac{\Gamma_x}{L_e} & \frac{2\Gamma_x}{L_e} & -\frac{\Gamma_x}{L_e} & & & \\ & -\frac{\Gamma_x}{L_e} & \frac{2\Gamma_x}{L_e} & -\frac{\Gamma_x}{L_e} & & \\ & & -\frac{\Gamma_x}{L_e} & \frac{2\Gamma_x}{L_e} & & \\ \vdots & & & & & \\ & & & & \frac{2\Gamma_x}{L_e} & -\frac{\Gamma_x}{L_e} \\ & & & & -\frac{\Gamma_x}{L_e} & \frac{\Gamma_x}{L_e}+h_{\mathrm{rs}} \end{bmatrix}$$

= global characteristics matrix

(13.73)

$$\{\Phi\}=\begin{Bmatrix} \phi_1 \\ \phi_2 \\ \phi_3 \\ \vdots \\ \vdots \\ \phi_n \\ \phi_{n+1} \end{Bmatrix} = \text{Vector for unknown nodal values in the solution domain} \tag{13.74}$$

$$\{F\}=\begin{Bmatrix} \frac{\bar{S}L_e}{2}+h_{\mathrm{ls}}\phi_{b,\mathrm{ls}} \\ \bar{S}L_e \\ \bar{S}L_e \\ \vdots \\ \vdots \\ \bar{S}L_e \\ \frac{\bar{S}L_e}{2}+h_{\mathrm{rs}}\phi_{b,\mathrm{rs}} \end{Bmatrix} = \text{Source or force vector} \tag{13.75}$$

The contribution of $\left[\kappa_b^{(e)}\right]$ to global characteristics matrix $[K]$ and $\left\{f_b^{(e)}\right\}$ to the global force vector occurs for boundary nodes with mixed boundary conditions or with non-zero convection coefficient, h.

Example 13.4: One-dimensional Steady State Conduction in Composite Wall

Consider a building wall made of four different homogeneous material layers. The thicknesses of these layers are: $l_A = 15\,\text{cm}$, $l_B = 20\,\text{cm}$, $l_C = 10\,\text{cm}$ and $l_D = 5\,\text{cm}$. The thermal conductivities of the corresponding layers are: k_A =1.3 W/m C, $k_B = 0.043$ W/m C, $k_C = 0.079$ W/m C and $k_D = 0.48$ W/m C respectively. The inside and outside convection heat transfer coefficient are $h_i = 10/\text{m}^2\,\text{C}$ and $h_o = 40/\text{m}^2\,\text{C}$. The inside temperature is $T_i = 25°\text{C}$, and the outside temperature is $T_0 = 10°\text{C}$. Determine the temperatures at the internal and the external surfaces as well temperatures at the interfaces of different materials using Galerkin-based finite element formulation and the mesh size distribution shown.

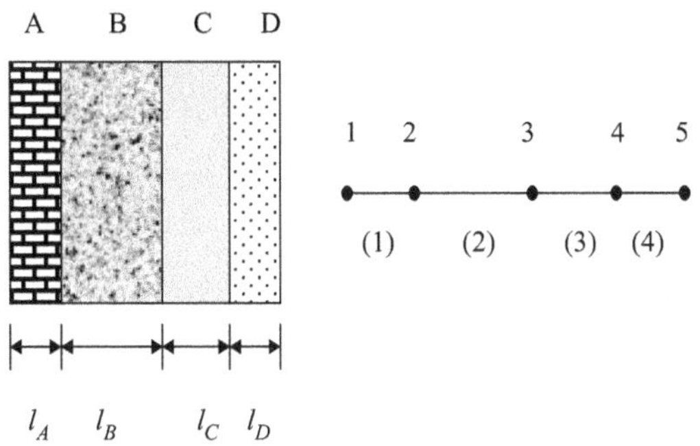

The mathematical statement of the problem is given as:

Governing Equation:

$$\frac{d}{dx}\left(kA\frac{dT_I}{dx}\right) = 0 \tag{E13.4.1}$$

Where subscript **I** is the index for different material layers.

Boundary Conditions:

$$1.\ x = 0,\ k\frac{dT}{dx} = h_{in}\left(T - T_{in}\right) \tag{E.13.4.2}$$

$$2.\ x = L,\ -k\frac{dT}{dx} = h_{out}\left(T - T_{out}\right) \tag{E13.4.3}$$

Solution:

For simplicity, let us discretize the solution domain with a non uniform mesh size distribution with one element in each layer of material as shown in the figure. Based on the formulation presented in Section 13.3.2, we select the element characteristic equations as follows:

Element #1: Material A with convective boundary on left side. From Equations 13.61, 13.62, 13.64, 13.65 and 13.68 and 13.69a

$$\left(\left[\kappa_A^{(1)}\right] + \left[\kappa_{b,ls}^{(1)}\right]\right)\left\{T^{(1)}\right\} = -\left\{I_j^{(1)}\right\} + \left\{f_{b,ls}^{(1)}\right\}$$

Where

$$\left[\kappa_A^{(1)}\right]=\frac{k_A}{l_A}\begin{bmatrix} 1 & -1 \\ -1 & 1 \end{bmatrix},\left[\kappa_{b,ls}^{(1)}\right]=\begin{bmatrix} h_{\text{In}} & 0 \\ 0 & 0 \end{bmatrix},\left\{f_{b,ls}^{n)}\right\}=\begin{Bmatrix} h_{in}T_{in} \\ 0 \end{Bmatrix}$$

$$\left\{I_j^{(1)}\right\}=\begin{Bmatrix} 0 \\ \left(-k_A\dfrac{dT}{dx}\right)^{(1)}\Bigg|_{xj} \end{Bmatrix},\left\{T^{(1)}\right\}=\begin{Bmatrix} T_1 \\ T_2 \end{Bmatrix}$$

Substituting, we get the characteristic equation for element #1 as

$$\begin{bmatrix} \dfrac{k_A}{l_A}+h_{in} & -\dfrac{k_A}{l_A} \\ -\dfrac{k_A}{l_A} & \dfrac{k_A}{l_A} \end{bmatrix}\begin{Bmatrix} T_1^{(1)} \\ T_2^{(1)} \end{Bmatrix}=\begin{Bmatrix} 0 \\ \left(k_A\dfrac{dT}{dx}\right)^{(1)}\Bigg|_{xj} \end{Bmatrix}+\begin{Bmatrix} h_{in}T_{in} \\ 0 \end{Bmatrix} \tag{E13.4.4a}$$

We can use Equations 13.16 and 13.17 for internal elements 2 and 3 as follows:

Element #2: Material B

$$\left[\kappa_B^{(2)}\right]\left\{T^{(2)}\right\}=\left\{I^{(2)}\right\}$$

Where

$$\left[\kappa_B^{(2)}\right]=\frac{k_B}{l_B}\begin{bmatrix} 1 & -1 \\ -1 & 1 \end{bmatrix},\left\{I^{(2)}\right\}=\begin{Bmatrix} -k_B\dfrac{dT}{dx}\Bigg|_{xi} \\ k_B\dfrac{dT}{dx}\Bigg|_{xj} \end{Bmatrix}=\begin{Bmatrix} I_i^{(e)} \\ I_j^{(e)} \end{Bmatrix},\left\{\phi^{(2)}\right\}=\begin{Bmatrix} T_2 \\ T_3 \end{Bmatrix}$$

Substituting, we get the characteristic equation for element #2 as

$$\begin{bmatrix} \dfrac{k_B}{l_B} & -\dfrac{k_B}{l_B} \\ -\dfrac{k_B}{l_B} & \dfrac{k_B}{lB} \end{bmatrix}\begin{Bmatrix} T_2^{(2)} \\ T_3^{(2)} \end{Bmatrix}=\begin{Bmatrix} -k_B\dfrac{dT}{dx}\Bigg|_{xi} \\ k_B\dfrac{dT}{dx}\Bigg|_{xj} \end{Bmatrix} \tag{E13.4.4b}$$

Element #3: Material C

$$\left[\kappa_C^{(3)}\right]\left\{T^{(3)}\right\}=\left\{I^{(3)}\right\}$$

Where

$$\left[\kappa_C^{(3)}\right]=\frac{k_C}{l_C}\begin{bmatrix} 1 & -1 \\ -1 & 1 \end{bmatrix},\left\{I^{(3)}\right\}=\begin{Bmatrix} -k_C\dfrac{dT}{dx}\Bigg|_{xi} \\ k_C\dfrac{dT}{dx}\Bigg|_{xj} \end{Bmatrix}=\begin{Bmatrix} I_i^{(3)} \\ I_j^{(3)} \end{Bmatrix},\left\{\phi^{(3)}\right\}=\begin{Bmatrix} T_3 \\ T_4 \end{Bmatrix}$$

Substituting, we get the characteristic equation for element #3 as

$$\begin{bmatrix} \dfrac{k_C}{l_C} & -\dfrac{k_C}{l_B} \\ -\dfrac{k_C}{l_C} & \dfrac{k_C}{l_C} \end{bmatrix} \begin{Bmatrix} T_3^{(3)} \\ T_4^{(3)} \end{Bmatrix} = \begin{Bmatrix} -k_C \left.\dfrac{dT}{dx}\right|_{x_i} \\ k_C \left.\dfrac{dT}{dx}\right|_{x_j} \end{Bmatrix} \tag{E13.4.4c}$$

Element #4: Material D with convective boundary on right side. From Equations 13.61, 13.63, 13.66, 13.67, and 13.69b

$$\left(\left[\kappa_D^{(4)}\right]+\left[\kappa_{b,rs}^{(4)}\right]\right)\left\{T^{(4)}\right\} = -\left\{I_i^{(4)}\right\}+\left\{f_{b,ls}^{(4)}\right\}$$

Where

$$\left[\kappa_A^{(4)}\right] = \frac{k_D}{l_D}\begin{bmatrix} 1 & -1 \\ -1 & 1 \end{bmatrix}, \left[\kappa_{b,rs}^{(4)}\right] = \begin{bmatrix} 0 & 0 \\ 0 & h_{out} \end{bmatrix}, \left\{f_{b,rs}^{4)}\right\} = \begin{Bmatrix} 0 \\ h_{out}T_{out} \end{Bmatrix}$$

$$\left\{I_i^{(\$)}\right\} = \begin{Bmatrix} \left.\left(K_D \dfrac{dT}{dx}\right)^{(4)}\right|_{x_i} \\ 0 \end{Bmatrix}, \left\{T^{(4)}\right\} = \begin{Bmatrix} T_4 \\ T_5 \end{Bmatrix}$$

Substituting we get the characteristic equation for the element #4 as

$$\begin{bmatrix} \dfrac{k_D}{l_D} & -\dfrac{k_D}{l_D} \\ -\dfrac{k_D}{l_D} & \dfrac{k_D}{l_D}+h_{out} \end{bmatrix} \begin{Bmatrix} T_4^{(4)} \\ T_5^{(4)} \end{Bmatrix} = \begin{Bmatrix} -\left.\left(K_D \dfrac{dT}{dx}\right)^{(4)}\right|_{x_i} \\ 0 \end{Bmatrix} + \begin{Bmatrix} 0 \\ h_{out}T_{out} \end{Bmatrix} \tag{E13.4.4d}$$

Assembly of all elements leads to the global system as

$$\begin{bmatrix} \dfrac{k_A}{l_A}+h_{in} & -\dfrac{k_A}{l_A} & & & \\ -\dfrac{k_A}{l_A} & \dfrac{k_A}{l_A}+\dfrac{k_B}{l_B} & -\dfrac{k_B}{l_B} & & \\ & -\dfrac{k_B}{l_B} & \dfrac{k_B}{l_B}+\dfrac{k_C}{l_C} & -\dfrac{k_C}{l_C} & \\ & & -\dfrac{k_C}{l_C} & \dfrac{k_C}{l_C}+\dfrac{k_D}{l_D} & -\dfrac{k_D}{l_D} \\ & & & -\dfrac{k_D}{l_D} & \dfrac{k_D}{l_D}+h_{out} \end{bmatrix} \begin{Bmatrix} T_1 \\ T_2 \\ T_3 \\ T_4 \\ T_5 \end{Bmatrix} = \begin{Bmatrix} h_{in}T_{in} \\ 0 \\ 0 \\ 0 \\ h_{out}T_{out} \end{Bmatrix} \tag{E13.4.5}$$

The numerical parameters for each elements are:

Element #A:

$$\frac{k_A}{l_A} = \frac{1.3}{0.15} = 8.666\,\mathrm{w/m^2 {}^\circ C},\ h_{in} = 10\,\mathrm{W/m^2C},\ h_{in}T_{in} = 10\times 25 = 250\,\mathrm{W/m^2}$$

Element #B:

$$\frac{k_B}{l_B} = \frac{0.043}{0.2} = 0.215\,\text{w/m}^2\,°\text{C}$$

Element #C:

$$\frac{k_C}{l_C} = \frac{0.079}{0.1} = 0.79\,\text{w/m}^2\,°\text{C}$$

Element #D:

$$\frac{k_D}{l_D} = \frac{0.48}{0.05} = 9.6\,\text{w/m}^2\,°\text{C},\; h_{out} = 40\,\text{W/m}^2\text{C},\; h_{out}T_{out} = 40 \times 10 = 400\,\text{W/m}^2$$

Substituting all numerical parameters into Equation E13.4.5, we get the global system as

$$\begin{bmatrix} 18.666 & -8.666 & & & \\ -8.666 & 8.881 & -0.215 & & \\ & -0.215 & 1.005 & -0.79 & \\ & & -0.79 & 10.39 & -9.6 \\ & & & -9.6 & 49.6 \end{bmatrix} \begin{Bmatrix} T_1 \\ T_2 \\ T_3 \\ T_4 \\ T_5 \end{Bmatrix} = \begin{Bmatrix} 250 \\ 0 \\ 0 \\ 0 \\ 400 \end{Bmatrix} \tag{E13.4.6}$$

Solution to the global system is

$$\begin{Bmatrix} T_1 \\ T_2 \\ T_3 \\ T_4 \\ T_5 \end{Bmatrix} = \begin{Bmatrix} 24.76 \\ 24.48 \\ 13.34 \\ 10.309 \\ 10.06 \end{Bmatrix}$$

13.4 VARIABLE SOURCE TERM

Let us now consider the case where the source term is assumed to be a linear function of the dependent variable as

$$S(\phi) = S_0 + S_1\phi \tag{13.76}$$

With this variable heat source, the general form of the governing equation can be written as

$$\frac{d}{dx}\left(\Gamma_x \frac{d\phi}{dx}\right) + S_1\phi + S_0 = 0 \tag{13.77}$$

Boundary Conditions:

$$3.\quad x = 0, T(0) = T_L \tag{13.78a}$$

$$4.\quad x = L, T(0) = T_R \tag{13.78b}$$

Let us demonstrate the application of finite element method to this problem using the Galerkin-based the finite element method as outlined in Section 13.1. Based on this procedure, we get the weak form of the problem as

$$-\left(\Gamma_x \frac{d\phi^*}{dx}\right)\Bigg|_{x_i} - \int_{x_i}^{x_j} \Gamma_x \frac{dN_i}{dx}\frac{d\phi^*}{dx} dx + \int_{x_i}^{x_j} S_0 N_i(x) dx + \int_{x_i}^{x_j} S_1 N_i(x)\phi^* \, dx = 0 \tag{13.79a}$$

and

$$\left(\Gamma_x \frac{d\phi^*}{dx}\right)\Bigg|_{x_j} - \int_{x_i}^{x_j} \Gamma_x \frac{dN_j}{dx}\frac{d\phi^*}{dx} dx + \int_{x_i}^{x_j} S_0 N_j(x) dx + \int_{x_i}^{x_j} S_1 N_j(x)\phi^* \, dx = 0 \tag{13.79b}$$

In the next step the element characteristics equation is obtained by evaluating the integrals and by substituting the approximate solution for the chosen element type. The first three terms of the Equation 13.79 are evaluated in Section 13.1 and leads to the element characteristic equations as

$$\left\{I^{(e)}\right\} + \left[\kappa_\Gamma^{(e)}\right]\left\{\phi^{(e)}\right\} - \left(f^{(e)}\right) = 0 \tag{13.80}$$

Combining Equations 13.79 and 13.80, we get

$$\left\{I^{(e)}\right\} + \left[\kappa_\Gamma^{(e)}\right]\left\{\phi^{(e)}\right\} - \left(f^{(e)}\right) + \int_{x_i}^{x_j} S_1 [N]^T \phi^* dx \tag{13.81}$$

The fourth integral term will have additional contribution to the element equation. Let us evaluate the term as follows:

$$\int_{x_i}^{x_j} S_1 N_i \phi^* \, dx = \int_{x_i}^{x_j} S_1 N_i \left(N_i \phi_i + N_j \phi_j\right) dx = \int_{x_i}^{x_j} S_1 N_i [N] \begin{Bmatrix} \phi_i \\ \phi_j \end{Bmatrix} dx \tag{13.82a}$$

$$= \int_{x_i}^{x_j} S_1 \left[N_i^2 \phi_i + N_i N_j \phi_j\right] dx \tag{13.82b}$$

and

$$\int_{x_i}^{x_j} S_1 N_j \phi^* \, dx = \int_{x_i}^{x_j} S_1 N_j \left(N_i \phi_i + N_j \phi_j\right) dx = \int_{x_i}^{x_j} S_1 N_j [N] \begin{Bmatrix} \phi_i \\ \phi_j \end{Bmatrix} dx \tag{13.83a}$$

$$= \int_{x_i} S_1 \left[N_j N_i \phi_i + N_j^2\right] dx \tag{13.83b}$$

Equations 13.82a and 13.83 can also be written in a matrix form as

$$\int_{x_i}^{x_j} S_1 [N]^T \phi^* dx = \left(\int_{x_i}^{x_j} S_1 \begin{bmatrix} N_i^2 & N_i N_j \\ N_j N_i & N_j^2 \end{bmatrix}^{(e)} dx \right) \begin{Bmatrix} \phi_i \\ \phi_j \end{Bmatrix} \tag{13.84a}$$

or

$$\int S_1 [N]^T [N] dx = \left[\kappa_S^{(e)}\right] \begin{Bmatrix} \phi_i \\ \phi_j \end{Bmatrix} \tag{13.84b}$$

where

$$\left[\kappa_S^{(e)}\right]=\int_{x_i}^{x_j} S_1\begin{bmatrix} N_i^2 & N_iN_j \\ N_jN_i & N_j^2 \end{bmatrix}^{(e)} \mathrm{d}x \tag{13.85}$$

The integrals appearing as the elements of the matrix $\left[k_S^{(e)}\right]$ can be evaluated either by numerical **quadrature formulas** as discussed in Chapter 4 or analytically, for some cases, as discussed in Chapter 12. As we have demonstrated in Chapter 12, for two-points line element, these integrals can be evaluated analytically using the natural coordinate systems and expresses as

$$\int_{x_i}^{x_j} N_i^2\,\mathrm{d}x=\frac{L_e}{3},\quad \int_{x_i}^{x_j} N_iN_j\,\mathrm{d}x=\frac{L_e}{6},\quad \int_{x_i}^{x_j} N_j^2\,\mathrm{d}x=\frac{L_e}{3},\quad \int_{x_i}^{x_j} N_jN_i\,\mathrm{d}x=\frac{L_e}{6} \tag{13.86}$$

Substituting Equation 13.86 into Equations 13.82 and 13.83, we have

$$\int_{x_i}^{x_j} S_1N_i\phi^{(*)}\mathrm{d}x=\frac{S_1L_e}{6}\left(2\phi_i+\phi_j\right) \tag{13.87a}$$

and

$$\int_{x_i}^{x_j} S_1N_j\phi^{(*)}\mathrm{d}x=\frac{S_1L_e}{3}\left(\phi_i+2\phi_j\right) \tag{13.87b}$$

Or in matrix form as

$$\left(\int_{x_i}^{x_j} S_1[N]^T[N]\mathrm{d}x\right)\left\{\phi^{(e)}\right\}=\left[k_S^{(e)}\right]\left\{\phi^{(e)}\right\}=\frac{S_1L_e}{6}\begin{bmatrix} 2 & 1 \\ 1 & 2 \end{bmatrix}^{(e)}\begin{Bmatrix} \phi_i \\ \phi_j \end{Bmatrix} \tag{13.87c}$$

Substituting Equation 13.87 into Equation 13.81, we obtain the element equation for the problem as

$$\left(\left[\kappa_\Gamma^{(e)}\right]+\left[\kappa_S^{(e)}\right]\right)\left\{\phi^{(e)}\right\}=-\left\{I^{(e)}\right\}+\left(f^{(e)}\right) \tag{13.88}$$

where

$$\left[\kappa_\Gamma^{(e)}\right]=\frac{\Gamma_x}{L_e}\begin{bmatrix} 1 & -1 \\ -1 & 1 \end{bmatrix}=\text{component of element characteristic matrix due to the diffusion term} \tag{13.89a}$$

$$\left[\kappa_\Gamma^{(e)}\right]=\frac{\Gamma_x}{L_e}\begin{bmatrix} 1 & -1 \\ -1 & 1 \end{bmatrix}=\text{component of element characteristic matrix due to the source term} \tag{13.89b}$$

$$\left\{f^{(e)}\right\}=\begin{Bmatrix} \dfrac{S_0L_e}{2} \\ \dfrac{S_0L_e}{2} \end{Bmatrix}=\begin{Bmatrix} f_i^{(e)} \\ f_j^{(e)} \end{Bmatrix}=\text{element source or force vector} \tag{13.89c}$$

$$\{I^{(e)}\} = \left\{ \begin{array}{c} -\Gamma_x \left.\dfrac{d\phi}{dx}\right|_{x_i} \\ \Gamma_x \left.\dfrac{d\phi}{dx}\right|_{x_j} \end{array} \right\} = \left\{ \begin{array}{c} I_i^{(e)} \\ I_j^{(e)} \end{array} \right\} \tag{13.89d}$$

Example 13.5: One-Dimensional Steady State Conduction in a Circular Pin Fin

An aluminum pin fin having a diameter D = 1 cm and length L = 6 cm is exposed to surrounding fluid with temperature, $T_\infty = 20°\text{C}$ and convection heat transfer coefficient, $h = 8.0\dfrac{\text{W}}{\text{m}^2 \cdot \text{C}}$. The fin base temperature is $T_0 = 100°\text{C}$. Determine the temperature distribution and fin heat loss. Assume thermal conductivity of the material as $k = 200\dfrac{\text{W}}{\text{m} \cdot °\text{C}}$.

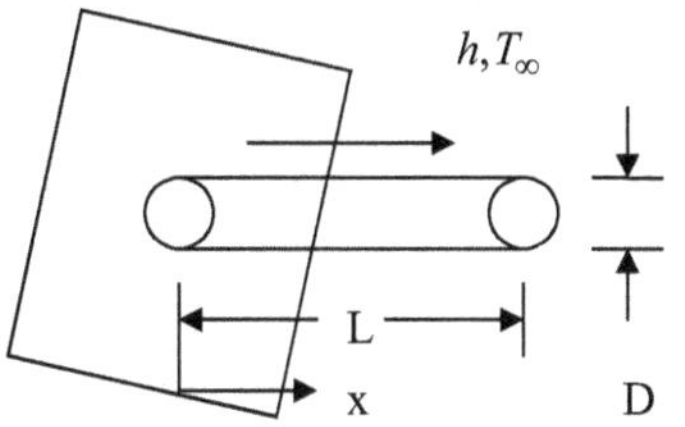

The mathematical statement of the problem is given as
Governing Equation:

$$\frac{d}{dx}\left(kA\frac{dT}{dx}\right) - hp(T - T_\infty) = 0 \tag{E13.5.1}$$

Boundary Conditions:

$$1.\ x = 0,\ T = T_o \tag{E13.5.2}$$

$$2.\ x = 0,\ -k\frac{dT}{dx} = h(T - T_\infty) \tag{E13.5.3}$$

Solution:

Let us rearrange the governing Equation E13.5.1 as follows:

$$\frac{d}{dx}\left(kA\frac{dT}{dx}\right) - hpT + hpT_\infty = 0$$

Or

$$\frac{d}{dx}\left(kA\frac{dT}{dx}\right) + S_1T + S_0 = 0$$

Where

$$S_0 = hpT_\infty = 8.0 \times \pi \times 0.01\,\text{m} \times 20 = 5.024\frac{\text{W}}{\text{m}},\ S_1 = -hp = -8.0 \times \pi \times 0.01 = -0.2512\frac{\text{W}}{\text{m°C}}$$

$$\Gamma_x = kA = 200 \times \frac{\pi \times (0.01)^2}{4} = 0.0157 \frac{\text{W} \cdot \text{m}}{\text{C}},\ L_e = \frac{0.06\,\text{m}}{4} = 0.015\,\text{m}$$

$$\frac{kA}{L_e} = \frac{0.0157}{0.015} = 1.0466,\ \frac{S_1 L_e}{3} = \frac{-0.2512 \times 0.015}{3} = -0.001256,$$

$$\frac{S_1 L_e}{6} = \frac{-0.2512 \times 0.015}{6} = -0.000628,\ \frac{S_o L_e}{2} = \frac{5.024 \times 0.015}{2} = 0.03768$$

$$S_0 L_e = 5.024 \times 0.015 = 0.07536$$

Let us discretize the solution domain by four equal size line element as shown below

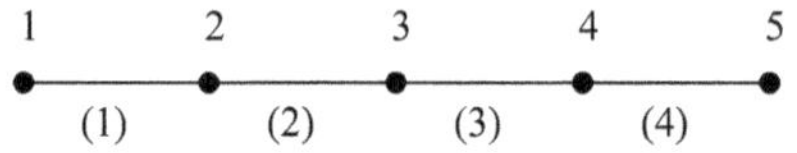

Following Equation 13.88, we can write the element characteristic matrix for elements 1–3 as

$$\left(\left[K_\Gamma^{(e)}\right] + \left[K_S^{(e)}\right]\right)\left\{T^{(e)}\right\} = -\left\{I^{(e)}\right\} + \left(f^{(e)}\right)$$

$$\left(\frac{kA}{L_e}\begin{bmatrix} 1 & -1 \\ -1 & 1 \end{bmatrix} + \frac{S_1 L_e}{6}\begin{bmatrix} 2 & 1 \\ 1 & 2 \end{bmatrix}\right)\begin{Bmatrix} \phi_i \\ \phi_j \end{Bmatrix} = -\begin{Bmatrix} -k\left.\dfrac{d\phi}{dx}\right|_{x_i} \\ k\left.\dfrac{d\phi}{dx}\right|_{x_j} \end{Bmatrix} + \begin{Bmatrix} \dfrac{S_0 L_e}{2} \\ \dfrac{S_0 L_e}{2} \end{Bmatrix}$$

$$\begin{bmatrix} \dfrac{kA}{L_e} + \dfrac{S_1 L_e}{3} & -\dfrac{kA}{L_e} + \dfrac{S_1 L_e}{6} \\ -\dfrac{kA}{L_e} + \dfrac{S_1 L_e}{6} & \dfrac{kA}{L_e} + \dfrac{S_1 L_e}{3} \end{bmatrix}\begin{Bmatrix} \phi_i \\ \phi_j \end{Bmatrix} = -\begin{Bmatrix} -k\left.\dfrac{d\phi}{dx}\right|_{x_i} \\ k\left.\dfrac{d\phi}{dx}\right|_{x_j} \end{Bmatrix} + \begin{Bmatrix} \dfrac{S_0 L_e}{2} \\ \dfrac{S_0 L_e}{2} \end{Bmatrix}$$

Applying this equation successively to element 1–3, we get

Element #1

$$\begin{bmatrix} \dfrac{kA}{L_e} + \dfrac{S_1 L_e}{3} & -\dfrac{kA}{L_e} + \dfrac{S_1 L_e}{6} \\ -\dfrac{kA}{L_e} + \dfrac{S_1 L_e}{6} & \dfrac{kA}{L_e} + \dfrac{S_1 L_e}{3} \end{bmatrix}\begin{Bmatrix} T_1^{(1)} \\ T_2^{(1)} \end{Bmatrix} = -\begin{Bmatrix} -k\left.\dfrac{d\phi}{dx}\right|_{x_1} \\ k\left.\dfrac{d\phi}{dx}\right|_{x_2} \end{Bmatrix} + \begin{Bmatrix} \dfrac{S_0 L_e}{2} \\ \dfrac{S_0 L_e}{2} \end{Bmatrix}$$

Element #2

$$\begin{bmatrix} \dfrac{kA}{L_e} + \dfrac{S_1 L_e}{3} & -\dfrac{kA}{L_e} + \dfrac{S_1 L_e}{6} \\ -\dfrac{kA}{L_e} + \dfrac{S_1 L_e}{6} & \dfrac{kA}{L_e} + \dfrac{S_1 L_e}{3} \end{bmatrix}\begin{Bmatrix} T_2^{(2)} \\ T_3^{(2)} \end{Bmatrix} = -\begin{Bmatrix} -k\left.\dfrac{d\phi}{dx}\right|_{x_2} \\ k\left.\dfrac{d\phi}{dx}\right|_{x_3} \end{Bmatrix} + \begin{Bmatrix} \dfrac{S_0 L_e}{2} \\ \dfrac{S_0 L_e}{2} \end{Bmatrix}$$

Element #3

$$\begin{bmatrix} \frac{\Gamma_x}{L_e}+\frac{S_1L_e}{3} & -\frac{\Gamma_x}{L_e}+\frac{S_1L_e}{6} \\ -\frac{\Gamma_x}{L_e}+\frac{S_1L_e}{6} & \frac{\Gamma_x}{L_e}+\frac{S_1L_e}{3} \end{bmatrix} \begin{Bmatrix} T_3^{(3)} \\ T_4^{(3)} \end{Bmatrix} = -\begin{Bmatrix} -k\left.\frac{d\phi}{dx}\right|_{x3} \\ k\left.\frac{d\phi}{dx}\right|_{x4} \end{Bmatrix} + \begin{Bmatrix} \frac{S_0L_e}{2} \\ \frac{S_0L_e}{2} \end{Bmatrix}$$

For **element #4** with convective boundary on right side, the element characteristic equation is

$$\left(\left[\kappa_\Gamma^{(5)}\right]+\left[\kappa_S^{(5)}\right]+\left[\kappa_b^{(5)}\right]\right)\left\{T^{(5)}\right\} = -\left\{I_i^{(4)}\right\}+\left(f^{(e)}+f_{b,rs}^4\right)$$

$$\begin{bmatrix} \frac{kA}{L_e}+\frac{S_1L_e}{3} & -\frac{kA}{L_e}+\frac{S_1L_e}{6} \\ -\frac{kA}{L_e}+\frac{S_1L_e}{6} & \frac{kA}{L_e}+\frac{S_1L_e}{3}+h \end{bmatrix} \begin{Bmatrix} T_4^{(4)} \\ T_5^{(4)} \end{Bmatrix} = -\begin{Bmatrix} -k_x\left.\frac{d\phi}{dx}\right|_{x4} \\ 0 \end{Bmatrix} + \begin{Bmatrix} \frac{S_0L_e}{2} \\ \frac{S_0L_e}{2}+hT_\infty \end{Bmatrix}$$

Assembly of all element equations leads to the global system as

$$\begin{bmatrix} \frac{kA}{L_e}+\frac{S_1L_e}{3} & -\frac{kA}{L_e}+\frac{S_1L_e}{6} & & & \\ -\frac{kA}{L_e}+\frac{S_1L_e}{6} & \frac{2kA}{L_e}+\frac{2S_1L_e}{3} & -\frac{kA}{L_e}+\frac{S_1L_e}{6} & & \\ & -\frac{kA}{L_e}+\frac{S_1L_e}{6} & \frac{2kA}{L_e}+\frac{2S_1L_e}{3} & -\frac{kA}{L_e}+\frac{S_1L_e}{6} & \\ & & -\frac{kA}{L_e}+\frac{S_1L_e}{6} & \frac{2kA}{L_e}+\frac{2S_1L_e}{3} & -\frac{kA}{L_e}+\frac{S_1L_e}{6} \\ & & & -\frac{kA}{L_e}+\frac{S_1L_e}{6} & \frac{2kA}{L_e}+\frac{2S_1L_e}{3}+h \end{bmatrix} \begin{Bmatrix} T_1 \\ T_2 \\ T_3 \\ T_4 \\ T_5 \end{Bmatrix}$$

$$= \begin{Bmatrix} k\left.\frac{\partial T}{\partial x}\right|_{x1}+\frac{S_0L_e}{2} \\ S_0L_e \\ S_0L_e \\ S_0L_e \\ \frac{S_0L_e}{2}+hT_\infty \end{Bmatrix}$$

With the application of the boundary condition at the base of the fin, i.e. $T_1 = T_o$, and rearranging, we get the global system as

$$\begin{bmatrix} k & -\frac{kA}{L_e}+\frac{S_1L_e}{6} & & & \\ 0 & \frac{2kA}{L_e}+\frac{2S_1L_e}{3} & -\frac{kA}{L_e}+\frac{S_1L_e}{6} & & \\ & -\frac{kA}{L_e}+\frac{S_1L_e}{6} & \frac{2kA}{L_e}+\frac{2S_1L_e}{3} & -\frac{kA}{L_e}+\frac{S_1L_e}{6} & \\ & & -\frac{kA}{L_e}+\frac{S_1L_e}{6} & \frac{2kA}{L_e}+\frac{2S_1L_e}{3} & -\frac{kA}{L_e}+\frac{S_1L_e}{6} \\ & & & -\frac{kA}{L_e}+\frac{S_1L_e}{6} & \frac{2kA}{L_e}+\frac{2S_1L_e}{3}+h \end{bmatrix} \begin{Bmatrix} \left.\frac{dT}{dx}\right|_{x_1} \\ T_2 \\ T_3 \\ T_4 \\ T_5 \end{Bmatrix}$$

$$= \begin{Bmatrix} -\left(\frac{kA}{L_e}+\frac{S_1L_e}{3}\right)T_o+\frac{S_0L_e}{2} \\ \left(\frac{kA}{L_e}-\frac{S_1L_e}{3}\right)T_0+S_0L_e \\ S_0L_e \\ S_0L_e \\ \frac{S_0L_e}{2}+hT_\infty \end{Bmatrix}$$

Substituting the numerical data, we get

$$\begin{bmatrix} 200 & -1.04728 & & & \\ 0 & 2.090688 & -1.04728 & & \\ & -1.04728 & 2.090688 & -1.04728 & \\ & & -1.04728 & 2.090688 & -1.04728 \\ & & & -1.04728 & 10.090688 \end{bmatrix} \begin{Bmatrix} \left.\frac{dT}{dx}\right|_{x_1} \\ T_2 \\ T_3 \\ T_4 \\ T_5 \end{Bmatrix} = \begin{Bmatrix} -104.57208 \\ 104.86096 \\ 0.07536 \\ 0.07536 \\ 800.03768 \end{Bmatrix}$$

13.5 AXISYMMETRIC PROBLEMS

There are many problems in radial co-ordinate systems such as in pipes and spheres that have symmetry about the axis of rotation. This symmetry arises not only because of the geometry but also due to the boundary conditions around the circumference or the circular directions. In this section, we will consider formulation and application of finite element method using a one-dimensional steady state equation for such an axisymmetric problem. Let us consider one-dimensional steady state diffusion along with volumetric generation process in a long circular hollow rod as shown. The boundary conditions are assumed to be of the first kind i.e. constant surface values ϕ_i and ϕ_j below:

The mathematical statement of this transport process is given as:

$$\frac{1}{r}\frac{d}{dr}\left(\Gamma_r r\frac{d\phi}{dr}\right)+S=0 \tag{13.90}$$

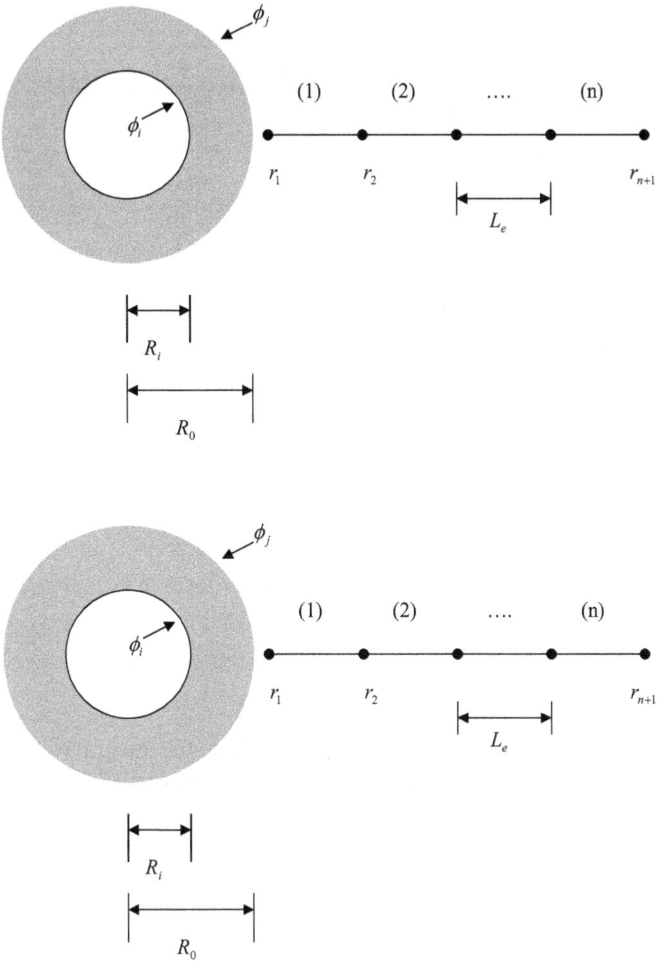

FIGURE 13.7 Axisymmetric problem

Boundary Conditions:

$$1.\ r = R_i,\ \ \phi = \phi_i \tag{13.91a}$$

$$2.\ r = R_0,\ \ \phi = \phi_0 \tag{13.91b}$$

Let us demonstrate the application of finite element method to this problem through detail description of several steps that constitute the finite element method.

Using **Galerkin's-based Weighted Residual Method** we obtain the integral statement of the problem as

$$\left\{R^{(e)}\right\} = \int_{r_i}^{r_j} [N]^T \left(\frac{d}{dr}\left(\Gamma_r r \frac{d\phi^*}{dr} \right) + Sr \right) dr = 0 \tag{13.92}$$

Evaluating the first term of the equation by integration-by-parts and using the properties of the shape functions for a linear line element, i.e. $N_i(r_i) = 1$, $N_i(r_j) = 0$and $N_j(r_i) = 0$, $N_j(r_j) = 1$, we can transform the Equation 13.92 into

$$\left\{\begin{array}{c} -\Gamma_r r \left.\dfrac{d\phi}{dx}\right|_{x_i} \\ \Gamma_r r \left.\dfrac{d\phi}{dx}\right|_{x_j} \end{array}\right\} - \int_{r_i}^{r_j} \Gamma_r r \frac{d[N]^T}{dx}\frac{d\phi^*}{dx} dr + \int_{r_i}^{r_j} Sr[N]^T dr = 0 \tag{13.93}$$

In the next step the element characteristics equation is obtained by evaluating the integrals by substituting the approximate solution for linear line element and following the procedure outline in Section 13.2. Only difference is in the treatment of the term $\Gamma_r r$ in evaluating the second integral. This is assumed as the product of a step-wise constant value for the transport property, Γ_r in the media and a mean value for r as follow:

$$\Gamma_r r = \Gamma_r \bar{r} \tag{13.94}$$

Where mean value of r is given as

$$\bar{r} = \frac{r_i + r_j}{2} \tag{13.95}$$

Following this procedure, we can evaluate the first integrals term in Equation 13.93 as follows:

$$\int_{r_i}^{r_j} \Gamma_r r \frac{d[N]^T}{dr}\frac{d\phi^*}{dr} dr = \frac{\Gamma_r \bar{r}}{L_e}\begin{bmatrix} 1 & -1 \\ -1 & 1 \end{bmatrix}\begin{Bmatrix} \phi_i \\ \phi_j \end{Bmatrix} \tag{13.96}$$

For the second **integral term**, it is more convenient to evaluate the integral using a local co-ordinate system. Let us use the shape function expression given by Equation 13.3 for the line element in a local co-ordinate system and substitute $r = r_i + r$ to change from the global co-ordinate system to the local co-ordinate system. Also, let us assume a step-wise constant value $S = \bar{S}$ for the source in the media for this demonstration, and evaluate the integrals as follows:

For $I = i$,

$$\int_{r_i}^{r_j} SrN_i(r)dr = \int_0^{L_e} \bar{S}\left(1 - \frac{r}{L_e}\right)(r_i + r)dr = \bar{S}\int_0^{L_e}\left(r_i\left(1 - \frac{r}{L_e}\right) + r\left(1 - \frac{r}{L_e}\right)\right)dr$$

$$= \bar{S}\left(r_i r\Big|_0^{L_e} - \frac{r_i}{L_e}\frac{r^2}{2}\Big|_0^{L_e} + \frac{r^2}{2}\Big|_0^{L_e} - \frac{r^3}{2L}\Big|_0^{L_e}\right)$$

or

$$\int_{r_i}^{r_j} SrN_i(r)dx = \frac{S\bar{L}_e}{6}(3r_i + L_e) \tag{13.97a}$$

For $I = j$,

$$\int_{r_i}^{r_j} SrN_j(r)dr = \int_0^{L_e} \bar{S}\frac{r}{L_e}(r_i + r)dr = \frac{\bar{S}}{L_e}\left(r_i\frac{r^2}{2}\Big|_0^{L_e} + \frac{r^3}{3}\Big|_0^{L_e}\right)$$

$$= \frac{\bar{S}}{L_e}\left(r_i\frac{L_e^2}{2} + \frac{L_e^2}{3}\right)$$

or

$$\int_{r_i}^{r_j} SrN_i(r)dx = \frac{S\bar{L}_e}{6}(3r_i + 2L_e) \tag{13.97b}$$

Equations 13.97a and 13.97b can be written in vector form as

$$\int_{x_i}^{x_j} Sr[N]^T dx = \frac{S\bar{L}_e}{6}\begin{Bmatrix} 3r_i + L_e \\ 3r_i + 2L_e \end{Bmatrix} \tag{13.97c}$$

Substituting Equations 13.96 and 13.97 into Equation 13.93, we obtain the element characteristics equations as follows:

$$\begin{Bmatrix} \Gamma_r r \left.\frac{d\phi}{dr}\right|_{r_i} \\ -\Gamma_r r \left.\frac{d\phi}{dr}\right|_{r_j} \end{Bmatrix} + \frac{\Gamma_r \bar{r}^{(e)}}{L_e}\begin{bmatrix} 1 & -1 \\ -1 & 1 \end{bmatrix}\begin{Bmatrix} \phi_i \\ \phi_j \end{Bmatrix} - \frac{S\bar{L}_e}{6}\begin{Bmatrix} 3r_i + L_e \\ 3r_i + 2L_e \end{Bmatrix} = 0 \tag{13.98a}$$

Or

$$\left[K_\Gamma^{(e)}\right]\left\{\phi^{(e)}\right\} = \left\{I^{(e)}\right\} + \left\{f^{(e)}\right\} = 0 \tag{13.98b}$$

Where

$$\left\{\phi^{(e)}\right\} = \begin{Bmatrix} \phi_i \\ \phi_j \end{Bmatrix} = \text{vector of element nodal values} \tag{13.99}$$

$$\left[K_\Gamma^{(e)}\right] = \frac{\Gamma_r \bar{r}}{L_e}\begin{bmatrix} 1 & -1 \\ -1 & 1 \end{bmatrix} = \text{element characteristic or stiffness matrix} \tag{13.100}$$

$$\left\{f^{(e)}\right\} = \frac{S\bar{L}_e}{6}\begin{Bmatrix} 3r_i + L_e \\ 3r_i + 2L_e \end{Bmatrix} = \begin{Bmatrix} f_i^{(e)} \\ f_j^{(e)} \end{Bmatrix} = \text{element source or force vector} \tag{13.101}$$

$$\left\{I^{(e)}\right\} = \begin{Bmatrix} -\Gamma_r r \left.\frac{d\phi}{dr}\right|_{r_i} \\ \Gamma_r r \left.\frac{d\phi}{dr}\right|_{r_j} \end{Bmatrix} = \begin{Bmatrix} I_i^{(e)} \\ I_j^{(e)} \end{Bmatrix} \tag{13.102}$$

= vector for element contribution to inter-element conditions

We can now apply Equation 13.98 repeatedly to all elements to obtain corresponding element equations, which are then assembled to form a global system of equations as follows:

$$
\begin{bmatrix}
\frac{\Gamma_r \bar{r}^{(1)}}{L_e} & -\frac{\Gamma_r \bar{r}^{(1)}}{L_e} & & & & \\
-\frac{\Gamma_r \bar{r}^{(1)}}{L_e} & \frac{\Gamma_r \bar{r}^{(1)}}{L_e}+\frac{\Gamma_r \bar{r}^{(2)}}{L_e} & -\frac{\Gamma_r \bar{r}^{(2)}}{L_e} & & & \\
 & -\frac{\Gamma_r \bar{r}^{(2)}}{L_e} & \frac{\Gamma_r \bar{r}^{(2)}}{L_e}+\frac{\Gamma_r \bar{r}^{(3)}}{L_e} & -\frac{\Gamma_r \bar{r}^{(3)}}{L_e} & & \\
 & & -\frac{\Gamma_r \bar{r}^{(3)}}{L_e} & \frac{\Gamma_r \bar{r}^{(3)}}{L_e}+\frac{\Gamma_r \bar{r}^{(4)}}{L_e} & & \\
\vdots & & & & & \\
 & & & & \frac{\Gamma_r \bar{r}^{(n-1)}}{L_e}+\frac{\Gamma_r \bar{r}^{(n)}}{L_e} & -\frac{\Gamma_r \bar{r}^{(n)}}{L_e} \\
 & & & & -\frac{\Gamma_r \bar{r}^{(n)}}{L_e} & \frac{\Gamma_r \bar{r}^{(n)}}{L_e}
\end{bmatrix}
\begin{Bmatrix} \phi_1 \\ \phi_2 \\ \phi_3 \\ \vdots \\ \vdots \\ \phi_{n-1} \\ \phi_n \end{Bmatrix}
$$

$$
= \begin{Bmatrix} +\Gamma_r \bar{r}^{(1)} \left.\frac{d\phi}{dr}\right|_{R_i} \\ 0 \\ 0 \\ \vdots \\ \vdots \\ 0 \\ -\Gamma_r \bar{r}^{(n)} \left.\frac{d\phi}{dx}\right|_{R_o} \end{Bmatrix} + \frac{\bar{S}}{3} \begin{Bmatrix} 3r_1 + L_1 \\ 3r_1 + 2L_1 + 3r_2 + L_2 \\ 3r_2 + 2L_2 + 3r_3 + L_3 \\ \vdots \\ \vdots \\ 3r_{n-1} + 2L_{n-1} + 3r_n + L_n \\ 3r_n + 2L_n \end{Bmatrix} \tag{13.103}
$$

The global system can be written in a compact form as follows:

$$
[\mathrm{K}]\{\Phi\} = \{F\} \tag{13.104}
$$

Where

$$
[\mathrm{K}]=\begin{bmatrix}
\frac{\Gamma_r \bar{r}^{(1)}}{L_e} & -\frac{\Gamma_r \bar{r}^{(1)}}{L_e} & 0 & 0 & 0 & 0 & 0 \\
-\frac{\Gamma_r \bar{r}^{(1)}}{L_e} & \frac{\Gamma_r \bar{r}^{(1)}}{L_e}+\frac{\Gamma_r \bar{r}^{(2)}}{L_e} & -\frac{\Gamma_r \bar{r}^{(2)}}{L_e} & 0 & 0 & 0 & 0 \\
 & -\frac{\Gamma_r \bar{r}^{(2)}}{L_e} & \frac{\Gamma_r \bar{r}^{(2)}}{L_e}+\frac{\Gamma_r \bar{r}^{(3)}}{L_e} & -\frac{\Gamma_r \bar{r}^{(3)}}{L_e} & 0 & 0 & 0 \\
 & & -\frac{\Gamma_r \bar{r}^{(3)}}{L_e} & \frac{\Gamma_r \bar{r}^{(3)}}{L_e}+\frac{\Gamma_r \bar{r}^{(4)}}{L_e} & & & \\
\vdots & & & & & & \\
0 & 0 & 0 & 0 & & \frac{\Gamma_r \bar{r}^{(n-1)}}{L_e}+\frac{\Gamma_r \bar{r}^{(n)}}{L_e} & -\frac{\Gamma_r \bar{r}^{(n)}}{L_e} \\
0 & 0 & 0 & 0 & 0 & -\frac{\Gamma_r \bar{r}^{(n)}}{L_e} & \frac{\Gamma_r \bar{r}^{(n)}}{L_e}
\end{bmatrix} \tag{13.105}
$$

$$
\{\Phi\}=\begin{Bmatrix} \phi_1 \\ \phi_2 \\ \phi_3 \\ \vdots \\ \vdots \\ \phi_{n-1} \\ \phi_n \end{Bmatrix} \tag{13.106}
$$

$$
\{F\}=\begin{Bmatrix} +\Gamma_r \bar{r}^{(1)} \left.\frac{d\phi}{dr}\right|_{R_i} \\ 0 \\ 0 \\ \vdots \\ \vdots \\ 0 \\ -\Gamma_r \bar{r}^{(n)} \left.\frac{d\phi}{dx}\right|_{R_o} \end{Bmatrix} + \frac{\bar{S}L_e}{6}\begin{Bmatrix} 3r_1+L_1 \\ 3r_1+2L_1+3r_2+L_2 \\ 3r_2+2L_2+3r_3+L_3 \\ \vdots \\ 3r_{n-1}+2L_{n-1}+3r_n+L_n \\ 3r_n+2L_n \end{Bmatrix} \tag{13.107}
$$

In the case of a uniform mesh size distribution, i.e. for $L_1 = L_2 = \cdots = L_n = L_e$, we can set $r_1 = R_i$, $r_2 = R_i + L_e$, $r_3 = R_i + 2L_e>$, $\ldots r_n = R_i + (n-1)L_e$, and $r_{n+1} = R_o$, and the Equation 13.107 can be simplified as

$$\{F\} = \left\{ \begin{array}{c} +\Gamma_r \bar{r}^{(1)} \left.\dfrac{d\phi}{dr}\right|_{R_i} \\ 0 \\ 0 \\ \vdots \\ \vdots \\ 0 \\ -\Gamma_r \bar{r}^{(n)} \left.\dfrac{d\phi}{dx}\right|_{R_o} \end{array} \right\} + \frac{\bar{S}L_e}{6} \left\{ \begin{array}{c} 3R_i + L_e \\ 6R_i + 6L_e \\ 6R_i + 12L_e \\ \vdots \\ \vdots \\ 6R_i + 6(n-2)L_e \\ 6R_i + 6(n-1)L_e \end{array} \right\} \tag{13.108}$$

In the next step, we apply the boundary conditions directly into the global systems. For boundary condition of the first kind or of constant surface values as defined by Equations 13.91, we directly assign the given values to the boundary nodes as $\phi_1 = \phi_i$ and $\phi_{n+1} = \phi_o$, and rearrange the global system as follows:

$$\begin{bmatrix} -\Gamma_r \bar{r}^{(1)} & -\dfrac{\Gamma_r \bar{r}^{(1)}}{L_e} & 0 & 0 & 0 & 0 & 0 \\ 0 & \dfrac{\Gamma_r \bar{r}^{(1)}}{L_e} + \dfrac{\Gamma_r \bar{r}^{(2)}}{L_e} & -\dfrac{\Gamma_r \bar{r}^{(2)}}{L_e} & 0 & 0 & 0 & 0 \\ 0 & -\dfrac{\Gamma_r \bar{r}^{(2)}}{L_e} & \dfrac{\Gamma_r \bar{r}^{(2)}}{L_e} + \dfrac{\Gamma_r \bar{r}^{(3)}}{L_e} & -\dfrac{\Gamma_r \bar{r}^{(3)}}{L_e} & 0 & 0 & 0 \\ 0 & 0 & -\dfrac{\Gamma_r \bar{r}^{(3)}}{L_e} & \dfrac{\Gamma_r \bar{r}^{(3)}}{L_e} + \dfrac{\Gamma_r \bar{r}^{(4)}}{L_e} & 0 & 0 & 0 \\ \vdots & & & & & & \\ & & & & \dfrac{\Gamma_r \bar{r}^{(n-1)}}{L_e} + \dfrac{\Gamma_r \bar{r}^{(n)}}{L_e} & & 0 \\ & & & & -\dfrac{\Gamma_r \bar{r}^{(n)}}{L_e} & & \Gamma_r \bar{r}^{n)} \end{bmatrix} \left\{ \begin{array}{c} \left.\dfrac{d\phi}{dr}\right|_{R_i} \\ \phi_2 \\ \phi_3 \\ \vdots \\ \vdots \\ \phi_n \\ \left.\dfrac{d\phi}{dr}\right|_{R_o} \end{array} \right\}$$

$$
= \left\{ \begin{array}{c}
-\dfrac{\Gamma_r \bar{r}^{(1)}}{L_e}\phi_i + \dfrac{\bar{S}L_e}{6}(3R_i + L_e) \\
\dfrac{\Gamma_r \bar{r}^{(1)}}{L_e}\phi_i + \dfrac{\bar{S}L_e}{6}(6R_i + 6L_e) \\
\dfrac{\bar{S}L_e}{6}(6R_i + 12L_e) \\
\vdots \\
\vdots \\
\dfrac{\Gamma_r \bar{r}^{(n)}}{L_e}\phi_o + \dfrac{\bar{S}L_e}{6}(6R_i + 6(n-2)L_e) \\
-\dfrac{\Gamma_r \bar{r}^{(n)}}{L_e}\phi_o + \dfrac{\bar{S}L_e}{6}6R_i{+}6(n-1)L_e
\end{array} \right\} \tag{13.109}
$$

PROBLEMS

13.1 The exposed surface ($x=0$) of a plane wall of thermal conductivity k is subjected to thermal radiation that causes volumetric heat generation in the media to vary as

$$\dot{q}(x) = \dot{q}_0\left(1 - \frac{x}{L}\right)$$

where $\dot{q}_0\left(W/m^3\right)$ is a constant. The boundary at $x=L$ is convectively cooled, while the exposed surface is maintained at a constant temperature T_0.

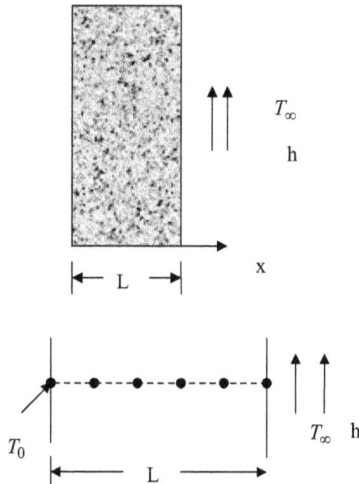

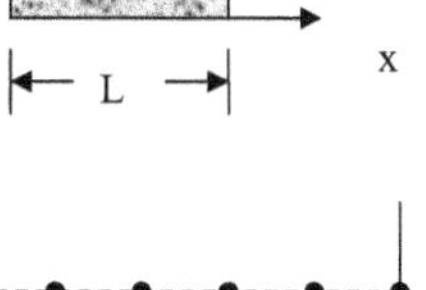

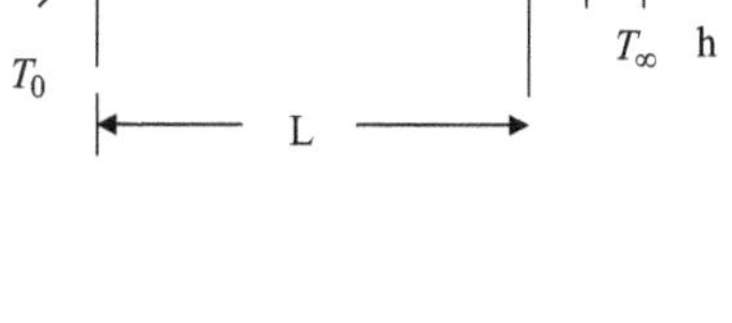

Governing Equation

$$\frac{\partial}{\partial x}\left(k\frac{\partial T}{\partial x}\right)+\dot{q}=0$$

Boundary Conditions

1. $x=0, T(x,0)=\mathrm{T}_0$, 2. $x=L, -k\dfrac{\partial T}{\partial x}=h(T-T_\infty)$

Consider mesh size distribution shown with five elements and determine the temperature distribution in the slab. Use following data for computation:

$$\dot{q}=1.0\times10^5\ \mathrm{W/m^3}, L=12\,\mathrm{cm}, T=30°\mathrm{C}, h=50\frac{W}{\mathrm{m}^2\cdot\mathrm{C}}, T_\infty=25°\mathrm{C}$$

13.2 Consider a straight fin of uniform cross-sectional area, A, length $L=2.0\,\mathrm{cm}$, width $Z=1.0\,\mathrm{cm}$ and a thickness $t=1.4\,\mathrm{mm}$. The fin thermal conductivity is $k=60$ W/m·C and it is exposed to a convection environment at $T_\infty=20°\mathrm{C}$ and $h=500\,\mathrm{W/n}$. The base of the fin is at a constant temperature $T_0=150°\mathrm{C}$ and the tip of the fin is assumed to be convective.

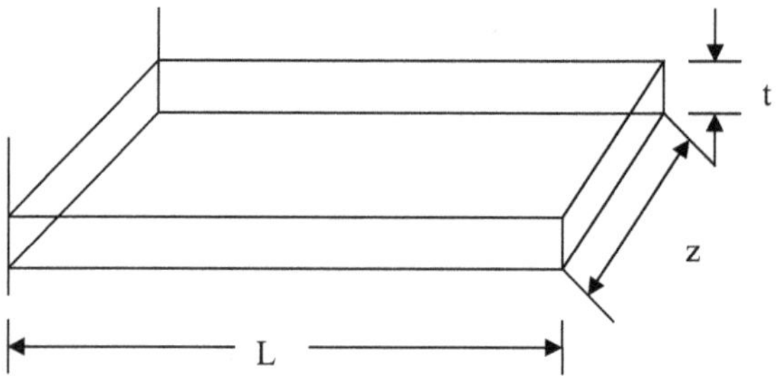

The mathematical statement of the problem is

Governing Equation:

$$\frac{\mathrm{d}}{\mathrm{d}x}\left(k\frac{\mathrm{d}T}{\mathrm{d}x}\right)+\frac{hP}{A}(T_\infty-T)=0$$

Boundary Conditions:

1. $x=0, T=T_0$
2. $x=L, -k\dfrac{\mathrm{d}T}{\mathrm{d}x}=h(T-T_\infty)$
   a. Use five linear line elements and determine the element characteristic equation for all elements.
   b. Assemble all element equations and derive the global system of equations.
   c. Determine the temperature distribution in the fin and fin heat loss.

13.3 A composite wall made of four homogeneous material layers as shown below:

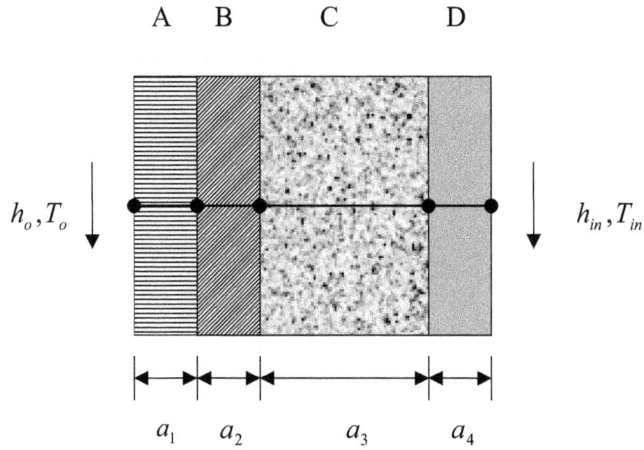

Calculate the nodal temperature values within the composite wall and evaluate the heat flow per unit width based assuming indoor and outdoor conditions as $T_{in} = 25°C$, $h_{in} = 10\,W/m^2 \cdot K$, and $T_{out} = -20°C$, $h_0 = 10\,W/m\cdot$ Consider four linear line elements and use the following data: $a_1=2\,cm$, $a_2=1\,cm$, $a_3=5\,cm$, $a_4=1\,cm$, $k_A = 0.72\,W/m \cdot K$, $k_C = 0.026\,W/m \cdot K$, and $k_D = 0.22\,W/m \cdot K$.

13.4 Consider the electrical heat generation in chip that is encapsulated with an aluminum cover plate of thickness $a_{al} = 2.5\,mm$ and thermal conductivity $k_b = 230\,W/m \cdot K$. The chip is mounted on a board of thickness $a_b = 4\,mm$ and thermal conductivity $k_b = 0.8\,W/m \cdot K$. The aluminum cover plate and the chip are cooled on the left by liquid with $h_l = 1200\,W/m^2 \cdot K$ and $T_i = 20°C$. The board on the right is cooled by air with $h_r = 20\,W/m^2 \cdot K$ and $T_r = 20°C$.

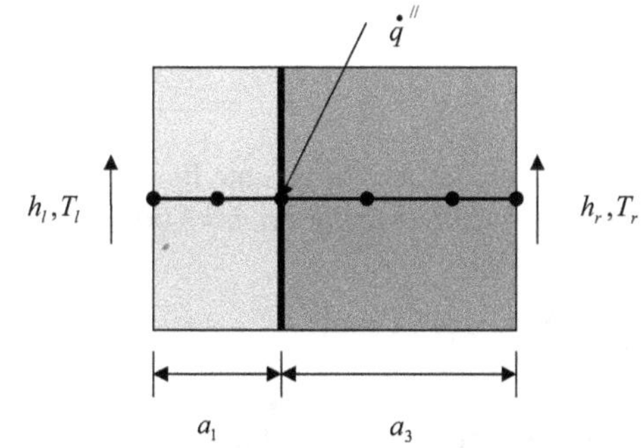

Considering a thin chip and assuming a uniform electric heat generation of $\dot{q}'' = 5000\,W/m^2$ at the interface of aluminum and board materials, determine the temperature distribution and heat transfer rates at two exterior surfaces using five linear elements.

13.5 Semitransparent wall is irradiated at the surface at $x=0$ by an incident radiation such that such absorbed radiation in the media results in an internal heat generation given as

$$\dot{q} = q_o'' a e^{-ax}$$

where a is the coefficient absorption of the material and $q_0^{//}$ is the incident radiation flux. The wall at x=0 is also exposed to a convective environment of h_0 and T_0, and the surface at $x=L$ is maintained at constant temperature T_L. Derive discretization equations for interior nodes and for the convective boundary node for the nonuniform mesh system shown. Assuming the length of the elements as l_{e1}, l_{e2} ... l_{e7}, derive the system of equations for the solution of nodal temperatures and heat transfer rates at inner and outer surfaces.

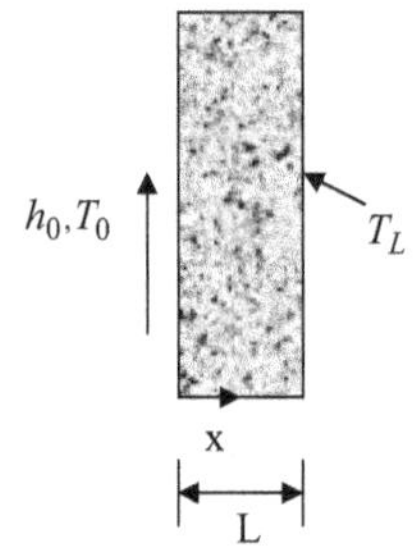

13.6 Consider the fully developed laminar steady flow of Newtonian viscous fluid through a two-dimensional channel formed by two infinite parallel plates as shown below. The lower plate is moving in the negative x-direction with a constant speed U_{w1} and the upper plate is moving in the positive x-direction with a constant speed, U_{w2}, in the positive x-direction.

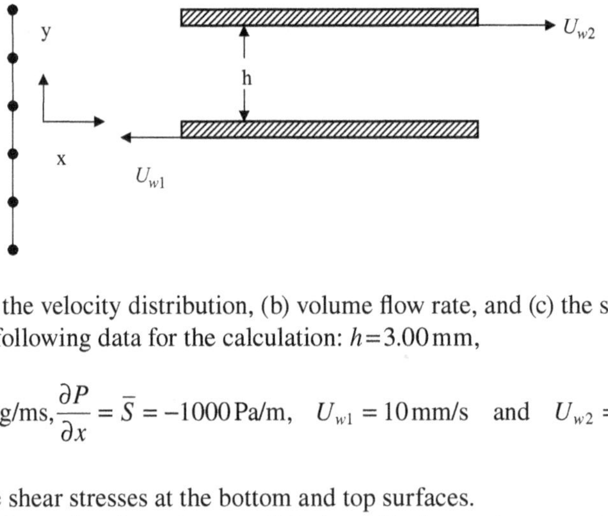

Determine (a) the velocity distribution, (b) volume flow rate, and (c) the shear stress at the plate and use following data for the calculation: $h=3.00\,\text{mm}$,

$$\mu = 0.02\,\text{kg/ms}, \frac{\partial P}{\partial x} = \bar{S} = -1000\,\text{Pa/m}, \quad U_{w1} = 10\,\text{mm/s} \quad \text{and} \quad U_{w2} = 20\,\text{mm/s}$$

Determine the shear stresses at the bottom and top surfaces.

13.7 Consider one-dimensional steady state diffusion of hydrogen in an anode electrode made of porous diffusion layer with a thin catalyst layer placed at the right side as shown. The catalyst layer provides a heterogeneous chemical reaction for the consumption of hydrogen at a rate $m_{H_2}^{//}$, defined as the consumption per unit area of the catalyst. Assume the reaction rate as given by a **first-order reaction**, i.e. $m_{H_2}^{//} = -k_1^{//}C(0)$, where $k_1^{//}$ is the reaction rate constant. Gas flow channel maintains a constant hydrogen concentration of C_0 at the left surface of the electrode.

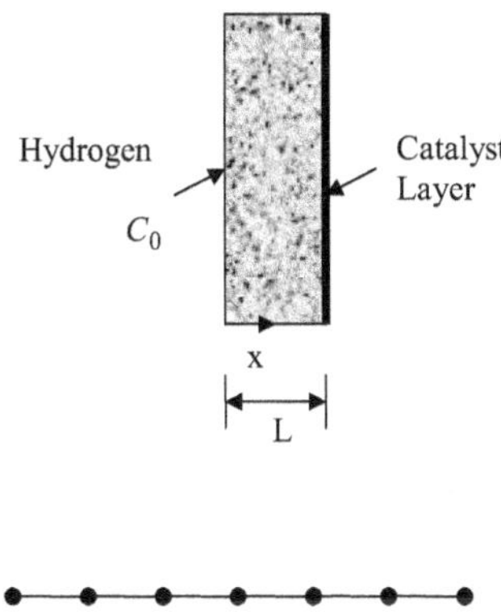

The mathematical statement of the problem is given as:

Governing Equation:

$$\frac{d}{dx}\left(D\frac{dC}{dx}\right)=0$$

Boundary Conditions:

$$1.\ x = 0,\ C(0)=C_0$$

$$2.\ x=L,\ -D\frac{dC}{dx}\bigg|_{x=L} = k_1^{\prime\prime}C(L)$$

Use a uniform mesh size distribution shown to derive the system of equations for the nodal hydrogen concentration.

13.8 Derive the element characteristic equation starting from Equation 13.10 and using quadratic line element defined by Equation 10.42.

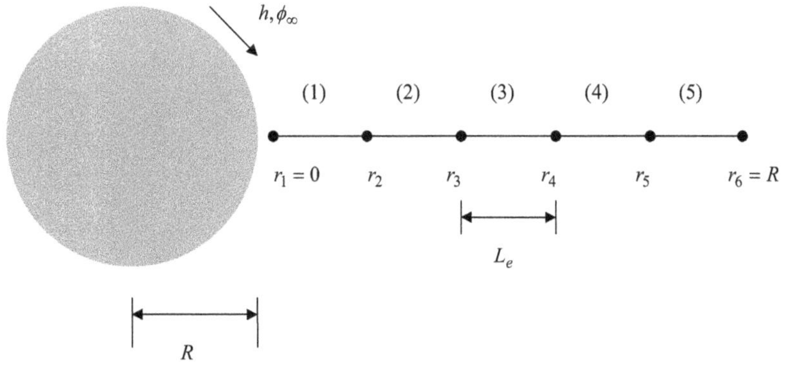

13.9 Consider the one-dimensional steady state transport process in a circular rod with an exponentially decaying source term.

The mathematical statement of the problem is given as follows:

Governing Equation:

$$\frac{1}{r}\frac{d}{dr}\left(\Gamma_r r\frac{d\phi}{dr}\right)+aI_0e^{-ar^2}=0$$

Boundary Conditions:

$$1.\ r = 0, \frac{d\phi}{dr} = 0$$

$$2.\ r = R, -\Gamma_r \frac{d\phi}{dr} = h(\phi - \phi_\infty)$$

Derive element characteristic equations for all elements. Assemble all elements to obtain the global system of equations.

13.10 Develop a computer code for the solution of one-dimensional diffuse equation in a plane slab of thickness L using the pseudo code based on finite element method as given in Appendix C. The pseudo code generates the element characteristic equation and forms the global system of equations based on the linear line element. Solve the system of equations using Gauss-elimination solver. Use the code for the solution of the problem given in Problem 13.1.

14 Finite Element Method
Multidimensional Steady-State Problems

In this chapter, we considered finite element procedures for solving two-dimensional and three-dimensional problems involving diffusion and source terms. Even though the basic steps and procedure used for one-dimensional problems are applicable to multidimensional problems, the solution of two- and three-dimensional problems poses considerably more difficulties in mesh generation, formulation of element characteristics equations, and in the assembly of element equations. Galerkin's weighted residual integral formulation is used in presenting this procedure.

14.1 TWO-DIMENSIONAL STEADY-STATE DIFFUSION EQUATION

In this section, we will consider the formulation and application of finite element method for solving two-dimensional steady-state problems. Let us consider two-dimensional steady-state transport process in a two-dimensional domain as show in Figure 14.1.

A general mathematical statement of the problem can be assumed as

$$\frac{\partial}{\partial x}\left(\Gamma_x \frac{\partial \phi}{\partial x}\right)+\frac{\partial}{\partial y}\left(\Gamma_y \frac{\partial \phi}{\partial y}\right)+S_1\phi+S_0=0 \tag{14.1}$$

The partial differential Equation 14.1 is like the Equation 13.1 except for the addition of an additional diffusion term in the y-direction.

14.1.1 Step 1: Mesh Generation or Discretization of the Solution Domain

As we have mentioned before that the **domain discretization** or **mesh generation** involves representing the solution region into a set of elements, referred to as the mesh. The solution domain could be discretized into a mesh using two-dimensional elements such as triangular or rectangular or any other two-dimensional elements discussed in Chapter 10. A mesh of rectangular elements is created easily by constructing all elements in a horizontal row of uniform height and all elements in vertical column of uniform length. Rectangular elements are well suited for regions consisting of straight edges. However, a combination of rectangular and triangular elements is used in irregular geometries with triangular elements being used to approximate the irregular or curved edges. For the irregular or curved boundaries, the sides of the element should closely approximate

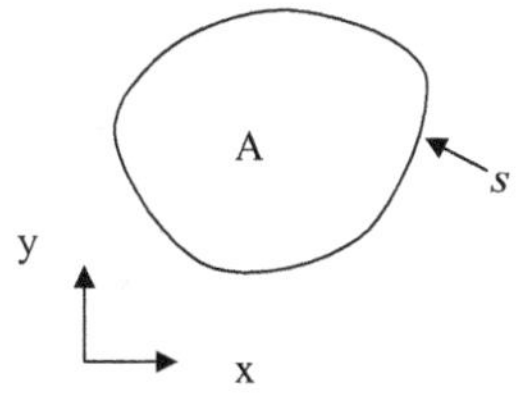

FIGURE 14.1 Two-dimensional domain.

boundary to reduce the error. For example, many linear straight-sided triangular elements could be used to approximate the irregular boundaries with good accuracies. A better approximation could be achieved by using higher-order or nonlinear curved-sided triangular elements. So, the choice of the element type, number of elements, and element density depends on the geometry of the domain, problem under investigation, and accuracy needed. The final selection is established by carrying out a comprehensive numerical experimentation.

Some general practices that are normally being used while creating a mesh are:

1. For simplicity, linear elements are always the initial elements of choice. The main advantage is that the linear shape functions and their derivatives are simpler to evaluate analytically. However, it requires large number of elements to keep the error down to acceptable level. Another disadvantage is that the derivative of the unknown function within an element is constant and this results in a greater discontinuity at interelement boundaries, and cause instability and/or slower rate convergence. Higher order or nonlinear curved-sided elements approximate curved or complex boundaries more accurately and with higher accuracy and rate of convergence, but at the expense of additional complexities in the computation process.
2. While creating a mesh, a solution domain is first divided into several subregions using simpler geometries such as quadrilaterals or triangles and then progressively divided into smaller subdomains. A mesh of triangular elements is generated by first dividing the region into number of quadrilateral elements and then dividing these into triangular regions by joining the nodes on opposite sides or inserting a diagonal and placing nodes at intersection points. To improve the accuracy, it is desirable to join the opposite nodes to insert the shortest diagonals. For example, a square plate with a hole at the center is first divided into subregions using quadrilateral elements as shown in Figure 14.2a.

   This is followed by dividing the quadrilateral elements into triangular elements as shown in Figure 14.2b. A more refined mesh near the circular hole can be created by

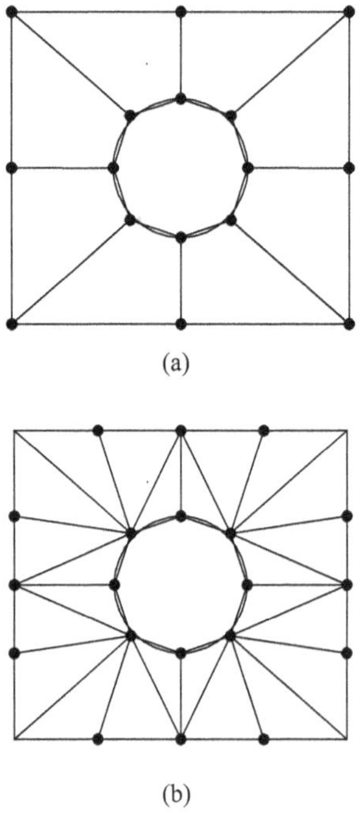

FIGURE 14.2 Mesh generation using (a) quadrilateral and (b) triangular geometries.

dividing the triangular elements into number of smaller triangles. A triangular region is divided into smaller triangular regions by locating same number of nodes on each sides and then joining the nodes in an orderly manner depicted in Figure 14.3.

3. A more refined mesh distribution is used in regions that involve large gradients or variation of dependent variable as shown in Figure 14.4.
4. For better numerical stability and convergence, mesh refinement should vary gradually from region with finer mesh distribution to region with coarse mesh distribution.
5. Elements must be compatible and complete. In order to ensure the continuity in dependent variables and continuity in the flux quantities at the interface of two elements, the elements at the interface must have the same number of nodes and have same location at the common face or edge, and defined by the same interpolation function. An example of compatible and noncompatible adjacent elements in a mesh is demonstrated in Figure 14.5.
6. Transition elements that connect lower-order element to higher-order element are used in regions away from region with steeper gradient of the dependent variables.

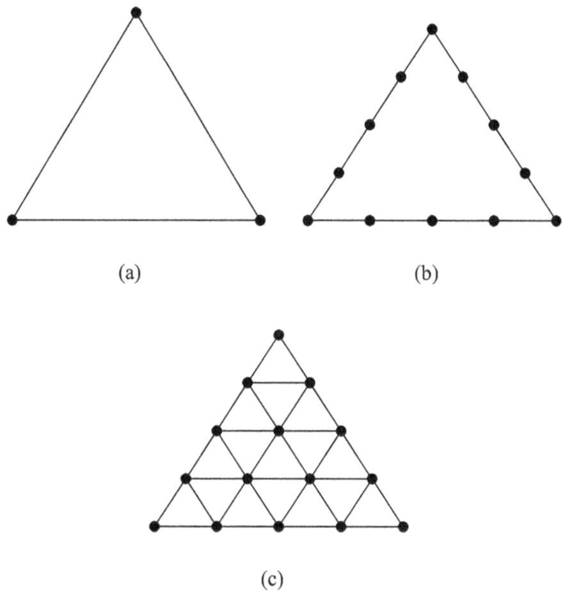

FIGURE 14.3 Division of a triangular region into smaller triangular elements. (a) A large triangular mesh, (b) multiple nodes on each side of the triangle to form refined mesh and (c) a refined mesh region formed.

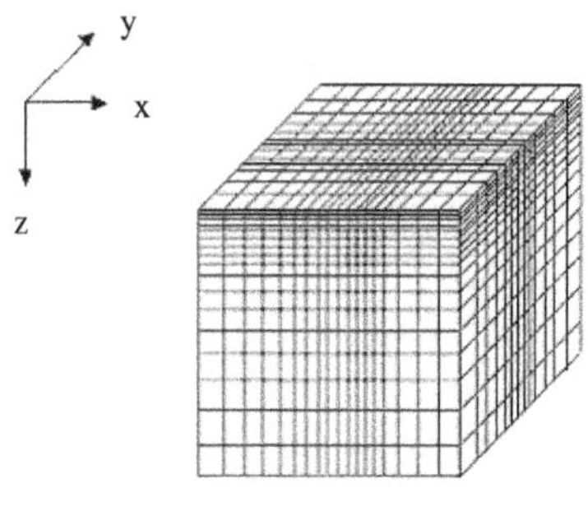

FIGURE 14.4 Refined mesh distributions.

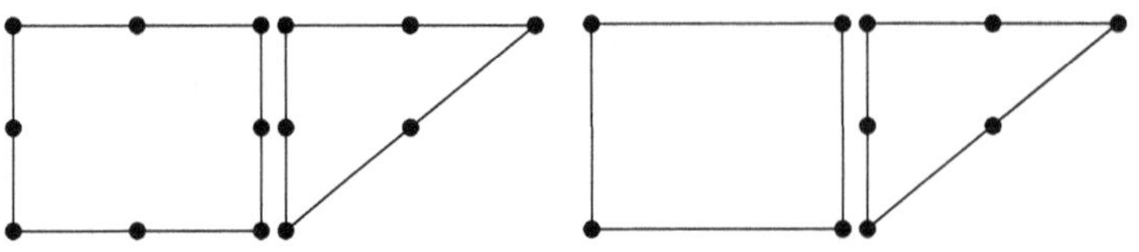

FIGURE 14.5 Nodes at the interface of two elements. (a) Compatible. (b) Noncompatible.

14.1.2 Step 2: Element and Node Numbering

The elements as well as the nodes in a mesh can be numbered sequentially in any arbitrary manner. However, the node numbering scheme will have a significant effect on the type of global matrix system formed after the assembly of all element characteristic matrix equations. As we have mentioned before that a banded system is always preferred as it only requires storing nonzero matrix elements in the band around the principal diagonal, and hence requiring less computer storage and computational time. Additionally, a more efficient solver for the solution of the system of linear algebraic equations could be selected as discussed in Chapter 2. So, a preferred numbering system should be the one that leads to a banded global system with band width (**bw**) as small as possible, preferably a band width (**bw**) of three as in a tridiagonal global matrix system. A random numbering system may increase the difference between node numbers in an element, and thus would increase the band width (**bw**). As we have noticed for one-dimensional problems that when the nodes are numbered sequentially from left to right, the maximum difference between node numbers in an element was only one, and assembly of all element characteristic matrices leads to a tridiagonal global system. A random numbering system would increase the bandwidth of the global characteristic matrix. Let us consider two different numbering schemes in a two-dimensional region as shown in Figure 14.6.

Comparing these two numbering schemes, we notice that there is considerable difference in node numbers in the second scheme (Figure 14.6b) compared to first scheme (Figure 14.6a). For example, the maximum difference between node numbers in any element was 3 as compared to 12 in the second scheme (Figure 14.6b). An assembly of all element characteristic matrices will lead to a much bigger band width in the global system for the second scheme.

14.1.3 Step 3: Selection of Approximate Solution Function

The approximate solution for the element is given by the general expression of the form given by Equation 14.2.

$$\phi^* = \sum_{I=1}^{n_p} N_I \phi_I = [N]\{\phi^{(e)}\} \tag{14.2}$$

where N_I = shape function, ϕ_I = nodal values, and n_p = number of nodes in an element.

14.1.4 Step 4: Formulation of an Integral Statement Using Galerkin's Approach

The integral form of the equation over an element is obtained by applying **Galerkin's weighted residual method** as

$$\{R^{(e)}\} = -\int_A [N]^T \left[\frac{\partial}{\partial x}\left(\Gamma_x \frac{\partial \phi^*}{\partial x}\right) + \frac{\partial}{\partial y}\left(\Gamma_y \frac{\partial \phi^*}{\partial y}\right) + S_1\phi^* + S_0 \right] dA = 0 \tag{14.3}$$

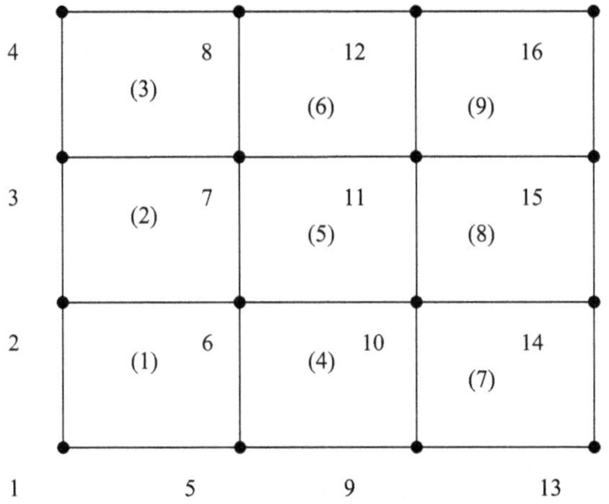

(a)

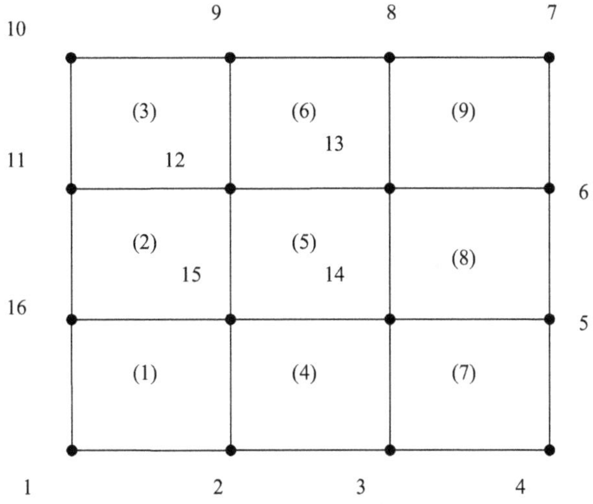

(b)

FIGURE 14.6 Node numbering scheme: (a) Node numbering scheme - 1 and (b) Node numbering scheme - 2.

The first two integral terms are evaluated as follows:

$$\int_A \frac{\partial}{\partial x}\left([N]^T \Gamma_x \frac{\partial \phi^*}{\partial x}\right) \mathrm{d}A - \int_A \Gamma_x \frac{\partial \phi^*}{\partial x}\frac{\partial [N]^T}{\partial x}\,\mathrm{d}A \tag{14.4a}$$

and

$$\int_A [N]^T \frac{\partial}{\partial y}\left(\Gamma_y \frac{\partial \phi^*}{\partial y}\right) \mathrm{d}A = \int_A \frac{\partial}{\partial y}\left([N]^T \Gamma_y \frac{\partial \phi^*}{\partial y}\right) \mathrm{d}A - \int_A \Gamma_y \frac{\partial \phi^*}{\partial y}\frac{\partial [N]^T}{\partial y}\,\mathrm{d}A \tag{14.4b}$$

The first area integral term on the right-hand side of Equation 14.4 can be transformed into a line integral using the Green's theorem as follows:

$$\int_A \frac{\partial}{\partial x}\left([N]^T \Gamma_x \frac{\partial \phi^*}{\partial x}\right) \mathrm{d}A = \oint_s [N]^T \Gamma_x \frac{\partial \phi^*}{\partial x} \hat{n}_x d \tag{14.5a}$$

and

$$\int_A \frac{\partial}{\partial y}\left([N]^T \Gamma_y \frac{\partial \phi^*}{\partial y}\right) \mathrm{d}A = \oint_s [N]^T \Gamma_y \frac{\partial \phi^*}{\partial y} \hat{n}_y \mathrm{d}s \tag{14.5b}$$

Substituting (14.5) into (14.4), we get

$$\int_A [N]^T \frac{\partial}{\partial x}\left(\Gamma_x \frac{\partial \phi^*}{\partial x}\right) dA = \oint_s [N]^T \Gamma_x \frac{\partial \phi^*}{\partial x} \hat{n}_x \, ds - \int_A \frac{\partial [N]^T}{\partial x} \Gamma_x \frac{\partial \phi^*}{\partial x} dA \tag{14.6a}$$

and

$$\int_A [N]^T \frac{\partial}{\partial y}\left(\Gamma_y \frac{\partial \phi^*}{\partial y}\right) \mathrm{d}A = \oint_s [N]^T \Gamma_y \frac{\partial \phi^*}{\partial y} \hat{n}_y \mathrm{d}s - \int_A \frac{\partial [N]^T}{\partial y} \Gamma_y \frac{\partial \phi^*}{\partial y} \mathrm{d}A \tag{14.6b}$$

Substituting (14.6) into the integral form of the elemental given by (14.3), we have

$$\begin{aligned} \left\{R^{(e)}\right\} = & -\int_s [N]^T \left(\Gamma_x \frac{\partial \phi^*}{\partial x} \hat{n}_x + \Gamma_y \frac{\partial \phi^*}{\partial y} \hat{n}_y\right) \mathrm{d}s \\ & + \int_A \left(\frac{\partial [N]^T}{\partial x} \Gamma_x \frac{\partial \phi^*}{\partial x} + \frac{\partial [N]^T}{\partial y} \Gamma_y \frac{\partial \phi^*}{\partial y}\right) \mathrm{d}A \\ & + \int_A S_1 [N]^T \phi^* \, \mathrm{d}A - \int_A S_0 [N]^T \, \mathrm{d}A = 0 \end{aligned} \tag{14.7}$$

14.1.5 Step 5: Formation of Element Characteristics Equation

To obtain the element characteristics equation, substitute the approximate solution function given by Equation 14.2 into the integral

$$\begin{aligned} & -\int_s [N]^T \left(\Gamma_x \frac{\partial \phi^*}{\partial x} \cos\theta + \Gamma_y \frac{\partial \phi^*}{\partial y} \sin\theta\right) \mathrm{d}s \\ & + \left\{\int_A \left(\frac{\partial [N]^T}{\partial x} \Gamma_x \frac{\partial [N]}{\partial x} + \frac{\partial [N]^T}{\partial y} \Gamma_y \frac{\partial [N]}{\partial y}\right) \mathrm{d}A\right\} \left\{\phi^{(e)}\right\} \\ & + \left\{\int S_1 [N]^T [N] \mathrm{d}A\right\} \left\{\phi^{(e)}\right\} - \int_A S_0 [N]^T \, \mathrm{d}A = 0 \end{aligned} \tag{14.8}$$

The element characteristics equation can be written in matrix notation as

$$\{I^{(e)}\}+[k^{(e)}]\{\phi^{(e)}\}-\{f_S^{(e)}\}=0 \tag{14.9}$$

where

$\{I^{(e)}\}$=column vector for the element contribution to the interelement requirement

$$=-\int_S [N]^T\left(\Gamma_x\frac{\partial\phi^*}{\partial x}\hat{n}_x+\Gamma_y\frac{\partial\phi^*}{\partial y}\hat{n}_y\right)ds \tag{14.10}$$

$\{f_S^{(e)}\}$=element source vector due to the volumetric source term

$$=\int_A S_0[N]^T\,dA \tag{14.11}$$

$\{\phi^{(e)}\}$=column vector of unknown nodal values

$[k^{(e)}]$=element characteristics or stiffness matrix

$$=\iint_A\left(\frac{\partial[N]^T}{\partial x}\Gamma_x\frac{\partial[N]}{\partial x}+\frac{\partial[N]^T}{\partial y}\Gamma_y\frac{\partial[N]}{\partial y}\right)dA+\int_A S_1[N]^T[N]dA$$

$$=[k_\Gamma^{(e)}]+[k_S^{(e)}] \tag{14.12}$$

Note that the element stiffness matrix has two components: $[k_\Gamma^{(e)}]$ and $[k_S^{(e)}]$. The second term $[k_S^{(e)}]$ is contributed by the convection term of the governing Equation 14.1 given as

$$[k_S^{(e)}]=\int_A S_1[N]^T[N]dA \tag{14.13}$$

The first term $[k_\Gamma^{(e)}]$ is contributed by the diffusion terms, which is expressed in a compact form as follows:

$$[k_\Gamma^{(e)}]=\iint_A\left(\frac{\partial[N]^T}{\partial x}\Gamma_x\frac{\partial[N]}{\partial x}+\frac{\partial[N]^T}{\partial y}\Gamma_y\frac{\partial[N]}{\partial y}\right)dA$$

$$=\int_A\{D\}^T[\Gamma]\{D\}dA \tag{14.14}$$

where

$$[\Gamma]=\text{transport property matrix}=\begin{bmatrix}\Gamma_x & 0\\ 0 & \Gamma_y\end{bmatrix} \tag{14.15}$$

$$\{D\}=\text{column vector with derivatives of shape functions}=\begin{Bmatrix}\dfrac{\partial[N]}{\partial x}\\ \dfrac{\partial[N]}{\partial y}\end{Bmatrix} \tag{14.16}$$

$$\{D\}^T = \text{row vector with derivatives of shape functions} = \left\{ \frac{\partial[N]}{\partial x} \quad \frac{\partial[N]}{\partial y} \right\} \tag{14.17}$$

Implementation of Boundary Conditions

Let us divide the boundary surface **s** into different surface sections s_1, s_1, and s_3 as shown in Figure 14.7.

and consider different types of boundary conditions at different sections of the boundary surface as follows:

On the Boundary Surface s_1: Boundary condition of the first kind

$$\phi = \phi_s = \text{constant surface temperature} \tag{14.18a}$$

On the Boundary Surface s_2: Boundary condition of the second kind

$$\Gamma_x \frac{\partial \phi}{\partial x}\hat{n}_x + \Gamma_y \frac{\partial \phi}{\partial y}\hat{n}_y = \pm f_s^{//} = \text{constant surface flux} \tag{14.18b}$$

Note that for the special case with adiabatic conditions or for lines of symmetry, this boundary condition reduces to

$$\Gamma_x \frac{\partial \phi}{\partial x}\hat{n}_x + \Gamma_y \frac{\partial \phi}{\partial y}\hat{n}_y = 0 \tag{14.18c}$$

On the Boundary Surface s_2: Boundary condition of the third kind

$$\Gamma_x \frac{\partial \phi}{\partial x}\hat{n}_x + \Gamma_y \frac{\partial \phi}{\partial y}\hat{n}_y = \pm h(\phi - \phi_s) = \text{mixed conditions} \tag{14.18d}$$

In general, Equations 14.18b–14.18d can expressed in a general form as

$$\Gamma_x \frac{\partial \phi}{\partial x}\hat{n}_x + \Gamma_y \frac{\partial \phi}{\partial y}\hat{n}_y = \pm h(\phi - \phi_s) \pm f_s^{//} \tag{14.18e}$$

The implementation of the derivative boundary conditions into the finite element formulation takes place through the interelement vector $\{I^{(e)}\}$ as given by Equation 14.10.

$$\{I^{(e)}\} = -\int_s [N]^T \left(\Gamma_x \frac{\partial \phi^*}{\partial x}\hat{n}_x + \Gamma_y \frac{\partial \phi^*}{\partial y}\hat{n}_y \right) ds \tag{14.19}$$

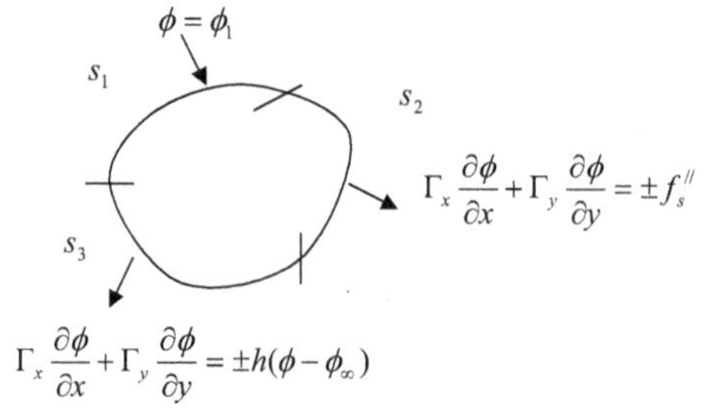

FIGURE 14.7 Different types of boundary conditions.

Note that this integral is the sum of all sides of an element including the side or sides that are subjected to the boundary conditions. To show the specific contribution from the boundary conditions, let us decompose the integral into two parts as

$$\{I^{(e)}\} = \{I_b^{(e)}\} + \{I_i^{(e)}\} \tag{14.20}$$

where $\{I_i^{(e)}\}$ is the integral contributions from all the other sides other than the boundary surface. This primarily constitutes the interelement requirements. The component $\{I_b^{(e)}\}$ is the integral contributions from sides that represent the boundary, and it is expressed as follows:

$$\{I_b^{(e)}\} = -\int_{s_b} [N]^T \left(\Gamma_x \frac{\partial \phi^*}{\partial x} \hat{n}_x + \Gamma_y \frac{\partial \phi^*}{\partial y} \hat{n}_y \right) ds \tag{14.21}$$

Substituting the boundary condition given by Equation 14.18e, which represents boundary condition in a general form, we get

$$\{I_b^{(e)}\} = -\int_{s_b} [N]^T \left\{ \pm h\left(\phi^* - \phi_\infty\right) \pm f_s^{//} \right\} ds \tag{14.22a}$$

or

$$\{I_b^{(e)}\} = \pm \int_{s_b} h[N]^T \phi^* ds \pm \int_{s_b} [N]^T \left\{ \left(h\phi_\infty + f_s^{//} \right) \right\} ds \tag{14.22b}$$

Now, substituting the approximate solution given by (13.2), we get

$$\{I_b^{(e)}\} = \pm \left(\int_{s_b} h[N]^T [N] ds \right) \phi^{(e)} \pm \int_{s_b} [N]^T \left\{ \left(h\phi_\infty + f_s^{//} \right) \right\} ds \tag{14.23a}$$

or

$$\{I_b^{(e)}\} = \left[k_b^{(e)}\right]\{\phi^{(e)}\} - \{f_{bs}^{(e)}\} \tag{14.23b}$$

where

$$\left[k_b^{(e)}\right] = \pm \int_{s_b} h[N]^T [N] ds = \text{Element stiffness matrix component due to boundary conditions} \tag{14.24}$$

$$\{f_{bs}^{(e)}\} = \pm \int_{s_b} [N]^T \left\{ \left(h\phi_\infty + f_s^{//} \right) \right\} ds = \text{Element source vector component due to boundary condition} \tag{14.25}$$

Finally, we can substitute Equations 14.29 and 14.23 into (14.9) to obtain element characteristic equation

$$\left\{I_i^{(e)}\right\} + \left(k_\Gamma^{(e)} + k_S^{(e)} + +k_b^{(e)} \right)\left\{\phi^{(e)}\right\} - \left\{ f_S^{(e)} + f_{bs}^{(e)} \right\} = 0 \tag{13.26}$$

or

$$\left\{I_i^{(e)}\right\} + k^{(e)}\left\{\phi^{(e)}\right\} - f^{//} = 0 \tag{14.27}$$

where

$$k^{(e)} = \left(k_\Gamma^{(e)} + k_S^{(e)} + +k_b^{(e)} \right) = \text{Element Characteristics Matrix} \tag{14.28}$$

and

$$f^{(e)} = \left\{ \left\{ f_S^{(e)} + f_{bs}^{(e)} \right\} \right\} = \text{Element source vector} \tag{14.29}$$

Let us demonstrate the formation of this element characteristics equation for different types of elements in the following sections:

Case: Element Characteristics Equation Using Linear Triangular Element

Let us consider a linear three-node triangular element as shown in Figure 10.13 and with the approximate solution and the corresponding shape functions are given by Equations 12.59 and 12.60, respectively. The vectors with derivatives of shape functions are derived by substituting the shape functions given as follows:

$$\{D\} = \begin{bmatrix} \dfrac{\partial N_i}{\partial x} & \dfrac{\partial N_j}{\partial x} & \dfrac{\partial N_k}{\partial x} \\ \dfrac{\partial N_i}{\partial y} & \dfrac{\partial N_j}{\partial y} & \dfrac{\partial N_k}{\partial y} \end{bmatrix} = \frac{1}{2A} \begin{bmatrix} \beta_i & \beta_j & \beta_k \\ \gamma_i & \gamma_j & \gamma_k \end{bmatrix} \tag{14.30a}$$

and

$$\{D\}^T = \begin{bmatrix} \dfrac{\partial N_i}{\partial x} & \dfrac{\partial N_i}{\partial y} \\ \dfrac{\partial N_j}{\partial x} & \dfrac{\partial N_j}{\partial y} \\ \dfrac{\partial N_k}{\partial x} & \dfrac{\partial N_k}{\partial y} \end{bmatrix} = \frac{1}{2A} \begin{bmatrix} \beta_i & \gamma_i \\ \beta_j & \gamma_j \\ \beta_k & \gamma_k \end{bmatrix} \tag{14.30b}$$

The component of element stiffness matrix due to the diffusion term $\left[k_\Gamma^{(e)} \right]$ can now be computed as

$$\left[k_{\Gamma}^{(e)}\right]=\int_A \{D\}^T[\Gamma]\{D\}\mathrm{d}A=\{D\}^T[\Gamma]\{D\}\int_A \mathrm{d}A$$

$$=\frac{1}{2A}\begin{bmatrix} b_i & c_i \\ b_j & c_j \\ b_k & c_k \end{bmatrix}\begin{bmatrix} \Gamma_x & 0 \\ 0 & \Gamma_y \end{bmatrix}\frac{1}{2A}\begin{bmatrix} b_i & b_j & b_k \\ c_i & c_j & c_k \end{bmatrix}A$$

$$=\frac{1}{4A^2}\begin{bmatrix} \beta_i & \gamma_i \\ \beta_j & \gamma_j \\ \beta_k & \gamma_k \end{bmatrix}\begin{bmatrix} \Gamma_x\beta_i & \Gamma_x\beta_j & \Gamma_x\beta_k \\ \Gamma_y\gamma_i & \Gamma_y\gamma_j & \Gamma_y\gamma_k \end{bmatrix}A$$

$$\left[k_{\Gamma}^{(e)}\right]=\frac{1}{4A}\begin{bmatrix} \Gamma_x\beta_i^2+\Gamma_y\gamma_i^2 & \Gamma_x\beta_i\beta_j+\Gamma_y\gamma_i\gamma_j & \Gamma_x\beta_i\beta_k+\Gamma_y\gamma_i\gamma_k \\ \Gamma_x\beta_j\beta_i+\Gamma_y\gamma_j\gamma_i & \Gamma_x\beta_j^2+\Gamma_y\gamma_j^2 & \Gamma_x\beta_j\beta_k+\Gamma_y\gamma_j\gamma_k \\ \Gamma_x\beta_k\beta_i+\Gamma_y\gamma_k\gamma_i & \Gamma_x\beta_k\beta_j+\Gamma_y\gamma_k\gamma_j & \Gamma_x\beta_k^2+\Gamma_y\gamma_k^2 \end{bmatrix} \tag{14.31a}$$

or

$$\left[k_{\Gamma}^{(e)}\right]=\frac{\Gamma_x}{4A}\begin{bmatrix} \beta_i^2 & \beta_i\beta_j & \beta_i\beta_k \\ \beta_j\beta_i & \beta_j^2 & \beta_j\beta_k \\ \beta_k\beta_i & \beta_k\beta_j & \beta_k^2 \end{bmatrix}+\frac{\Gamma_y}{4A}\begin{bmatrix} \gamma_i^2 & \gamma_i\gamma_j & \gamma_i\gamma_k \\ \gamma_j\gamma_i & \gamma_j^2 & \gamma_j\gamma_k \\ \gamma_k\gamma_i & \gamma_k\gamma_j & \gamma_k^2 \end{bmatrix} \tag{14.31b}$$

The component of element stiffness matrix due to the convection term, $\left[k_S^{(e)}\right]$, can now be computed as

$$\left[k_S^{(e)}\right]=\int_A S_1[N]^T[N]\mathrm{d}A$$

$$=S_1\int_A \begin{Bmatrix} N_i \\ N_j \\ N_k \end{Bmatrix}\begin{Bmatrix} N_i & N_j & N_k \end{Bmatrix}\mathrm{d}A$$

$$\left[k_{\beta}^{(e)}\right]=S_1\int_A \begin{bmatrix} N_i^2 & N_iN_j & N_iN_k \\ N_jN_i & N_j^2 & N_jN_k \\ N_kN_i & N_kN_j & N_k^2 \end{bmatrix}\mathrm{d}A \tag{14.32}$$

This integral can either be evaluated by direct integral formula or by Gauss-Legendre numerical formulas. As direct integral formulas are applicable, let us continue evaluating this integral analytically using natural co-ordinate system. From the expression of natural co-ordinate system (Equation 12.68) for the triangular element:

$$N_i=L_i,\quad N_j=L_j \quad \text{and} \quad N_k=L_k \tag{12.68}$$

Using (12.68) into (14.32), we get

$$\left[k_S^{(e)}\right] = S_1 \int_A \begin{bmatrix} L_i^2 & L_iL_j & L_iL_k \\ L_jL_i & L_j^2 & L_jL_k \\ L_kL_i & L_kL_j & L_k^2 \end{bmatrix} \tag{14.33}$$

Each of the element integral can now be evaluated using the integral formula given by Equation 12.71. For example, the integral elements in the first row are evaluated as follows:

First Integral Element (I_{11}):

$$I_{11} = \int_A L_i^2 \mathrm{d}A = \int L_i^2 L_j^0 L_k^0 \,\mathrm{d}A$$

$$= \frac{1!1!0!}{(1+1+0+2)} 2A$$

$$I_{11} = \frac{A}{6} \tag{14.34a}$$

Second Integral Element (I_{12}):

$$\int_A L_iL_j \mathrm{d}A = \int L_i^1 L_j^1 L_k^0 \mathrm{d}A$$

$$= \frac{1!1!0!}{(1+1+0+2)} 2A$$

$$= \frac{A}{12} \tag{14.34b}$$

Third Integral Element (I_{13}):

$$\int_A L_iL_k \mathrm{d}A = \int L_i^1 L_j^0 L_k^1 \mathrm{d}A$$

$$= \frac{1!1!0!}{(1+1+0+2)} 2A$$

$$= \frac{A}{12} \tag{14.34c}$$

Similarly, the integral elements in other rows can be evaluated and the integral expression (14.33) becomes

$$\left[k_S^{(e)}\right] = \frac{S_1 A}{12} \begin{bmatrix} 2 & 1 & 1 \\ 1 & 1 & 1 \\ 1 & 1 & 2 \end{bmatrix} \tag{14.34}$$

Substituting Equations 14.31 and 14.34 into 14.28, we have the expression for the element characteristic equation for an interior element as

$$\left[k^{(e)}\right]=\frac{\Gamma_x}{4A}\begin{bmatrix}\beta_i^2 & \beta_i\beta_j & \beta_i\beta_k\\ \beta_j\beta_i & \beta_j^2 & \beta_j\beta_k\\ \beta_k\beta_i & \beta_k\beta_j & \beta_k^2\end{bmatrix}+\frac{\Gamma_y}{4A}\begin{bmatrix}\gamma_i^2 & \gamma_i\gamma_j & \gamma_i\gamma_k\\ \gamma_j\gamma_i & \gamma_j^2 & \gamma_j\gamma_k\\ \gamma_k\gamma_i & \gamma_k\gamma_j & \gamma_k^2\end{bmatrix}+\frac{S_1A}{12}\begin{bmatrix}2 & 1 & 1\\ 1 & 1 & 1\\ 1 & 1 & 2\end{bmatrix}$$

(14.35)

The component of element characteristic matrix due to convective boundary conditions as given by Equation 14.24 is computed as follows:

$$\left[k_b^{(e)}\right]=\pm\int_{S_b} h[N]^T[N]ds$$

$$=\pm h\int_{S_b}\begin{Bmatrix}N_i\\ N_j\\ N_k\end{Bmatrix}\begin{Bmatrix}N_i & N_j & N_k\end{Bmatrix}ds$$

$$=\pm h\int_{S_b}\begin{bmatrix}N_i^2 & N_iN_j & N_iN_k\\ N_jN_i & N_j^2 & N_jN_k\\ N_kN_i & N_kN_j & N_k^2\end{bmatrix}ds \tag{14.36}$$

The integral is further evaluated by considering only the sides that are subjected to convection boundary conditions. For example, if convection is specified over side *i-j* of the element, then $N_k = 0$ and Equation 14.36 becomes

$$\left[k_b^{(e)}\right]=\pm h\int_{S_b}\begin{bmatrix}N_i^2 & N_iN_j & 0\\ N_jN_i & N_j^2 & 0\\ 0 & 0 & 0\end{bmatrix}ds \tag{14.37}$$

Again, using natural coordinate system (Equation 12.68) and integral formula (Equation (12.71), Equation 14.37 is transformed into

$$\left[k_b^{(e)}\right]=\pm\frac{hL_{ij}}{6}\begin{bmatrix}2 & 1 & 0\\ 1 & 2 & 0\\ 0 & 0 & 0\end{bmatrix} \tag{14.38a}$$

Similarly, for convection condition specified on side *j-k* and *k-i* sides, we get

$$\left[k_b^{(e)}\right]=\pm h\int_{S_b}\begin{bmatrix}0 & 0 & 0\\ 0 & N_j^2 & N_jN_k\\ 0 & N_kN_j & N_k^2\end{bmatrix}ds=\pm\frac{hL_{jk}}{6}\begin{bmatrix}0 & 0 & 0\\ 0 & 2 & 1\\ 0 & 1 & 2\end{bmatrix} \tag{14.38b}$$

and

$$\left[k_b^{(e)}\right] = \pm h \int_{S_b} \begin{bmatrix} N_i^2 & 0 & N_i N_k \\ 0 & 0 & 0 \\ N_k N_i & 0 & N_k^2 \end{bmatrix} ds = \pm \frac{hL_{ki}}{6} \begin{bmatrix} 2 & 0 & 1 \\ 0 & 0 & 0 \\ 1 & 0 & 2 \end{bmatrix} \tag{14.38c}$$

The element characteristic matrix for the element with external boundary specified at the edge i-j is obtained by substituting Equations 14.31, 14.34, and 14.38a into 14.28

$$\frac{\Gamma_x}{4A} \begin{bmatrix} \beta_i^2 & \beta_i \beta_j & \beta_i \beta_k \\ \beta_j \beta_i & \beta_j^2 & \beta_j \beta_k \\ \beta_k \beta_i & \beta_k \beta_j & \beta_k^2 \end{bmatrix}$$

$$\left[k^{(e)}\right] = \frac{\Gamma_x}{4A} \begin{bmatrix} \beta_i^2 & \beta_i \beta_j & \beta_i \beta_k \\ \beta_j \beta_i & \beta_j^2 & \beta_j \beta_k \\ \beta_k \beta_i & \beta_k \beta_j & \beta_k^2 \end{bmatrix} + \frac{\Gamma_y}{4A} \begin{bmatrix} \gamma_i^2 & \gamma_i \gamma_j & \gamma_i \gamma_k \\ \gamma_j \gamma_i & \gamma_j^2 & \gamma_j \gamma_k \\ \gamma_k \gamma_i & \gamma_k \gamma_j & \gamma_k^2 \end{bmatrix}$$

$$+ \frac{S_1 A}{12} \begin{bmatrix} 2 & 1 & 1 \\ 1 & 1 & 1 \\ 1 & 1 & 2 \end{bmatrix} \pm \frac{hL_{ij}}{6} \begin{bmatrix} 2 & 1 & 0 \\ 1 & 2 & 0 \\ 0 & 0 & 0 \end{bmatrix} \tag{14.39}$$

Let us now evaluate the force vectors given by Equations 14.11 and 14.25.

The element force vector due to a constant source term is computed using natural co-ordinate system for a triangular element as follows:

$$\left\{f_S^{(e)}\right\} = \int_A S_0 [N]^T \, dA \tag{14.11}$$

$$= S_0 \int_A \begin{bmatrix} N_i \\ N_j \\ N_k \end{bmatrix} dA$$

$$= S_0 \int_A \begin{bmatrix} L_i \\ L_j \\ L_k \end{bmatrix} dA \tag{14.40}$$

Computing the integral elements using integration formula (Equation 12.71), we get

$$\left\{f_S^{(e)}\right\} = \frac{S_0 A}{3} \begin{Bmatrix} 1 \\ 1 \\ 1 \end{Bmatrix} \tag{14.41}$$

Similarly, the element force vector component due to the boundary condition is computed from Equation 14.25 as

$$\left\{f_{\text{bs}}^{(e)}\right\} = \pm\int_{S_b} [N]^T \left\{\left(h\phi_\infty + f_s^{//}\right)\right\} \mathrm{d}s \tag{14.25}$$

$$= \left(h\phi_\infty + f_s^{//}\right)\int_{S_b} \begin{Bmatrix} N_i \\ N_j \\ N_k \end{Bmatrix} \mathrm{d}s \tag{14.42}$$

For boundary condition specified on *i-j* side of the element, $N_k = 0$, and we use natural co-ordinate system for a triangle and use integral formula (Equation 12.28), we get

$$\left\{f_{bs}^{(e)}\right\} = \pm\left(h\phi_\infty + f_s^{//}\right)\int_{S_b} \begin{Bmatrix} N_i \\ N_j \\ 0 \end{Bmatrix} \mathrm{d}s = \pm\left(h\phi_\infty + f_s^{//}\right)\int_{S_b} \begin{Bmatrix} L_i \\ L_j \\ 0 \end{Bmatrix} \mathrm{d}s$$

$$\left\{f_{bs}^{(e)}\right\} = \pm\frac{\left(h\phi_\infty + f_s^{//}\right)L_{ij}}{2}\begin{Bmatrix} 1 \\ 1 \\ 0 \end{Bmatrix} \tag{14.43a}$$

Similarly, for boundary conditions specified on the sides *j-k*, $N_i = 0$, and Equation 14.42 transforms into

$$\left\{f_{\text{bs}}^{(e)}\right\} = \left(h\phi_\infty + f_s^{//}\right)\int_{S_b} \begin{Bmatrix} 0 \\ N_j \\ N_k \end{Bmatrix} \mathrm{d}s \pm \frac{\left(h\phi_\infty + f_s^{//}\right)L_{jk}}{2}\begin{Bmatrix} 0 \\ 1 \\ 1 \end{Bmatrix} \tag{14.43b}$$

and for boundary conditions specified on the side *k-i*, $N_j = 0$, and we get

$$\left\{f_{\text{bs}}^{(e)}\right\} = \left(h\phi_\infty + f_s^{//}\right)\int_{S_b} \begin{Bmatrix} N_i \\ 0 \\ N_k \end{Bmatrix} \mathrm{d}s = \pm\frac{\left(h\phi_\infty + f_s^{//}\right)L_{ki}}{2}\begin{Bmatrix} 1 \\ 0 \\ 1 \end{Bmatrix} \tag{14.43c}$$

Example 14.1 Two-Dimensional Problem Using Triangular Element

Determine the element characteristic equation for a triangular element in the finite element mesh used to solve the temperature distribution and heat transfer rate a two-dimensional fin shown in the figure. The convective conditions are $h = 20.0\frac{\mathrm{W}}{\mathrm{m^2K}}$ and $T_\infty = 25°\mathrm{C}$. The fin is made of aluminum with thermal conductivity, $k = 202\frac{\mathrm{W}}{\mathrm{m\cdot °C}}$.

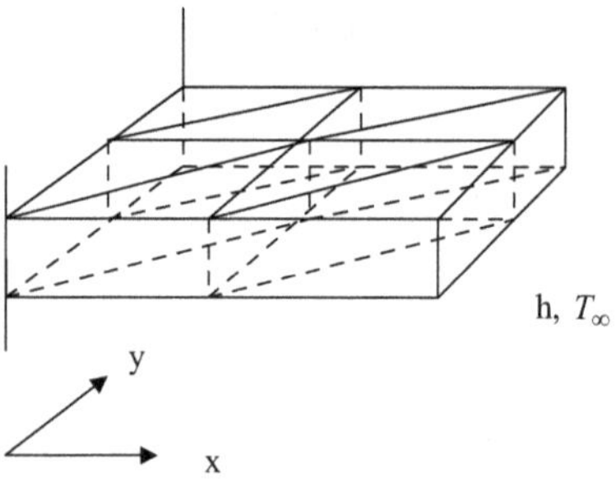

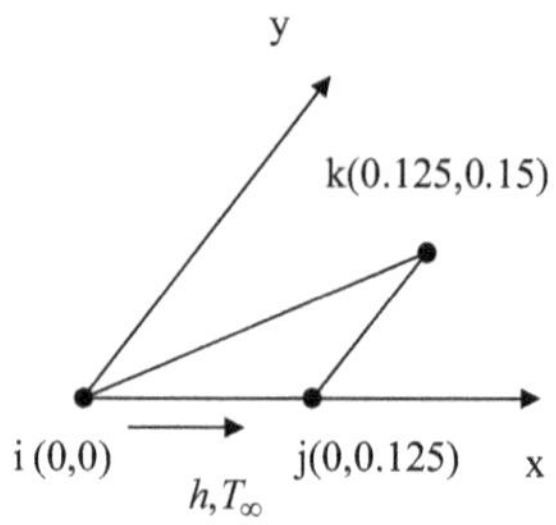

The length and width of the fin are $L=25$ cm and $W=30$ cm, respectively. The thickness of the fin is $t=5$ mm. A typical triangular element in the mesh is shown in the figure. The co-ordinates of the nodes are *i*(0,0), *j*(0.125), and *k*(0.125,0.15). Determine the element characteristic equation for this element.

Solution:

The governing heat equation for a two-dimensional fine is

$$\frac{\partial}{\partial x}\left(k_x A_x \frac{\partial T}{\partial x}\right)+\frac{\partial}{\partial y}\left(k_y A_y \frac{\partial T}{\partial y}\right)-hdA_s(T-T_\infty)=0 \tag{E.14.1.1}$$

Substituting $A_x = t \cdot \mathrm{d}y$, $A_y = t \cdot \mathrm{d}x$, and $\mathrm{d}A_s = 2\mathrm{d}x\mathrm{d}y$, Equation E.14.1.1 reduces to

$$k_x \frac{\partial^2 T}{\partial x^2}+k_y \frac{\partial^2 T}{\partial y^2}-\frac{2h}{t}T+\frac{2hT_\infty}{t}=0$$

This is of the same form as the general Equation 14.1 with

$$\Gamma_x = k_x = k = 202\frac{\mathrm{W}}{\mathrm{m}\cdot{}^\circ\mathrm{C}}, \quad \Gamma_y = k_y = k = 202\frac{W}{\mathrm{m}\cdot{}^\circ\mathrm{C}}$$

$$S_1 = \frac{2h}{t} = \frac{2\times 20}{0.005} = 8000\frac{\mathrm{W}}{\mathrm{m}\cdot{}^\circ\mathrm{C}}$$

and

$$S_0 = \frac{2hT_\infty}{t} = \frac{2\times 20\times 25}{0.005} = 2\times 10^5\,\mathrm{W/m^3}$$

The element characteristic matrix is evaluated based on Equation 14.28

$$k^{(e)} = \left(k_\Gamma^{(\mathrm{e})} + k_S^{(\mathrm{e})} + k_\mathrm{b}^{(\mathrm{e})}\right)$$

$$\left[k_\Gamma^{(e)}\right]=\frac{\Gamma_x}{4A}\begin{bmatrix}\beta_i^2 & \beta_i\beta_j & \beta_i\beta_k\\ \beta_j\beta_i & \beta_j^2 & \beta_j\beta_k\\ \beta_k\beta_i & \beta_k\beta_j & \beta_k^2\end{bmatrix}+\frac{\Gamma_y}{4A}\begin{bmatrix}\gamma_i^2 & \gamma_i\gamma_j & \gamma_i\gamma_k\\ \gamma_j\gamma_i & \gamma_j^2 & \gamma_j\gamma_k\\ \gamma_k\gamma_i & \gamma_k\gamma_j & \gamma_k^2\end{bmatrix} \tag{14.31b}$$

$$\left[k_S^{(e)}\right]=\frac{S_1A}{12}\begin{bmatrix}2 & 1 & 1\\ 1 & 1 & 1\\ 1 & 1 & 2\end{bmatrix},\quad k_b^{(e)}=\int_{Sb} h[N]^T[N]ds$$

The element source vector is given by Equation 14.29 as

$$f^{(e)}=\left\{f_S^{(e)}+f_{bs}^{(e)}\right\}$$

The co-ordinates of the three nodes of the triangular element are

$$x_i=0,\quad x_j=0.125,\quad x_k=0.125$$

$$y_i=0,\quad y_j=0,\quad y_k=0.15$$

Let us first determine all coefficients and shape function for the triangular element based on these coordinate points and using Equation 12.62 as follows:

$$\alpha_i=x_jy_k-x_ky_j=0.125(0.15)-0.125(0.0)=0.01875$$

$$\alpha_j=x_ky_i-x_iy_k=0.125(0.0)-0.0(0.15)=0.0$$

$$\alpha_k=x_iy_j-x_jy_i=0.0(0.0)-0.125(0)=0.0$$

$$\beta_i=y_j-y_k=0.0-0.15=-0.15$$

$$\beta_j=y_k-y_i=0.15-0.0=0.15$$

$$\beta_k=y_i-y_j=0.0-0.0=0.0$$

$$\gamma_i=x_k-x_j=0.125-0.125=0.0$$

$$\gamma_j=x_i-x_k=0.0-0.125=-0.125$$

$$\gamma_k=x_j-x_i=0.125-0.0=0.125$$

$$A=\frac{1}{2}\begin{vmatrix}1 & x_i & y_i\\ 1 & x_j & y_j\\ 1 & x_k & y_k\end{vmatrix}=\frac{1}{2}\begin{vmatrix}1 & 0 & 0\\ 1 & 0.125 & 0\\ 1 & 0.125 & 0.15\end{vmatrix}=0.009375$$

The component of element characteristic due to the diffusion terms is estimated as

$$\left[k_\Gamma^{(e)}\right]=\frac{\Gamma_x}{4A}\begin{bmatrix}\beta_i^2 & \beta_i\beta_j & \beta_i\beta_k\\ \beta_j\beta_i & \beta_j^2 & \beta_j\beta_k\\ \beta_k\beta_i & \beta_k\beta_j & \beta_k^2\end{bmatrix}+\frac{\Gamma_y}{4A}\begin{bmatrix}\gamma_i^2 & \gamma_i\gamma_j & \gamma_i\gamma_k\\ \gamma_j\gamma_i & \gamma_j^2 & \gamma_j\gamma_k\\ \gamma_k\gamma_i & \gamma_k\gamma_j & \gamma_k^2\end{bmatrix}$$

$$=\frac{202}{4\times 0.009375}\begin{bmatrix}(-0.15)^2 & (-0.15)(0.15) & (-0.15)(0.0)\\ (0.15)(-0.15) & (0.15)^2 & (0.15)(0.0)\\ (0.0)(-0.15) & (0.0)(0.15) & (0.0)^2\end{bmatrix}$$

$$+\frac{202}{4\times 0.009375}\begin{bmatrix}(0.0)^2 & (0.0)(-0.125) & (0.0)(0.125)\\ (-0.125)(0.0) & (-0.125)^2 & (-0.125)(0.125)\\ (0.125)(0.0) & (0.125)(-0.125) & (0.125)^2\end{bmatrix}$$

$$=5386.6666\begin{bmatrix}0.0225 & -0.0225 & 0.0\\ -0.0225 & 0.0225 & 0.0\\ 0.0 & 0.0 & 0.0\end{bmatrix}+5386.6666\begin{bmatrix}0.0 & 0.0 & 0.0\\ 0.0 & 0.015625 & -0.015625\\ 0.0 & -0.015625 & 0.015625\end{bmatrix}$$

$$=121.1999\begin{bmatrix}1 & -1 & 0\\ -1 & 1 & 0\\ 0 & 0 & 0\end{bmatrix}+84.1666\begin{bmatrix}0 & 0 & 0\\ 0 & 1 & -1\\ 0 & -1 & 1\end{bmatrix}$$

$$=\begin{bmatrix}121.1999 & -121.1999 & 0.0\\ -121.1999 & 121.1999 & 0.0\\ 0.0 & 0.0 & 0.0\end{bmatrix}+\begin{bmatrix}0.0 & 0.0 & 0.0\\ 0.0 & 84.1666 & -84.1666\\ 0.0 & -84.1666 & 84.1666\end{bmatrix}$$

$$\left[k_\Gamma^{(e)}\right]=\begin{bmatrix}121.1999 & -121.1999 & 0.0\\ -121.1999 & 205.3665 & -84.1666\\ 0.0 & -84.1666 & 84.1666\end{bmatrix} \tag{E14.1.2}$$

$$\left[k_\beta^{(e)}\right]=\frac{\beta A}{12}\begin{bmatrix}2 & 1 & 1\\ 1 & 1 & 1\\ 1 & 1 & 2\end{bmatrix}$$

$$=\frac{8000\times 0.01875}{12}\begin{bmatrix}2 & 1 & 1\\ 1 & 1 & 1\\ 1 & 1 & 2\end{bmatrix}=12.5\begin{bmatrix}2 & 1 & 1\\ 1 & 1 & 1\\ 1 & 1 & 2\end{bmatrix}$$

$$\left[k_\beta^{(e)}\right]=\begin{bmatrix}25 & 12.5 & 12.5\\ 12.5 & 12.5 & 12.5\\ 12.5 & 12.5 & 25\end{bmatrix} \tag{E.14.1.3}$$

The component of element characteristic matrix due to boundary conditions as given by Equation 14.24 is computed as follows:

$$\left[k_b^{(e)}\right] = \pm\int_{s_b} h[N]^T[N]ds$$

$$= \pm\int_{s_b} h\begin{Bmatrix} N_i \\ N_j \\ N_k \end{Bmatrix}\begin{Bmatrix} N_i & N_j & N_k \end{Bmatrix}ds$$

$$= \pm h\int_{S_b}\begin{bmatrix} N_i^2 & N_iN_j & N_iN_k \\ N_jN_i & N_j^2 & N_jN_k \\ N_kN_i & N_kN_j & N_k^2 \end{bmatrix}ds$$

Since convection boundary conditions are specified over side *i-j* of the element, then $N_k = 0$ and Equation 14.36 transformed into

$$\left[k_b^{(e)}\right] = \frac{hL_{ij}}{6}\begin{bmatrix} 2 & 1 & 0 \\ 1 & 2 & 0 \\ 0 & 0 & 0 \end{bmatrix} = \frac{20\times 0.125}{6}\begin{bmatrix} 2 & 1 & 0 \\ 1 & 2 & 0 \\ 0 & 0 & 0 \end{bmatrix}$$

$$= \begin{bmatrix} 0.8333 & 0.4166 & 0 \\ 0.4166 & 0.8333 & 0 \\ 0 & 0 & 0 \end{bmatrix} \tag{E.14.1.4}$$

Combining Equations E.14.1.1–E.14.1.3, we get the element characteristics equation as

$$k^{(e)} = \left(k_\Gamma^{(e)} + k_S^{(e)} + k_b^{(e)}\right)$$

$$= \begin{bmatrix} 121.1999 & -121.1999 & 0.0 \\ -121.1999 & 205.3665 & -84.1666 \\ 0.0 & -84.1666 & 84.1666 \end{bmatrix} + \begin{bmatrix} 25 & 12.5 & 12.5 \\ 12.5 & 12.5 & 12.5 \\ 12.5 & 12.5 & 25 \end{bmatrix} + \begin{bmatrix} 0.8333 & 0.4166 & 0 \\ 0.4166 & 0.8333 & 0 \\ 0 & 0 & 0 \end{bmatrix}$$

$$k^{(e)} = \begin{bmatrix} 146.1999 & -108.6999 & 12.5 \\ -108.6999 & 217.8665 & -71.6666 \\ 12.5 & -71.6666 & 109.1666 \end{bmatrix} + \begin{bmatrix} 0.8333 & 0.4166 & 0 \\ 0.4166 & 0.8333 & 0 \\ 0 & 0 & 0 \end{bmatrix}$$

$$k^{(e)} = \begin{bmatrix} 147.0332 & -108.2833 & 12.5 \\ -108.2833 & 218.6998 & -71.6666 \\ 12.5 & -71.6666 & 109.1666 \end{bmatrix} \tag{E.14.1.5}$$

The element force vector due to a constant source term is computed by Equation 14.41 as

$$\left\{f_S^{(e)}\right\} = \frac{SA}{3}\begin{Bmatrix} 1 \\ 1 \\ 1 \end{Bmatrix} = \frac{2\times 10^5 \times 0.009375}{3}\begin{Bmatrix} 1 \\ 1 \\ 1 \end{Bmatrix}$$

$$= \begin{Bmatrix} 625 \\ 625 \\ 625 \end{Bmatrix} \tag{E.14.1.6}$$

The element source vector component due to the boundary condition is computed from Equation 14.42 with convective boundary condition and $f_s^{//}=0$ as

$$\{f_{bs}^{(e)}\}=(hT_\infty)\int_{S_b}\begin{Bmatrix}N_i\\N_j\\N_k\end{Bmatrix}ds$$

Since convective boundary condition is specified on *i-j* side of the element, $N_k=0$. Along with the use natural co-ordinate system and integral formula given by Equation 14.27a, we get

$$\{f_{bs}^{(e)}\}=\frac{(hT_\infty)L_{ij}}{2}\begin{Bmatrix}1\\1\\0\end{Bmatrix}=\frac{(25\times 20)0.125}{2}\begin{Bmatrix}1\\1\\0\end{Bmatrix}$$

$$\{f_{bs}^{(e)}\}=\frac{(20\times 25)0.125}{2}\begin{Bmatrix}1\\1\\0\end{Bmatrix}$$

$$\{f_{bs}^{(e)}\}=\begin{Bmatrix}31.25\\31.25\\0\end{Bmatrix}\tag{E.14.1.7}$$

Combining Equations E.14.1.6 and E.14.1.7, we get the element source vector as

$$f^{(e)}=\{f_S^{(e)}+f_{bs}^{(e)}\}$$

$$=\begin{Bmatrix}625\\625\\625\end{Bmatrix}+\begin{Bmatrix}31.25\\31.25\\0\end{Bmatrix}$$

$$f^{(e)}=\begin{Bmatrix}656.5\\656.5\\625.0\end{Bmatrix}\tag{E.14.1.8}$$

The element characteristic matrix is given by Equation 12.27 as

$$k^{(e)}\{\phi^{(e)}\}=-\{I_i^{(e)}\}+f^{(e)}$$

$$\begin{bmatrix}147.0332 & -108.2833 & 12.5\\ -108.2833 & 218.6998 & -71.6666\\ 12.5 & -71.6666 & 109.1666\end{bmatrix}\begin{Bmatrix}\phi_i^{(e)}\\ \phi_j^{(e)}\\ \phi_k^{(e)}\end{Bmatrix}=-\begin{Bmatrix}I_i^{(e)}\\ I_j^{(e)}\\ I_k^{(e)}\end{Bmatrix}+\begin{Bmatrix}656.5\\656.5\\625\end{Bmatrix}\tag{E.14.1.9}$$

Case: Element Characteristics Equation Using Bilinear Rectangular Element

For a bilinear rectangular element (Figure 12.19), the approximate solution and the corresponding shape functions in the local co-ordinate system are given by Equations 12.87 and 12.88, respectively:

$$\phi(\tilde{x},\tilde{y})=N_i\phi_i+N_j\phi_j+N_k\phi_k+N_l\phi_l=\sum_{i=i,j,k,l}N_I\phi_I=[N]\{\phi^{(e)}\}\tag{12.87}$$

with shape functions given as

$$N_i = \left(1-\frac{\tilde{x}}{2l_e}\right)\left(1-\frac{\tilde{y}}{2h_e}\right), \quad N_j = \frac{\tilde{x}}{2l_e}\left(1-\frac{\tilde{y}}{2h_e}\right)$$

$$N_k = \frac{\tilde{x}\tilde{y}}{4l_e h_e} \qquad N_l = \frac{\tilde{y}}{2h_e}\left((1-\frac{\tilde{x}}{2l_e}\right) \tag{12.88}$$

The vectors with derivatives of shape functions are derived by substituting the shape functions given by Equations 12.88 for bilinear rectangular element as follows:

$$\{D\} = \begin{bmatrix} \dfrac{\partial N_i}{\partial x} & \dfrac{\partial N_j}{\partial x} & \dfrac{\partial N_k}{\partial x} & \dfrac{\partial N_l}{\partial x} \\ \dfrac{\partial N_i}{\partial y} & \dfrac{\partial N_j}{\partial y} & \dfrac{\partial N_k}{\partial y} & \dfrac{\partial N_l}{\partial y} \end{bmatrix} = \frac{1}{4l_e h_e}\begin{bmatrix} -(2h_e-\tilde{y}) & (2h_e-\tilde{y}) & \tilde{y} & -\tilde{y} \\ -(2l_e-\tilde{x}) & -\tilde{x} & \tilde{x} & (2l_e-\tilde{x}) \end{bmatrix} \tag{14.44}$$

$$\left\{D^T\right\} = \begin{bmatrix} \dfrac{\partial N_i}{\partial x} & \dfrac{\partial N_i}{\partial y} \\ \dfrac{\partial N_j}{\partial x} & \dfrac{\partial N_j}{\partial y} \\ \dfrac{\partial N_k}{\partial x} & \dfrac{\partial N_k}{\partial y} \\ \dfrac{\partial N_l}{\partial x} & \dfrac{\partial N_l}{\partial y} \end{bmatrix} = \begin{bmatrix} -(2h_e-\tilde{y}) & -(2l_e-\tilde{x}) \\ (2h_e-\tilde{y}) & \tilde{x} \\ \tilde{y} & \tilde{x} \\ -\tilde{y} & (2l_e-\tilde{x}) \end{bmatrix} \tag{14.45}$$

The component of element characteristic matrix due to the diffusion term $\left[k_\Gamma^{(e)}\right]$ can now be computed from Equation 14.14 as

$$\left[k_\Gamma^{(e)}\right] = \int_A \{D\}^T[\Gamma]\{D\}\mathrm{d}A = \{D\}^T[\Gamma]\{D\}\int_A \mathrm{d}A$$

$$= \int_A \begin{bmatrix} \dfrac{\partial N_i}{\partial x} & \dfrac{\partial N_i}{\partial y} \\ \dfrac{\partial N_j}{\partial x} & \dfrac{\partial N_j}{\partial y} \\ \dfrac{\partial N_k}{\partial x} & \dfrac{\partial N_k}{\partial y} \\ \dfrac{\partial N_l}{\partial x} & \dfrac{\partial N_l}{\partial y} \end{bmatrix} \begin{bmatrix} \Gamma_x & 0 \\ 0 & \Gamma_y \end{bmatrix} \begin{bmatrix} \dfrac{\partial N_i}{\partial x} & \dfrac{\partial N_j}{\partial x} & \dfrac{\partial N_k}{\partial x} & \dfrac{\partial N_l}{\partial x} \\ \dfrac{\partial N_i}{\partial y} & \dfrac{\partial N_i}{\partial y} & \dfrac{\partial N_i}{\partial y} & \dfrac{\partial N_i}{\partial y} \end{bmatrix}$$

$$
= \int_A \begin{bmatrix} \dfrac{\partial N_i}{\partial \tilde{x}} & \dfrac{\partial N_i}{\partial \tilde{y}} \\ \dfrac{\partial N_j}{\partial \tilde{x}} & \dfrac{\partial N_j}{\partial \tilde{y}} \\ \dfrac{\partial N_k}{\partial \tilde{x}} & \dfrac{\partial N_k}{\partial \tilde{y}} \\ \dfrac{\partial N_l}{\partial \tilde{x}} & \dfrac{\partial N_l}{\partial \tilde{y}} \end{bmatrix} \begin{bmatrix} \Gamma_x & 0 \\ 0 & \Gamma_y \end{bmatrix} \begin{bmatrix} \dfrac{\partial N_i}{\partial \tilde{x}} & \dfrac{\partial N_j}{\partial \tilde{x}} & \dfrac{\partial N_k}{\partial \tilde{x}} & \dfrac{\partial N_l}{\partial \tilde{x}} \\ \dfrac{\partial N_i}{\partial \tilde{y}} & \dfrac{\partial N_i}{\partial \tilde{y}} & \dfrac{\partial N_i}{\partial \tilde{y}} & \dfrac{\partial N_i}{\partial \tilde{y}} \end{bmatrix}
$$

$$
= \int_0^{2l_e} \int_0^{2h_e} \frac{1}{4l_e h_e} \begin{bmatrix} -(2h_e - \tilde{y}) & -(2l_e - \tilde{x}) \\ (2h_e - \tilde{y}) & -\tilde{x} \\ \tilde{y} & \tilde{x} \\ -\tilde{y} & (2l_e - \tilde{x}) \end{bmatrix} \begin{bmatrix} \Gamma_x & \\ & \Gamma_y \end{bmatrix}
$$

$$
\times \frac{1}{4l_e h_e} \begin{bmatrix} -(2h_e - \tilde{y}) & (2h_e - \tilde{y}) & \tilde{y} & -\tilde{y} \\ -(2l_e - \tilde{x}) & -\tilde{x} & \tilde{x} & (2l_e - \tilde{x}) \end{bmatrix} \mathrm{d}\tilde{x}\mathrm{d}\tilde{y}
$$

$$
= \int_0^{2l_e} \int_0^{2h_e} \frac{1}{16 l_e^2 h_e^2} \begin{bmatrix} -(2h_e - \tilde{y}) & -(2l_e - \tilde{x}) \\ (2h_e - \tilde{y}) & -\tilde{x} \\ \tilde{y} & \tilde{x} \\ -\tilde{y} & (2l_e - \tilde{x}) \end{bmatrix}
$$

$$
\times \begin{bmatrix} -\Gamma_x (2h_e - \tilde{y}) & \Gamma_y (2h_e - \tilde{y}) & \Gamma_x \tilde{y} & -\Gamma_y \tilde{y} \\ -\Gamma_y (2l_e - \tilde{x}) & -\Gamma_y \tilde{x} & \Gamma_y \tilde{x} & \Gamma_y (2l_e - \tilde{x}) \end{bmatrix} \mathrm{d}\tilde{x}\mathrm{d}\tilde{y}
$$

$$
= \int_0^{2l_e} \int_0^{2h_e} \frac{1}{16 l_e^2 h_e^2} \begin{bmatrix} -(2h_e - \tilde{y}) & -(2l_e - \tilde{x}) \\ (2h_e - \tilde{y}) & -\tilde{x} \\ \tilde{y} & \tilde{x} \\ -\tilde{y} & (2l_e - \tilde{x}) \end{bmatrix}
$$

$$
\times \begin{bmatrix} -\Gamma_x (2h_e - \tilde{y}) & \Gamma_y (2h_e - \tilde{y}) & \Gamma_x \tilde{y} & -\Gamma_y \tilde{y} \\ -\Gamma_y (2l_e - \tilde{x}) & -\Gamma_y \tilde{x} & \Gamma_y \tilde{x} & \Gamma_y (2l_e - \tilde{x}) \end{bmatrix} \mathrm{d}\tilde{x}\mathrm{d}\tilde{y}
$$

$$
= \int_0^{2l_e}\int_0^{2h_e} \frac{1}{16l_e^2h_e^2}\begin{bmatrix} \Gamma_x(2h_e-\tilde{y})^2+\Gamma_e(2l_e-\tilde{x})^2 & -\Gamma_x(2h_e-\tilde{y})^2+\Gamma_e(2l_e-\tilde{x})\tilde{x} & -\Gamma_x(2h_e-\tilde{y})\tilde{y}-\Gamma_e(2l_e-\tilde{x})\tilde{x} & \Gamma_x(2h_e-\tilde{y})\tilde{y}-\Gamma_e(2l_e-\tilde{x})^2 \\ -\Gamma_x(2h_e-\tilde{y})^2+\Gamma_e(2l_e-\tilde{x})^2\tilde{x} & -\Gamma_x(2h_e-\tilde{y})^2+\Gamma_e(2l_e-\tilde{x})\tilde{x} & \Gamma_x(2h_e-\tilde{y})\tilde{y}-\Gamma_e\tilde{x}^2 & -\Gamma_x(2h_e-\tilde{y})\tilde{y}-\Gamma_e(2l_e-\tilde{x})\tilde{x} \\ -\Gamma_x(2h_e-\tilde{y})\tilde{y}-\Gamma_e(2l_e-\tilde{x})\tilde{x} & -\Gamma_x(2h_e-\tilde{y})^2+\Gamma_e(2l_e-\tilde{x})\tilde{x} & \Gamma_x\tilde{y}^2-\Gamma_e\tilde{x}^2 & -\Gamma_x\tilde{y}^2+\Gamma_e(2l_e-\tilde{x})\tilde{x} \\ \Gamma_x(2h_e-\tilde{y})\tilde{y}-\Gamma_e(2l_e-\tilde{x})^2 & -\Gamma_x(2h_e-\tilde{y})^2+\Gamma_e(2l_e-\tilde{x})\tilde{x} & -\Gamma_x\tilde{y}^2+\Gamma_e(2l_e-\tilde{x})\tilde{x} & \Gamma_x\tilde{y}^2+\Gamma_e(2l_e-\tilde{x})^2 \end{bmatrix} d\tilde{x}d\tilde{y} \tag{14.46}
$$

Each integral element of the equation can be integrated one at a time. For example, the first integral element is evaluated as follows:

$$
I_{11} = \int_0^{2l_e}\int_0^{2h_e} \frac{1}{16l_e^2h_e^2}\left(\Gamma_x(2h_e-\tilde{y})^2+\Gamma_e(2l_e-\tilde{x})^2\right)d\tilde{x}d\tilde{y}
$$

$$
= \int_0^{2l_e}\int_0^{2h_e} \frac{1}{16l_e^2h_e^2}\Gamma_x(2h_e-\tilde{y})^2 d\tilde{x}d\tilde{y} + \int_0^{2l_e}\int_0^{2h_e} \frac{1}{16l_e^2h_e^2}\Gamma_e(2l_e-\tilde{x})^2 d\tilde{x}d\tilde{y} \tag{14.47a}
$$

$$
I_{11} = \frac{\Gamma_x h_e}{3l_e} + \frac{\Gamma_y l_e}{3h_e}
$$

In a similarly manner, all other elements are evaluated and the expression for element characteristic matrix due to the diffusion terms is given as

$$
\left[k_\Gamma^{(e)}\right] = \frac{\Gamma_x h_e}{6l_e}\begin{bmatrix} 2 & -2 & -1 & 1 \\ -2 & 2 & 1 & -1 \\ -1 & 1 & 2 & -2 \\ 1 & -1 & -2 & 2 \end{bmatrix} + \frac{\Gamma_y l_e}{6h_e}\begin{bmatrix} 2 & 1 & -1 & -2 \\ 1 & 2 & -2 & -1 \\ -1 & -2 & 2 & 1 \\ -2 & -1 & 1 & 2 \end{bmatrix} \tag{14.47}
$$

The component of element stiffness matrix due to the convection term $\left[k_S^{(e)}\right]$ can now be computed as

$$
\left[k_S^{(e)}\right] = \int_A S_1[N]^T[N]dA
$$

$$
= \int_A S_1 \begin{Bmatrix} N_i \\ N_j \\ N_k \\ N_l \end{Bmatrix}\begin{Bmatrix} N_i & N_j & N_k & N_l \end{Bmatrix} dA
$$

$$
\left[k_S^{(e)}\right] = \int_A S_1 \begin{bmatrix} N_i^2 & N_iN_j & N_iN_k & N_iN_l \\ N_jN_i & N_j^2 & N_jN_k & N_jN_l \\ N_kN_i & N_kN_j & N_k^2 & N_kN_l \\ N_lN_i & N_lN_j & N_lN_k & N_l^2 \end{bmatrix} dA \tag{14.48}
$$

Each element integral can be evaluated after substituting the shape function expressions given by Equation 12.88. For cases with constant coefficient β, the integrals can be evaluated directly as follows:

$$I_{33} = \int_A S_1 N_k^2 \mathrm{d}A = S_1 \int_0^{2l_e} \int_0^{2h_e} N_k^2 \mathrm{d}\tilde{x}\mathrm{d}\tilde{y}$$

$$= S_1 \int_0^{2l_e} \int_0^{2h_e} \left(\frac{\tilde{x}\tilde{y}}{4l_e h_e} \right)^2 \mathrm{d}\tilde{x}\mathrm{d}\tilde{y} = S_1 \int_0^{2l_e} \frac{\tilde{x}^2}{16 l_e^2 h_e^2} \frac{\tilde{y}^3}{3} \Bigg|_0^{2h_e} \mathrm{d}\tilde{x}$$

$$= S_1 \int_0^{2l_e} \frac{\tilde{x}^2 h_e}{6 l_e^2} \mathrm{d}\tilde{x} = S_1 \frac{h_e}{18 l_e^2} \tilde{x}^3 \Big|_0^{2l_e} = S_1 \frac{4 l_e h_e}{9}$$

$$I_{33} = \frac{S_1 A}{9} \tag{14.49a}$$

Similarly, the rest of the integral elements are evaluated, and the element stiffness matrix due to the convection term is expressed as

$$\left[k_\beta^{(e)}\right] = \frac{S_1 A}{36} \begin{bmatrix} 4 & 2 & 1 & 2 \\ 2 & 4 & 2 & 1 \\ 1 & 2 & 4 & 2 \\ 2 & 1 & 2 & 4 \end{bmatrix} \tag{14.49b}$$

The element stiffness matrix component due to boundary conditions as given by Equation (14.24) is computed as follows:

$$\left[k_b^{(e)}\right] = \pm \int_{S_b} h [N]^T [N] \mathrm{d}s$$

$$= \pm \int_{S_b} h \begin{Bmatrix} N_i \\ N_j \\ N_k \\ N_l \end{Bmatrix} \begin{Bmatrix} N_i & N_j & N_k & N_l \end{Bmatrix} \mathrm{d}s$$

$$\left[k_b^{(e)}\right] = \pm \int_{S_b} h \begin{bmatrix} N_i^2 & N_i N_j & N_i N_k & N_i N_l \\ N_j N_i & N_j^2 & N_j N_k & N_j N_l \\ N_k N_i & N_k N_j & N_k^2 & N_k N_l \\ N_l N_i & N_l N_j & N_l N_k & N_l^2 \end{bmatrix} \mathrm{d}s \tag{14.50}$$

The integral is further evaluated by considering only the sides that is subjected to convection boundary conditions. For example, if convection is specified over side i-j of the element, then $N_k = 0$ and $N_l = 0$, and Equation 14.50 becomes

$$\left[k_b^{(e)}\right] = \pm\int_{S_b} h \begin{bmatrix} N_i^2 & N_iN_j & 0 & 0 \\ N_jN_i & N_j^2 & 0 & 0 \\ 0 & 0 & 0 & 0 \\ 0 & 0 & 0 & 0 \end{bmatrix} ds \tag{14.51}$$

Again, using natural coordinate system (Equation 12.27) and integral formula (Equation 12.28), the integral elements can be evaluated. For example, the first integral element is evaluated

$$\int_{S_b} hN_i^2 ds = \frac{2l_e}{3} = \frac{L_{ij}}{3} \tag{14.52}$$

Using these integral elements, the Equation 14.37 is transformed into

$$\left[k_b^{(e)}\right] = \pm\frac{hL_{ij}}{6}\begin{bmatrix} 2 & 1 & 0 & 0 \\ 1 & 2 & 0 & 0 \\ 0 & 0 & 0 & 0 \\ 0 & 0 & 0 & 0 \end{bmatrix} \text{for convective boundary on the side } i\text{-}j \tag{14.53a}$$

Similarly,

For convective boundary on the side j-k

$$\left[k_b^{(e)}\right] = \pm\int_{S_b} h \begin{bmatrix} 0 & 0 & 0 & 0 \\ 0 & N_j^2 & N_jN_k & 0 \\ 0 & N_kN_j & N_k^2 & 0 \\ 0 & 0 & 0 & 0 \end{bmatrix} ds = \pm\frac{hL_{jk}}{6}\begin{bmatrix} 0 & 0 & 0 & 0 \\ 0 & 2 & 1 & 0 \\ 0 & 1 & 2 & 0 \\ 0 & 0 & 0 & 0 \end{bmatrix} \tag{14.53b}$$

For convective boundary on the side k-l

$$\left[k_b^{(e)}\right] = \pm\int_{S_b} h \begin{bmatrix} 0 & 0 & 0 & 0 \\ 0 & 0 & 0 & 0 \\ 0 & 0 & N_k^2 & N_kN_l \\ 0 & 0 & N_lN_k & N_l^2 \end{bmatrix} ds = \pm\frac{hL_{kl}}{6}\begin{bmatrix} 0 & 0 & 0 & 0 \\ 0 & 0 & 0 & 0 \\ 0 & 0 & 2 & 1 \\ 0 & 0 & 1 & 2 \end{bmatrix} \tag{14.53c}$$

For convective boundary on the side l-i

$$\left[k_b^{(e)}\right] = \pm\int_{S_b} h \begin{bmatrix} N_i^2 & 0 & 0 & N_iN_l \\ 0 & 0 & 0 & 0 \\ 0 & 0 & 0 & 0 \\ N_lN_i & 0 & 0 & N_l^2 \end{bmatrix} ds = \pm\frac{hL_{li}}{6}\begin{bmatrix} 2 & 0 & 0 & 1 \\ 0 & 0 & 0 & 0 \\ 0 & 0 & 0 & 0 \\ 1 & 0 & 0 & 2 \end{bmatrix} \tag{14.53d}$$

The element force vector due to a constant source term is computed using natural co-ordinate system for a triangular element as follows:

$$\left\{f_S^{(e)}\right\} = \int_A S_0 [N]^T \,\mathrm{d}A \tag{14.11}$$

$$= \int_A S_0 \begin{Bmatrix} N_i \\ N_j \\ N_k \\ N_l \end{Bmatrix} \mathrm{d}A \tag{14.54}$$

$$\left\{f_S^{(e)}\right\} = \int_0^{2l_e}\int_0^{2h_e} S_0 \begin{Bmatrix} 1 - \dfrac{\tilde{x}}{2l_e} - \dfrac{\tilde{y}}{2h_e} + \dfrac{\tilde{x}\tilde{y}}{4l_e h_e} \\ \dfrac{\tilde{x}}{2l_e} - \dfrac{\tilde{x}\tilde{y}}{4l_e h_e} \\ \dfrac{\tilde{x}\tilde{y}}{4l_e h_e} \\ \dfrac{\tilde{y}}{2h_e} - \dfrac{\tilde{x}\tilde{y}}{4l_e h_e} \end{Bmatrix} \mathrm{d}\tilde{x}\mathrm{d}\tilde{y} \tag{14.55}$$

Each element integral in the force vector can be evaluated analytically for cases with source term. For example, the third element is evaluated by integral rules as follows:

$$f_{S_3}^{(e)} = S_0 \int_0^{2l_e}\int_0^{h_e} \frac{\tilde{x}\tilde{y}}{4l_e h_e} \mathrm{d}\tilde{x}\mathrm{d}\tilde{y}$$

$$= S_0 \int_0^{2l_e} \frac{\tilde{x}}{4l_e h_e} \frac{\tilde{y}^2}{2}\bigg|_0^{2h_e} = S_0 \frac{h_e}{2l_e} \frac{\tilde{x}^2}{2}\bigg|_0^{2l_e}$$

$$= S_0 l_e h_e = \frac{S_0 4 l_e h_e}{4}$$

$$f_{S_3}^{(e)} = \frac{S_0 A}{4} \tag{14.56}$$

Evaluating all elements in a similarly manner, the element force vector due to the source term is expressed as

$$f_S^{(e)} = \frac{S_0 A}{4} \begin{Bmatrix} 1 \\ 1 \\ 1 \\ 1 \end{Bmatrix} \tag{14.57}$$

Similarly, the element force vector component due to the boundary condition is computed as

$$\left\{f_{\mathrm{bs}}^{(e)}\right\} = \pm \int_{s_b} [N]^T \left\{\left(h\phi_\infty + f_s^{//}\right)\right\} \mathrm{d}s \tag{14.25}$$

$$= \left(h\phi_\infty + f_s^{//}\right)\int_{S_b}\begin{Bmatrix} N_i \\ N_j \\ N_k \\ N_l \end{Bmatrix} ds \tag{14.58}$$

For boundary condition specified on *i-j* side of the element, $N_k = 0$ and $N_l = 0$. We use natural coordinate system for a triangle and use integral formula (Equation 12.28), we get

$$\left\{f_{bs}^{(e)}\right\} = \pm\left(h\phi_\infty + f_s^{//}\right)\int_{S_b}\begin{Bmatrix} N_i \\ N_j \\ 0 \\ 0 \end{Bmatrix} ds = \pm\left(h\phi_\infty + f_s^{//}\right)\int_{S_b}\begin{Bmatrix} L_i \\ L_j \\ 0 \\ 0 \end{Bmatrix} ds$$

$$\left\{f_{bs}^{(e)}\right\} = \pm\frac{\left(h\phi_\infty + f_s^{//}\right)L_{ij}}{2}\begin{Bmatrix} 1 \\ 1 \\ 0 \\ 0 \end{Bmatrix} \tag{14.59a}$$

Similarly, for boundary conditions specified on the sides *j-k*, $N_i = 0$ and $N_l = 0$, and Equation 14.26 transforms into

$$\left\{f_{bs}^{(e)}\right\} = \left(h\phi_\infty + f_s^{//}\right)\int_{S_b}\begin{Bmatrix} 0 \\ N_j \\ N_k \\ \end{Bmatrix} ds = \pm\frac{\left(h\phi_\infty + f_s^{//}\right)L_{jk}}{2}\begin{Bmatrix} 0 \\ 1 \\ 1 \\ 0 \end{Bmatrix} \tag{14.59b}$$

For boundary conditions specified on the side *k-l*, $N_i = 0$, and $N_j = 0$, and we get

$$\left\{f_{bs}^{(e)}\right\} = \left(h\phi_\infty + f_s^{//}\right)\int_{S_b}\begin{Bmatrix} 0 \\ 0 \\ N_k \\ N_l \end{Bmatrix} ds = \pm\frac{\left(h\phi_\infty + f_s^{//}\right)L_{kl}}{2}\begin{Bmatrix} 0 \\ 0 \\ 1 \\ 1 \end{Bmatrix} \tag{14.59c}$$

For boundary conditions specified on the side *l-i*, $N_j = 0$, and $N_k = 0$, and we get

$$\left\{f_{bs}^{(e)}\right\} = \left(h\phi_\infty + f_s^{//}\right)\int_{S_b}\begin{Bmatrix} N_i \\ 0 \\ 0 \\ 1 \end{Bmatrix} ds = \pm\frac{\left(h\phi_\infty + f_s^{//}\right)L_{li}}{2}\begin{Bmatrix} 1 \\ 0 \\ 0 \\ 1 \end{Bmatrix} \tag{14.59d}$$

Example 14.2 Two-Dimensional Problem Using Bilinear Rectangular Element

Determine the element characteristic equation for the bilinear rectangular element (Element #5) in the finite element mesh used to solve the steady-state temperature distribution and heat transfer rate in plane carbon steel rectangular irradiated by constant heat flux at the top surface as shown. All surfaces of the work piece are convectively cooled with $h = 50.0\frac{\text{W}}{\text{m}^2\,\text{K}}$ and $T_\infty = 20°\text{C}$. The workpiece is made of plane carbon steel with thermal conductivity, $k = 43\frac{\text{W}}{\text{m}\cdot°\text{C}}$. The constant heat flux at the top surface of the element #5 is $q_s^{//} = 5\times10^5\,\text{W/m}^2$. The length and height of the element are l_e=0.05 m and h_e=0.05 m, respectively. The co-ordinates of the nodes are: *i*(0.0,0.0), *j*(0.05,0.0), *k*(0.05,0.05), and *l*(0.0,0.05).

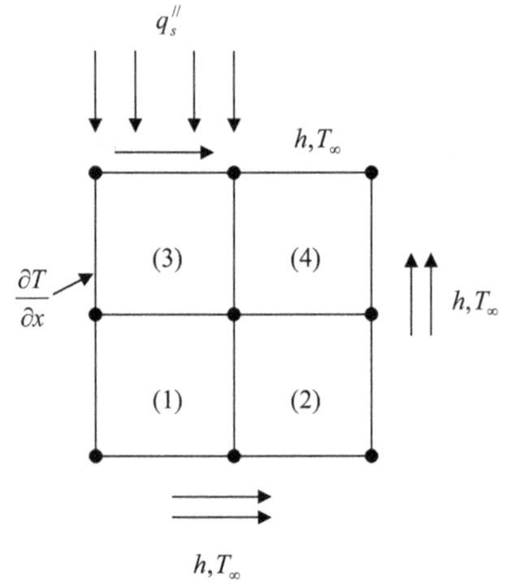

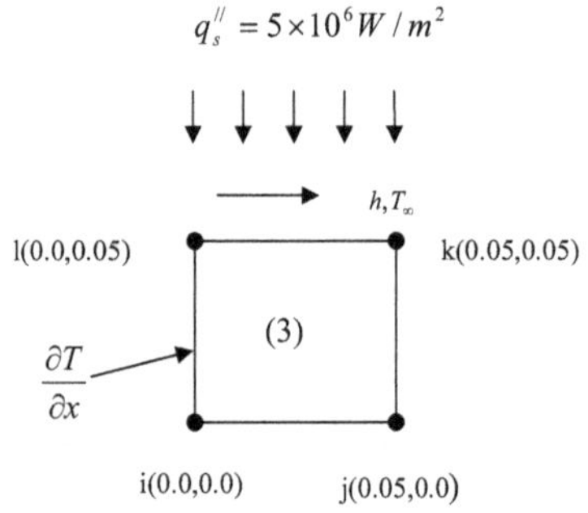

Solution

The governing heat equation for a two-dimensional fine is

$$\frac{\partial}{\partial x}\left(k_x\frac{\partial T}{\partial x}\right)+\frac{\partial}{\partial y}\left(k_y\frac{\partial T}{\partial y}\right)=0 \tag{E14.2.1}$$

Comparing this equation with the general Equation 14.1, we get

$$\Gamma_x = k_x = k = 43\frac{\text{W}}{\text{m}\cdot°\text{C}},\quad \Gamma_y = k_y,\quad k = 43\frac{\text{W}}{\text{m}\cdot°\text{C}},\quad S_1 = 0,\quad S_0 = 0$$

The boundary condition on the top surface is

$$k\frac{\partial T}{\partial y} = h(T - T_\infty) - f_s^{//} \tag{E.14.2.2}$$

Rest of the sides of the element is internal.

The element characteristic matrix is evaluated based on the Equation 14.2.8

$$k^{(e)} = \left(k_\Gamma^{(e)} + k_S^{(e)} + k_b^{(e)}\right)$$

where the element characteristic matrix component due the diffusion terms is given for a bilinear rectangular element is given by Equation 14.47b as

$$\left[k_\Gamma^{(e)}\right] = \frac{kh_e}{6l_e}\begin{bmatrix} 2 & -2 & -1 & 1 \\ -2 & 2 & 1 & -1 \\ -1 & 1 & 2 & -2 \\ 1 & -1 & -2 & 2 \end{bmatrix} + \frac{k_y l_e}{6h_e}\begin{bmatrix} 2 & 1 & -1 & -2 \\ 1 & 2 & -2 & -1 \\ -1 & -2 & 2 & 1 \\ -2 & -1 & 1 & 2 \end{bmatrix} \quad \text{(E.14.2.3)}$$

For the square element with $l_e = h_e$, the equation reduces to

$$\left[k_\Gamma^{(e)}\right] = \frac{43}{6}\begin{bmatrix} 4 & -1 & -2 & -1 \\ -1 & 4 & -1 & -2 \\ -2 & -1 & 4 & -1 \\ -1 & -2 & -1 & 4 \end{bmatrix} = 7.166\begin{bmatrix} 4 & -1 & -2 & -1 \\ -1 & 4 & -1 & -2 \\ -2 & -1 & 4 & -1 \\ -1 & -2 & -1 & 4 \end{bmatrix}$$

$$\left[k_\Gamma^{(e)}\right] = \begin{bmatrix} 28.666 & -7.166 & -14.333 & -7.166 \\ -7.166 & 28.666 & -7.166 & -14.333 \\ -14.333 & -7.166 & 28.666 & -7.166 \\ -7.166 & -14.333 & -7.166 & 28.666 \end{bmatrix} \quad \text{(E.14.2.4)}$$

Since $S_1 = 0$, the component of characteristic matrix due to the convection term is

$$\left[k_S^{(e)}\right] = 0 \quad \text{(E.14.2.5)}$$

The component of element characteristic matrix due to boundary conditions is given by Equation 14.24 and is computed as follows:

$$\left[k_b^{(e)}\right] = \pm\int_{S_b} h[N]^T[N]ds$$

$$\left[k_b^{(e)}\right] = \pm\int_{S_b} h\begin{bmatrix} N_i^2 & N_iN_j & N_iN_k & N_iN_l \\ N_jN_i & N_j^2 & N_jN_k & N_jN_l \\ N_kN_i & N_kN_j & N_k^2 & N_kN_l \\ N_lN_i & N_lN_j & N_lN_k & N_l^2 \end{bmatrix} ds$$

The integral is further evaluated by considering only the sides that is subjected to convection boundary conditions. Since the side *k-l* of the element is subjected to convective condition, the matrix is evaluated as

$$\left[k_b^{(e)}\right] = \pm\int_{S_b} h\begin{bmatrix} 0 & 0 & 0 & 0 \\ 0 & 0 & 0 & 0 \\ 0 & 0 & N_k^2 & N_kN_l \\ 0 & 0 & N_lN_k & N_l^2 \end{bmatrix} ds = \frac{hL_{kl}}{6}\begin{bmatrix} 0 & 0 & 0 & 0 \\ 0 & 0 & 0 & 0 \\ 0 & 0 & 2 & 1 \\ 0 & 0 & 1 & 2 \end{bmatrix}$$

$$= \frac{50 \times 0.05}{6}\begin{bmatrix} 0 & 0 & 0 & 0 \\ 0 & 0 & 0 & 0 \\ 0 & 0 & 2 & 1 \\ 0 & 0 & 1 & 2 \end{bmatrix} = \begin{bmatrix} 0 & 0 & 0 & 0 \\ 0 & 0 & 0 & 0 \\ 0 & 0 & 0.8333 & 0.4166 \\ 0 & 0 & 0.4166 & 0.8333 \end{bmatrix} \quad \text{(E.14.2.6)}$$

The element characteristic matrix is now obtained by combining Equations E14.2.4–E.14.2.6

$$k^{(e)} = \left(k_{\Gamma}^{(e)} + k_S^{(e)} + k_b^{(e)}\right)$$

$$= \begin{bmatrix} 28.666 & -7.166 & -14.333 & -7.166 \\ -7.166 & 28.666 & -7.166 & -14.333 \\ -14.333 & -7.166 & 28.666 & -7.166 \\ -7.166 & -14.333 & -7.166 & 28.666 \end{bmatrix} + \begin{bmatrix} 0 & 0 & 0 & 0 \\ 0 & 0 & 0 & 0 \\ 0 & 0 & 0.8333 & 0.4166 \\ 0 & 0 & 0.4166 & 0.8333 \end{bmatrix}$$

$$k^{(e)} = \begin{bmatrix} 28.666 & -7.166 & -14.333 & -7.166 \\ -7.166 & 28.666 & -7.166 & -14.333 \\ -14.333 & -7.166 & 29.4999 & -6.75 \\ -7.166 & -14.333 & -6.75 & 29.4999 \end{bmatrix} \quad \text{(E.14.2.7)}$$

The element source vector is given as

$$f^{(e)} = \left\{ f_S^{(e)} + f_{bs}^{(e)} \right\} \quad \text{(E.14.2.8)}$$

where $\left\{ f_S^{(e)} \right\}$ = element source vector due to the volumetric source term, which for the case of no volumetric heat generation is given as

$$\left\{ f_S^{(e)} \right\} = \int_A \dot{q}[N]^T \, dA = 0 \quad \text{(E.14.2.9)}$$

For the bilinear rectangular element with the mixed boundary condition is specified on the side *k-l*, $N_i = 0$, and $N_j = 0$, and we get the source vector due to the boundary condition for the element as

$$\left\{ f_{bs}^{(e)} \right\} = \left(hT_\infty + q_s^{//}\right) \int_{S_b} \begin{Bmatrix} 0 \\ 0 \\ N_k \\ N_l \end{Bmatrix} ds = \frac{\left(h\varphi_\infty + q_s^{//}\right) L_{jk}}{2} \begin{Bmatrix} 0 \\ 0 \\ 1 \\ 1 \end{Bmatrix}$$

$$\left\{ f_{bs}^{(e)} \right\} = \frac{\left(hT_\infty + q_s^{//}\right) 2l_e}{2} \begin{Bmatrix} 0 \\ 0 \\ 1 \\ 1 \end{Bmatrix} \quad \text{(E.14.2.10)}$$

Substituting Equations E.14.2.9 and E.14.2.10 into Equation E.14.2.8, we get the element source vector as

$$f^{(e)} = \frac{\left(hT_\infty + q_s^{//}\right) 2l_e}{2} \begin{Bmatrix} 0 \\ 0 \\ 1 \\ 1 \end{Bmatrix} = \frac{\left(50 \times 20 + 5 \times 10^5\right) 2 \times 0.05}{2} \begin{Bmatrix} 0 \\ 0 \\ 1 \\ 1 \end{Bmatrix}$$

$$f^{(e)} = \begin{Bmatrix} 0 \\ 0 \\ 50 + 0.25 \times 10^5 \\ 50 + 0.25 \times 10^5 \end{Bmatrix} \quad \text{(E.14.2.11)}$$

The element characteristic matrix for the bilinear rectangular element is

$$k^{(e)} \left\{ \varphi^{(e)} \right\} = -\left\{ I_i^{(e)} \right\} + f^{(e)}$$

$$\begin{bmatrix} 28.666 & -7.166 & -14.333 & -7.166 \\ -7.166 & 28.666 & -7.166 & -14.333 \\ -14.333 & -7.166 & 29.4999 & -6.75 \\ -7.166 & -14.333 & -6.75 & 29.4999 \end{bmatrix} \begin{Bmatrix} \varphi_i \\ \varphi_j \\ \varphi_k \\ \varphi_l \end{Bmatrix} = \begin{Bmatrix} I_i \\ I_j \\ I_k \\ I_l \end{Bmatrix} + \begin{Bmatrix} 0 \\ 0 \\ 50+0.25\times10^5 \\ 50+0.25\times10^5 \end{Bmatrix}$$

(E.14.2.12)

Example 14.3:

Consider a square rod with uniform volumetric heat generation $q = 3.0\times10^4 W/m^3$. The rod is convectively cooled at the surface with $h = 200\frac{W}{m^2C}$ and 10° C.

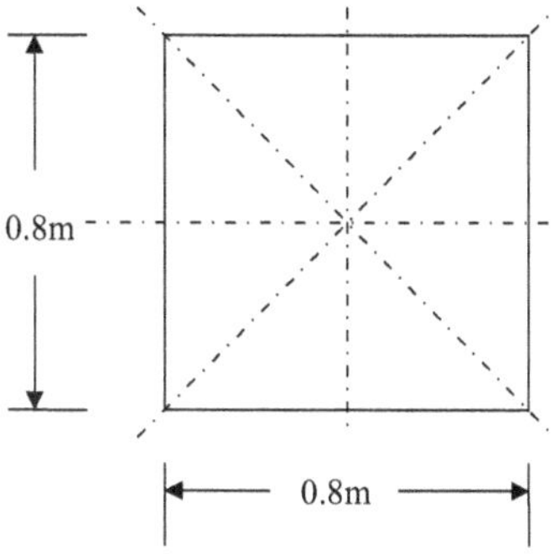

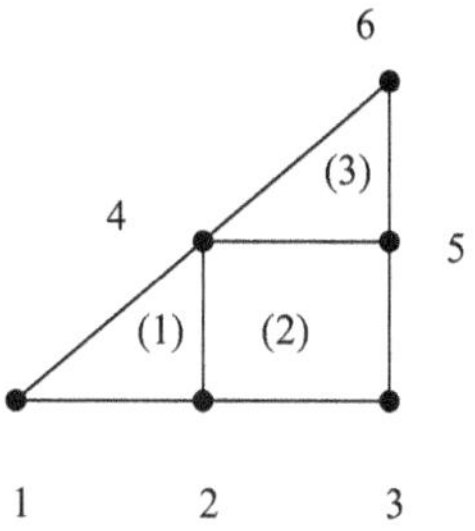

14.1.6 Step 6: Assembly of Element Equations and Formation of Global System

The assembly of all element characteristic system to form the global system is done by using a procedure known as the direct stiffness method. Let us demonstrate the procedure by considering two triangular elements with both local and global numbers shown in Figure 14.8.

A connectivity array between the local and global numbering of the elements needs to be constructed as shown in Table 14.1.

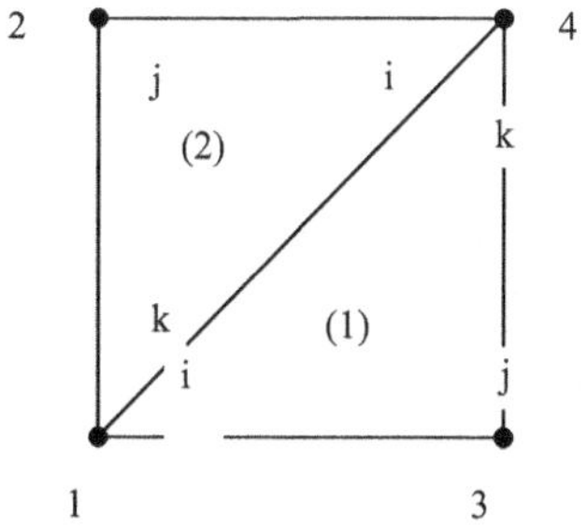

FIGURE 14.8 Assembly of elements.

TABLE 14.1
Connectivity Array

Element Number (e)	Global Node Numbers		
	Local Node (i)	Local Node (j)	Local Node (k)
1	1	3	4
2	4	2	1

Let us consider a general expression for these elements as follows:

Element #1

1 3 4

$$\begin{bmatrix} k_{ii}^{(1)} & k_{ij}^{(1)} & k_{ik}^{(1)} \\ k_{ji}^{(1)} & k_{jj}^{(1)} & k_{jk}^{(1)} \\ k_{ki}^{(1)} & k_{kj}^{(1)} & k_{kk}^{(1)} \end{bmatrix} \begin{Bmatrix} \phi_i \\ \phi_j \\ \phi_k \end{Bmatrix} = \begin{Bmatrix} f_i^{(1)} \\ f_j^{(1)} \\ f_k^{(1)} \end{Bmatrix} \begin{matrix} 1 \\ 3 \\ 4 \end{matrix} \tag{14.60a}$$

Element #2

4 2 1

$$\begin{bmatrix} k_{ii}^{(2)} & k_{ij}^{(2)} & k_{ik}^{(2)} \\ k_{ji}^{(2)} & k_{jj}^{(2)} & k_{jk}^{(2)} \\ k_{ki}^{(2)} & k_{kj}^{(2)} & k_{kk}^{(2)} \end{bmatrix} \begin{Bmatrix} \phi_i \\ \phi_j \\ \phi_k \end{Bmatrix} = \begin{Bmatrix} f_i^{(1)} \\ f_j^{(1)} \\ f_k^{(1)} \end{Bmatrix} \begin{matrix} 4 \\ 2 \\ 1 \end{matrix} \tag{14.60b}$$

Note that the global node numbers are written over the columns and along rows of the element characteristic matrix $\left[k^{(e)}\right]$ and along rows of the force vector $\left\{f^{(e)}\right\}$.

Let us start with the initial global characteristic equation as

1 2 3 4

$$\begin{bmatrix} 0 & 0 & 0 & 0 \\ 0 & 0 & 0 & 0 \\ 0 & 0 & 0 & 0 \\ 0 & 0 & 0 & 0 \end{bmatrix} \begin{Bmatrix} \phi_1 \\ \phi_2 \\ \phi_3 \\ \phi_4 \end{Bmatrix} = \begin{Bmatrix} 0 \\ 0 \\ 0 \\ 0 \end{Bmatrix} \begin{matrix} 1 \\ 2 \\ 3 \\ 4 \end{matrix} \tag{14.61}$$

To form the global system, add the coefficients of the element characteristic matrix and force vector one at a time. For example, the coefficients of the element characteristics equation for the element #1 can be added as follows:

Add $k_{ii}^{(1)}$ to K_{11}, Add $k_{ij}^{(1)}$ to K_{13}, and Add $k_{ik}^{(1)}$ to K_{14}
Add $k_{ji}^{(1)}$ to K_{31}, Add $k_{jj}^{(1)}$ to K_{33}, and Add $k_{jk}^{(1)}$ to K_{34}
Add $k_{ki}^{(1)}$ to K_{41}, Add $k_{kj}^{(1)}$ to, K_{43} and Add $k_{kk}^{(1)}$ to K_{44}
$f_i^{(1)}$ to F_1, $f_j^{(1)}$ to F_3, and $f_k^{(1)}$ to F_4

With these additions of the **element #1**, the global system becomes

1 2 3 4

$$\begin{bmatrix} k_{ii}^{(1)} & 0 & k_{ij}^{(1)} & k_{ik}^{(1)} \\ 0 & 0 & 0 & 0 \\ k_{ji}^{(1)} & 0 & k_{jj}^{(1)} & k_{jk}^{(1)} \\ k_{ki}^{(1)} & 0 & k_{kj}^{(1)} & k_{kk}^{(1)} \end{bmatrix} \begin{Bmatrix} \phi_1 \\ \phi_2 \\ \phi_3 \\ \phi_4 \end{Bmatrix} = \begin{Bmatrix} f_i^{(1)} \\ 0 \\ f_j^{(1)} \\ f_k^{(1)} \end{Bmatrix} \begin{matrix} 1 \\ 2 \\ 3 \\ 4 \end{matrix} \tag{14.62}$$

TABLE 14.2
Pseudo Code for the Assembly of Element Characteristic Equations

```
do for i 1 = ne                    : ne = number of element
   do for j = 1, np                : np = number of node in the element
      i1 = cr (j)                  : cr = connectivity array
      F(i1) = F (i1) + f(j)        : F,f = global and element force vectors
      do for k = 1, np
         j1=cr (k)
         K(i1,j1) = K(i1,j1)+k(j,k)
   end do
  end do
end do
```

In a similar manner, coefficients of the element characteristic matrix and force vector of the **element # 2** are added to the coefficient positions of the global system, and the global system takes the form

1 2 3 4

$$\begin{bmatrix} k_{ii}^{(1)}+k_{kk}^{(2)} & k_{kj}^{(2)} & k_{ij}^{(1)} & k_{ik}^{(1)}+k_{ki}^{(2)} \\ k_{jk}^{(2)} & k_{jj}^{(2)} & 0 & k_{ji}^{(2)} \\ k_{ji}^{(1)} & 0 & k_{jj}^{(1)} & k_{jk}^{(1)} \\ k_{ki}^{(1)}+k_{ik}^{(2)} & k_{ij}^{(2)} & k_{kj}^{(1)} & k_{kk}^{(1)}+k_{ii}^{(2)} \end{bmatrix} \begin{Bmatrix} \phi_1 \\ \phi_2 \\ \phi_3 \\ \phi_4 \end{Bmatrix} = \begin{Bmatrix} f_i^{(1)}+f_k^{(1)} \\ f_j^{(2)} \\ f_j^{(1)} \\ f_k^{(1)}+f_i^{(2)} \end{Bmatrix} \begin{matrix} 1 \\ 2 \\ 3 \\ 4 \end{matrix} \tag{14.63}$$

A pseudo code for the formation of the global system by assembling all element characteristic equation using the connectivity array is given in Table 14.2.

14.2 THREE-DIMENSIONAL PROBLEMS

In this section, we will consider the formulation and application finite element method for solving three-dimensional steady-state problems. The three-dimensional formulation is like those discussed for two- and one-dimensional problems. Let us consider three-dimensional steady-state transport process in a three-dimensional domain as shown in the figure below (Figure 14.9).

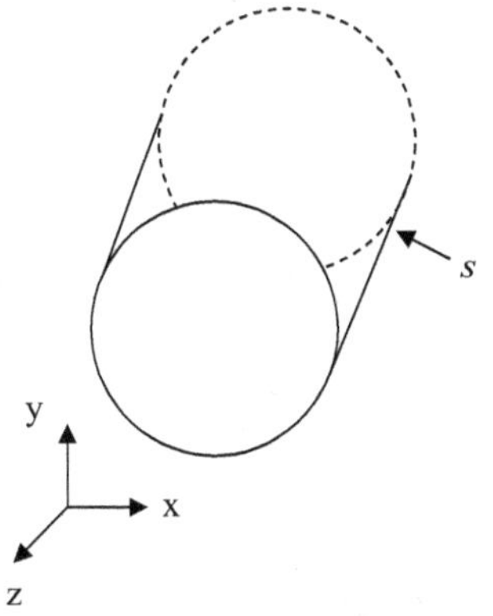

FIGURE 14.9 Three-dimensional domain.

A general mathematical statement of the problem can be assumed as

$$\frac{\partial}{\partial x}\left(\Gamma_x \frac{\partial \phi}{\partial x}\right)+\frac{\partial}{\partial y}\left(\Gamma_y \frac{\partial \phi}{\partial y}\right)+\frac{\partial}{\partial z}\left(\Gamma_z \frac{\partial \phi}{\partial z}\right)+S=0 \tag{14.64}$$

Subject to the associate general boundary conditions as

On the Boundary Surface A_b Boundary condition of the third kind

$$\Gamma_x \frac{\partial \phi}{\partial x} n_x+\Gamma_y \frac{\partial \phi}{\partial y} n_y+\Gamma_z \frac{\partial \phi}{\partial z} n_z=\pm h(\phi-\phi_s) \pm f_s^{//} \tag{14.65}$$

Approximation Function:

The approximate solution for the element is given by the similar expression as in the case of two-dimensional problem, i.e.,

$$\phi^*=\sum_{I=1}^{n_p} N_I \phi_I=[N]\left\{\phi^{(e)}\right\} \tag{14.2}$$

where N_I=shape functions of the three-dimensional element, ϕ_I=nodal values, and n_p=number of nodes in the three-dimensional element.

Integral Formulation:

The integral form of the equation over an element is obtained by applying **Galerkin's weighted residual method** as

$$\left\{R^{(e)}\right\}=-\int_V [N]^T\left[\frac{\partial}{\partial x}\left(\Gamma_x \frac{\partial \phi^*}{\partial x}\right)+\frac{\partial}{\partial y}\left(\Gamma_y \frac{\partial \phi^*}{\partial y}\right)+\frac{\partial}{\partial z}\left(\Gamma_z \frac{\partial \phi^*}{\partial z}\right)+S\right] dV=0 \tag{14.66}$$

The first three integral terms are evaluated as follows:

$$\int_V [N]^T \frac{\partial}{\partial x}\left(\Gamma_x \frac{\partial \phi^*}{\partial x}\right) dV=\int_V \frac{\partial}{\partial x}\left([N]^T \Gamma_x \frac{\partial \phi^*}{\partial x}\right) dV-\int_V \Gamma_x \frac{\partial \phi^*}{\partial x} \frac{\partial [N]^T}{\partial x} dV \tag{14.67a}$$

$$\int_V [N]^T \frac{\partial}{\partial y}\left(\Gamma_y \frac{\partial \phi^*}{\partial y}\right) dV=\int_V \frac{\partial}{\partial y}\left([N]^T \Gamma_y \frac{\partial \phi^*}{\partial y}\right) dV-\int_V \Gamma_y \frac{\partial \phi^*}{\partial y} \frac{\partial [N]^T}{\partial y} dV \tag{14.67b}$$

and

$$\int_V [N]^T \frac{\partial}{\partial z}\left(\Gamma_z \frac{\partial \phi^*}{\partial z}\right) dV=\int_V \frac{\partial}{\partial z}\left([N]^T \Gamma_z \frac{\partial \phi^*}{\partial z}\right) dV-\int_V \Gamma_z \frac{\partial \phi^*}{\partial z} \frac{\partial [N]^T}{\partial z} dV \tag{14.67c}$$

The first volume integral term on right-hand side of Equation 14.67 can be transformed into an area integral using the Gauss's divergence theorem as follows:

$$\int_V \frac{\partial}{\partial x}\left([N]^T \Gamma_x \frac{\partial \phi^*}{\partial x}\right) dV=\oint_s [N]^T \Gamma_x \frac{\partial \phi^*}{\partial x} n_x dA \tag{14.68a}$$

$$\int_V \frac{\partial}{\partial y}\left([N]^T \Gamma_y \frac{\partial \phi^*}{\partial y}\right) dV = \oint_A [N]^T \Gamma_y \frac{\partial \phi^*}{\partial y} n_y dA \tag{14.68b}$$

and

$$\int_V \frac{\partial}{\partial z}\left([N]^T \Gamma_z \frac{\partial \phi^*}{\partial z}\right) dV = \oint_A [N]^T \Gamma_y \frac{\partial \phi^*}{\partial y} n_z dA \tag{14.68c}$$

Substituting (14.68) into (14.67), we get

$$\int_V [N]^T \frac{\partial}{\partial x}\left(\Gamma_x \frac{\partial \phi^*}{\partial x}\right) dV = \oint_A [N]^T \Gamma_x \frac{\partial \phi^*}{\partial x} n_x dA - \int_V \frac{\partial [N]^T}{\partial x} \Gamma_x \frac{\partial \phi^*}{\partial x} dV \tag{14.69a}$$

$$\int_V [N]^T \frac{\partial}{\partial y}\left(\Gamma_y \frac{\partial \phi^*}{\partial y}\right) dV = \oint_A [N]^T \Gamma_y \frac{\partial \phi^*}{\partial y} n_y dA - \int_V \frac{\partial [N]^T}{\partial y} \Gamma_y \frac{\partial \phi^*}{\partial y} dV \tag{14.69b}$$

and

$$\int_V [\mathrm{N}]^T \frac{\partial}{\partial Z}\left(\Gamma_z \frac{\partial \phi^*}{\partial Z}\right) dV = \oint_A [N]^T \Gamma_z \frac{\partial \phi^*}{\partial Z} n_z\, dA - \int_V \frac{\partial [N]^T}{\partial z} \Gamma_z \frac{\partial \varphi^*}{\partial z} dV \tag{14.69c}$$

Substituting (14.169) into the integral form of the equation over an element given by (14.66), we have

$$\begin{aligned}\{R^{(e)}\} = &-\int_A [N]^T \left(\Gamma_x \frac{\partial \phi^*}{\partial x} n_x + \Gamma_y \frac{\partial \phi^*}{\partial y} n_y + \Gamma_z \frac{\partial \phi^*}{\partial z} n_z\right) dA \\ &+ \int_V \left(\frac{\partial [N]^T}{\partial x} \Gamma_x \frac{\partial \phi^*}{\partial x} + \frac{\partial [N]^T}{\partial y} \Gamma_y \frac{\partial \phi^*}{\partial y} + \frac{\partial [N]^T}{\partial z} \Gamma_z \frac{\partial \phi^*}{\partial z}\right) dV \\ &- \int_V S[N]^T dV = 0\end{aligned} \tag{14.70}$$

Formation of Element Characteristics Equation:

To obtain the element characteristics equation, substitute the approximate solution function given by Equation 14.2 into the integral Equation 14.109

$$\begin{aligned}&-\int_A [N]^T \left(\Gamma_x \frac{\partial \phi^*}{\partial x} n_x + \Gamma_y \frac{\partial \phi^*}{\partial y} n_y + \Gamma_z \frac{\partial \phi^*}{\partial z} n_z\right) dA \\ &+ \left\{\int_V \left(\frac{\partial [N]^T}{\partial x} \Gamma_x \frac{\partial [N]}{\partial x} + \frac{\partial [N]^T}{\partial y} \Gamma_y \frac{\partial [N]}{\partial y} + \frac{\partial [N]^T}{\partial z} \Gamma_z \frac{\partial [N]}{\partial z}\right) dV\right\} \{\phi^{(e)}\} \\ &- \int_V S[N]^T dV = 0\end{aligned} \tag{14.71}$$

The element characteristics equation can be written in matrix notation as

$$\{I^{(e)}\}+[k^{(e)}]\{\phi^{(e)}\}-\{f_S^{(e)}\}=0 \tag{14.72}$$

where
$\{I^{(e)}\}$=column vector for the element contribution to the interelement requirement

$$=-\int_A [N]^T\left(\Gamma_x\frac{\partial\phi^*}{\partial x}n_x+\Gamma_y\frac{\partial\phi^*}{\partial y}n_y+\Gamma_z\frac{\partial\phi^*}{\partial z}n_z\right)dA \tag{14.73}$$

$\{f_S^{(e)}\}$=element source vector due to the volumetric source term

$$=\int_V S[N]^T\,dV \tag{14.74}$$

$\{\phi^{(e)}\}$=column vector of unknown nodal values
$[k^{(e)}]$=element characteristics matrix

$$=\int_V\left(\frac{\partial[N]^T}{\partial x}\Gamma_x\frac{\partial[N]}{\partial x}+\frac{\partial[N]^T}{\partial y}\Gamma_y\frac{\partial[N]}{\partial y}+\frac{\partial[N]^T}{\partial z}\Gamma_z\frac{\partial[N]}{\partial z}\right)dV$$

$$=[k_\Gamma^{(e)}]=[k_{\Gamma x}^{(e)}]+[k_{\Gamma x}^{(e)}]+[k_{\Gamma z}^{(e)}] \tag{14.75}$$

The first term $[k_\Gamma^{(e)}]$ is contributed by the three diffusion terms, which is expressed in a compact form as follows:

$$[k_\Gamma^{(e)}]=\int_V\left(\frac{\partial[N]^T}{\partial x}\Gamma_x\frac{\partial[N]}{\partial x}+\frac{\partial[N]^T}{\partial y}\Gamma_y\frac{\partial[N]}{\partial y}+\frac{\partial[N]^T}{\partial z}\Gamma_z\frac{\partial[N]}{\partial z}\right)dV$$

$$=\int_V\{D\}^T[\Gamma]\{D\}dV \tag{14.76}$$

where

$$[\Gamma]=\text{transport property matrix}=\begin{bmatrix}\Gamma_x & 0 & 0\\ 0 & \Gamma_y & 0\\ 0 & 0 & \Gamma_z\end{bmatrix} \tag{14.77a}$$

$$\{D\}=\text{column vector with derivatives of shape functions}=\begin{Bmatrix}\dfrac{\partial[N]}{\partial x}\\ \dfrac{\partial[N]}{\partial y}\\ \dfrac{\partial[N]}{\partial z}\end{Bmatrix} \tag{14.77b}$$

$$\{D\}^T = \text{row vector with derivatives of shape functions} = \left\{ \frac{\partial[N]}{\partial x} \quad \frac{\partial[N]}{\partial y} \quad \frac{\partial[N]}{\partial z} \right\} \tag{14.78}$$

The implementation of the derivative boundary conditions into the finite element formulation takes place in the same manner as in the case of one- and two-dimensional problems through the interelement vector $\{I^{(e)}\}$ as given by Equation 14.10.

$$\{I^{(e)}\} = -\int_A [N]^T \left(\Gamma_x \frac{\partial\phi^*}{\partial x} n_x + \Gamma_y \frac{\partial\phi^*}{\partial y} n_y + \Gamma_y \frac{\partial\phi^*}{\partial z} n_z \right) dA \tag{14.79}$$

To show the specific contribution from the boundary conditions, let us decompose the integral into two parts as

$$\{I^{(e)}\} = \{I_b^{(e)}\} + \{I_i^{(e)}\} \tag{14.80}$$

where $\{I_i^{(e)}\}$ is the integral contribution from all the other sides other than the boundary surface, constituting the interelement requirements. The component $\{I_b^{(e)}\}$ is the integral contributions from the surfaces that represent the external boundary surfaces, and it is expressed as follows:

$$\{I_b^{(e)}\} = -\int_{A_b} [N]^T \left(\Gamma_x \frac{\partial\phi^*}{\partial x} n_x + \Gamma_y \frac{\partial\phi^*}{\partial y} n_y + \Gamma_z \frac{\partial\phi^*}{\partial z} n_z \right) dA \tag{14.81}$$

Substituting the boundary condition given by Eq. (14.18e), we get

$$\{I_b^{(e)}\} = -\int_{A_b} [N]^T \left\{ \pm h(\phi^* - \phi_\infty) \pm f_s^{//} \right\} dA \tag{14.82a}$$

or

$$\{I_b^{(e)}\} = \pm \int_{A_b} h[N]^T \phi^* dA \pm \int_{s_b} [N]^T \left\{ (h\phi_\infty + f_s^{//}) \right\} dA \tag{14.82b}$$

Now, substituting the approximate solution given by (14.2), we get

$$\{I_b^{(e)}\} = \pm \left(\int_{A_b} h[N]^T [N] ds \right) \phi^{(e)} \pm \int_{A_b} [N]^T \left\{ (h\phi_\infty + f_s^{//}) \right\} dA \tag{14.83a}$$

or

$$\{I_b^{(e)}\} = [k_b^{(e)}]\{\phi^{(e)}\} - \{f_{bs}^{(e)}\} \tag{14.83b}$$

where

$$[k_b^{(e)}] = \pm \int_{A_b} h[N]^T [N] dA = \text{Element stiffness matrix component due to boundary conditions} \tag{14.84a}$$

$$\left\{f_{\text{bs}}^{(e)}\right\} = \pm\int_{s_b}[N]^T\left\{\left(h\phi_\infty + f_s^{//}\right)\right\}\mathrm{d}A = \text{Element source vector component due to boundary condition} \tag{14.84b}$$

Finally, we can substitute equations

$$\left\{I_i^{(e)}\right\} + \left(k_\Gamma^{(e)} + k_b^{(e)}\right)\left\{\phi^{(e)}\right\} - \left\{f_S^{(e)} + f_{\text{bs}}^{(e)}\right\} = 0 \tag{14.85}$$

or

$$\left\{I_i^{(e)}\right\} + k^{(e)}\left\{\phi^{(e)}\right\} - f^{//} = 0 \tag{14.86}$$

where

$$k^{(e)} = \left(k_\Gamma^{(e)} + k_b^{(e)}\right) = \text{Element Characteristics Matrix} \tag{14.87a}$$

and

$$f^{(e)} = \left\{f_S^{(e)} + f_{\text{bs}}^{(e)}\right\} = \text{Element source vector} \tag{14.87b}$$

Case: Element Characteristics Equation Using Linear Tetrahedron Element

Let us consider a linear four-node tetrahedron element as shown in Figure 10.22a, and with the approximate solution, the corresponding shape functions are given by Equations 12.112 and 12.114, respectively (Figure 14.10).

The vectors with derivatives of shape functions are derived by substituting the shape functions given as follows:

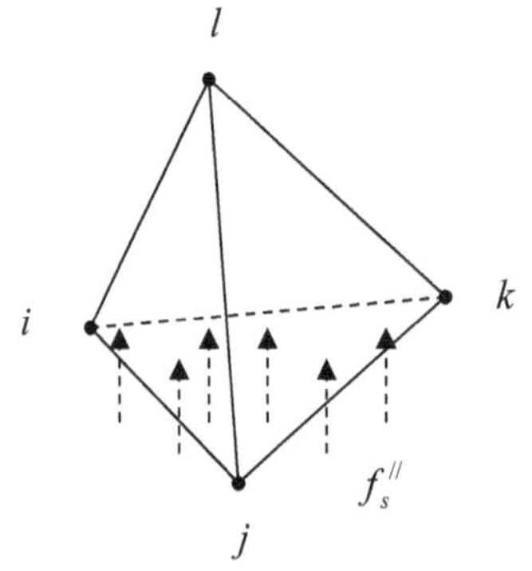

FIGURE 14.10 Tetrahedron element with mixed boundary condition at the area surface *i-j-k*.

$$\{D\} = \left\{ \begin{matrix} \dfrac{\partial[N]}{\partial x} \\ \dfrac{\partial[N]}{\partial y} \\ \dfrac{\partial[N]}{\partial z} \end{matrix} \right\}$$

$$\{D\} = \begin{bmatrix} \dfrac{\partial N_i}{\partial x} & \dfrac{\partial N_j}{\partial x} & \dfrac{\partial N_k}{\partial x} & \dfrac{\partial N_l}{\partial x} \\ \dfrac{\partial N_i}{\partial y} & \dfrac{\partial N_j}{\partial y} & \dfrac{\partial N_k}{\partial y} & \dfrac{\partial N_l}{\partial y} \\ \dfrac{\partial N_i}{\partial z} & \dfrac{\partial N_j}{\partial z} & \dfrac{\partial N_k}{\partial z} & \dfrac{\partial N_l}{\partial z} \end{bmatrix} \tag{14.88}$$

Evaluating the derivatives from Equation 12.114, we can get

$$\{D\} = \frac{1}{6V} \begin{bmatrix} \beta_i & \beta_j & \beta_k & \beta_l \\ \gamma_i & \gamma_j & \gamma_k & \gamma_l \\ \delta_i & \delta_j & \delta_k & \delta_l \end{bmatrix}$$

Similarly, we get

$$\{D\}^T = \left\{ \begin{matrix} \dfrac{\partial[N]}{\partial x} & \dfrac{\partial[N]}{\partial y} & \dfrac{\partial[N]}{\partial z} \end{matrix} \right\}$$

$$\{D\}^T = \begin{bmatrix} \dfrac{\partial N_i}{\partial x} & \dfrac{\partial N_i}{\partial y} & \dfrac{\partial N_i}{\partial z} \\ \dfrac{\partial N_j}{\partial x} & \dfrac{\partial N_j}{\partial y} & \dfrac{\partial N_j}{\partial z} \\ \dfrac{\partial N_k}{\partial x} & \dfrac{\partial N_k}{\partial y} & \dfrac{\partial N_k}{\partial z} \\ \dfrac{\partial N_l}{\partial x} & \dfrac{\partial N_l}{\partial y} & \dfrac{\partial N_l}{\partial z} \end{bmatrix} = \frac{1}{6V} \begin{bmatrix} \beta_i & \gamma_i & \delta_i \\ \beta_j & \gamma_j & \delta_j \\ \beta_k & \gamma_k & \delta_k \\ \beta_l & \gamma_l & \delta_l \end{bmatrix} \tag{14.89}$$

The component of element characteristic matrix due to the diffusion term $\left[k_\Gamma^{(e)}\right]$ can now be computed as

$$\left[k_{\Gamma}^{(e)}\right] = \int_V \{D\}^T [\Gamma]\{D\} \mathrm{d}V = \{D\}^T [\Gamma]\{D\} \int_V \mathrm{d}V$$

$$= \frac{1}{6V}\begin{bmatrix} \beta_i & \gamma_i & \delta_i \\ \beta_j & \gamma_j & \delta_j \\ \beta_k & \gamma_k & \delta_k \\ \beta_l & \gamma_l & \delta_l \end{bmatrix}\begin{bmatrix} \Gamma_x & 0 & 0 \\ 0 & \Gamma_y & 0 \\ 0 & 0 & \Gamma_z \end{bmatrix}\frac{1}{6V}\begin{bmatrix} \beta_i & \beta_j & \beta_k & \beta_l \\ \gamma_i & \gamma_j & \gamma_k & \gamma_l \\ \delta_i & \delta_j & \delta_k & \delta_l \end{bmatrix} V$$

$$= \frac{1}{36V^2}\begin{bmatrix} \beta_i & \gamma_i & \delta_i \\ \beta_j & \gamma_j & \delta_j \\ \beta_k & \gamma_k & \delta_k \\ \beta_l & \gamma_l & \delta_l \end{bmatrix}\begin{bmatrix} \Gamma_x\beta_i & \Gamma_x\beta_j & \Gamma_x\beta_k & \Gamma_x\beta_l \\ \Gamma_y\gamma_i & \Gamma_y\gamma_j & \Gamma_y\gamma_k & \Gamma_z\gamma_l \\ \Gamma_z\delta_i & \Gamma_z\delta_j & \Gamma_z\delta_k & \Gamma_z\delta_l \end{bmatrix}$$

$$= \frac{1}{36V^2}\begin{bmatrix} \Gamma_x\beta_i^2 + \Gamma_y\gamma_i^2 + \Gamma_z\delta_i^2 & \Gamma_x\beta_i\beta_j + \Gamma_y\gamma_i\gamma_j + \Gamma_z\delta_i\delta_j & \Gamma_x\beta_i\beta_k + \Gamma_y\gamma_i\gamma_k + \Gamma_z\delta_i\delta_k & \Gamma_x\beta_i\beta_l + \Gamma_y\gamma_i\gamma_l + \Gamma_z\delta_i\delta_l \\ \Gamma_x\beta_i\beta_j + \Gamma_y\gamma_i\gamma_j + \Gamma_z\delta_i\delta_j & \Gamma_x\beta_j^2 + \Gamma_y\gamma_j^2 + \Gamma_z\delta_j^2 & \Gamma_x\beta_j\beta_k + \Gamma_y\gamma_j\gamma_k + \Gamma_z\delta_j\delta_k & \Gamma_x\beta_j\beta_l + \Gamma_y\gamma_j\gamma_l + \Gamma_z\delta_j\delta_l \\ \Gamma_x\beta_i\beta_k + \Gamma_y\gamma_i\gamma_k + \Gamma_z\delta_i\delta_k & \Gamma_x\beta_j\beta_k + \Gamma_y\gamma_j\gamma_k + \Gamma_z\delta_j\delta_k & \Gamma_x\beta_k^2 + \Gamma_y\gamma_k^2 + \Gamma_z\delta_k^2 & \Gamma_x\beta_l\beta_k + \Gamma_y\gamma_l\gamma_k + \Gamma_z\delta_l\delta_k \\ \Gamma_x\beta_i\beta_l + \Gamma_y\gamma_i\gamma_l + \Gamma_z\delta_i\delta_l & \Gamma_x\beta_j\beta_l + \Gamma_y\gamma_j\gamma_l + \Gamma_z\delta_j\delta_l & \Gamma_x\beta_k\beta_l + \Gamma_y\gamma_k\gamma_l + \Gamma_z\delta_k\delta_l & \Gamma_x\beta_l^2 + \Gamma_y\gamma_l^2 + \Gamma_z\delta_l^2 \end{bmatrix} \tag{14.90a}$$

or

$$\left[k_{\Gamma}^{(e)}\right] = \frac{\Gamma_x}{36V^2}\begin{bmatrix} \beta_i^2 & \beta_i\beta_j & \beta_i\beta_k & \beta_i\beta_l \\ \beta_i\beta_j & \beta_j^2 & \beta_j\beta_k & \beta_j\beta_l \\ \beta_i\beta_k & \beta_j\beta_k & \beta_k^2 & \beta_l\beta_k \\ \beta_i\beta_l & \beta_j\beta_l & \beta_k\beta_l & \beta_l^2 \end{bmatrix} + \frac{\Gamma_y}{36V^2}\begin{bmatrix} \gamma_i^2 & \gamma_i\gamma_j & \gamma_i\gamma_k & \gamma_i\gamma_l \\ \gamma_i\gamma_j & \gamma_j^2 & \gamma_j\gamma_k & \gamma_j\gamma_l \\ \gamma_i\gamma_k & \gamma_j\gamma_k & \gamma_k^2 & \gamma_l\gamma_k \\ \gamma_i\gamma_l & \gamma_j\gamma_l & \gamma_k\gamma_l & \gamma_l^2 \end{bmatrix} + \frac{\Gamma_z}{36V^2}\begin{bmatrix} \delta_i^2 & \delta_i\delta_j & \delta_i\delta_k & \delta_i\delta_l \\ \delta_i\delta_j & \delta_j^2 & \delta_j\delta_k & \delta_j\delta_l \\ \delta_i\delta_k & \delta_j\delta_k & \delta_k^2 & \delta_l\delta_k \\ \delta_i\delta_l & \delta_j\delta_l & \delta_k\delta_l & \delta_l^2 \end{bmatrix} \tag{14.90b}$$

where β 's, γ 's, and δ 's are given by Equations 12.108b, 12.109b, and 12.110b.

The component of element characteristic matrix due to the diffusion term, $\left[k_\Gamma^{(e)}\right]$ can now be computed as

$$\left[k_b^{(e)}\right] = \pm \int_{A_b} h[N]^T [N] dA$$

$$= \int_{A_b} h \left[\begin{Bmatrix} N_i \\ N_j \\ N_k \\ N_l \end{Bmatrix} \begin{Bmatrix} N_i & N_j & N_k & N_l \end{Bmatrix} \right] dA$$

$$= \int_{A_b} h \begin{bmatrix} N_i^2 & N_iN_j & N_iN_k & N_iN_l \\ N_jN_i & N_j^2 & N_jN_k & N_jN_l \\ N_kN_i & N_kN_j & N_k^2 & N_kN_l \\ N_lN_i & N_lN_j & N_lN_k & N_l^2 \end{bmatrix} dA \tag{14.91}$$

The integral is further evaluated by considering only the sides that are subjected to external boundary conditions. For example, if the boundary condition given by Equation 14.65 is specified over the area surface side *i-j-k* of the element, then we set $N_l = 0$ for the area integral, and Equation 14.91 becomes

$$\left[k_b^{(e)}\right] = \int_{A_{ijk}} h \begin{bmatrix} N_i^2 & N_iN_j & N_iN_k & 0 \\ N_jN_i & N_j^2 & N_jN_k & 0 \\ N_kN_i & N_kN_j & N_k^2 & 0 \\ N_lN_i & N_lN_j & N_lN_k & 0 \end{bmatrix} dA \tag{14.92}$$

Again, using natural volume coordinate system (Equation 12.116), the integral elements can be evaluated as

$$\left[k_b^{(e)}\right] = \pm \int_{A_{ijk}} h \begin{bmatrix} L_i^2 & L_iL_j & L_iL_k & 0 \\ L_jL_i & L_j^2 & L_jL_k & 0 \\ L_kL_i & L_kL_j & L_k^2 & 0 \\ 0 & 0 & 0 & 0 \end{bmatrix} dA \tag{14.93}$$

$$\left[k_b^{(e)}\right] = \pm h \begin{bmatrix} \int\limits_{A_{ijk}} L_i^2\,\mathrm{d}A & \int\limits_{A_{ijk}} L_i L_j\,\mathrm{d}A & \int\limits_{A_{ijk}} L_i L_k\,\mathrm{d}A & 0 \\ \int\limits_{A_{ijk}} L_j L_i\,\mathrm{d}A & \int\limits_{A_{ijk}} L_j^2\,\mathrm{d}A & \int\limits_{A_{ijk}} L_j L_k\,\mathrm{d}A & 0 \\ \int\limits_{A_{ijk}} L_k L_i\,\mathrm{d}A & \int\limits_{A_{ijk}} L_k L_j\,\mathrm{d}A & \int\limits_{A_{ijk}} L_k^2\,\mathrm{d}A & 0 \\ 0 & 0 & 0 & 0 \end{bmatrix}$$

$$= \pm h \begin{bmatrix} \int\limits_{A_{ijk}} L_i^2 L_j^0 L_k^0\,\mathrm{d}A & \int\limits_{A_{ijk}} L_i^1 L_j^1 L_k^0\,\mathrm{d}A & \int\limits_{A_{ijk}} L_i^1 L_j^0 L_k^1\,\mathrm{d}A & 0 \\ \int\limits_{A_{ijk}} L_i^1 L_j^1 L_k^0\,\mathrm{d}A & \int\limits_{A_{ijk}} L_i^0 L_j^2 L_k^0\,\mathrm{d}A & \int\limits_{A_{ijk}} L_i^0 L_j^1 L_k^1\,\mathrm{d}A & 0 \\ \int\limits_{A_{ijk}} L_i^1 L_j^0 L_k^1\,\mathrm{d}A & \int\limits_{A_{ijk}} L_i^0 L_j^1 L_k^1\,\mathrm{d}A & \int\limits_{A_{ijk}} L_i^0 L_j^0 L_k^2\,\mathrm{d}A & 0 \\ 0 & 0 & 0 & 0 \end{bmatrix}$$

Using analytical integral formulas involving area natural co-ordinate system, we get

$$\left[k_b^{(e)}\right] = \pm h \begin{bmatrix} \frac{2!0!0!}{(2+0+0+2)!}2A_{ijk} & \frac{1!1!0!}{(1+1+0+2)!}2A_{ijk} & \frac{1!0!1!}{(1+0+1+2)!}2A_{ijk} & 0 \\ \frac{1!1!0!}{(1+1+0+2)!}2A_{ijk} & \frac{0!2!0!}{(0+2+0+2)!}2A_{ijk} & \frac{0!1!1!}{(0+1+1+2)}2A_{ijk} & 0 \\ \frac{1!1!0!}{(1+1+0+2)!}2A_{ijk} & \frac{0!1!1!}{(0+1+1+2)}2A_{ijk} & \frac{0!0!2!}{(0+0+2+2)!}2A_{ijk} & 0 \\ 0 & 0 & 0 & 0 \end{bmatrix}$$

$$= \pm h \begin{bmatrix} \frac{A_{ijk}}{6} & \frac{A_{ijk}}{12} & \frac{A_{ijk}}{12} & 0 \\ \frac{A_{ijk}}{12} & \frac{A_{ijk}}{6} & \frac{A_{ijk}}{12} & 0 \\ \frac{A_{ijk}}{12} & \frac{A_{ijk}}{12} & \frac{A_{ijk}}{6} & 0 \\ 0 & 0 & 0 & 0 \end{bmatrix}$$

$$\left[k_b^{(e)}\right] = \pm \frac{hA_{ijk}}{12} \begin{bmatrix} 2 & 1 & 1 & 0 \\ 1 & 2 & 1 & 0 \\ 1 & 1 & 2 & 0 \\ 0 & 0 & 0 & 0 \end{bmatrix} \tag{14.94a}$$

Similarly,

For boundary condition specified on the side j-k-l

$$\left[k_b^{(e)}\right] = \pm\int_{A_b} h \begin{bmatrix} 0 & 0 & 0 & 0 \\ 0 & N_j^2 & N_jN_k & N_jN_l \\ 0 & N_kN_j & N_k^2 & N_kN_l \\ 0 & N_lN_j & N_lN_k & N_l^2 \end{bmatrix} \mathrm{d}A$$

or

$$= \pm\frac{hA_{jkl}}{12}\begin{bmatrix} 0 & 0 & 0 & 0 \\ 0 & 2 & 1 & 1 \\ 0 & 1 & 2 & 1 \\ 0 & 1 & 1 & 2 \end{bmatrix} \tag{14.94b}$$

For boundary condition specified on the side *i-j-l*

$$\left[k_b^{(e)}\right] = \int_{A_b} h \begin{bmatrix} N_i^2 & N_iN_j & 0 & N_iN_l \\ N_jN_i & N_j^2 & 0 & N_jN_l \\ 0 & 0 & 0 & 0 \\ N_lN_i & N_lN_j & 0 & N_l^2 \end{bmatrix} \mathrm{d}A$$

or,

$$\left[k_b^{(e)}\right] = \pm\frac{hA_{ijl}}{12}\begin{bmatrix} 2 & 1 & 0 & 1 \\ 1 & 2 & 0 & 1 \\ 0 & 0 & 0 & 0 \\ 1 & 1 & 0 & 2 \end{bmatrix} \tag{14.94c}$$

For boundary condition specified on the side *i-k-l*

$$\left[k_b^{(e)}\right] = \int_{A_b} h \begin{bmatrix} N_i^2 & 0 & N_iN_k & N_iN_l \\ 0 & 0 & 0 & 0 \\ N_kN_i & 0 & N_k^2 & N_kN_l \\ N_lN_i & 0 & N_lN_k & N_l^2 \end{bmatrix} \mathrm{d}A$$

or

$$\left[k_b^{(e)}\right] = \pm\frac{hA_{ikl}}{12}\begin{bmatrix} 2 & 0 & 1 & 1 \\ 0 & 0 & 0 & 0 \\ 1 & 0 & 2 & 1 \\ 1 & 0 & 1 & 2 \end{bmatrix} \tag{14.94d}$$

The source term is evaluated as follows:

$$\{f_S^{(e)}\} = \int_V S[N]^T \, dV$$

$$= \int_V S \begin{Bmatrix} N_i \\ N_j \\ N_k \\ N_l \end{Bmatrix} dV \tag{14.95}$$

Using the natural volume co-ordinate system given by Equation 12.116 and integral formula in natural volume co-ordinate system as given by Equation 12.119, we get

$$\{f_S^{(e)}\} = \int_V S \begin{Bmatrix} N_i \\ N_j \\ N_k \\ N_l \end{Bmatrix} dV = \int_V S \begin{Bmatrix} L_i \\ L_j \\ L_k \\ L_l \end{Bmatrix} dV \tag{14.96}$$

$$= S \begin{Bmatrix} \int_V L_i^1 L_j^0 L_k^0 L_l^0 \, dV \\ \int_V L_i^0 L_j^1 L_k^0 L_l^0 \, dV \\ \int_V L_i^0 L_j^0 L_k^1 L_l^0 \, dV \\ \int_V L_i^0 L_j^0 L_k^0 L_l^1 \, dV \end{Bmatrix}$$

$$= S \begin{Bmatrix} \dfrac{1!0!0!0!}{(1+0+0+0+3)!} 6V \\[2mm] \dfrac{0!1!0!0!}{(0+1+0+0+3)!} 6V \\[2mm] \dfrac{0!0!1!0!}{(0+0+1+0+3)!} 6V \\[2mm] \dfrac{0!0!0!1!}{(0+0+0+1+3)!} 6V \end{Bmatrix} \tag{14.97}$$

$$\{f_S^{(e)}\} = \frac{SV}{4} \begin{Bmatrix} 1 \\ 1 \\ 1 \\ 1 \end{Bmatrix}$$

Element source vector component due to boundary condition is evaluated as follows:

$$\{f_{bs}^{(e)}\} = \pm \int_{s_b} [N]^T \left\{ \left(h\phi_\infty + f_s'' \right) \right\} dA$$

$$\left(h\phi_\infty + f_s^{//}\right)\int_{A_b}\begin{Bmatrix} N_i \\ N_j \\ N_k \\ N_l \end{Bmatrix} dA \tag{14.98}$$

For boundary condition specified on *i-j-k* side of the element, $N_l = 0$, and we use natural co-ordinate system and use integral formula (Equation 12.71), we get,

$$\left\{f_{\text{bs}}^{(e)}\right\} = \left(h\phi_\infty + f_s^{//}\right)\int_{A_b}\begin{Bmatrix} N_i \\ N_j \\ N_k \\ 0 \end{Bmatrix} dA = \left(h\phi_\infty + f_s^{//}\right)\int_{A_b}\begin{Bmatrix} L_i \\ L_j \\ L_k \\ 0 \end{Bmatrix} dA$$

$$\left\{f_{\text{bs}}^{(e)}\right\} = \pm\frac{\left(h\phi_\infty + f_s^{//}\right)A_{ijk}}{3}\begin{Bmatrix} 1 \\ 1 \\ 1 \\ 0 \end{Bmatrix} \tag{14.99a}$$

Similarly, for boundary condition specified on the surface area *j-k-l*, we get

$$\left\{f_{\text{bs}}^{(e)}\right\} = \left(h\phi_\infty + f_s^{//}\right)\int_{A_{jkl}}\begin{Bmatrix} 0 \\ N_j \\ N_k \\ N_l \end{Bmatrix} dA = \pm\frac{\left(h\phi_\infty + f_s^{//}\right)A_{jkl}}{3}\begin{Bmatrix} 0 \\ 1 \\ 1 \\ 1 \end{Bmatrix} \tag{14. 99b}$$

For boundary condition specified on the surface area *i-j-l*, we get

$$\left\{f_{\text{bs}}^{(e)}\right\} = \left(h\phi_\infty + f_s^{//}\right)\int_{A_{ijl}}\begin{Bmatrix} N_i \\ N_j \\ 0 \\ N_l \end{Bmatrix} dA = \pm\frac{\left(h\phi_\infty + f_s^{//}\right)A_{ijl}}{3}\begin{Bmatrix} 1 \\ 1 \\ 0 \\ 1 \end{Bmatrix} \tag{14.99c}$$

For boundary condition specified on the surface area *i-k-l*, we get

$$\left\{f_{\text{bs}}^{(e)}\right\} = \left(h\phi_\infty + f_s^{//}\right)\int_{A_{ikl}}\begin{Bmatrix} N_i \\ 0 \\ N_k \\ N_l \end{Bmatrix} dA = \pm\frac{\left(h\phi_\infty + f_s^{//}\right)A_{ikl}}{3}\begin{Bmatrix} 1 \\ 0 \\ 1 \\ 1 \end{Bmatrix} \tag{14.99d}$$

The element characteristic equation for the tetrahedron element with area surface *i-j-k* as the external boundary surface with mixed boundary condition is obtained by substituting Equations 14.130a, 14.134a, 14.137, and 14.139a into 14.124:

$$\left\{ \frac{1}{36V^2}\begin{bmatrix} \Gamma_x\beta_i^2+\Gamma_y\gamma_i^2+\Gamma_z\delta_i^2 & \Gamma_x\beta_i\beta_j+\Gamma_y\gamma_i\gamma_j+\Gamma_z\delta_i\delta_j & \Gamma_x\beta_i\beta_k+\Gamma_y\gamma_i\gamma_k+\Gamma_z\delta_i\delta_k & \Gamma_x\beta_i\beta_l+\Gamma_y\gamma_i\gamma_l+\Gamma_z\delta_i\delta_l \\ \Gamma_x\beta_i\beta_j+\Gamma_y\gamma_i\gamma_j+\Gamma_z\delta_i\delta_j & \Gamma_x\beta_j^2+\Gamma_y\gamma_j^2+\Gamma_z\delta_j^2 & \Gamma_x\beta_j\beta_k+\Gamma_y\gamma_j\gamma_k+\Gamma_z\delta_j\delta_k & \Gamma_x\beta_j\beta_l+\Gamma_y\gamma_j\gamma_l+\Gamma_z\delta_j\delta_l \\ \Gamma_x\beta_i\beta_k+\Gamma_y\gamma_i\gamma_k+\Gamma_z\delta_i\delta_k & \Gamma_x\beta_j\beta_k+\Gamma_y\gamma_j\gamma_k+\Gamma_z\delta_j\delta_k & \Gamma_x\beta_k^2+\Gamma_y\gamma_k^2+\Gamma_z\delta_k^2 & \Gamma_x\beta_l\beta_k+\Gamma_y\gamma_l\gamma_k+\Gamma_z\delta_l\delta_k \\ \Gamma_x\beta_i\beta_l+\Gamma_y\gamma_i\gamma_l+\Gamma_z\delta_i\delta_l & \Gamma_x\beta_j\beta_l+\Gamma_y\gamma_j\gamma_l+\Gamma_z\delta_j\delta_l & \Gamma_x\beta_k\beta_l+\Gamma_y\gamma_k\gamma_l+\Gamma_z\delta_k\delta_l & \Gamma_x\beta_l^2+\Gamma_y\gamma_l^2+\Gamma_z\delta_l^2 \end{bmatrix} \pm \frac{hA_{ijk}}{12}\begin{bmatrix} 2 & 1 & 1 & 0 \\ 1 & 2 & 1 & 0 \\ 1 & 1 & 2 & 0 \\ 0 & 0 & 0 & 0 \end{bmatrix} \right\}$$

$$\begin{Bmatrix} \varphi_i \\ \varphi_j \\ \varphi_k \\ \varphi_l \end{Bmatrix} = \frac{SV}{4}\begin{Bmatrix} 1 \\ 1 \\ 1 \\ 1 \end{Bmatrix} + \pm\frac{\left(h\varphi_\infty + f_s^{//}\right)A_{ijk}}{6}\begin{Bmatrix} 1 \\ 1 \\ 1 \\ 0 \end{Bmatrix} \tag{14.100}$$

14.4 POINT SOURCE

In many cases, the source term may not be distributed over the entire domain but applied at some discrete points. Let us consider a three-dimensional tetrahedron element with source located a point (x_0, y_0, z_0) (Figure 14.11).

The source term now can be expressed as a delta source function over the element as

$$S = S_0\delta(x-x_0)\delta(y-y_0)\delta(z-z_0) \tag{14.101}$$

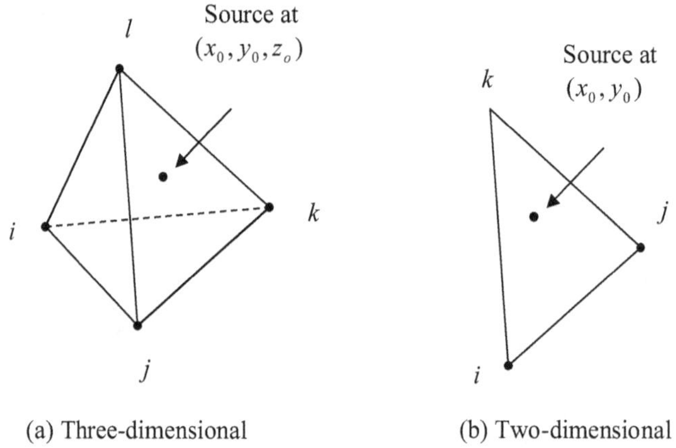

FIGURE 14.11 Source located at a discrete point.

The element force vector for a four nodded three-dimensional element can be estimates as follows:

$$\left\{f_S^{(e)}\right\} = \int_V S[N]^T \, dV$$

$$= S_0 \int_A \begin{bmatrix} N_i \\ N_j \\ N_k \\ N_l \end{bmatrix} \delta(x - x_0)\delta(y - y_0)\delta(z - z_0)dxdydz \tag{14.102}$$

$$\left\{f_S^{(e)}\right\} = S_0 \begin{Bmatrix} N_i(x_0, y_0, z_0) \\ N_j(x_0, y_0, z_0) \\ N_k(x_0, y_0, z_0) \\ N_l(x_0, y_0, z_0) \end{Bmatrix}$$

Basically, the source term at a point is distributed among the nodes of the element with weights given by the relative values of shape functions calculated at the co-ordinates of point.

In a similar manner, the element force vector for a three-noded two-dimensional element can be estimates as follows:

$$\left\{f_S^{(e)}\right\} = \int_A S[N]^T \, dA$$

$$= S_0 \int_A \begin{bmatrix} N_i \\ N_j \\ N_k \end{bmatrix} \delta(x - x_0)\delta(y - y_0)dxdy$$

$$\left\{f_S^{(e)}\right\} = S_0 \begin{Bmatrix} N_i(x_0, y_0) \\ N_j(x_0, y_0) \\ N_k(x_0, y_0) \end{Bmatrix}$$

PROBLEMS

14.1 Consider two-dimensional steady-state conduction in a 1.5 m by 1.5 m square slab with the top and the left surfaces maintained at a high temperature, T_H = 600 °C, and a low temperature, T_C = 20 °C, respectively. The right and the bottom surfaces are subjected to convection condition with h_c = 30 w/m^2°C and T_∞ = 25°C. The material conductivity is k = 30 w/m · °C. Apply finite element method using six bilinear rectangular elements to calculate the two-dimensional temperature distribution in the slab and heat transfer rates at all surfaces and show the overall energy balance.

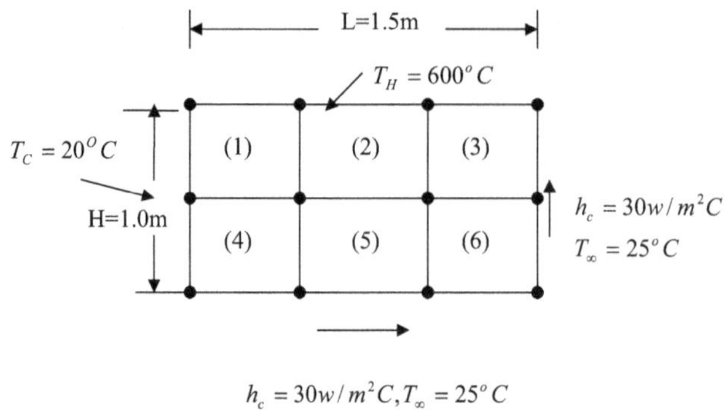

14.3 Repeat Problem 14.1 using linear triangular elements using the mesh shown in the figure.

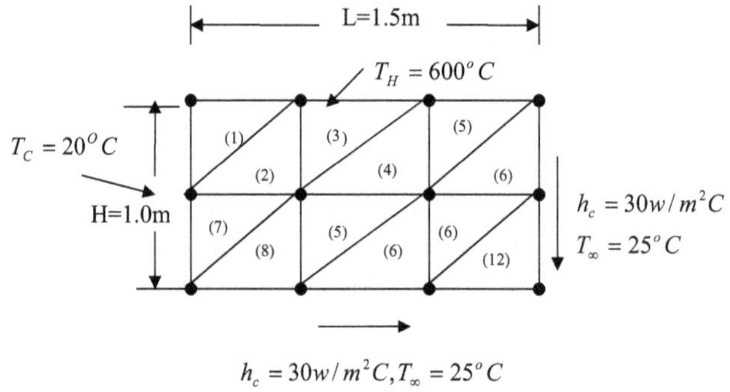

14.3 The fin shown has a base maintained at 300°C and is exposed to the convection environment shown in the figure. Consider the mesh consisting of four-node rectangular elements and write the characteristics equation for each element.

Assemble them to obtain the global system of equation.

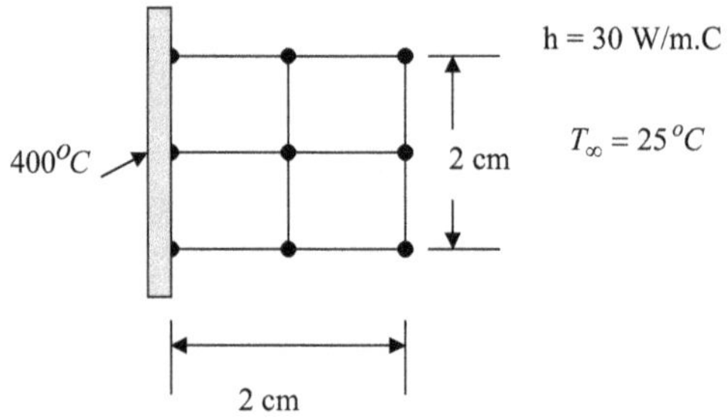

14.4 Consider temperature rise in a metal cutting tool due to heat is generation in the shear zones of a metallic workpiece and metal chips during metal cutting processes.

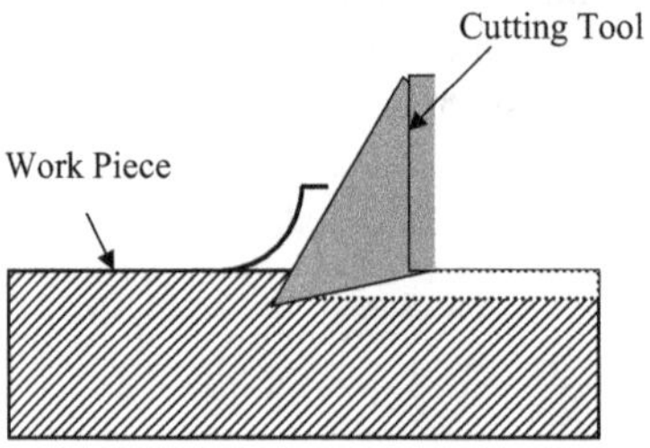

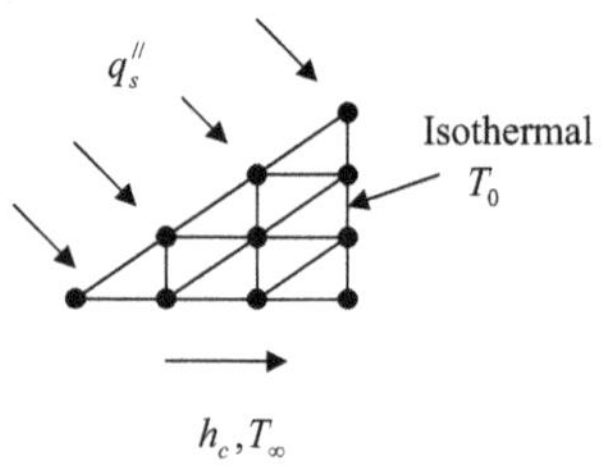

A simplified heating condition is considered in this problem by considering a constant surface heat flux at the top rake surface and forced convective cooling at the bottom surface. The further interior regions, i.e., the right surface is assumed to be isothermal or at a constant temperature T_o. Consider the mesh size distribution with linear triangular elements and derive the element characteristic equation and the global system of equations for the nodal temperatures.

14.5 Repeat Problem 14.4 with a mesh size distribution consisting of triangular and rectangular elements as shown.

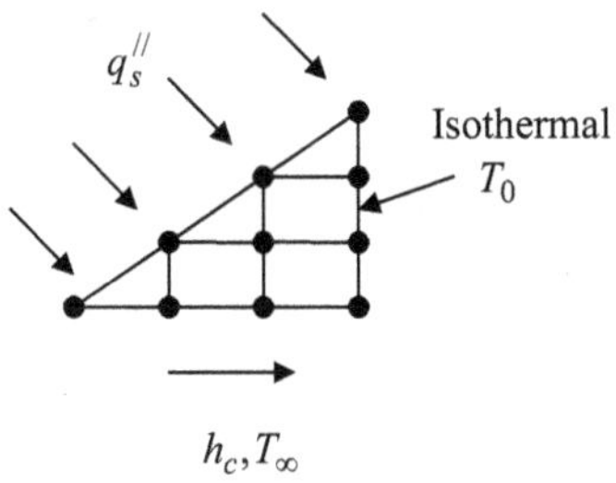

14.6 Use finite element method to solve two-dimensional steady-state conduction in a rectangular carbon steel (k = 200 W/m°C) slab subjected to a constant surface heat flux irradiated by a high-energy laser beam at the top surface. For simplicity, assume the heat flux distribution to be a constant average value $q_0^{//} = 2 \times 10^8$ W/m^2, acting over a section of the surface. The top surface is also subjected to forced convection with h_c = 500 W/m^2°C. The left surface is assumed as adiabatic as a line of symmetry. All other surfaces are assumed to be subjected to free convection with T_∞ = 25°C and $h = 40\dfrac{\text{W}}{\text{m}^2\text{C}}$. Derive the finite element equations for all elements and derive the global system of equations.

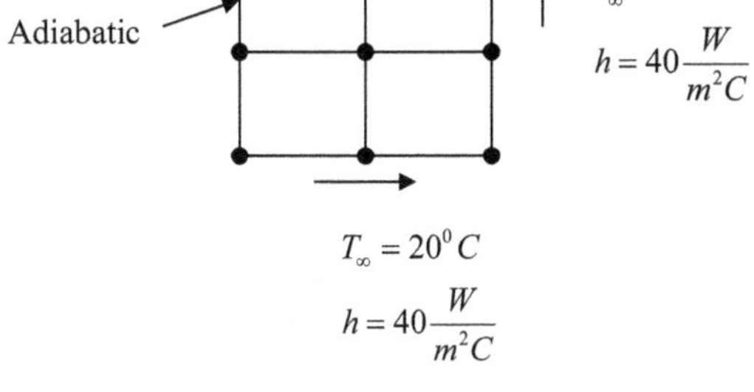

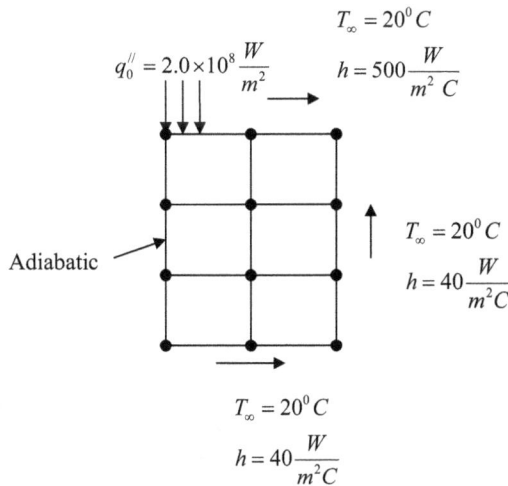

Solve the system of equations using Gauss elimination solver and check for the overall energy balance.

14.7 Consider three-dimensional steady-state heat conduction within a tetrahedron element with convective boundary condition shown in the figure. Derive the element characteristic equation for the element with conditions shown in the figure.

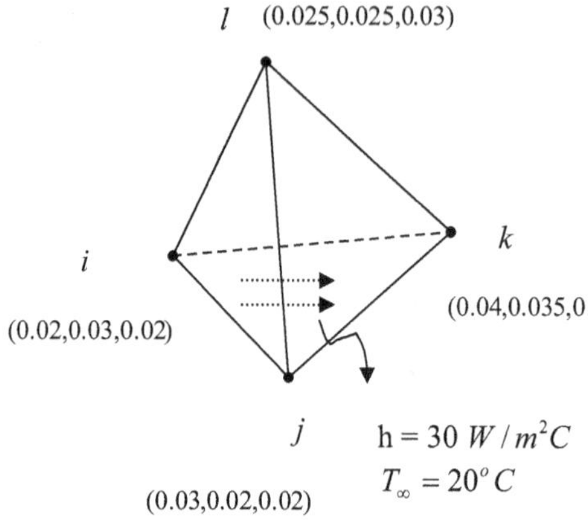

14.8 Use bilinear rectangular element to derive the element characteristic equation for the two-dimensional axisymmetric steady-state problem without heat generation.

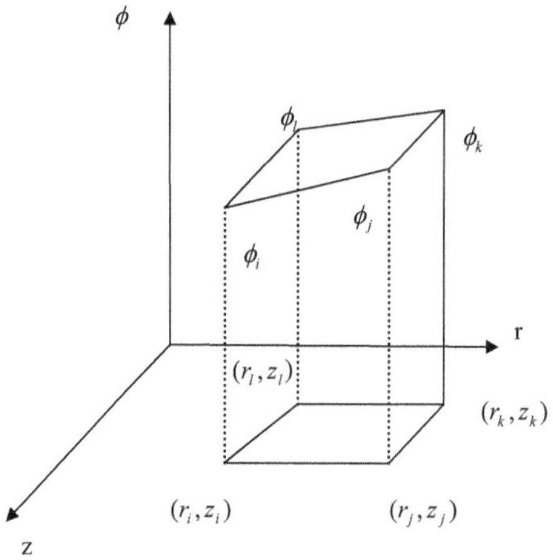

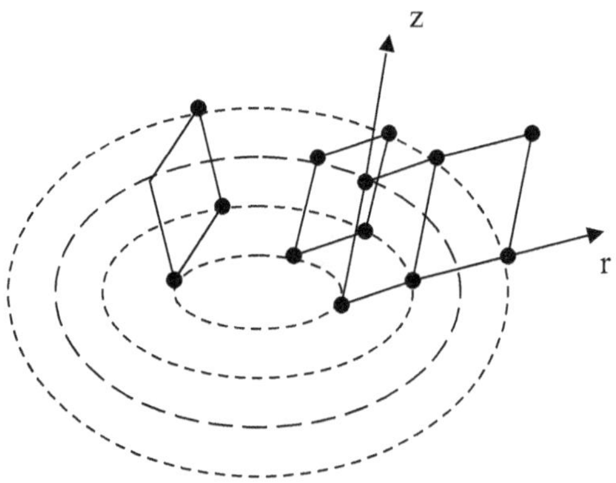

14.9 Use trilinear eight-node hexahedron element to derive the three-dimensional element characteristic equation starting from Equations 14.125 to 14.127.

15 Finite Element Method
Unsteady-State Problems

In this chapter we will consider development of finite element solution for unsteady-state problems. The treatment of the diffusion terms, source term, and the boundary condition will remain the same as outlined for the steady-state problems, and we will primarily focus on the treatment of the first-order time derivative term. We will first demonstrate the development of finite element solution for a general unsteady-state one-dimensional problem. The procedure will then be extended to two- and three-dimensional unsteady-state problems.

15.1 DISCRETIZATION SCHEME

The basic steps for the finite element formulation of unsteady-state problems are identical to those described in Chapters 13 and 14 except for the treatment of unsteady-state term. There are two major changes in the basic steps: (a) **spatial discretization scheme** in selecting the element approximation solution and (b) **time approximation scheme** for the unsteady term.

In the **spatial discretization scheme**, the approximate solution for the element is assumed as

$$\phi^* = \sum_{I=1}^{n_p} N_I(\tilde{x})\phi_I(t) = [N(\tilde{x})]\{\phi^{(e)}(t)\} \tag{15.1}$$

where N_I is the shape function, which is assumed to be a function of space variable only, $\phi_I(t)$ = nodal values as a function time, and n_p = number of nodes in an element. Basically, it is assumed that the spatial and time variations of the dependent variable are separable, and this is appropriate for linear problems and in nonlinear problems with small time steps. Next, the finite element formulation is carried out in the same manner as in the case of a steady-state problem but retaining all time-dependent terms in the equation. This procedure leads to a system of ordinary equations in time for the nodal values $\phi^{(e)}(t)$. This formulation is also referred to as **semi-discrete finite element formulation**.

In the **next step**, the system of ordinary differential equations is solved numerically. In the numerical approach, the time derivatives are approximated by finite difference schemes such as the one discussed in Chapters 5 and 6.

15.2 ONE-DIMENSIONAL UNSTEADY-STATE PROBLEM

Let us consider a general one-dimensional unsteady-state problem of the form

$$C\frac{\partial \phi}{\partial t} = \frac{\partial}{\partial x}\left(\Gamma_x \frac{\partial \phi}{\partial x}\right) + S_1\phi + S_0 \tag{15.2}$$

15.2.1 Semi-Discrete Finite Element Formulation

The Galerkin's-based weighted integral statement of Equation 15.2 over an element is

$$R_i^{(e)} = \int_{x_i}^{x_j} [N]^T\left(\frac{\partial}{\partial x}\left(\Gamma_x \frac{\partial \phi^*}{\partial x}\right) + S_1\phi^* + S_0 - C\frac{\partial \phi^*}{\partial t}\right)dx = 0 \tag{15.3}$$

Substituting the approximate solution function given by Equation 15.1 for one-dimensional space co-ordinate, we get the integral statement of the problem as

$$\left\{R^{(e)}\right\}=\left\{\begin{array}{c}-\Gamma_x \left.\dfrac{\mathrm{d}\phi}{\mathrm{d}x}\right|_{x_i} \\ \Gamma_x \left.\dfrac{\mathrm{d}\phi}{\mathrm{d}x}\right|_{x_j}\end{array}\right\}-\int_{x_i}^{x_j}\Gamma_x \frac{\mathrm{d}[N]^T}{\mathrm{d}x}\frac{\mathrm{d}\phi^*}{\mathrm{d}x}\mathrm{d}x+\int_{x_i}^{x_j}[N]^T\left(S_1\phi^*\right)\mathrm{d}x$$

$$+\int_{x_i}^{x_j} S_0[N]^T\mathrm{d}x-\int_{x_i}^{x_j} C[N]^T\frac{\partial\left\{\phi^{(e)}\right\}}{\partial t}\mathrm{d}x=0 \tag{15.4}$$

The first four terms in the Equation 15.2 were evaluated in Chapter 11 for one-dimensional two-point line element and are given as follows:

$$\int_{x_i}^{x_j}\Gamma_x \frac{\mathrm{d}[N]^T}{\mathrm{d}x}\frac{\mathrm{d}\phi^*}{\mathrm{d}x}\mathrm{d}x=[k_\Gamma]\left\{\phi^{(e)}\right\} \tag{15.5a}$$

where

$$[k_\Gamma]=\int_{x_i}^{x_j}\Gamma_x \frac{\mathrm{d}[N]^T}{\mathrm{d}x}\frac{\mathrm{d}[N]}{\mathrm{d}x}\mathrm{d}x = \text{Element characteristic matrix due to the diffusion term}$$

$$=\frac{\Gamma_x}{L_e}\begin{bmatrix}1 & -1\\ -1 & 1\end{bmatrix} \tag{15.5b}$$

$$\int_{x_i}^{x_j}[N]^T\left(S_1\phi^*\right)\mathrm{d}x=\left[k_S^{(e)}\right]\left\{\phi^{(e)}\right\} \tag{15.6a}$$

where

$$\left[k_S^{(e)}\right]=\int_{x_i}^{x_j}S_1[N]^T[N]\mathrm{d}x = \text{Element characteristic matrix due to the source term}$$

$$=\frac{S_1L_e}{6}\begin{bmatrix}2 & 1\\ 1 & 2\end{bmatrix} \tag{15.6b}$$

$$\int_{x_i}^{x_j}[N]^T(S_0)\mathrm{d}x=\left\{\begin{array}{c}f_i^{(e)}\\ f_j^{(e)}\end{array}\right\}=\left\{f^{(e)}\right\}=\text{element force vector}$$

$$=\left\{\begin{array}{c}\dfrac{S_0L_e}{2}\\ \dfrac{S_0L_e}{2}\end{array}\right\} \tag{15.7}$$

and

$$\begin{Bmatrix} -\Gamma_x \left.\dfrac{d\phi}{dx}\right|_{x_i} \\ \Gamma_x \left.\dfrac{d\phi}{dx}\right|_{x_j} \end{Bmatrix} = \begin{Bmatrix} I_i^{(e)} \\ I_j^{(e)} \end{Bmatrix} = \left\{ I^{(e)} \right\} = \text{inter} - \text{element vector} \tag{15.8}$$

The unsteady-state term is evaluated in the following manner:

$$\left(\int_{x_i}^{x_j} C[N]^T [N] \right) \frac{\partial \{\phi^{(e)}\}}{\partial t} dx = \left[c^{(e)} \right] \left\{ \dot{\phi}^{(e)} \right\} \tag{15.9a}$$

where

$$\left[c^{(e)} \right] = \left(\int_{x_i}^{x_j} C[N]^T [N] \right) dx = \text{local capacitance matrix} \tag{15.9b}$$

and

$$\left\{ \dot{\phi}^{(e)} \right\} = \begin{Bmatrix} \dfrac{d\phi_i}{dt} \\ \dfrac{d\phi_j}{dt} \end{Bmatrix} \tag{15.9c}$$

The capacitance matrix $c^{(e)}$ can be evaluated by selecting appropriate element type and associated shape functions. For example, for **one-dimensional linear element** Equation 15.9b can be evaluated in the same manner as the $\left[k_S^{(e)} \right]$ given by (13.87) as

$$\left[c^{(e)} \right] = \int_{x_i}^{x_j} C[N]^T [N] dx = \frac{C l_e}{6} \begin{bmatrix} 2 & 1 \\ 1 & 2 \end{bmatrix} \tag{15.10}$$

Substituting Equations 15.5–15.9 into Equation 15.4, we get the **element characteristic equation** as

$$\left[c^{(e)} \right] \frac{d\{\phi^{(e)}\}}{dt} = \left\{ I^{(e)} \right\} - \left[k^{(e)} \right] \left\{ \phi^{(e)} \right\} + \left\{ f^{(e)} \right\} \tag{15.11a}$$

or

$$\left[c^{(e)} \right] \left\{ \dot{\phi}^{(e)} \right\} = \left\{ I^{(e)} \right\} - \left[k^{(e)} \right] \left\{ \phi^{(e)} \right\} + \left\{ f^{(e)} \right\} \tag{15.11b}$$

The element characteristic Equation 15.11 represents a system of first-order ordinary differential equations.

15.2.2 Time Approximation

The system of first-order differential equation in time can be solved numerically by approximating the first-order time derivatives by finite difference formulas presented in Chapter 5. For example, for a continuous variation of the dependent variable $\phi(t)$ as shown in Figure 15.1, the time derivative at a time step $m+\frac{1}{2}$ can be estimated based on the **central difference scheme** using nodal values at $m+1$ and m as follows:

$$\frac{\mathrm{d}\{\phi\}}{\mathrm{d}t} = \frac{\{\phi\}^{m+1} - \{\phi\}^{m}}{\Delta t} + O\left(\Delta t^2\right) \tag{15.12}$$

Let us write the element characteristic equation at the time step $m+\frac{1}{2}$ as

$$\left[c^{(e)}\right]\frac{d\left\{\phi^{(e)}\right\}}{dt}\Bigg|^{m+\frac{1}{2}} = \left\{I^{(e)}\right\}^{m+\frac{1}{2}} - \left[k^{(e)}\right]\left\{\phi^{(e)}\right\}^{m+\frac{1}{2}} + \left\{f^{(e)}\right\}^{m+\frac{1}{2}} \tag{15.13}$$

Using the central difference scheme for the time derivative and a linear piece-wise profile for ϕ, f and I in the time interval, the element characteristic equation (15.13) becomes

$$\left[c^{(e)}\right]\frac{\{\phi\}^{m+1} - \{\phi\}^{m}}{\Delta t} = \frac{\{I\}^{m+1} + \{I\}^{m}}{2} - [k]\frac{\{\phi\}^{m+1} + \{\phi\}^{m}}{2} + \frac{\{f\}^{m+1} + \{f\}^{m}}{2} \tag{15.14}$$

Rearranging,

$$\left(\left[c^{(e)}\right] + \frac{1}{2}\Delta t\left[k^{(e)}\right]\right)\{\phi\}^{m+1} = \left(\left[c^{(e)}\right] - \frac{1}{2}\Delta t\left[k^{(e)}\right]\right)\{\phi\}^{m} + \frac{\Delta t}{2}\left(\{I\}^{m+1} + \{I\}^{m}\right) + \frac{\Delta t}{2}\left(\{f\}^{m+1} + \{f\}^{m}\right) \tag{15.15}$$

The element characteristic equation given by Equation 15.15 referred to as **Crank-Nicolson or semi-implicit time integration scheme**, which is obtained by using **central difference scheme**. In a similar manner, we can derive the element characteristic equation based on forward difference scheme and backward difference scheme.

Forward Difference Scheme

In this scheme, it is assumed that the old values of $\left\{\phi^{(e)}\right\}^m$, $\left\{f^{(e)}\right\}^m$, and $\left\{I^{(e)}\right\}^m$ prevail throughout the time interval $(t, \Delta t)$ except at the time $t+\Delta t$ as shown in Figure 7.2. The time derivative at the time step i is approximated by the forward difference scheme in terms of nodal values at steps m and $m+1$ as shown in Figure 15.2.

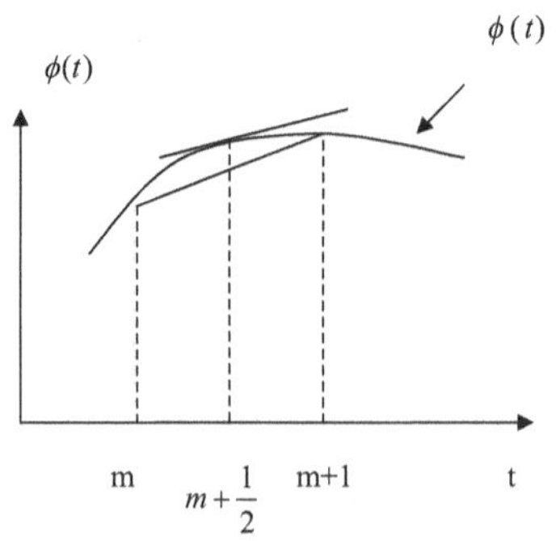

FIGURE 15.1 Central difference time approximations.

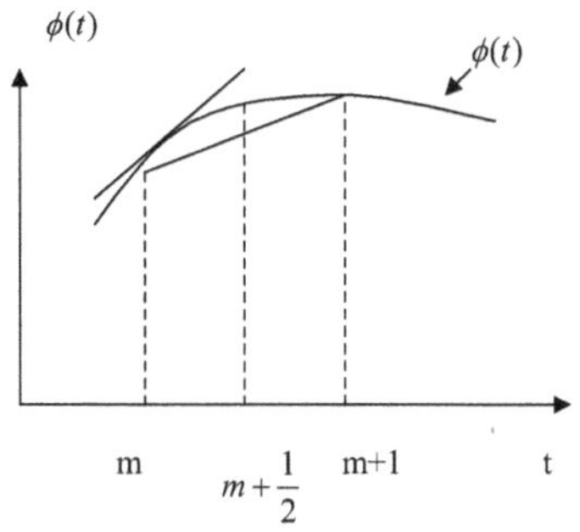

FIGURE 15.2 forward difference time approximation.

and by the equation

$$\frac{\mathrm{d}\{\phi\}^m}{\mathrm{d}t}=\frac{\{\phi\}^{m+1}-\{\phi\}^m}{\Delta t}+O(\Delta t) \tag{15.16}$$

Let us write the element characteristic equation at the time step m as

$$\left[c^{(e)}\right]\frac{\mathrm{d}\{\phi^{(e)}\}}{\mathrm{d}t}\bigg|^m=\{I^{(e)}\}^m-\left[k^{(e)}\right]\{\phi^{(e)}\}^m+\{f^{(e)}\}^m \tag{15.17}$$

Now, substituting the forward difference scheme for the first-time derivative given by Equation 15.16, we get

$$\left[c^{(e)}\right]\frac{\{\phi\}^{m+1}-\{\phi\}^m}{\Delta t}=\{I^{(e)}\}^m-\left[k^{(e)}\right]\{\phi^{(e)}\}^m+\{f^{(e)}\}^m \tag{15.18}$$

Rearranging,

$$\left[c^{(e)}\right]\{\phi^{(e)}\}^{m+1}=\left(\left[c^{(e)}\right]-\Delta t\left[k^{(e)}\right]\right)\{\phi^{(e)}\}^m+\Delta t\{I^{(e)}\}^m+\Delta t\{f^{(e)}\}^m \tag{15.19}$$

The element characteristic equation given by Equation 15.16 is referred to as **explicit time integration scheme**, which is obtained by using forward difference scheme.

Backward Difference Scheme

In this scheme, it is assumed that the new values of $\{\phi^{(e)}\}^{m+1}$, $\{f^{(e)}\}^{m+1}$, and $\{I^{(e)}\}^{m+1}$ prevail throughout the time interval $(t,\Delta t)$ except at the time t as shown in Figure 7.2. The time derivative at the time step $m+1$ is approximated by the backward difference scheme in terms of nodal values at steps m and $m+1$ as depicted in Figure 15.3.

and by the equation

$$\frac{\mathrm{d}\{\phi\}^{m+1}}{\mathrm{d}t}=\frac{\{\phi\}^{m+1}-\{\phi\}^m}{\Delta t}+O(\Delta t) \tag{15.20}$$

Let us write the element characteristic equation that can be at the time step $m+1$ as

$$\left[c^{(e)}\right]\frac{\mathrm{d}\{\phi^{(e)}\}}{\mathrm{d}t}\bigg|^{m+1}=\{I^{(e)}\}^{m+1}-\left[k^{(e)}\right]\{\phi^{(e)}\}^{m+1}+\{f^{(e)}\}^{m+1} \tag{15.21}$$

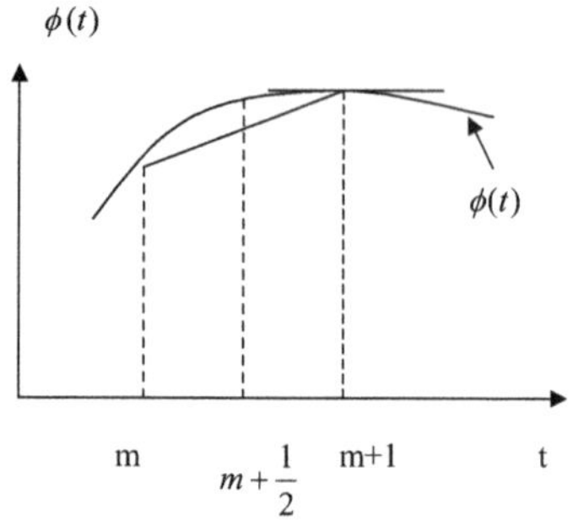

FIGURE 15.3 backward difference time approximations.

Substituting the backward difference scheme for the first-time derivative given by Equation 15.20, we get

$$\left[c^{(e)}\right]\frac{\{\phi\}^{m+1}-\{\phi\}^{m}}{\Delta t}=\left\{I^{(e)}\right\}^{m+1}-\left[k^{(e)}\right]\left\{\phi^{(e)}\right\}^{m+1}+\left\{f^{(e)}\right\}^{m+1} \tag{15.22}$$

Rearranging,

$$\left(\left[c^{(e)}\right]+\Delta t\left[k^{(e)}\right]\right)\left\{\phi^{(e)}\right\}^{m+1}=\left[c^{(e)}\right]\left\{\phi^{(e)}\right\}^{m}+\Delta t\left\{I^{(e)}\right\}^{m+1}+\Delta t\left\{f^{(e)}\right\}^{m+1} \tag{15.23}$$

The element characteristic equation given by Equation 15.20 is referred to as **implicit time integration scheme**, which is obtained by using backward difference scheme.

Equations 15.15, 15.19, and 15.23 can be expressed in a general form

$$\begin{aligned}\left(\left[c^{(e)}\right]+\lambda\Delta t\left[k^{(e)}\right]\right)\{\phi\}^{m+1}&=\left(\left[c^{(e)}\right]-(1-\lambda)\Delta t\left[k^{(e)}\right]\right)\{\phi\}^{m}\\&+\Delta t\left(\lambda\left\{I^{(e)}\right\}^{m+1}+(1-\lambda)\left\{I^{(e)}\right\}^{m}\right)+\Delta t\left(\lambda\left\{f^{(e)}\right\}^{m+1}+(1-\lambda)\left\{f^{(e)}\right\}^{m}\right)\end{aligned} \tag{15.24a}$$

Equation 15.24a can further be written in a compact form as

$$A\{\phi\}^{m+1}=B\{\phi\}^{m}+C \tag{15.24b}$$

where

$$A=\left[c^{(e)}\right]+\lambda\Delta t\left[k^{(e)}\right] \tag{15.24c}$$

$$B=\left[c^{(e)}\right]+(1-\lambda)\Delta t\left[k^{(e)}\right] \tag{15.24d}$$

$$C=\Delta t\left(\lambda\left\{I^{(e)}\right\}^{m+1}+(1-\lambda)\left\{I^{(e)}\right\}^{m}\right)-\Delta t\left(\lambda\left\{f^{(e)}\right\}^{m+1}+(1-\lambda)\left\{f^{(e)}\right\}^{m}\right) \tag{15.24e}$$

where λ is the weighting factor for the time approximation scheme. We can derive different time approximation scheme by selecting different values for the weighting factor. This weighting factor is usually specified in the range of $0\le\lambda\le 1$. For example,

$\lambda = 0$ leads to the **explicit scheme** (Equation 15.19) based on the **forward difference scheme**

$\lambda = 1$ leads to the fully **implicit scheme** (Equation 15.23) based on the **backward difference scheme**

$\lambda = \frac{1}{2}$ leads to the semi-implicit or **Crank-Nicolson scheme** based on the **central difference scheme** (Equation 15.15).

With the known initial values of $\{\varphi\}^0$, Equation 15.24 is solved with marching in time. For example, with the known values of $\{\varphi\}^0$ at time $t=0$ and a time step Δt, we can solve Equation 15.24 for $\{\varphi\}^1$ at time $t=\Delta t$. Now with the known value of $\{\varphi\}^1$, we can proceed to determine $\{\varphi\}^2$ at time $t=2\ \Delta t$ and so on. So, at each time step, the solution is equivalent to solving a steady-state problem. It can also be noticed that for constant material properties and constant time step Δt, matrices A and B are same for all time steps. For variable material properties as in a nonlinear problem and for variable time steps, the matrices A and B are reevaluated at each time step.

15.2.3 Stability Consideration

As we have discussed in Chapter 7, the advantage of a fully implicit scheme (based on the backward difference approximation ($\lambda = 1$)) is that it is unconditionally stable. Whereas an explicit scheme ($\lambda = 0$) based on the forward difference approximation is associated with the stability problem, where the temporal approximation error grows with time and the estimated solution oscillates when the selected time step Δt exceeds certain minimum value. A convergence for such a scheme is achieved only if the time steps satisfy the stability criterion. A stability analysis for such a system shows (Segerlind, 1984) that for $\lambda < \frac{1}{2}$ the largest Δt is given by

$$\Delta t = \frac{\beta_{\min}}{1-\lambda} \tag{15.25}$$

where $\beta_{\min}$ is the smallest eigenvalue of the equation given by

$$\left|\left(\left[c^{(e)}\right] - \beta\left[k^{(e)}\right]\right)\right| = 0 \tag{15.26}$$

Example 15.1: One-Dimensional Unsteady-State Conduction

Consider one-dimensional unsteady-state conduction without heat generation in a plane slab as shown. Initially the slab is at uniform temperature $T_I = 25°\text{C}$. The boundary surface at x=0 is subjected to a constant surface heat flux, $f_{ls}'' = 1\times 10^3\,\text{W/m}^2$, and the boundary surface x=L is maintained at constant surface temperature $T_R = 25°\text{C}$, respectively. Determine the temperature distribution in the slab and heat transfer rates at the right surface as a function of time using implicit time approximation and time step $\Delta t = 1\,\text{s}$. Use $k = 60.5\frac{\text{W}}{\text{m}\cdot\text{C}}$, $\rho = 7854\,\text{kg/m}^3$, $c_p = 434\,\text{J/kg}\cdot\text{K}$, L=0.08 m.

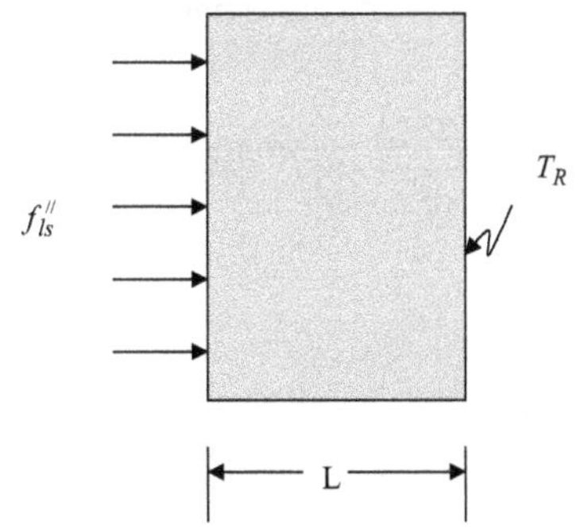

The mathematical statement of the problem is

Governing equation:

$$\rho c_p \frac{\partial T}{\partial t} = \frac{\mathrm{d}}{\mathrm{d}x}\left(k\frac{\mathrm{d}T}{\mathrm{d}x}\right) \tag{E.15.1.1}$$

Boundary condition:

$$1.\quad x = 0, -k\frac{\partial T}{\partial x} = f_{\mathrm{ls}}^{//} \tag{E.15.1.2a}$$

$$2.\quad x = L, T = T_R \tag{E.15.1.2b}$$

Solution

In the **first step**, let us discretize the domain using four uniform line elements as shown in the figure.

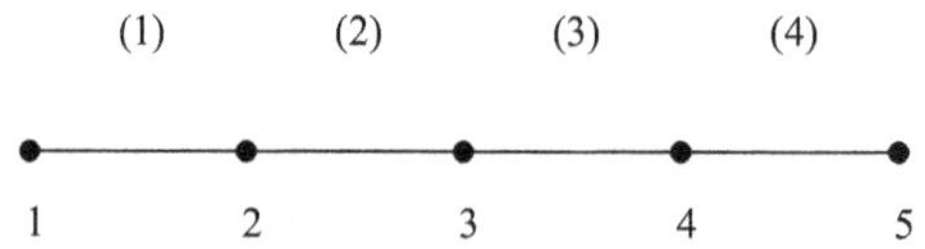

With this mesh size distribution, we have $l_e = l = 0.02\,\mathrm{m}$ and number of grid points N=5 with node i=1 and i =5 located at x=0 and x=0.08 m.

In the **second step**, we assume that the temperature distribution in each element be represented by the approximate function

$$T^* = N_i T_i + N_j T_j \tag{E.15.1.3}$$

In the **third step**, we write the weak form of the governing equation following Equation 13.9, which is obtained by Galerkin's weighted residual method as

$$\left(\int_{x_i}^{x_j} \rho c_p [N]^T [N]\right)\frac{\partial\{T^{(e)}\}}{\partial t}\mathrm{d}x = \begin{Bmatrix} -k\left.\dfrac{\mathrm{d}T}{\mathrm{d}x}\right|_{x_i} \\ k\left.\dfrac{\mathrm{d}T}{\mathrm{d}x}\right|_{x_j} \end{Bmatrix} - \int_{x}^{x_j} k\frac{\mathrm{d}[N]^T}{\mathrm{d}x}\frac{\mathrm{d}T^*}{\mathrm{d}x}\mathrm{d}x = 0 \tag{E.15.1.4}$$

Comparing with Equation 15.11, we can write the **element characteristic equation** as

$$\left[c^{(e)}\right]\frac{\mathrm{d}\{T^{(e)}\}}{\mathrm{d}t} = \{I^{(e)}\} - \left[k_k^{(e)}\right]\{T^{(e)}\} \tag{15.8a}$$

or

$$\left[c^{(e)}\right]\left\{\dot{T}^{(e)}\right\}=\left\{I^{(e)}\right\}-\left[k_k^{(e)}\right]\left\{T^{(e)}\right\}$$

Using fully implicit time approximation scheme as given by Equation 15.24 with $\lambda = 1$, we get

$$\left(c^{(e)}+\Delta t\left[k^{(e)}\right]\right)\{T\}^{m+1}=\left[c^{(e)}\right]\{T\}^{m}+\Delta t\left\{I^{(e)}\right\}^{m+1} \tag{E.15.1.5}$$

where

$$\left[c^{(e)}\right]=\frac{Cl_e}{6}\begin{bmatrix} 2 & 1 \\ 1 & 2 \end{bmatrix}, C=\rho c_p \tag{E.15.1.6a}$$

$$\left[k_k\right]=\frac{k}{l_e}\begin{bmatrix} 1 & -1 \\ -1 & 1 \end{bmatrix} \tag{E.15.1.6b}$$

$$\left\{I^{(e)}\right\}=\begin{Bmatrix} -k\left.\dfrac{dT}{dx}\right|_{xi} \\ k\left.\dfrac{dT}{dx}\right|_{xj} \end{Bmatrix} \tag{E.15.1.6c}$$

$$\left\{T^{(e)}\right\}=\begin{Bmatrix} T_i \\ T_j \end{Bmatrix} \tag{E.15.1.6d}$$

Based on these expressions, the element characteristics equation (E.15.1.5) is written as

$$\left(\frac{Cl_e}{6}\begin{bmatrix} 2 & 1 \\ 1 & 2 \end{bmatrix}+\frac{k\Delta t}{l_e}\begin{bmatrix} 1 & -1 \\ -1 & 1 \end{bmatrix}\right)\begin{Bmatrix} T_i^{m+1} \\ T_j^{m+1} \end{Bmatrix}=\frac{Cl_e}{6}\begin{bmatrix} 2 & 1 \\ 1 & 2 \end{bmatrix}\begin{Bmatrix} T_i^{m} \\ T_j^{m} \end{Bmatrix}+\Delta t\begin{Bmatrix} -k\left.\dfrac{dT}{dx}\right|_{xi}^{m+1} \\ k\left.\dfrac{dT}{dx}\right|_{xj}^{m+1} \end{Bmatrix}$$

$$\left(\begin{bmatrix} \dfrac{Cl_e}{3}+\dfrac{k\Delta t}{l_e} & \dfrac{Cl_e}{6}-\dfrac{k\Delta t}{l_e} \\ \dfrac{Cl_e}{6}-\dfrac{k\Delta t}{l_e} & \dfrac{Cl_e}{3}+\dfrac{k\Delta t}{l_e} \end{bmatrix}\right)\begin{Bmatrix} T_i^{m+1} \\ T_j^{m+1} \end{Bmatrix}=\begin{bmatrix} \dfrac{Cl_e}{3} & \dfrac{Cl_e}{6} \\ \dfrac{Cl_e}{6} & \dfrac{Cl_e}{3} \end{bmatrix}\begin{Bmatrix} T_i^{m} \\ T_j^{m} \end{Bmatrix}+\Delta t\begin{Bmatrix} -k\left.\dfrac{dT}{dx}\right|_{xi}^{m+1} \\ k\left.\dfrac{dT}{dx}\right|_{xj}^{m+1} \end{Bmatrix} \tag{E.15.1.7}$$

Since the material is homogeneous and the mesh is uniform, the element characteristics equation is identical for all elements.

In the **fifth step**, we assemble all element equations and obtain the global system following Equation 15.21 through the connectivity relations. For example, addition of the **element #1** into the global system results

$$\begin{bmatrix} \frac{Cl_e}{3}+\frac{k\Delta t}{l_e} & \frac{Cl_e}{6}-\frac{k\Delta t}{l_e} & 0 & 0 & 0 \\ \frac{Cl_e}{6}-\frac{k\Delta t}{l_e} & \frac{Cl_e}{3}+\frac{k\Delta t}{l_e} & 0 & 0 & 0 \\ 0 & 0 & 0 & 0 & 0 \\ 0 & 0 & 0 & 0 & 0 \\ 0 & 0 & 0 & 0 & 0 \end{bmatrix} \begin{Bmatrix} T_1^{m+1} \\ T_2^{m+1} \\ T_3^{m+1} \\ T_4^{m+1} \\ T_5^{m+1} \end{Bmatrix} = \begin{bmatrix} \frac{Cl_e}{3} & \frac{Cl_e}{6} & 0 & 0 & 0 \\ \frac{Cl_e}{6} & \frac{Cl_e}{3} & 0 & 0 & 0 \\ 0 & 0 & 0 & 0 & 0 \\ 0 & 0 & 0 & 0 & 0 \\ 0 & 0 & 0 & 0 & 0 \end{bmatrix} \begin{Bmatrix} T_1^{m} \\ T_2^{m} \\ T_3^{m} \\ T_4^{m} \\ T_5^{m} \end{Bmatrix} + \Delta t \begin{Bmatrix} -k\left.\frac{dT}{dx}\right|_{x1}^{m+1} \\ k\left.\frac{dT}{dx}\right|_{x2}^{m+1} \\ 0 \\ 0 \\ 0 \end{Bmatrix} \quad \text{(E.15.1.8a)}$$

Similarly, addition of **element #2** gives

$$\begin{bmatrix} \frac{Cl_e}{3}+\frac{k\Delta t}{l_e} & \frac{Cl_e}{6}-\frac{k\Delta t}{l_e} & 0 & 0 & 0 \\ \frac{Cl_e}{6}-\frac{k\Delta t}{l_e} & \frac{2Cl_e}{3}+\frac{2k\Delta t}{l_e} & \frac{Cl_e}{6}-\frac{k\Delta t}{l_e} & 0 & 0 \\ 0 & \frac{Cl_e}{6}-\frac{k\Delta t}{l_e} & \frac{Cl_e}{3}+\frac{k\Delta t}{l_e} & 0 & 0 \\ 0 & 0 & 0 & 0 & 0 \\ 0 & 0 & 0 & 0 & 0 \end{bmatrix} \begin{Bmatrix} T_1^{m+1} \\ T_2^{m+1} \\ T_3^{m+1} \\ T_4^{m+1} \\ T_5^{m+1} \end{Bmatrix}$$

$$= \begin{bmatrix} \frac{Cl_e}{3} & \frac{Cl_e}{6} & 0 & 0 & 0 \\ \frac{Cl_e}{6} & \frac{2Cl_e}{3} & \frac{Cl_e}{6} & 0 & 0 \\ 0 & \frac{Cl_e}{6} & \frac{Cl_e}{3} & 0 & 0 \\ 0 & 0 & 0 & 0 & 0 \\ 0 & 0 & 0 & 0 & 0 \end{bmatrix} \begin{Bmatrix} T_1^{m} \\ T_2^{m} \\ T_3^{m} \\ T_4^{m} \\ T_5^{m} \end{Bmatrix} + \Delta t \begin{Bmatrix} -k\left.\frac{dT}{dx}\right|_{x1}^{m+1} \\ k\left.\frac{dT}{dx}\right|_{x2}^{m+1} - k\left.\frac{dT}{dx}\right|_{x2}^{m+1} \\ k\left.\frac{dT}{dx}\right|_{x3}^{m+1} \\ 0 \\ 0 \end{Bmatrix} \quad \text{(E.15.1.8b)}$$

Addition of **Element #3** gives

$$
\begin{bmatrix}
\frac{Cl_e}{3}+\frac{k\Delta t}{l_e} & \frac{Cl_e}{6}-\frac{k\Delta t}{l_e} & 0 & 0 & 0 \\
\frac{Cl_e}{6}-\frac{k\Delta t}{l_e} & \frac{2Cl_e}{3}+\frac{2k\Delta t}{l_e} & \frac{Cl_e}{6}-\frac{k\Delta t}{l_e} & 0 & 0 \\
0 & \frac{Cl_e}{6}-\frac{k\Delta t}{l_e} & \frac{2Cl_e}{3}+\frac{2k\Delta t}{l_e} & \frac{Cl_e}{6}-\frac{k\Delta t}{l_e} & 0 \\
0 & 0 & \frac{Cl_e}{6}-\frac{k\Delta t}{l_e} & \frac{Cl_e}{3}+\frac{k\Delta t}{l_e} & 0 \\
0 & 0 & 0 & 0 & 0
\end{bmatrix}
\begin{Bmatrix} T_1^{m+1} \\ T_2^{m+1} \\ T_3^{m+1} \\ T_4^{m+1} \\ T_5^{m+1} \end{Bmatrix}
$$

$$
=\begin{bmatrix}
\frac{Cl_e}{3} & \frac{Cl_e}{6} & 0 & 0 & 0 \\
\frac{Cl_e}{6} & \frac{2Cl_e}{3} & \frac{Cl_e}{6} & 0 & 0 \\
0 & \frac{Cl_e}{6} & \frac{2Cl_e}{3} & \frac{Cl_e}{6} & 0 \\
0 & 0 & \frac{Cl_e}{6} & \frac{Cl_e}{3} & 0 \\
0 & 0 & 0 & 0 & 0
\end{bmatrix}
\begin{Bmatrix} T_1^{m} \\ T_2^{m} \\ T_3^{m} \\ T_4^{m} \\ T_5^{m} \end{Bmatrix}
+\Delta t
\begin{Bmatrix}
-k\left.\frac{dT}{dx}\right|_{x1}^{m+1} \\
k\left.\frac{dT}{dx}\right|_{x2}^{m+1}-k\left.\frac{dT}{dx}\right|_{x2}^{m+1} \\
k\left.\frac{dT}{dx}\right|_{x3}^{m+1}-k\left.\frac{dT}{dx}\right|_{x3}^{m+1} \\
k\left.\frac{dT}{dx}\right|_{x3}^{m+1} \\
0
\end{Bmatrix}
\tag{E.15.1.8c}
$$

Finally, addition **element #4** gives the final global system as

$$
\begin{bmatrix}
\frac{Cl_e}{3}+\frac{k\Delta t}{l_e} & \frac{Cl_e}{6}-\frac{k\Delta t}{l_e} & 0 & 0 & 0 \\
\frac{Cl_e}{6}-\frac{k\Delta t}{l_e} & \frac{2Cl_e}{3}+\frac{2k\Delta t}{l_e} & \frac{Cl_e}{6}-\frac{k\Delta t}{l_e} & 0 & 0 \\
0 & \frac{Cl_e}{6}-\frac{k\Delta t}{l_e} & \frac{2Cl_e}{3}+\frac{2k\Delta t}{l_e} & \frac{Cl_e}{6}-\frac{k\Delta t}{l_e} & 0 \\
0 & 0 & \frac{Cl_e}{6}-\frac{k\Delta t}{l_e} & \frac{2Cl_e}{3}+\frac{2k\Delta t}{l_e} & \frac{Cl_e}{6}-\frac{k\Delta t}{l_e} \\
0 & 0 & 0 & \frac{Cl_e}{6}-\frac{k\Delta t}{l_e} & \frac{Cl_e}{3}+\frac{k\Delta t}{l_e}
\end{bmatrix}
\begin{Bmatrix} T_1^{m+1} \\ T_2^{m+1} \\ T_3^{m+1} \\ T_4^{m+1} \\ T_5^{m+1} \end{Bmatrix}
$$

$$
=\begin{bmatrix}
\frac{Cl_e}{3} & \frac{Cl_e}{6} & 0 & 0 & 0 \\
\frac{Cl_e}{6} & \frac{2Cl_e}{3} & \frac{Cl_e}{6} & 0 & 0 \\
0 & \frac{Cl_e}{6} & \frac{2Cl_e}{3} & \frac{Cl_e}{6} & 0 \\
0 & 0 & \frac{Cl_e}{6} & \frac{2Cl_e}{3} & \frac{Cl_e}{6} \\
0 & 0 & 0 & \frac{Cl_e}{6} & \frac{Cl_e}{3}
\end{bmatrix}
\begin{Bmatrix} T_1^{m} \\ T_2^{m} \\ T_3^{m} \\ T_4^{m} \\ T_5^{m} \end{Bmatrix}
+\Delta t
\begin{Bmatrix}
-k\left.\frac{dT}{dx}\right|_{x1}^{m+1} \\
k\left.\frac{dT}{dx}\right|_{x2}^{m+1}-k\left.\frac{dT}{dx}\right|_{x2}^{m+1} \\
k\left.\frac{dT}{dx}\right|_{x3}^{m+1}-k\left.\frac{dT}{dx}\right|_{x3}^{m+1} \\
k\left.\frac{dT}{dx}\right|_{x4}^{m+1}-k\left.\frac{dT}{dx}\right|_{x4}^{m+1} \\
k\left.\frac{dT}{dx}\right|_{x5}^{m+1}
\end{Bmatrix}
\tag{E.15.1.8d}
$$

In the **sixth step**, we incorporate the interelement continuity conditions and apply the boundary conditions at the left boundary node as $-k\left.\frac{\partial T}{\partial x}\right|_{x1} = f_{1s}^{//}$ and at the right boundary as $T_5 = T_R$, we get

$$\begin{bmatrix} \frac{Cl_e}{3}+\frac{k\Delta t}{l_e} & \frac{Cl_e}{6}-\frac{k\Delta t}{l_e} & 0 & 0 & 0 \\ \frac{Cl_e}{6}-\frac{k\Delta t}{l_e} & \frac{2Cl_e}{3}+\frac{2k\Delta t}{l_e} & \frac{Cl_e}{6}-\frac{k\Delta t}{l_e} & 0 & 0 \\ 0 & \frac{Cl_e}{6}-\frac{k\Delta t}{l_e} & \frac{2Cl_e}{3}+\frac{2k\Delta t}{l_e} & \frac{Cl_e}{6}-\frac{k\Delta t}{l_e} & 0 \\ 0 & 0 & \frac{Cl_e}{6}-\frac{k\Delta t}{l_e} & \frac{2Cl_e}{3}+\frac{2k\Delta t}{l_e} & \frac{Cl_e}{6}-\frac{k\Delta t}{l_e} \\ 0 & 0 & 0 & \frac{Cl_e}{6}-\frac{k\Delta t}{l_e} & \frac{Cl_e}{3}+\frac{k\Delta t}{l_e} \end{bmatrix} \begin{Bmatrix} T_1^{m+1} \\ T_2^{m+1} \\ T_3^{m+1} \\ T_4^{m+1} \\ T_R \end{Bmatrix}$$

$$= \begin{bmatrix} \frac{Cl_e}{3} & \frac{Cl_e}{6} & 0 & 0 & 0 \\ \frac{Cl_e}{6} & \frac{2Cl_e}{3} & \frac{Cl_e}{6} & 0 & 0 \\ 0 & \frac{Cl_e}{6} & \frac{2Cl_e}{3} & \frac{Cl_e}{6} & 0 \\ 0 & 0 & \frac{Cl_e}{6} & \frac{2Cl_e}{3} & \frac{Cl_e}{6} \\ 0 & 0 & 0 & \frac{Cl_e}{6} & \frac{Cl_e}{3} \end{bmatrix} \begin{Bmatrix} T_1^{m} \\ T_2^{m} \\ T_3^{m} \\ T_4^{m} \\ T_R \end{Bmatrix} + \Delta t \begin{Bmatrix} f_{1s}^{//} \\ 0 \\ 0 \\ 0 \\ k\left.\frac{dT}{dx}\right|_{x5}^{m+1} \end{Bmatrix} \tag{E.15.1.9}$$

Since the boundary conditions at the right surface are of the first kind, we rearrange

$$\begin{bmatrix} \frac{cl_e}{3}+\frac{k\Delta t}{l_e} & \frac{cl_e}{6}-\frac{k\Delta t}{l_e} & 0 & 0 & 0 \\ \frac{cl_e}{6}-\frac{k\Delta t}{l_e} & \frac{2cl_e}{3}+\frac{2k\Delta t}{l_e} & \frac{cl_e}{6}-\frac{k\Delta t}{l_e} & 0 & 0 \\ 0 & \frac{cl_e}{6}-\frac{k\Delta t}{l_e} & \frac{2cl_e}{3}+\frac{2k\Delta t}{l_e} & \frac{cl_e}{6}-\frac{k\Delta t}{l_e} & 0 \\ 0 & 0 & \frac{cl_e}{6}-\frac{k\Delta t}{l_e} & \frac{2cl_e}{3}+\frac{2k\Delta t}{l_e} & \Delta t \\ 0 & 0 & 0 & \frac{cl_e}{6}-\frac{k\Delta t}{l_e} & \Delta t \end{bmatrix} \begin{Bmatrix} T_1^{m+1} \\ T_2^{m+1} \\ T_3^{m+1} \\ T_4^{m+1} \\ -k\left.\frac{dT}{dx}\right|_{x_s}^{m+1} \end{Bmatrix}$$

$$= \begin{bmatrix} \frac{Cl_e}{3} & \frac{Cl_e}{6} & 0 & 0 & 0 \\ \frac{Cl_e}{6} & \frac{2Cl_e}{3} & \frac{Cl_e}{6} & 0 & 0 \\ 0 & \frac{Cl_e}{6} & \frac{2Cl_e}{3} & \frac{Cl_e}{6} & 0 \\ 0 & 0 & \frac{Cl_e}{6} & \frac{2Cl_e}{3} & 0 \\ 0 & 0 & 0 & \frac{Cl_e}{6} & 0 \end{bmatrix} \begin{Bmatrix} T_1^{l} \\ T_2^{l} \\ T_3^{l} \\ T_4^{l} \\ -k\left.\frac{dT}{dx}\right|_{x5}^{m} \end{Bmatrix} + \begin{Bmatrix} f_{1s}^{//}\Delta t \\ 0 \\ 0 \\ \frac{k\Delta t T_R}{l_e} \\ -\frac{k\Delta t T_R}{l_e} \end{Bmatrix} \tag{E.15.1.10}$$

For the problem with the given data, we get

$$C = \rho c_p = 7854 \times 434 \frac{\text{kg}}{\text{m}^3\text{K}} = 7854, \quad k = 60.5 \frac{\text{W}}{\text{m} \cdot \text{K}}, \quad l_e = 0.02\,\text{m}$$

The element matrices and interelement vector are

$$\left[K_k^{(e)}\right] = \frac{k}{l_e}\begin{bmatrix} 1 & -1 \\ -1 & 1 \end{bmatrix} = \frac{60.5}{0.02}\begin{bmatrix} 1 & -1 \\ -1 & 1 \end{bmatrix} = 3025\begin{bmatrix} 1 & -1 \\ -1 & 1 \end{bmatrix}$$

$$\frac{k\Delta t T_R}{l_e} = 7.5625 \times 10^4$$

$$\left|c^{(e)}\right| = \int_{x_i}^{x_j} \rho c_p [N]^T [N] dx = \frac{\rho c_p l_e}{6}\begin{bmatrix} 2 & 1 \\ 1 & 2 \end{bmatrix} = \frac{7854 \times 434 \times 0.02}{6}\begin{bmatrix} 2 & 1 \\ 1 & 2 \end{bmatrix}$$

$$= 1.1362.12 \times 10^4 \begin{bmatrix} 2 & 1 \\ 1 & 2 \end{bmatrix}, \quad \frac{Cl_e}{6} = 1.1362 \times 10^4 \frac{\text{J}}{\text{m}^2\text{K}}$$

$$\frac{Cl_e}{3} + \frac{k\Delta t}{l_e} = 22724 + 3025 = 2.5749 \times 10^4$$

$$\frac{2Cl_e}{3} + \frac{2k\Delta t}{l_e} = 5.1498 \times 10^4$$

$$\frac{Cl_e}{6} - \frac{k\Delta t}{l_e} = 11362 - 3025 = 0.8337 \times 10^4$$

Substituting the numerical values, we obtain the global system as

$$\begin{bmatrix} 2.5749\times10^4 & 0.8337\times10^4 & 0 & 0 & 0 \\ 0.8337\times10^4 & 5.1498\times10^4 & 0.8337\times10^4 & 0 & 0 \\ 0 & 0.8337\times10^4 & 5.1498\times10^4 & 0.8337\times10^4 & 0 \\ 0 & 0 & 0.8337\times10^4 & 5.1498\times10^4 & 1 \\ 0 & 0 & 0 & 0.8337\times10^4 & 1 \end{bmatrix} \begin{Bmatrix} T_1^{m+1} \\ T_2^{m+1} \\ T_3^{m+1} \\ T_4^{m+1} \\ -k\left.\dfrac{dT}{dx}\right|_{x_5}^{m+1} \end{Bmatrix}$$

$$= \begin{bmatrix} 2.2724\times10^4 & 1.1362\times10^4 & 0 & 0 & 0 \\ 1.1362\times10^4 & 4.5448\times10^4 & 1.1362\times10^4 & 0 & 0 \\ 0 & 1.1362\times10^4 & 4.5448\times10^4 & 1.1362\times10^4 & 0 \\ 0 & 0 & 1.1362\times10^4 & 4.5448\times10^4 & 0 \\ 0 & 0 & 0 & 1.1362\times10^4 & 0 \end{bmatrix} \begin{Bmatrix} T_1^{m} \\ T_2^{m} \\ T_3^{m} \\ T_4^{m} \\ k\left.\dfrac{dT}{dx}\right|_{x_5}^{m} \end{Bmatrix}$$

$$+ \begin{Bmatrix} -0.1\times10^4 \\ 0 \\ 0 \\ 7.5625\times10^4 \\ -7.5625\times10^4 \end{Bmatrix} \qquad \text{(E.15.1.11)}$$

Considering the initial values at $t=0$ as

$$\begin{Bmatrix} T_1^0 \\ T_2^0 \\ T_3^0 \\ T_4^0 \\ k\dfrac{\mathrm{d}T}{\mathrm{d}x}\bigg|_{x5}^{0} \end{Bmatrix} = \begin{Bmatrix} 25 \\ 25 \\ 25 \\ 25 \\ 0 \end{Bmatrix}$$

We can solve the system of Equation E.15.1.11 for solution at $t = t + \Delta t = 1\,\mathrm{s}$ as

$$\begin{Bmatrix} T_1^1 \\ T_2^1 \\ T_3^1 \\ T_4^1 \\ k\dfrac{\mathrm{d}T}{\mathrm{d}x}\bigg|_{x5}^{1} \end{Bmatrix} = \begin{Bmatrix} 29.1048 \\ 24.3170 \\ 25.1141 \\ 24.9780 \\ 183.80 \end{Bmatrix}$$

The heat transfer rate at $x=L$ is estimated as

$$q''_{x=L} = -k\frac{\mathrm{d}T}{\mathrm{d}x}\bigg|_{x5} = 183.80\frac{\mathrm{W}}{\mathrm{m}^2}$$

Example 15.2: One-dimensional Un-steady State Conduction with Volume heat generation

Consider one dimensional unsteady state bio-heat equation [Ravi and Majumdar] in a plane slab of biological tissue irradiated by a laser beam as shown. The boundary surface at x = 0 is subjected to a convective cooling constant and the boundary surface x = H is maintained at constant surface temperature T_b.

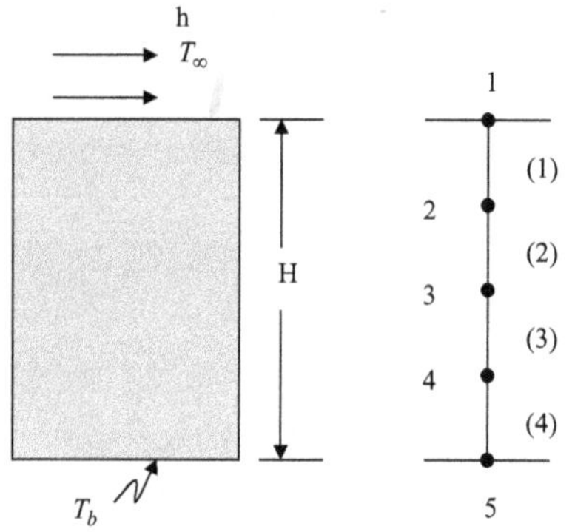

15.3 TWO-DIMENSIONAL UNSTEADY-STATE DIFFUSION EQUATION

In this section we will consider the formulation and application finite element method for solving two-dimensional unsteady-state problems. Let us consider two-dimensional steady-state transport process in a two-dimensional domain as shown in Figure 15.4 below:

A general mathematical statement of the problem can be assumed as

$$C\frac{\partial \phi}{\partial t} = \frac{\partial}{\partial x}\left(\Gamma_x \frac{\partial \phi}{\partial x}\right) + \frac{\partial}{\partial y}\left(\Gamma_y \frac{\partial \phi}{\partial y}\right) + S_1\phi + S_0 \qquad (15.27)$$

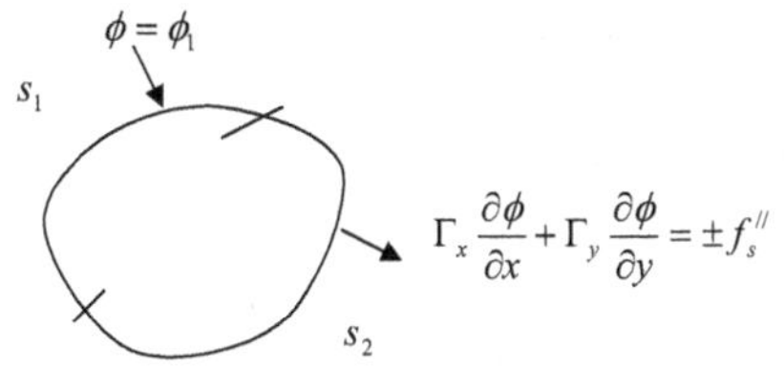

FIGURE 15.4 Two-dimensional domain and boundary conditions.

The partial differential equation (15.27) is like Equation 14.1 except for a storage term on the left-hand side. The associated boundary conditions at different sections of the boundary surface are assumed as follows:

On the boundary surface s_1: Boundary condition of the first kind

$$\phi = \phi_s = \text{constant surface temperature} \tag{15.28a}$$

On the boundary surface s_2:

$$\Gamma_x \frac{\partial \phi}{\partial x} \hat{n}_x + \Gamma_y \frac{\partial \phi}{\partial y} \hat{n}_y = \pm h(\phi - \phi_s) \pm f_s^{\prime\prime} \tag{15.28b}$$

Initial condition

$$\phi(x, y, z, 0) = \phi_I \tag{15.29}$$

In the semidiscrete finite element formulation, the **spatial discretization scheme** assumes the approximate solution for the element as

$$\phi^* = \sum_{I=1}^{n_p} N_I(\tilde{x}, \tilde{y}) \phi_I(t) = [N(\tilde{x}, \tilde{y})]\{\phi^{(e)}(t)\} \tag{15.30}$$

where N_I is a two-dimensional shape function, which is assumed to be a function of space variable only, ϕ_I (t) = nodal values as a function time, and n_p = number of nodes in an element.

Galerkin's integral procedure can be applied for obtaining the integral form of the equation over an element as

$$\{R^{(e)}\} = -\int_A [N]^T \left[\frac{\partial}{\partial x}\left(\Gamma_x \frac{\partial \phi^*}{\partial x} \right) + \frac{\partial}{\partial y}\left(\Gamma_y \frac{\partial \phi^*}{\partial y} \right) + S_1 \phi^* + S_0 - C \frac{\partial \phi^*}{\partial t} \right] \mathrm{d}A = 0 \tag{15.31}$$

Following the procedure outlined in Chapter 12, Equation 15.31 reduces to

$$\begin{aligned} \{R^{(e)}\} = &-\int_s [N]^T \left(\Gamma_x \frac{\partial \phi^*}{\partial x} \hat{n}_x + \Gamma_y \frac{\partial \phi^*}{\partial y} \hat{n}_y \right) \mathrm{d}s \\ &+ \int_A \left(\frac{\partial [N]^T}{\partial x} \Gamma_x \frac{\partial \phi^*}{\partial x} + \frac{\partial [N]^T}{\partial y} \Gamma_y \frac{\partial \phi^*}{\partial y} \right) \mathrm{d}A \\ &- \int_A S_1 [N]^T \phi^* \, \mathrm{d}A - \int_A S_0 [N]^T \, \mathrm{d}A - -\int_A [N]^T \left[C \frac{\partial \phi^*}{\partial t} \right] \mathrm{d}A = 0 \end{aligned} \tag{15.32}$$

Substituting the approximate solution function given by Equation 15.28 into the integral

$$-\int_{s}[N]^{T}\left(\Gamma_{x}\frac{\partial\phi^{*}}{\partial x}\cos\theta+\Gamma_{y}\frac{\partial\phi^{*}}{\partial y}\sin\theta\right)\mathrm{d}s$$

$$+\left\{\int_{A}\left(\frac{\partial[N]^{T}}{\partial x}\Gamma_{x}\frac{\partial[N]}{\partial x}+\frac{\partial[N]^{T}}{\partial y}\Gamma_{y}\frac{\partial[N]}{\partial y}\right)\mathrm{d}A\right\}\{\phi^{(e)}\}$$

$$-\left\{\int S_{1}[N]^{T}[N]\mathrm{d}A\right\}\{\phi^{(e)}\}-\int_{A}S_{0}[N]^{T}\mathrm{d}A+\int_{A}C[N]^{T}[N]\left[\frac{\partial\phi^{(e)}}{\partial t}\right]\mathrm{d}A \tag{15.33}$$

The element characteristics equation can be written in matrix notation as

$$\left[c^{(e)}\right]\frac{\mathrm{d}\{\phi^{(e)}\}}{\mathrm{d}t}=\left\{I^{(e)}\right\}+\left[k^{(e)}\right]\{\phi^{(e)}\}-\left\{f_{S}^{(e)}\right\}=0 \tag{15.34}$$

where

$$\left\{I^{(e)}\right\}=-\int_{s}[N]^{T}\left(\Gamma_{x}\frac{\partial\phi^{*}}{\partial x}\hat{n}_{x}+\Gamma_{y}\frac{\partial\phi^{*}}{\partial y}\hat{n}_{y}\right)\mathrm{d}s \tag{15.35a}$$

$$\left\{f_{S}^{(e)}\right\}=\int_{A}S_{0}[N]^{T}\mathrm{d}A \tag{15.35b}$$

$$\left[k^{(e)}\right]=\int_{A}\left(\frac{\partial[N]^{T}}{\partial x}\Gamma_{x}\frac{\partial[N]}{\partial x}+\frac{\partial[N]^{T}}{\partial y}\Gamma_{y}\frac{\partial[N]}{\partial y}\right)\mathrm{d}A-\int_{A}S_{1}[N]^{T}[N]\mathrm{d}A$$

$$=\left[k_{\Gamma}^{(e)}\right]-\left[k_{S}^{(e)}\right] \tag{15.35c}$$

$$\left[k_{S}^{(e)}\right]=\int_{A}S_{1}[N]^{T}[N]\mathrm{d}A \tag{15.35d}$$

$$\left[k_{\Gamma}^{(e)}\right]=\int_{A}\left(\frac{\partial[N]^{T}}{\partial x}\Gamma_{x}\frac{\partial[N]}{\partial x}+\frac{\partial[N]^{T}}{\partial y}\Gamma_{y}\frac{\partial[N]}{\partial y}\right)\mathrm{d}A$$

$$=\int_{A}\{D\}^{T}[\Gamma]\{D\}\mathrm{d}A \tag{15.35e}$$

$$\left[c^{(e)}\right]=\int_{A}C[N]^{T}[N]\mathrm{d}A \tag{15.35f}$$

The implementation of the derivative boundary conditions into the finite element formulation takes place through the interelement vector $\left\{I^{(e)}\right\}$ and results into following element characteristic equation

$$\left[c^{(e)}\right]\frac{\mathrm{d}\{\phi^{(e)}\}}{\mathrm{d}t}=\left\{I_{i}^{(e)}\right\}+\left(\mathrm{k}_{\Gamma}^{(e)}-\mathrm{k}_{S}^{(e)}-k_{b}^{(e)}\right)\{\phi^{(e)}\}-\left\{f_{S}^{(e)}+f_{\mathrm{bs}}^{(e)}\right\}=0 \tag{15.36a}$$

or

$$\left[c^{(e)}\right]\frac{\mathrm{d}\{\phi^{(e)}\}}{\mathrm{d}t}=\left\{I_{i}^{(e)}\right\}+k^{(e)}\left\{\phi^{(e)}\right\}-f^{//} \tag{15.36b}$$

where

$$k^{(e)} = \left(k_{\Gamma}^{(e)} - k_S^{(e)} - k_b^{(e)}\right) = \text{Element Characteristic Matrix} \tag{15.36a}$$

$$f^{(e)} = \left\{f_S^{(e)} + f_{bs}^{(e)}\right\} = \text{Element source vector} \tag{15.36b}$$

$$\left[k_b^{(e)}\right] = \pm\int_{s_b} h[N]^T[N]\mathrm{d}s \tag{15.36c}$$

$$\left\{f_{bs}^{(e)}\right\} = \pm\int_{s_b} [N]^T\left\{\left(h\varphi_\infty + f_s^{//}\right)\right\}\mathrm{d}s \tag{15.36d}$$

$$\left[c^{(e)}\right] = \int_A C[N]^T[N]\mathrm{d}A = \text{Element Characteristic Matrix} \tag{15.36e}$$

The element characteristic matrix, force vector, and capacitance matrix can now be evaluated for a particular element type. For example, we have seen the development of element characteristic matrix and force vector for **linear triangular element** and by **bilinear rectangular element** in Chapter 14 for steady-state problems. The capacitance matrix $\left[c^{(e)}\right]$ can also be evaluated by selecting appropriate element type and associated shape functions in the same manner as $\left[k_S^{(e)}\right]$ given by Equation (14.32). For **linear triangular element** as

$$\left[c^{(e)}\right] = \int_A C[N]^T[N]\mathrm{d}A$$

$$= C\int_A \begin{Bmatrix} N_i \\ N_j \\ N_k \end{Bmatrix}\begin{Bmatrix} N_i & N_j & N_k \end{Bmatrix}\mathrm{dA}$$

$$\left[c^{(e)}\right] = C\int_A \begin{bmatrix} N_i^2 & N_iN_j & N_iN_k \\ N_jN_i & N_j^2 & N_jN_k \\ N_kN_i & N_kN_j & N_k^2 \end{bmatrix}\mathrm{dA} \tag{15.37}$$

Using natural co-ordinate system (Equation 12.68) for the triangular element:

$$N_i = L_i, \quad N_j = L_j \quad \text{and} \quad N_k = L_k \tag{12. 68}$$

we get

$$\left[c^{(e)}\right] = C\int_A \begin{bmatrix} L_i^2 & L_iL_j & L_iL_k \\ L_jL_i & L_j^2 & L_jL_k \\ L_kL_i & L_kL_j & L_k^2 \end{bmatrix} \tag{15.38}$$

Using the integral formula given by Equation 12.71, we can evaluate the integral elements and the integral expression (15.38) becomes

$$\left[c^{(e)}\right] = \frac{CA}{12}\begin{bmatrix} 2 & 1 & 1 \\ 1 & 1 & 1 \\ 1 & 1 & 2 \end{bmatrix} \quad \text{for triangular element} \tag{15.39}$$

Similarly, for **bilinear rectangular element**, the capacitance matrix is given as

$$\left[c^{(e)}\right]=\int_A C[N]^T[N]\mathrm{d}A$$

$$=\int_A C\begin{Bmatrix} N_i \\ N_j \\ N_k \\ N_l \end{Bmatrix}\left\{ N_i \quad N_j \quad N_k \quad N_l \right\}\mathrm{d}A$$

$$\left[c^{(e)}\right]=\int_A C\begin{bmatrix} N_i^2 & N_iN_j & N_iN_k & N_iN_l \\ N_jN_i & N_j^2 & N_jN_k & N_jN_l \\ N_kN_i & N_kN_j & N_k^2 & N_kN_l \\ N_lN_i & N_lN_j & N_lN_k & N_l^2 \end{bmatrix}\mathrm{d}A \tag{15.40}$$

For cases with constant capacitance value, the integrals can be evaluated directly by substituting the shape function expressions in local co-ordinate system given by Equation 12.88, and the element capacitance matrix is expressed as

$$\left[k_S^{(e)}\right]=\frac{S_1 A}{36}\begin{bmatrix} 4 & 2 & 1 & 2 \\ 2 & 4 & 2 & 1 \\ 1 & 2 & 4 & 2 \\ 2 & 1 & 2 & 4 \end{bmatrix} \quad \text{for rectangular element} \tag{15.41}$$

In the next step, the element characteristic equation (15.36), which represents a set of first-order differential equation, can be solved using the time approximation scheme used for one-dimensional problem in Section 15.2.2.

15.4 THREE-DIMENSIONAL UNSTEADY-STATE DIFFUSION EQUATION

Solution of three-dimensional unsteady-state problems follows the same procedure as outlined for one- and two-dimensional problems. Let us consider three-dimensional steady-state transport process in a three-dimensional domain with the mathematical statement given as

$$C\frac{\partial\phi}{\partial t}=\frac{\partial}{\partial x}\left(\Gamma_x\frac{\partial\phi}{\partial x}\right)+\frac{\partial}{\partial y}\left(\Gamma_y\frac{\partial\phi}{\partial y}\right)+\frac{\partial}{\partial y}\left(\Gamma_y\frac{\partial\phi}{\partial y}\right)+S \tag{15.41}$$

with associated general boundary conditions given as

On the boundary surface A_b: Boundary condition of the third kind

$$\Gamma_x\frac{\partial\phi}{\partial x}n_x+\Gamma_y\frac{\partial\phi}{\partial y}n_y+\Gamma_z\frac{\partial\phi}{\partial z}n_z=\pm h(\phi-\phi_s)\pm f_s^{//} \tag{15.42a}$$

$$\text{On the remaining boundary surface: } \phi=\phi_s \tag{15.42b}$$

and with initial condition

$$\phi(x,y,z,0)=\phi_I \tag{15.43}$$

In the semidiscrete finite element formulation, the **spatial discretization scheme** assumes the approximate solution for the element as

$$\phi^* = \sum_{I=1}^{n_p} N_I(x,y,z)\phi_I(t) = [N(x,y,z)]\{\phi^{(e)}(t)\} \tag{15.44}$$

where N_I is a three-dimensional shape function, which is assumed to be a function of space variable only, $\phi_I(t)$ = nodal values as a function time, and n_p = number of nodes in an element.

Galerkin's integral procedure can be applied for obtaining the integral form of the equation over an element as

$$\{R^{(e)}\} = -\int_V [N]^T \left[\frac{\partial}{\partial x}\left(\Gamma_x \frac{\partial \phi^*}{\partial x}\right) + \frac{\partial}{\partial y}\left(\Gamma_y \frac{\partial \phi^*}{\partial y}\right) + S_1\phi^* + S_0 - C\frac{\partial \phi^*}{\partial t}\right] \mathrm{d}A = 0 \tag{15.45}$$

Evaluating the integral we, get

$$\int_V C[N]^T[N]\mathrm{d}V = -\int_A [N]^T\left(\Gamma_x \frac{\partial \phi^*}{\partial x}n_x + \Gamma_y \frac{\partial \phi^*}{\partial y}n_y + \Gamma_z \frac{\partial \phi^*}{\partial z}n_z\right)\mathrm{d}A$$
$$+\int_V \left(\frac{\partial [N]^T}{\partial x}\Gamma_x \frac{\partial \phi^*}{\partial x} + \frac{\partial [N]^T}{\partial y}\Gamma_y \frac{\partial \phi^*}{\partial y} + \frac{\partial [N]^T}{\partial z}\Gamma_z \frac{\partial \phi^*}{\partial z}\right)\mathrm{d}V$$
$$-\int_V S[N]^T\,\mathrm{d}V = 0 \tag{15.46}$$

To obtain the element characteristics equation, substitute the approximate solution function given by Equation 15.28 into the integral equation (15.42)

$$\int_V C[N]^T[N]\left[\frac{\partial \phi^{(e)}}{\partial t}\right]\mathrm{d}V = -\int_A [N]^T\left(\Gamma_x \frac{\partial \phi^*}{\partial x}n_x + \Gamma_y \frac{\partial \phi^*}{\partial y}n_y + \Gamma_z \frac{\partial \phi^*}{\partial z}n_z\right)\mathrm{d}A$$
$$+\left\{\int_V \left(\frac{\partial [N]^T}{\partial x}\Gamma_x \frac{\partial [N]}{\partial x} + \frac{\partial [N]^T}{\partial y}\Gamma_y \frac{\partial [N]}{\partial y} + \frac{\partial [N]^T}{\partial z}\Gamma_z \frac{\partial [N]}{\partial z}\right)\mathrm{d}V\right\}\{\phi^{(e)}\}$$
$$-\int_V S[N]^T\,\mathrm{d}V \tag{15.47}$$

The terms on the right-hand side of Equation 15.43 have already been evaluated in Chapter 14 for three-dimensional steady-state problems along with the implementation of the boundary conditions. With the addition of the new storage term on the left-hand side, the element characteristic equation takes the form as

$$\left[c^{(e)}\right]\frac{\mathrm{d}\{\phi^{(e)}\}}{\mathrm{d}t}=\left\{I_i^{(e)}\right\}+k^{(e)}\left\{\phi^{(e)}\right\}-f'' \tag{15.48}$$

where

$$\left[c^{(e)}\right]=\int_V C[N]^T[N]\mathrm{d}V \tag{15.49a}$$

$$k^{(e)}=\left(\mathrm{k}_\Gamma^{(e)}+k_b^{(e)}\right) \tag{15.49b}$$

$$\left[k^{(e)}\right]=\int_V\left(\frac{\partial[N]^T}{\partial x}\Gamma_x\frac{\partial[N]}{\partial x}+\frac{\partial[N]^T}{\partial y}\Gamma_y\frac{\partial[N]}{\partial y}+\frac{\partial[N]^T}{\partial z}\Gamma_z\frac{\partial[N]}{\partial z}\right)\mathrm{d}V$$

$$=\int_V\{D\}^T[\Gamma]\{D\}\mathrm{d}V \tag{15.49c}$$

$$[\Gamma]=\text{transport property matrix}=\begin{bmatrix}\Gamma_x & 0 & 0\\ 0 & \Gamma_y & 0\\ 0 & 0 & \Gamma_z\end{bmatrix} \tag{15.49d}$$

$$\{D\}=\begin{Bmatrix}\dfrac{\partial[N]}{\partial x}\\ \dfrac{\partial[N]}{\partial y}\\ \dfrac{\partial[N]}{\partial z}\end{Bmatrix} \tag{15.49e}$$

$$\left[k_b^{(e)}\right]=\pm\int_{A_b}h[N]^T[N]\mathrm{d}A \tag{15.49f}$$

$$f^{(e)}=\left\{f_S^{(e)}+f_{\mathrm{bs}}^{(e)}\right\} \tag{15.49g}$$

$$\left\{f_S^{(e)}\right\}=\int_V S[N]^T\,\mathrm{d}V \tag{15.49h}$$

$$\left\{f_{\mathrm{bs}}^{(e)}\right\}=\pm\int_{S_b}[N]^T\left\{\left(h\phi_\infty+f_s''\right)\right\}\mathrm{d}A \tag{15.49i}$$

The element characteristic matrix, force vector, and capacitance matrix can now be evaluated for a particular element type. For tetrahedron element, we have already derived the element characteristic matrix and force vector in Chapter 14 for steady-state problems. The capacitance matrix $\left[c^{(e)}\right]$ can now be evaluated for the three-dimensional such as the tetrahedron element as follows:

$$\left[c^{(e)}\right]=\int_V C[N]^T[N]\mathrm{d}V$$

$$=\int_V C\left\{\begin{matrix} N_i \\ N_j \\ N_k \\ N_l \end{matrix}\left\{ N_i \quad N_j \quad N_k \quad N_l \right\}\right\}\mathrm{d}V$$

$$=\int_V C\begin{bmatrix} N_i^2 & N_iN_j & N_iN_k & N_iN_l \\ N_jN_i & N_j^2 & N_jN_k & N_jN_l \\ N_kN_i & N_kN_j & N_k^2 & N_kN_l \\ N_lN_i & N_lN_j & N_lN_k & N_l^2 \end{bmatrix}\mathrm{d}V \tag{15.50}$$

Using natural volume coordinate system (Equation 12.116), the integral elements can be evaluated as

$$\left[c^{(e)}\right]=\int_V C\begin{bmatrix} L_i^2 & L_iL_j & L_iL_k & L_iL_l \\ L_jL_i & L_j^2 & L_jL_k & L_jL_l \\ L_kL_i & L_kL_j & L_k^2 & L_kL_l \\ L_lL_i & L_lL_j & L_lL_k & L_l^2 \end{bmatrix}\mathrm{d}V \tag{15.51}$$

$$c^{(e)}=C\begin{bmatrix} \int_V L_i^2\,\mathrm{d}V & \int_V L_iL_j\,\mathrm{d}V & \int_V L_iL_k\,\mathrm{d}V & \int_V L_iL_l\,\mathrm{d}V \\ \int_V L_jL_i\,\mathrm{d}V & \int_V L_j^2\,\mathrm{d}V & \int_V L_jL_k\,\mathrm{d}V & \int_V L_jL_l\,\mathrm{d}V \\ \int_V L_kL_i\,\mathrm{d}V & \int_V L_kL_j\,\mathrm{d}V & \int_V L_k^2\,\mathrm{d}V & \int_V L_kL_l\,\mathrm{d}V \\ \int_V L_lL_i\,\mathrm{d}V & \int_V L_lL_j\,\mathrm{d}V & \int_V L_lL_k\,\mathrm{d}V & \int_V L_l^2\,\mathrm{d}V \end{bmatrix}$$

$$=C\begin{bmatrix} \int_V L_i^2L_j^0L_k^0L_l^0\,\mathrm{d}V & \int_V L_i^1L_j^1L_k^0L_l^0\,\mathrm{d}V & \int_V L_i^1L_j^0L_k^1L_l^0\,\mathrm{d}V & \int_V L_i^1L_j^0L_k^0L_l^1\,\mathrm{d}V \\ \int_V L_i^1L_j^1L_k^0L_l^0\,\mathrm{d}V & \int_V L_i^0L_j^2L_k^0L_l^0\,\mathrm{d}V & \int_V L_i^0L_j^1L_k^1L_l^0\,\mathrm{d}V & \int_V L_i^0L_j^1L_k^0L_l^1\,\mathrm{d}V \\ \int_V L_i^1L_j^0L_k^1L_l^0\,\mathrm{d}V & \int_V L_i^0L_j^1L_k^1L_l^0\,\mathrm{d}V & \int_V L_i^0L_j^0L_k^2L_l^0\,\mathrm{d}V & \int_V L_i^0L_j^0L_k^1L_l^1\,\mathrm{d}V \\ \int_V L_i^1L_j^0L_k^0L_l^1\,\mathrm{d}V & \int_V L_i^0L_j^1L_k^0L_l^1\,\mathrm{d}V & \int_V L_i^0L_j^0L_k^1L_l^1\,\mathrm{d}V & \int_V L_i^0L_j^0L_k^0L_l^2\,\mathrm{d}V \end{bmatrix} \tag{15.52}$$

Using analytical integral formulas involving in natural volume co-ordinate system given by Equation 12.119, we get,

$$\left[c^{(e)}\right]=6VC\begin{bmatrix} \dfrac{2!0!0!0}{(2+0+0+0+3)!} & \dfrac{1!1!0!0!}{(1+1+0+0+3)!} & \dfrac{1!0!1!0!}{(1+0+1+0+3)!} & \dfrac{1!0!0!1!}{(1+0+0+1+3)!} \\ \dfrac{1!1!0!0!}{(1+1+0+0+3)!} & \dfrac{0!2!0!0!}{(0+2+0+0+3)!} & \dfrac{0!1!1!0!}{(0+1+1+0+3)} & \dfrac{0!1!0!1!}{(0+1+0+1+3)!} \\ \dfrac{1!0!1!0!}{(1+0+1+0+2)!} & \dfrac{0!1!10!!}{(0+1+1+0+2)} & \dfrac{0!0!2!0!}{(0+0+2+0)!} & \dfrac{1!0!1!}{(0+0+1+1+3)!} \\ \dfrac{1!0!0!1!}{(1+0+0+1+3)!} & \dfrac{0!1!0!1!}{(0+1+0+1+3)!} & \dfrac{0!0!1!1!}{(0+0+1+1+3)!} & \dfrac{0!0!0!2!}{(0+0+0+2+3)!} \end{bmatrix}$$

or

$$\left[c^{(e)}\right]=\frac{CV}{20}\begin{bmatrix} 2 & 1 & 1 & 1 \\ 1 & 2 & 1 & 1 \\ 1 & 1 & 2 & 1 \\ 1 & 1 & 1 & 2 \end{bmatrix} \tag{15.53}$$

PROBLEMS

15.1 Consider one-dimensional unsteady-state conduction in a plane slab [$k=50$ W/m·C, $\alpha=2\times 10^{-5}$m²/s] of thickness $L=4$ mm. Initially the slab is at a 20°C. The right side of the plate is suddenly brought to a temperature of 200°C. The left side is maintained at 20°C.

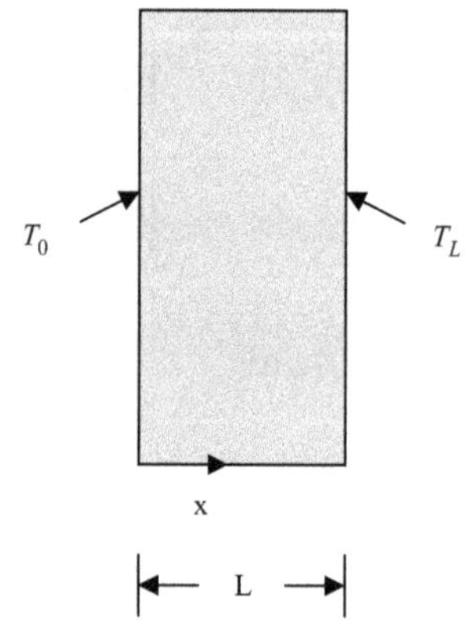

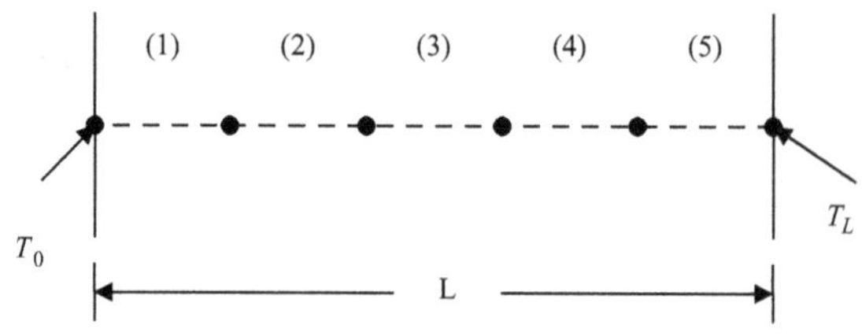

Use finite element method and fully implicit time approximation to derive the global system of equations. Carry out the computation for five consecutive time steps using a time step $\Delta t = 0.2$ s. Plot heat flux at the right surface as a function of time.

15.2 Redo Problem 15.1 using the Crank-Nicolson time approximation scheme.

15.3 Derive the global system of equation in Problem 15.1 using the explicit time approximation scheme. Estimate the largest $\Delta t_{\text{largest}}$ based on Equations 15.25 and 15.26. Solve the global system of equations using tome step $\Delta t < \Delta t_{\text{largest}}$ for temperature distribution and heat transfer rates. Plot heat transfer rate at the right surface as a function of time.

15.4 Derive the global system of equations in Problem 15.1 using quadratic line element.

15.5 Consider one-dimensional unsteady-state conduction in a plane slab [$k = 50$ W/m·C, $\alpha = 2 \times 10^{-5}$ m²/s] of thickness $L = 4$ mm. Initially the slab is at a temperature of 20°C. The right side of the plate is suddenly brought to a convective environment with a temperature of $T_\infty = 200$°C and $h = 80$ W/m·C. The left side is maintained at 20°C. Consider the same mesh size distribution as in Problem 15.1 and determine the transient temperature distribution in the slab. Plot heat transfer rate at the right surface as a function of time.

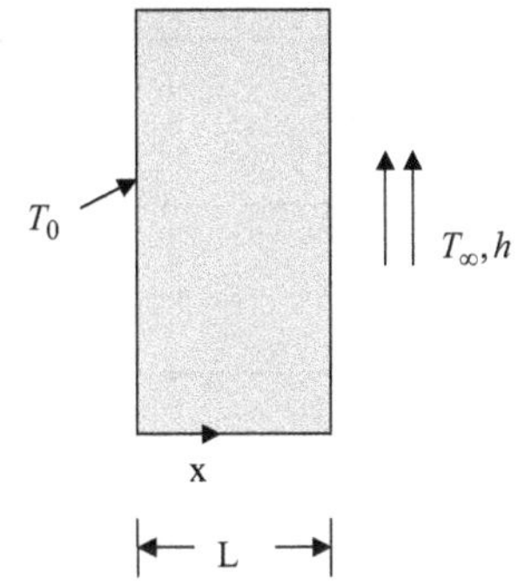

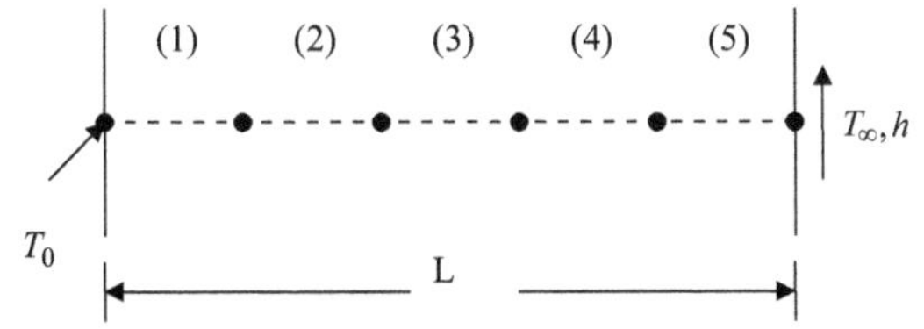

15.6 Consider one-dimensional unsteady-state heat conduction in a plane slab of thickness $L = 10$ cm. Initially the slab is at a uniform temperature of $T_I = 20$°C. At time $t > 0$, the left side of the slab is subjected to uniform constant surface heat flux $q_s^{//} = 2.0 \times 10^4$ W/cm^2 and the right surface is maintained at $T_R = 20$°C.

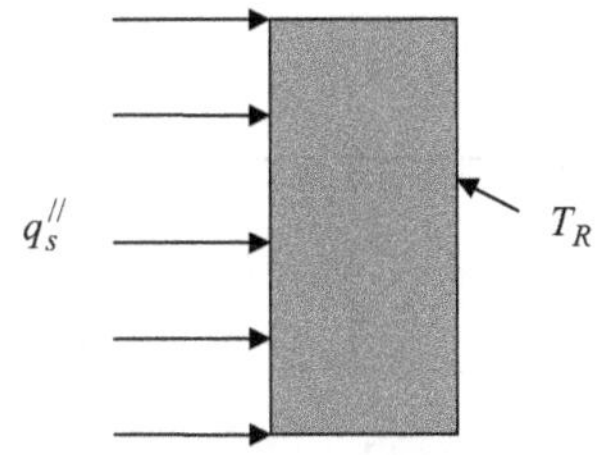

The slab is carbon steel with $\alpha = 0.835$ cm²/s. Use fully implicit scheme to obtain the global system of equations with six linear line elements and a time step of $\Delta t = 0.2$ s. Show computations for two-time steps.

15.7 Develop a computer program using a fully implicit scheme and Gauss elimination solver to study the transient temperature distribution and rate of heat transfer for Problem 15.3. Presents results in terms of temperature distribution and heat transfer rates at left and right surfaces with a time step of $\Delta t = 0.2$ s up to 5 s. Perform a mesh refinement study by using the number of elements as 10, 15, and 20. Presents results for the variation of heat transfer rate at $x = L$ with increase in time for different number of elements.

15.8 Redo Problem 7.4 based on developing a computer code that uses a matrix inversion procedure with Gauss-Jordon solver.

15.9 Let us consider a two-dimensional unsteady-state problem with the bilinear rectangular mesh distribution system and boundary conditions as shown in the figure.

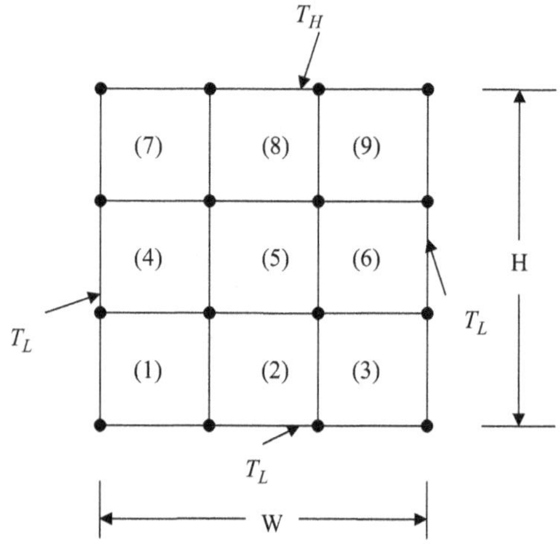

Derive the system of equations using fully implicit scheme and using following data: $W = 8$ cm, $H = 8$ cm, $T_I = 25°C$, $T_H = 400°C$, $T_L = 20°C$. Solve the system for two-time steps with $\Delta t = 0.2$ s.**15.10** A two-dimensional 8 cm × 8 cm square slab, initially at a uniform of 300°C, is suddenly exposed to a convection environment at a temperature of 20°C and $h = 40\,\text{W/m}^2 \cdot \text{C}$ all around. Consider two-dimensional unsteady-state conduction in the slab and determine temperature distribution and rate of heat loss as a function of time.

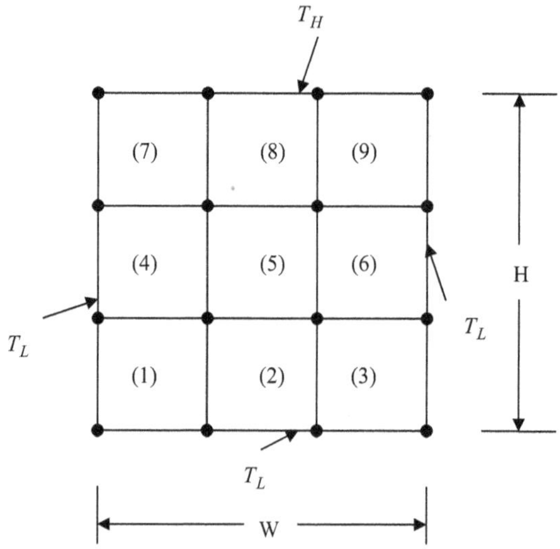

a. Use the 3×3 mesh and derive the system of equations for grid temperatures using fully implicit scheme.
b. Select an appropriate Δt and calculate grid points temperature for five-time increments. Material thermo-physical properties are $\rho = 7900\,\text{kg/m}^3$, $c = 477\,\text{J/kg·K}$, and $k = 14.9\,\text{W/m·K}$.

15.11 Repeat Problem 15.2 using Crank-Nicolson scheme.

15.12 Develop a two-dimensional finite element program using fully implicit time approximation scheme and bilinear rectangular elements, and using input data given for Problem 15.9. The computer program should incorporate the following features: (a) construct element capacitance and matrix $\lfloor c^{(e)} \rfloor$, the element characteristic matrix $\lfloor k^{(e)} \rfloor$, and element force vector $\{f^{(e)}\}$ based on the bilinear rectangular element; (b) assemble the global system of equations following the pseudo code given in Table 14.2; and (c) solve the global system using Gauss elimination solver. Redo Problem 15.9 with refined grid size distributions as 4×4, 6×6, and 8×8 using this code.

Present temperature distribution along with percent relative error for the vertical center line with refined grid size distributions at different time steps.

15.13 Use finite element method to solve two-dimensional unsteady state conduction in a rectangular carbon steel ($k = 200$ W/m·°C) slab subjected to a constant surface heat flux irradiated by a continuous high-energy laser beam at the top surface.

Assume the heat flux distribution to be a constant average value, $q_s^{\prime\prime} = 2 \times 10^8\ \text{W/cm}^2$, acting over a section of the surface. The top surface is also subjected to forced convection with $h_c = 500$ W/m²·°C. The left surface is assumed as adiabatic as a line of symmetry. All other surfaces are assumed to be subjected to free convection with $T_\infty = 25°\text{C}$ and $h = 40 \dfrac{\text{W}}{\text{m}^2\text{C}}$.

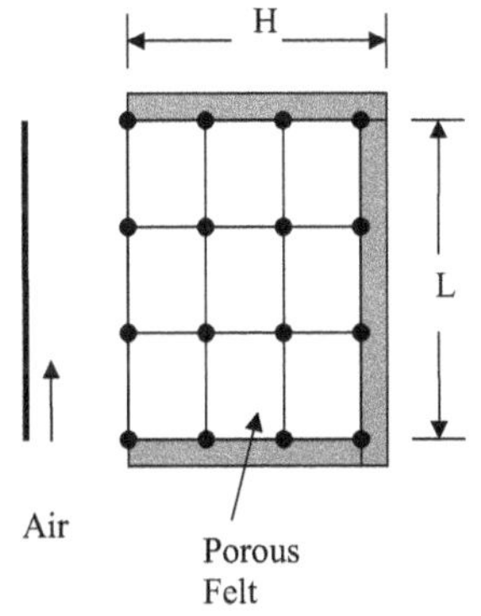

a. Derive the element characteristic equations for all elements using bilinear rectangular elements and fully implicit scheme. Assemble all element equations to derive the global system in matrix form.
b. Solve the system of equations using Gauss elimination solver and present results for temperature distribution as a function of time.

15.15 Use bilinear rectangular element to derive the element characteristic equation for the two-dimensional axisymmetric unsteady state problem with heat generation.

15.16 Use trilinear eight-nodded hexahedron element to derive the three-dimensional element characteristic equation staring from Equations 15.48 to 15.49.

16 Finite Element Method
Convection Problems

As we have mentioned in Chapter 8 that the convection problems involve distribution and dissipation of scalar transport quantities such as heat and mass species by molecular diffusion as well as by bulk motion of the fluid. Since the convection problems are primarily influenced by the associated flow field, the solution of convection problems starts with the solution of flow field governed by the conservation laws of mass and momentum equations such as the **Navier-Stokes Equations**, followed by the solution of energy or mass concentration equations. These governing equations constitute a set of coupled nonlinear partial differential equations for three velocity components, pressure, temperature, and/or mass concentration. Major computational difficulties arise due to the presence of the nonlinear convective terms as well as for pressure term for which there is no distinct equation. In many problems, the inertial forces or the convective terms are significantly small compared to viscous and pressure force terms as in the case of large Reynolds number laminar flow problems, and hence, the nonlinear convective terms are dropped as an approximation giving rise to so-called **Stokes equations**. Solution of the **Stokes equations** is computationally similar to the solution of diffusion equations as discussed in the previous chapters. In another class of problems, simplification is achieved due to a constant velocity of the fluid body, making the problem as linear. However, solution of this class of problem also exhibits considerable difficulties when the convective term becomes significantly greater in magnitude compared to the diffusion term as in large **Peclet number** flows. A discussion of this class of problems was given in Chapter 8 while discussing finite difference methods for nonlinear convection problems involving Navier-Stokes equations and scalar transport equations. We will start our discussion of finite element method for the solution of convection problems by first considering the linear convection problems.

Also, as we have discussed before that there are two major approaches for solving the flow field: (a) direct solution method involving the primitive variable and (b) stream function-/vorticity-based method. The stream function-/vorticity-based method is computationally less convenient as discussed in Chapter 8. The direct solution method is increasingly becoming more popular and is the method of choice for most commercial finite element codes. We will concentrate our discussion of finite element methods based on the direct solution method involving the primitive variables

16.1 CLASSIFICATION OF FINITE ELEMENT METHODS FOR CONVECTION PROBLEMS

There are primarily two finite element formulation methods for convections problems, namely, **velocity-pressure formulation or mixed formulation** and **penalty finite element formulation.** *The* **velocity-pressure or mixed formulation** starts with the weak form of the governing equations retaining both velocity and pressure variables. The continuity equation leads to the formation of a pressure equation. In the **penalty formulation**, the pressure term is effectively eliminated by satisfying the continuity equation in a least square sense. More discussions on penalty method are given in the books by Reddy (1993), Reddy and Gartling (1994), and Bathe (1996). In this book, we restrict our discussion to only **velocity-pressure formulation or mixed formulation**.

16.2 VELOCITY-PRESSURE OR MIXED FORMULATION

This is a direct formulation in terms of the primitive variables velocity and pressure given by the original mass and momentum Equations. The procedure for developing the finite element model is similar to the basic steps outlined for the diffusion equation presented in the previous chapters with the exception of the additional convective terms. In order to keep the derivation simple, we will first demonstrate the procedure by first considering one- and two-dimensional linear convections problems. We will subsequently consider two-dimensional steady-state nonlinear flow problems. This will be followed by inclusion of the unsteady-state term, and considerations of scalar transport Equations for convective heat and mass transfer problems.

16.2.1 One-Dimensional Convection-Diffusion Problem

Let us consider a one-dimensional steady-state problem involving convection and diffusion terms as below

$$\frac{\mathrm{d}}{\mathrm{d}x}(Cu\phi)=\frac{\mathrm{d}}{\mathrm{d}x}\left(\Gamma_x\frac{\mathrm{d}\phi}{\mathrm{d}x}\right)+S \tag{16.1}$$

In this problem, it is assumed that the value of the convective velocity u is known as constant value. Let us demonstrate the application of finite element method to this problem through brief description of several main steps that constitute the finite element method.

Step – 1 Discretization of the Solution Domain

The solution domain is discretized into a mesh size distribution using one-dimensional line elements as shown in Figure 16.1.

Step – 2 Formation of Integral Statement of the Problem

The Galerkin's-based weighted integral statement of the Equation 16.1 over an element is given as

$$R_i^{(e)}=\int_{x_i}^{x_j}[N]^T\left(\frac{\partial}{\partial x}\left(\Gamma_x\frac{\partial\phi^*}{\partial x}\right)+Cu\frac{\partial\phi}{\partial x}+S\right)\mathrm{d}x=0 \tag{16.2}$$

Evaluating the first integral, we get the integral weak statement of the problem as

$$\left\{\begin{matrix}-\Gamma_x\left.\frac{\mathrm{d}\phi}{\mathrm{d}x}\right|_{x_i}\\ \Gamma_x\left.\frac{\mathrm{d}\phi}{\mathrm{d}x}\right|_{x_j}\end{matrix}\right\}-\int_{x_i}^{x_j}\Gamma_x\frac{\mathrm{d}[N]^T}{\mathrm{d}x}\frac{\mathrm{d}\phi^*}{\mathrm{d}x}\mathrm{d}x+\int_{x_i}^{x_j}[N]^T Cu\frac{\mathrm{d}\phi^*}{\mathrm{d}x}\mathrm{d}x+\int_{x_i}^{x_j}S[N]^T\mathrm{d}x=0 \tag{16.3}$$

Step – 3 Approximation Solution Function

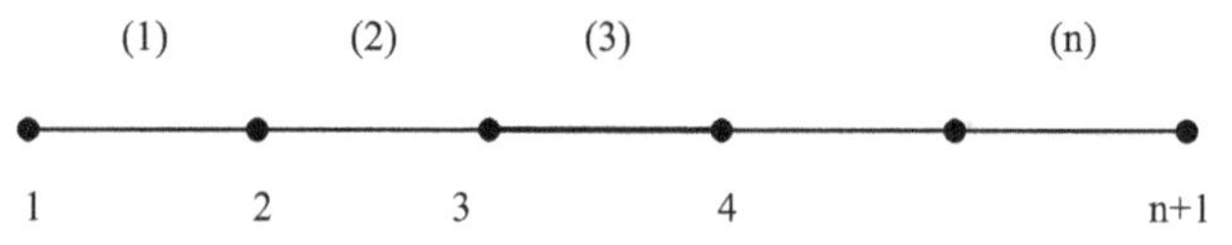

FIGURE 16.1 Mesh size distributions.

Let us consider the elements as linear two-point line elements with a typical element shown in Figure 16.1. The approximate solution function, $\phi \approx \phi^*$, over this element is given as

$$\phi^* = \sum N_I \phi_I^{(e)} = [N]\{\phi^{(e)}\} \tag{16.4}$$

Step-4 Formation of Element Characteristics Equation

Substituting the approximate solution function given by Equation 16.4 for one-dimensional space co-ordinate and evaluating the integrals for one-dimensional two-points line element, we get the **element characteristic equation** as

$$\begin{Bmatrix} -\Gamma_x \left.\dfrac{d\phi}{dx}\right|_{x_i} \\ \Gamma_x \left.\dfrac{d\phi}{dx}\right|_{x_j} \end{Bmatrix} - \int_{x_i}^{x_j} \Gamma_x \frac{d[N]^T}{dx}\frac{d[N]}{dx}\{\phi^{(e)}\}dx + \int_{x_i}^{x_j} [N]^T Cu \frac{d[N]}{dx}\{\phi^{(e)}\}dx + \int_{x_i}^{x_j} S[N]^T dx = 0 \tag{16.5}$$

or

$$\{I^{(e)}\} - [k^{(e)}]\{\phi^{(e)}\} + \{f^{(e)}\} = 0 \tag{16.6}$$

where

$$[k] = [k_\Gamma] + [k_I] = \text{Element characteristic matrix}$$

$$= \frac{\Gamma_x}{L_e}\begin{bmatrix} 1 & -1 \\ -1 & 1 \end{bmatrix} + \frac{Cu}{2}\begin{bmatrix} -1 & 1 \\ -1 & 1 \end{bmatrix} \tag{16.7a}$$

$$[k_\Gamma] = \int_{x_i}^{x_j} \Gamma_x \frac{d[N]^T}{dx}\frac{d[N]}{dx}dx = \text{Element characteristic matrix due to the diffusion term}$$

$$= \frac{\Gamma_x}{L_e}\begin{bmatrix} 1 & -1 \\ -1 & 1 \end{bmatrix} \tag{16.7b}$$

$$[k_I] = \int_{x_i}^{x_j} Cu[N]^T \frac{d[N]}{dx}dx = \text{Element characteristic matrix due to the inertia term}$$

$$= \frac{Cu}{2}\begin{bmatrix} -1 & 1 \\ -1 & 1 \end{bmatrix} \tag{16.7c}$$

$$\{f^{(e)}\} = \int_{x_i}^{x_j} [N]^T (S)dx = \text{element force vector}$$

$$= \begin{Bmatrix} \dfrac{SL_e}{2} \\ \dfrac{SL_e}{2} \end{Bmatrix} \tag{16.7d}$$

and

$$\left\{I^{(e)}\right\} = \begin{Bmatrix} -\Gamma_x \left.\dfrac{d\phi}{dx}\right|_{x_i} \\ \Gamma_x \left.\dfrac{d\phi}{dx}\right|_{x_j} \end{Bmatrix} = \text{inter-element vector} \tag{16.7e}$$

Step – 5 Assembly of Element Equations to Form the Global System

Equation 16.6 is applied repeatedly to all elements to obtain corresponding element equations and are assembled through use of the connectivity scheme to form a global system of equations

$$[K]\{\Phi\} = \{F\} \tag{16.8a}$$

or

$$\begin{bmatrix} \dfrac{\Gamma_x}{L_e}-\dfrac{Cu}{2} & -\dfrac{\Gamma_x}{L_e}+\dfrac{Cu}{2} & & & & \\ -\dfrac{\Gamma_x}{L_e}-\dfrac{Cu}{2} & \dfrac{2\Gamma_x}{L_e} & -\dfrac{\Gamma_x}{L_e}+\dfrac{Cu}{2} & & & \\ & -\dfrac{\Gamma_x}{L_e}-\dfrac{Cu}{2} & \dfrac{2\Gamma_x}{L_e} & -\dfrac{\Gamma_x}{L_e}+\dfrac{Cu}{2} & & \\ & & -\dfrac{\Gamma_x}{L_e}-\dfrac{Cu}{2} & \dfrac{2\Gamma_x}{L_e} & & \\ & & & & \dfrac{2\Gamma_x}{L_e} & -\dfrac{\Gamma_x}{L_e}+\dfrac{Cu}{2} \\ & & & & -\dfrac{\Gamma_x}{L_e}-\dfrac{Cu}{2} & \dfrac{\Gamma_x}{L_e}+\dfrac{Cu}{2} \end{bmatrix} \begin{Bmatrix} \phi_1 \\ \phi_2 \\ \phi_3 \\ \vdots \\ \vdots \\ \phi_{n-1} \\ \phi_n \end{Bmatrix}$$

$$= \begin{Bmatrix} -\Gamma_x \left.\dfrac{d\phi}{dx}\right|_{x_1} \\ 0 \\ 0 \\ \vdots \\ \vdots \\ 0 \\ \Gamma_x \left.\dfrac{d\phi}{dx}\right|_{x_{n+1}} \end{Bmatrix} + \begin{Bmatrix} \dfrac{SL_e}{2} \\ SL_e \\ SL_e \\ \vdots \\ \vdots \\ SL_e \\ \dfrac{SL_e}{2} \end{Bmatrix} \tag{16.8b}$$

Step – 6 Implementation of Boundary Conditions

In this step, the boundary conditions are introduced directly into the global system of equations. For boundary conditions of the *first kind*, i.e., of constant surface value, we simply assign the specified values to the boundary nodes as $\phi_1 = \phi_l$ and $\phi_{n+1} = \phi_r$, and rearranged the system so that the natural boundary conditions $\left.\frac{d\phi}{dx}\right|_{x_1}$ and $\left.\frac{d\phi}{dx}\right|_{x_{n+1}}$ appear as one of the unknowns as follows:

$$[\mathrm{K}]\{\Phi\} = \{F\}$$

or

$$\begin{pmatrix} \frac{\Gamma_x}{L_e} - \frac{Cu}{2} & -\frac{\Gamma_x}{L_e} + \frac{Cu}{2} & & & & \\ -\frac{\Gamma_x}{L_e} - \frac{Cu}{2} & \frac{2\Gamma_x}{L_e} & -\frac{\Gamma_x}{L_e} + \frac{Cu}{2} & & & \\ & -\frac{\Gamma_x}{L_e} - \frac{Cu}{2} & \frac{2\Gamma_x}{L_e} & -\frac{\Gamma_x}{L_e} + \frac{Cu}{2} & & \\ & & -\frac{\Gamma_x}{L_e} - \frac{Cu}{2} & \frac{\Gamma_x}{L_e} + \frac{\Gamma_x}{L_e} & & \\ \vdots & & & & & \\ & & & & \frac{2\Gamma_x}{L_e} & -\frac{\Gamma_x}{L_e} + \frac{Cu}{2} \\ & & & & -\frac{\Gamma_x}{L_e} - \frac{Cu}{2} & \frac{\Gamma_x}{L_e} + \frac{Cu}{2} \end{pmatrix} \left\{ \begin{bmatrix} \left.\frac{d\phi}{dx}\right|_{x_1} \\ \phi_2 \\ \phi_3 \\ \vdots \\ \vdots \\ \phi_n \\ \left.\frac{d\phi}{dx}\right|_{x_{n+1}} \end{bmatrix} \right\}$$

$$= \left\{ \begin{bmatrix} -\frac{\Gamma_x \phi_l}{L_e} + \frac{\bar{S}L_e}{2} \\ \bar{S}L_e \\ \bar{S}L_e \\ \vdots \\ \vdots \\ \frac{\Gamma_x \phi_r}{L_e} + \bar{S}L_e \\ -\frac{\Gamma_x \phi_r}{L_e} + \frac{\bar{S}L_e}{2} \end{bmatrix} \right\} \tag{16.9}$$

Step – 7 Solution of the Global System of Equations

Before deciding an appropriate solver for solving the global system of equations, let us examine the nature of the element characteristic matrix *K*. It can be noticed that due to the presence of the convective term, the characteristic matrix is nonsymmetric. Also, as the inertia term becomes greater in magnitude compared to the diffusion term, i.e., $\frac{Cu}{2} > \frac{\Gamma_x}{L_e}$ or in dimensionless form, $\frac{uL_e}{\Gamma / C} > 2$, the characteristic matrix no longer remains diagonally dominant. So, the use of an iterative solver may experience instability and may not lead to a converged solution. In heat transfer problem, this dimensionless number is named as Peclet number, $\left(\text{Pe} = \frac{uL_e}{\alpha} \right)$, and the criterion for a diagonally dominant characteristic matrix is given as $\text{Pe} \leq 2$. Severity of this difficulty caused by the convection terms grows as local element Peclet number becomes greater and greater.

This difficulty is also experienced in the use of central difference scheme in **finite difference/control volume methods** as discussed in Chapter 8. As we have discussed in Chapter 8, this difficulty was resolved through use of alternate schemes such as **upwind scheme**, **exponential scheme**, **hybrid scheme,** and **power law scheme**. Similar concepts are also utilized in developing finite element techniques as an alternate to Galerkin finite element methods. The alternate finite element techniques that have shown improved accuracy in one-dimensional problems and increasingly being experimented in two- and three-dimensional fluid flow and heat transfer are **Petrov-Galerkin Method** and **Galerkin Least Squares Method**. More detailed discussions of these techniques are given by Bathe (1996) and Pepper and Heinrich (1992).

16.2.2 Two-Dimensional Viscous Incompressible Flow

The governing equations for two-dimensional Newtonian viscous incompressible flow over a flow domain A are given as

Continuity

$$\frac{\partial u}{\partial x} + \frac{\partial v}{\partial y} = 0 \tag{16.10a}$$

***x*-Momentum**

$$\rho\left(u\frac{\partial u}{\partial x} + v\frac{\partial u}{\partial y} \right) = b_x - \frac{\partial p}{\partial x} + \mu\left(\frac{\partial^2 u}{\partial x^2} + \frac{\partial^2 u}{\partial y^2} \right) \tag{16.10b}$$

***y*-Momentum**

$$\rho\left(u\frac{\partial v}{\partial x} + v\frac{\partial v}{\partial y} \right) = b_y - \frac{\partial p}{\partial y} + \mu\left(\frac{\partial^2 v}{\partial x^2} + \frac{\partial^2 v}{\partial y^2} \right) \tag{16.10c}$$

With boundary conditions given in a general form as

$$1.\, u = u_s, \quad \text{and} \quad v = v_s \quad \text{on the boundary} \quad s_1 \tag{16.11a}$$

$$\begin{aligned} 2.\quad & \mu\left(\frac{\partial u}{\partial x}\hat{n}_x + \frac{\partial u}{\partial y}\hat{n}_y \right) - p\hat{n}_x = f_{sx} \\ & \mu\left(\frac{\partial v}{\partial x}\hat{n}_x + \frac{\partial v}{\partial y}\hat{n}_y \right) - p\hat{n}_y = f_{sy} \text{ on the boundary } s_2 \end{aligned} \tag{16.11b}$$

Notice that the boundary surface s is divided into two sections s_1 and s_2. Boundary condition of the first kind, i.e., of constant surface value, is specified on the boundary surface s_1. Boundary condition of the second kind, i.e., constant stress, is specified on the boundary surface s_2. Notice that this constant stress boundary condition given by Equation 8.2b is a general boundary condition for the total stress that includes the normal hydrostatic pressure and the viscous boundary shear stress on the boundary.

Let us now demonstrate the finite element formulation following the basic steps outlined in Chapters 13–15.

Step – 1 Discretization of the Solution Domain

As a first step, the solution domain is discretized into a finite element mesh following the procedure outlined for the diffusion problems involving only a single variable as in Chapter 14. However, solution of the flow field involves two different kinds of variables such as velocity, $\vec{V}$, and pressure, p. This requires the need for two different finite element meshes: one for the velocity and other for the pressure. In general the mesh for the pressure is a subset of the primary mesh for the velocity, and so, only one mesh is displayed along with the nodal degrees of freedom associated with the nodes of the element.

Step – 2 Formulation of an Integral or Weak Form Using Galerkin's Method

In order to obtain the weak forms of the governing set of equations using weighted residual method, we substitute the approximate solutions in the governing equations. The weighted residual method involves finding a minimum for the residuals over an element $A^{(e)}$ by multiplying the residuals of Equations 8.1a–8.1c by the weighting functions W_p, W_u, and W_v, and integrating over the element in the following manner:

$$\left\{R_P^{(e)}\right\} = -\int_{A^{(e)}} W_p \left\{\frac{\partial u^*}{\partial x} + \frac{\partial v^*}{\partial y}\right\} dA = 0 \tag{16.12a}$$

$$\left\{R_u^{(e)}\right\} = \int_{A^{(e)}} W_u \left\{\rho\left(u^*\frac{\partial u^*}{\partial x} + v\frac{\partial u^*}{\partial y}\right) - b_x + \frac{\partial p^*}{\partial x} - \mu\left(\frac{\partial^2 u^*}{\partial x^2} + \frac{\partial^2 u^*}{\partial y^2}\right)\right\} dA = 0 \tag{16.12b}$$

$$\left\{R_v^{(e)}\right\} = \int_{A^{(e)}} W_v \left\{\rho\left(u\frac{\partial v^*}{\partial x} + v\frac{\partial v^*}{\partial y}\right) - b_y + \frac{\partial p^*}{\partial y} - \mu\left(\frac{\partial^2 v^*}{\partial x^2} + \frac{\partial^2 v^*}{\partial y^2}\right)\right\} dA = 0 \tag{16.12c}$$

In the **Galerkin finite element method** the weighting functions are assumed to be same as the shape functions used for the approximation functions given by Equation 16.4, i.e.,

$$W_u = [N]^T, \quad W_v = [N]^T \quad \text{and} \quad W_p = [N_P]^T \tag{16.13}$$

where $[N]$ and $[N_p]$ are the shape functions associated with velocity and pressure approximation functions, respectively.

The integral equations are transformed into a weak form by evaluating the integral in the following manner:

***x*–Momentum**

$$\int_{A^{(e)}} [N^T]\rho\left(u^*\frac{\partial u^*}{\partial x} + v\frac{\partial u^*}{\partial y}\right) dA - \int_{A^{(e)}} [N]^T b_x dA + \int_{A^{(e)}} [N]^T \frac{\partial p^*}{\partial x} dA$$

$$-\int_{A^{(e)}} [N]^T \mu \frac{\partial^2 u^*}{\partial x^2} dA + \int_{A^{(e)}} [N]^T \mu \frac{\partial^2 u^*}{\partial y^2} dA = 0$$

or

$$\int_{A^{(e)}} [N^T]\rho\left(u^* \frac{\partial u^*}{\partial x} + v\frac{\partial u^*}{\partial y}\right) dA - \int_{A^{(e)}} [N]^T b_x dA$$
$$+ \int_{A^{(e)}} \frac{\partial}{\partial x}\left([N]^T p^*\right) dA - \int_{A^{(e)}} p^* \frac{\partial [N]^T}{\partial x} dA$$
$$- \int_{A^{(e)}} \frac{\partial}{\partial x}\left(\mu [N]^T \frac{\partial u^*}{\partial x}\right) dA + \int_{A^{(e)}} \mu \frac{\partial u^*}{\partial x}\frac{\partial [N]^T}{\partial x} dA$$
$$- \int_{A^{(e)}} \frac{\partial}{\partial y}\left(\mu [N]^T \frac{\partial u^*}{\partial y}\right) dA + \int_{A^{(e)}} \mu \frac{\partial u^*}{\partial y}\frac{\partial [N]^T}{\partial y} dA = 0$$

or

$$\int_{A^{(e)}} [N^T]\rho\left(u^* \frac{\partial u^*}{\partial x} + v\frac{\partial u^*}{\partial y}\right) dA - \int_{A^{(e)}} [N]^T b_x dA$$
$$+ \oint_s [N]^T p^* \hat{n}_x ds - \int_{A^{(e)}} p^* \frac{\partial [N]^T}{\partial x} dA$$
$$- \oint_s \mu [N]^T \frac{\partial u^*}{\partial x} \hat{n}_x ds + \int_{A^{(e)}} \mu \frac{\partial u^*}{\partial x}\frac{\partial [N]^T}{\partial x} dA$$
$$- \oint_s \mu [N]^T \frac{\partial u^*}{\partial y} \hat{n}_y ds + \int_{A^{(e)}} \mu \frac{\partial u^*}{\partial y}\frac{\partial [N]^T}{\partial y} dA = 0$$

or

$$\int_{A^{(e)}} [N^T]\rho\left(u^* \frac{\partial u^*}{\partial x} + v\frac{\partial u^*}{\partial y}\right) dA - \int_{A^{(e)}} [N]^T b_x dA - \int_{A^{(e)}} p^* \frac{\partial [N]^T}{\partial x} dA$$
$$+ \int_{A^{(e)}} \mu \frac{\partial u^*}{\partial x}\frac{\partial [N]^T}{\partial x} dA + \int_{A^{(e)}} \mu \frac{\partial u^*}{\partial y}\frac{\partial [N]^T}{\partial y} dA \tag{16.14}$$
$$+ \oint_s [N]^T \left(p^* n_x - \mu\left(\frac{\partial u^*}{\partial x}\hat{n}_x + \frac{\partial u^*}{\partial y}\hat{n}_y\right)\right) ds = 0$$

Similarly, the y-momentum can be transformed in a weak form as

$$\int_{A^{(e)}} [N^T]\rho\left(u^*\frac{\partial v^*}{\partial x}+v\frac{\partial v^*}{\partial y}\right)\mathrm{d}A-\int_{A^{(e)}}[N]^T b_y \mathrm{d}A-\int_{A^{(e)}} p^*\frac{\partial[N]^T}{\partial y}\mathrm{d}A$$

$$+\int_{A^{(e)}}\mu\frac{\partial v^*}{\partial x}\frac{\partial[N]^T}{\partial x}\mathrm{d}A+\int_{A^{(e)}}\mu\frac{\partial v^*}{\partial y}\frac{\partial[N]^T}{\partial y}\mathrm{d}A \tag{16.15}$$

$$+\oint_s [N]^T\left(p^* n_y-\mu\left(\frac{\partial v^*}{\partial x}\hat{n}_x+\frac{\partial v^*}{\partial y}\hat{n}_y\right)\right)\mathrm{d}s=0$$

Step – 3 Selection of Approximate Solution Function

The approximate solutions for velocity and pressure fields can be assumed as follows:

$$u^*=\sum_{I=1}^{n_p} N_I u_I=[N]\{u^{(e)}\} \tag{16.16a}$$

$$v^*=\sum_{I=1}^{n_p} N_I v_I=[N]\{v^{(e)}\} \tag{16.16b}$$

$$p^*=\sum_{I=1}^{n_p} N_{pI} p_I=[N_p]\{p^{(e)}\} \tag{16.16c}$$

Note that that two different shape functions or interpolation functions are used for velocity and pressure fields. This is due to the fact that the order of the derivative for pressure in the governing equations as well as in its weak form is one order less than that of the velocity components. In general the order of the shape function N_p used for pressure is one order less than the order for the shape function N for the velocity components. Additionally, the pressure need not be approximated as continuous function across interelement boundaries. So, we could use quadratic shape function for the velocity components and discontinuous linear shape function for the pressure. For example, we could use linear triangular element with pressure calculated at three nodes located at the vertices of the triangle and use quadratic triangular element with velocity components calculated at six nodes of the triangle as shown in Figure 16.2a.

Figure 16.2 shows quadratic rectangular element for velocity and bilinear rectangular element for pressure. Figure 16.3 shows that the velocity components are calculated at three nodes in a linear triangular element whereas pressure is calculated at a node located at the center of the triangle approximated by a constant discontinuous shape function. The pressure is assumed as constant over the entire element.

So, in essence, there are two different finite element meshes: one for the velocity components and other for the pressure. However, since the mesh for the pressure appears as a subset of the general mesh for the velocity, a common practice is to display the general mesh and indicate the total degrees of freedom at each node.

Step – 4 Formation of Element Characteristic Equation

The element characteristic equation can be derived by substituting the approximate solution functions given by Equation 16.2 in the integral form of the governing Equations 16.12a, 16.14, and 16.15. Let us demonstrate this by considering the x-momentum Equation 16.14 as follows:

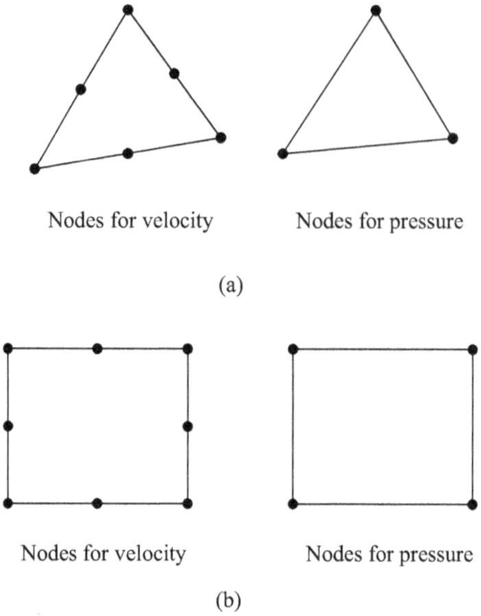

FIGURE 16.2 Different element types for velocity and pressure: (a) quadratic triangular element for velocity and linear triangular element for pressure; (b) quadratic rectangular element for velocity and bilinear rectangular element for pressure.

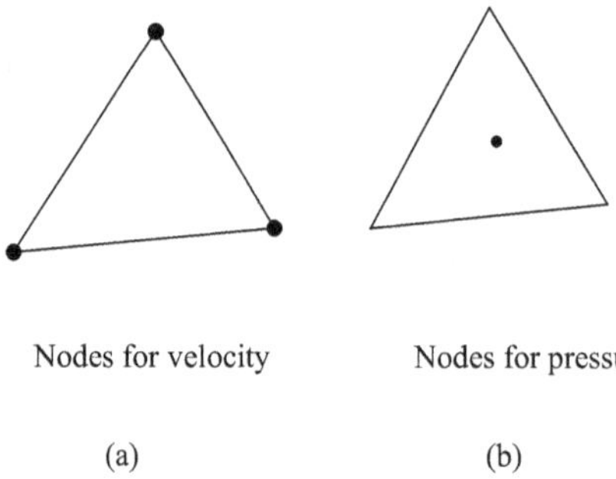

FIGURE 16.3 Different element types for velocity and pressure: (a) three-noded linear triangular element for velocity and (b) discontinuous-constant pressure at the center of the triangle.

$$\int_{A^{(e)}} [N^T]\rho\left([N]\{u^{(e)}\}\frac{\partial[N]}{\partial x}\{u^{(e)}\}+[N]\{v^{(e)}\}\frac{\partial[N]}{\partial y}\{u^{(e)}\}\right)\mathrm{d}A-\int_{A^{(e)}}[N]^T b_x \mathrm{d}A-$$

$$\int_{A^{(e)}}[N_p]\{p^{(e)}\}\frac{\partial[N]^T}{\partial x}\mathrm{d}A+\int_{A^{(e)}}\mu\frac{\partial[N]}{\partial x}\{u^{(e)}\}\frac{\partial[N]^T}{\partial x}\mathrm{d}A+\int_{A^{(e)}}\mu\frac{\partial[N]}{\partial y}\{u^{(e)}\}\frac{\partial[N]^T}{\partial y}\mathrm{d}A$$

$$+\oint_s [N]^T\left(p^* n_x-\mu\left(\frac{\partial u^*}{\partial x}\hat{n}_x+\frac{\partial u^*}{\partial y}\hat{n}_y\right)\right)\mathrm{d}s=0$$

or

$$\left\{\int_{A^{(e)}} \rho[N^T][N]\left(\{u^{(e)}\}\frac{\partial[N]}{\partial x}+\{v^{(e)}\}\frac{\partial[N]}{\partial y}\right)\mathrm{d}A\right\}\{u^{(e)}\}$$

$$+\left[\left\{\int_{A^{(e)}} \mu\frac{\partial[N]}{\partial x}\frac{\partial[N]^T}{\partial x}\mathrm{d}A\right\}+\left\{\int_{A^{(e)}} \mu\frac{\partial[N]}{\partial y}\frac{\partial[N]^T}{\partial y}\mathrm{d}A\right\}\right]\{u^{(e)}\}-\left\{\int_{A^{(e)}}[N_p]\frac{\partial[N]^T}{\partial x}\mathrm{d}A\right\}\{p^{(e)}\} \quad (16.17a)$$

$$+\oint_s [N]^T\left(p^* n_x-\mu\left(\frac{\partial u^*}{\partial x}\hat{n}_x+\frac{\partial u^*}{\partial y}\hat{n}_y\right)\right)\mathrm{d}s-\int_{A^{(e)}}[N]^T b_x\,\mathrm{d}A=0$$

Writing the above equation in matrix notation, we get

$$\left[k_I^{(e)}(u,v)\right]\{u^{(e)}\}+\left[k_\mu^{(e)}\right]\{u^{(e)}\}-\left[k_{\mathrm{px}}^{(e)}\right]\{p^{(e)}\}+\{I_x^{(e)}\}-\{f_{\mathrm{bx}}^{(e)}\}=0 \quad (16.17b)$$

where

$$\left[k_I^{(e)}(u,v)\right]=\int_{A^{(e)}}\rho[N^T][N]\left(\{u^{(e)}\}\frac{\partial[N]}{\partial x}+\{v^{(e)}\}\frac{\partial[N]}{\partial y}\right)\mathrm{d}A \quad (16.18a)$$

= element characteristic matrix due to inertia forces

$$k_\mu^{(e)}=k_{xx}^{(e)}+k_{yy}^{(e)} = \text{element characteristic matrix due to viscous forces} \quad (16.18b)$$

$$k_{xx}^{(e)}=\int_{A^{(e)}}\mu\frac{\partial[N]}{\partial x}\frac{\partial[N]^T}{\partial x}\mathrm{d}A \quad (16.18c)$$

$$k_{yy}^{(e)}=\int_{A^{(e)}}\mu\frac{\partial[N]}{\partial y}\frac{\partial[N]^T}{\partial y}\mathrm{d}A \quad (16.18d)$$

$$\left[k_{\mathrm{px}}^{(e)}\right]=\int_{A^{(e)}}[N_p]\frac{\partial[N]^T}{\partial x}\mathrm{d}A = \text{element characteristic matrix due to the pressure force in } x\text{-direction}$$

(16.18e)

$\{f_{\mathrm{bx}}^{(e)}\}$ = element source vector due to the x-component body force term

$$=\int_{A^{(e)}}[N]^T b_x\mathrm{d}A \quad (16.18f)$$

$\{I_x^{(e)}\}$ = column vector for the element contribution to the inter-element requirement

$$=\oint_s [N]^T\left(p^* n_x-\mu\left(\frac{\partial u^*}{\partial x}\hat{n}_x+\frac{\partial u^*}{\partial y}\hat{n}_y\right)\right)\mathrm{d}s \quad (16.18g)$$

Similarly, the element characteristic equation corresponding to y-momentum is derived from Equation 16.15 as follows:

$$\int_{A^{(e)}} [N^T]\rho\left([N]\{u^{(e)}\}\frac{\partial[N]}{\partial x}\{v^{(e)}\}+[N]\{v^{(e)}\}\frac{\partial[N]}{\partial y}\{v^{(e)}\}\right)\mathrm{d}A-\int_{A^{(e)}}[N]^T b_y \mathrm{d}A$$

$$-\int_{A^{(e)}}[N_P]\{p^{(e)}\}\frac{\partial[N]^T}{\partial y}\mathrm{d}A+\int_{A^{(e)}}\mu\frac{\partial[N]\{v^{(e)}\}}{\partial x}\frac{\partial[N]^T}{\partial x}\mathrm{d}A+\int_{A^{(e)}}\mu\frac{\partial[N]\{v^{(e)}\}}{\partial x}\frac{\partial[N]^T}{\partial x}\mathrm{d}A$$

$$+\oint_s [N]^T\left(p^* n_y-\mu\left(\frac{\partial v^*}{\partial x}\hat{n}_x+\frac{\partial v^*}{\partial y}\hat{n}_y\right)\right)\mathrm{d}s=0$$

or

$$\left\{\int_{A^{(e)}}\rho[N^T][N]\left(\{u^{(e)}\}\frac{\partial[N]}{\partial x}+\{v^{(e)}\}\frac{\partial[N]}{\partial y}\right)\mathrm{d}A\right\}\{v^{(e)}\}$$

$$+\left\{\int_{A^{(e)}}\mu\frac{\partial[N]}{\partial x}\frac{\partial[N]^T}{\partial x}\mathrm{d}A+\int_{A^{(e)}}\mu\frac{\partial[N]}{\partial y}\frac{\partial[N]^T}{\partial y}\mathrm{d}A\right\}\{v^{(e)}\}-\left\{\int_{A^{(e)}}[N_p]\frac{\partial[N]^T}{\partial y}\mathrm{d}A\right\}\{p^{(e)}\} \quad (16.19a)$$

$$+\oint_s [N]^T\left(p^* n_y-\mu\left(\frac{\partial v^*}{\partial x}\hat{n}_x+\frac{\partial v^*}{\partial y}\hat{n}_y\right)\right)\mathrm{d}s-\int_{A^{(e)}}[N]^T b_y\,\mathrm{d}A=0$$

In matrix notation, we get

$$\left[k_I^{(e)}(u,v)\right]\{v^{(e)}\}+\left[k_\mu^{(e)}\right]\{v^{(e)}\}-\left[k_{\mathrm{py}}^{(e)}\right]\{p^{(e)}\}+\{I_y^{(e)}\}-\{f_{\mathrm{by}}^{(e)}\}=0 \quad (16.19b)$$

where

$$\left[k_I^{(e)}(u,v)\right]=\int_{A^{(e)}}\rho[N^T][N]\left(\{u^{(e)}\}\frac{\partial[N]}{\partial x}+\{v^{(e)}\}\frac{\partial[N]}{\partial y}\right)\mathrm{d}A \quad (16.20a)$$

$$k_\mu^{(e)}=k_{xx}^{(e)}+k_{yy}^{(e)} \quad (16.20b)$$

$$k_{xx}^{(e)}=\int_{A^{(e)}}\mu\frac{\partial[N]}{\partial x}\frac{\partial[N]^T}{\partial x}\mathrm{d}A \quad (16.20c)$$

$$k_{yy}^{(e)}=\int_{A^{(e)}}\mu\frac{\partial[N]}{\partial y}\frac{\partial[N]^T}{\partial y}dA \quad (16.20d)$$

$\left[k_{\mathrm{py}}\right]=\int_{A^{(e)}}[N_p]\frac{\partial[N]^T}{\partial y}\mathrm{d}A$ = element characteristic matrix due to the pressure force in y-direction

(16.20e)

$$\left\{f_{by}^{(e)}\right\} = \text{element source vector due to the body force term}$$

$$= \int_{A^{(e)}} [N]^T b_y \mathrm{d}A \tag{16.20f}$$

$$\left\{I_y^{(e)}\right\} = \text{column vector for the element contribution to the inter-element requirement}$$

$$= \oint_s [N]^T \left(p^* n_y - \mu \left(\frac{\partial u^*}{\partial x} \hat{n}_x + \frac{\partial u^*}{\partial y} \hat{n}_y \right) \right) \mathrm{d}s \tag{16.20g}$$

In a similar manner, we can reduce the integral form of the continuity equation given by Equation 16.12a to the following element characteristic equation in the following manner:

$$-\int_{A^{(e)}} [N_p]^T \left\{ \frac{\partial [N]}{\partial x} \left\{u^{(e)}\right\} + \frac{\partial [N]}{\partial y} \left\{v^{(e)}\right\} \right\} \mathrm{d}A = 0$$

or

$$-\left\{ \int_{A^{(e)}} [N_p]^T \frac{\partial [N]}{\partial x} \mathrm{d}A \right\} \left\{u^{(e)}\right\} - \left\{ \int_{A^{(e)}} [N_p]^T \frac{\partial [N]}{\partial y} \mathrm{d}A \right\} \left\{v^{(e)}\right\} = 0 \tag{16.21a}$$

In the matrix notation, the above equation is written as

$$-\left[k_{px}^{(e)}\right]^T \left\{u^{(e)}\right\} - \left[k_{px}^{(e)}\right]^T \left\{v^{(e)}\right\} = 0 \tag{16.21b}$$

Finally, the set of element characteristic equations for u, v, and p are assembled by collecting Equations 16.17, 16.19, and 16.21 as follows:

$$\begin{aligned} &\left[k_I^{(e)}(u,v)\right]\left\{u^{(e)}\right\} + \left[k_\mu^{(e)}\right]\left\{u^{(e)}\right\} - \left[k_{px}^{(e)}\right]\left\{p^{(e)}\right\} + \left\{I_x^{(e)}\right\} - \left\{f_{bx}^{(e)}\right\} = 0 \quad \text{(a)} \\ &\left[k_I^{(e)}(u,v)\right]\left\{v^{(e)}\right\} + \left[k_\mu^{(e)}\right]\left\{v^{(e)}\right\} - \left[k_{py}^{(e)}\right]\left\{p^{(e)}\right\} + \left\{I_y^{(e)}\right\} - \left\{f_{by}^{(e)}\right\} = 0 \quad \text{(b)} \\ &-\left[k_{px}^{(e)}\right]^T \left\{u^{(e)}\right\} - \left[k_{py}^{(e))}\right]^T \left\{v^{(e)}\right\} \text{(c)} \end{aligned} \tag{16.22}$$

In the matrix form, the element characteristic equations are written as

$$\begin{bmatrix} k_I^{(e)}(u,v) & 0 & 0 \\ 0 & k_I^{(e)}(u,v) & 0 \\ 0 & 0 & 0 \end{bmatrix} \begin{Bmatrix} \left\{u^{(e)}\right\} \\ \left\{v^{(e)}\right\} \\ \left\{p^{(e)}\right\} \end{Bmatrix} + \begin{bmatrix} k_\mu^{(e)} & 0 & -k_{px}^{(e)} \\ 0 & k_\mu^{(e)} & -k_{yy}^{(e)} \\ -k_{px}^{(e)T} & -k_{py}^{(e)T} & 0 \end{bmatrix} \begin{Bmatrix} \left\{u^{(e)}\right\} \\ \left\{v^{(e)}\right\} \\ \left\{p^{(e)}\right\} \end{Bmatrix}$$

$$+ \begin{Bmatrix} \left\{I_x^{(e)}\right\} \\ \left\{I_y^{(e)}\right\} \\ 0 \end{Bmatrix} - \begin{Bmatrix} \left\{f_{bx}^{(e)}\right\} \\ f_{by}^{(e)} \\ 0 \end{Bmatrix} = 0 \tag{16.23a}$$

The above equation can be further simplified as

$$\begin{bmatrix} k_I^{(e)}(u,v)+k_\mu^{(e)} & 0 & -k_{px}^{(e)} \\ 0 & k_I^{(e)}(u,v)+k_\mu^{(e)} & -k_{py}^{(e)} \\ -k_{px}^{(e)T} & -k_{py}^{(e)T} & 0 \end{bmatrix} \begin{Bmatrix} \{u^{(e)}\} \\ \{v^{(e)}\} \\ \{p^{(e)}\} \end{Bmatrix} + \begin{Bmatrix} \{I_x^{(e)}\} \\ \{I_y^{(e)}\} \\ 0 \end{Bmatrix} - \begin{Bmatrix} \{f_{bx}^{(e)}\} \\ f_{by}^{(e)} \\ 0 \end{Bmatrix} = 0 \tag{16.23b}$$

Step – 5 Implementation of the Boundary Conditions

The implementation of the boundary conditions of the second kind into the finite element formulation takes place through the interelement vector $\{I_x^{(e)}\}$ given by Equation 16.18g. In order to show the specific contribution from the boundary condition, the interelement vector can be decomposed into two parts:

$$\{I_x^{(e)}\} = \{I_{sx}^{(e)}\} + \{I_{ix}^{(e)}\} \tag{16.24}$$

The vector $\{I_{ix}^{(e)}\}$ is the integral contributions from all interior sides of the element other than the boundary sides and constitutes the interelement requirements. The vector $\{I_{sx}^{(e)}\}$ constitutes the integral contributions from the boundary forces, and it is expressed as

$$\{I_{sx}^{(e)}\} = \oint_s [N]^T \left(p^* n_x - \mu\left(\frac{\partial u^*}{\partial x}\hat{n}_x + \frac{\partial u^*}{\partial y}\hat{n}_y \right)\right) \mathrm{d}s \tag{16.25}$$

Substituting the boundary conditions given by Equation 16.11, we get

$$\{I_{sx}^{(e)}\} = \oint_s [N]^T f_{sx} \mathrm{d}s = \{f_{bsx}^{(e)}\} = \text{Element source vector due to boundary conditions} \tag{16.26}$$

Substituting Equations 16.24 and 16.26 into 16.17b, we get the x-momentum element characteristics equation as

$$[k_I^{(e)}(u,v)]\{u^{(e)}\} + [k_\mu^{(e)}]\{u^{(e)}\} - [k_{px}^{(e)}]\{p^{(e)}\} + \{I_{ix}^{(e)}\} - \{f_{bx}^{(e)} - f_{bsx}^{e}\} = 0$$

or

$$[k_I^{(e)}(u,v)]\{u^{(e)}\} + [k_\mu^{(e)}]\{u^{(e)}\} - [k_{px}^{(e)}]\{p^{(e)}\} + \{I_{ix}^{(e)}\} - \{f_x^{(e)}\} = 0 \tag{16.27}$$

In a similar manner, the boundary condition for y-momentum equation can be implemented in Equation 16.19b, and the set of element characteristic equations for u, v, and p are assembled as follows:

$$\begin{aligned} &[k_I^{(e)}(u,v) + [k_\mu^{(e)}]]\{u^{(e)}\} - [k_{px}^{(e)}]\{p^{(e)}\} + \{I_{ix}^{(e)}\} - \{f_x^{(e)}\} = 0 \quad \text{(a)} \\ &[k_I^{(e)}(u,v) + [k_\mu^{(e)}]]\{v^{(e)}\} - [k_{py}^{(e)}]\{p^{(e)}\} + \{I_{iy}^{(e)}\} - \{f_y^{(e)}\} = 0 \quad \text{(b)} \\ &-[k_{px}^{(e)}]^T\{u^{(e)}\} - [k_{py}^{(e))}]^T\{v^{(e)}\} = 0 \quad \text{(c)} \end{aligned} \tag{16.28}$$

and in the matrix form as

$$\begin{bmatrix} k_I^{(e)}(u,v)+k_\mu^{(e)} & 0 & -k_{px}^{(e)} \\ 0 & k_I^{(e)}(u,v)+k_\mu^{(e)} & -k_{py}^{(e)} \\ -k_{px}^{(e)T} & -k_{py}^{(e)T} & 0 \end{bmatrix} \begin{Bmatrix} \{u^{(e)}\} \\ \{v^{(e)}\} \\ \{p^{(e)}\} \end{Bmatrix} + \begin{Bmatrix} \{I_{ix}^{(e)}\} \\ \{I_{iy}^{(e)}\} \\ 0 \end{Bmatrix} - \begin{Bmatrix} \{f_x^{(e)}\} \\ f_y^{(e)} \\ 0 \end{Bmatrix} = 0 \tag{16.29a}$$

and in a more compact form as

$$\left[k^{(e)}\right]\left\{U^{(e)}\right\}+\left\{I^{(e)}\right\}-\left\{F^{(e)}\right\}=0 \tag{16.29b}$$

where

$$\left[k_I^{(e)}(u,v)\right]=\int_{A^{(e)}} \rho\left[N^T\right][N]\left(\left\{u^{(e)}\right\}\frac{\partial[N]}{\partial x}+\left\{v^{(e)}\right\}\frac{\partial[N]}{\partial y}\right)\mathrm{d}A \tag{16.30a}$$

$$k_\mu^{(e)} = k_{xx}^{(e)} + k_{yy}^{(e)} \tag{16.30b}$$

$$k_{xx}^{(e)} = \int_{A^{(e)}} \mu\frac{\partial[N]}{\partial x}\frac{\partial[N]^T}{\partial x}\mathrm{d}A, \quad k_{yy}^{(e)} = \int_{A^{(e)}} \mu\frac{\partial[N]}{\partial y}\frac{\partial[N]^T}{\partial y}\mathrm{d}A \tag{16.20c}$$

$$\left[k_{px}^{(e)}\right]=\int_{A^{(e)}}\left[N_p\right]\frac{\partial[N]^T}{\partial x}\mathrm{d}A, \quad \left[k_{py}^{(e)}\right]=\int_{A^{(e)}}\left[N_p\right]\frac{\partial[N]^T}{\partial y}\mathrm{d}A \tag{16.30d}$$

$$\left\{f_x^{(e)}\right\}=\left\{f_{bx}^{(e)}\right\}-\left\{f_{bsx}^{(e)}\right\} \tag{16.30e}$$

$$\left\{f_{bx}^{(e)}\right\}=\int_{A^{(e)}}[N]^T b_x \mathrm{d}A, \quad \int_{A^{(e)}}[N]^T b_x \mathrm{d}A=\oint_s [N]^T f_{sx}\mathrm{d}s \tag{16.30f}$$

$$\left\{f_y^{(e)}\right\}=\left\{f_{by}^{(e)}\right\}-\left\{f_{bsy}^{(e)}\right\}, \quad \left\{f_{by}^{(e)}\right\}=\int_{A^{(e)}}[N]^T b_y \mathrm{d}A \tag{16.30g}$$

$$\left\{f_{bsy}^{(e)}\right\}=\oint_s [N]^T f_{sy}\mathrm{d}s \tag{16.30i}$$

$$\left\{U^{(e)}\right\}=\begin{Bmatrix} u^{(e)} \\ v^{(e)} \\ p^{(e)} \end{Bmatrix} \quad \text{and} \quad \left\{F^{(e)}\right\}=\begin{Bmatrix} f_x^{(e)} \\ f_y^{(e)} \\ 0 \end{Bmatrix} \tag{16.30j}$$

An examination of the matrices in Equation 16.30a clearly shows that characteristics matrix $k_I^{(e)}(u,v)$ is nonlinear, and that makes Equation 16.29 also nonlinear.

Case: Element Characteristic Equations Using Linear Triangular Element

Let us consider a linear three-node triangular element for velocity components as shown in Figure 12.14 or Figure 16.3a, and with the approximate solution, the corresponding shape functions are given by Equations 12.59 and 12.60, respectively. For pressure, we assume discontinuous constant pressure at the center of the triangle as shown in Figure 16.3b. The vectors with derivatives of shape functions are derived by substituting the shape functions given as follows:

$$\frac{\partial[N]}{\partial x}=\begin{bmatrix} \frac{\partial N_i}{\partial x} & \frac{\partial N_j}{\partial x} & \frac{\partial N_k}{\partial x} \end{bmatrix}=\frac{1}{2A}\begin{bmatrix} \beta_i & \beta_j & \beta_k \end{bmatrix} \quad (16.31a)$$

and

$$\frac{\partial[N]}{\partial y}=\begin{bmatrix} \frac{\partial N_i}{\partial y} & \frac{\partial N_j}{\partial y} & \frac{\partial N_k}{\partial y} \end{bmatrix}=\frac{1}{2A}\begin{bmatrix} \gamma_i & \gamma_j & \gamma_k \end{bmatrix} \quad (16.31b)$$

Using these expressions, we can evaluate the element stiffness matrices given by Equations 16.31a and 16.31b in the following manner:

The element characteristic matrix due to inertia forces is given as

$$\left[k_I^{(e)}(u,v)\right]=\int_{A^{(e)}}\rho\left[N^T\right][N]\left(\left\{u^{(e)}\right\}\frac{\partial[N]}{\partial x}+\left\{v^{(e)}\right\}\frac{\partial[N]}{\partial y}\right)dA$$

$$=\int_{A^{(e)}}\rho\begin{Bmatrix} N_i \\ N_j \\ N_k \end{Bmatrix}\begin{Bmatrix} N_i & N_j & N_k \end{Bmatrix}\begin{Bmatrix} u_i^{(e)} \\ u_j^{(e)} \\ u_k^{(e)} \end{Bmatrix}\begin{Bmatrix} \frac{\partial N_i}{\partial x} & \frac{\partial N_j}{\partial x} & \frac{\partial N_k}{\partial x} \end{Bmatrix}dA$$

$$+\int_{A^{(e)}}\rho\begin{Bmatrix} N_i \\ N_j \\ N_k \end{Bmatrix}\begin{Bmatrix} N_i & N_j & N_k \end{Bmatrix}\begin{Bmatrix} v_i^{(e)} \\ v_j^{(e)} \\ v_k^{(e)} \end{Bmatrix}\begin{Bmatrix} \frac{\partial N_i}{\partial y} & \frac{\partial N_j}{\partial y} & \frac{\partial N_k}{\partial y} \end{Bmatrix}dA$$

$$=\rho\int_{A^{(e)}}\begin{bmatrix} N_i^2 & N_iN_j & N_iN_k \\ N_jN_i & N_j^2 & N_jN_k \\ N_kN_i & N_kN_j & N_k^2 \end{bmatrix}\begin{Bmatrix} u_i^{(e)} \\ u_j^{(e)} \\ u_k^{(e)} \end{Bmatrix}\frac{1}{2A}\begin{Bmatrix} \beta_i & \beta_j & \beta_k \end{Bmatrix}dA$$

$$+\rho\int_{A^{(e)}}\begin{bmatrix} N_i^2 & N_iN_j & N_iN_k \\ N_jN_i & N_j^2 & N_jN_k \\ N_kN_i & N_kN_j & N_k^2 \end{bmatrix}\begin{Bmatrix} v_i^{(e)} \\ v_j^{(e)} \\ v_k^{(e)} \end{Bmatrix}\frac{1}{2A}\begin{Bmatrix} \gamma_i & \gamma_j & \gamma_k \end{Bmatrix}dA$$

$$=\frac{\rho}{2A}\begin{Bmatrix} u_i^{(e)} \\ u_j^{(e)} \\ u_k^{(e)} \end{Bmatrix}\begin{Bmatrix} \beta_i & \beta_j & \beta_k \end{Bmatrix}\int_{A^{(e)}}\begin{bmatrix} N_i^2 & N_iN_j & N_iN_k \\ N_jN_i & N_j^2 & N_jN_k \\ N_kN_i & N_kN_j & N_k^2 \end{bmatrix}dA$$

$$+\frac{\rho}{2A}\begin{Bmatrix} v_i^{(e)} \\ v_j^{(e)} \\ v_k^{(e)} \end{Bmatrix}\begin{Bmatrix} \gamma_i & \gamma_j & \gamma_k \end{Bmatrix}\int_{A^{(e)}}\begin{bmatrix} N_i^2 & N_iN_j & N_iN_k \\ N_jN_i & N_j^2 & N_jN_k \\ N_kN_i & N_kN_j & N_k^2 \end{bmatrix}dA$$

$$=\frac{\rho}{24}\begin{bmatrix} 2 & 1 & 1 \\ 1 & 2 & 1 \\ 1 & 1 & 2 \end{bmatrix}\begin{bmatrix} u_i^{(e)}\beta_i & u_i^{(e)}\beta_j & u_i^{(e)}\beta_k \\ u_j^{(e)}\beta_i & u_j^{(e)}\beta_j & u_j^{(e)}\beta_k \\ u_k^{(e)}\beta_i & u_k^{(e)}\beta_j & u_k^{(e)}\beta_k \end{bmatrix}$$

$$+\frac{\rho}{24}\begin{bmatrix} 2 & 1 & 1 \\ 1 & 2 & 1 \\ 1 & 1 & 2 \end{bmatrix}\begin{bmatrix} v_i^{(e)}\gamma_i & v_i^{(e)}\gamma_j & v_i^{(e)}\gamma_k \\ v_j^{(e)}\gamma_i & v_j^{(e)}\gamma_j & v_j^{(e)}\gamma_k \\ v_k^{(e)}\gamma_i & v_k^{(e)}\gamma_j & v_k^{(e)}\gamma_k \end{bmatrix} \quad (16.32)$$

The element characteristics matrix due to viscous forces is given as

$$k_{11}^{(e)} = \int_{A^{(e)}} \mu \frac{\partial[N]}{\partial x}\frac{\partial[N]^T}{\partial x}\mathrm{d}A$$

$$= \int_{A^{(e)}} \mu \begin{Bmatrix} \frac{\partial N_i}{\partial x} & \frac{\partial N_j}{\partial x} & \frac{\partial N_k}{\partial x} \end{Bmatrix}\begin{Bmatrix} \frac{\partial N_i}{\partial x} \\ \frac{\partial N_j}{\partial x} \\ \frac{\partial N_k}{\partial x} \end{Bmatrix}dA \quad (16.33a)$$

$$= \frac{\mu}{4A^2}\begin{Bmatrix} \beta_i & \beta_j & \beta_k \end{Bmatrix}\begin{Bmatrix} \beta_i \\ \beta_j \\ \beta_k \end{Bmatrix}A$$

$$k_{11}^{(e)} = \frac{\mu}{4A}\begin{bmatrix} \beta_i^2 & \beta_i\beta_j & \beta_i\beta_k \\ \beta_j\beta_i & \beta_j^2 & \beta_j\beta_i \\ \beta_k\beta_i & \beta_k\beta_j & \beta_k^2 \end{bmatrix}$$

Similarly,

$$k_{12}^{(e)} = \int_{A^{(e)}} \mu \frac{\partial[N]}{\partial y}\frac{\partial[N]^T}{\partial y}\mathrm{d}A = \frac{\mu}{4A}\begin{bmatrix} \gamma_i^2 & \gamma_i\gamma_j & \gamma_i\gamma_k \\ \gamma_j\gamma_i & \gamma_j^2 & \gamma_j\gamma_i \\ \gamma_k\gamma_i & \gamma_k\gamma_j & \gamma_k^2 \end{bmatrix} \quad (16.33b)$$

Combining Equations 16.33a and 16.33b, we get the element characteristics matrix as

$$\left[k_{\mu}^{(e)}\right]=\frac{\mu}{4A}\begin{bmatrix} \beta_i^2+\gamma_i^2 & \beta_i\beta_j+\gamma_i\gamma_j & \beta_i\beta_k+\gamma_i\gamma_k \\ \beta_j\beta_i+\gamma_j\gamma_i & \beta_j^2+\gamma_j^2 & \beta_j\beta_k+\gamma_j\gamma_k \\ \beta_k\beta_i+\gamma_k\gamma_i & \beta_k\beta_j+\gamma_k\gamma_j & \beta_k^2+\gamma_k^2 \end{bmatrix} \tag{16.34}$$

The component of element characteristics matrix due to the pressure force term can now be computed as

$$\left[k_{px}^{(e)}\right]=\int_{A^{(e)}}\left[N_p\right]\frac{\partial[N]^T}{\partial x}\mathrm{d}A=\int_A N_p\begin{Bmatrix} \dfrac{\partial N_i}{\partial x} \\ \dfrac{\partial N_j}{\partial x} \\ \dfrac{\partial N_k}{\partial x} \end{Bmatrix}\mathrm{d}A=N_p\int_A\frac{1}{2A}\begin{Bmatrix} \beta_i \\ \beta_j \\ \beta_k \end{Bmatrix}\mathrm{d}A$$

$$\left[k_{px}^{(e)}\right]=\frac{N_p}{2}\begin{Bmatrix} \beta_i \\ \beta_j \\ \beta_k \end{Bmatrix} \tag{16.35a}$$

Similarly, for the y-component pressure force

$$\left[k_{py}^{(e)}\right]=\int_{A^{(e)}}\left[N_p\right]\frac{\partial[N]^T}{\partial y}\mathrm{d}A=\int_A N_p\begin{Bmatrix} \dfrac{\partial N_i}{\partial y} \\ \dfrac{\partial N_j}{\partial y} \\ \dfrac{\partial N_k}{\partial y} \end{Bmatrix}\mathrm{d}A=N_p\int_A\frac{1}{2A}\begin{Bmatrix} \gamma_i \\ \gamma_j \\ \gamma_k \end{Bmatrix}\mathrm{d}A$$

$$\left[k_{py}^{(e)}\right]=\frac{N_p}{2}\begin{Bmatrix} \gamma_i \\ \gamma_j \\ \gamma_k \end{Bmatrix} \tag{16.35b}$$

Elements source vector due to the body forces can be evaluated as follows:

$$\left\{f_{bx}^{(e)}\right\}=\int_{A^{(e)}}[N]^T b_x\mathrm{d}A$$

$$=b_x\int_A\begin{Bmatrix} N_i \\ N_j \\ N_k \end{Bmatrix}\mathrm{d}A=b_x\int_A\begin{Bmatrix} L_i \\ L_j \\ L_k \end{Bmatrix}\mathrm{d}A$$

Using natural coordinate system for a triangle and integral formula (Equation 12.71), we get

$$\left\{f_{bx}^{(e)}\right\}=\frac{b_xA}{3}\begin{Bmatrix}1\\1\\1\end{Bmatrix} \tag{16.36a}$$

Similarly, for the y-component body force term,

$$\left\{f_{by}^{(e)}\right\}=\int_{A^{(e)}}[N]^T b_y\mathrm{d}A=\frac{b_yA}{3}\begin{Bmatrix}1\\1\\1\end{Bmatrix} \tag{16.36b}$$

Element source vector due to x-component boundary surface forces

$$\left\{f_{bsx}^{(e)}\right\}=\oint_s[N]^T f_{sx}\mathrm{d}s$$

$$=f_{sx}\oint_s\begin{Bmatrix}N_i\\N_j\\N_k\end{Bmatrix}\mathrm{d}s \tag{16.37}$$

For boundary condition specified on i-j side of the element, $N_k=0$. Using natural co-ordinate system for a triangle and integral formula (Equation 12.28), we get,

$$\left\{f_{bsx}^{(e)}\right\}=f_{sx}\int_{S_b}\begin{Bmatrix}N_i\\N_j\\0\end{Bmatrix}\mathrm{d}s=f_{sx}\int_{S_b}\begin{Bmatrix}L_i\\L_j\\0\end{Bmatrix}\mathrm{d}s$$

$$\left\{f_{bsx}^{(e)}\right\}=\frac{f_{sx}L_{ij}}{2}\begin{Bmatrix}1\\1\\0\end{Bmatrix} \tag{16.38a}$$

Similarly, for boundary conditions specified on the sides j-k, $N_i=0$, and Equation 16.37 transforms into

$$\left\{f_{bsx}^{(e)}\right\}=f_{sx}\int_{S_b}\begin{Bmatrix}0\\N_j\\N_k\end{Bmatrix}\mathrm{d}s=\frac{f_{sx}L_{jk}}{2}\begin{Bmatrix}0\\1\\1\end{Bmatrix} \tag{16.39a}$$

and for boundary conditions specified on the side *k-i*, $N_j = 0$, and we get

$$\left\{f_{bsx}^{(e)}\right\} = f_{sx}\int_{S_b}\begin{Bmatrix} N_i \\ 0 \\ N_k \end{Bmatrix} ds = \frac{f_{sx}L_{ki}}{2}\begin{Bmatrix} 1 \\ 0 \\ 1 \end{Bmatrix} \tag{16.39b}$$

Element source vector due to *y*-component boundary surface forces

$$\begin{aligned}\left\{f_{bsy}^{(e)}\right\} &= \oint_s [N]^T f_{sy} ds \\ &= f_{sy}\oint_s \begin{Bmatrix} N_i \\ N_j \\ N_k \end{Bmatrix} ds\end{aligned} \tag{16.40}$$

For boundary condition specified on *i-j* side of the element, Equation 16.40 reduces to

$$\left\{f_{bsy}^{(e)}\right\} = \frac{f_{sy}L_{ij}}{2}\begin{Bmatrix} 1 \\ 1 \\ 0 \end{Bmatrix} \tag{16.41a}$$

For boundary conditions specified on the side *j-k*, Equation 16.40 reduces to

$$\left\{f_{bsy}^{(e)}\right\} = \frac{f_{sy}L_{jk}}{2}\begin{Bmatrix} 0 \\ 1 \\ 1 \end{Bmatrix} \tag{16.41b}$$

For boundary conditions specified on the side *k-i*, Equation 16.40 reduces to

$$\left\{f_{bsy}^{(e)}\right\} = \frac{f_{sy}L_{ki}}{2}\begin{Bmatrix} 1 \\ 0 \\ 1 \end{Bmatrix} \tag{16.41c}$$

Substituting Equations 16.32, 16.34, 16.35a, and 14.36a into 16.28a, we have the expression for the element characteristic equation for an **interior element** with $f_{bsx}^{(e)} = 0$ as

$$\left[k_I^{(e)}(u,v) + \left[k_\mu^{(e)}\right]\right]\left\{u^{(e)}\right\} - \left[k_{px}^{(e)}\right]\left\{p^{(e)}\right\} + \left\{I_{ix}^{(e)}\right\} - \left\{f_{bx}^{(e)}\right\} = 0$$

$$\left\{\frac{\rho}{24}\begin{bmatrix}2&1&1\\1&2&1\\1&1&2\end{bmatrix}\left[\begin{bmatrix}u_i^{(e)}\beta_i & u_i^{(e)}\beta_j & u_i^{(e)}\beta_k\\ u_j^{(e)}\beta_i & u_j^{(e)}\beta_j & u_j^{(e)}\beta_k\\ u_k^{(e)}\beta_i & u_k^{(e)}\beta_j & u_k^{(e)}\beta_k\end{bmatrix}+\begin{bmatrix}v_i^{(e)}\gamma_i & v_i^{(e)}\gamma_j & v_i^{(e)}\gamma_k\\ v_j^{(e)}\gamma_i & v_j^{(e)}\gamma_j & v_j^{(e)}\gamma_k\\ v_k^{(e)}\gamma_i & v_k^{(e)}\gamma_j & v_k^{(e)}\gamma_k\end{bmatrix}\right]\right\}\begin{Bmatrix}u_i^{(e)}\\u_j^{(e)}\\u_k^{(e)}\end{Bmatrix}+$$

$$\frac{\mu}{4A}\begin{bmatrix}\beta_i^2+\gamma_i^2 & \beta_i\beta_j+\gamma_i\gamma_j & \beta_i\beta_k+\gamma_i\gamma_k\\ \beta_j\beta_i+\gamma_j\gamma_i & \beta_j^2+\gamma_j^2 & \beta_j\beta_k+\gamma_j\gamma_k\\ \beta_k\beta_i+\gamma_k\gamma_i & \beta_k\beta_j+\gamma_k\gamma_j & \beta_k^2+\gamma_k^2\end{bmatrix}\begin{Bmatrix}u_i^{(e)}\\u_j^{(e)}\\u_k^{(e)}\end{Bmatrix}-\frac{N_p}{2}\begin{Bmatrix}\beta_i\\\beta_j\\\beta_k\end{Bmatrix}\left\{p^{(e)}\right\}+\left\{I_{ix}^{(e)}\right\}$$

$$-\frac{b_xA}{3}\begin{Bmatrix}1\\1\\1\end{Bmatrix} \tag{16.42}$$

The element characteristic matrix for a **boundary element** with boundary surface force specified at the edge *i-j* is obtained by substituting Equations 16.32, 16.34, 16.35a, 16.36a, and 16.38a into 16.28a as

$$\left\{\frac{\rho}{24}\begin{bmatrix}2&1&1\\1&2&1\\1&1&2\end{bmatrix}\left[\begin{bmatrix}u_i^{(e)}\beta_i & u_i^{(e)}\beta_j & u_i^{(e)}\beta_k\\ u_j^{(e)}\beta_i & u_j^{(e)}\beta_j & u_j^{(e)}\beta_k\\ u_k^{(e)}\beta_i & u_k^{(e)}\beta_j & u_k^{(e)}\beta_k\end{bmatrix}+\begin{bmatrix}v_i^{(e)}\gamma_i & v_i^{(e)}\gamma_j & v_i^{(e)}\gamma_k\\ v_j^{(e)}\gamma_i & v_j^{(e)}\gamma_j & v_j^{(e)}\gamma_k\\ v_k^{(e)}\gamma_i & v_k^{(e)}\gamma_j & v_k^{(e)}\gamma_k\end{bmatrix}\right]\right\}\begin{Bmatrix}u_i^{(e)}\\u_j^{(e)}\\u_k^{(e)}\end{Bmatrix}$$

$$+\frac{\mu}{4A}\begin{bmatrix}\beta_i^2+\gamma_i^2 & \beta_i\beta_j+\gamma_i\gamma_j & \beta_i\beta_k+\gamma_i\gamma_k\\ \beta_j\beta_i+\gamma_j\gamma_i & \beta_j^2+\gamma_j^2 & \beta_j\beta_k+\gamma_j\gamma_k\\ \beta_k\beta_i+\gamma_k\gamma_i & \beta_k\beta_j+\gamma_k\gamma_j & \beta_k^2+\gamma_k^2\end{bmatrix}\begin{Bmatrix}u_i^{(e)}\\u_j^{(e)}\\u_k^{(e)}\end{Bmatrix}$$

$$-\frac{N_p}{2}\begin{Bmatrix}\beta_i\\\beta_j\\\beta_k\end{Bmatrix}\left\{p^{(e)}\right\}+\left\{I_{ix}^{(e)}\right\}-\frac{b_xA}{3}\begin{Bmatrix}1\\1\\1\end{Bmatrix}+\frac{f_{sx}L_{ij}}{2}\begin{Bmatrix}1\\1\\0\end{Bmatrix}=0 \tag{16.43}$$

In a similar manner, we get to derive the element characteristic equation for the *y*-component velocity *v* and pressure *p* starting from Equations 16.28b and 16.28c, respectively.

Step – 6 Assembly of Elements and Formation of Global System

The assembly of all element characteristics equations to form the global system is done by following the procedure and the connectivity array presented in Chapter 14. In a compact form the global system of equations is written in a compact form as

$$K\{U\}=\{F\} \tag{16.44}$$

Example 16.1

Consider a two-dimensional viscous incompressible flow induced in rectangular cavity by the motion of upper surface at constant velocity. Consider **element #7** for the mesh consisting of linear triangular elements for velocity components and a constant pressure at the center of the triangle as shown in the figure and derive the element characteristic equation. The governing equations over a flow domain Ω are given as

Continuity Equation:

$$\frac{\partial u}{\partial x}+\frac{\partial v}{vy}=0$$

X-momentum:

$$\rho\left(u\frac{\partial u}{\partial x}+v\frac{\partial u}{\partial y}\right)=-\frac{\partial P}{\partial x}+\mu\left(\frac{\partial^2 u}{\partial x^2}+\frac{\partial^2 u}{\partial y^2}\right)$$

Y-momentum:

$$\rho\left(u\frac{\partial v}{\partial x}+v\frac{\partial v}{\partial y}\right)=-\frac{\partial P}{\partial y}+\mu\left(\frac{\partial^2 v}{\partial x^2}+\frac{\partial^2 v}{\partial y^2}\right)$$

with boundary conditions given as

1. $x=0$, $u=0$; $v=0$
2. $x=1$, $u=0$; $v=0$
3. at $y=0$, $u=0$; $v=0$
4. at $y=1$, $u=1$; $v=0$

Steps for the finite element formulation are:

Step 1 Discretization of the Solution Domain

Solution for flow field involves two different kinds of variables such as velocity, *V*, and pressure *P*. This requires the need for two different finite element meshes: one for the velocity and other for the pressure.

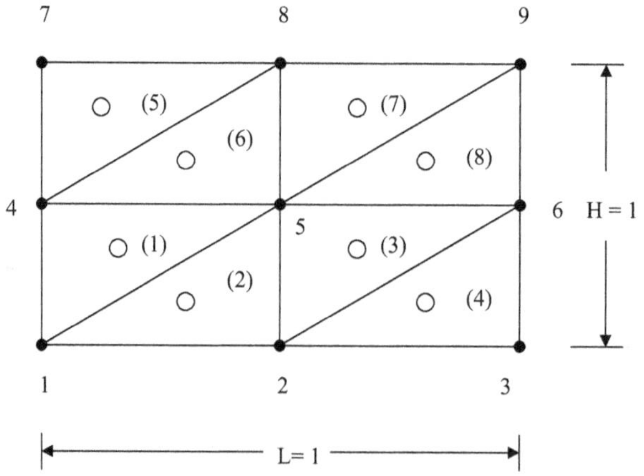

Step 2 Formulation of an Integral or Weak Form Using Galerkin's Method

Weighted residual involves finding a minimum for the residuals over an element $A^{(e)}$ by multiplying the residuals of governing equations by the weighted functions W_p, W_u, and W_v, and integrating over the element in the following manner:

$$\left\{R_p^{(e)}\right\} = -\int W_P\left\{\frac{\partial u^*}{\partial x} + \frac{\partial v^*}{\partial y}\right\}dA = 0$$

$$\left\{R_u^{(e)}\right\} = -\int_{A^{(e)}} W_u\left\{\rho\left(u^*\frac{\partial u^*}{\partial x} + v^*\frac{\partial u^*}{\partial y}\right) - b_x + \frac{\partial P^*}{\partial x} - \mu\left(\frac{\partial^2 u^*}{\partial x^2} + \frac{\partial^2 u^*}{\partial y^2}\right)\right\}dA = 0$$

$$\left\{R_v^{(e)}\right\} = -\int_{A^{(e)}} W_v\left\{\rho\left(u^*\frac{\partial v^*}{\partial x} + v^*\frac{\partial v^*}{\partial y}\right) - b_y + \frac{\partial P^*}{\partial y} - \mu\left(\frac{\partial^2 v^*}{\partial x^2} + \frac{\partial^2 v^*}{\partial y^2}\right)\right\}dA = 0$$

In the Galerkin finite element method, the weighting functions are assumed to be same as the shape functions used for the approximate functions

$$W_u = [N]^T, \quad W_y = [N]^T \quad \text{and} \quad W_p = \left[N_P\right]^T$$

where $[N]$ and $[N_p]$ are the shape functions associated with velocity and pressure approximation function, respectively.

The integral equations are transformed into a weak form

$$\int_{A^{(e)}}[N^T]\rho\left(u^*\frac{\partial u^*}{\partial x} + v\frac{\partial u^*}{\partial y}\right)dA - \int_{A^{(e)}}[N]^T b_x\,dA - \int_{A^{(e)}} p^*\frac{\partial[N]^T}{\partial x}dA + \int_{A^{(e)}}\mu\frac{\partial u^*}{\partial x}\frac{\partial[N]^T}{\partial x}dA$$
$$+\int_{A^{(e)}}\mu\frac{\partial u^*}{\partial y}\frac{\partial[N]^T}{\partial y}dA + \oint_s[N]^T\left(p^* n_x - \mu\left(\frac{\partial u^*}{\partial x}\hat{n}_x + \frac{\partial u^*}{\partial x}\hat{n}_y\right)\right)dS = 0$$

and

$$\int_{A^{(e)}}[N^T]\rho\left(u^*\frac{\partial v^*}{\partial x} + v\frac{\partial v^*}{\partial y}\right)dA - \int_{A^{(e)}}[N]^T b_y\,dA - \int_{A^{(e)}} p^*\frac{\partial[N]^T}{\partial y}dA + \int_{A^{(e)}}\mu\frac{\partial v^*}{\partial x}\frac{\partial[N]^T}{\partial x}dA$$
$$+\int_{A^{(e)}}\mu\frac{\partial v^*}{\partial y}\frac{\partial[N]^T}{\partial y}dA + \oint_s[N]^T\left(p^* n_x - \mu\left(\frac{\partial v^*}{\partial x}\hat{n}_x + \frac{\partial v^*}{\partial x}\hat{n}_y\right)\right)dS = 0$$

Step 3 Selection of Approximate Solution Functions

The approximate solutions for velocity and pressure fields can be assumed as follows:

$$u^* = \sum_{I=1}^{np} N_I u_I = [N]\left\{u^{(e)}\right\}, \quad v^* = \sum_{I=1}^{np} N_I v_I = [N]\left\{v^{(e)}\right\} \quad \text{and} \quad p^* = \sum_{I=1}^{np} N_I p_I = [N_p]\left\{p^{(e)}\right\}$$

Step 4 Formation of Element Characteristic Equation

The element characteristic equation can be derived by substituting the approximate solution functions in the integral form of the governing Equations. The set of element characteristic equations for *u*, *v*, and *p* is assembled as

$$\left[k_1^{(e)}(u,v)\right]\left\{u^{(e)}\right\}+\left[k_\mu^{(e)}\right]\left\{u^{(e)}\right\}-\left[k_{px}^{(e)}\right]\left\{p^{(e)}\right\}+\left\{I_x^{(e)}\right\}-\left\{f_{bx}^{(e)}\right\}=0$$

$$\left[k_1^{(e)}(u,v)\right]\left\{v^{(e)}\right\}+\left[k_\mu^{(e)}\right]\left\{v^{(e)}\right\}-\left[k_{py}^{(e)}\right]\left\{p^{(e)}\right\}+\left\{I_y^{(e)}\right\}-\left\{f_{by}^{(e)}\right\}=0$$

$$-\left[k_{px}^{(e)}\right]\left\{u^{(e)}\right\}-\left[k_{py}^{(e)}\right]\left\{v^{(e)}\right\}=0$$

This can be written in matrix form as

$$\begin{bmatrix} k_1^{(e)}(u,v)+k_\mu^{(e)} & 0 & -k_{px}^{(e)} \\ 0 & k_1^{(e)}(u,v)+k_\mu^{(e)} & -k_{py}^{(e)} \\ -k_{px}^{(e)} & -k_{py}^{(e)} & 0 \end{bmatrix}\begin{bmatrix} \{u^{(e)}\} \\ \{v^{(e)}\} \\ \{p^{(e)}\} \end{bmatrix}+\begin{Bmatrix} \{I_x^{(e)}\} \\ \{I_y^{(e)}\} \\ 0 \end{Bmatrix}-\begin{Bmatrix} \{f_{bx}^{(e)}\} \\ \{f_{by}^{(e)}\} \\ 0 \end{Bmatrix}=0$$

The matrices and vectors in the element characteristic equations are now evaluated using linear triangular element as follows:

$$\left[k_1^{(e)}(u,v)\right]=\int_{A^{(e)}}\rho\left[N^T\right][N]\left(\left\{u^{(e)}\right\}\frac{\partial[N]}{\partial x}+\left\{v^{(e)}\right\}\frac{\partial[N]}{\partial y}\right)\mathrm{d}A$$

$$=\frac{\rho}{24}\begin{bmatrix} 2 & 1 & 1 \\ 1 & 2 & 1 \\ 1 & 1 & 2 \end{bmatrix}\begin{bmatrix} u_i^{(e)}\beta_i+v_i^{(e)}\gamma_i & u_i^{(e)}\beta_j+v_i^{(e)}\gamma_j & u_i^{(e)}\beta_k+v_i^{(e)}\gamma_k \\ u_j^{(e)}\beta_i+v_j^{(e)}\gamma_i & u_j^{(e)}\beta_j+v_j^{(e)}\gamma_j & u_j^{(e)}\beta_k+v_j^{(e)}\gamma_k \\ u_k^{(e)}\beta_i+v_k^{(e)}\gamma_i & u_k^{(e)}\beta_j+v_k^{(e)}\gamma_j & u_k^{(e)}\beta_k+v_k^{(e)}\gamma_k \end{bmatrix}$$

Stiffness matrix due to viscous forces is given as

$$k_{xx}^{(e)}=\int_{A^{(e)}}\mu\frac{\partial[N]}{\partial x}\frac{\partial[N]^T}{\partial x}\mathrm{d}A$$

$$=\frac{\mu}{4A}\begin{bmatrix} \beta_i^2 & \beta_i\beta_j & \beta_i\beta_k \\ \beta_j\beta_j & \beta_j^2 & \beta_j\beta_k \\ \beta_k\beta_j & \beta_k\beta_j & \beta_k^2 \end{bmatrix}$$

$$k_{12}^{(e)}=\int_{A^{(e)}}\mu\frac{\partial[N]}{\partial y}\frac{\partial[N]^T}{\partial y}\mathrm{d}A$$

$$=\frac{\mu}{24}\begin{bmatrix} \gamma_i^2 & \gamma_i\gamma_j & \gamma_i\gamma_k \\ \gamma_j\gamma_i & \gamma_j^2 & \gamma_j\gamma_k \\ \gamma_k\gamma_i & \gamma_k\gamma_j & \gamma_k^2 \end{bmatrix}$$

The component of element stiffness matrix due to the pressure force term can now be computed as

$$\left[k_{px}^{(e)}\right] = \int_{A^{(e)}} \left[N_{px}\right] \frac{\partial [N]^T}{\partial x} dA = \frac{N_p}{2} \begin{Bmatrix} \beta_i \\ \beta_j \\ \beta_k \end{Bmatrix}$$

$$\left[k_{py}^{(e)}\right] = \int_{A^{(e)}} \left[N_{py}\right] \frac{\partial [N]^T}{\partial y} dA = \frac{N_p}{2} \begin{Bmatrix} \beta_i \\ \beta_j \\ \beta_k \end{Bmatrix}$$

Element source vector due to the body force can be evaluated as follows:

$$\left\{f_{bx}^{(e)}\right\} = \int_{A^{(e)}} [N]^T \, b_x \, dA = 0$$

Similarly, for y-component

$$\left\{f_{by}^{(e)}\right\} = \int_{A^{(e)}} [N]^T \, b_y \, dA = 0$$

Element source vector due to x-component boundary surface forces

$$\left\{f_{bsx}^{(e)}\right\} = \oint_S [N]^T \, f_{sx} \, ds = 0$$

For boundary condition specified on i-j side of the element, $N_k=0$. Using natural co-ordinate system for a triangle and integral formula, we get

$$\left\{f_{bsx}^{(e)}\right\} = \frac{f_{sx} L_{ij}}{2} \begin{Bmatrix} 1 \\ 1 \\ 0 \end{Bmatrix} = 0$$

Element source vector due to y-component boundary surface forces

$$\left\{f_{bsy}^{(e)}\right\} = \oint_S [N]^T \, f_{sy} \, ds$$

$$= f_{sy} \oint_s \begin{Bmatrix} N_i \\ N_j \\ N_k \end{Bmatrix} ds$$

For boundary condition specified on i-j side of the element, $N_k=0$. Using the natural co-ordinate system for a triangle and integral formula, we get

$$\left\{f_{bsy}^{(e)}\right\} = \frac{f_{sy} L_{ij}}{2} \begin{Bmatrix} 1 \\ 1 \\ 0 \end{Bmatrix} = 0$$

Model Calculation for Element #7

$$(x_i, y_i) = (0.5, 0.5); (x_j, y_j) = (1,1) \quad \text{and} \quad (x_k, y_k) = (0.5, 1)$$

$$\beta_i = (y_j - y_k) = 1 - 1 = 0, \quad \beta_j = (y_k - y_i) = 1 - 0.5 = 0.5, \quad \beta_i = (y_i - y_j) = 0.5 - 1 = -0.5$$

$$\gamma_i = -(x_j - x_k) = 0.5 - 1 = -0.5, \quad \gamma_j = -(x_k - x_i) = 0.5 - 0.5 = 0, \quad \gamma_i = -(x_i - x_j) = 1 - 0.5 = 0.5$$

$$k_1^{(7)}(u,v) = \frac{\rho}{24}\begin{bmatrix} 2 & 1 & 1 \\ 1 & 2 & 1 \\ 1 & 1 & 2 \end{bmatrix}\begin{bmatrix} u_i^{(e)}\beta_i + v_i^{(e)}\gamma_i & u_i^{(e)}\beta_j + v_i^{(e)}\gamma_j & u_i^{(e)}\beta_k + v_i^{(e)}\gamma_k \\ u_j^{(e)}\beta_i + v_j^{(e)}\gamma_i & u_j^{(e)}\beta_j + v_j^{(e)}\gamma_j & u_j^{(e)}\beta_k + v_j^{(e)}\gamma_k \\ u_k^{(e)}\beta_i + v_k^{(e)}\gamma_i & u_k^{(e)}\beta_j + v_k^{(e)}\gamma_j & u_k^{(e)}\beta_k + v_k^{(e)}\gamma_k \end{bmatrix}$$

$$= \frac{\rho}{24}\begin{bmatrix} 2 & 1 & 1 \\ 1 & 2 & 1 \\ 1 & 1 & 2 \end{bmatrix}\begin{bmatrix} (u_5 * 0) + (v_5 * -0.5) & (u_5 * 0.5) + (v_5 * 0) & (u_5 * -0.5) + (v_5 * 0.5) \\ (u_9 * 0) + (v_9 * -0.5) & (u_9 * 0.5) + (v_9 * 0) & (u_9 * -0.5) + (v_9 * 0.5) \\ (u_8 * 0) + (v_8 * -0.5) & (u_8 * 0.5) + (v_8 * 0) & (u_8 * -0.5) + (v_8 * 0.5) \end{bmatrix}$$

$$= \frac{\rho}{24}\begin{bmatrix} 2 & 1 & 1 \\ 1 & 2 & 1 \\ 1 & 1 & 2 \end{bmatrix}\begin{bmatrix} -0.5v_5 & 0.5u_5 & 0.5*(u_5 - v_5) \\ -0.5v_9 & 0.5u_9 & 0.5*(u_9 - v_9) \\ -0.5v_8 & 0.5u_8 & 0.5*(u_8 - v_8) \end{bmatrix}$$

$$k_\mu^{(7)} = \frac{\mu}{4A}\begin{bmatrix} \beta_i^2 + \gamma_i^2 & \beta_i\beta_j + \gamma_i\gamma_j & \beta_i\beta_k + \gamma_i\gamma_k \\ \beta_j\beta_i + \gamma_j\gamma_i & \beta_j^2 + \gamma_j^2 & \beta_j\beta_k + \gamma_j\gamma_k \\ \beta_k\beta_i + \gamma_k\gamma_i & \beta_k\beta_j + \gamma_k\gamma_j & \beta_k^2 + \gamma_k^2 \end{bmatrix}$$

$$= \frac{\mu}{4A}\begin{bmatrix} 0.25 & 0 & -0.25 \\ 0 & 0.25 & -0.25 \\ -0.25 & -0.25 & 0.5 \end{bmatrix}$$

$$k_{px}^{(7)} = \frac{N_p}{2}\begin{Bmatrix} \beta_i \\ \beta_j \\ \beta_k \end{Bmatrix} = \frac{N_p}{2}\begin{Bmatrix} 0 \\ 0.5 \\ -0.5 \end{Bmatrix}$$

$$k_{py}^{(7)} = \frac{N_p}{2}\begin{Bmatrix} \gamma_i \\ \gamma_j \\ \gamma_k \end{Bmatrix} = \frac{N_p}{2}\begin{Bmatrix} -0.5 \\ 0 \\ 0.5 \end{Bmatrix}$$

16.2.3 Unsteady Two-Dimensional Viscous Incompressible Flow

Let us consider the development of the transient problem by considering two-dimensional **Stokes flow** which is applicable for problems with low speed. As we have discussed in Chapter 1, appropriate governing equations for Stokes flow are derived by neglecting the nonlinear inertial or convective terms, and are given as follows:

Continuity

$$\frac{\partial u}{\partial x}+\frac{\partial v}{\partial y}=0 \tag{16.45a}$$

***x*-Momentum**

$$\rho\frac{\partial u}{\partial t}=b_x-\frac{\partial p}{\partial x}+\frac{\partial}{\partial x}\left[\mu\left(\frac{\partial u}{\partial x}\right)\right]+\frac{\partial}{\partial y}\left[\mu\left(\frac{\partial u}{\partial y}\right)\right] \tag{16.45b}$$

***y*-Momentum**

$$\rho\frac{\partial v}{\partial t}=b_y-\frac{\partial p}{\partial y}+\frac{\partial}{\partial x}\left[\mu\left(\frac{\partial v}{\partial x}\right)\right]+\frac{\partial}{\partial y}\left[\mu\left(\frac{\partial v}{\partial y}\right)\right] \tag{16.45c}$$

The treatment of the unsteady-state term is similar to that discussed in Chapter 15.

Following the procedure outlined in Section 16.1.1, we can obtain the weak for equations using Galerkin's weighted residual method as

$$\left\{R_P^{(e)}\right\}=-\int_{A^{(e)}}\left[N_p\right]^T\left\{\frac{\partial u^*}{\partial x}+\frac{\partial v^*}{\partial y}\right\}dA=0 \tag{16.46a}$$

$$\left\{R_u^{(e)}\right\}=\int_{A^{(e)}}[N]^T\left\{\rho\frac{\partial u^*}{\partial t}-b_x+\frac{\partial p^*}{\partial x}-\frac{\partial}{\partial x}\left(\mu\frac{\partial u^*}{\partial x}\right)-\frac{\partial}{\partial y}\left(\mu\frac{\partial u^*}{\partial y}\right)\right\}dA=0 \tag{16.46b}$$

$$\left\{R_v^{(e)}\right\}=\int_{A^{(e)}}[N]^T\left\{\rho\frac{\partial v^*}{\partial t}-b_y+\frac{\partial p^*}{\partial y}-\frac{\partial}{\partial x}\left(\mu\frac{\partial v^*}{\partial x}\right)-\frac{\partial}{\partial y}\left(\mu\frac{\partial v^*}{\partial y}\right)\right\}dA=0 \tag{16.46c}$$

In the **semi-discrete finite element formulation as outlined in Chapter 15**, the approximate solutions for velocity and pressure fields can be assumed as follows:

$$u^*=\sum_{I=1}^{n_p}N_I(x)u_I(t)=\left[N(x)\right]\left\{u^{(e)}(t)\right\} \tag{16.47a}$$

$$v^*=\sum_{I=1}^{n_p}N_I(x)v_I(t)=\left[N(x)\right]\left\{v^{(e)}(t)\right\} \tag{16.47b}$$

$$p^*=\sum_{I=1}^{n_p}N_{pI}(x)p_I(t)=\left[N_p(x)\right]\left\{p^{(e)}(t)\right\} \tag{16.47c}$$

where the shape functions N_I and N_{pI} are assumed to be a function of space variable only, and the nodal values u_I, v_I, and p_I as a function time. With the substitution of the approximation solutions and evaluation of integrals, we get the system of element characteristics equations as

$$\begin{aligned}
&\left[c^{(e)}\right]\frac{d\left\{u^{(e)}\right\}}{dt}+\left[k_{\mu}^{(e)}\right]\left\{u^{(e)}\right\}-\left[k_{px}^{(e)}\right]\left\{p^{(e)}\right\}+\left\{I_{ix}^{(e)}\right\}-\left\{f_{x}^{(e)}\right\}=0 \quad \text{(a)}\\
&\left[c^{(e)}\right]\frac{d\left\{v^{(e)}\right\}}{dt}+\left[k_{\mu}^{(e)}\right]\left\{v^{(e)}\right\}-\left[k_{py}^{(e)}\right]\left\{p^{(e)}\right\}+\left\{I_{iy}^{(e)}\right\}-\left\{f_{y}^{(e)}\right\}=0 \quad \text{(b)}\\
&-\left[k_{px}^{(e)}\right]^{T}\left\{u^{(e)}\right\}-\left[k_{py}^{(e))}\right]^{T}\left\{v^{(e)}\right\}=0 \quad \text{(c)}
\end{aligned} \tag{16.48}$$

and in the matrix form as

$$\begin{bmatrix} c^{(e)} & 0 & 0 \\ 0 & c^{(e)} & 0 \\ 0 & 0 & 0 \end{bmatrix}\begin{Bmatrix} \frac{\partial u^{(e)}}{\partial t} \\ \frac{\partial v^{(e)}}{\partial t} \\ \frac{\partial p}{\partial t} \end{Bmatrix}+\begin{bmatrix} k_{\mu}^{(e)} & 0 & -k_{px}^{(e)} \\ 0 & k_{\mu}^{(e)} & -k_{py}^{(e)} \\ -k_{px}^{(e)T} & -k_{py}^{(e)T} & 0 \end{bmatrix}\begin{Bmatrix} \left\{u^{(e)}\right\} \\ \left\{v^{(e)}\right\} \\ \left\{p^{(e)}\right\} \end{Bmatrix}$$

$$+\begin{Bmatrix} \left\{I_{ix}^{(e)}\right\} \\ \left\{I_{iy}^{(e)}\right\} \\ 0 \end{Bmatrix}-\begin{Bmatrix} \left\{f_{x}^{(e)}\right\} \\ f_{y}^{(e)} \\ 0 \end{Bmatrix}=0 \tag{16.49a}$$

and in a compact form as

$$C^{(e)}\left\{\frac{\partial U^{(e)}}{\partial t}\right\}+k^{(e)}\left\{U^{(e)}\right\}+\left\{I^{(e)}\right\}-\left\{F^{(e)}\right\} \tag{16.49b}$$

where

$$c^{(e)}=\int_{A}\rho[N]^{T}[N]dA \tag{16.50a}$$

$$C^{(e)}=\begin{bmatrix} c^{(e)} & 0 & 0 \\ 0 & c^{(e)} & 0 \\ 0 & 0 & 0 \end{bmatrix} \tag{16.50b}$$

All other terms are same as those defined by Equation 16.21.

16.2.4 Unsteady Three-Dimensional Viscous Incompressible Flow

The procedure described for two-dimensional problems can easily be extended to a three-dimensional problem. Let us consider the three-dimensional incompressible viscous flow problem given as

Continuity:

$$\frac{\partial u}{\partial x}+\frac{\partial v}{\partial y}+\frac{\partial w}{\partial y}=0 \tag{16.51a}$$

***x*-Momentum:**

$$\rho\left(\frac{\partial u}{\partial t}+u\frac{\partial u}{\partial x}+v\frac{\partial u}{\partial y}+w\frac{\partial u}{\partial z}\right)=b_x-\frac{\partial p}{\partial x}+\mu\left(\frac{\partial^2 u}{\partial x^2}+\frac{\partial^2 u}{\partial y^2}+\frac{\partial^2 u}{\partial z^2}\right) \tag{16.51b}$$

***y*-Momentum:**

$$\rho\left(\frac{\partial v}{\partial t}+u\frac{\partial v}{\partial x}+v\frac{\partial v}{\partial y}+w\frac{\partial v}{\partial z}\right)=b_y-\frac{\partial p}{\partial y}+\mu\left(\frac{\partial^2 v}{\partial x^2}+\frac{\partial^2 v}{\partial y^2}+\frac{\partial^2 v}{\partial z^2}\right) \tag{16.51c}$$

***z*-Momentum:**

$$\rho\left(\frac{\partial w}{\partial t}+u\frac{\partial w}{\partial x}+v\frac{\partial w}{\partial y}+ww\right)=b_z-\frac{\partial p}{\partial z}+\mu\left(\frac{\partial^2 w}{\partial x^2}+\frac{\partial^2 w}{\partial y^2}+\frac{\partial^2 w}{\partial z^2}\right) \tag{16.51d}$$

With boundary conditions given in a general form as

$$1.\quad u=u_s,\quad v=v_s\quad \text{and}\quad w=w_s\quad \text{on the boundary } s_1 \tag{16.52a}$$

$$
\begin{aligned}
2.\quad &\mu\left(\frac{\partial u}{\partial x}\hat{n}_x+\frac{\partial u}{\partial y}\hat{n}_y+\frac{\partial u}{\partial z}\hat{n}_z\right)-p\hat{n}_x=f_{sx}\\
&\mu\left(\frac{\partial v}{\partial x}\hat{n}_x+\frac{\partial v}{\partial y}\hat{n}_y+\frac{\partial v}{\partial z}\hat{n}_z\right)-p\hat{n}_y=f_{sy}\\
&\mu\left(\frac{\partial w}{\partial x}\hat{n}_x+\frac{\partial w}{\partial y}\hat{n}_y+\frac{\partial w}{\partial z}\hat{n}_z\right)-p\hat{n}_z=f_{sz}\quad \text{on the boundary } s_2
\end{aligned} \tag{16.52b}
$$

Following the procedure outlined in Sections 16.1.1 and 16.1.2, the set of element characteristic equations for u, v, w, and p is assembled as follows:

$$
\begin{aligned}
&\left[c^{(e)}\right]\frac{\mathrm{d}\left\{u^{(e)}\right\}}{\mathrm{d}t}+\left[k_I^{(e)}(u,v,w)+\left[k_\mu^{(e)}\right]\right]\left\{u^{(e)}\right\}-\left[k_{px}^{(e)}\right]\left\{p^{(e)}\right\}+\left\{I_{ix}^{(e)}\right\}-\left\{f_x^{(e)}\right\}=0 \quad \text{(a)}\\
&\left[c^{(e)}\right]\frac{\mathrm{d}\left\{v^{(e)}\right\}}{\mathrm{d}t}+\left[k_I^{(e)}(u,v,w)+\left[k_\mu^{(e)}\right]\right]\left\{v^{(e)}\right\}-\left[k_{py}^{(e)}\right]\left\{p^{(e)}\right\}+\left\{I_{iy}^{(e)}\right\}-\left\{f_y^{(e)}\right\}=0 \quad \text{(b)}\\
&\left[c^{(e)}\right]\frac{\mathrm{d}\left\{w^{(e)}\right\}}{\mathrm{d}t}+\left[k_I^{(e)}(u,v,w)+\left[k_\mu^{(e)}\right]\right]\left\{w^{(e)}\right\}-\left[k_{pz}^{(e)}\right]\left\{p^{(e)}\right\}+\left\{I_{iz}^{(e)}\right\}-\left\{f_z^{(e)}\right\}=0 \quad \text{(c)}\\
&-\left[k_{px}^{(e)}\right]^T\left\{u^{(e)}\right\}-\left[k_{py}^{(e))}\right]^T\left\{v^{(e)}\right\}-\left[k_{pz}^{(e))}\right]^T\left\{w^{(e)}\right\}=0 \quad \text{(d)}
\end{aligned} \tag{16.53}
$$

and in the matrix form as

$$\begin{bmatrix} c^{(e)} & 0 & 0 & 0 \\ 0 & c^{(e)} & 0 & 0 \\ 0 & 0 & c^{(e)} & 0 \\ 0 & 0 & 0 & 0 \end{bmatrix} \begin{Bmatrix} \frac{\partial u}{\partial t} \\ \frac{\partial v}{\partial t} \\ \frac{\partial w}{\partial t} \\ \frac{\partial p}{\partial t} \end{Bmatrix} +$$

$$\begin{bmatrix} k_I^{(e)}(u,v,w)+k_\mu^{(e)} & 0 & 0 & -k_{px}^{(e)} \\ 0 & k_I^{(e)}(u,v,w)+k_\mu^{(e)} & 0 & -k_{py}^{(e)} \\ 0 & 0 & k_I^{(e)}(u,v,w)+k_\mu^{(e)} & -k_{pz}^{(e)} \\ -k_{px}^{(e)T} & -k_{py}^{(e)T} & -k_{pz}^{(e)T} & 0 \end{bmatrix} \begin{Bmatrix} u^{(e)} \\ v^{(e)} \\ w^{(e)} \\ p^{(e)} \end{Bmatrix} + \tag{16.54a}$$

$$\begin{Bmatrix} I_{ix}^{(e)} \\ I_{iy}^{(e)} \\ I_{iz}^{(e)} \\ 0 \end{Bmatrix} - \begin{Bmatrix} f_{sx}^{(e)} \\ f_{sy}^{(e)} \\ f_{sz}^{(e)} \\ 0 \end{Bmatrix} = 0$$

and in compact notation as

$$C^{(e)}\left\{\frac{\partial U^{(e)}}{\partial t}\right\} + k^{(e)}\left\{U^{(e)}\right\} + \left\{I^{(e)}\right\} = \left\{F^{(e)}\right\} \tag{16.54b}$$

Assembly of all element characteristic equations gives the global system as

$$C\left\{\frac{\partial U}{\partial t}\right\} + K\{U\} = \{F\} \tag{16.54c}$$

16.2.5 Convective Heat and Mass Transfer

For convection heat and mass transfer, we need to consider the solution of flow field followed by the solution of the energy or mass concentration equations.

Let us consider the following general set of governing equations and boundary conditions for forced convection problems involving three-dimensional incompressible viscous flow as:

Continuity:

$$\frac{\partial u}{\partial x} + \frac{\partial v}{\partial y} + \frac{\partial w}{\partial y} = 0 \tag{16.55a}$$

***x*-Momentum**

$$\rho\left(\frac{\partial u}{\partial t} + u\frac{\partial u}{\partial x} + v\frac{\partial u}{\partial y} + w\frac{\partial u}{\partial z}\right) = b_x - \frac{\partial p}{\partial x} + \mu\left(\frac{\partial^2 u}{\partial x^2} + \frac{\partial^2 u}{\partial y^2} + \frac{\partial^2 u}{\partial z^2}\right) \tag{16.55b}$$

***y*-Momentum:**

$$\rho\left(\frac{\partial v}{\partial t}+u\frac{\partial v}{\partial x}+v\frac{\partial v}{\partial y}+w\frac{\partial v}{\partial z}\right)=b_y-\frac{\partial p}{\partial y}+\mu\left(\frac{\partial^2 v}{\partial x^2}+\frac{\partial^2 v}{\partial y^2}+\frac{\partial^2 v}{\partial z^2}\right) \tag{16.55c}$$

***z*-Momentum:**

$$\rho\left(\frac{\partial w}{\partial t}+u\frac{\partial w}{\partial x}+v\frac{\partial w}{\partial y}+w\frac{\partial w}{\partial z}\right)=b_z-\frac{\partial p}{\partial z}+\mu\left(\frac{\partial^2 w}{\partial x^2}+\frac{\partial^2 w}{\partial y^2}+\frac{\partial^2 w}{\partial z^2}\right) \tag{16.55d}$$

Energy Equation:

$$C_\phi\left(\frac{\partial \phi}{\partial t}+u\frac{\partial \phi}{\partial x}+v\frac{\partial \phi}{\partial y}+w\frac{\partial \phi}{\partial z}\right)=\Gamma\left(\frac{\partial^2 \phi}{\partial x^2}+\frac{\partial^2 \phi}{\partial y^2}+\frac{\partial^2 \phi}{\partial z^2}\right)+S \tag{16.55e}$$

where ϕ represents a scalar transport quantity such as temperature, T, in the energy equation or concentration, c, in the mass transport equation. S represents a generation term such as volumetric heat generation, Q''', and/or viscous dissipation term, Φ, in the energy equation, or mass source or sink term in mass transport equation.

Boundary conditions are given in a general form as

$$1.\quad u=u_s,\quad v=v_s \quad \text{and} \quad w=w_s \quad \text{on the boundary } s_1 \tag{16.56a}$$

$$2.\quad \mu\left(\frac{\partial u}{\partial x}\hat{n}_x+\frac{\partial u}{\partial y}\hat{n}_y+\frac{\partial u}{\partial z}\hat{n}_z\right)-p\hat{n}_x=f_{sx}$$

$$\mu\left(\frac{\partial v}{\partial x}\hat{n}_x+\frac{\partial v}{\partial y}\hat{n}_y+\frac{\partial v}{\partial z}\hat{n}_z\right)-p\hat{n}_y=f_{sy} \tag{16.56b}$$

$$\mu\left(\frac{\partial w}{\partial x}\hat{n}_x+\frac{\partial w}{\partial y}\hat{n}_y+\frac{\partial w}{\partial z}\hat{n}_z\right)-p\hat{n}_z=f_{sz} \quad \text{on the boundary } s_2$$

$$3.\quad \phi=\phi_s \quad \text{on the boundary } s_1 \tag{16.56c}$$

$$4.\quad \Gamma\left(\frac{\partial \phi}{\partial x}\hat{n}_x+\frac{\partial \phi}{\partial y}\hat{n}_y+\frac{\partial \phi}{\partial z}\hat{n}_z\right)=f''_{\phi s} \tag{16.56d}$$

where $f''_{\phi s}$ represents a surface flux quantity such surface heat flux, q''_s, or surface mass flux, m''_s.

Following the procedure outlined for the **velocity-pressure** or **mixed finite element method**, the weighted residual statement of the equations over an element $A^{(e)}$

$$-\int_{A^{(e)}} W_p\left\{\frac{\partial u^*}{\partial x}+\frac{\partial v^*}{\partial y}+\frac{\partial w^*}{\partial z}\right\}dA=0 \tag{16.57a}$$

$$\int_{A^{(e)}} W_u\left\{\rho\left(\frac{\partial u^*}{\partial t}+u\frac{\partial u^*}{\partial x}+v\frac{\partial u^*}{\partial y}+w\frac{\partial u^*}{\partial z}\right)-b_x+\frac{\partial p^*}{\partial x}-\mu\left(\frac{\partial^2 u^*}{\partial x^2}+\frac{\partial^2 u^*}{\partial y^2}+\frac{\partial^2 u^*}{\partial z^2}\right)\right\}dA=0 \quad (16.57b)$$

$$\int_{A^{(e)}} W_v\left\{\rho\left(\frac{\partial v^*}{\partial t}+u\frac{\partial v^*}{\partial x}+v\frac{\partial v^*}{\partial y}+w\frac{\partial v^*}{\partial z}\right)-b_y+\frac{\partial p^*}{\partial y}-\mu\left(\frac{\partial^2 v^*}{\partial x^2}+\frac{\partial^2 v^*}{\partial y^2}+\frac{\partial^2 v^*}{\partial z^2}\right)\right\}dA=0 \quad (16.57c)$$

$$\int_{A^{(e)}} W_w\left\{\rho\left(\frac{\partial w^*}{\partial t}+u\frac{\partial w^*}{\partial x}+v\frac{\partial w^*}{\partial y}+w\frac{\partial w^*}{\partial z}\right)-b_z+\frac{\partial p^*}{\partial z}-\mu\left(\frac{\partial^2 w^*}{\partial x^2}+\frac{\partial^2 w^*}{\partial y^2}+\frac{\partial^2 w^*}{\partial z^2}\right)\right\}dA=0 \quad (16.57d)$$

$$\int_{A^{(e)}} W_\phi\left\{C\left(\frac{\partial \phi}{\partial t}+u\frac{\partial \phi^*}{\partial x}+v\frac{\partial \phi^*}{\partial y}+w\frac{\partial \phi^*}{\partial z}\right)-S-\mu\left(\frac{\partial^2 \phi^*}{\partial x^2}+\frac{\partial^2 \phi^*}{\partial y^2}+\frac{\partial^2 \phi^*}{\partial z^2}\right)\right\}dA=0 \quad (16.57e)$$

The approximate solutions for velocity, pressure, and transport quantity are assumed as follows:

$$u^*=\sum_{I=1}^{n_p} N_I(x)u_I(t)=[N(x)]\{u^{(e)}(t)\} \quad (16.58a)$$

$$v^*=\sum_{I=1}^{n_p} N_I(x)v_I(t)=[N(x)]\{v^{(e)}(t)\} \quad (16.58b)$$

$$p^*=\sum_{I=1}^{n_p} N_{pI}(x)p_I(t)=[N_p(x)]\{p^{(e)}(t)\} \quad (16.58c)$$

$$\phi^*=\sum_{I=1}^{n_p} N_{\varphi I}(x)\phi_I(t)=[N_\varphi(x)]\{\phi^{(e)}(t)\} \quad (16.58d)$$

It can be seen that three different shape functions are used for velocity, pressure, and scalar transport quantity requiring three different finite element meshes. However, since the forms of the momentum equations and scalar transport equations are of the same form, we could also use the same type of shape functions for the velocity (u, v, w) and scalar transport ϕ.

In the **Galerkin finite element method**, the weighting functions are assumed to be same as the shape functions used for the approximations functions given by Equation 16.58, i.e.,

$$W_u=W_v=W_w=[N]^T,\quad W_p=[N_P]^T,\quad W_\phi=[N_\phi]^T \quad (16.59)$$

where $[N]$, $[N_p]$ and $[N_\phi]$ are the shape functions associated with velocity, pressure, and scalar transport approximation functions, respectively.

The weighted residual statements given by Equation 16.57 are transformed into weak forms by integrating the integrals. This is followed by substitution of approximation functions given by Equations 16.58 and formation of element characteristics equations in the following form:

$$\left[c^{(e)}\right]\frac{\mathrm{d}\{u^{(e)}\}}{\mathrm{d}t}+\left[k_I^{(e)}(u,v,w)+\left[k_\mu^{(e)}\right]\right]\{u^{(e)}\}-\left[k_{px}^{(e)}\right]\{p^{(e)}\}+\{I_{ix}^{(e)}\}-\{f_x^{(e)}\}=0 \quad \text{(a)}$$

$$\left[c^{(e)}\right]\frac{\mathrm{d}\{v^{(e)}\}}{\mathrm{d}t}+\left[k_I^{(e)}(u,v,w)+\left[k_\mu^{(e)}\right]\right]\{v^{(e)}\}-\left[k_{py}^{(e)}\right]\{p^{(e)}\}+\{I_{iy}^{(e)}\}-\{f_y^{(e)}\}=0 \quad \text{(b)}$$

$$\left[c^{(e)}\right]\frac{\mathrm{d}\{w^{(e)}\}}{\mathrm{d}t}+\left[k_I^{(e)}(u,v,w)+\left[k_\mu^{(e)}\right]\right]\{w^{(e)}\}-\left[k_{pz}^{(e)}\right]\{p^{(e)}\}+\{I_{iz}^{(e)}\}-\{f_z^{(e)}\}=0 \quad \text{(c)} \qquad (16.60)$$

$$-\left[k_{px}^{(e)}\right]^T\{u^{(e)}\}-\left[k_{py}^{(e))}\right]^T\{v^{(e)}\}-\left[k_{pz}^{(e))}\right]^T\{w^{(e)}\}=0 \quad \text{(d)}$$

$$\left[d^{(e)}\right]\frac{\mathrm{d}\{\phi^{(e)}\}}{\mathrm{d}t}+\left[k_{\phi I}^{(e)}(u,v,w)+\left[k_\Gamma^{(e)}\right]\right]\{w^{(e)}\}+\{I_{i\phi}^{(e)}\}-\{f_\phi^{(e)}\}=0 \quad \text{(e)}$$

and in a

$$\begin{bmatrix} c^{(e)} & 0 & 0 & 0 & 0 \\ 0 & c^{(e)} & 0 & 0 & 0 \\ 0 & 0 & c^{(e)} & 0 & 0 \\ 0 & 0 & 0 & 0 & 0 \\ 0 & 0 & 0 & 0 & d^{(e)} \end{bmatrix}\begin{Bmatrix} \frac{\partial u}{\partial t} \\ \frac{\partial v}{\partial t} \\ \frac{\partial w}{\partial t} \\ \frac{\partial p}{\partial t} \\ \frac{\partial \phi}{\partial t} \end{Bmatrix}+\begin{bmatrix} k_I+k_\mu & 0 & 0 & 0 & -k_{px}{}^T \\ 0 & k_I+k_\mu & 0 & 0 & -k_{py}{}^T \\ 0 & 0 & k_I+k_\mu & 0 & -k_{pz}{}^T \\ -k_{px}{}^T & -k_{py}{}^T & -k_{pz}{}^T & 0 & 0 \\ 0 & 0 & 0 & 0 & k_I+k_\mu \end{bmatrix} 0\begin{Bmatrix} u \\ v \\ w \\ p \\ \phi \end{Bmatrix}$$

$$+\begin{Bmatrix} I_{ix} \\ I_{iy} \\ I_{iz} \\ 0 \\ I_{i\phi} \end{Bmatrix}=\begin{Bmatrix} f_x \\ f_y \\ f_z \\ 0 \\ f_\phi \end{Bmatrix} \qquad (16.61a)$$

and in compact notation as

$$C^{(e)}\left\{\frac{\partial U^{(e)}}{\partial t}\right\}+k^{(e)}\{U^{(e)}\}+\{I^{(e)}\}=\{F^{(e)}\} \qquad (16.61b)$$

where

$$c^{(e)}=\int_A \rho[N]^T[N]\mathrm{d}A, d^{(e)}=\int_A C_\phi[N]^T[N]\mathrm{d}A$$

$$\left[k_I^{(e)}(u,v)\right]=\int_{A^{(e)}} \rho\left[N^T\right][N]\left(\{u^{(e)}\}\frac{\partial[N]}{\partial x}+\{v^{(e)}\}\frac{\partial[N]}{\partial y}\right)\mathrm{d}A$$

$$\left[k_{\phi I}^{(e)}(u,v)\right]=\int_{A^{(e)}} C\left[N_\phi^T\right]\left[N_\phi\right]\left(\left\{u^{(e)}\right\}\frac{\partial\left[N_\phi\right]}{\partial x}+\left\{v^{(e)}\right\}\frac{\partial\left[N_\phi\right]}{\partial y}\right)\mathrm{d}A$$

$$k_\mu^{(e)}=\int_{A^{(e)}}\mu\frac{\partial[N]}{\partial x}\frac{\partial[N]^T}{\partial x}\mathrm{d}A+\int_{A^{(e)}}\mu\frac{\partial[N]}{\partial y}\frac{\partial[N]^T}{\partial y}\mathrm{d}A+\int_{A^{(e)}}\mu\frac{\partial[N]}{\partial z}\frac{\partial[N]^T}{\partial z}\mathrm{d}A$$

$$\left[k_{px}^{(e)}\right]=\int_{A^{(e)}}\left[N_p\right]\frac{\partial[N]^T}{\partial x}\mathrm{d}A,\left[k_{py}^{(e)}\right]=\int_{A^{(e)}}\left[N_p\right]\frac{\partial[N]^T}{\partial y}\mathrm{d}A$$

$$\left\{f_x^{(e)}\right\}=\int_{A^{(e)}}[N]^T b_x\mathrm{d}A-\oint_s[N]^T f_{sx}\mathrm{d}s,\left\{f_y^{(e)}\right\}=\int_{A^{(e)}}[N]^T b_y\mathrm{d}A-\oint_s[N]^T f_{sy}\mathrm{d}s$$

$$\left\{f_z^{(e)}\right\}=\int_{A^{(e)}}[N]^T b_z\mathrm{d}A-\oint_s[N]^T f_{sz}\mathrm{d}s,\left\{f_\phi^{(e)}\right\}=\int_{A^{(e)}}\left[N_\phi\right]^T S\mathrm{d}A-\oint_s\left[N_\phi\right]^T f_{\phi s}''\mathrm{d}s$$

$$C^{(e)}=\begin{bmatrix} c^{(e)} & 0 & 0 & 0 & 0\\ 0 & c^{(e)} & 0 & 0 & 0\\ 0 & 0 & c^{(e)} & 0 & 0\\ 0 & 0 & 0 & 0 & 0\\ 0 & 0 & 0 & 0 & d^{(e)}\end{bmatrix},\left\{U^{(e)}\right\}=\begin{Bmatrix} u^{(e)}\\ v^{(e)}\\ w^{(e)}\\ p^{(e)}\\ \phi^{(e)}\end{Bmatrix},\left\{F^{(e)}\right\}=\begin{Bmatrix} f_x\\ f_y\\ f_z\\ 0\\ f_\phi\end{Bmatrix} \quad (16.61c)$$

Assembly of all element characteristic equations gives the global system as

$$C\left\{\frac{\partial U}{\partial t}\right\}+K\{U\}=\{F\} \quad (16.62)$$

16.3 SOLUTION METHODS

Before selecting any particular solution techniques, let us first examine the nature and type of the element and global system of equations given by Equation 16.62. For unsteady-state problems, Equation 16.62 represents a set of first-order nonlinear differential equations. The nonlinearity is caused due to the nonlinear characteristics of the matrices, $k_I^{(e)}(u,v,w)$ and $k_{\phi I}^{(e)}(u,v,w)$, and by variable material properties such as Γ, μ, ρ, and C. Also, the system is nonsymmetric because the element characteristic matrix due to convection is also nonsymmetric. For constant material properties and very low fluid velocities, the system becomes linear and symmetric. The choice of the solution technique differs depending on whether the system is linear or nonlinear, degree of nonlinearity, and symmetric or nonsymmetric. For unsteady problems, the nonlinear differential equations are reduced to a system of nonlinear algebraic equation by using the time approximation scheme. The general solution techniques can be categorized into two basic kinds: (a) linear system of equations that are solved either by direct solver or by iterative methods. (b) For nonlinear problems, the system of nonlinear equations is first linearized or reduced to a system of linear algebraic equations, and then an iterative technique is used to solve the linear system. Let us begin with the discussion of these methods for steady-state problems first.

16.3.1 Steady-State Problems

For steady-state problems, the global system of equations is given as

$$K\{U\}=\{F\} \tag{16.63}$$

and component form as

$$\left[K_I(u,v,w)+K_\mu\right]\{u\}-\left[K_{px}\right]\{p\}-\{F_x\}=0 \tag{16.64a}$$

$$\left[K_I(u,v,w)+K_\mu\right]\{v\}-\left[K_{py}\right]\{p\}-\{F_y\}=0 \tag{16.64b}$$

$$\left[K_I(u,v,w)+K_\mu\right]\{w\}-\left[K_{pz}\right]\{p\}-\{F_z\}=0 \tag{16.64c}$$

$$-\left[k_{px}\right]^T\{u\}-\left[k_{py}\right]^T\{v\}-\left[k_{pz}\right]^T\{w\}=0 \tag{16.64d}$$

Solution to this set of nonlinear equations is usually solved by methods such as **Picard method, Newton-Raphson method,** and **velocity-pressure correction method**.

16.3.1.1 Picard Method

The Picard method for solution of Equations is based on the ***fixed-point iteration method*** *or* ***successive substitution method*** used for the solution of nonlinear algebraic equations (Chapra and Canale, 2002; Rao, 2002). The iterative algorithm for the **Picard method** is given as

$$K(U^i)\left\{U^{i+1}\right\}=\left\{F(U^i\right\} \tag{16.65}$$

where U^i is the unknown vector at previous iteration step and U^{i+1} is the unknown vector at present iteration step. The iteration process is carried out until the convergence criteria given as $\left|\dfrac{U^{i+1}-U^i}{U^{i+1}}\right|\le\varepsilon_s$ are satisfied. In order to enhance the convergence rate and/or eliminate any oscillation in the convergence process, a relaxation scheme is used in the following manner:

$$U^{i+1}=\lambda U^i+(1-\lambda)U^{i+1} \tag{16.66}$$

The inherent slow convergence rate in the method allows a poor initial guess vector. For example, an initial starting vector $U^0=0$ can always be used if no better guess can be made. However, a better initial guess will always enhance the convergence rate.

Also note that Equation 16.65 basically represents a system of linear algebraic equations, which can be solved by using methods used for linear system of Equations, i.e., either by a direct solver such as Gauss elimination or by an iterative solver such as Gauss-Seidel method.

16.3.1.2 Newton-Raphson Method

The Newton-Raphson method is one of the most efficient solution algorithms for solving nonlinear equations (Chapra and Canale, 2002). The algorithm is derived by using a truncated Taylor series expansion. In order to derive the algorithm, let us rewrite Equation 16.63 in the following manner:

$$g(U)=K\{U\}-\{F\}=0 \tag{16.67}$$

and component form as

$$\begin{aligned} g_u(u,v,w,p) &= \left[K_I(u,v,w)+K_\mu\right]\{u\}-\left[K_{px}\right]\{p\}-\{F_x\}=0 \quad \text{(a)}\\ g_v(u,v,w,p) &= \left[K_I(u,v,w)+K_\mu\right]\{v\}-\left[K_{py}\right]\{p\}-\{F_y\}=0 \quad \text{(b)}\\ g_w(u,v,w,p) &= \left[K_I(u,v,w)+K_\mu\right]\{w\}-\left[K_{pz}\right]\{p\}-\{F_z\}=0 \quad \text{(c)}\\ g_p(u,v,w,p) &= -\left[k_{px}\right]^T\{u\}-\left[k_{py}\right]^T\{v\}-\left[k_{pz}\right]^T\{w\}=0 \quad \text{(d)} \end{aligned} \tag{16.68}$$

Truncated multivariable Taylor series expansion of the function $g\ (U)$ is given as

$$g\left(U^{i+1}\right)=g\left(U^i\right)+\left.\frac{\partial g}{\partial U}\right|^i\left(U^{i+1}-U^i\right)+\cdots \tag{16.69}$$

and in component form as

$$g_u\left(u^{i+1},v^{i+1},w^{i+1},p^{i+1}\right)=g_u\left(u^i,v^i,w^i\right)+\left.\frac{\partial g_u}{\partial u}\right|^i\left(u^{i+1}-u^i\right)+\left.\frac{\partial g_u}{\partial v}\right|^i\left(v^{i+1}-v^i\right)$$

$$+\left.\frac{\partial g_u}{\partial w}\right|^i\left(w^{i+1}-w^i\right)+\left.\frac{\partial g_u}{\partial p}\right|^i\left(p^{i+1}-p^i\right)+\cdots \tag{16.70a}$$

$$g_v\left(u^{i+1},v^{i+1},w^{i+1},p^{i+1}\right)=g_v\left(u^i,v^i,w^i\right)+\left.\frac{\partial g_v}{\partial u}\right|^i\left(u^{i+1}-u^i\right)+\left.\frac{\partial g_v}{\partial v}\right|^i\left(v^{i+1}-v^i\right)$$

$$+\left.\frac{\partial g_v}{\partial w}\right|^i\left(w^{i+1}-w^i\right)+\left.\frac{\partial g_v}{\partial p}\right|^i\left(p^{i+1}-p^i\right)+\cdots \tag{16.70b}$$

$$g_w\left(u^{i+1},v^{i+1},w^{i+1},p^{i+1}\right)=g_w\left(u^i,v^i,w^i\right)+\left.\frac{\partial g_w}{\partial u}\right|^i\left(u^{i+1}-u^i\right)+\left.\frac{\partial g_w}{\partial v}\right|^i\left(v^{i+1}-v^i\right)$$

$$+\left.\frac{\partial g_w}{\partial w}\right|^i\left(w^{i+1}-w^i\right)+\left.\frac{\partial g_w}{\partial p}\right|^i\left(p^{i+1}-p^i\right)\cdots \tag{16.70c}$$

$$g_p\left(u^{i+1},v^{i+1},w^{i+1},p^{i+1}\right)=g_p\left(u^i,v^i,w^i\right)+\left.\frac{\partial g_p}{\partial u}\right|^i\left(u^{i+1}-u^i\right)+\left.\frac{\partial g_p}{\partial v}\right|^i\left(v^{i+1}-v^i\right)$$

$$+\left.\frac{\partial g_p}{\partial w}\right|^i\left(w^{i+1}-w^i\right)+\left.\frac{\partial g_p}{\partial p}\right|^i\left(p^{i+1}-p^i\right)+\cdots \tag{16.70d}$$

Equation 16.70 constitutes a system of linear equations, which can be written in matrix form as

$$\begin{bmatrix} \frac{\partial g_u}{\partial u} & \frac{\partial g_u}{\partial v} & \frac{\partial g_u}{\partial w} & \frac{\partial g_u}{\partial p} \\ \frac{\partial g_v}{\partial u} & \frac{\partial g_v}{\partial v} & \frac{\partial g_v}{\partial w} & \frac{\partial g_v}{\partial p} \\ \frac{\partial g_w}{\partial u} & \frac{\partial g_w}{\partial v} & \frac{\partial g_w}{\partial w} & \frac{\partial g_w}{\partial p} \\ \frac{\partial g_p}{\partial u} & \frac{\partial g_p}{\partial u} & \frac{\partial g_p}{\partial u} & \frac{\partial g_p}{\partial u} \end{bmatrix} \begin{Bmatrix} (u^{i+1} - u^i) \\ (v^{i+1} - v^i) \\ (w^{i+1} - w^i) \\ (p^{i+1} - p^i) \end{Bmatrix} = \begin{Bmatrix} g_u(u^i, v^i, w^i, p^i) \\ g_v(u^i, v^i, w^i, p^i) \\ g_w(u^i, v^i, w^i, p^i) \\ g_p(u^i, v^i, w^i, p^i) \end{Bmatrix} \tag{16.71}$$

Defining the **Jacobian matrix** as

$$J = \begin{bmatrix} \frac{\partial g_u}{\partial u} & \frac{\partial g_u}{\partial v} & \frac{\partial g_u}{\partial w} & \frac{\partial g_u}{\partial p} \\ \frac{\partial g_v}{\partial u} & \frac{\partial g_v}{\partial v} & \frac{\partial g_v}{\partial w} & \frac{\partial g_v}{\partial p} \\ \frac{\partial g_w}{\partial u} & \frac{\partial g_w}{\partial v} & \frac{\partial g_w}{\partial w} & \frac{\partial g_w}{\partial p} \\ \frac{\partial g_p}{\partial u} & \frac{\partial g_p}{\partial u} & \frac{\partial g_p}{\partial u} & \frac{\partial g_p}{\partial u} \end{bmatrix} \tag{16.72}$$

and solving, we get Newton-Raphson scheme as

$$U^{i+1} = U^i - J^{-1}(U^i)g(U^i) \tag{16.73}$$

and in component form as

$$u^{i+1} = u^i - J^{-1}(u^i, v^i, w^i, p^i)g_u(u^i, v^i, w^i, p^i) \tag{16.74a}$$

$$v^{i+1} = v^i - J^{-1}(u^i, v^i, w^i, p^i)g_v(u^i, v^i, w^i, p^i) \tag{16.74b}$$

$$w^{i+1} = w^i - J^{-1}(u^i, v^i, w^i, p^i)g_w(u^i, v^i, w^i, p^i) \tag{16.74c}$$

$$p^{i+1} = p^i - J^{-1}(u^i, v^i, w^i, p^i)g_p(u^i, v^i, w^i, p^i) \tag{16.74d}$$

The Jacobian for the problem under consideration can be evaluated from Equation 16.68 as follows:

$$J = \begin{bmatrix} K_I(u^i, v^i, w^i) + K_\mu & 0 & 0 & -K_{px} \\ 0 & K_I(u^i, v^i, w^i) + K_\mu & 0 & -K_{py} \\ 0 & 0 & K_I(u^i, v^i, w^i) + K_\mu & -K_{pz} \\ -K_{px}^T & -K_{py}^T & -K_{pz}^T & 0 \end{bmatrix} \tag{16.75}$$

16.3.1.3 Velocity-Pressure Correction Method

The concept of velocity-pressure correction method in Finite Element Method (FEM) originated from the algorithms such as **SIMPLE** and **SIMPLER** (Patankar, 1980) that are successfully used in finite difference solution of fluid flow and heat transfer problems as discussed in Chapter 8. In order to describe this method, let us consider the global system of equations for three-dimensional steady-state fluid flow Equation 16.63 given as

$$K\{U\}=\{F\} \tag{16.63}$$

and in component form as

$$\left[K_I(u,v,w)+K_\mu\right]\{u\}-\left[K_{px}\right]\{p\}-\{F_x\}=0 \tag{16.76a}$$

$$\left[K_I(u,v,w)+K_\mu\right]\{v\}-\left[K_{py}\right]\{p\}-\{F_y\}=0 \tag{16.76b}$$

$$\left[K_I(u,v,w)+K_\mu\right]\{w\}-\left[K_{pz}\right]\{p\}-\{F_z\}=0 \tag{16.76c}$$

$$-\left[K_{px}\right]^T\{u\}-\left[K_{py}\right]^T\{v\}-\left[K_{pz}\right]^T\{w\}=0 \tag{16.76d}$$

Equations 16.76a–16.76c are written explicitly for u, v, and w in the following manner:

$$\{u\}=\left[K_I(u,v,w)+K_\mu\right]^{-1}\left(\left[K_{px}\right]\{p\}-\{F_x\}\right) \tag{16.77a}$$

$$\{v\}=\left[K_I(u,v,w)+K_\mu\right]^{-1}\left(\left[K_{py}\right]\{p\}-\{F_y\}\right) \tag{16.77b}$$

$$\{w\}=\left[K_I(u,v,w)+K_\mu\right]^{-1}\left(\left[K_{pz}\right]\{p\}-\{F_z\}\right) \tag{16.77c}$$

Substituting Equation 16.77 into Equation 16.76d, we get

$$\begin{aligned}&-\left[K_{px}\right]^T\left[K_I(u,v,w)+K_\mu\right]^{-1}\left(\left[K_{px}\right]\{p\}-\{F_x\}\right)\\&-\left[K_{py}\right]^T\left[K_I(u,v,w)+K_\mu\right]^{-1}\left(\left[K_{py}\right]\{p\}-\{F_y\}\right)\\&-\left[K_{pz}\right]^T\left[K_I(u,v,w)+K_\mu\right]^{-1}\left(\left[K_{pz}\right]\{p\}-\{F_z\}\right)=0\end{aligned}$$

or

$$\begin{aligned}&\left(\left[K_{px}^T\right]\left[K_I(u,v,w)+K_\mu\right]^{-1}+\left[K_{px}^T\right]+\left[K_{py}^T\right]\left[K_I(u,v,w)+K_\mu\right]^{-1}\left[K_{pu}^T\right]\right.\\&\left.+\left[K_{pz}^T\right]\left[K_I(u,v,w)+K_\mu\right]^{-1}\left[K_{pz}^T\right]\right)p=-\left[K_{px}^T\right]\left[K_I(u,v,w)+K_\mu\right]^{-1}\{F_x\}\\&-\left[K_{py}^T\right]\left[K_I(u,v,w)+K_\mu)\right]^{-1}\{F_y\}\\&-\left[K_{pz}^T\right]\left[K_I(u,v,w)+K_\mu\right]^{-1}\{F_z\}\end{aligned} \tag{16.78}$$

Equation 16.78 represents a direct form of the pressure discretization equation. However, an approximation is often made by approximating the $\left[K_I + K_\mu\right]$ by matrix $\left[\hat{K}\right]$ by retaining only the diagonal elements. The approximate pressure discretization equation is written as

$$\left(\left[K_{px}^T\right]\left[\hat{K}\right]^{-1} + \left[K_{px}^T\right] + \left[K_{py}^T\right]\left[\hat{K}\right]^{-1}\left[K_{py}\right] + \left[K_{pz}^T\right]\left[\hat{K}\right]^{-1}\left[K_{pz}\right]\right)p^* = -\left[K_{px}^T\right]\left[\hat{K}\right]^{-1}\{F_x\} - \left[K_{py}^T\right]\left[\hat{K}\right]^{-1}\{F_y\} - \left[K_{pz}^T\right]\left[\hat{K}\right]^{-1}\{F_z\} \tag{16.79}$$

where superscript * is used to indicate an approximate quantity. The matrices $\hat{K}$ and force vector F are evaluated based on velocities u^i, v^i and w^i at previous iteration step. A relaxation scheme is also performed at this time step to establish the pressure estimate at the present iteration step as follows:

$$p^{i+1} = \lambda_p p^i + \left(1 - \lambda_p\right) p^* \tag{16.80}$$

With the known pressure values, an estimate of the velocity components is now given by Equation 16.76 as

$$\left[K_I(u,v,w) + K_\mu\right]\left\{u^*\right\} = \left[K_{px}\right]\left\{p^{i+1}\right\} + \left\{F_x\right\} \tag{16.81a}$$

$$\left[K_I(u,v,w) + K_\mu\right]\left\{v^*\right\} = \left[K_{py}\right]\left\{p^{i+1}\right\} + \left\{F_y\right\} = 0 \tag{16.81b}$$

$$\left[K_I(u,v,w) + K_\mu\right]\left\{w^*\right\} = \left[K_{pz}\right]\left\{p^{i+1}\right\} + \left\{F_z\right\} \tag{16.81c}$$

In the next step, a discretization equation for the corrected pressure is given as

$$\left(\left[K_{px}^T\right]\left[\hat{K}\right]^{-1}\left[K_{px}\right] + \left[K_{px}^T\right] + \left[K_{py}^T\right]\left[\hat{K}\right]^{-1}\left[K_{py}\right] + \left[K_{pz}^T\right]\left[\hat{K}\right]^{-1}\left[K_{pz}\right]\right)p = -\left[K_{px}^T\right]u^* - \left[K_{py}^T\right]v^* - \left[K_{pz}^T\right]w^* \tag{16.82}$$

Finally, the velocity components are corrected by equations

$$\begin{aligned} u^{i+1} &= u^* + \left[\hat{K}\right]^{-1}\left[K_{px}\right]p \quad \text{(a)} \\ v^{i+1} &= v^* + \left[\hat{K}\right]^{-1}\left[K_{py}\right]p \quad \text{(b)} \\ w^{i+1} &= w^* + \left[\hat{K}\right]^{-1}\left[K_{pz}\right]p \quad \text{(c)} \end{aligned} \tag{16.83}$$

Equations 16.79–16.83 are solved iteratively until convergence is achieved. An algorithm based on these equations and concepts is given as follows:

1. Guess the velocity field u, v, and w.
2. Solve the pressure equation (16.79) for the pressure field, p^*
3. Use Equation 16.80 to update pressure estimate at present iteration using relaxation
4. Use this pressure estimate to solve discretization Equations 16.81 for u^*, v^* and w^*.

5. Solve for corrected pressure using (16.82)
6. Calculate velocity components u, v, and w from their starred components and corrected pressure using Equation 16.83
7. Solve the discretization equation for other scalar quantities such as temperature or mass concentration.
8. Repeat Steps 2–7 with the new velocity estimates until solution converges within a specified tolerance limit.

16.3.2 Unsteady-State Problem

For unsteady-state problems, the governing global system of equations is given by Equation 16.62 as

$$C\left\{\frac{\partial U}{\partial t}\right\}+K\{U\}=\{F\} \tag{16.62}$$

and in component form as

$$[c]\frac{\mathrm{d}\{u\}}{\mathrm{d}t}+\left[K_I(u,v,w)+\left[K_\mu\right]\right]\{u\}-\left[K_{px}\right]\{p\}-\{F_x\}=0 \tag{16.84a}$$

$$[c]\frac{\mathrm{d}\{u\}}{\mathrm{d}t}+\left[K_I(u,v,w)+\left[K_\mu\right]\right]\{v\}-\left[K_{py}\right]\{p\}-\{F_y\}=0 \tag{16.84b}$$

$$[c]\frac{\mathrm{d}\{w\}}{\mathrm{d}t}+\left[K_I(u,v,w)+\left[K_\mu\right]\right]\{w\}-\left[K_{pz}\right]\{p\}-\{F_z\}=0 \tag{16.84c}$$

$$-\left[K_{px}\right]^T\{u\}-\left[K_{py}\right]^T\{v\}-\left[K_{pz}\right]^T\{w\}=0 \tag{16.84d}$$

The first-order ordinary differential equations representing the system of global characteristics equations are of the same form as for the characteristic Equation 15.11 for the diffusion problem, and so the equations can be discretized using the similar time approximation schemes discussed in Chapter 15. Since the equation does not involve any time derivative for pressure, complete explicit time integration is not possible. If we use an implicit time approximation, the system of equations given by Equation 16.84 reduces to

$$\left(\left[c^{(e)}\right]+\Delta t\left[k^{(e)}\right]\right)\left\{u^{(e)}\right\}^{m+1}=\left[c^{(e)}\right]\left\{u^{(e)}\right\}^{m}-\left[k_{px}^{(e)}\right]\left\{p^{(e)}\right\}^{m+1}+\Delta t\{I_{ix}\}^{m+1}+\Delta t\left\{f_x^{(e)}\right\}^{m+1} \tag{16.85a}$$

$$\left(\left[c^{(e)}\right]+\Delta t\left[k^{(e)}\right]\right)\left\{v^{(e)}\right\}^{m+1}=\left[c^{(e)}\right]\left\{v^{(e)}\right\}^{m}-\left[k_{py}^{(e)}\right]\left\{p^{(e)}\right\}^{m+1}+\Delta t\{I_{iy}\}^{m+1}+\Delta t\left\{f_y^{(e)}\right\}^{m+1} \tag{16.85b}$$

$$\left(\left[c^{(e)}\right]+\Delta t\left[k^{(e)}\right]\right)\left\{w^{(e)}\right\}^{m+1}=\left[c^{(e)}\right]\left\{w^{(e)}\right\}^{m}-\left[k_{pz}^{(e)}\right]\left\{p^{(e)}\right\}^{m+1}+\Delta t\{I_{iz}\}^{m+1}+\Delta t\left\{f_z^{(e)}\right\}^{m+1} \tag{16.85c}$$

Notice that with use of the time approximation, the equations transformed into a set nonlinear algebraic equations. Procedure discussed for solving nonlinear and linear system of equations now can be used to solve Equation 16.84. More discussions on time approximation schemes including an explicit time integration scheme are given in Reddy and Garting (1994).

PROBLEMS

16.1 Derive the element characteristic equation for all elements in Example 16.1. Show the global system of equations for the mesh shown.

16.2 Derive the element characteristic equation for u, v, and pressure p using bilinear rectangular element for velocity components and constant pressure at the center of the rectangle based on Equations 16.28 and 16.30.

16.3 Consider fully developed and heat transfer in a duct of rectangular cross-section as shown in the figure.

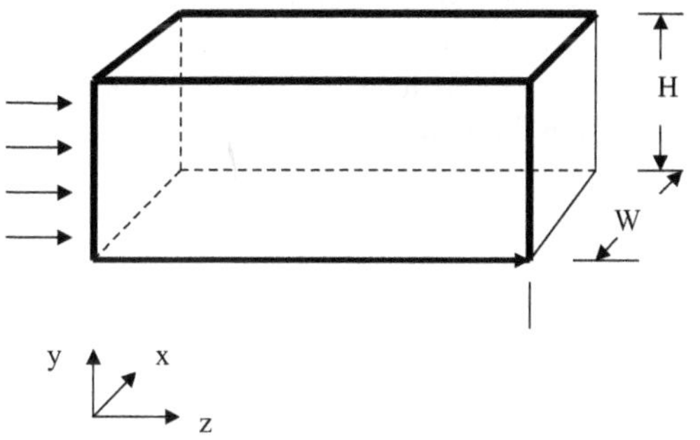

The top surface is maintained at a constant surface heat flux q''_w and all other three surfaces are adiabatic. The associated governing equations and boundary conditions for the fully developed flow and heat transfer with constant surface heat flux are given as follows:

Momentum:

$$\nu\left(\frac{\partial^2 w}{\partial x^2}+\frac{\partial^2 w}{\partial y^2}\right)=-\frac{1}{\rho}\frac{\partial \bar{p}}{\partial z}$$

Energy:

$$\frac{k}{\rho c_p}\left(\frac{\partial^2 T}{\partial x^2}+\frac{\partial^2 T}{\partial y^2}\right)=w\frac{\partial \bar{T}}{\partial z}$$

The pressure gradient $\frac{\mathrm{d}\bar{p}}{\mathrm{d}z}$ is assumed to be known. The temperature gradient $\frac{\partial \bar{T}}{\partial x}$ is constant and given based on an energy balance as

$$\frac{\partial \bar{T}}{\partial x}=\frac{q''_w}{\rho c_p \bar{w}}$$

where $\bar{w}$ and $\bar{T}$ are the average velocity and average temperature given as

$$\bar{w}=\frac{1}{WH}\int_A w\,\mathrm{d}A \quad \text{and} \quad \bar{T}=\frac{1}{WH\bar{w}}\int_A wT\,\mathrm{d}A$$

Boundary Conditions:

No-slip velocity on all wall, i.e., $w=0$ on all walls

Constant surface heat flux on top wall, i.e., at $y=H$, $k\dfrac{\partial T}{\partial x}=q''_w$

Rest of the walls are adiabatic, i.e., $\dfrac{\partial T}{\partial n}=0$

Derive the element characteristic equations for w and T using Galerkin's weighted residual formulation and using a mesh consisting of bilinear rectangular elements. Present the sequence of steps to be used for calculating velocity and temperature field and convection heat transfer coefficient.

16.4 Consider free-convection motion and heat transfer in two-dimensional rectangular chamber as shown. The free convection motion is created by maintaining the left wall at a high-temperature T_H and right surface at a low-temperature T_C. The top and bottom assumed as insulated. The governing equations for mass, momentum, and energy with Boussinesq assumption and for Newtonian fluid are given as:

Continuity:

$$\frac{\partial u}{\partial x}+\frac{\partial v}{\partial y}=0$$

x-Momentum:

$$u\frac{\partial u}{\partial x}+v\frac{\partial u}{\partial y}=-\frac{1}{\rho}\frac{\partial p}{\partial x}+\nu\left(\frac{\partial^2 u}{\partial x^2}+\frac{\partial^2 u}{\partial y^2}\right)$$

y-Momentum:

$$u\frac{\partial v}{\partial x}+v\frac{\partial v}{\partial y}=-\frac{1}{\rho}\frac{\partial p}{\partial y}+\nu\left(\frac{\partial^2 v}{\partial x^2}+\frac{\partial^2 v}{\partial y^2}\right)-g\left[1-\beta(T-T_0)\right]$$

Energy:

$$u\frac{\partial T}{\partial x}+v\frac{\partial T}{\partial y}=\alpha\left(\frac{\partial^2 T}{\partial x^2}+\frac{\partial^2 T}{\partial y^2}\right)$$

Boundary Conditions

Velocity: No slip condition on all walls

Temperature

1. at $x=0$, $0<y<H$, $T(0,y)=T_H$
2. at $x=W$, $0<y<H$, $T(0,W)=T_C$
3. at $y=0$, $0<x<W$, $\left.\dfrac{\partial T}{\partial y}\right|_{y=0}=0$
4. at $y=H$, $0<x<W$, $\left.\dfrac{\partial T}{\partial y}\right|_{y=H}=0$

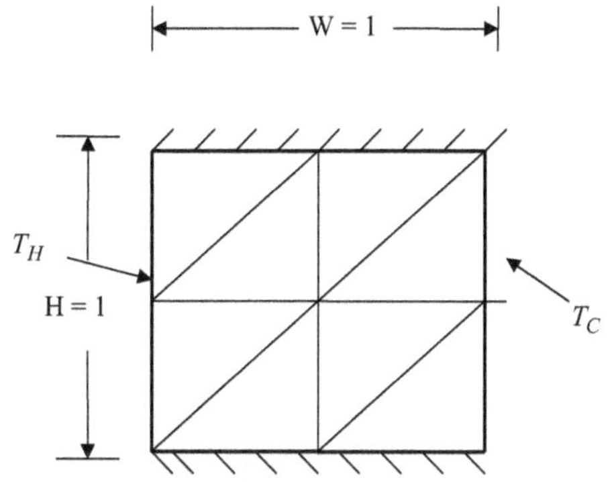

a. Derive the element characteristic equations for u, v, p, and T using a mesh consisting of linear triangular elements as shown. Show the system of equations for each variable.
b. Derive the correction equations for u, v, and p
c. Develop a computer program implementing the SIMPLE algorithm for velocity and pressure coupling and a direct solver for solving the system of equation for each variable.

16.5 Derive the element characteristic equation for an interior element in a two-dimensional channel assuming linear triangular elements. Consider the flow as unsteady Stokes flow with governing equations given by Equations 16.45a–16.45c.

Appendix A
Review of Vectors and Matrices

A.1 VECTORS

A.1.1 Symbols

A vector is written in terms of its components as

$$\tilde{a} = \hat{i}a_x + \hat{j}a_y + \hat{k}a_z \tag{A.1a}$$

where $\hat{i}$, $\hat{j}$, and $\hat{k}$ are unit vectors in x-, y-, and z–direction, respectively.

The unit vector in the direction of $\tilde{a}$ is given as

$$\hat{i}_a = \frac{\tilde{a}}{|\tilde{a}|} \tag{A.1b}$$

Another notation for the unit vectors in x-, y-, and z-directions are given as n_x, n_y, and n_z, which are simply the components of a unit vector $\hat{n}$ normal to an arbitrary surface. Physically, these components are represented by the direction cosine of the unit vector $\hat{n}$ in x-, y-, and z-directions, and written as

$$n_x = \cos\theta_{xn}, \quad n_y = \cos\theta_{yn} \quad \text{and} \quad n_z = \cos\theta_{zn}$$

A.1.2 Vector Norm

A **norm** is a single-valued quantity that represents the size of vectors or and matrices. For example, for a vector $\{x\}$, *Euclidean norm* is defined as

$$\|x\|_e = \sqrt{\sum_{i=1}^{n} x_i^2} = \sqrt{\left((x_1)^2 + (x_2)^2 \cdots + (x_n)\right)^2} \tag{A.2}$$

An alternative to *Euclidean norm* is given in terms of **uniform vector norm**, which is defined as

$$\|x\|_\infty = \max_{1 \le i \le n} |x_i| \tag{A.3}$$

With a *uniform vector norm*, the element with the largest absolute value is used to represent the size of the vector.

A.1.3 Dot or Scalar Product

Dot product of two vectors $\tilde{a}$ and $\tilde{b}$ is defined as

$$\tilde{a} \cdot \tilde{b} = |\tilde{a}||\tilde{b}|\cos\theta_{ab} \tag{A.4}$$

where θ_{ab} is the included angle between two vectors $\tilde{a}$ and $\tilde{b}$. Physically, a dot product represents the product of $|\tilde{a}|$ with the component of $\tilde{b}$ in the direction of $\tilde{a}$.

Example:

Volume flow rate through a differential area vector $\tilde{\mathrm{d}}A$ is given by the product of the component of velocity $\tilde{V}$ normal to an area and the area and expressed as

$$d\forall = \left|\tilde{V}\right| \cos\theta_{vdA} \left|\tilde{\mathrm{d}}A\right| = \tilde{V} \cdot \tilde{\mathrm{d}}A = \hat{V} \cdot \hat{n}\mathrm{d}A \tag{A.5}$$

A.1.3 Cross or Vector Product or Curl

Cross product or curl of two vectors $\tilde{a}$ and $\tilde{b}$ is a vector $\tilde{c}$, and it is written as

$$\tilde{c} = \tilde{a} \times \tilde{b} \tag{A.6}$$

Physically, its magnitude is given as

$$\tilde{c} = \tilde{a} \times \tilde{b} = |\tilde{a}||\tilde{b}| \sin\theta_{ab} \tag{A.7}$$

and its sense is given by the right-hand rule, i.e., as $\tilde{a}$ is rotated into $\tilde{b}$, the vector $\tilde{c} = \tilde{a} \times \tilde{b}$ points vertically upward and perpendicular to both $\tilde{a}$ and $\tilde{b}$.

The cross-product vector $\tilde{c} = \tilde{a} \times \tilde{b}$ is evaluated as follows:

$$\tilde{c} = \tilde{a} \times \tilde{b} = \begin{vmatrix} \hat{i} & \hat{j} & \hat{k} \\ a_x & a_y & a_z \\ b_x & b_y & b_z \end{vmatrix} \tag{A.8}$$

$$= \hat{i}\left(a_y b_z - a_z b_y\right) - \hat{j}\left(a_x b_z - a_z b_x\right) + \vec{k}\left(a_x b_y - a_y b_x\right)$$

Example

Torque given by a force vector $\tilde{F}$ acting at a distance $\tilde{x}$ is given as

$$\tilde{T} = |\tilde{r}||\tilde{F}| \sin\theta_{Fr} = \tilde{r} \times \tilde{F} \tag{A.9}$$

VECTOR OPERATOR

Del, ∇

Cartesian Co-ordinate

$$\nabla = \hat{i}\frac{\partial}{\partial x} + \hat{j}\frac{\partial}{\partial y} + \hat{k}\frac{\partial}{\partial z} \tag{A.10a}$$

or

$$\nabla = n_x \frac{\partial}{\partial x} + n_y \frac{\partial}{\partial y} + n_z \frac{\partial}{\partial z} = \hat{n} \cdot \nabla \tag{A.10b}$$

Cylindrical Co-ordinate

$$\nabla = \hat{i}_r \frac{\partial}{\partial r} + \hat{i}_\theta \frac{1}{r}\frac{\partial}{\partial \theta} + \hat{i}_z \frac{\partial}{\partial z} \tag{A.10c}$$

Gradient

Gradient is the operation of ∇ on a differentiable scalar function such as pressure or temperature. For example, gradient of a scalar field, ϕ, is written as

$$\nabla\phi = \hat{i}\frac{\partial\phi}{\partial x} + \hat{j}\frac{\partial\phi}{\partial y} + \hat{k}\frac{\partial\phi}{\partial z} \tag{A.11}$$

Divergence

Divergence is the dot product of ∇ and a vector function such as velocity field. For example, the divergence of the velocity field is given as

$$\nabla.\vec{V} = \left(\hat{i}\frac{\partial}{\partial x} + \hat{j}\frac{\partial}{\partial y} + \hat{k}\frac{\partial}{\partial z}\right).\left(\hat{i}u + \hat{j}v + \hat{k}w\right)$$

$$= \left(\frac{\partial u}{\partial x} + \frac{\partial v}{\partial y} + \frac{\partial w}{\partial z}\right) \tag{A.12}$$

Laplacian Operator

It is the dot product of two Del operators:

Cartesian Co-ordinate

$$\nabla\cdot\nabla = \nabla^2 = \frac{\partial^2}{\partial x^2} + \frac{\partial^2}{\partial y^2} + \frac{\partial^2}{\partial z^2} \tag{A.13a}$$

Laplacian of a function ϕ is given as

$$\nabla^2\phi = \frac{\partial^2\phi}{\partial x^2} + \frac{\partial^2\phi}{\partial y^2} + \frac{\partial^2\phi}{\partial z^2} \tag{A.13b}$$

Cylindrical Co-ordinate

$$\nabla\cdot\nabla = \nabla^2 = \frac{\partial^2}{\partial r^2} + \frac{1}{r}\frac{\partial}{\partial r} + \frac{1}{r^2}\frac{\partial^2}{\partial\theta^2} + \frac{\partial^2}{\partial z^2} \tag{A.14a}$$

Laplacian of the function ϕ is

$$\nabla^2\phi = \frac{\partial^2\phi}{\partial r^2} + \frac{1}{r}\frac{\partial\phi}{\partial r} + \frac{1}{r^2}\frac{\partial^2\phi}{\partial\theta^2} + \frac{\partial^2\phi}{\partial z^2} \tag{A.14b}$$

A.2 MATRICES

Coefficient Matrix

A co-efficient matrix is a matrix whose elements are the coefficients of the unknowns in a system of equations. For example, the coefficient matrix for the system of equation given by Equation 3.1 is

$$A = \begin{bmatrix} a_{11} & a_{12} & \cdots & \cdots & \cdots & a_{1n} \\ a_{21} & a_{22} & \cdots & \cdots & \cdots & a_{2n} \\ \vdots & & & & & \\ \vdots & & & & & \\ \vdots & & & & & \\ a_{n1} & a_{n2} & \cdots & \cdots & \cdots & a_{nn} \end{bmatrix} \tag{A-15}$$

The elements of the matrix A are specified a_{ij}, where index i represents row number and varies from 1,2, … n. The index j represents column number and varies from 1, 2…n.

If matrix A is symmetric, then

$$a_{ij} = a_{ji} \tag{A.16}$$

Augmented Matrix

Augmented matrix is formed by adding the constant vector of the system to the coefficient matrix as a column of elements

$$A = \begin{bmatrix} a_{11} & a_{11} & & a_{11} & c_1 \\ a_{21} & a_{21} & & a_{21} & c_2 \\ \vdots & & & & \\ \vdots & & & & \\ \vdots & & & & \\ a_{n1} & a_{n1} & & a_{n1} & c_n \end{bmatrix}$$

$$= \begin{bmatrix} a_{11} & a_{11} & & a_{11} & a_{1m} \\ a_{21} & a_{21} & & a_{21} & a_{2m} \\ \vdots & & & & \\ \vdots & & & & \\ \vdots & & & & \\ a_{n1} & a_{n2} & & a_{nn} & c_{nm} \end{bmatrix} = a_{i,j}$$

$$= \begin{bmatrix} a_{11} & a_{11} & & a_{11} & a_{1m} \\ a_{21} & a_{21} & & a_{21} & a_{2m} \\ \vdots & & & & \\ \vdots & & & & \\ \vdots & & & & \\ a_{n1} & a_{n2} & & a_{nn} & c_{nm} \end{bmatrix} = a_{i,j} \tag{A.17}$$

where index i represents the row number and varies from 1, 2…n. The index j represents the column number and varies from 1,2…m, and $m = n + 1$, respectively.

A **square matrix** is a matrix with number of rows is same as number of columns, i.e., $m = n$. A **diagonal matrix** is a square matrix with all elements off the main diagonal are equal to zero.

$$\begin{bmatrix} a_{11} & & & & & \\ & a_{22} & & & & \\ & & a_{33} & & & \\ & & & \ddots & & \\ & & & & \ddots & \\ & & & & & a_{nn} \end{bmatrix} \tag{A.18}$$

Notice that all zero elements are left blank.

Lower Triangular Matrix

A matrix is termed as a **lower triangular matrix** if all elements above the main diagonal are zero. For example,

$$L = A = \begin{bmatrix} a_{11} & & & & & \\ a_{21} & a_{22} & & & & \\ a_{31} & a_{32} & a_{33} & & & \\ \vdots & & & \ddots & & \\ \vdots & & & & \ddots & \\ a_{n1} & a_{n2} & a_{n3} & \cdots & \cdots & a_{nn} \end{bmatrix} \tag{A.19}$$

Upper Triangular Matrix

A matrix is termed as an **upper triangular matrix** if all elements below the main diagonal are zero. For example,

$$U = A = \begin{bmatrix} a_{11} & a_{12} & a_{13} & \cdots & \cdots & a_{1n} \\ & a_{22} & a_{23} & \cdots & \cdots & a_{2n} \\ & & a_{33} & & & a_{3n} \\ & & & \ddots & & \vdots \\ & & & & \ddots & \vdots \\ & & & & & a_{nn} \end{bmatrix} \tag{A.20}$$

Banded Matrix

A matrix that has all elements equal to zero except for a band centered around the main diagonal. For example,

$$A = \begin{bmatrix} a_{11} & a_{12} & a_{12} & & & \\ a_{21} & a_{22} & a_{23} & a_{24} & & \\ a_{31} & a_{32} & a_{33} & a_{34} & \ddots & \\ & a_{42} & a_{43} & \ddots & \ddots & \\ & & \ddots & \ddots & \ddots & a_{n-1,n} \\ & & & a_{n,n-2} & a_{n,n-1} & a_{nn} \end{bmatrix} \tag{A.21}$$

Tridiagonal Matrix

A tridiagonal matrix is banded matrix with a bandwidth of three. For example:

$$A = \begin{bmatrix} a_{11} & a_{12} & & & & \\ a_{21} & a_{22} & a_{23} & & & \\ & a_{32} & a_{33} & a_{34} & & \\ & & a_{43} & \ddots & \ddots & \\ & & & \ddots & \ddots & a_{n-1,n} \\ & & & & a_{n,n-1} & a_{nn} \end{bmatrix} \tag{A.22}$$

Equality of Matrices

$$A = B, \text{ if } a_{ij} = b_{ij},\ i = 1, 2, \ldots n;\ j = 1,2, \ldots n$$

Summation of Matrices

$$A + B = C, \text{ if } a_{ij} + b_{ij} = c_{ij},\ i = 1, 2 \ldots n;\ j = 1, 2, \ldots n$$

Difference of Matrices

$$A - B = C, \text{ if } a_{ij} - b_{ij} = c_{ij},\ i = 1, 2 \ldots n;\ j = 1, 2, \ldots n$$

Matrix Multiplication

Product of two matrices $A = \left[a_{ij}\right]$ and $B = \left[b_{ij}\right]$ is written as

$$\left[a_{ij}\right]\left[b_{ij}\right] = \left[c_{ij}\right] \tag{A.23}$$

or

$A\,B = C$

where the elements of the product matrix are given by

$$c_{ij} = \sum_{k=1}^{n} a_{ik} b_{kj} \tag{A.24}$$

Properties of Matrix Multiplication

1. Matrix multiplication is not commutative, i.e., $AB \neq BA$
2. Product of two matrices A and B is defined if the number of columns in A is equal to number of rows in B.
3. Multiplication of a matrix A $(n \times n)$ by a column matrix B $(n \times 1)$ results in a column matrix C $(n \times 1)$.
4. Matrix multiplication is associative. For example,

   $ABC = (AB)\,C = A\,(BC)$

5. Premultiplication of the product AB by C means

   $CAB = D$

6. Postmultiplication of AB by C means

   $ABC = E$

Identity Matrix or Unit Matrix

It is a matrix with 1's on the main diagonal and all off-diagonal elements as zero. So, basically it is diagonal matrix with all elements on the main diagonal are 1's.

$$I = \begin{bmatrix} 1 & & & & & \\ & 1 & & & & \\ & & 1 & & & \\ & & & \ddots & & \\ & & & & \ddots & \\ & & & & & 1 \end{bmatrix} \tag{A.25}$$

It gives following matrix properties:

$$I\,A = A\,I = A \tag{A.26}$$

Zero Matrix

A matrix with all elements as zero. For example,

$$\begin{bmatrix} 0 & 0 & 0 & \cdots & \cdots & 0 \\ 0 & 0 & 0 & \cdots & \cdots & 0 \\ 0 & 0 & 0 & \cdots & \cdots & 0 \\ \vdots & & & \ddots & & \\ \vdots & & & & \ddots & \\ 0 & 0 & 0 & \cdots & \cdots & 0 \end{bmatrix} \tag{A.27}$$

Transpose of a Matrix

Transpose of a matrix is obtained by interchanging the row and columns. For example, the transpose of the matrix A as given by Equation A.1 is

$$A^T = \begin{bmatrix} a_{11} & a_{21} & \cdots & \cdots & \cdots & a_{n1} \\ a_{12} & a_{22} & \cdots & \cdots & \cdots & a_{n2} \\ \vdots & & & & & \\ \vdots & & & & & \\ a_{1n} & a_{2n} & \cdots & \cdots & \cdots & a_{nn} \end{bmatrix} \tag{A.28}$$

If A is a symmetric matrix, then

$$A = A^T$$

One important property of transpose can be demonstrated as

$$[BA]^T = B^T A^T \tag{A.29}$$

Minor (M_{ij})

The minor of an element a_{ij} in a coefficient matrix A is defined as the determinant of the square array obtained by deleting the ith row and j^{th} column. For example, the minor of the element a_{11} of the matrix

$$A = \begin{bmatrix} a_{11} & a_{12} & a_{13} \\ a_{21} & a_{22} & a_{23} \\ a_{31} & a_{32} & a_{33} \end{bmatrix} \tag{A.30a}$$

is

$$M_{11} = \begin{vmatrix} a_{22} & a_{23} \\ a_{32} & a_{33} \end{vmatrix} \tag{A.30b}$$

Co-Factor (F_{ij})

The co-factor F_{ij} is defined as the result of changing the sign of the minor by $(-1)^{i+j}$ as

$$F_{ij} = (-1)^{i+j} M_{ij} \tag{A.31}$$

Determinant $(|A|)$

It is the sum of the products of each element by its co-factor in any row of the matrix. For example, the determinant of the matrix A is given as

$$|A| = \sum_{\substack{i=1 \\ j=1,2\ldots n}} a_{ij} F_{ij} = \sum_{\substack{i=1 \\ j=1,2\ldots n}} (-1)^{i+j} a_{ij} M_{ij} \tag{A.32}$$

Examples

Determinant of the 2×2 matrix

$$\mathrm{A} = \begin{bmatrix} a_{11} & a_{12} \\ a_{21} & a_{22} \end{bmatrix}$$

is

$$|A| = \begin{vmatrix} a_{22} & a_{23} \\ a_{32} & a_{33} \end{vmatrix} = a_{11}a_{22} - a_{12}a_{21}$$

Determinant of the 3×3 matrix

$$A = \begin{bmatrix} a_{11} & a_{12} & a_{13} \\ a_{21} & a_{22} & a_{23} \\ a_{31} & a_{32} & a_{33} \end{bmatrix}$$

is

$$|A| = \begin{vmatrix} a_{11} & a_{12} & a_{13} \\ a_{21} & a_{22} & a_{23} \\ a_{31} & a_{32} & a_{33} \end{vmatrix}$$

Using Equation A.30

$$|A| = a_{11}\begin{vmatrix} a_{22} & a_{23} \\ a_{32} & a_{33-} \end{vmatrix} - a_{12}\begin{vmatrix} a_{21} & a_{23} \\ a_{31} & a_{33} \end{vmatrix} + a_{13}\begin{vmatrix} a_{21} & a_{22} \\ a_{31} & a_{32} \end{vmatrix}$$

$$|A| = a_{11}\left(a_{22}a_{33} - a_{23}a_{32}\right) - a_{12}\left(a_{21}a_{33} - a_{23}a_{31}\right) + a_{13}\left(a_{21}a_{32} - a_{22}a_{31}\right)$$

Inverse of a Matrix

The inverse of a square matrix A is written as A^{-1} and it is defined by the relationship

$$A^{-1}A = AA^{-1} = I \tag{A.33}$$

The inverse of a matrix is determined by

$$A^{-1} = \frac{adj[A]}{|A|} \tag{A.34}$$

where $adj[A]$ is defined as the adjoint of a matrix A and it is obtained by replacing each element a_{ij} of matrix by its co-factor F_{ij} and then taking a transpose of the resulting matrix. So, adjoint of a matrix is written as

$$\text{Adj}[A] = \begin{bmatrix} F_{11} & F_{12} & \cdots & \cdots & F_{1n} \\ F_{21} & F_{22} & & & F_{2n} \\ \vdots & & & & \\ \vdots & & & & \\ F_{n1} & F_{n2} & \cdots & \cdots & F_{nn} \end{bmatrix}^T$$

$$= \begin{bmatrix} F_{11} & F_{21} & \cdots & \cdots & F_{n1} \\ F_{12} & F_{22} & & & F_{n2} \\ \vdots & & & & \\ \vdots & & & & \\ F_{1n} & F_{2n} & \cdots & \cdots & F_{nn} \end{bmatrix} \tag{A.35}$$

Orthogonal Matrix

A square matrix A is said to be orthogonal if

$$A^T = A^{-1} \tag{A.36}$$

Dense Matrix

A matrix is termed as dense if it contains few zero elements and the size is relatively small.

Sparse Matrix

A large matrix with many zero elements is termed as sparse matrix.

Well-Conditioned System

A system is termed as **well-conditioned** if a small relative change in one or more of the coefficients of a system of equations results in a small change in the solution.

Ill-Conditioned System

A system is ill-conditioned if a small change in one or more of the coefficients results in a large change in the solution of the system.

Singular Matrix

A matrix is termed as singular if the determinant of the matrix $|A|$ is zero. A system that is close to singular poses considerable difficulties in resulting into a solution.

Quadratic Form

For a $n \times n$ matrix A and a $n \times 1$ vector, the scalar quantity $x^T Ax$ is called a quadratic form. The quadratic form can be shown by expanding $x^T Ax$ as follows:

$$x^T Ax = \left\{ \begin{matrix} x_1 & x_2 & x_3 & \cdots & x_n \end{matrix} \right\} \begin{bmatrix} a_{11} & a_{12} & a_{13} & \cdots & a_{1n} \\ a_{21} & a_{22} & a_{23} & \cdots & a_{2n} \\ a_{31} & a_{32} & a_{33} & \cdots & a_{3n} \\ \vdots & & & & \\ a_{n1} & a_{n2} & a_{n3} & \cdots & a_{nn} \end{bmatrix} \begin{Bmatrix} x_1 \\ x_2 \\ x_3 \\ \vdots \\ x_n \end{Bmatrix}$$

$$\begin{aligned} &= x_1 a_{11} x_1 + x_1 a_{12} x_2 + x_1 a_{13} x_3 + \cdots + x_1 a_{1n} x_n \\ &+ x_2 a_{21} x_1 + x_2 a_{22} x_2 + x_2 a_{23} x_3 + \cdots + x_2 a_{2n} x_n \\ &+ x_3 a_{31} x_1 + x_3 a_{32} x_2 + x_3 a_{33} x_3 + \cdots + x_3 a_{3n} x_n \\ &\vdots \\ &+ x_n a_{n1} x_1 + x_n a_{n2} x_2 + x_n a_{n3} x_3 + \cdots + x_n a_{nn} x_n \end{aligned} \quad \text{(A.37)}$$

Eigenvalues and Eigenvectors

For a nontrivial solution to exist for a homogeneous system of equations such as $Ax = c$, an **eigenvalue problem** is defined as

$$Ax = \lambda x \quad \text{(A.38a)}$$

and the matrix $A - \lambda I$ is singular, i.e.,

$$|A - \lambda I| = 0 \quad \text{(A.38b)}$$

Equation A.38b is referred to as the characteristic equation and its n roots $\lambda_1, \lambda_2 \ldots \lambda_n$ are the eigenvalues. For each eigenvalue λ_i, the corresponding eigenvector x is obtained from

$$(A - \lambda_i I)x = 0 \quad \text{(A.39)}$$

Positive Definite Matrix

A real symmetric matrix A is positive definite if the real quadric form $x^T Ax$ is nonnegative for all real values of the vector, x, i.e.,

$$x^T Ax > 0 \quad \text{(A.40)}$$

Note that the quadratic form vanishes only if each of those vector elements is zero.

Matrix Norm

As in case of a one-dimensional vector, **a matrix norm** is a single-valued quantity that represents the size of a two-dimensional vector array or matrix array $a_{i,j}$. A **Frobenius norm** is defined as

$$\left\|a_{ij}\right\|_e = \sqrt{\sum_{i=1}^{n}\sum_{j=1}^{n}(a_{ij})^2} \tag{A.41}$$

A **uniform matrix norm** or **row sum norm** is defined as

$$\left\|a_{ij}\right\|_\infty = \max_{1\le i\le n}\sum_{j=1}^{n}\left|a_{ij}\right| \tag{A.42}$$

Matrix Condition Number

The matrix condition number is used to assess the condition of a system and it is defined as

$$\mathrm{Cond}[\mathrm{A}] = \|A\|\cdot\left\|A^{-1}\right\| \tag{A.43}$$

The condition number is equal to or greater than one. Greater the condition number of a system compared to one, more ill-conditioned the system is.

Appendix B
Integral Theorems

B.1 GRADIENT THEOREM

If $\forall$ is volume bounded by a surface A and φ is a scalar, then, the gradient theorem states that

$$\iiint_{\forall} \nabla\phi \mathrm{d}\forall = \iint_{A} \hat{n}\phi \mathrm{d}A = \iint_{A} \phi \cdot \mathrm{d}\tilde{A} \tag{B.1}$$

Similarly, for an area A bounded by a contour s,

$$\iint_{A} \nabla\phi \mathrm{d}A = \oint_{s} \hat{n}\phi \mathrm{d}s \tag{B.2}$$

Basically, the gradient theorem changes the volume integral into a surface integral and area integral into a line integral involving a scalar, and vice versa.

B.1.1 Divergence Theorem

If $\forall$ is volume bounded by a surface A and $\tilde{\phi}$ is a vector, then, the divergence theorem states that

$$\iiint_{\forall} \nabla \cdot \tilde{\phi} \mathrm{d}\forall = \iint_{A} \hat{n} \cdot \tilde{\phi} \mathrm{d}A = \iint_{A} \tilde{\phi} \cdot \mathrm{d}\tilde{A} \tag{B.3}$$

where dA is the differential area vector on the surface A.

Similarly, for an area A bounded by a contour s,

$$\iint_{A} \nabla \cdot \tilde{\phi} dA = \oint_{s} \hat{n} \cdot \tilde{\phi} \mathrm{d}s = \oint_{s} \tilde{\phi} \cdot \mathrm{d}\tilde{s} \tag{B.4}$$

Basically, the divergence theorem changes the volume integral into a surface integral and area integral into a line integral involving a vector, and vice versa.

Bibliography

Abdel-Wahed, R. M., S. V. Patankar, and E. M. Sparrow, *Letters in Heat and Mass Transfer*, Vol. 3, pp. 355–364, 1976.

Akai, T. J., *Applied Numerical Methods for Engineers, Z*, John Wiley & Sons, 1994.

Anderson, D. A., J. C. Tannehill, and R. H. Pletcher, *Computational Fluid Mechanics and Heat Transfer*, McGraw-Hill, New York, 1984.

Angirasa, D., M. J. B. Pourquie, and F. T. M. Nieuwstadt, "Numerical Study of Transient and Steady Laminar Buoyancy-Driven Flows and Heat Transfer in a Square Open Cavity," *Journal of Numerical Heat Transfer, Part A*, Vol. 22, pp. 223–239, 1992.

Arnoldi, W., "The Principle of Minimized Iteration in the Solution of the Matrix Eigenvalue Problem, Quarterly," *Applied Mathematics*, 9, pp. 17–29, 1951.

Arpaci, V., *Conduction Heat Transfer*, Addison-Wesley, Menlo Park, CA, 1966.

Baker, A. J., *Finite Element Computational Fluid Mechanics*, Taylor & Francis, New York, 1991.

Baker, A. J., and D. W. Pepper, *Finite Elements 1–2–3*, McGraw-Hill, New York, 1991.

Batchelor, G. K., *Fluid Dynamics*, Cambridge University Press, Cambridge, 1967.

Bathe, K.-J., *Finite Element Procedures*, Prentice Hall, Englewood Cliffs, NJ, 1996.

Bathe, K.-J., *Finite Element Procedures in Engineering Analysis*, Prentice Hall, Englewood Cliffs, NJ, 1982.

Bathe, K.-J., and E. L. Wilson, *Numerical Methods in Finite Element Analysis*, Prentice Hall, Englewood Cliffs, NJ, 1976.

Bayazittoglu, Y., and M. N. Izisik, *Elements of Heat Transfer*, McGraw-Hill, New York, 1988.

Becker, M., *The Principles and Applications of Variational Methods*, MIT Press, Cambridge, MA, 1964.

Bejan, A., *Convection Heat Transfer*, John Wiley, New York, 1984.

Bejan, A., *Heat Transfer*, John Wiley, New York, 1993.

Benim, A., and W. Zinser, "A Segregated Formulation of Navier–Stokes Equations with Finite Elements," *Computer-Methods in Applied Mechanics and Engineering*, Vol. 57, pp. 223–237, 1986.

Bennett, C. O., and J. E. Myers, *Momentum, Heat, and Mass Transfer*, McGraw-Hill, New York, 1982.

Bickford, W. B., *A First Course in the Finite Element Method*, Eichard D. Irwin, Inc., Burr Ridge, IL, 1994.

Biot, M. A., Variational Principles in Heat Transfer,

Bird, R. B., W. E. Stewart, and E. N. Lightfoot, *Transport Phenomena*, Second Edition, Wiley, New York, 2002.

Bliss, G. A., *Calculus of Variations*, Mathematical Association of America, The Open Court Publishing Company, La sale, IL, 1925.

Bliss, G. A., *Lectures on the Calculus of Variations*, University of Chicago Press, Chicago, IL, 1946.

Bolza, O., *Lectures on the Calculus of Variations*, Stechert-Hafner, Inc., New York, 1946.

Brandt, A., J. Brannick, K. Karsten and I. Livshits, "Bootstrap Algebraic Multigrid: Status Report, Open Problems, and Outlook," *Numerical Mathematics: Theory, Methods and Applications*, Vol. 8, No. 1, pp. 112–135, 2015.

Burden, R. L., J. D. Faires, and A. C. Reynolds, *Numerical Analysis*, PWS Publishers, Boston, MA, 1981.

Carslaw, H. S., and J. C. Jaeger, *Conduction of Heat in Solids*, Oxford Science Publications, Oxford, 1959.

Cengel, Y. A., *Heat Transfer – A Practical Approach*, McGraw-Hill, New York, 2003.

Chandrupatla, T. R., and A. D. Belegundu, *Introduction to Finite Elements in Engineering*, Prentice Hall, Englewood Cliffs, NJ, 1997.

Chapra, S. C., and R. P. Canale, *Numerical Methods for Engineers*, McGraw-Hill, New York, 2002.

Chung, T. J., *Computational Fluid Dynamics*, Cambridge University press, New York, 2002.

Churchill, R. V., and J. W. Brown, *Fourier Series and Boundary Value problems*, Third Edition, McGraw Hill, New York, 1978.

Constantinides, A., *Applied Numerical Methods with Personal Computers*, McGraw-Hill, New York, 1987.

Courant, R., and D. Hilbert, *Methods of Mathematical Physics*, Vol. I and II, Interscience, New York, 1962.

CUDA: NVIDIA, https://developer.nvidia.com/about-cuda.

Date, A. W., *Introduction to Computational Fluid Dynamics*, Cambridge University Press, New York, 2005.

Datta, B. N., *Numerical Linear Algebra and Applications*, Brooks/Cole Publishing Company, Pacific Grove, CA, 1995.

Deb, P. and P. Majumdar, Direct Numerical Simulation of Mixing of a Passive Scalar Decaying Turbulence, *Proceedings of the 1999 International Mechanical Engineering Congress and Exhibition (IMECE)*, Nov. HTD. Vol. 364–3, pp. 299–306, 1999.

Desai, C. S., *Elementary Finite Element Method*, Prentice Hall Inc., Englewood Cliffs, NJ, 1979.

Desai, C. S., and J. F. Abel, *Introduction to the Finite Element Method*, Van Nostrand Reinhold, New York, 1972.

Eckert, E. R. G., and R. M. Drake, Jr. *Analysis of Heat and Mass Transfer*, McGraw-Hill, New York, 1972.

Elsgolc, E.-L., *Calculus of Variations*, Addition-Wesley, Reading, MA, 1962.

Eswaran, V. and S. B. Pope, "Direct Numerical Simulations of the Turbulent Mixing of a Passive Scalar," *Physics of Fluids*, Vol. 31, p. 506, 1988.

Ewing, G. M., *Calculus of Variations with Applications*, Dover Publications, New York, 1985.

Falgout, R. D., and J. E. Jones, "Conceptual interfaces in HYPRE," *Future Generation Computing Systems*, Vol. 22, No. 1–2, 2006, pp. 239–251.

Finlayson, B. A., *The Method of Weighted residuals and Variational Principles*, Academic Press, New York, 1972.

Fletcher, A. J., *Computational Techniques for Fluid Dynamics*, Vol. I, Second Edition, Springer- Verlag, Berlin, Germany, 1991.

Foray, M. J., *Variational Calculus in Science and Engineering*, McGraw-Hill, New York, 1968.

Forsythe, G. E., M. A. Malcolm, and C. B. Moler, *Computer Methods for Mathematical Computations*, Prentice Hall, Englewood Cliffs, NJ, 1977.

Fox, C., *An Introduction to the Calculus of Variations*, Oxford University Press, Inc., New York, 1950.

Fox, R. W., and A. T. McDonald, *Introduction to Fluid Mechanics*, Fifth Edition, John Wiley, New York, NY, 1998.

Gallagher, R. H., *Finite Element Analysis Fundamentals*, Prentice Hall, Inc., Englewood Cliffs, NJ, 1975.

Gelfand, I. M., and S. V. Fomin, *Calculus of Variations*, Prentice Hall, Inc., Englewood Cliffs, NJ, 1963.

Gerald, C. F., and P. O. Wheatley, *Applied Numerical Analysis*, Fifth Edition, Addison Wesley, New York, 1994.

Golub, G. H., and C. F. Van Loan, *Matrix Computations*, Johns Hopkins University Press, Baltimore, MD, 1989.

Grandin, H., *Fundamentals of the Finite Element Method*, Macmillan Publishing Company, New York, 1986.

Grinstein, F. F., L. G. Margolin, and W. J. Rider, Implicit Large Eddy Simulation, Cambridge University Press, New York, 2007.

Harlow, F, H., and J. E. Welch, "Numerical Calculation of Time-Dependent Viscous Incompressible Flow of Fluid with Free Surface," *Journal of the Physics of Fluid*, Vol. 8, No. 12, p. 2182, 1965.

Haroutunian, V., M. S. Engelman, and I. Hasbani, "Segregated Finite Element Algorithms for the Numerical Solution of Large-Scale Incompressible Flow Problems," *International Journal for Numerical Methods in Fluids*, Vol. 17, pp. 323–348, 1993.

Hilderbrand, F. B., *Introduction to Numerical Analysis*, Second Edition, McGraw-Hill, New York, 1956.

Hilderbrand, F. B., *Methods of Applied Mathematics*, Prentice-Hall, Englewood Cliffs, NJ, 1965.

Hinze, J. O., *Turbulence*, Second Edition, McGraw-Hill, New York, 1975.

Hirt, C. W., and J. L. Cook, "Calculating Three-Dimensional Flows around Structures and over Rough Terrain," *Journal of Computational Physics*, Vol. 10, pp. 324–340, 1972.

Holman, J. P., *Heat Transfer*, 7th Edition, McGraw-Hill, New York, 1990.

Holman, J. P., *Heat Transfer*, Eighth Edition, McGraw-Hill, New York, 1997.

Howel, J. R., and R. O. Buckius., *Fundamentals of Engineering Thermodynamics*, McGraw-Hill, New York, 1987.

Incropera, F. P., and D. P. DeWitt, *Fundamentals of Heat and Mass Transfer*, Fifth Edition, John Wiley, New, NY, 2002.

Jakob, M., *Heat Transfer*, John Wiley, New York, 1949.

Jaluria, Y., and K. E. Torrance, *Computational Heat Transfer*, Hemisphere Publishing Corp, New York, 1986.

James, M. L., G. M. Smith, and J. C. Wolford, *Applied Numerical Methods for Digital Computation*, Third Edition, Harper Collins Publishers, New York, 1985.

Jones, W. P and B. F. Launder, The Prediction of Laminarization with Two Equation Model of Turbulence, Vol. 15, 2, pp. 301–314 (1972).

Kolmogorov, A. N., Equations of Turbulent Motion of an Incompressible Fluid, Izv. Akad. Nauk SSR, Ser. fiz. 6, pp. 56–58, 1942.

Kays, W. M., and M. E. Crawford, *Convective Heat and Mass Transfer*, McGraw-Hill, New York, 1993.

Keenan, J. H., *Thermodynamics*, The MIT Press, Cambridge, MA, 1970.

Kim, S. F and and Choudhury, D., A Near–Wall treatment using Wall Function Sensitized to Pressure Gradient, ASME FED – Vol. 217, Separated Complex Flows, ASME 1995. Flows, ASME, 1995.
Kim, J., P. Moin., and R. Moser, Turbulence Statistics in Fully Developed Channel Flow at Low Reynolds Number, J. Fluid Mech., Vol. 177, pp. 133–166, 1987
Kincaid, D., and W. Cheney, *Numerical Analysis*, Brooks/Cole Publishing, Pacific Grove, CA, 1991.
Klebanoff, P. S., Characteristics of Influence in a Boundary Layer with Zero Pressure Gradient, NACA-R-1247, 1955.
Kreith, F. K., and M. S. Bohn, *Principles of Heat Transfer*, Sixth Edition, Brooks/Cole, Pacific Grove, CA, 2001.
Lam, C. K. G., and Bremhorst, K., A Modified form of the $\kappa-\varepsilon$ Model for Predicting Wall Turbulence, J. Fluids Engineering, Vol. 103, pp. 456–460, 1981.
Lamb, H., *Hydrodynamics*, Cambridge University Press, Cambridge, 1945.
Lardner, T., "Biot's Variational Principle in Heat Conduction," *AIAA Journal*, Vol. 1, No.1, pp. 196–206, 2012.
Lanczos, C., *The Variational Principles of Mechanics*, The University of Toronto Press, Toronto, 1964.
Launder, B. E., and D. B. Spalding, Numerical Computation of Turbulent Flows, Computer Methods in Applied Mechanics and Engineering, Vol. 3, pp. 269–289, 1974.
Launder, B. E and B. I Sharma, Application of the Energy-Dissipation Model of Turbulence to the Calculation Flow and Near Spinning Disc, Vol. 1, 2, pp. 131–137, 1974.
Lax, P. D., "Weak Solutions of Non-linear Hyperbolic Equations and Their Numerical Computation," *Communications on Pure and Applied Mathematics*, Vol. VII, pp. 159–193, 1954.
Leipholz, H., *Direct Variational Methods and Eigenvalue Problems in Engineering*, Noordhoff, Leyden, IL, 1977.
Leonard, B. P., "Order of Accuracy of QUICK and related Convective-diffusive Schemes," *Applied Mathematical Modeling*, Vol. 19, pp. 640–653, 1995.
Lippmann, H., *Extremum and Variational Principles in Mechanics*, Springer-Verlag, New York, 1972.
Logan, D. L., *A First Course in the Finite Element Method*, Brooks/Cole Publishers, Pacific Grove, CA, 2000.
Majumdar, P. and P. Deb, "Computational Analysis of Turbulent Fluid Flow and Heat Transfer Over an Array of Heated Modules Using Turbulence Models," *Journal of Numerical Heat Transfer*, Vol. 43, No. 7, pp. 669–692, 2003.
Mikhlin, S. G., *Variational Methods in Mathematical Physics*, Pergamon Press, New York, 1964.
Mills, A. F., *Basic Heat and Mass Transfer*, Second Edition, Prentice Hall, Englewood Cliffs, NJ, 1999.
Minkowycz, W. J., E. M. Sparrow, G. E. Schneider, and R. H. Pletcher, *Handbook of Numerical Heat Transfer*, John Wiley, New York, 1988.
Mitchell, A. R., *Computational Methods in Partial Differential Equations*, Wiley, New York, 1969.
Montgomery, S. B., and P. Wibulswas, "Laminar Flow and Heat Transfer in Ducts of Rectangular Cross-Section," *Proceedings of the International Heat Transfer Conference*, Vol. 1, pp. 104–112, 1966.
Moran, M. J., H. N. Shapiro, D. D. Boettner and M. B. Bailey, *Fundamentals of Engineering Thermodynamics*, 9th Edition, John Wiley, New York, 2018.
Morse, R. M., and H. Feshbach, *Methods of Theoretical Physics*, McGraw-Hill, New York, 1953.
Myers, G. E., *Analytical Methods in Conduction Heat Transfer*, AMCHT Publications, Madison, WI, 1998.
Noye, B. J., *Numerical Solution of Differential Equations*, North-Holland Publishing Company, Amsterdam, 1983.
Ortega, J. M., and W. G. Poole, *An Introduction to Numerical Methods for Differential Equations*, Pitman Publishers Inc., Marshfield, MA, 1981.
Ozisik, M. N., *Heat Conduction*, John Wiley, New York, 1980.
Panton, R. L., *Incompressible Flow*, John Wiley, New York, 1984.
Patankar, S. V., *Numerical Heat Transfer and Fluid Flow*, Taylor & Francis, Washington, DC, 1980.
Patankar, S. V., "Parabolic Systems: Finite Difference Method – I," *Handbook of Numerical Heat Transfer*, Edited by Minkowycz, W. J., E. M. Sparrow, G. E. Schneider, and R. H. Pletcher, Chapter – 2, pp. 89–115, Wiley, New York, 1988a.
Patankar, S. V., "Elliptic Systems: Finite Difference Method – I," *Handbook of Numerical Heat Transfer*, Edited by Minkowycz, W. J., E. M. Sparrow, G. E. Schneider, and R. H. Pletcher, pp. 215–240, Wiley, New York, 1988b.
Patankar, S. V., *Computation of Conduction and Duct Flow Heat Transfer*, Innovative Research, Inc., Maple Grove, MN, 1991.
Patankar, S. V., and D. B. Spalding, "A Calculation Procedure for Heat, Mass and Momentum Transfer in Three-Dimensional Parabolic Flows," *International Journal of Heat and Mass Transfer*, Vol. 15, pp. 1787–1806, 1972.

Payvar, P., and P. Majumdar, "Developing Flow and Heat Transfer in a Rectangular Duct with a Moving Wall," *Journal of Numerical Heat Transfer, Part A*, Vol. 26, pp. 17–30, 1994.

Peaceman, D. W., and H. H. Rachford, "Numerical Solution of Parabolic and Elliptic Differential Equations," *Journal of the Society for Industrial and Applied Mathematics*, Vol. 3, pp. 28–41, 1955.

Pepper, D. W., and J. C. Heinrich, *The Finite Element Method – Basic Concepts and Applications*, Hemisphere Publishing Corp, New York, 1992.

Plybon, B. F., *An Introduction to Applied Numerical Analysis*, PWS-Kent Publishing Co., Mishawaka, IN, 1992.

Poulikakos, D., *Conduction Heat Transfer*, Prentice Hall, Englewood Cliffs, NJ, 1994.

Rai, M. M. and P. Moin, "Direct Simulation of Turbulent Flow Using Finite Difference Schemes," *Journal of Computational Physics*, Vol. 96, p. 15, 1991.

Rao, S. S, *The Finite Element Method in Engineering*, Butterworth and Heinemann, Boston, MA, 1999.

Rao, S. S., *Applied Numerical Methods for Engineers and Scientists*, Prentice Hall, Englewood Cliffs, NJ, 2002.

Recktenwald, G., *Numerical Methods with MatLab – Implementation and Application*, Prentice Hall, Englewood Cliffs, NJ, 2000.

Reddy, J. N., *Energy and Variational Methods in Applied Mechanics*, John Wiley, New York, 1984.

Reddy, J. N., *Applied Functional Analysis and Variational Methods in Engineering*, McGraw-Hill, New York, 1986.

Reddy, J. N., *An Introduction to the Finite Element Method*, Second Edition, McGraw-Hill, Inc., New York, 1994.

Reddy, J. N., and D. K. Gartling, *The Finite Element Method in Heat Transfer and Fluid Dynamics*, CRC Press, Boca Raton, FL, 1994.

Rektorys, K., *Variational Methods in Mathematics, Science and Engineering*, Reidel, Boston, MA, 1977.

Roaches, P. J., *Computational Fluid Dynamics*, Hermosa Publishers, Albuquerque, NM, 1982.

Robertson, J. A., D. F. Elger, and C. T. Crowe, *Engineering Fluid Mechanics*, 12th Edition, John Wiley, New York, 2019.

Rogallo, R. S., Numerical Experiments in homogeneous Turbulence, NASA Report NO: N81-31508, 1981.

Ruge, J., and K. Stuben, "Algebraic Multigrid", *Multigrid Methods*, Edited by S. McCormick (Ed.), pp. 73–130, SIAM, Philadelphia, PA, 1987.

Saad, Y., *Iterative Methods for Sparse Linear Systems*, PWS Publishers, Boston, MA, 1996.

Saad, Y., "Variations on Arnoldi's Method for Computing Eigenelements of Large Unsymmetric Matrices," *Linear Algebra and Its Applications*, Vol. 29, pp. 269–295, 1980.

Saad, Y., and M. H. Schultz, "GMRES: A Generalized Minimal Residual Algorithm for Solving Nonsymmetric Linear Systems," *SIAM Journal on Scientific Computing*, Vol. 7, No. 3, pp. 856–869, 1986.

Schechter, R. S., *The Variational Methods in Engineering*, McGraw-Hill, New York, 1967.

Schilling, R. J., and S. L. Harris, *Applied Numerical Methods for Engineers Using MATLAB and C*, Brooks/Cole Publishing, Pacific Grove, CA, 1999.

Schlichting, H., *Boundary Layer Theory*, McGraw-Hill, New York, 1951.

Schlichting, H and K. Gersten, Boundary Layer Theory, Springer-Verlag, 1996.

Schlichting, H., and K. Gersten, *Boundary Layer Theory*, Eighth Revised and Enlarged Edition, Springer-Verlag, New York, 2000

Segerlind, L. J. *Applied Finite Element Analysis*, John Wiley, New York, 1984.

Sentker, A., and W. Riess, Experimental Investigation of Turbulent Wake-Blade Interaction in Axial Compressors, *Proceedings of the 4th International Symposium on Engineering Turbulence Modeling and Measurement*, France, pp. 731–740, 1999.

Sherman, F. S., *Viscous Flow*, McGraw-Hill, New York, 1990.

Sherwood, T. K., R. L. Pigford, and C. R. Wilke, *Mass Transfer*, McGraw-Hill, New York, 1975.

Shih, T. and Liou, W. (1995). A New Eddy- Viscosity Model for High Reynolds Number Turbulent Flows – Model Development and Validation. *Computers Fluids*. pp. 227–238.

Shih, T. M., *Numerical Heat Transfer*, Hemisphere Publishing Corp, New York, 1984.

Siegel, R., and J. R. Howell, *Thermal Radiation Heat Transfer*, Hemisphere Publishing Corp, New York, 1992.

Skeel, R. D., and J. B. Keiper, *Elementary Numerical Computing with Mathematica*, McGraw-Hill, New York, 1993.

Smith, G. D., *Numerical Solution of Partial Differential Equation*, Oxford University Press, London, 1978.

Sonntag, R. E., C. Borgnakke, and G. J. Van Wylen., *Fundamentals of Thermodynamics*, John Wiley, New York, 1998.

Sparrow, E. M., and R. D. Cess, *Radiation Heat Transfer*, McGraw-Hill, New York, 1978.

Stakgold, I., *Green's Functions and Boundary Value Problems*, John Wiley, New York, 1979.
Suga, K., "Chapter 9: Analytical Wall-Functions of Turbulence for Complex Surface Flow Phenomena," *Transactions on State of the Art in Science and Engineering*, Vol. 41, WIT Press, England, 2010.
Thomas, L. C., *Heat Transfer*, Prentice-Hall, Inc, Englewood Cliffs, NJ, 1992.
Thompson, P. A., *Compressible Fluid Dynamics*, McGraw-Hill, New York, 1972.
Tu, J., G. H. Yeah, and C. Liu, *Computational Fluid Dynamics – A Practical Approach*, Elsevier, Oxford, UK, 2008.
Tzou, D. Y., *Macro- to Microscale Heat Transfer*, Taylor & Francis, Washington, DC, 1997.
Van Doormaal, J. P., and G. D. Rathby, "Enhancements of the SIMPLE Method for Predicting Incompressible Fluid Flows," *Journal of Numerical Heat Transfer*, Vol. 7, pp. 147–163, 1984.
van Leer, B., "Towards the Ultimate Conservative Difference Scheme - A Second Order Sequel to Godunov's Method," *Journal of Computational Physics*, Vol. 32, pp. 101–136, 1979.
Van Loan, C. F., *Introduction to Scientific Computing: A Matrix-Vector Approach using Matlab*, Prentice-Hall, Inc., Englewood Cliffs, NJ, 1997.
Weinstock, R., *Calculus of Variation with Applications to Physics and Engineering*, McGraw-Hill, New York, 1952.
White, F. M., *Heat and Mass Transfer*, Addison-Wesley, Menlo Park, CA, 1988.
White, F. M., *Viscous Fluid Flow*, Second Edition, McGraw Hill, New York, 1991
Wilcox, D., *Turbulence Modeling for CFD*, DCW Industries, Inc., La Canada, CA, 1993.
Xin, N., and L. Li, A comparison of Low Reynolds Number k-ε Models, *4th International Conference on Computer Mechatronics, Control and Electronics Engineering (ICCMCEE)*, pp. 1334–1339, 2015.
Yakhot, V., Thangam, S., Gatski, T. B., Orszag, S. A., and Speziale, C. G., Development of Turbulence Models for Shear Flows By a Double Expansion Technique, *Physics of Fluids A: Fluid Dynamics*, Vol. 4, 1992.
Yuan, S. W., Foundations of Fluid Mechanics, Prentice-Hall, Englewood Cliffs, New Jersey, 1967.
Zienkiewicz, O. C. *The Finite Element Method in Engineering Science*, McGraw-Hill, London, England, 1971.
Zienkiewicz, O. C., and R. L. Taylor, *The Finite Element Method*, Vol – 1, McGraw-Hill, 1986.
Zienkiewicz, O. C., and R. L. Taylor, *The Finite Element Method*, Vol – 2, McGraw-Hill, 1986.

Index

G

H

I

J

K

L

M

N

O

P

Q

R

S

T

For Product Safety Concerns and Information please contact our EU
representative GPSR@taylorandfrancis.com
Taylor & Francis Verlag GmbH, Kaufingerstraße 24, 80331 München, Germany

www.ingramcontent.com/pod-product-compliance
Lightning Source LLC
Chambersburg PA
CBHW080343220326
41598CB00030B/4588
* 9 7 8 1 0 3 2 0 4 0 9 4 3 *